Springer Aerospace Technology

The series explores the technology and the science related to the aircraft and spacecraft including concept, design, assembly, control and maintenance. The topics cover aircraft, missiles, space vehicles, aircraft engines and propulsion units. The volumes of the series present the fundamentals, the applications and the advances in all the fields related to aerospace engineering, including:

- structural analysis,
- aerodynamics,
- aeroelasticity,
- aeroacoustics,
- flight mechanics and dynamics
- orbital maneuvers,
- avionics,
- systems design,
- materials technology,
- launch technology,
- payload and satellite technology,
- space industry, medicine and biology.

The series' scope includes monographs, professional books, advanced textbooks, as well as selected contributions from specialized conferences and workshops.

The volumes of the series are single-blind peer-reviewed.

To submit a proposal or request further information, please contact: Mr. Pierpaolo Riva at pierpaolo.riva@springer.com (Europe and Americas) Mr. Mengchu Huang at mengchu.huang@springer.com (China)

The series is indexed in Scopus and Compendex

Mikhail A. Pogosyan
Editor

Aircraft Design

Springer

Editor
Mikhail A. Pogosyan
Moscow Aviation Institute (National
Research University)
Moscow, Russia

ISSN 1869-1730 ISSN 1869-1749 (electronic)
Springer Aerospace Technology
ISBN 978-981-96-4598-5 ISBN 978-981-96-4599-2 (eBook)
https://doi.org/10.1007/978-981-96-4599-2

This Springer imprint is published by the registered company Springer Nature Singapore Pte Ltd.
The registered company address is: 152 Beach Road, #21-01/04 Gateway East, Singapore 189721, Singapore

Preface

This is a textbook for the course "Aircraft Design", which is a part of the curriculum of the specialty "Aircraft and Helicopter Engineering" in Russian aerospace universities and colleges. It was conceived as the new edition of the popular textbook of the same name, edited by S. M. Yeger, published in 1983 and reprinted in 2005. For a quarter of a century, students and postgraduates not only from Russia but also from a number of other countries have been using it to master the basics of aircraft design science.

The authors of the new edition sought, while preserving the undoubted advantages of the previous one, to reflect in the textbook recent achievements in the theory and methods of aircraft design.

The aircraft industry at its current stage is characterized by a high level of cooperation between countries involved in the development of aircraft systems. Such cooperation requires the unification and standardization of the language and procedures throughout all stages of the aircraft life cycle, from initial concept to disposal, which became possible through the use of modern information technologies. It is clear that a modern specialist must master these technologies.

As in the previous edition, the textbook focuses on the application of scientific foundations of aircraft design, which are not subject to rapid aging. The textbook explores ways and methods to form a rational aircraft configuration that meets the imposed requirements. It also outlines specific features in the design of the main classes of airplanes, with the exception of light propeller-driven aircraft. The design of light aircraft is the subject of the textbook by A. N. Arepyev, published in 2005.

For the first time, the textbook includes chapters on the design of seaplanes, in the development and operation of which Russia holds a leading position, as well as unmanned aircraft. The latter are becoming increasingly significant competitors of manned aircraft in solving a wide variety of tasks.

The structure of the textbook has also undergone changes. It consists of two parts. In the first part (Chapters "Design as a Key Stage in the Development of a New Aircraft and Its Modifications"–"Unmanned Aircraft"), as before, the scientific and methodological foundations of design common to all aircraft are considered, as well as specific design features of various classes of aircraft. Methods for determining aircraft parameters are now discussed in a separate chapter. Separate chapters are also devoted to the relationships between aircraft parameters and characteristics, as well as parametric analysis and optimization.

The second part (Chapters “Aircraft Wing Design”–“Landing Gear Design”) presents the materials related to the design of individual aircraft parts and retaines the structure of the previous edition.

In order to save the volume of the textbook, the third part, devoted to the basics of design automation, is excluded from it. This is justified by the fact that the principles of building and the structure of computer-aided design systems are described in a separate course, and the methodology of computer-aided design, as prevailing in modern practice, is described in the relevant chapters of the first part.

To facilitate the work on the graduation thesis, the Annexes have been expanded and upgraded.

The general editing and preparation for publication of the textbook was carried out by the editorial board consisting of M. A. Poghosyan—Chairman, N. K. Liseitsev—editor—compiler, D. Y. Strelets—member of the board.

The textbook was written by leading experts from the industry (Sukhoi Design Bureau, MiG, Molniya, G. M. Beriev Design Bureau), as well as universities (Bauman Moscow State Technical University and Zhukovsky Air Technical University).

Meanwhile, in order to preserve continuity, the editorial board of this publication sought to use non-outdated materials from the authors of the previous edition, N. A. Fomin (Chapter “Relationships Between Aircraft Parameters and Performance. Ways to Improve Performance”), S. M. Yeger (Chapters “Design Input Data. Conditions and Relationships Defining the Feasibility of a Project”, “Aircraft Layout”, “Design Features of Commercial Airliners and Cargo Aircraft”, “Aircraft Wing Design”), B. F. Mishin (Chapters “Analysis and Selection of Aicraft Configuration and Type of Power Plant”, “Air-Space and Hypersonic Aircraft Design”, “Aircraft Power Plant Design”), F. I. Sklyansky (Chapters “Aircraft Layout”, “Empennage Design”, “Aircraft Control System Design”), A. A. Badyagin (Chapter “About Aircraft Weight Analysis (Weight Design)”), V. E. Rotin (Chapter “Aircraft Wing Design”), and N. A. Kandrashov (Chapter “Landing Gear Design”).

The responsibilities for writing the textbook were distributed as follows.

The Preface, Introduction, and Chapter “Design as a Key Stage in the Development of a New Aircraft and Its Modifications” were written jointly by M. A. Pogosyan and N. K. Liseitsev. N. K. Liseitsev wrote Chapters “Theoretical and Methodological Foundations of Aircraft Design”, “Design Input Data. Conditions and Relationships Defining the Feasibility of a Project”, “Determination of Aircraft Design Parameters”, “About Aircraft Weight Analysis (Weight Design)”, “Relationships Between Aircraft Parameters and Performance. Ways to Improve Performance”, with the exception of Sect. 8.6, written by V. A. Klyagin, as well as Chapters “Optimization in Design. Methods of Solving Optimization Problems” and “Design Features of Short and Vertical Take-Off and Landing Aircraft”.

Chapter “Analysis and Selection of Aircraft Configuration and Type of Power Plant” was written jointly by N. K. Liseitsev (Sect. 4.1), A. Z. Tarasov (Sect. 4.2), and E. A. Penkov (Sect. 4.5). Sections 4.3, 4.4, and 4.6 are materials modified by N. K. Liseytsev from the previous edition of the textbook.

Chapter "Aircraft Layout" was written by N. K. Liseitsev using materials from the previous edition, as well as by M. Y. Kuprikov (Sects. 7.3.1, 7.3.2, 7.5.2), V. V. Malchevsky (Sects. 7.2.4, 7.3.3, 7.4.1, 7.5.3), and E. A. Penkov (Sect. 7.4.4).

V. A. Kiselev wrote Chapters "Design Features of Commercial Airliners and Cargo Aircraft" and "Fuselage Design", with the exception of Sect. 17.6, written by E. A. Penkov.

Chapter "Design Features of Combat Maneuvering Aircraft" was written by M. A. Pogosyan and D. Y. Strelets.

Chapter "Air-Space and Hypersonic Aircraft Design" was written by E. N. Dudar and V. P. Timoshenko.

Chapter "Design Features of Hydro-Aviation Aircraft" was written by L. G. Fortinov and A. M. Hairov.

Chapter "Unmanned Aircraft" was written by N. A. Eremin.

Chapter "Aircraft Wing Design" was written by M. A. Pogosyan.

Chapters "Aircraft Power Plant Design" and "Empennage Design" are based on materials modified and supplemented by N. K. Liseitsev from the previous edition of the textbook.

Chapter Aircraft Control System Design was written by O. S. Dolgov using materials by F. I. Sklyansky, as well as materials kindly provided by professor of Bauman Moscow State Technical University I. S. Shumilov on modern electric remote control systems.

Appendices to the textbook were prepared by E. I. Plykin.

The Editorial Board expresses its gratitude to the leading specialists of Sukhoi Company and Sukhoi Civil Aircraft, who reviewed the manuscript of the textbook and made a number of valuable suggestions for its improvement, as well as M. V. Shkurin and E. A. Karpovich for preparing the manuscript for publication.

Moscow, Russia

Mikhail A. Pogosyan
kaf101@mai.ru

Nikolai K. Liseitsev

Introduction

"Aircraft design" is a scientific and technical discipline studied at the final stage of training for an aeronautical engineer, after completion of all previous special courses (aerodynamics and flight mechanics, structural design, power plants, special equipment and instruments, production technology, economics and efficiency, etc.). In this discipline, previously acquired knowledge is consolidated to analyze and synthesize the design of a new aircraft. Thus, the peculiarity of this discipline is its complex, interdisciplinary nature. At the same time, it is an independent discipline that has its own object of research, its own tasks, and methods.

The object of aircraft design is an airplane as an element of a certain system, in which it performs specific functions.

Airplanes belong to the class of airborne vehicles employing the aerodynamic principle of flight. They use the energy of the stored fuel, as well as ambient air, to produce propulsive, lifting, and control forces. Currently, this is the most common type of aircraft, with huge potential for further development, since the achieved ranges of speed and altitude are very small, and the demand of society for such aircraft is constantly increasing.

The subject of the discipline is the study of the relationships between aircraft parameters and characteristics, the search for ways to improve its technical and economic performance, as well as to improve design methods. In addition, the discipline is aimed at:

- aircraft concept development;
- selection of the aircraft layout, composition, and structure of its systems and elements;
- determination of aircraft parameters;
- parametric analysis of aircraft characteristics and optimization of its parameters according to project quality criteria;
- development of design documentation for the manufacture, testing and operation of the new aircraft.

The approaches to solving the above tasks in the most effective way form the basis of design methods.

For more than a century of aircraft construction, the aircraft design process has undergone significant changes. The aircraft have gone from the first shakily floating "bookcases" of the Wright brothers, Farman, Fokker, Voisin, Grizodubov, Gakkel to modern high-speed, comfortable airliners carrying hundreds of passengers over intercontinental distances, as well as highly effective fifth-generation combat systems.

In parallel with a changing aircraft as a design object, the design goals and methods also change. At the dawn of the aircraft industry, the goal and basic design principle was to create a reliably flying aircraft. Based on the maturing science of aerodynamics, by copying birds, empirically, through trial and error, the pioneers of the aircraft industry created and improved the first aircraft. After the First World War, when aviation proved its viability and demonstrated great practical benefits, this principle was replaced by a new one, which dictated the need to develop an aircraft with flight, technical, and economic characteristics exceeding those of already existing aircraft. A popular slogan of this period was "Fly the highest, furthest, fastest." The question of the price of implementing this slogan has not yet been raised. It was the golden age of aviation, which lasted about half a century. The beginning of this period (the late 1930s–early 40s) saw the emergence of design as an independent discipline, driven by the objective needs of aircraft development. Based on the generalization and systematization of accumulated experience and knowledge in aircraft design, the subject and objectives of the discipline were determined, its conceptual framework was formulated, and the first guidelines and textbooks on aircraft design appeared (by K. Wood, P. D. Samsonov, N. N. Fadeev, N. N. Polikarpov, and many others). The statistical design method developed during this period was based on a certain scientific approach. Its essence was that during the evolutionary development of aircraft, the processing of past experience made it possible to establish empirical patterns reflecting the main relationships between aircraft shape, size, and flight performance. By extrapolating these patterns to a new project, under certain conditions, it was possible to reliably determine the parameters of the newly designed aircraft.

The development of methods for estimating the weight and aerodynamic characteristics, stability and controllability of aircraft, fuel consumption and performance of engines formed the prerequisites for the development of analytical methods for determining the design parameters of the aircraft. These methods were based on the solution of a system of equations describing the essential relationships between aircraft parameters and characteristics. As a result of such a solution, a set of design parameters was sought, in which the required characteristics were realized and, at the same time, the design constraints were satisfied.

The unity equation of aircraft, proposed by V. F. Bolkhovitinov and N. A. Fomin, played a special role in the development of these methods. Professor N. A. Fomin's textbook *Aircraft Design*, published in 1961, most fully reflected the development of the scientific foundations of design for the period under review. He played a significant role in shaping the professional training of designers, many of whom still work in the aviation industry.

Within the framework of analytical methods, approximate methods based on the so-called growth coefficients (weight gradients) were developed for evaluating the behavior of the selected project quality indicator (most often take-off weight). These methods have become widely used in the practice of design bureaus in solving various tasks of preliminary design of the aircraft, as well as in the development of its modifications. The works of G. Nertel, V. Herbst, A. A. Badyagin, and V. M. Sheinin made a significant contribution to the development of these methods.

The development of analytical methods has opened up opportunities for setting and solving optimization problems, as well as developing optimal design methods. The practical application of these methods in design began with solving particular problems of optimizing the parameters of individual aircraft assemblies and sections of the flight path, and then logically developed into attempts to solve the problem of choosing rational parameters from the standpoint of the effectiveness of its use within a specific combat or transport system. In the second edition of the textbook *Aircraft Design*, published in 1972, this shift in the design process was adequately reflected. In particular, the textbook explored the design of new classes of aircraft (hypersonic and aerospace, as well as vertical and short take-off and landing), developed optimal design concepts, and reflected the initial level of computer application in design.

By about the end of the 1960s, objective prerequisites had developed for the further development of the theory of aircraft design. The most important of them were:

- development of the general theory of complex technical systems (system engineering);
- advances in mathematics, primarily optimization theory and numerical methods;
- the advent of computers.

With the accumulation of experience in solving various optimization problems of individual aircraft elements, and the increasing use of computers in solving such problems, the number of parameters involved in the simultaneous optimization process gradually increased. The criteria for evaluating design solutions were improved in the direction of taking into account the multipurpose nature of the aircraft and the dynamics of changes in requirements for it over the period of operation.

Along with this, a general theory of large-scale system design was developing, on the basis of which the practice and theory of aircraft design were gradually gaining logical completeness, providing a truly scientific approach to predicting the parameters and characteristics of a future aircraft. It was on the basis of these achievements that the prerequisites appeared for overcoming the crisis in aircraft design that was brewing by the end of the 1960s and early 70s. The essence of the crisis was that, on the one hand, the development of increasingly complex aircraft systems required more and more resources (human, material, financial), critical even for major aviation powers. On the other hand, the development time for new aircraft increased dramatically. This led to the fact that by the time the new aircraft was released, it sometimes became obsolete and lost its relevance.

The need to overcome this crisis in the design led to the replacement of the principle of "design what the achievements of science and technology allow" with the principle of "design what is minimally necessary to solve the current tasks" (the principle of reasonable sufficiency).

The desire to reduce the cost and shorten the design time stimulated the development and increasing use of automation methods.

The third edition of the textbook *Aircraft Design*, published in 1983, as well as the textbook *Fundamentals of Computer-Aided Aircraft Design*, published in 1986, played an important role in the training of specialists who knew the basics of the new design methodology. In these books, the discipline "Aircraft Design" was considered for the first time as a branch of "system engineering", from which it drew design principles common to technical systems, methods for constructing and researching models, making decisions, and optimizing them.

A modern aircraft is a complex technical system, the creation and operation of which require a powerful scientific, technical, technological and financial potential of the country.

Paying tribute to the contribution of France, Great Britain, Germany, the USA and other countries to the development of the aircraft industry, we can be proud that Russia is rightfully one of the leading aviation powers in the world. The aircraft built in Russia at the beginning of the 20th century: the "Russian Knight" and "Ilya Muromets" designed by I. I. Sikorsky, the "Svyatogor" designed by V. A. Slesarev, and the aircraft by Ya. M. Gakkel had unique characteristics for their time.

During the Soviet period, in conditions of competition and often confrontation between the two economic systems, our country gave priority to the development of aircraft industry.

Despite the incredibly difficult conditions caused by the consequences of the First World War and the Civil Wars, Soviet aircraft of original design appeared as early as 1923–1924. Among them: the I-1 designed by N. N. Polikarpov, the world's first cantilever monoplane, and the ANT—2 designed by A. N. Tupolev, the first Soviet all-metal aircraft. In 1925, tests began on the ANT-4 (TB-1) aircraft, an all—metal cantilever monoplane with engines located on the wing, which became the standard for all aircraft of the same class.

During the first five-year plan, reconstruction of existing and construction of new aviation enterprises continued. Scientific and research institutions were strengthened. During this period, well-known aircraft were created: U-2, I-3, I-4, I-5, R-5, TB-3, ANT-9, ANT-14, K-5, Stal-2, Stal-3, Sh-2 and others. These aircraft formed the backbone of the country's air fleet in the late 1920s and the first half of the 1930s. Record-breaking flights were carried out on them, which showed that Soviet aviation was confidently moving to the forefront in the world. So, the R-5 aircraft designed by N. N. Polikarpov took first place at the native reconnaissance aircraft competition in Tehran in 1930.

In the 1930s, based on the achievements of the first five-year plan, which ensured the creation of Soviet aircraft engines of various capacities, the production of high-strength aviation materials, as well as successes in the development of aviation science, the I-15, I-16, ANT-25, Maxim Gorky (ANT-20), Rodina (ANT-37 BIS), DB-3 (IL-4), UT-2, Pe-8 (ANT-42), and I-153 Chaika. They set international speed, altitude, and range records.

In the pre-war years and the years of the Great Patriotic War, the famous Pe-2, Il-2, Yak-1, Yak-3, La-5, La-7, LaGG-3, MiG-3, Tu-2 aircraft were created, which ensured the superiority of Soviet aviation over enemy aircraft.

In the post-war years, the Soviet aviation made the transition to jet engines as soon as possible. Airplanes with variable wing geometry, short and vertical takeoff and landing are appearing.

The world's first jet passenger aircraft, the Tu-104, and the Tu-114 and An-22 Antey turboprop giants, which had no analogues, were created in the Soviet Union. In December 1968, the world's first supersonic passenger aircraft, the Tu-144, took off from a Soviet airfield.

The An-124 Ruslan and AN-225 Mriya heavy transport aircraft, the A-40 amphibious aircraft, the Tu-160 strategic bomber, the famous Sukhoi and Mikoyan Su 27 and MiG-29 and their modifications, the vertical-take-off combat aircraft of the Yakovlev Design Bureau Yak-38 and later the supersonic Yak-141, created in the USSR in the 70s and 80s, will undoubtedly become bright pages in the history of aircraft development.

In the early 90s, the Soviet Union collapsed. This disrupted the established system of cooperation and negatively affected the development of the aircraft industry as a whole.

The current leadership of the country, realizing the role of aviation in the development of the economy and ensuring the country's defense capability, is taking measures aimed at restoring Russia's status as a major aviation power and the best traditions of the Soviet school of aviation designers.

The characteristic features of these traditions, created in the world-famous design teams of A. N. Tupolev, N. N. Polikarpov, A. S. Yakovlev, S. V. Ilyushin, A. I. Mikoyan, P. O. Sukhoi, S. L. Lavochkin, O. K. Antonov, V. M. Myasishchev, G. M. Beriev, are:

- constant striving for simplicity, reliability, and high manufacturability of the designs being developed;
- innovative, creative approach to solving emerging problems;
- the ability to rationally combine technical risk with rigorous calculations, and constant reliance on scientific achievements;
- focus on solving the most complex engineering and technical problems at minimal cost.

The success of these schools is largely determined by a well-thought-out system of training aviation designers. The course "Aircraft Design" plays a special role in the formation of an aviation engineer's education. As part of this course, students not only study the theoretical foundations and methods of aircraft design, but also gain practical skills in solving design tasks, including those offered by the leading Russian aircraft design bureau.

The authors hope that the proposed textbook, as well as popular books on aircraft design by foreign colleagues such as E. Torenbik, J. Roskam, D. Reimer and many others, will allow the student to acquire the necessary knowledge in their chosen specialty. However, as Kuhemann rightly writes in the introduction to his fundamental work "Aerodynamic Design of Aircraft", "... it would be a dangerous mistake to consider our knowledge of aircraft design approaching the top and final...". It follows from his words, firstly, that, in addition to textbooks, the future engineer must constantly consult publications on various problems of the aviation and related industries. And, secondly, the future engineer should be involved as early as possible in the practical design work.

We wish you success on the difficult but fascinating path of becoming an aircraft designer!

Moscow, Russia

Mikhail A. Pogosyan
kaf101@mai.ru

Nikolai K. Liseitsev

Contents

General Aircraft Design

Design as a Key Stage in the Development of a New Aircraft and Its Modifications

Mikhail A. Pogosyan and Nikolai K. Liseitsev

Abstract This chapter discusses the definition of aircraft design, the aircraft design goals and objectives, as well as the role of innovation and creativity in the design process. In addition, the chapter lists out some design limitations, including the physical laws of aircraft flight mechanics, the level of advances in aircraft engineering and related industries, some production and technological aspects, and the available design time. The chapter also provides an insight into the aircraft as a complex system. Furthermore, the essence and goals of each stage of aircraft design, the relationship between the designer's freedom and the maturity of the project, as well as the cost of design errors, are explored. Finally, in this chapter, the reader will learn about the role of technology demonstrators, as well as the concept of aircraft families and design for modifications.

Keywords Aircraft general design · Limitations · Innovation · System engineering · Aircraft family · Modification

1 Design Goals and Objectives. Aircraft as a Design Object

Design is a part of the process of developing a new aircraft, which together with its production, operation, and disposal constitute the life cycle of an aircraft.

Among the attempts to define the term "design", in our opinion, the most complete idea of the meaning of the word is derived from Dahl's [1] definition: "to design means to conceive, to set, to propose for execution something new, previously non-existing". This proposal is implemented in the form of an orderly set of technical documentation, which ensures the possibility of commercial production of a new aircraft that meets the specified requirements and allows its proper operation under specified conditions. In accordance with The Unified System for Design Documentation requirements, technical documentation is divided into 3 groups. The first group includes schematic, design, assembly, and text documentation. It reflects the ideas

M. A. Pogosyan (✉) · N. K. Liseitsev
Moscow Aviation Institute (National Research University), Moscow, Russia
e-mail: kaf101@mai.ru

M. A. Pogosyan (ed.), *Aircraft Design*, Springer Aerospace Technology,
https://doi.org/10.1007/978-981-96-4599-2_1

and principles embedded in the aircraft being designed, answering the question of what is to be produced. The second group is the process documentation, which regulates the methods and means of manufacturing the aircraft. Finally, the third group is the operational documentation, which ensures the correct operation of the aircraft.

The development of the mentioned documentation is an important part of the design process, and the documentation itself is the design of the future aircraft. However, an equally essential component of the design process derives from the words 'conceive, envision'. It is in the process of their implementation that ways and means of achieving design goals are sought and proposed.

At the present stage of the aircraft engineering, the design objective is to create a new aircraft with unique characteristics that ensure more effective performance of either traditional functions for this aircraft type, or new national economic, defense, or other functions, whose performance in other ways is either impossible or ineffective.

Although this goal can be achieved by using well-known elements, methods, or principles in the project, a creative search for such combinations in a new, original way, that would lead to new quantitative or qualitative results is always required.

As a result, an important conclusion can be drawn: designing is a creative process, not a routine one.

The design problem is to develop such a project, the implementation of which, within certain constraints, would ensure the most effective achievement of the design objectives. The solution to this problem requires a clear definition of design objectives and the establishment of criteria for evaluating design results, since achieving design objectives "at any" cost is, of course, unacceptable.

The above-mentioned limitations are determined primarily by the physical laws of aircraft flight mechanics, using the aerodynamic principle of flight. These restrictions are objective and cannot be cancelled.

The second group of constraints stems from the level of scientific and technological advances in aircraft engineering and related industries available to the project developer. It is clear that this level is constantly developing and improving.

The development of the mentioned documentation is an important part of the design process, and the documentation itself is the design of the future aircraft. However, an equally essential component of the design process derives from the words 'conceive, envision'. It is in the process of their implementation that ways and means of achieving design goals are sought and proposed.

Equally essential design constraints are dictated by production and technological aspects. They are determined by the availability of necessary materials and components and their capabilities, laboratory and production equipment, as well as the qualification level of the aviation enterprise's personnel. The experience in the development of fundamentally new aircraft indicates that the success of their creation was due to radical reconstruction of the experimental and production and technological base of aviation enterprises. Moreover, the creation of such aircraft sometimes required the reconstruction of entire industries of the country, being a powerful incentive for its scientific and technological progress.

Design time, i.e. the time available to designers to develop designs, should also be noted among the constraints.

As an object of design, a modern aircraft is a complex technical system with a developed hierarchical structure and many elements and internal connections, the number of which increases approximately in proportion to the number of elements squared. For example, the airframe of a modern wide-body aircraft includes more than a million parts. It is clear that taking into account all the interconnections between the elements of such an aeroplane turns into a complex problem.

A number of functional subsystems can be distinguished in any aircraft, which together determine its useful properties. These include subsystems for generating lift and propulsion forces, ensuring stability and controllability of the aircraft on a given flight path, performing target functions, providing power supply and life support, providing control and navigation in various flight conditions, etc. Each of these subsystems may include a set of simple and complex systems consisting of individual elements.

The partitioning of an aircraft into subsystems, which is convenient for the purposes of study and analysis, does not mean that they are completely autonomous. They are interconnected and interdependent. At the same time, an ever tighter integration of aircraft elements and subsystems is a characteristic trend of aviation development. For example, for supersonic manoeuvrable aircraft, configurations based on the integration of the wing, fuselage, power plant, control, and stability systems in order to achieve maximum efficiency are widely used. At the same time, an aircraft is an element of a more complex structure—an aircraft system including other aircraft, flight crews, technical personnel, and means to prepare the aircraft for flight and its support—and the aircraft in question can perform certain tasks.

The aviation system is understood as an organic combination of human and material resources, the actions of which produce a certain useful effect for society. The functional and structural diagram of the aircraft as an element of the aviation complex is shown in Fig. 1.

The aircraft system, in turn, is an element of the transportation system, national defence system, or other organizational and technical systems.

Such specifics of an aeroplane dictate the need for a system approach to its design.

2 Organization and Directions for Improvement in the Design Process

As noted above, design is an integral part of the complex process of developing a new aircraft. The main features of this process, as applied to the development of military aircraft under state order, are shown in Table 1.

It can be seen that this is a multi-level hierarchical process in which each level has its own object of study, the tasks to be performed at this level, and a certain range of persons (organizations or units) responsible for their solution.

Nevertheless, these levels are interrelated, because they have one ultimate goal—the creation of a combat aviation complex (CAF), providing an increase in the

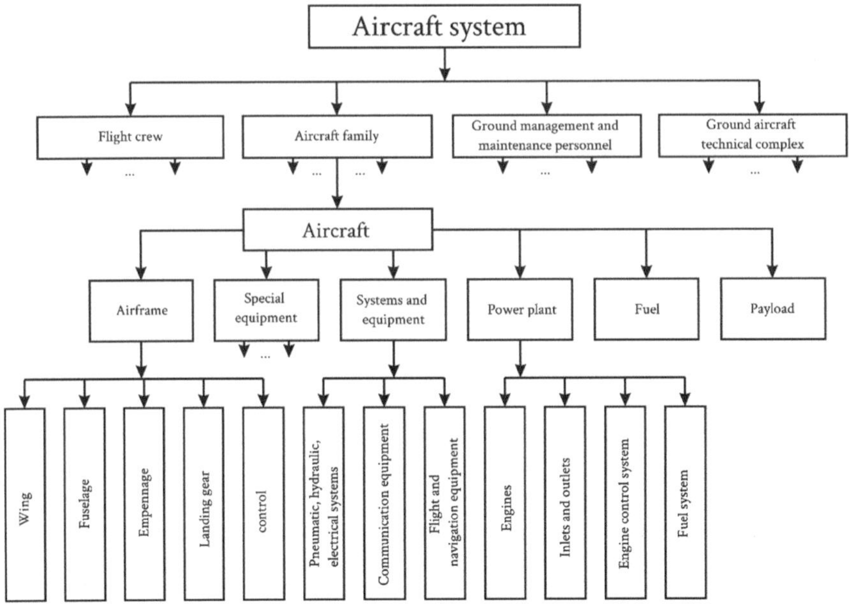

Fig. 1 Functional and structural diagram of an aircraft system

country's defence capability at minimal cost for its development. It is clear that the implementation of this goal is possible only with the rational organization of this process.

Note that the structure of the new aircraft development process described above is not the only possible one. There are examples of successful development of new aircraft at the initiative of aircraft design bureaus.

The structure of the process of developing a new civil aircraft in a market economy is similar to the structure described above. However, it has a number of differences, mainly concerning the first two hierarchical levels in terms of the object of their research and the range of persons (organizations) involved in the solution. Here the models of cargo-passenger transportation (transport operations) by air transport, their nomenclature, and the desirable characteristics of promising aircraft for certain airlines are the object of research.

As with military aircraft development, tasks are solved to answer questions such as: what should be designed and built, in what time frame, in what quantities and for whom, who will finance the work, and many others.

Civil aircraft are generally developed on a commercial basis, and their development is therefore co-financed by aircraft and component designers and manufacturers, airlines transporting passengers and cargo, and banks and various funds interested in efficiently investing their money.

Table 1 Organization of the military aircraft design process

Level	Tasks to be solved	Organizations responsible for solving tasks	Object of study
I	Preparation of proposals for the development of new aviation facilities, terms, and volumes of financing	Security Council, Ministry of defence, general staff and their institutions	Models of a country's defence capability system
II	Formation of the aircraft concept, preparation of baseline data for development tender, including preliminary technical requirements specification	Specialized research institutes of the defense industry, research institutes and design bureaus involved in the development of the combat air system and their components	Models of the combat air system as an element of air force, air defense, Navy, etc.
III	Assessment of possibilities and ways of implementation of the technical requirements specifications, development of aircraft design options and selection of the preferred ones (pre-project)	Aircraft design bureaus (parent organizations), industry research institutes, developers of the combat air system components	Models of aircraft as an element of the combat air system and its constituent components
IV	Clarification of the appearance of the aircraft, its structural and general layout, determination of the composition and parameters of its subsystems (conceptual design)	Subject divisions of experimental design bureau and subcontractor organizations	
V	Development of units, components and parts of the aircraft, preparation of documentation for the production and operation of the aircraft (full-scale design)	Functional departments of experimental design bureaus	Structural and functional models of aircraft elements
VI	Development of technological processes and tooling for manufacturing of aircraft components and its assembly	Technological divisions of experimental design bureau, preproduction, and subcontractor organizations	Ways to materialize an airplane project

(continued)

Table 1 (continued)

Level	Tasks to be solved	Organizations responsible for solving tasks	Object of study
VII	Component fabrication, unit assembly, system assembly and general assembly	Pilot plant, organizations-suppliers of materials and components	Aircraft, its components and systems
VIII	Ground and flight testing of the aircraft and its components to determine the degree of compliance with the specified characteristics and possible ways to eliminate non-compliance	Flying research institute and development base and other structural subdivisions of experimental design bureau	

Further, we will limit ourselves to considering only levels III…IV related to the design process itself, as well as partially, in accordance with the requirements of the system approach, the levels adjacent to them.

The complexity of the aircraft as a design object determines the specificity of the design process organization. The implementation of the aircraft design process required the creation of specialized design organizations—design bureaus—including many specialists from various technical and other fields and complex laboratory and production units. Design bureaus are large organizational and technical systems that evolve and improve as the design object and the means for its design become more complex. In their activities, they rely on the work of industry research institutes engaged in scientific and technical research in the interests of aircraft engineering, the assessment of aviation development prospects in its various directions (the Central Institute of Aerohydrodynamics, the State Research Institute of Aviation Systems, the State Research Institute of Civil Aviation, the Central Institute of Aviation Motors, the All-Russian Research Institute of Aviation Materials, the All-Russian Institute of Light Alloys, etc.), and the experience of aircraft production and operation.

Designing is a complex process that requires an enormous amount of calculation, graphic work, and research of all kinds.

An enlarged step-by-step structure of this process, developed by the middle of the last century, is shown in Fig. 2.

Let us briefly describe the tasks to be solved at the individual stages. Direct design is preceded by the requirements development phase, which is carried out jointly by the customer's R&D institute, industry institutes, and design bureaus. At this stage (sometimes known as "external design"), a concept of the future aeroplane is generated based on parametric studies of prospective aeroplanes as components of some kind of transport or combat system and analysis of their interactions with

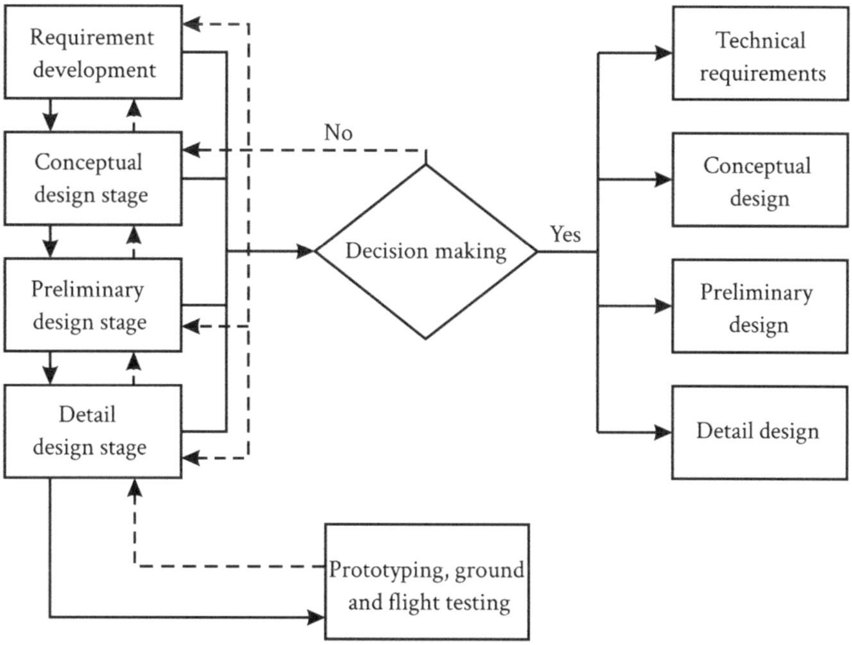

Fig. 2 Main stages of aircraft design process

the components of the complex they are to operate within. The design objectives are quantified as a list of required usable properties of the future aeroplane to be achieved under specific operating conditions. These objectives are finally formalized in the form of design briefs. There is a certain procedure for developing the Technical Requirements Specification and its typical structure.

The purpose of the next design stage, called the development of a technical proposal (conceptual design, pre-project), is to select the layout and determine the most advantageous combination of the basic parameters of the aircraft and the composition and structure of its systems, ensuring either the fulfilment of the specified requirements or justification of the need for their adjustment.

At this stage, based on the analysis of technical specifications, the chief designer's ideas, the experience of the design bureau, and the recommendations of the research institute, the previously formulated concept of the aircraft is implemented in the form of a pre-project. In the first approximation, the basic geometric, mass, and energy characteristics of the designed aircraft are determined, as are the laws of its control at various sections of the trajectory for a variety of flight profiles provided by the requirements. A feasibility study and risk assessment are prepared.

This phase brings together various aspects of aircraft design related to the selection and coordination of its geometry, mass, aerodynamic characteristics, altitude-velocity and engine performance, equipment and gear structure, flight data, and combat and transport performance characteristics.

The result of the pre-project is drawings of the general views of a rational variant of the aircraft and its layout, as well as documentation of its flight, economic, and operational characteristics. Based on the analysis of this documentation, the competent authorities decide whether to proceed with the project or to adjust the Technical Requirements Specification.

At the conceptual design stage, previously obtained geometric, mass, and energy parameters of the aircraft are embodied in a specific design layout that meets different, very contradictory requirements, including operational and technological. In the process of configuration, the aircraft alignment is clarified, which calculation is preceded by a weight summary based on strength and weight calculations of airframe and power plant units, lists of equipment, cargo, etc. During the conceptual design process, extensive theoretical and experimental studies of the aircraft's assemblies and systems are typically conducted. Models of the aircraft and its individual assemblies are manufactured and blown through wind tunnels. Based on the results of the studies, aerodynamic calculations, stability and controllability calculations, and aeroelasticity characteristics are refined.

Based on these calculations, appropriate corrections are made to the aircraft layout, and weight calculations are refined. A mock-up of the airplane is built, which allows mutual spatial coordination of airplane units and systems, placement of equipment, evaluation of crew convenience, and target load placement.

The result of this stage is a conceptual design. It provides information on the specified characteristics of the aircraft as well as on the shape, dimensions, and mutual arrangement of the main units and functional elements of the aircraft. The conceptual design is submitted to the state mock-up committee, which performs a comprehensive evaluation of the design necessary to make a decision to develop a detailed design and implement it. As a rule, such a decision is made after the remarks of the mock-up commission have been eliminated.

The second and third stages are sometimes referred to as "general design," meaning that the object of study here is the aircraft as a whole.

The final fourth stage, "detailed engineering," is aimed at producing all the technical documentation required for the manufacturing, assembly, and installation of individual assemblies and systems as well as the aircraft as a whole.

At this stage, drawings of general views of aircraft assemblies and assembly and detail drawings of individual aircraft parts are produced on the basis of the design and engineering study. This process (design or structural engineering) is aimed at the material implementation of the project in production. At this stage, as a rule, the following are carried out:

Experimental and research works related to the introduction of

- new materials and types of structures;
- static and dynamic tests of structures for strength;
- vibration, resource;
- bench tests of equipment systems, control;
- life support systems.

The information obtained at this stage allows one to clarify and specify the project data and, if necessary, make appropriate adjustments to the weight and strength calculations made in the previous stages.

The subsequent development phases, including prototype production and ground and flight testing, are organically linked to the design phase. In the process of prototype production, technical documentation and aircraft manufacturing technology are fine-tuned. In the course of factory and then state (certification testing for civil aircraft) tests, the actual characteristics of the aircraft and the degree of its compliance with the design specification are determined, as are its compliance with the requirements and limitations of the regulatory documents governing the development and acceptance of the new aircraft established in individual countries or groups of countries. Among the documents in force in our country are the Air Force's Technical Requirements Specification for military aircraft, the Aviation Rules for civil aviation, state, and industry standards in force at the time of the aircraft's development, etc. Similar documents exist in all countries involved in aircraft development.

Based on the test results, modifications are made to the design, if necessary. The result of this stage is a decision on the further fate of the developed aircraft.

Analysis of the design process shows that it is a complex, multi-step process of preparation and decision-making. The essential thing is that it is iterative. And iterations are carried out not only between the main design stages but also within each of them.

Let us analyze the role and "specific weight" of different stages of aircraft development. Figure 3 shows the trend in aircraft development costs and the percentage of decisions made at individual stages. Conceptual design is the most critical stage of aircraft development, since spending a maximum of 20–25% of time and no more than 5–10% of allocated funds means that at least 75–80% of the main design decisions (both technical and organizational) are made here. And it is here that the probability of errors is the highest, which is explained by the nature of design.

The division of the design process into stages, which facilitates the problem of solving individual tasks, requires taking into account all the essential links between adjacent stages. These links are divided into direct and inverse links (see Fig. 2). Direct links (solid lines) are output information (design result) for the upper level and input information (constraints) for the lower level. Inverse links (dashed lines) are the opposite. They reflect the technical and economic characteristics of the elements of the system of the lower hierarchical level, modeling the functioning of this element within the system of the higher hierarchical level.

The design at each subsequent stage seeks to confirm the characteristics stated in the previous stage. Since it is usually conducted from the top down, the information corresponding to the feedback is largely predictive in nature. This circumstance, as well as ignoring some structural and functional relationships, carries the risk of making unjustified and even erroneous decisions. The aforementioned neglect may be due not only to subjective but also to objective reasons, caused by the inability to take certain factors into account due to the lack of reliable information at the stage under consideration.

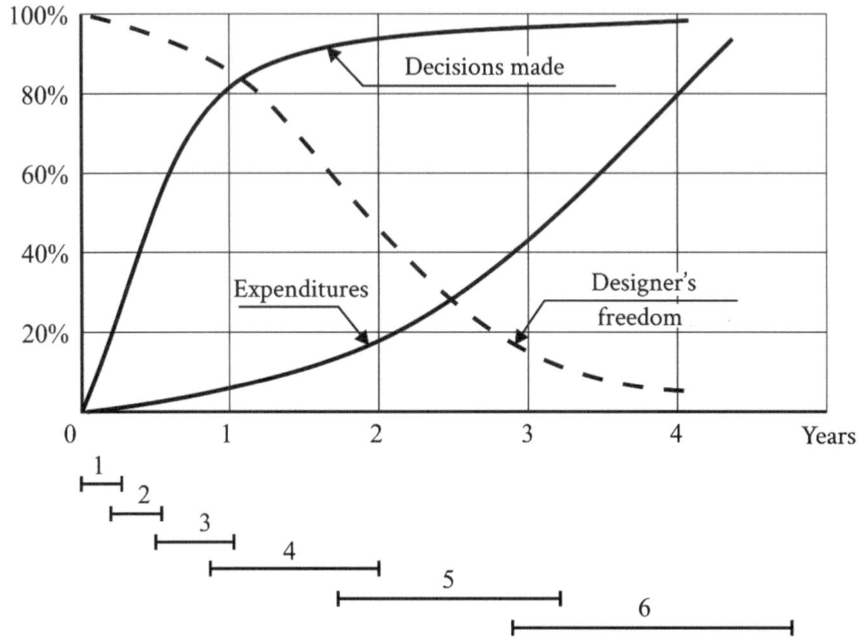

Fig. 3 Changes in costs, decisions made and degree of freedom of the designer at different stages of aircraft development: 1—Research of fields of application, analysis of technical requirements; 2—Development of technical proposal; 3—Development of conceptual design; 4—Development of detailed design; 5—Aircraft construction, bench testing; 6—Flight tests and aircraft refinement

Thus, at the stage of developing a technical proposal, it is necessary to make decisions on the shape, size, design, and mutual arrangement of the main elements of the airframe and its power plant based on the condition of ensuring the specified performance. These characteristics are determined by the weight and aerodynamic characteristics of the aircraft, as well as the characteristics of the power plant, which are determined with relatively low accuracy at this stage, particularly in the design of original aircraft.

The cost of such errors will be illustrated by examples from passenger aircraft design experience. For a supersonic passenger airplane with an estimated takeoff weight of 272 t and a payload of 20 t (200 passengers), with an average statistical value of the structural mass ratio of 0.25, its value will be 68 t. The error in the estimate of design mass of only 3% toward its increase, which is considered quite admissible in the initial stages of design, means a 10% reduction in payload. A 30% error is equivalent to the calculated value of the entire payload. No less significant is the cost of errors in the calculation of aerodynamic characteristics. For example, for a passenger aircraft with a design takeoff weight of 180 t, a 10% miscalculation of the maximum lift coefficient while keeping the payload fixed will lead to a 10% increase in runway length. If the runway length remains the same, the flight range is reduced by 25%, which also significantly changes the capabilities of the aircraft and, consequently, its level of competitiveness.

The predictive nature of the information on the basis of which decisions often have to be made requires iterative cycles between phases. However, in practice, due to the time and cost constraints of project development, it is not possible to implement the number of iterations required for the convergence of the result.

To this we should add the paradoxical mismatch between the trends in the amount of information about the design object and the degree of freedom of designers in selecting design variables (the dotted curve in Fig. 3). In the sequential decision-making technology, the paradox noted reduces the possibility of making the most profitable decisions and sometimes even eliminates the possibility of implementing the project itself. This occurs when the "first button effect" is triggered, which, when incorrectly buttoned, deprives us of the opportunity to "button up the jacket"—to design the aircraft.

Thus, in aircraft design, there are a number of objective factors, the impact of which, worsening as the design objects become more complex, makes it difficult to find rational solutions in an acceptable time frame, leading to an increase in the cost and timing of the design. It is proved by practice that the time required for wind tunnel studies, human labor costs, and the required amount of calculations to create promising aircraft increases exponentially.

The duration of aircraft development (up to the first prototype flight) is 3–5 years, and when developing brand new aircraft that have no analogues or prototypes (XB-70, Concord, Harrier, F-35), this time increases up to 8–10 years.

The time to first flight of a serial airplane ranges from 3 to 8 years, and in some cases longer (Mirage F1, Harrier—10 to 11 years, Concorde—14 years).

It is also characteristic that the planned and actual costs and development times tend to differ significantly, following the upward trend shown in Fig. 4.

The analysis of the curves in Fig. 3 reveals the main directions and tasks for improving the process of new aircraft development. At the early stages of design, efforts should be focused on improving the quality of decisions made and increasing forecasting accuracy (even sometimes at the cost of increased costs and time). At later stages, the task of increasing productivity is more relevant.

At present, there are two directions to take to improve the quality of design. One of them is to increase the forecasting accuracy, reduce the degree of risk due to making unreasonable decisions, and, consequently, reduce the cost and timing of development through the use of the results of the scientific and technological backlog. This is because in order to create an aircraft that is competitive in the world market, it is necessary to use a considerable amount (according to published data, over 50) of new technical (innovative) solutions in the aircraft project. At the same time, at least two-thirds of them must have been tested before the start of design. For example, it is possible to test a new engine or some other system by installing it on an existing aircraft. This is all the more relevant given that the cycles for creating individual aircraft subsystems, such as the airframe and engine, are significantly different.

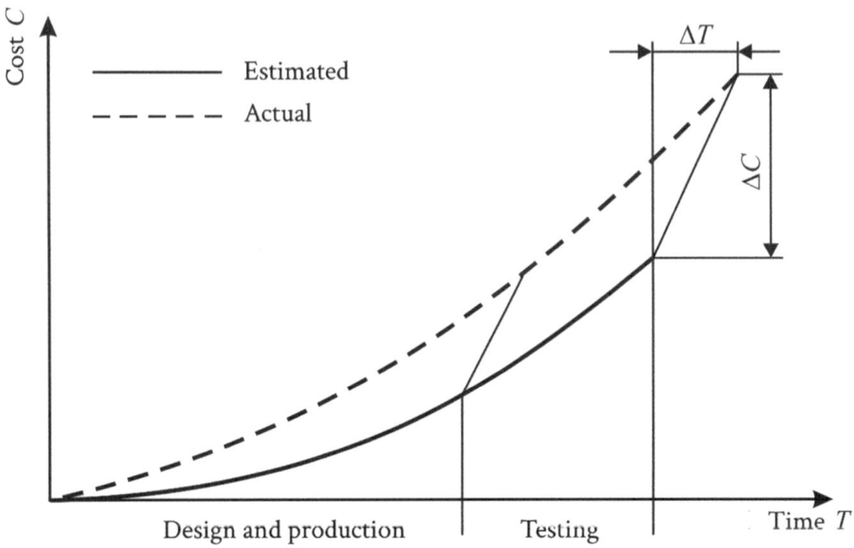

Fig. 4 Planned and actual costs by stage of aircraft development

A logical development of this direction is the concept of designing new aircraft systems based on the experience of the development and testing of so-called demonstrators, aircraft on which new technical solutions are tested and intended for use on future aircraft. It is according to this concept that the F-35 multi-purpose attack aircraft, as well as the X-45 and X-47 unmanned combat aircraft, were developed on a competitive basis in the USA.

Another example of such an approach to the development of advanced combat aircraft is the Sukhoi Design Bureau's Berkut S-37 program.

The second direction is related to the improvement of the design process by using CAD (see Chap. 2). This direction is being implemented both in terms of improving the organization of the aircraft development process and in terms of improving the methods of solving design and engineering and other tasks on the basis of modern information technologies [1]. The development of this direction was the creation and introduction into practice of the so-called CALS-technologies, which provide information support of the entire life cycle of products up to their decommissioning and disposal [2].

Implementation of the above-mentioned technologies became especially relevant when, in conditions of excessive cost for new aircraft development programs, there was a tendency to unite the efforts of the aviation industries of different countries. It was expressed in the creation of international associations for the development of advanced aircraft systems.

An important factor in reducing the cost of developing new aircraft and better tracking changing market requirements was the widespread introduction of the concept of designing a family of aircraft based on a certain basic version. This concept, successfully implemented over the past decades by leading foreign and domestic firms involved in the development of passenger and transport aircraft, is used in the development of advanced combat aircraft. As an example, let us consider an international development program for the mentioned fifth-generation aircraft, the F-35, under U.S. auspices. Under this program, the U.S. plans to create on a common basis (about 70% of the overall design and engineering solutions) several versions of the aircraft that meet the requirements of the Air Force, Navy, and Marine Corps. It is also envisaged that the aircraft supplied to the countries participating in the program will be further developed to meet their specific requirements. It is clear that the implementation of such programs is possible only within the framework of modern information technology for aircraft development.

New technologies appeared and were used by the leading aerospace companies as an adequate response to overcoming the above-mentioned "crisis" for the development of prospective airplanes that require the solution of a complex of interrelated tasks for a rather long period of time (5–6 years), large investments (from 5 to 8 billion dollars), and wide international cooperation in their implementation.

The goals of such technologies were to ensure that the development of new aircraft takes customer requirements more thoroughly into account, reduces the time from the beginning of development to the appearance of the aircraft on the market, improves production manufacturability, and reduces product life cycle costs.

Some of these technologies that Boeing has refined and implemented in creating the first fully digitally engineered Boeing 777 aircraft include:

- digital representation of the object of development (DPD, digital product definition);
- computer-integrated manufacturing (CIM);
- lean manufacturing (LM) technologies;
- total quality management (TQM);
- structured quality functions (QFD, or "quality function deployment").

In order to achieve the greatest effect from the use of these technologies, a systematic approach was required to organize the process of new aircraft development as an important, but not the only, part of the product life cycle. It was called concurrent engineering. This term means organization of development based on the close cooperation and coordinated activities of interdisciplinary integrated product teams (IPTs), which include not only design engineers, tooling developers, and process developers of components and aircraft assembly, but also specialists in quality management, aircraft operation, marketing specialists, financiers, etc.

The use of this organization made it possible to significantly reduce the time of design and the issue of design documentation. Thus, with the traditional technology (Boeing 767 aircraft), it was 40 months, and with the new technology (Boeing 777 aircraft), it was reduced to 27 months [3].

An example of how the Russian aviation industry applies new technologies in the development of promising aircraft is the development program for the SuperJet 100 short-range passenger aircraft, which is being implemented by the Sukhoi Aviation Holding Company.

References

1. Liseitsev NK (1982) A new stage in the development of design methods. In: Methods of research and development of aircraft designs and structures, MAI, Moscow
2. Pogosyan MA, Bratukhin AG, Savelyevskikh EP et al (2004) Proektirovanie aviacionnyh kompleksov s primeneniem informacionnyh tehnologij (Design of aviation complexes using information technologies) Aviac Prom 2:14–18
3. Funke C (1997) Concurrent engineering in the aircraft industry and it's relationship to the development process. SAE Tech Paper. https://doi.org/10.4271/972245

Theoretical and Methodological Foundations of Aircraft Design

Nikolai K. Liseitsev

Abstract The chapter explores the nature of design and analysis in aircraft engineering, as well as the role of modeling, optimization, and digital technologies in aircraft design. The chapter shows how the decision-making process can be formalized and explains some scientific foundations of computer-aided design technology. In addition, the chapter presents some defining features of a system and introduces the concepts of systematic approach, system decomposition, and optimization. Finally, in this chapter, the reader will learn what differentiates parameters from characteristics and how both can be used in aircraft modeling and analysis at different levels.

Keywords Aircraft modeling · Optimization · Formalization · Computer-aided design technology · Design parameters and characteristics

1 The Nature of the Design Process

As noted, engineering design is a specific type of creative activity aimed at developing previously non-existing objects, systems, and processes. The tasks and connection of engineering design with other types of creative activity can be explained with the help of the scheme presented in Table 1 [1].

Let us illustrate this scheme with examples of wing design and analysis, taking into account the wing's flutter characteristics. In the first case (analysis), the question sought is whether or not the flutter of the selected design will occur at known input data (speed, altitude). In the second case, on the contrary, we determine the conditions under which this wing will collapse due to flutter.

In the third case, a scientific research is conducted to determine the causes of the phenomenon known as "flutter" and the conditions under which it occurs.

Finally, engineering design, based on the results of scientific research on the theory of flutter, makes it possible to create structures that do not collapse under certain conditions due to flutter.

N. K. Liseitsev (✉)
Moscow Aviation Institute (National Research University), Moscow, Russia
e-mail: nikolai.liseitsev@yandex.ru

M. A. Pogosyan (ed.), *Aircraft Design*, Springer Aerospace Technology,
https://doi.org/10.1007/978-981-96-4599-2_2

Table 1 The place of engineering design among other scientific and technical activities

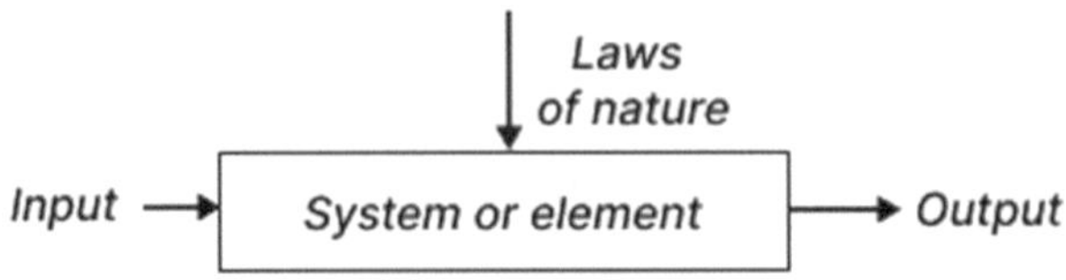

Given	Find	Process
Input, laws of nature, system or element	Output	Analysis (deduction)
Exit, laws of nature, system or element	Input	Inverse analysis
Input, output, system or element	Laws of nature	Scientific research (induction)
Input, output, laws of nature	System or element	Engineering design

Noting the usefulness of the considered scheme for understanding the nature of the design process, it should be highlighted that the real design process of such a complex object as an airplane is carried out on a collective basis by specialists in various scientific and technical disciplines. Practically, it combines all the listed types of creative activities, including scientific research, if the known technical solutions do not ensure the achievement of the design objectives. At the same time, it cannot be implemented without such important components as synthesis and decision-making.

Indeed, before analyzing something, it is necessary to propose (synthesize) some hypotheses or options (alternatives) to solve the problem. For example, in the case of flutter, such options may be the installation of anti-flutter weight, structural reinforcement, changing the structural layout, etc. After assessing the feasibility of each option, the designer should choose the best of them if there are alternatives. This process of selecting one alternative out of several defines the essence of the third component of the design process: decision-making. It is these three components, intertwining and complementing each other, that implement the design process.

Let us briefly characterize the named components of the design process and assess the possible level of application of formal apparatus (formalization) in their implementation.

The scheme of the design process algorithm for a complex technical object is shown in Fig. 1. After determining the design goal, relying on his or her creative abilities, the designer forms the main idea, the concept of the future object, outlining possible variants (alternatives) of solving the design problem. This process is often based on invention, i.e., a person's ability to synthesize new useful ideas and principles to solve engineering problems.

Invention, which is based on a person's ability to think abstractly, requires an engineer to have extensive expertise, be able to connect various facts and phenomena, and be able to overcome psychological inertia. Although there are some tricks and methods for improving inventiveness, inventing belongs almost entirely to the field of

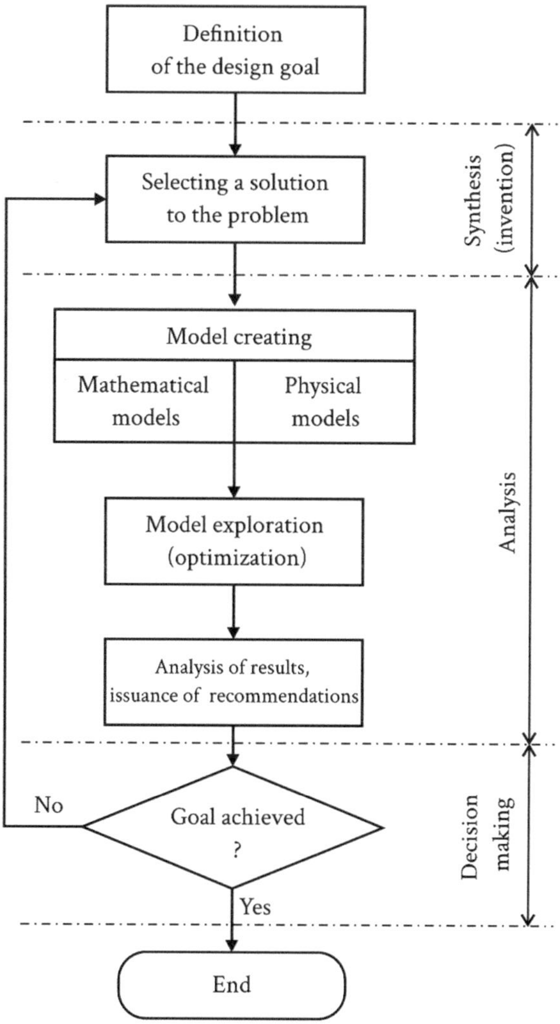

Fig. 1 Design process algorithm for a complex technical object

heuristics. Thanks to invention, it is possible to achieve fundamentally new technical solutions that significantly improve the characteristics of the designed object.

Invention accompanies the entire design process, but it is most pronounced at its initial stages. When designing airplanes, it is the inventive ideas of the chief designer that often determine the overall concept of the future airplane. At the same time, the experience of aircraft construction shows that only very few aircraft designs can be recognized as completely new. But even in them, a significant number of technical solutions are taken on the basis of past experience. In practice, aviation progress is based on both evolutionary development and revolutionary, radical new technical solutions. In order to reduce the degree of risk associated with the implementation of such solutions obtained as a result of inventive activity or from scientific and technical groundwork, each of these solutions is thoroughly tested on stands, mock-ups, experimental samples, etc.

The next component of the design process, engineering analysis, is aimed at detailing the intended variants of the problem solution (defining the structure and dimensional parameters of the designed object) and checking its compliance with physical laws and other constraints arising from the specific requirements for the designed object.

Engineering analysis is based on modeling, i.e., the study of the design object with the help of a model, which, by displaying or reproducing the object of study, is able to provide the necessary information about it. Both physical (materially realized) and mathematical (abstract) models are widely used in aircraft design. Examples of physical models are airplane mock-ups, airplane scale models, various test benches, etc. Simulation on physical models gives the most complete and reliable picture of the phenomena under study. At the same time, the application of physical modeling is often associated with significant consumption of time and material resources. Nevertheless, physical modeling is practically the only tool for investigating new concepts that are either unpredictable theoretically or require experimental verification and confirmation of some hypotheses. It is clear that the role of physical modeling increases when creating fundamentally new aircraft based on new technical concepts.

Mathematical modeling is based on known physical laws describing certain aspects of aircraft design and calculation. Mathematical modeling, in turn, can be divided into analytical and numerical modeling. Analytical modeling allows conducting research in the most general form and obtaining results in a clear, convenient form for analysis. However, the construction of analytical models is often associated with the need to significantly simplify the phenomenon in question, which reduces the reliability of the results obtained. Examples of successful analytical models are the beam model of a large aspect ratio wing, the lifting line theory, etc. Numerical simulation using computers is currently becoming one of the main methods for studying complex objects and processes, ensuring high accuracy and reliability of the results obtained. Thus, on the basis of finite-element models, highly effective numerical methods for calculating the aerodynamic characteristics of aircraft as well as methods for analyzing the stress–strain state of aircraft structure and synthesising rational design and force schemes were developed.

The information obtained as a result of the analysis allows us to implement the third component of the design process: decision making. This is a very important component, the purpose of which is to choose the only one among the possible variants of solving the problem. This choice, as a rule, is a compromise since each variant has certain advantages and disadvantages. In other words, the designer faces the necessity of searching for a constrained optimum. Decision-making is a multi-step process in which each subsequent step narrows the field of search, limiting both the number of alternatives and the number of factors to be taken into account.

A brief analysis of the design process reveals the possibilities of formalizing the individual components of this process. The first component of the design process—the generation of problem solutions—lends itself to formalization to the least extent. Currently, attempts are being made to solve this problem with the help of special heuristic methods for searching for rational technical solutions [2]. However, the possibility of using such methods in designing complex technical objects is rather limited. At the same time, if the solution is based on known elements or principles, then the task of finding a rational combination of these elements (a combinatorial problem) can be formalized. Thus, on the basis of the matrix of features of various layouts, it is possible to formalize the process of generating variants of aircraft configurations, which allows us to automate this process and, on this basis, significantly expand the field of search for a rational variant [3].

The correlation between mathematical and physical modeling determines, to a large extent, the possibilities of formalizing the analysis stage. At the same time, when a fundamentally new aircraft based on new technical concepts is developed, the role and scope of physical modeling increase, and consequently, the possibility of formalization and, automation of this stage decrease.

There are certain possibilities for formalizing the decision-making process. These opportunities are due to the achievements of the discipline "Decision Theory", which is based on the apparatus of mathematical programming. So, for example, if the design goal can be quantitatively expressed through a certain criterion, which is a function of the design parameters, then the decision-making problem about the specific values of these parameters can be reduced to the problem of finding a combination of parameters in which the criterion reaches an extreme value.

Thus, we see that there is a significant proportion of design procedures in the overall aircraft design process that can be formalized.

The possible level of formalization of design process components (in percent), determined on the basis of expert evaluations, is presented in Fig. 2.

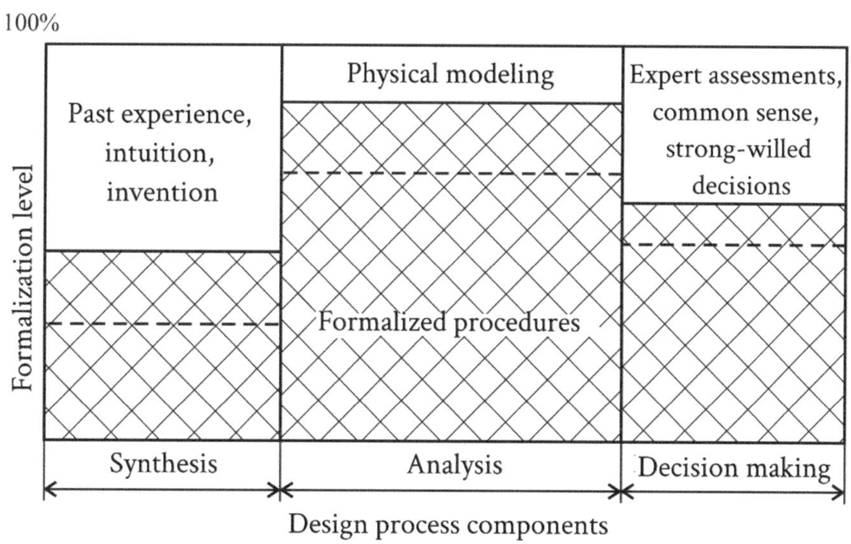

Fig. 2 Expert evaluation of the possible level of formalization of the design process components (shaded area)

The dotted line in the figure reflects the fact that when designing an unconventional aircraft or an aircraft with characteristics significantly exceeding the achieved performance, the value of past experience and available scientific and technical knowledge decreases, and with them, the level of formalization. In a broad sense, the correlation between formal and informal procedures of the design process reflects the correlation between the science and art of design. It is the shaded area that shows the possible level of design process automation. At the same time, it defines the subject area of the theoretical part of special education for aircraft designers and aviation engineers.

2 Scientific Foundations of Computer-Aided Design Methodology

The study of the scientific foundations of the design process, revealing the ways to improve the organization and methods of design, plays an important role in solving the problem of training and retraining specialists in the field of design, specialists able not only to fit quickly enough into the team of qualified designers, possessing a modern high-tech arsenal of methods and means of design, but also to contribute to improving the quality and reducing the time and cost of design. In this connection, let us consider the achieved level and directions of improvement of organization and methods of design work performance within the framework of modern methodologies for designing complex technical systems.

In a broad sense, methodology investigates the structure, logical organization, methods, and means of activity. In our case, methodology is a set of principles, methods, and means for solving scientific, technical, and organizational problems associated with the development of a new aircraft.

As noted above, by the mid-60s of the last century, there was an urgent need and, at the same time, objective prerequisites for further development of the scientific foundations of aircraft design and improvement of design methods on this basis. These prerequisites were caused, first of all, by the emergence and intensive development of electronic computing machines (ECM), the development of applied computational mathematics, and optimization theory. The successes in the development of applied aviation disciplines, which provided opportunities for the numerical solution of various problems in the aerodynamics and strength of aircraft as well as their flight dynamics and efficiency, also played an important role.

The first attempts to introduce computers into design date back to the late fifties of the last century, but practical results were not obtained until the second half of the sixties. During this period, computers with input and output means for graphic information as well as mathematical support systems appeared. It made it possible to work with the computer in an interactive mode, which significantly increased the efficiency of the computer-aided design process.

On the basis of the obtained results, the concept of computer-aided design (CAD) systems was formed, and their practical implementation was carried out.

When the first generation of CAD was created, the task was to combine all available industry programs into an integrated system. This task was largely solved only in the 1970s, when second-generation CAD emerged. They had a well-developed system of control and auxiliary programs, which significantly increased the efficiency of their use.

The creation and ever-wider application of CAD in the development of new aircraft meant the emergence of new automated technologies for design work, the implementation of which promised to improve the quality and the technical and economic level of designed objects, increasing productivity and reducing the time, cost, and labor intensity of design.

Prospects for improving the quality of CAD-designed projects are based on:

- possibilities of improving design methods, in particular on the use of multi-variant design and optimization methods for finding rational options and making decisions;
- increasing the amount of creative work of designers by freeing them from routine work that does not require high qualifications;
- improving the quality of project documentation;
- improving the management of the project development process.

Increased productivity is achieved by:

- significant acceleration of calculation and graphic information processing operations;
- increasing the speed of information transfer between divisions of the project organization and reducing the time of its approval;
- speeding up the preparation of project documentation;
- partial replacement of in-situ experiments and modeling by simulation on analog-digital computer complexes;
- improving the organization of the work of specialists engaged in routine work (calculations, processing of text and graphic information, as well as specialists of information and reference services);
- reducing the volume of testing and development of aircraft prototypes as a result of improved design quality and the level of anticipation of expected design results.

Less obvious advantages of computer-aided design technology can be realized by: combining tasks of related stages (sequentially-parallel design technology); more complete consideration of all significant relationships (application of more complex models); simultaneous consideration of as many design factors as possible in optimizing the appearance of the aircraft; increasing (if necessary) the number of iterations to convergence; and using a single source of information by all divisions.

However, realization of the advantages of new design technology turned out to be a very complex, multifaceted problem whose solution required the development of CAD methodology, the creation of compliant software and hardware complexes, the training of personnel proficient in the new technology, as well as solving a number of organizational issues related to changes in the structure of design organizations and the functions of its individual links.

The theoretical foundations of designing complex technical systems were formed within the framework of systems engineering, a scientific direction that studies the most general patterns of functioning and methods of analysis and synthesis of complex systems. The methodological and mathematical apparatus developed within the framework of systems engineering and related disciplines (systems analysis and operations research) is rightfully considered one of the most important achievements of modern science.

A great contribution to this methodology was made by scientists such as Wiener N., Mesarovich M., Kweid E., and others. In our country, academicians N.N. Moiseev and V.M. Glushkov and scientists of the scientific schools headed by them worked fruitfully on this problem. It was on the basis of this methodology, that it became possible to fully realize the advantages of new (automated) design technology.

Let us briefly consider the main components of the CAD methodology and identify the problems associated with its implementation in relation to aircraft.

The need for a systems approach in aircraft design is dictated by two circumstances. First, the peculiarities of the aircraft as a design object. The second circumstance is that the design process of modern aircraft is implemented within the framework of organizational and technical systems, the complexity of which is higher than the complexity of the designed objects.

Let us list out some general features of a system [4].

Any system is an element of a higher-order system, and the elements of this system, in turn, act as lower-order systems (system communicability).

The properties of a system (as a whole) are not the sum of the properties of its constituent elements; new qualities appear. At the same time, a change in the characteristics of at least one component of the system causes a change in all its other parts and the system as a whole (integrity of systems).

Every system has a threshold, a limit of viability and integrity (the feasibility and potential efficiency of systems). Taking into account this regularity allows avoiding, on the one hand, gigantomania and, on the other hand, primitivism when setting project tasks.

When designing a system, its structure is usually represented in the form of several hierarchical levels that characterize the complex relationships between the elements of the structure. The relationship between the components of the system (subsystems) is represented in the form of set-theoretic descriptions, matrices, graphs, and other structure description languages.

The systemic approach involves the study of the system and its behavior as a whole, as a single object that performs certain functions under specific conditions. At the same time, it is based on the assumption that complex systems and processes can be broken down into relatively independent subsystems (subprocesses) in order to facilitate their study. The essence of the system approach is the ability to see the particular in the general and, conversely, the general in the particular.

In the light of the above, the central problem of implementing the system approach in design is to reasonably divide (decompose) the system into subsystems, the design process into subprocesses (stages), and determine the essential structural and functional relationships between the dissected elements (the definition of "inputs" and "outputs").

The following points should guide the choice of the object of a systematic study:

- the object should have sufficient scale so that the results of its study have a significant effect compared to the consideration of the system elements;
- the object should allow description (at least approximate) of its structure and functioning by mathematical methods;
- the scale of the object under study should take into account the computational capabilities of modern computers.

Let us note, as a positive trend, the enlargement of the objects of system research in aircraft design. It is based on the achievements of applied aeronautical sciences, which are increasingly revealing multifaceted relationships between parameters and characteristics of aircraft systems, as well as on the continuous improvement of computational methods. The concept of shaping an aircraft while taking into account the parameters of the control system and matching aircraft and engine parameters are clear examples of the successful use of this trend in design practice.

Since computer-aided design, by definition, is based on extensive use of mathematical modeling on the computer, the next important aspect of implementing the methodology of computer-aided design is the development of acceptable mathematical models of the systems and processes under study. In accordance with the principles of the system approach, each stage of the design process has its own models that differ in the degree of detail and nomenclature of submodels. At each hierarchical level, the model is a set of submodels reflecting those or other aspects of the structure and functioning of the system and subsystems. To build a model of a system means to describe it quantitatively, using a system of equations linking parameters (arguments) and characteristics (functions) of the system. The totality of these values, united by a single system of equations, to one degree or another reflects the structure, relationships, and processes occurring in the system under study. It should be taken into account that in hierarchical systems the terms "parameter" and "characteristic" are relative because the parameters of the upper level become the characteristics of the lower one. For example, such parameters as an aircraft's payload capacity and its flight range, which are determined in the study of a prospective transport system, at the stage of forming the shape of the aircraft element of this system, are transformed into characteristics.

Model creation begins with identifying the list of variables that can have a noticeable impact on the simulation results. The main problem here is the identification, for each hierarchical level of design, of the minimum required set of variables and relationships between them that would provide a solution to the problem in question, out of the entire set of relationships between the parameters and characteristics of the object under study.

The next stage of constructing a model is to establish relationships between parameters and between parameters and characteristics. At the same time, it is necessary to select the most significant relationships out of the variety of such relationships. If, at the stage under consideration, some parameters cannot be associated with characteristics and other parameters, they are excluded from the formalized description of the model, leaving them for consideration at later stages.

System design is always aimed at finding the optimal parameters of the system in the sense of fulfilling its functional purpose, taking into account its possible multifunctionality and interrelation with other subsystems. On this basis, the important problem of realizing the systems approach is the development of principles and methods for finding the most profitable, optimal parameters of the system and its elements. In this case, it is necessary to be guided by the following basic provisions about the optimality of systems:

In general, a system consisting of optimal elements (subsystems) will not necessarily be optimal; it should be optimized as a whole, as a single object with a given purpose. This does not mean, however, that optimization by parts does not make sense at all.

The system should be optimized by a quantitatively defined criterion reflecting in mathematical form the design goal. The absence of such a criterion indicates that the problem is underdetermined.

Since the system, as a rule, is optimized under quantitatively defined constraints, its optimality is always relative and conditional.

In the light of the above, there is a problem of determining an interrelated set of criteria that would allow at each hierarchical level of system decomposition to objectively assess the design results and to find, using formal methods for each element of a subsystem, such parameters that would ensure high efficiency of the system as a whole. When solving it, it is necessary to be guided by the principle of optimality. It states: "If the objects of elements and subsystems of all levels are optimal in the sense of criteria, corresponding to the higher-level systems, then the whole system is optimal". This means that although there may be a different evaluation criterion for each design stage and each subsystem, all of them must be consistent, meeting the overall design goals.

The problems of criterion selection as applied to the problems of optimal aircraft design are considered in [5, 6].

With a set of models that reflect the real object or process to a certain degree of accuracy, you can begin to solve the design problem.

At the same time, the development of a machine-learning algorithm for its solution is perhaps the most important task in the implementation of computer-aided design technology.

The first results in the development and practical implementation of the new methodology as applied to the design of aerospace systems in our country were obtained when solving "external" design problems. Theoretical foundations and practical methods of aerospace system design based on mathematical modeling of combat and transportation operations using computers were created.

The successes in the development of practical methods for implementing the principles of systems engineering in solving the problems of aircraft design are more modest. This is explained by the fact that, as has been shown, a significant share of the design procedures at this stage are not amenable to formalization, especially when creating fundamentally new aircraft. In addition, the peculiarities of "internal" design problems from a mathematical point of view make it extremely difficult to use the existing mathematical apparatus for solving such problems, especially optimization ones. The fact that aircraft design is a collaborative process and that its successful implementation requires taking into account not only technical but also organizational and psychological factors is also important.

Let us consider the main directions of computer-aided design methodology on the example of a transport aircraft.

3 Structural and Functional Diagram of an Aircraft as a Design Object. Types of Design Models

Taking into account the existing structure of design organizations and the systematization of design objects and tasks, an aircraft as a design object can be represented as an eight-level hierarchical structure (see Fig. 3).

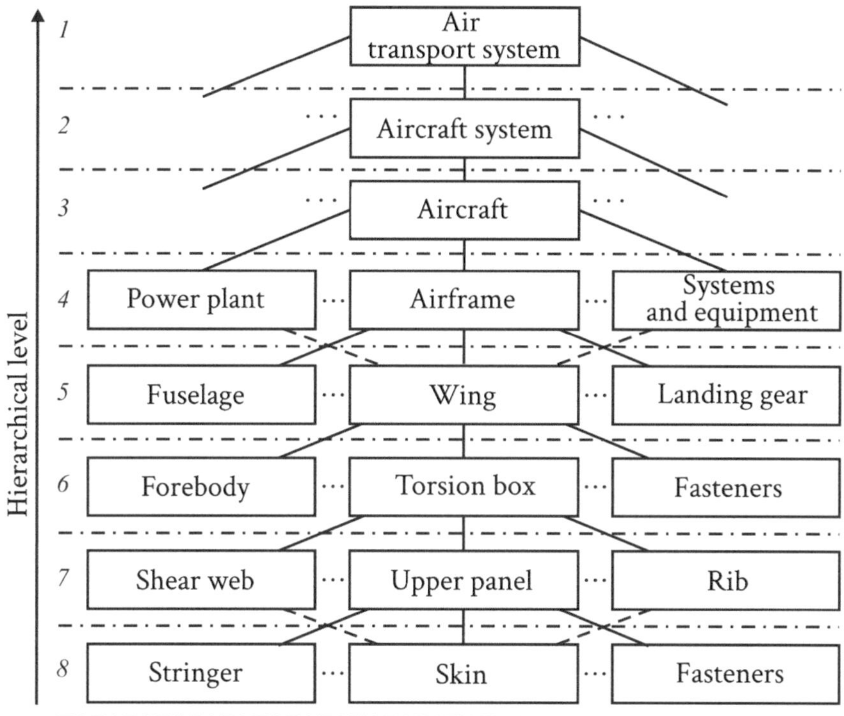

Fig. 3 Fragments of aircraft structural and functional diagram

Hierarchical levels are linked by two types of relationships. One type characterizes the structure of the system, ordering the composition of its elements (blocks, parts, etc.) and the relationship of parts to each other (solid lines). At the same time, every structure is created to perform certain functions. For instance, the wing structure provides functions such as load transfer, creation of lift, placement of fuel, etc. Thus, elements of the system and its individual hierarchical levels are also connected by functional relationships, another type (dashed lines).

Having decomposed the system, we can analyze the relationships between the levels. In doing so, it is necessary to identify all relationships that can have a significant impact on the results of solving the problem at a particular level. Ignoring even one such relationship can lead to results that are very far from what is expected.

In accordance with the principles of the systems approach, when solving the problem at the i-level, it is necessary to consider, at a minimum, the links between the $i + 1$ and $i - 1$ levels. Since we are interested in the aircraft as a design object, let us consider the links at hierarchical levels 1–5, which correspond to the development of the technical requirements specification, conceptual design, and project design.

From the information point of view, these relationships are divided into direct and inverse ones. Direct relationships between the levels of development of the Technical Requirements Specification and the conceptual design are performance characteristics regulated by the Technical Requirements Specification. The direct relationships between conceptual and project design reflect those design decisions that must be made by the time project design begins. These decisions are informally expressed in the form of numerical, graphical, and verbal descriptions of the project that allow evaluating the possibility or degree of fulfillment of the Technical Requirements Specification.

For the level of conceptual design, the inverse links reflect the predicted technical and economic characteristics of promising aircraft, expressed through such generalized indicators as the empty-operational-weight fraction (weight perfection), cruising lift-to-drag ratio (aerodynamic perfection), and specific fuel consumption (perfection of the power plant). The outputs of the project design stage are the aerodynamic characteristics of various aircraft layouts, the characteristics of engines, equipment, and systems, the physical and mechanical properties of structural materials, etc.

The analysis of input and output information allows specifying the problem statement at a certain hierarchical level. The input information determines the initial data for solving the problem, and the output information dictates in what form the answer is to be obtained. Of course, the answer itself, i.e., the specific values of the parameters of the designed object, can be obtained as a result of solving the problem.

Each level of design has its own models, which differ in the number of factors and relationships taken into account in them as well as the nomenclature of submodels, which depends on the nature of the processes studied at the level under consideration and reflects those or other aspects of the structure and functioning of the system and its subsystems.

When developing models, it is necessary to be guided by the following important proposition, which follows from the principles of the systems approach: The efficiency of a system at a higher hierarchical level depends on the generalized characteristics of subsystems at a lower hierarchical level and, if these characteristics are equal, does not depend on how exactly they are provided. This allows us to significantly reduce the dimensionality of the models at the corresponding level.

Let us consider the structure of the aircraft model at the stage of technical proposal development. Its main purpose is to obtain reliable information on the possibility of implementing certain design solutions to meet the set design objectives as well as to assess the comparative effectiveness of these solutions in order to select a rational one. The aircraft model is based on relationships between its parameters and characteristics: flight and technical, operational and production manufacturability, reliability, etc. These relationships in the development of the model are combined into separate blocks, which represent submodels of the aircraft.

The block principle of model construction has a number of advantages. In particular, there is a possibility to develop and improve submodels by specialists competent in a certain field of knowledge. There is a certain flexibility in building computational models for various formulations of the design problem, as well as some advantages in the software implementation of the models.

Let us name the main submodels. One of the most important submodels is the geometric model. Its importance is determined by the following:

- it is used to form a visual image of the designed object;
- for an airplane, its external shape and dimensions largely determine its performance.

The geometric model describes the relationship between the aircraft parameters and its shape and dimensions. It is used to define the shapes, areas, volumes, and cross-sections of the aircraft and its assemblies.

Data from this model is used for mass, aerodynamic, and strength calculations, aircraft layout, graphical representation of design results, as well as the development of tooling and computer codes.

The weight model is used to estimate the aircraft mass and its components in accordance with a weight specification, the degree of detail of which is determined by the stage of design development. It is based on the relationships between the geometry of the aircraft, the loads acting on it, the features of the design and structural layout, the payload, the composition and placement of equipment and gear, on the one hand, and the mass of the aircraft and its components, on the other.

The aerodynamic model is used to calculate the aerodynamic characteristics of the aircraft in flight and in takeoff and landing configurations. It is based on the relationships between the shape and size parameters of the aircraft, as well as between the flight mode and the nature of the forces and moments acting on the aircraft (their magnitude and variation).

The propulsion system model is used to calculate the geometric, mass, and altitude-velocity characteristics of engines. This model is based on the relationships between geometric and gasdynamic parameters of different types of engines, their mass, thrust, and fuel consumption in various flight modes.

The weight and aerodynamic models, as well as the power plant model, are used to predict the loads acting on the aircraft. This makes it possible to solve important tasks such as determining the general indicators of the aircraft's maneuverability, flight trajectories, and field performance, as well as analyzing its stability and controllability. These tasks are solved by a set of dynamic models describing the relations between the above-mentioned characteristics and the forces and moments acting on the aircraft.

When forming the aircraft configuration, the center of gravity model plays an important role in ensuring the mutual spatial relationship of the main components of the aircraft to meet the conflicting requirements of aerodynamics and durability, stability and controllability, serviceability and manufacturability, etc.

Various indicators of their technical and economic perfection are used to assess variants of design and engineering solutions. Efficiency models are used for their calculation. They allow calculating the performance indicators of an aircraft (or its individual subsystems) depending on its parameters. Thus, economic models of transport aircraft reflect the links between the transport performance of the aircraft, determined by the level of its technical perfection, and the costs of its design, manufacture, and operation.

An enlarged block diagram of an airplane model is shown in Fig. 4.

The blocks shown here are the submodels described above. Each block, in turn, consists of a number of analysis modules. The partitioning of a block into modules is determined by the list of tasks for the stage under consideration.

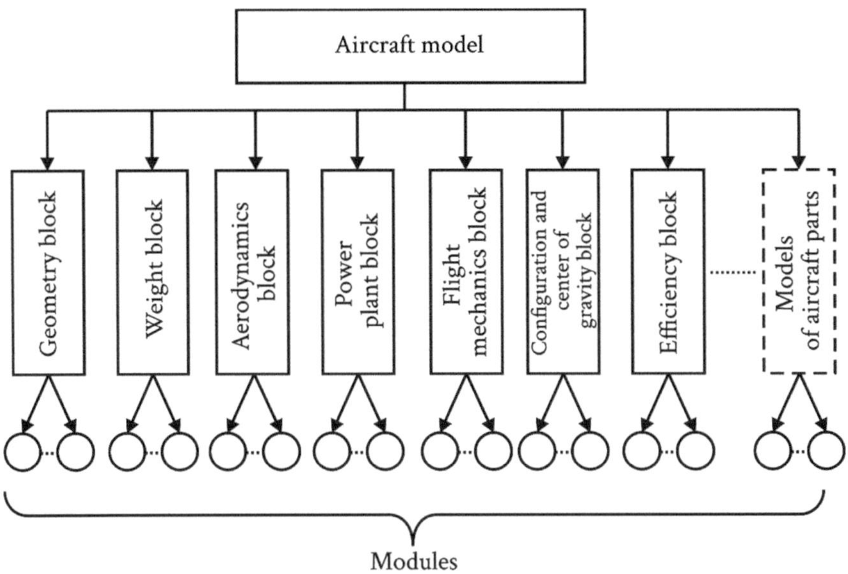

Fig. 4 High-level structural diagram of an aircraft model

Depending on the project development stage and the type of aircraft being designed, the composition and structure of the model may vary. In particular, during the formation of the aircraft design, in addition to the above models constructed according to the interdisciplinary principle, the so-called parts models describing individual structural components of the aircraft (equipment units, payload, etc.) are used.

The mathematical description of aircraft submodels is based on various methods of weight and aerodynamic analysis, strength calculations and analyses of stability and controllability, and power plant performance analysis.

The critical factors determining the choice of one or another method for the model are, on the one hand, the availability of input information, and, on the other hand, the expedient dimensionality of the model. Since the information about the designed object expands as the design elaboration is deepened and detailed, aircraft design models are developed and refined both at different design stages and within each stage.

4 Types of Design Problems and Structure of Algorithms for Their Solution. Optimization in Design

The construction of formalized procedures, which implement the process of aircraft design within the framework of computer-aided design technology, is a very complex problem. The complexity of its solution is determined both by the variety of design problems actually encountered in practice, each of which is characterized by its initial data, required variables, constraints, and criteria, and by the fact that design, as noted, is a creative process.

In this regard, it is necessary to systematize design problems, analyze the algorithms of their solutions, determine the invariant fragments in them, and determine the possibility of applying mathematical apparatus for their solution, as well as ways of combining formal and informal procedures.

Let us consider these issues in relation to the conceptual design stage, which is key in the aircraft design process. As already noted, its purpose is to determine the feasibility and ways to implement an aircraft with the technical and economic characteristics specified in the requirements specification. In the process of solving a number of tasks, input information about the design object is transformed into output information. These tasks can be conditionally divided into two classes, differing in the degree of freedom to vary the design variables.

The first class includes design problems "based on the technical requirements specifications," in which the majority of design variables, except for the parameters and characteristics stipulated in the technical requirements specifications, are subject to selection and optimization in the process of design. Let us list the main ones:

(a) selection of the reference aircraft and its scaling, i.e., determination of the basic design parameters, layout, and center of gravity of the aircraft meeting the requirements;
(b) analysis of the project's sensitivity to changes in design factors, i.e., assessment of the impact of different design options (schematic, parametric) on the degree of satisfaction of the technical requirements specifications and the technical and economic characteristics of the aircraft;
(c) optimization of aircraft design parameters according to the selected criteria, taking into account the limitations imposed by the level of performance specified in the Technical Requirements Specifications as well as the requirements of regulatory documents;
(d) prediction of performance and technical–economic characteristics of the aircraft with the selected parameters.

The second class includes tasks associated with the design of various versions (modifications) of the aircraft based on the existing basic one.

They include tasks of developing modifications associated with the replacement of individual units and systems (engine, wing, equipment), the introduction of new materials, and the improvement of the aerodynamics of aircraft parts. A special group consists of design modification tasks for different variants of its application with maximum use of parts, systems, and compartments of the base variant (the principle of modularity in design). In such tasks, the designer's freedom in the selection of design variables is considerably restricted by the previously adopted solutions. In nomenclature, the tasks of this class are similar to tasks (a) and (d).

Let's analyze the mentioned tasks and define the structure of their solution algorithms.

Let us denote the vector of parameters uniquely describing the structure, shape and dimensions of the designed object by $X = (x_1, x_2,..., x_n)$.

We denote the characteristics of the object by the vector $Y = (y_1, y_2,..., y_n)$.

Characteristics are a function of parameters and, in general, other characteristics.

The relationships between the parameters and characteristics represent the basis of the object model. The model can be used to solve the following three types of problems, which differ in the structure of their algorithms:

1. Parametric analysis is associated with determining the characteristics of the designed object Y at the selected parameters $Y = (X)$. Examples are the prediction of aircraft performance and technical–economic characteristics, the analysis of the mass and strength of the aircraft and its parts, etc., i.e., tasks (b) and (d). These problems are solved on the basis of linear and branching algorithms.
2. Parametric synthesis is related to the definition of design parameters, in which the values of characteristics satisfy the predetermined conditions (constraints), i.e., are in the feasible area Y_{allow}:

$$Y = \Phi'(X) \tag{1}$$

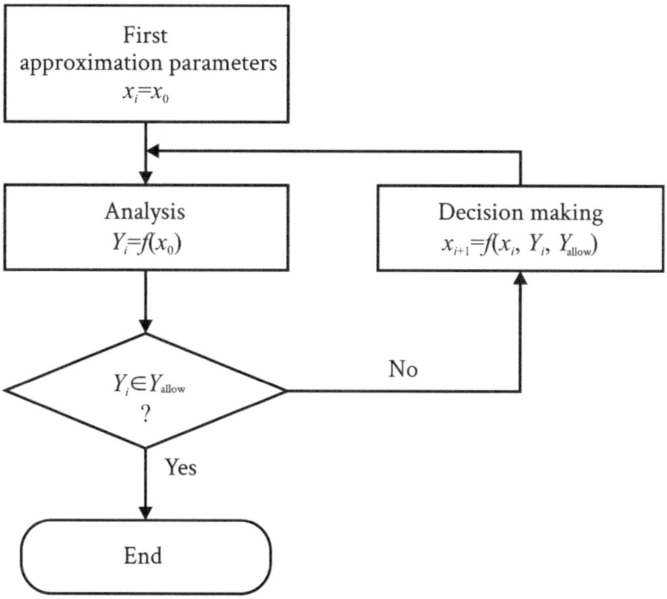

Fig. 5 A generalized algorithm for solving design problems

This type includes a wide class of problems associated with determining the rational structure and parameters of the aircraft and its parts, i.e., problems (a).

The solution to these two types of problems can be implemented on the basis of a generalized algorithm, the scheme of which is presented in Fig. 5.

It can be seen that synthesis includes analysis.

3. Optimization problems arise when, when solving a structural-parametric synthesis problem, the goal is to reach the extremum of the target function, which is the criterion of project optimality. This category includes problems of type (c).

   Their solution requires the quantitative expression of an optimality criterion or measure of merit F depending on parameters and, in general, on characteristics, and then the extremum of this function is searched for in the area of allowable changes of design parameters X_{allow}:

$$extrF(X, Y(X)), X \in X \in X_{allow} \tag{2}$$

The general scheme of solving optimization problems is presented in Fig. 6.

The role of optimization in design and methods for solving optimization problems are discussed in Chap. 9.

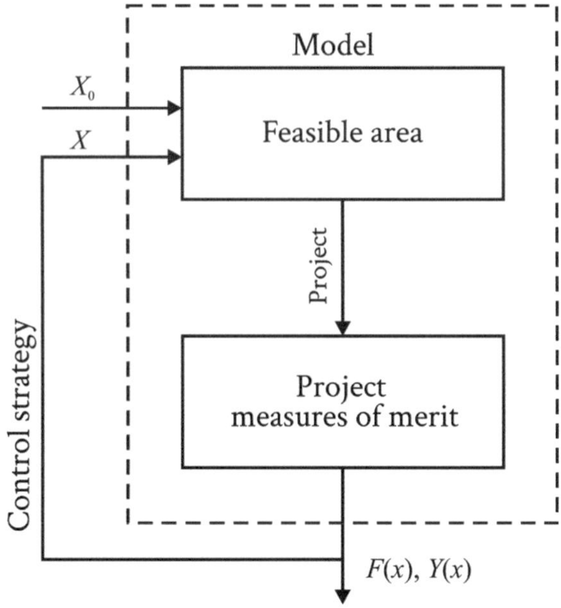

Fig. 6 Scheme of solving optimization problems

5 Integrated Automation of Design Work

The modern stage of aircraft engineering is characterized by the widespread introduction of powerful integrated software and hardware systems for the automation of engineering calculations (CAE, or computer-aided engineering), computer-aided manufacturing (CAM), as well as computer-aided design (CAD, or computer-aided design). Among them, CATIA and Unigraphics are best known. Unigraphics is a multimodule system of geometric modeling and engineering design, strength calculations, preparation of design documentation, and technological production preparation. Its CAD subsystem has tools for solid-state design, spline-based modeling, the creation of drawings on the basis of 3D models, and the design of assemblies. The CAM subsystem develops control programs for various technological processes realized on CNC production centers. The CAE subsystem solves problems of strength analysis using the finite element method, kinematic and dynamic analysis of mechanisms, etc.

Nevertheless, the problem of the development and implementation in practice of complex systems of design work automation in aircraft design is still relevant.

The experience of developing a number of such domestic systems showed that the reason for their insufficient efficiency was the neglect of organizational problems during their development. They were created as an additional structural subdivision of a design bureau rather than as its organic part, penetrating all subdivisions involved in project development. This caused an insurmountable psychological barrier to their introduction, although within the framework of research institutes and universities, they worked quite effectively.

The principles of the systems approach require the construction of such a CAD as a hierarchical system consisting of separate subsystems that perform the functions of structural subdivisions of the design organization. These subsystems should be based on a single or compatible hardware, information, and software base, which will ensure their operation both in stand-alone mode and within the whole system functioning under the control of the Chief Designer and Project Manager. Implementation of the technology of distributed parallel calculations on a computer network allows effective teamwork of design bureau subdivisions within the framework of such systems to achieve a single goal. At the same time, decision makers at each hierarchical level are able to evaluate the impact of these decisions on other interrelated subsystems, and the project manager, using generalized information, can conduct a directed search for optimal aircraft parameters.

A fragment of the structure of such a system is shown in Fig. 7.

Informationally, it is connected to other subsystems implementing the aircraft development process: ASTPP (automated system of technological production preparation), APMS (automated project management system), etc., which corresponds to the modern concept of information support of all stages of the product life cycle within the CALS technologies.

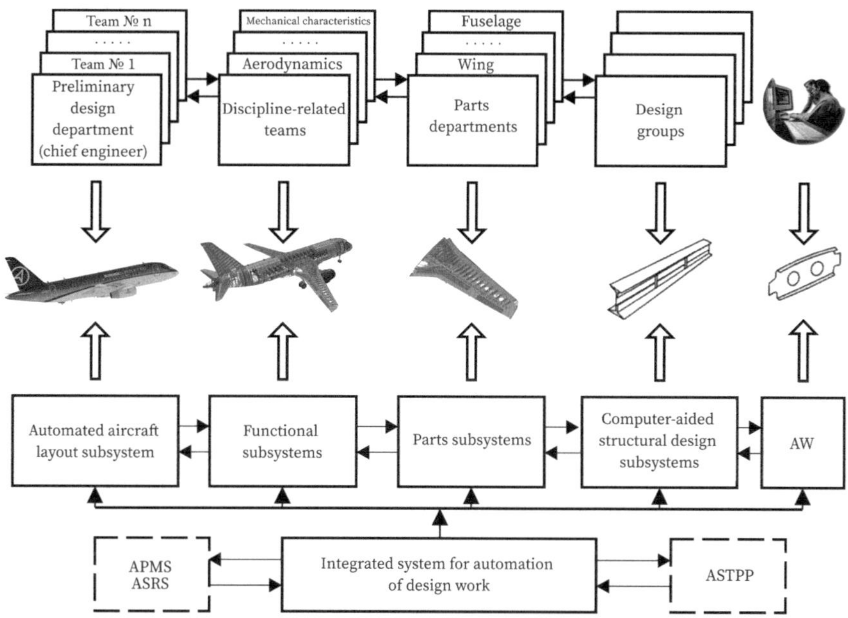

Fig. 7 Enlarged structure of an integrated system for automation of design and development work. AW = automated workplace; APMS = automated project management system; ASRS = automated scientific research system; ASTPP = automated system of technological production preparation

In conclusion, let us dwell on the directions and prospects of development of scientific foundations of design. Undoubtedly, research aimed at expanding the field of formalized procedures of the design process is relevant. There are two ways to achieve this goal. The first one is to create new methods of describing difficult to formalize tasks, for example, on the basis of information-semantic models and expert systems reflecting the experience of solving such tasks by qualified specialists.

The second way is more traditional. It is associated with the expansion of the knowledge in the field of aeronautical engineering through research and development of unconventional aircraft layouts, expansion of the range of speeds and flight altitudes, the use of alternative energy sources and ways to create lift, improvement of the hardware based on the achievements of nanotechnology, etc.

References

1. Dixon JR (1966) Design engineering: inventiveness, analysis and decision making. McGraw-Hill, New York
2. Dvoryankin AM, Polovinkin AI, Sobolev AN (1977) Metody sinteza tekhnicheskih reshenij (Methods of synthesis of technical solutions) Nauka, Moscow
3. Eger SM, Liseitsev NK, Samoilovich OS (1986) Osnovy avtomatizirovannogo proektirovaniya samoletov (Fundamentals of Computer Aided Aircraft Design). Mashinostroenie, Moscow
4. Quade ES (1964) Analysis for military decisions. Rand McNally and Co., Chicago
5. Vyazgin VA (1983) On the synthesis of quality criteria for the problem of aircraft parameter optimization. In: Methods for selecting rational design solutions in the process of creating an aircraft. MAI, Moscow
6. Krasnoshchekov PS, Morozov VV, Fedorov VV (1979) Decomposition in design problems. Izvestiia Akademii nauk SSSR. Tekhnicheskaia kibernetika 2:7–17

Design Input Data. Conditions and Relationships Defining the Feasibility of a Project

Nikolai K. Liseitsev

Abstract The chapter presents a detailed categorization and explanation of the input data required to launch the aircraft design process. In addition, the chapter offers the reader an algorithm for sizing and laying out an aircraft at the conceptual design stage. Finally, in this chapter, the reader will find an example of a three-view drawing of a commercial airplane.

Keywords Conceptual design · Input data · Aircraft sizing · Aircraft layout

1 Introductory Remarks

Having studied the first two chapters of this textbook, the reader already has a fairly complete understanding of the goals, objectives, and organization of the design process, as well as the theoretical and methodological foundations on which the solution of the entire range of design problems is based. The content of the following chapters of the textbook is the presentation of practical methods, techniques, and recommendations for solving these tasks in relation to a wide class of aircraft with a degree of detailing that enables the student to independently develop a preliminary aircraft design. Their sequence mainly corresponds to the technological chain of the aircraft design process.

This chapter outlines the general issues of objective selection, preparation of input data, and analyzes various kinds of constraints affecting the choice of aircraft design, parameters, and flight modes.

The need to develop a new aircraft, and hence design it in the most general case, arises for two reasons.

The first reason is the emergence of socially conscious tasks related to the scope of aviation, whose solution by the existing types of aircraft is either impossible or economically unreasonable. For example, it may turn out that the existing types of aircraft do not meet the newly emerging requirements in terms of dimensions and/

N. K. Liseitsev (✉)
Moscow Aviation Institute (National Research University), Moscow, Russia
e-mail: nikolai.liseitsev@yandex.ru

M. A. Pogosyan (ed.), *Aircraft Design*, Springer Aerospace Technology,
https://doi.org/10.1007/978-981-96-4599-2_3

or mass of transported cargo, range of transportation, operating conditions, etc. The same type of prerequisites include the need to find ways to ensure the country's defense capacity.

The second reason is that, on the one hand, existing aircraft are gradually becoming obsolete, while on the other hand, as a result of scientific and technological progress, new opportunities arise, the implementation of which allows to offer the market aircraft that have the required competitive performance indicators.

It is clear that the success of the design is largely determined by the extent to which the project developers were able to forecast the future and, on this basis, correctly define the goals and set design objectives. The solution of this range of problems, as it has been shown, is an important part of the "external" design, which belongs to the competence of special organizations and departments of the Design Bureau.

Nevertheless, the chief designer and graduate student defending their project should clearly answer the questions: why did they design their aircraft, how is this aircraft better than the existing ones, and then, how did they achieve this?

To answer the first question, it is necessary to have an understanding of global and domestic experience in the development and operation of aircraft of the type and purpose of interest, the achieved level of their technical, economic, and flight performance characteristics, the prospects and trends in aviation and related branches of science and technology, as well as the requirements that potential customers have for future aircraft types. On the basis of this information, design goals and objectives are formed, which are presented in the form of a Technical Requirements Specification (TRS) (see below). The degree of mastery of the author of the project of the basic provisions of the discipline "Aircraft Design" determines the ability to propose and justify effective ways to achieve the set design objectives.

2 Preparation of Design Input Data

The initial phase of the design process is the preparation of baseline data, which largely determines and streamlines the design process.

They typically include the following groups of data:

- a design specification that contains formalized data defining the design objectives and the basic constraints within which it is carried out;
- systematized data reflecting the global level of achievements in the design of aircraft of a particular class;
- the list, characteristics, and areas of application of new technical solutions from science and technology that can be used to improve the performance of newly designed aircraft.

The structure and content of the TRS are set by regulatory documents. Let us list the sections usually included in this document.

The "Mission" section specifies the class to which the projected aircraft belongs and its intended use.

The "General Provisions" section includes a list of regulatory documents, the implementation of which is mandatory for the aviation industry in the construction of new aircraft. Such documents include the Standards for Airworthiness of Civil Aircraft, the Standards for Aviation Noise and Environmental Pollution, as well as general technical requirements regulated by industry standards and restrictions, such as "General technical requirements to passenger cabin layouts and component passenger, household, and auxiliary equipment", "Standards for establishing landing minimums for civil aviation aircraft", etc. [1, 2].

Similar documents exist for military aircraft as well (Regulation on Military Aircraft Engineering, State Standards, etc.).

The "Flight Specifications" section contains a list of required output characteristics of the aircraft, such as maximum speed, range, service ceiling, payload and nomenclature of transported cargo, takeoff and maneuvering characteristics, etc. The values of the required aircraft performance characteristics are a determining factor when selecting the design parameters of an aircraft.

The section "Characteristics of the technical level of serial aircraft and its cost performance" includes generalized performance indicators of the aircraft (transport, combat, fuel, etc.) to assess its competitiveness. It also includes an indicative aircraft price. The information in this section makes it possible to establish the criteria for project optimization.

The section "Further Development of the Aircraft and its Main Systems" contains information on the directions of development of the designed aircraft and options for its use. The significance of this section is particularly important in the context of the modern concept of developing a family of aircraft based on the base variant.

The section "Aircraft Structures, Systems, and Equipment" contains requirements for the service life of the airframe and its primary structural members in flight hours and landings, the total service life in years (total and to first repair), and the durability of components and equipment.

The section "Power Plant and its Systems" contains the main characteristics of the engine used or developed (modified) for the designed aircraft, including the long-term performance, noise, and emission levels, as well as the requirements for systems and assemblies that support the engine. A special technical specification is developed for newly created engines.

The "Special Equipment" section includes a list of the main tasks to be solved by such equipment on the designed aircraft (flight navigation support, automatic flight control, aircraft communication with ground air traffic control systems, intra-aircraft communication, power supply of special equipment and onboard systems, as well as equipment for flight data collection and proccssing).

The section "Requirements for Reliability, Maintenance, and Repair System" contains quantitative values of probability of occurrence of special situations in flight, levels of reliability indicators of the aircraft fleet in operation, indicators of operational manufacturability, probability of timely departure, as well as the maintenance and repair program. The implementation of this section of requirements largely determines the level of aircraft competitiveness.

The section "Level of Standardization and Unification" sets the degree of unification of engines, components, and equipment.

The information contained in the TRS not only imposes significant restrictions on the choice of design solutions, but often determines some of them. For example, requirements for safety in emergency situations and passenger comfort largely determine the parameters of passenger compartment layout.

A simplified version of the design specification for a passenger airplane as part of the diploma project is given in Appendix B.

For military aircraft, the content of the TRS sections is determined by the project developer depending on the specifics of the aircraft system being developed (upgraded), its application, and its operation conditions. In general, the TRS shall include the following sections:

- the name of the project and the reason for its implementation;
- the name of the aircraft system, and its purpose;
- the composition of the aircraft system;
- operational requirements for the aircraft system;
- technical and economic requirements;
- requirements for ground support;
- requirements for raw materials and component parts for cross-industry applications;
- requirements for preservation, packaging, and labeling;
- special requirements;
- requirements to ensure state secrecy during project implementation,
- project milestones;
- the procedure for implementing and accepting the project stages.

The second group of data includes selected and systematized data on prototype aircraft. More than a century of practical aircraft manufacturing experience, appropriately systematized and processed, serves as a valuable source of information for the designer in finding and making decisions in the process of designing a new aircraft. This information is especially important at the initial design stages, since it makes it possible to reasonably narrow the field of search when choosing an aircraft layout, type and number of engines, and a range of design parameters. In the framework of computer-aided design technology, such information is systematized, processed, stored, and replenished in the CAD data bank.

Prototype aircraft include aircraft of similar purpose whose performance is as close as possible to that of the designed aircraft. The development time of the aircraft, i.e. its generation, which determines the level of scientific and technological achievements used in its creation, is taken into account.

For each of the prototype aircraft, information on the aircraft's layout and design, its geometry, mass, and aerodynamic characteristics, power plant, on-board equipment and systems, as well as its performance characteristics, is selected, processed, and systematized.

An example of the design input data is presented in Table 1.

Table 1 Specifications of prototype aircraft

Aircraft properties	Units	Prototype aircraft		
General				
Aircraft name				
Country of manufacture				
Mission				
Year of serial production (or certification)				
Mass specifications				
Take-off mass	kg			
Operational empty mass	kg			
Fuel mass	kg			
Payload mass (maximum)	kg			
Landing mass (maximum)	kg			
Geometric characteristics				
Wing				
Area	m^2			
Aspect ratio				
Taper ratio				
Sweep angle	°			
Aileron-to-wing area ratio				
Flaps (slats) area ratio				
Fuselage				
Length	m			
Width	m			
Height	m			
Midsection area	m^2			
Wetted area	m^2			
Payload compartment dimensions	m × m × m			
Horizontal tail				
Area	m^2			
Aspect ratio				
Taper ratio				
Sweep angle	°			
Moment arm	m			
Volume coefficient				
Elevator-to-horizontal tail area ratio				
Vertical tail				
Area	m^2			
Aspect ratio				

(continued)

Table 1 (continued)

Aircraft properties	Units	Prototype aircraft		
Taper ratio				
Sweep angle	°			
Moment arm	m			
Volume coefficient				
Rudder-to-vertical tail area ratio				
Landing gear				
Number of main supports				
Track	m			
Base	m			
Tyre pressure	Pa			
Tail clearance angle	°			
Propulsion				
Engine name				
Country of manufacture				
Motor type				
Bypass ratio				
Number of engines on the aircraft				
Take-off thrust	daN			
Take-off specific fuel consumption	kg/daN × h			
Cruising specific fuel consumption	kg/daN × h			
Motor mass	kg			
Flight performance				
Maximum speed/at altitude	(km/h)/m/N			
At maximum payload				
– Range	km			
– Speed	km/h			
– Altitude	m			
At maximum fuel reserve				
– Range	km			
– Speed	km/h			
– Altitude	m			
Safe take-off speed (lift-off speed)	km/h			
Landing approach speed (landing speed)	km/h			
Take-off distance (take-off ground roll)	m			
Landing distance (landing ground roll)	m			
General characteristics				
Take-off wing loading	daN/m^2			

(continued)

Table 1 (continued)

Aircraft properties	Units	Prototype aircraft		
Take-off thrust loading (power-to-mass)	– (kWh/kg)			
Fuel efficiency	g/passenger/km (g/t.km)			
Aerodynamic characteristics				
Cruise lift-to-drag ratio				
Take-off (landing) lift-to-drag ratio				
Cruise lift coefficient				
Take-off (landing) lift coefficient				

The success of creating a new aircraft is largely determined by the level of application in its design of new progressive scientific and technological solutions (innovations) accumulated by the time of aircraft development in various branches of science and technology. Such innovations may include new structural materials, new methods of manufacturing and joining structural elements, various improvements aimed at enhancing aerodynamic performance, new engines with improved specific parameters, more advanced onboard equipment and systems, etc. Information on these new solutions and technologies is included in the third group of input data.

In the actual design, the input data can also be:

- of the chief designer's ideas;
- industry research institute recommendations;
- limitations on operating conditions and on the specifics of the aircraft application.

Atmosphere. The International Standard Atmosphere (ISA) gives the air temperature (t_H), pressure (p_H), specific mass (γ), density (ρ), the ratio of air density at a given altitude to that at sea level (Δ), the speed of sound (a) and other physical characteristics of air as a function of height above ocean level (see Annex C).

Three characteristic atmospheric conditions are commonly used in aircraft design (see Table 2):

Table 2 Basic atmospheric conditions

Atmospheric parameters	Designation	Units	Normal conditions	Design conditions	Stratospheric conditions
Height	H	m	0	0	11000
Temperature	t_H	°C	+15	+30	−56.5
Pressure	p_H	mm Hg	760	730	–
	p_H	GPa	1013.25	973.35	–
Air density	ρ	kg/m^3	1.225	1.120	–
Speed of sound	a	m/s	340.294	350	295,069

- normal conditions (NC);
- design conditions (DC), which are used to estimate an aircraft's field performance;
- stratospheric conditions.

Climatic conditions (limit), see Table 3.

The designer must take into account that:

- in tropical conditions, non-metallic materials are affected by mould and insects (termites, maggots), which should be particularly taken into account when insulating electrical wiring, as damage to this insulation leads to short circuits and failure of equipment and controls;
- in tropical conditions, a humidity of 10% is specified at high outdoor temperatures (+40 °C). This defines the design conditions for all cooling systems.

Wind. To calculate field characteristics, the cross-wind speed is assumed to be 15 m/s.

Turbulent flight loads are calculated at equivalent vertical wind gust speeds. There are design constraints that are imposed by the design office itself. An example of such limitations is the choice of design conditions, mainly the design dynamic pressure and the design Mach number, which largely determine the appearance of the aircraft.

Let us consider an example of the selection of design conditions for a long-haul passenger airplane with a maximum airspeed of 950 km/h. Figure 1 shows the variation of possible flight speeds depending on the flight altitude for a given power plant and selected aerodynamic layout. In this case, the maximum airspeed of flight (significantly higher than the given one) is obtained at sea level.

However, the flight altitude of a long-haul passenger aircraft is mostly greater than 8000 m. The aircraft flies at lower altitudes either during the climb or during the descent. Therefore, it is advisable to introduce a limitation on the flight speed at altitudes less than 8000 m. To do this, we take *A* as a design point. In this case, the flight speed in the region of aircraft normal operation altitudes (8000–12000 m) is equal to or greater than the set speed.

Table 3 Basic data on climatic conditions

Climate	Properties	Temperature, °C		Relative humidity
		max	min	
Desert	Dry heat, intense solar radiation, sandy dust	Air up to +60, ground up to +75	At night, down to −10	5
Tropical	Damp heat, high relative humidity, mould growth, destructive effect of insects	Air up to +40, ground up to +70	At night, up to + 25	100
Marine	Wet salt dust	+50	−40	100
Arctic	Low temperature, icy dust	−25	Down to −70	–
Great heights	Low temperature, low pressure, water vapour condensation due to rapid temperature changes	–	Down to −90	–

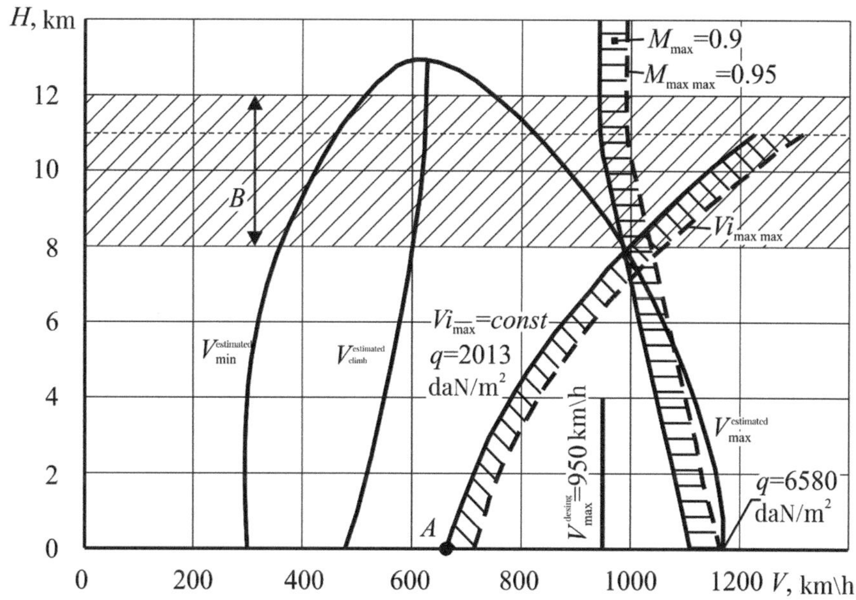

Fig. 1 Example of selection of design conditions for a passenger aircraft: *B* denotes the altitude of normal passenger aircraft operation

At point *A*, $q = \frac{\rho_H V^2}{2} \cong 2013\,\text{daN/m}^2$, which corresponds to the equivalent flight speed $V_i = 655\,\text{km/h}$.

The airworthiness standards set the values of $V_{\max\max}$ and $M_{\max\max}$:

$$V_{i\ \max\ \max} \approx V_{i\ \max} + 50;$$

$$\text{M}_{\max\ \max} \approx \text{M}_{\max} + 0,05.$$

In the example under consideration, $V_{i\max\max} \approx 700\,\text{km/h}$ and $M_{\max\max} = 0.95$.

A similar approach to selecting design conditions, taking into account the specifics of aircraft operation, can be performed when establishing initial data for the design of any aircraft.

Another example of limitations imposed by the design bureau itself is the choice of the maximum design landing mass of the aircraft, determined by considerations of safe landing of the aircraft.

3 Logical Procedure for Aircraft Conceptual Design. The Main Constraints Defining a Feasible Project

Using the initial data, the designer has the opportunity to begin solving the main task of general design: the formation of the shape of the future aircraft. This task is formulated as follows: find the parameters characterizing the shape, structure, and dimensions of the aircraft that would ensure that the requirements and constraints imposed on the future aircraft are met (an acceptable design) and that the extremum of the objective function, the design measure of merit, is achieved (an optimal design).

By parameters, we mean design variables (continuous, discrete), the choice of values of which allows us to unambiguously describe the model of the future aircraft and, later, in the production process, to materialize this model.

An enlarged diagram of the iterative process of aircraft design is presented in Fig. 2.

This process starts with the selection of the aircraft layout and the type and number of engines, which largely determines the appearance of the aircraft. These issues are addressed in Chap. 4. The scaling of the appearance is carried out in the process of determining the dimensional parameters of the aircraft, which allow drawing the general view of the aircraft (see Chap. 5). An example of such a drawing made as part of the graduation project for a passenger aircraft is shown in Fig. 3.

An important role in shaping an aircraft is played by the issues of ensuring a rational aircraft interior, i.e., its internal arrangement. The results of this process are formalized in the form of layout drawings. A fragment of such drawings of the aircraft is shown in Fig. 4.

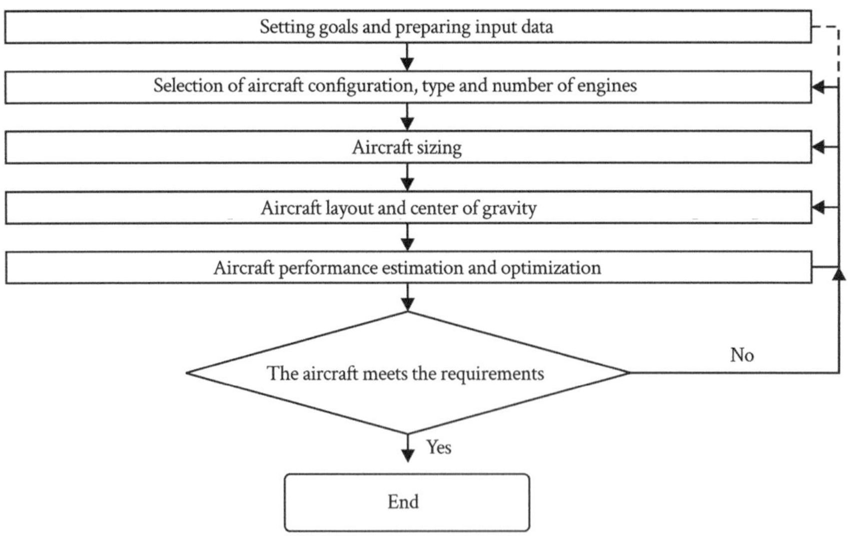

Fig. 2 The main stages of the aircraft design process

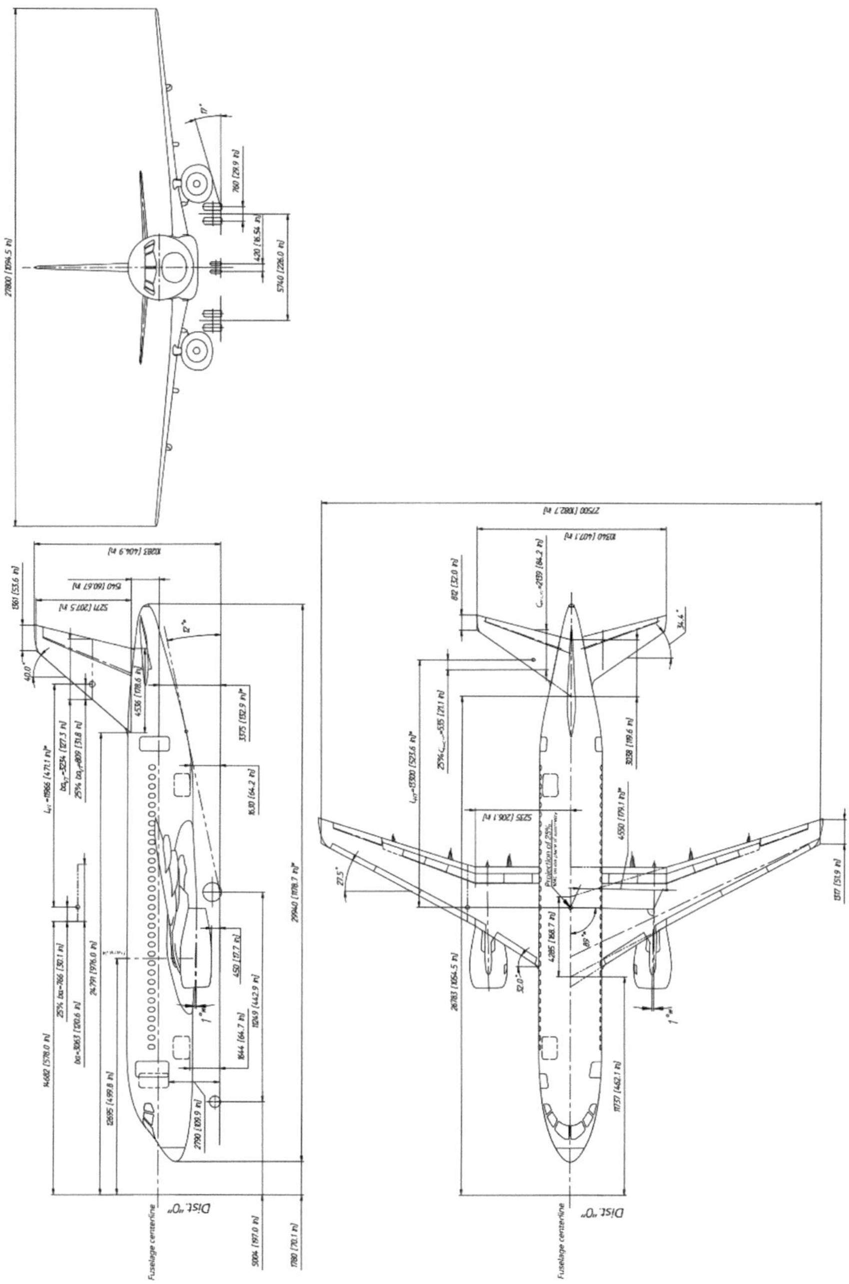

Fig. 3 Three-view drawing of a commercial airplane

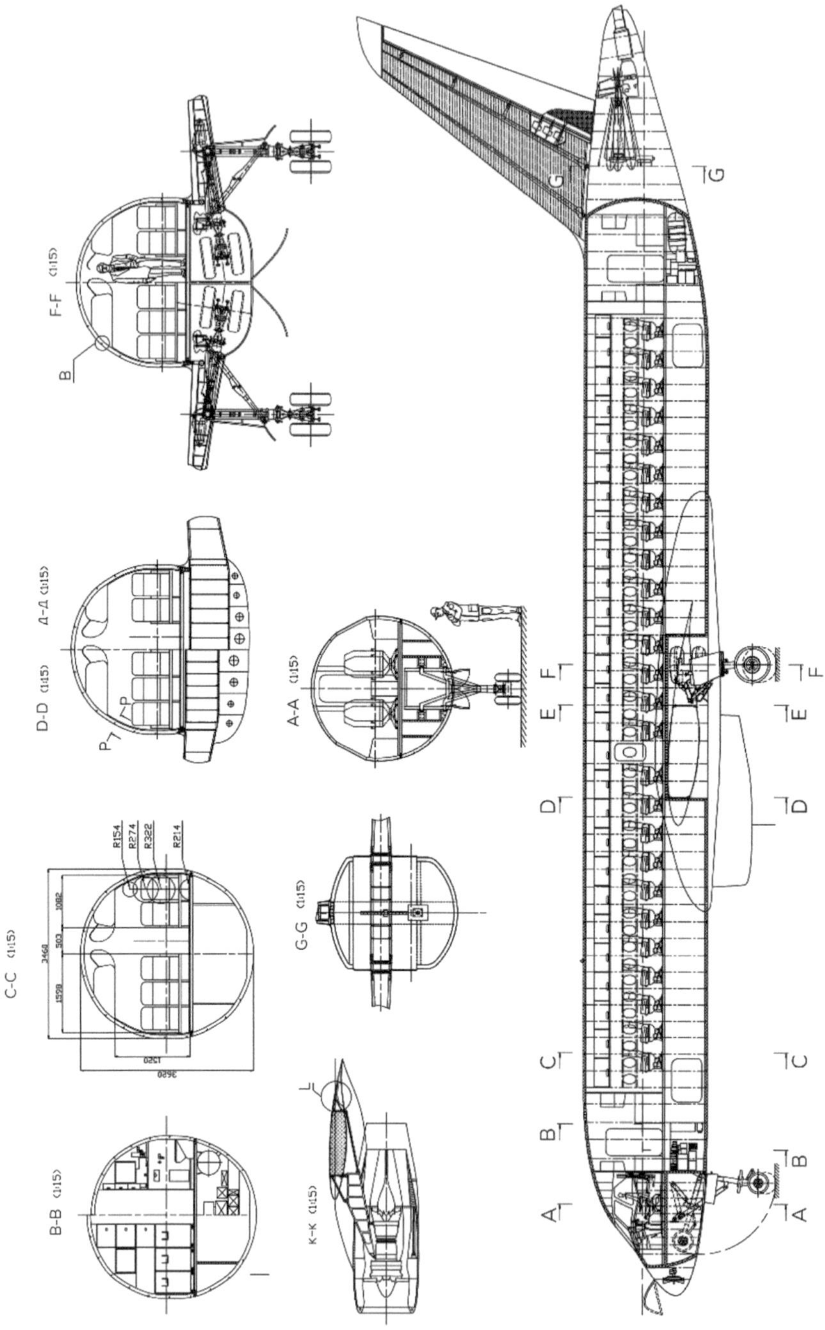

Fig. 4 Fragment of a passenger aircraft layout

Note that the final three-view drawing is produced only after all layout and balance problems have been solved.

Both permissible and optimal designs are identified within a feasibility region, the boundaries of which are formed by different conditions and relations (design constraints).

Three groups of such restrictions can be distinguished. The first group includes limitations common to any aircraft, regardless of its purpose. These are constraints dictated by the physical feasibility of the aircraft as an aerodynamic flight vehicle. They are described by mathematical relationships between the aircraft's parameters and characteristics derived from the laws of flight mechanics. Fulfillment of these constraints ensures sustainable flight of the aircraft, and, in this sense, they can be called the equations of the existence of the aircraft, regardless of whether it is good or bad.

The main of these equations are written in the following form:

$$\sum m_i - 1 = 0,$$

$$n_z mg - L = 0,$$

$$n_x mg + D - T = 0,$$

$$x_{aft} - x_{AC} + C_{m,\min}^{C_L} = 0,$$

$$C_m = 0.$$

Here:

m_i are the relative masses of the aircraft components,

m is the aircraft mass,

L, D are the aircraft lift and drag,

T is the engine thrust,

x_{aft} is the aircraft aft center of gravity limit,

x_{AC} is the X-coordinate of the aircraft's aerodynamic center,

C_m, $C_{m,\min}^{C_L}$ are the aircraft's pitching moment coefficient and its derivative with respect to the lift coefficient.

The first of these equations is the mass balance equation, the second is the weight balance equation, and the third is the energy balance equation. The fourth and fifth equations represent the conditions for ensuring the aircraft's longitudinal static stability and trim.

The way in which the constraints arising from these equations are taken into account depends on the formulation of the design problem and the available input data.

The second group of constraints derives from the need to meet a number of the previously listed requirements.

The third group of constraints determining the choice of design parameters includes the characteristics of the designed aircraft, specific to each newly developed aircraft.

However, these data are usually insufficient to unambiguously determine the aircraft parameters and estimate its performance. Some additional data required for passenger aircraft design are presented in Table 4.

Restrictions can also be imposed on aircraft parameters such as its overall dimensions, take-off mass, etc. Such restrictions are typical for heavy transport aircraft, which operate from existing airfields, and for military ship-based aircraft.

The listed conditions and relationships, which largely determine the project, are often insufficient to unambiguously define its feasibility boundaries. The satisfaction of requirements for placement of payload, equipment, and systems, taking into account their interaction, ensuring operational and production manufacturability, survivability, etc., affect not only the airframe but also its shape and dimensions. This determines the key role of the hard-to-formalize layout process in solving the problems of aircraft conceptual design.

Note that, depending on the formulation of the design problem, certain parameters may pass into the category of constraints, and vice versa, constraints may pass into parameters. Nevertheless, categorizing all variables is a necessary step in formalizing the design problem and developing an algorithm for its solution.

Table 4 Additional input data

	Additional aircraft properties	Designation	Units
Operational and economic data			
1	Fuel price	C_F	RUB/L
2	Load factor	n	–
3	Annual usage time	T_A	hour/year
4	Depreciation period	T_D	Years
5	Net book value	C_{NBV}	%
6	Number of aircraft in the series	N_S	Units
7	Ratio of the number of passengers in the 1st class cabins to the total number of passengers	$\frac{N_{pas}^1}{N_{pas}}$	–
Other data			
1	Maximum dynamic pressure	q_{max}	daPa
2	Equipment mass – Radio equipment – Flight and navigation – Test facility	m_R $m_{F\&N}$ $m_{C\&T}$	kg kg kg
3	Capacity ratio of APU	N_{APU}	–

An important stage in formalization of the design problem is the selection of criteria allowing objective evaluation of the design results. Direct operating costs (DOC), cost per tonne-kilometer, fuel efficiency, etc., may be used as measures of merit for the passenger aircraft design. The choice of one or another of them is determined by the purpose of the designed aircraft and the formulation of the design problem [3, 4].

In the design of other airplanes, the measures of merit will be different. For example, for military aircraft, the criteria may be the effectiveness of the typical combat operation (see Chap. 11).

Variables

Symbol	Description	Units
H_{cruise}	Cruising altitude	m
$\left(L/D\right)_{\max}$	Maximum lift-to-drag ratio	–
C_L	Lift coefficient	–
$V_{\min}$	Minimum speed	m/s
L	Lift	N
m_0	Take-off mass	kg
g	Gravity acceleration	m/s^2
ρ	Density	kg/m^3
S	Wing area	m^2
V_2	Safety speed	m/s
C_m	Pitching moment coefficient	–
ρ_H	Density at altitude H	kg/m^3
V_2	Safe take-off speed	m/s
V_{st}	Stalling speed	m/s
n_{eng}	Number of engines	–
q	Dynamic pressure	N/m^2
S	Wing area	m^2
x_{AC}	Aerodynamic center X-position	% MAC
$C_m^{C_L}$	Pitching moment coefficient derivative with respect to the lift coefficient	–
x_{aft}	Aft center of gravity limit	% MAC
T	Thrust	N
m_i	Mass of i-component of aircraft	kg
n	Load factor	–
L_{HT}	Horizontal tail moment arm	m
L_{VT}	Vertical tail moment arm	m
c_{MAC}	Mean aerodynamic chord length	m
M	Mach number	–
p	Pressure	N/m^2

(continued)

(continued)

Variables		
Symbol	Description	Units
t	Temperature	°

References

1. Aviation Rules. Part 23—Light civil category airplanes
2. Aviation Rules. Part 25—Transport civil category airplanes
3. Sheinin VM (1962) Vesovaya i transportnaya effektivnost' passazhirskih samoletov (Weight and transport efficiency of passenger aircraft), Oborongiz Moscow
4. Badyagin AA, Ovrutsky EA (1964) Proektirovanie passazhirskih samoletov s uchetom ekonomiki ekspluatacii (Design of passenger aircraft taking into account operating economics). Mashinostroenie, Moscow

Analysis and Selection of Aircraft Configuration and Type of Power Plant

Nikolai K. Liseitsev, Alexey Z. Tarasov, and Evgeniy A. Pen'kov

Abstract In this chapter, the reader will familiarize themselves with a wide pallette of aircraft configurations, distinguished by location and main purpose of lifting surfaces. The chapter also presents the aircraft-related reference frames and introduces the concept of aircraft longitudinal static stability, as well as trim and balance, including how they can be measured and ensured by design. The reader will learn in detail how the pitching moment of the three basic aircraft aerodynamic configurations (wing-tail, canard, and tailless aircraft) can be estimated and what is meant by balancing an aircraft in flight. Aircraft stability is closely related to its aerodynamic center and the volume coefficients of the tail, and they are also thoroughly discussed in this chapter. In particular, the reader will find a series of crossplots intended to facilitate the correct sizing of the tail for the desired center of gravity envelope. Furthermore, the conditions for the minimum trim losses and a generalized drag polar for the three configurations of aircraft are presented. The chapter also discusses aircraft control, including the requirements imposed on the efficiency of control surfaces.

Keywords Aircraft configuration · Longitudinal static stability · Trim · Tail volume coefficient · Aerodynamic center · Center of gravity envelope

N. K. Liseitsev (✉)
Moscow Aviation Institute (National Research University), Moscow, Russia
e-mail: nikolai.liseitsev@yandex.ru

A. Z. Tarasov · E. A. Pen'kov
Sukhoi Design Bureau, Moscow, Russia

M. A. Pogosyan (ed.), *Aircraft Design*, Springer Aerospace Technology,
https://doi.org/10.1007/978-981-96-4599-2_4

1 Selection of Aircraft Configuration

The aircraft's general layout includes the following:

- crew and payload arrangements;
- arrangement of the aerodynamic load-bearing system for the main (cruising) flight mode, as well as the take-off and landing or other flight phases;
- propulsion system layout (type, number of engines or thrusters, their location on the aircraft, fuel, and system layout);
- landing gear layout;
- structural design of the aircraft and the arrangement of its structural members, taking into account production and operation considerations;
- selection of the on-board equipment and control systems required to meet the technical specifications;
- design of configuration breakdowns.

Figure 1 shows a wide, but far from complete, range of configurations used in aircraft designed for various purposes.

Layout selection begins with the generation of alternative layout versions. The generation procedure, like other creative activities, is very difficult to formalize. In real design, a number of alternatives are formed by highly qualified specialists in the design bureau's preliminary design department. At the same time, they rely on their own and the world's experience in designing aircraft of a given class, the results of scientific research conducted by research institutes and universities, as well as studies of patents for original aircraft configurations. One of the configurations of a passenger aircraft developed to Airbus specifications by scientists from the MAI, Zhukovsky Aircraft Research Institute, and TsAGI is presented in Fig. 2 [1].

The final choice of the aircraft configuration is usually made from a number of competing options based on the elaboration of each of them and their subsequent comprehensive analysis. The best option is selected on the basis of a comprehensive assessment of alternatives using the "cost-efficiency" or "efficiency-cost" criteria, with one side of the criterion acting as a measure of merit and the other as a constraint. At the same time, the degree of technical risk associated with the implementation of alternative configurations is assessed.

As an evaluation criterion (measure of merit) of most aircraft can be taken, for example, their take-off weight, and as constraints—flight and technical characteristics specified by technical specifications. In this case, the best configuration version will be the one with the lowest take-off weight, all other things being equal.

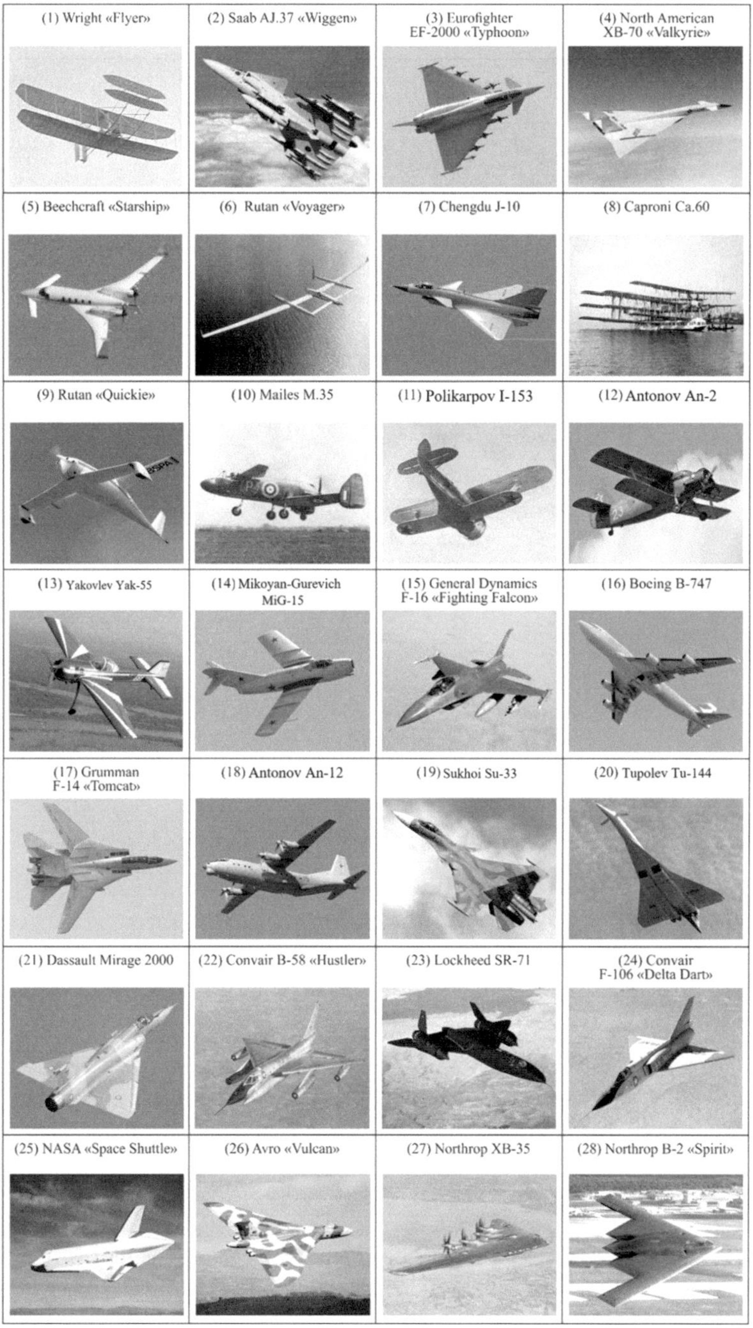

Fig. 1 Some aircraft configurations

Fig. 2 General view of a multiplane-type passenger aircraft

2 Selection of Aircraft Aerodynamic Configuration

2.1 Basic Aircraft Aerodynamics, Longitudinal Trim and Stability

The main feature characterizing the aerodynamic layout of an aircraft is the system of load-bearing and stabilizing surfaces, as well as the fuselage, which can be described by their mutual location, relative size, shape, and purpose. In the system of load-bearing surfaces, the main ones are the wings, which create the main share of lift. Other surfaces are designed to balance and control the aircraft (horizontal and vertical fins, stabilizers, etc.).

The configuration with one main bearing surface (wing) is called a monoplane (Fig. 1, configurations 2–7, 13–28). When there are several main bearing surfaces, we speak of a multiplane configuration. In a multiplane configuration, wings can be placed both vertically and horizontally, the latter option is often called "tandem" (Fig. 1, configurations 1, 8–12).

Depending on the arrangement of the auxiliary and control surfaces, the aerodynamic layouts are divided into the following main categories:

- wing-tail configuration, if the stabilizer and pitch control surfaces are located behind the wing (Fig. 1, configurations 11–19);
- canard configuration, if the stabilizer and pitch control surfaces are located ahead of the wing (Fig. 1, configurations 1–7);
- "tailless" or "flying wing" configurations, if the aerodynamic configuration includes only one lifting surface (Fig. 1, configurations 20–28).

If an airplane does not have an explicit fuselage, such an aerodynamic layout is called integral. Based on this feature, a "tail-less" and a "flying wing", which is a "tail-less" without an explicit fuselage, are usually distinguished (Fig. 1, configurations 26–28).

This categorization is not rigid, in particular, an airplane with fore and aft horizontal stabilizers is sometimes called a "triplane" (Fig. 1, configuration 19), and a canard aircraft with small fore horizontal stabilizer is sometimes called a tailless canard aircraft, especially if several surfaces are used for control purposes. It can also be difficult to distinguish between biplane-tandem and canard configurations with large area foreplanes, and a polyplane is referred to as a wing-tail or canard layout.

The main properties that make controlled flight of an aircraft possible are its ability to be balanced around three axes, and stability. Therefore, all three main configurations under consideration are called **balancing**. To analyze aircraft stability, the coordinate systems shown in Fig. 3.

Stability is defined as the aircraft's ability to maintain the prescribed flight path under the action of disturbances. An airplane is longitudinally stable, when some perturbation (increment) in the angle of attack $\Delta\alpha$ (and the corresponding increment in lift coefficient ΔC_L) occurs, the pitching moment coefficient increment ΔC_m decreases the angle of attack, i.e. $\frac{\Delta C_m}{\Delta C_L} < 0$. The derivative $\frac{dC_m}{dC_L} = C_{mC_L}$ is called the longitudinal static stability margin. The aircraft is longitudinally stable with respect to angle of attack, that is, it tends to return to the original angle of attack if the center of mass is located ahead of the point of application of the lifting force increment, the so-called aircraft aerodynamic center.

Stability can be provided by layout-related means (the selection of the CG position and envelope, the size and location of lifting surfaces) or by artificial means (provided by the pilot or automatic deflection of control surfaces). For example, the Wright Brothers' Flyer aircraft was longitudinally unstable and its stabilization was done manually by the pilot, while the stability of T-4 (Sukhoi), F-22 (Lockheed-Martin), Rafale (Dassault), Su-30MKI (Sukhoi) and some others, which have longitudinal instability at subsonic speeds, is ensured by means of automatic control systems.

An important parameter for the choice of aerodynamic layout is the hinge moment acting on a control surface and determining the loads and requirements for the control system. Different configurations have their own peculiarities in terms of magnitudes of hinge moments of control surfaces.

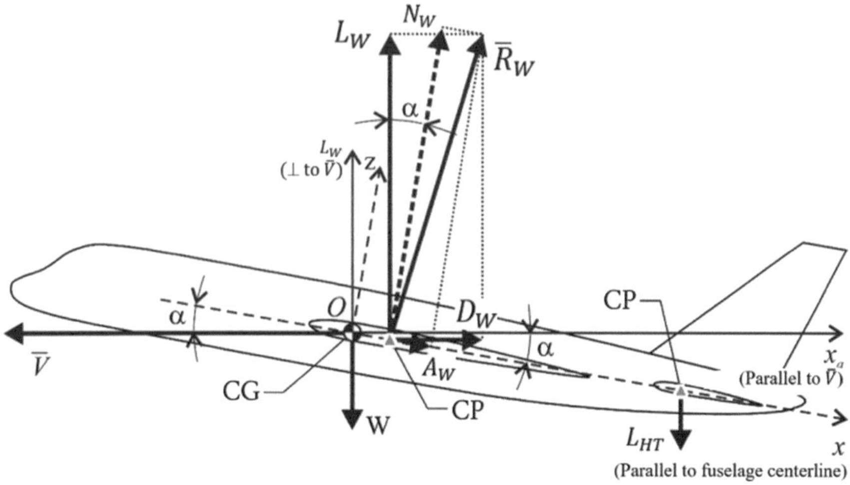

Fig. 3 Reference frames used in aircraft stability analysis: body reference frame $O_{X_B Z_B}$ and wind reference frame $O_{X_W Z_W}$. R_W is the resultant aerodynamic force. In the wind frame, the components of R_W are: L_W, the wing lift; D_W, the wing drag. In the wind frame, the components of R_W are: N_W, the wing normal force; A_W, the wing axial force

The load-bearing surfaces (wings, stabilizers) in the first approximation can be considered flat, i.e., having small thickness ratio $\bar{t}$. Depending on the aircraft type, the thickness ratio of the load-bearing surfaces can vary from $\bar{t} = 0.10 \ldots 0.15$ (subsonic transport, aerobatic aircraft) to $\bar{t} = 0.03 \ldots 0.05$ (supersonic aircraft with small aspect ratio wings). Real lifting surfaces tend to be twisted, which must be taken into account when considering aircraft trim capabilities.

Let's compare the wing-tail, canard, and tailless configurations with respect to their trim and stability properties, and hinge moment magnitudes.

2.2 Wing-Tail Configuration and Its Characteristics

Figure 3 shows the forces and moments acting in the vertical plane on a wing-tail airplane. For simplicity, we assume that the lifting surfaces—the wing and the horizontal tail (HT)—are flat, and at $\alpha = 0°$ the forces and moments are zero. At subsonic speeds, the lift generated by the wing + fuselage combination L_{W-B} is applied to its center of pressure located at the quarter of the mean aerodynamic chord of the wing c_{MAC}. The tail lift L_{HT} is applied to the quarter of the aerodynamic chord of the tail $c_{MAC,HT}$. Let us write down the equations for the longitudinal equilibrium of an aircraft with the weight W exposed to the given normal load factor n_z:

$$L = W \cdot n_z = L_{W-B} + L_{HT}, M = 0 = L_W \left(x_{CG} - x_{AC,wing}\right) + L_{HT}\left(x_{CG} - x_{AC,HT}\right). \tag{1}$$

Fig. 4 The wing downwash ε

The parameter $l_{HT} = x_{AC,HT} - x_{CG}$ is called the horizontal tail moment arm.

Figure 3 shows that if the center of mass is located ahead of the center of pressure of the wing + fuselage combination, then in a balanced flight the HT lift has the opposite direction to the wing lift, which referred to as the trim penalties. On the contrary, if the aircraft center of mass is located behind the wing center of pressure, the wing and tail lifts add up when the aircraft is in equilibrium.

For a wing-tail airplane, the span of the wing significantly exceeds the span of the horizontal tail, so it is possible to introduce the concept of flow downwash behind the wing ε, and the tail is considered as being exposed to the flow deflected at an angle with respect to the relative wind, Fig. 4.

The wing downwash is caused by the fact that the flow comes off the wing parallel to its trailing edge—this effect is called the Kutta-Jukovsky condition. With a distance from the trailing edge, the flow gradually returns to the original direction determined by the angle of attack of the wing, and the downwash decreases. As a result, the HT angle of attack is $\alpha_{HT} = \alpha - \varepsilon + \varphi_{HT}$, where φ_{HT} is the HT incidence angle relative to the aircraft longitudinal axis. In the transition to supersonic speeds, the downwash decreases (Fig. 5a), and practically disappears at a Mach number of 1.5. At subsonic flight speeds, before the onset of stall on the wing, the downwash angle ε is proportional to the angle of attack and can be described by two derivatives, $\varepsilon^{\alpha} = d\varepsilon/d\alpha$ and $\varepsilon^{\delta} = d\varepsilon/d\delta$: $\varepsilon \approx \varepsilon^{\alpha} \cdot \alpha + \varepsilon^{\delta} \cdot \delta$, where δ is the deflection angle of the high-lift devices (Fig. 5b). The typical value of ε^{α} for aircraft with the HT arranged at a distance 1…1.5 MAC from the wing aerodynamic center is 0.4…0.5 for $M < 1$, and the local angle of attack of the HT is $\alpha_{HT} \approx (1 - \varepsilon^{\alpha})\alpha$. For aircraft with the HT located very close to the wing (e.g., as in the F-22), the value of ε^{α} may be as high as is 0.7…0.9 at $M < 1$.

In addition to the downwash, the dynamic pressure at the tail q_{HT} is reduced in comparison with the free stream dynamic pressure due to the wing wake. The typical value of the tail efficiency factor is $k_{HT} = \frac{q_{HT}}{q} \approx 0.9$, and if the tail gets immersed in the wake produced by powerful high-lift devices (multislotted flaps

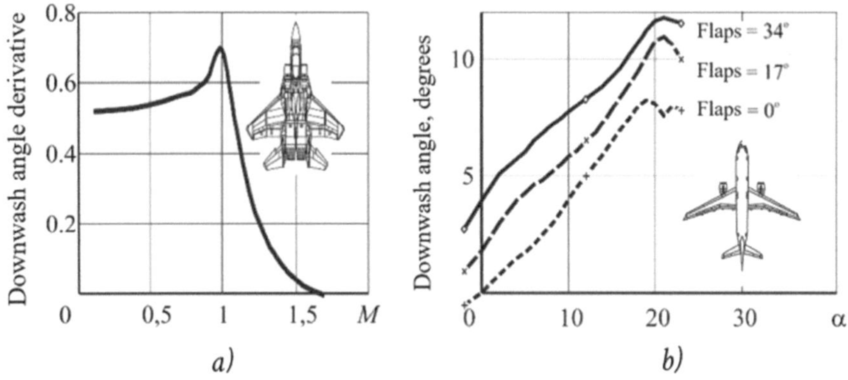

Fig. 5 Downwash angle variation: **a** with the Mach number; **b** with the angle of attack

deflected to a large angle), the tail efficiency factor may be even smaller. Hereafter, for simplification, we assume $k_{HT} = 1$.

To analyze the aircraft longitudinal stability, we write down the expressions for C_L and C_m, taking into account the parameter ε^α, the area ratio of the horizontal tail $\overline{S}_{HT} = \frac{S_{HT}}{S_w}$, and the distance ratio $\overline{x} = \frac{x}{c_{MAC}}$ expressed in fractions of the mean aerodynamic chord of the wing:

$$\begin{aligned} C_L &= C_{L,w_\alpha}\alpha + (1-\varepsilon^\alpha)\overline{S}_{HT}C_{L,\mathrm{HT}_\alpha}\alpha \\ C_m &= C_{L,w_\alpha}\big(\overline{x}_{CG} - \overline{x}_{AC,W}\big)\alpha + (1-\varepsilon^\alpha)\overline{S}_{HT}C_{L,\mathrm{HT}_\alpha}\alpha\big(\overline{x}_{CG} - \overline{x}_{AC,HT}\big). \end{aligned} \tag{2}$$

In these equations, the lift force coefficients of the wing and horizontal tail are referred to their areas S_w and S_{HT}.

From (2) it can be shown that the aircraft longitudinal static stability margin is defined as $C_{m_{C_L}} = \overline{x}_{CG} - \overline{x}_{AC}$; it can be represented as follows:

$$\begin{aligned} C_{m_{C_L}} &= \overline{x}_{CG} - \overline{x}_{AC,W} - \frac{C_{L,HT_\alpha}\overline{S}_{HT}(1-\varepsilon^\alpha)\big(\overline{x}_{AC,HT} - \overline{x}_{CG}\big)}{C_{L,W_\alpha} + C_{L,HT_\alpha}\overline{S}_{HT}(1-\varepsilon^\alpha)}, \\ \overline{x}_{CG} - \overline{x}_{AC,W} &= \big(C_{m_{C_L}}\big)_{W-B}, \\ C_{L,W_\alpha} + C_{L,HT_\alpha}\overline{S}_{HT}(1-\varepsilon^\alpha) &= \Delta C_{m_{C_L}} = -\Delta X_{AC,HT}. \end{aligned} \tag{3}$$

The center of mass location, the aerodynamic center of the wing-body combination, and the aerodynamic center displacement due to tail produce the following types of longitudinal stability of a wing-tail aircraft, shown in Table 1.

When deflecting the high-lift devices (flaps, flaperons on the trailing edge, slats, hinged leading edges), for example, during takeoff and landing, there are additional lift ΔL_{HL} and pitching moment ΔM_{HL}. They represent the sum of the additional force and moment applied to the wing and the additional force on the HT due to the change in the downwash when the high-lift devices are deflected, Fig. 5b.

Table 1 Longitudinal stability of a wing-tail aircraft

№	CG, wing AC, aircraft AC	Trim	Stability
1	CG is ahead of the wing + fuselage AC and the aircraft AC: $\left(C_{mC_L}\right)_{W-B} < 0$; $C_{mC_L} < 0$	There are trim penalties	The aircraft is longitudinally statically stable with respect to angle of attack With small stability margins, an (automatic) control system may be required to provide good control characteristics
2	The CG is behind the wing + fuselage AC and ahead of the aircraft AC: $\left(C_{mC_L}\right)_{W-B} > 0$; $C_{mC_L} < 0$	The tail lift adds up to the lift of the wing	
3	The CG is behind the wing + fuselage AC and the Aircraft AC: $\left(C_{mC_L}\right)_{W-B} > 0$; $C_{mC_L} > 0$		The aircraft is longitudinally statically unstable with respect to angle of attack The use of an (automatic) control system is necessary to ensure the stability of the aircraft

The additional force and the moment due to high-lift devices are included in the expressions for C_L and C_m:

$$\begin{aligned} C_L &= \Delta C_{L,HL} + C_{L,W}(\alpha) + \overline{S}_{HT} C_{LHT}, \\ C_m &= \Delta C_{mHL} + \left(\frac{dC_m}{dC_L}\right)_{W-B} C_{Lw}(\alpha) + \overline{S}_{HT}(\overline{x}_{CG} - \overline{x}_{ACHT}) C_{LHT}. \end{aligned} \tag{4}$$

If the aircraft is balanced in flight ($C_m = 0$), the value of the HT lift coefficient will be determined by the following expression:

$$C_{LHT} = \frac{\Delta C_{mHL} + \left(\frac{dC_m}{dC_L}\right)_{HToff} C_{Lw}(\alpha)}{\overline{S}_{HT} l_{HT}}. \tag{5}$$

The attainable maximum value of $C_{LHT\max}$ is limited, that is why the trim requirement at landing may restrict the center of gravity travel of aircraft $\left(\left(\frac{dC_m}{dC_L}\right)_{W-B}\right)$.

For a clean configuration, the wing lift coefficient is as follows:

$$C_{LHT} = \frac{\left(\frac{dC_m}{dC_L}\right)_{W-B} C_L}{\overline{S}_{HT} \overline{c}_{AC}}; \; C_{LW} = \frac{\overline{l}_{HT} C_L}{\overline{c}_{AC}}. \tag{6}$$

The aerodynamic drag of the wing and the HT are described by the polar, defined by the dependence $C_D = f(C_L)$:

$$C_D = C_{D0} + C_{Di,W}|C_{Lw} + \overline{S}_{HT}\left[C_{Di,HT}|C_{LHT} + \varepsilon C_{LHT}\right],$$
where $C_{D0} + C_{Di,W}|C_{Lw}$ is the wing polar,
$C_{Di,W}|C_{Lw}$ is the induced wing polar,
$C_{Di,HT}|C_{LHT}$ is the horizontal tail induced polar. (7)

Under equilibrium conditions, C_{LHT} and C_{LW} are defined by relations (6). The induced polars of the wing and HT for small α, when $C_L \approx C_N$, can be represented as $C_{Di,W} = K_2 \cdot C_{L,W}^2$; $C_{Di,HT} = k_2 \cdot C_{L,HT}^2$, and the downwash angle ε can be expressed via the wing lift coefficient C_{Lw} using the relation $\varepsilon \approx \varepsilon^{\alpha} \cdot \frac{C_{Lw}}{(C_{L\alpha})_W}$. As a result, from (7), it is possible to predict the aircraft induced drag coefficient $K_{2,trim} = \frac{C_{Di}}{C_L^2}$ for the equilibrium condition, which depends on the stability margin$C_{m_{C_L}}$. The optimal position of the center of mass provides a minimum $K_{2,trim}$ for the parameter $C_{m_{C_L}}$ and the following expression can be used to estimate $C_{m_{C_L},opt}$:

$$C_{m_{C_L},opt} \approx \frac{\varepsilon^{\alpha}}{2(1-\varepsilon^{\alpha})}\Delta X_{ACHT}. \quad (8)$$

For modern Su-27, MiG-29 and F-15/16/18 fighters, the optimum longitudinal static instability, which provides minimum trim drag for lift coefficients of 0…0.6, is $C_{m_{C_L}} = +0.05\ldots0.15$.

When the aircraft's center of gravity deviates from the optimum position, its trim drag for a constant value of lift coefficient increases according to the quadratic law, however, when the center of gravity deviates within (±5…10)% MAC, this increase is insignificant. Figure 6 shows, for a fighter aircraft, the trim drag for different C_L as a function of $\frac{dC_m}{dC_L}$ expressed as a percentage of the MAC.

The aerodynamic characteristics at large angles of attack are important for the selection of aerodynamic configuration and its parameters. For maneuvering fighter aircraft, the operating range of angles of attack is from −10° to 30°, and for non-maneuvering transport aircraft, the typical range of permissible angles of attack is from −10° to 20°.

The authority of the pitch control surfaces should be sufficient to trim the aircraft over the range of allowable angles of attack and to provide a control margin over a wider range of angles of attack to ensure its descent at low angles of attack. Maneuvering aircraft usually require a nose-down pitching moment ($C_{mdown} < -0.05\ldots-0.10$ (with some rudder deflection)) over an entire range of angles of attack up to 90°, Fig. 7a. Non-maneuvering aircraft usually require a nose-down pitching moment C_{mdown} over a range of angles of attack from 5° to 10°.

The pitching moment curve depends primarily on the wing shape and the location of the HT relative to the wing, as well as on the shape of the fuselage nose part. For large aspect ratio swept wings *(AR = 5…10)*, at angles of attack higher than 10…15°, due to flowing of the boundary layer along the wing, flow separation begins at the tip sections, which causes the onset of pitch instability—a zone of non-linearity in moment curve, where the aircraft can become unstable with respect to angle of

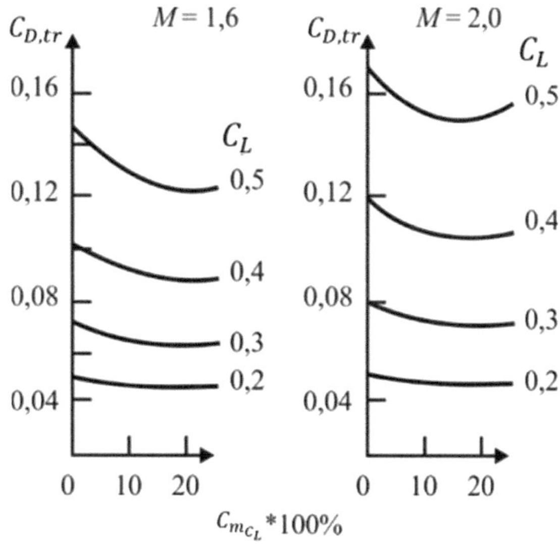

Fig. 6 Trim drag $C_{D,tr}$ of a wing-tail aircraft

attack, Fig. 7b. If an aircraft has a T-tail, at certain angles of attack (usually around 20°…25°), the HT gets into the stalled wake behind the wing, and as a result, the lift of the HT decreases and this also leads to the appearance of the pitch instability in the moment curve. For trapezoidal wings with an aspect ratio less than 5, typical for maneuvering fighter aircraft, the moment curve is usually smoother.

On a wing-tail aircraft, the HT can also be used for roll and yaw control by differential deflection of the right and left part of the HT:

- for roll control when flying at high speeds, at which the roll efficiency of ailerons and flaperons is greatly reduced due to elasticity effects;
- for yaw control at large angles of attack (20…25°), at which the effectiveness of the rudder is greatly reduced.

The choice of the HT area on a wing-tail aircraft is made taking into account the maximum possible number of design cases, which usually include the following:

(1) ensuring the specified longitudinal static stability margin;
(2) ensuring aircraft trimmability during landing with the high-lift devices extended;
(3) ensuring the required longitudinal control margin (e.g., the pitching down moment margin C_{mdown} at high angles of attack, trimming the aircraft at a given load factor at supersonic speed, trimming the aircraft at the maximum allowable angle of attack, etc.);
(4) nose wheel lift on takeoff.

For the first design case, the ratio for the aerodynamic center displacement due to the horizontal tail, derived from (3), is used:

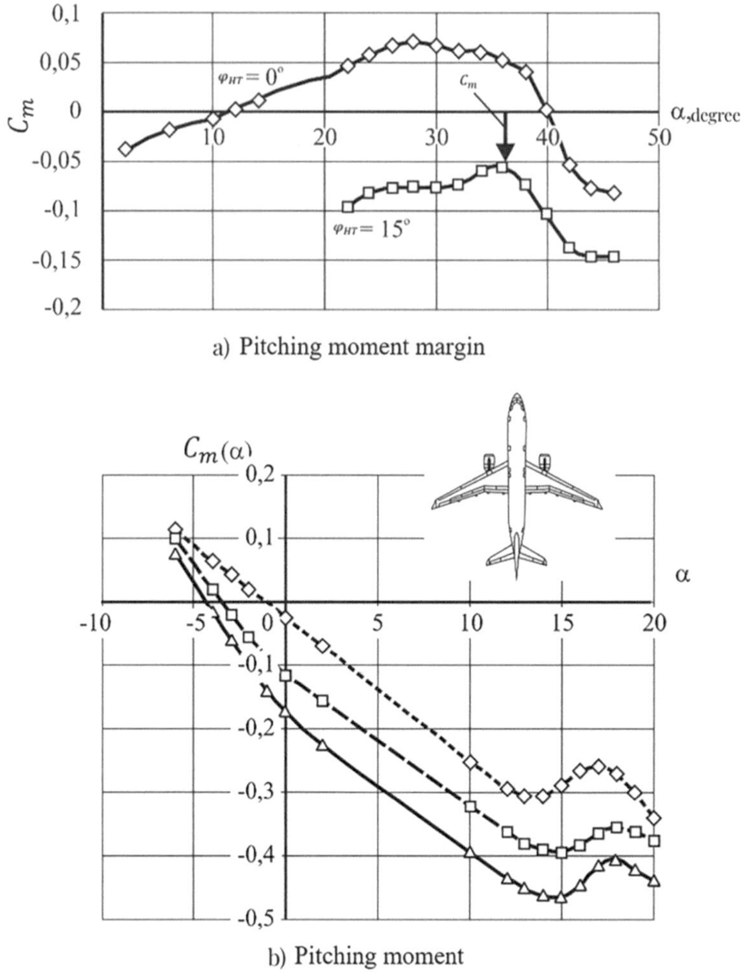

Fig. 7 Pitching moment and pitching moment margin as functions of angle of attack

$$\bar{x}_{AC} = \bar{x}_{Ac,W} \frac{C_{L_{a,HT}} \overline{S}_{HT}(1 - \varepsilon^{\alpha})\bar{c}_{AC}}{C_{L_{a,W}} + C_{L_{a,HT}} \overline{S}_{HT}(1 - \varepsilon^{\alpha})} \tag{9}$$

For the second design case, expression (5) for the balance factor of the tail lift with the high-lift device extended is used: using the moment coefficient due to high-lift devices ΔC_{mh-l} and the HT lift coefficient $C_{L,HT}$, expression (5) yields the HT area S_{HT} as a function of the center-of-gravity location $\bar{x}_{AC}$.

To calculate the pitching down moment as a function of the area (or volume coefficient) of the HT, the prototype data can be used. From wind tunnel tests of aircraft models at large angles of attack, the force and longitudinal moment coefficients of the wing-body combination, as well as those of the HT are derived. Taking into account that the authority of a HT is proportional to its area, the following relationship is used for $C_m\left(\alpha, \overline{S}_{HT0}, \overline{x}_{AC0}\right)$:

$$C_m\left(\alpha, \bar{S}_{HT}, \bar{x}_{AC}\right) = C_{m,\,W-B}\left(\alpha, \bar{x}_{AC\,0}\right) + \frac{\bar{S}_{HT}}{\bar{S}_{HT\,0}}\left\{\begin{array}{l}\left[C_m\left(\alpha, \bar{S}_{HT\,0}, \bar{x}_{AC\,0}\right) - C_{m,\,W-B}\left(\alpha, \bar{x}_{AC\,0}\right)\right] \\ +\left(\bar{x}_{AC} - \bar{x}_{AC\,0}\right)\left[C_L\left(\alpha, \bar{S}_{HT\,0}\right) - C_{L,\,W-B}(\alpha)\right]\end{array}\right\} \tag{10}$$

By setting the C_{mdown} parameter in this relation for several angles of attack, we can determine $\overline{S}_{HT} = f\left(\overline{x}_{AC}\right)$.

The results are plotted in terms of the HT area ratio $\overline{S}_{HT}$ and the distance reduced by the wing MAC (the so-called "cross plot") with the limits of constant stability margins, trim lines with the high-lift devices deployed and other constraints, for example, on C_{mdown}. Using this diagram, it is possible to determine the required area (volume coefficient) of the HT and/or the allowable center of gravity envelope of the aircraft. Figure 8 shows an example of selecting the center of gravity envelope for a passenger aircraft with a given HT area, a stability margin of at least 5% of the MAC and an angle of attack during landing up to 10°. Figure 9 also demonstrates an example of selection of the HT volume coefficient for a transport aircraft while ensuring the stability margin $C_{m_{C_L}} = -0.05$, a landing trim capability with the high-lift devices deployed at an angle of attack of 10°, and a given center of gravity travel $x_{CGforward} \ldots x_{CGaft}$. From the diagrams, it is seen that the reduction in stability margin from $C_{m_{C_L}} = -0.05$ to -0.025 (which is possible due to automation of control) in the first case allows to extend the CG envelope by 2.5% rearward, and in the second case allows a reduction of the HT area by ~5%.

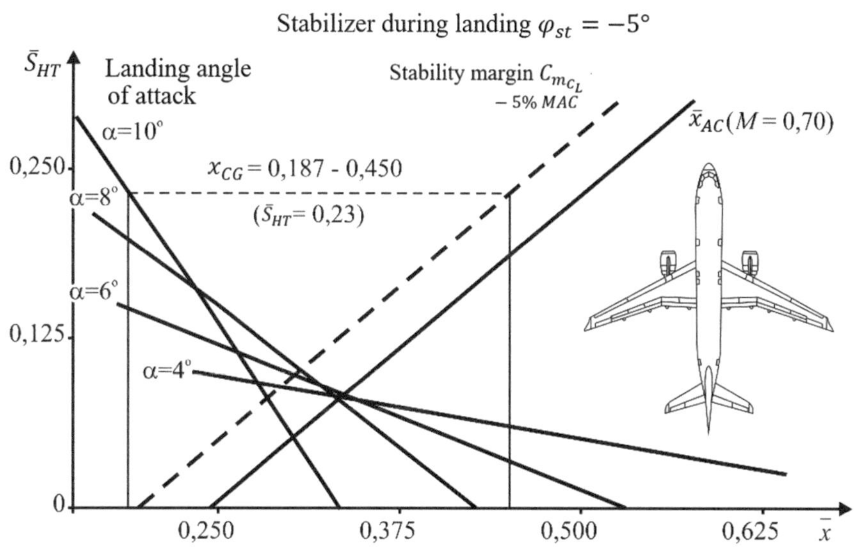

Fig. 8 Example of selection of the center of gravity envelope for a transport aircraft

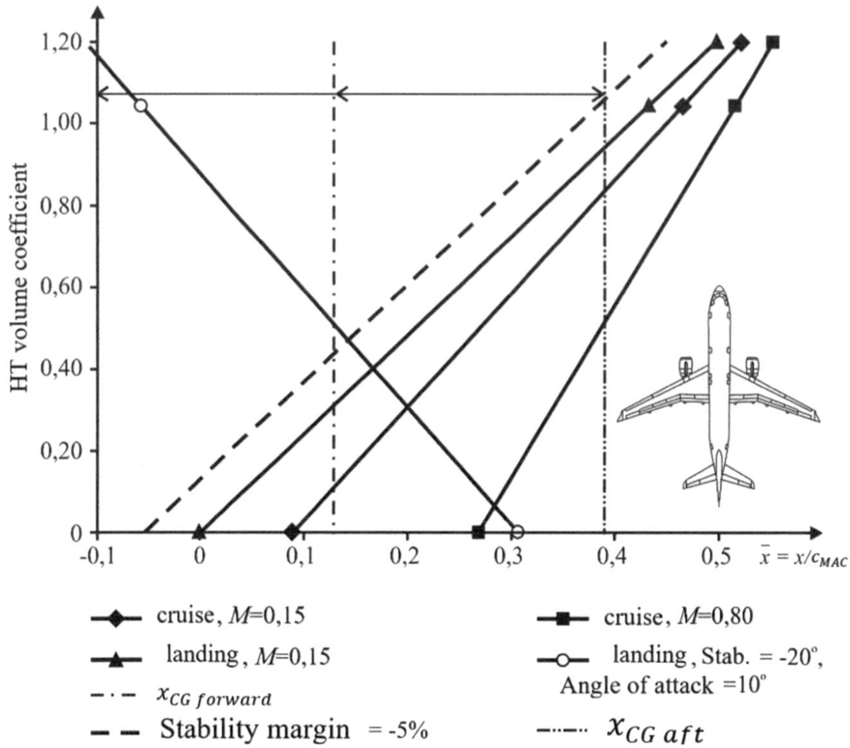

Fig. 9 Example of selection of the HT volume coefficient of a transport aircraft

Figure 10 shows the effect of the pitching moment margin requirement at high angles of attack on the HT area for a fighter aircraft—this requirement determines the lower limit for the HT area.

The aerodynamic hinge moment M_h acting on the HT determines the possibility of manual control of the aircraft, the choice of the HT drive when using boosters, and the loads on the HT. For a wing-tail airplane with an all-movable HT, the hinge moment M_h can be estimated using the following equation:

$$M_{h,HT} = L_{HT}\left(x_h - x_{AC,HT}\right) = qS_{HT}c_{MAC,HT}C_{L,HT}\left(\overline{x}_h - \overline{x}_{AC,HT}\right), \quad (11)$$

where $\overline{x}_{AC,HT}$ and $\overline{x}_h$ are the positions of the HT aerodynamic center and the HT hinge axis normalized by the HT MAC. From (6) and (11), using the relation $Gn_z = qSC_L$, one can obtain an estimate of the HT hinge moment:

$$M_{h,HT} = \frac{c_{MAC,HT}}{\overline{c}_{AC}(M)}\left[\overline{X}_h - \overline{X}_{AC,HT}(M)\right]\left(C_{m_{C_L}}\right)_{W-B}(M)Wn_z. \quad (12)$$

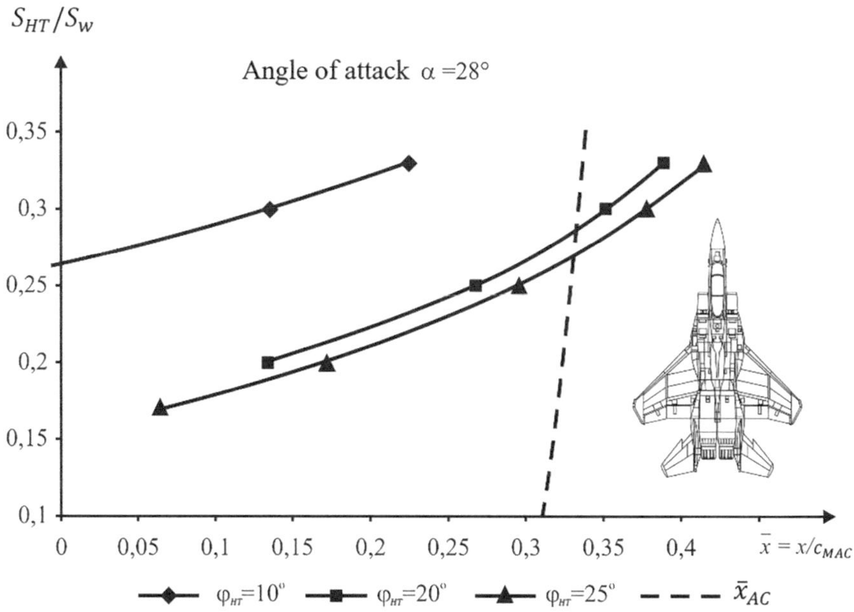

Fig. 10 Impact of the nose-down pitching moment coefficient C_{mdown} on the HT area at high angles of attack

Expression (12) shows that $M_{h,HT}$ does not depend directly on the dynamic pressure, is determined by Gn_z and M. Figure 11 shows variations of $\frac{M_{h,HT}}{Gn_{zy}c_{MAC,HT}}$ as a function of M for $C_{m_{C_L}} = -0.10 \cdots +0.15$. The HT hinge moments increase sharply at supersonic speeds on an airplane that is statically stable or close to neutral. The small values of the hinge moments at subsonic speeds are related to the fact that the hinge axis of the HT is usually slightly ahead of its subsonic aerodynamic center.

2.3 *Canard Configuration and Its Trimability*

Figure 12 shows the diagram of forces acting in the vertical plane on a canard aircraft with the front horizontal tail (canard) used to trim the aircraft. In contrast to the wing-tail aircraft, the canard has a much smaller span than that of the wing, and the canard downwash all falls on the wing. This induces some additional force on the wing L_{ind}, which is always directed opposite to the lift on the canard. Therefore, for the canard configuration, the interaction between the load bearing surfaces can be described using the parameter η_i, which characterizes the fraction of the canard lift that is dampened on the wing: $\eta_i = \frac{L_{ind}}{L_{canard}}$. As the angle of attack increases, this parameter decreases in magnitude, remaining approximately constant at $M < 1$ and decreasing at $M > 1$. Within the linear range of lift curve, the typical value of this parameter

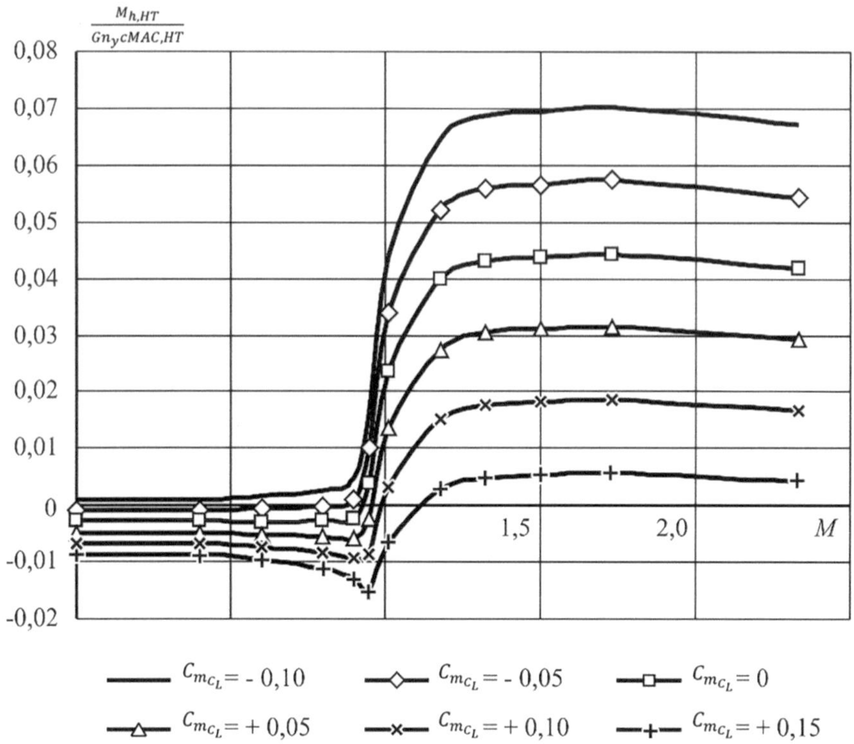

Fig. 11 HT hinge moment as a function of the Mach number and the static stability margin of an aircraft

for a closely-to-wing-located canard is $\eta_i = -(0.7...0.9)$. The point of application of this induced force x_{ind} is usually located near the wing MAC leading-edge.

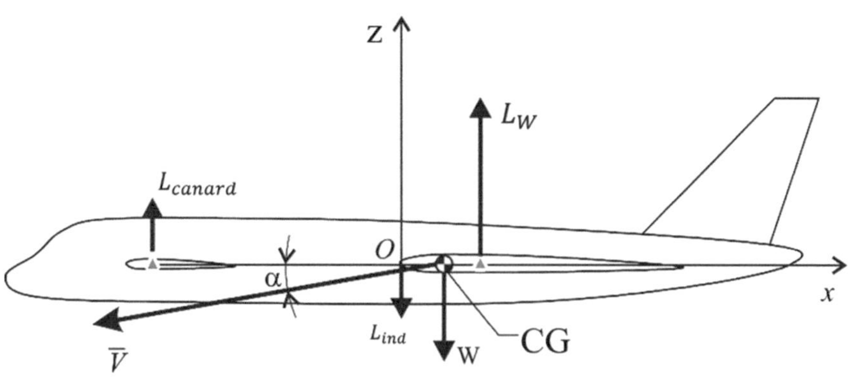

Fig. 12 Diagram of forces acting in the vertical plane on a canard aircraft

Taking the above into account, we can write down the equations for longitudinal trim of a canard aircraft:

$$\begin{aligned} L &= W \cdot n_z = L_W + L_{canard} + L_{ind}, \\ M &= 0 = L_W(x_{CG} - x_{ACW}) + L_{HT}(x_{CG} - x_{ACHT}) + L_{ind}(x_{CG} - x_{ind}). \end{aligned} \tag{13}$$

The positive value of $l_{canard} = -(x_{ACHT} - x_{CG})$ is called the canard moment arm.

The magnitude of the canard trim force will be determined by the following expression:

$$L_{canard} = -\frac{x_{CG} - x_{AC,W}}{l_{canard} + \eta_i (x_{CG} - x_{ind})} L_W. \tag{14}$$

If the center of mass is located ahead of the wing-body combination AC, then the trim lift of the canard is added to the wing lift and vice versa.

The expressions for the aerodynamic coefficients C_L and C_m are written as follows:

$$\begin{aligned} C_L &= C_{L,w_\alpha}\alpha + (1 - \eta_{ind})\bar{S}_{canard} C_{L,canard_\alpha}\alpha \\ C_m &= C_{L,w_\alpha}\left(\bar{x}_{CG} - \bar{x}_{AC,W}\right)\alpha + \bar{S}_{canard} C_{L,canard_\alpha}\alpha\left(\bar{x}_{CG} - \bar{x}_{AC,canard}\right) \\ &\quad + \eta_{ind}\bar{S}_{canard} C_{L,canard_\alpha}\alpha(\bar{x}_{CG} - \bar{x}_{induced}). \end{aligned} \tag{15}$$

From (15), it is possible to obtain an expression for the static stability margin of a canard aircraft:

$$\begin{aligned} C_{m_{C_L}} &= \bar{x}_{CG} - \bar{x}_{AC,W} - \frac{C_{L,canard_\alpha}\bar{S}_{canard}\left(\bar{x}_{AC,W} - \bar{x}_{AC,canard} + \eta_{ind}\left(\bar{x}_{AC,W} - \bar{x}_{ind}\right)\right)}{C_{L,W_\alpha} + C_{L,canard_\alpha}\bar{S}_{canard}(1 + \eta_{ind})}, \\ \bar{x}_{CG} - \bar{x}_{AC,W} &= \left(C_{m_{C_L}}\right)_{W-B}, \ \bar{x}_{AC,W} - \bar{x}_{AC,canard} = \bar{c}_{AC,y}, \\ \bar{x}_{AC,W} - \bar{x}_{ind} &= \bar{c}_{ind} C_{L,W_\alpha} + C_{L,canard_\alpha}\bar{S}_{canard}(1 + \eta_{ind}) = \Delta C_{m_{C_L}} = -\Delta x_{AC,canard}, \end{aligned} \tag{16}$$

where $\bar{c}_{AC,y}$ is the distance between the canard and wing AC, $\bar{c}_{ind}$ is the distance from the point of application of the induced force to the wing AC, both parameters are positive (in the adopted coordinate system $\bar{x}_{AC,canard} < 0$). Taking into account typical values of these parameters, we can show that due to canard installation, the aircraft aerodynamic center is moved forward by a value of $(1 \ldots 1.5)V_{canard}$, $V_{canard} = \bar{c}_{AC,y} \cdot \overline{S}_{canard}$ (canard volume coefficient). Thus, the canard shifts the aircraft AC forward approximately 2…3 times more per unit of volume coefficient than the HT backward in the wing-tail layout.

The center of mass, the wing-body AC and the AC shift due to canard lead to the following possible trimming and longitudinal stability options for a canard aircraft, shown in Table 2.

Table 2 Trimming and longitudinal stability options for a canard aircraft

№	Location of center of gravity, wing and aircraft aerodynamic center	Trim	Stability
1	The center of mass is located ahead of the wing-body AC and the aircraft AC: $\left(C_{mC_L}\right)_{W-B} < 0; C_{mC_L} < 0$	Canard generates positive lift	The aircraft is longitudinally statically stable with respect to angle of attack With small stability margins, a (automatic) control system may be required to provide good control characteristics
2	The center of mass is located ahead of the wing-body AC and behind the aircraft AC: $\left(C_{mC_L}\right)_{W-B} < 0; C_{mC_L} > 0$		The aircraft is longitudinally statically unstable with respect to angle of attack The use of an (automatic) control system is necessary to ensure the stability of the aircraft
3	The center of mass is located behind the wing-body AC and aircraft AC: $\left(C_{mC_L}\right)_{W-B} > 0, C_{mC_L} > 0$	There are trim penalties	

The deflection of high-lift devices (elevons, flaps) in the canard configuration always causes, along with an increase in lift by an amount of $\Delta C_{L,HL} > 0$, the nose-down pitching moment. From the equations for canard trim with the wing high-lift devices deployed, the trim lift coefficient is obtained:

$$C_{L,canard} = -\frac{\Delta C_{m,HL} + \left(C_{mC_L}\right)_{canard-off} C_{L,W}(\alpha)}{\bar{S}_{canard}\left(\bar{l}_{canard} + \eta_{ind}\bar{c}_{ind}\right)}. \tag{17}$$

From (17) it can be seen that to trim the deployed high-lift devices, it is required to produce additional lift on canard. Since the maximum value of the canard lift coefficient $C_{L,canard,max}$ is limited, this limits the trim capabilities for the high-lift configuration of a canard aircraft.

The choice of the canard area is made based on taking into account the same design cases as for the wing-tail layout.

The AC line for a canard airplane is given by the relation obtained from (16):

$$\bar{x}_{AC} = \bar{x}_{AC,W-B} - \frac{C_{L_{a,canard}}\bar{S}_{canard}\left(\bar{c}_{AC,z} + \eta_{ind}\bar{c}_{ind}\right)}{C_{L_{a,W}} + C_{L_{a,canard}}\bar{S}_{canard}(1 + \eta_{ind})} \approx \bar{x}_{AC,W-B} - \frac{C_{L_{a,canard}}}{C_{L_{a,W}}}\left(\bar{c}_{AC,z} + \eta_{ind}\bar{c}_{ind}\right)\bar{S}_{canard}. \tag{18}$$

The position of the centre of mass to satisfy the trim requirement for landing can be obtained from relation (17):

$$\bar{x}_{CG} = \bar{x}_{AC,W-B} - \frac{\Delta C_{m,HL}}{C_L - \Delta C_{L,HL}} - \frac{C_{L,canard}}{C_L - \Delta C_{L,HL}} \left(\bar{c}_{AC,y} + \eta_{ind} \bar{c}_{ind}\right) \bar{S}_{canard}. \quad (19)$$

As a result (Fig. 13), the AC lines and trim lines in terms of $\overline{S}_{canard}$ and $\bar{x}$ for a canard airplane essentially differ from a conventional wing-tail configuration. An increase in the canard area shifts the CG forward, and to ensure the required CG range Δx_{CG} for the given $\overline{S}_{canard}$, the slope of the trim line should be greater than the slope of the AC line, which can be expressed in the form of the following inequality:

$$\frac{C_{L,canard}}{C_L(\alpha)} > \frac{\left(C_{L_\alpha}\right)_{canard}}{\left(C_{L_\alpha}\right)_{wing}}, \quad (20)$$

where $C_L(\alpha) = C_L - \Delta C_{L,HL}$ denotes the lift coefficient due to angle of attack. This inequality can be represented in an equivalent form:

$$\frac{C_{L,canard}}{\left(C_{L_\alpha}\right)_{canard}} > \frac{C_L(\alpha)}{\left(C_{L_\alpha}\right)_{wing}}, \quad (21)$$

or otherwise $\alpha_{canard} > \alpha_w$.

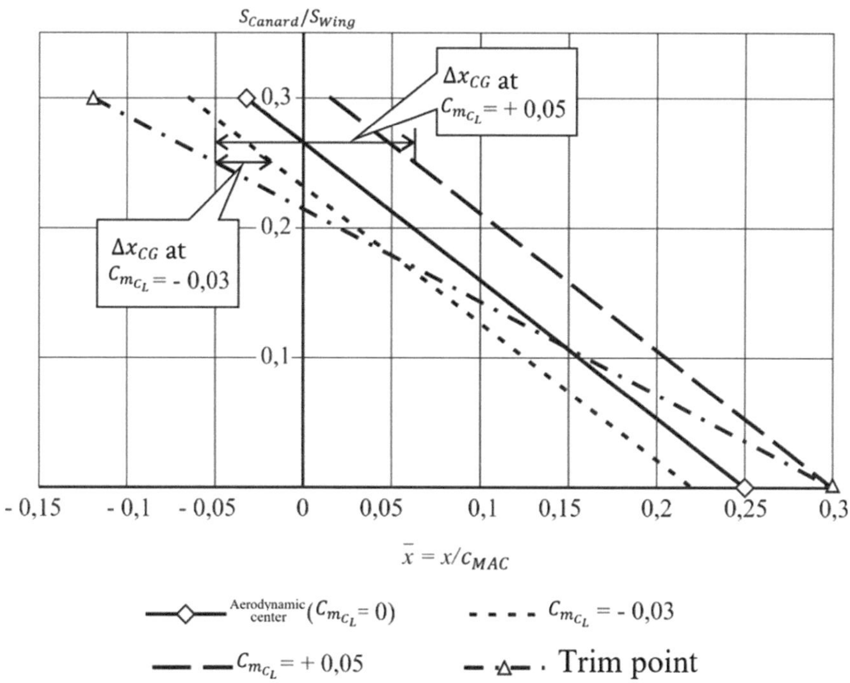

Fig. 13 Expanding the aircraft CG envelope by using an automated control system

In (21), α_{canard} and α_w are the local angles of attack of the canard and the wing.

Inequality (21) means that in order to achieve a given CG range on a canard airplane, the canard should provide as large lift as possible, with the lift slope as small as possible. Another interpretation of (21) is that the local angle of attack of the canard should be greater than the angle of attack of the wing ("longitudinal dihedral"). This is the reason for the selection of the canard geometry: it should be of a small aspect ratio (which contributes to the maximum lift coefficient) with a large leading edge sweep angle (which contributes to the reduction of the derivative $\left(C_{L_\alpha}\right)_{canard}$). The use of a rudder on a canard, as implemented in the Wiggen JAS-37 aircraft, contributes to the fulfillment of this condition. It is also possible to fulfill (21) by applying a feathering canard with a servo tab on the trailing edge, which creates virtually no lift due to angle of attack, i.e. $\left(C_{L_\alpha}\right)_{canardfeath} \approx 0$. However, such a solution is characterized by a significant technical complexity and has not found application on modern aircraft.

Another way to extend the CG range of a canard configuration is to allow static instability $C_{m_{C_L}} > 0$ by using an automated control system, Fig. 13.

Due to the complex canard-wing interaction, it is difficult to model the canard polar with a simple two-term approximation. Calculations for various canard aircraft show that the minimum induced drag due to trim is achieved when the aircraft is longitudinally unstable. An approximated expression for the optimal degree of instability for a canard aircraft aiming at the minimum trimming induced drag is following:

$$\begin{aligned} \left(C_{m_{C_L}}\right)_{W-B,opt} &\approx -\left(1+\frac{\eta_{ind}}{2}\right)\Delta\bar{x}_{AC,canard}, \\ \left(C_{m_{C_L}}\right)_{opt} &\approx -\frac{\eta_{ind}}{2}\Delta\overline{X}_{AC,canard}. \end{aligned} \quad (22)$$

It can be seen from (22) that at the optimum center of gravity, the aircraft is statically stable without canard and unstable with canard. The relative position of the AC at the optimum position of the center of mass is shown in Fig. 14a. The optimal degree of instability for a canard aircraft is greater than for the wing-tail configuration and amounts to 20…25% MAC (M < 1), Fig. 14b.

By analogy with the wing-tail configuration, one can obtain an expression for the estimation of the canard hinge moment $M_{h,canard}$ (analogous to expression (12)):

$$M_{h,canard} = \frac{c_{MAC,canard}}{\bar{c}_{AC}(M) + \eta_{ind}\bar{c}_{ind}(M)}\left[\bar{x}_h - \bar{x}_{AC,canard}(M)\right]\left(C_{m_{C_L}}\right)_{W-B}(M)Gn_z. \quad (23)$$

Figure 15 shows an example variation of the parameter $M_{h,canard}/\left(Gn_z c_{MAC,canard}\right)$ as a function of the Mach number for a range of $C_{m_{C_L}}$. As in a wing-tail aircraft, with an increasing static instability, the canard hinge moments decrease; however, comparing Figs. 11 and 15, it can be seen that, for the same static instability $C_{m_{C_L}}$, the canard hinge moments are about ~2 times larger than those of the HT. The same magnitude of hinge moments may be reached at different vales of static instability:

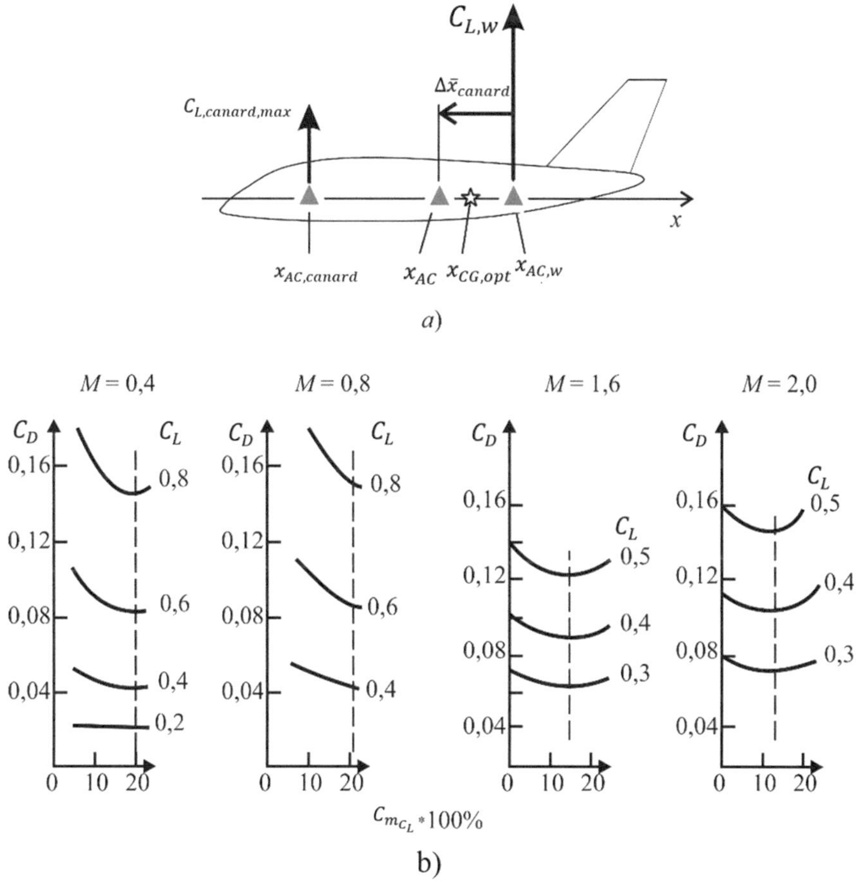

Fig. 14 Optimal degree of static instability for a canard aircraft

$C_{m_{C_L}} = 0 \ldots + 0.05$ for the wing-tail layout and $C_{m_{C_L}} = +0.25 \cdots + 0.30$ for a canard aircraft.

In contrast to the HT with a substantially reduced or eliminated authority at high angles of attack, the canard at high angles of attack retains its nose-down pitching authority, provided that the canard deflection angles are also high ($\varphi_{canard} = -50 \cdots - 70°$, nose down), see Fig. 16. In contrast, the canard pitching-up authority disappears very quickly with an increasing angle of attack. As a result, in a canard aircraft, the canard is usually deflected to negative angles $\varphi_{canard} < 0$, although positive lift can be generated on the wingtip. At these deflection angles, some of the canard authority is retained at all angles of attack, so if the wing-body combination has a nose-down pitching moment margin, then with the all-movable canard deflected at a high angle, the nose-down pitching moment margin can be maintained or even increased.

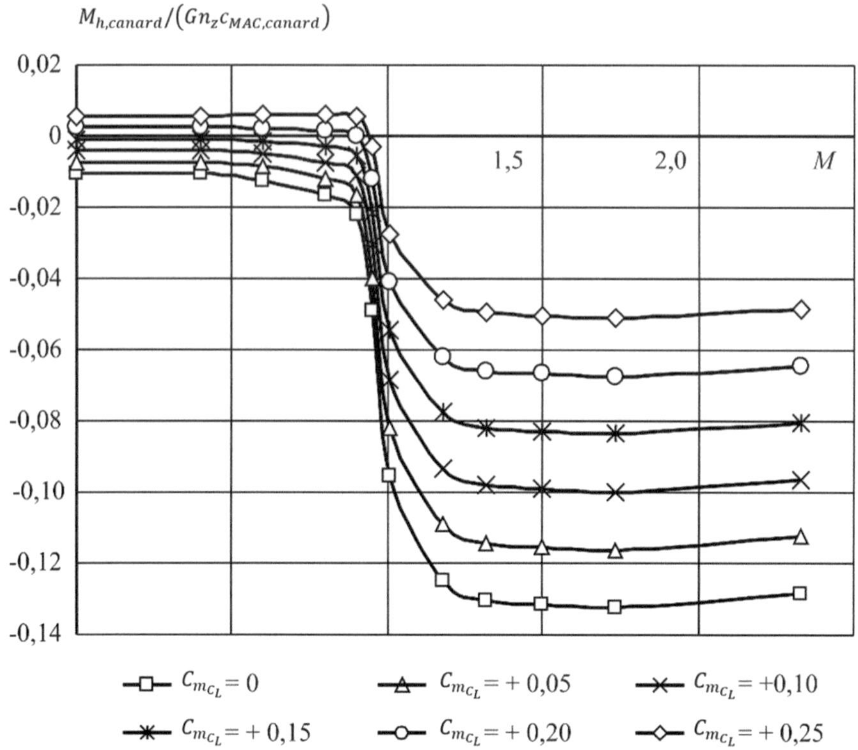

Fig. 15 $\frac{M_{h,canard}}{Gn_z c_{MAC,canard}}$ as a function of the Mach number for a range of C_{mC_L}

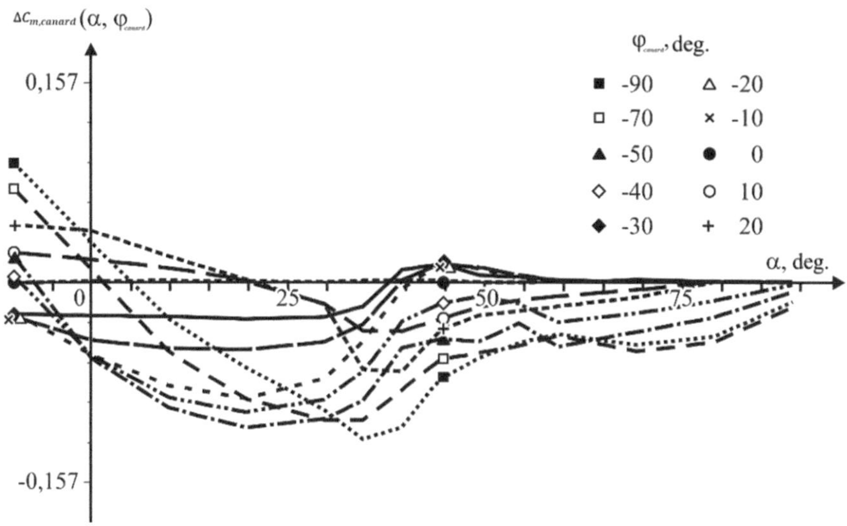

Fig. 16 Canard authority as a function of α and φ_{canard}

In contrast to the wing-tail configuration, the canard is ineffective for roll control at high flight speeds due to an almost complete damping of the lift due to canard by the wing. At the same time, a differential deflection of the canard halves can be effective for roll control at high angles of attack.

Thus, the following main features are characteristic of a canard aircraft as compared to the wing-tail layout:

- the trim authority for balancing the deployed high-lift devices in landing and for the required center-of-gravity range improves with an increasing degree of longitudinal instability;
- the trim penalties are minimized at a static instability of 20…30% MAC;
- the canard hinge moments are larger than the HT hinge moments and decrease with an increasing static instability.

These features make it difficult to design a multi-mode, maneuvering canard aircraft using only a fore tail for pitch balance. Therefore, most aircraft of this configuration use elevons on the wing trailing edge to provide trim capability in addition to the canard. All existing maneuvering canard fighters (Dassault "Rafale", "Eurofighter" EF-2000, "Grippen" JAS-39, IAI "Kfir" C-2, J-10 and J-14 of China, Mikoyan MFI "1.44") use this method. The simultaneous deflection of the canard and elevons on a statically unstable aircraft facilitates its trimming in all flight modes, extends the center of gravity envelope, reduces the value of the optimum longitudinal instability to an acceptable level ($C_{m_{C_L}} = +0.15$) without a significant drag penalty, and reduces canard hinge moments.

The canard and elevon deflection angles must be adjusted by the control system depending on the Mach number, angle of attack, and other flight parameters. Nowadays, remote-control systems ensure a stable flight for statically unstable aerodynamic configurations with $C_{m_{C_L}} = +0.15 \cdots + 0.20$. This could not be realized before, which explains the generally unsuccessful attempts to create maneuverable canard aircraft in the 40s and 60s and the appearance of a great number of fighters of this configuration in the 70s and 90s, following the introduction of remote-control systems technology. On the other hand, non-maneuvering and, as a rule, single-mode aircraft, such as the HV-70 "Valkyrie" bomber, T-4 (Sukhoi), "Starship" 1, and a number of others, were created long before that.

2.4 *Tailless Layout and Its Trim Characteristics*

Tailless aircraft have only one lifting surface, a wing, therefore, their pitch balance is achieved by the deflection of rudders located on the trailing edge of the wing, elevons, which, as a rule, consist of several sections. Independent deflection of different sections provides pitch control (symmetric deflection of elevons on the right and left wings) and roll (asymmetric, or differential, deflection of elevons on the right and left wings).

At M < 1, the deflected downward elevons generate positive lift and nose-down pitching moment. The lift due to elevons is applied to the elevon aerodynamic center $x_{AC,elevon}$(which is also called the second wing AC $x_{AC,2}$), located at low subsonic speeds at ~70…80% of the local chord of the wing. At supersonic speeds the elevon aerodynamic center shifts backwards up to ~50% of the MAC of the deflected part of the elevons.

The force diagram for a tailless aircraft is shown in Fig. 17.

The equations for a tailless aircraft equilibrium with respect to its center of gravity are as follows (the elevons are deflected at angle δ_{elevon}, which changes the lift by $\Delta L_{elevon}(\delta_{elevon})$:

$$\begin{aligned} L &= Wn_z = L_w + \Delta L_{elevon}(\delta_{elevon}), \\ M &= 0 = L_w\left(x_{CG} - x_{AC,w}\right) + \Delta L_{elevon}(\delta_{elevon})\left(x_{CG} - x_{AC,elevon}\right). \end{aligned} \tag{24}$$

For a tailless aircraft, the longitudinal stability will always be determined by the mutual location of the center of gravity x_{CG} and the wing AC with respect to the angle of attack $x_{AC,w}$, so the static longitudinal stability is always accompanied by trim penalties, as shown in Table 3.

The equilibrium Eq. (24) can be expressed in terms of aerodynamic coefficients, angles of attack α and elevon deflection angles δ_{elevon}:

$$\begin{aligned} &C_L = C_{L_\alpha}(M)\alpha + C_{L_{\delta elevon}}(M)\delta_{elevon}, \\ &0 = C_{L_\alpha}(M)\left[\overline{x}_{CG} - \overline{x}_{AC,w}(M)\right]\alpha + C_{L_{\delta elevon}}(M)\left[\overline{x}_{CG} - \overline{x}_{AC,elevon}(M)\right]\delta_{elevon}. \end{aligned} \tag{25}$$

From (25), we obtain expressions for elevon deflection angle and elevon share in total lift of the tailless configuration, trimmed at the lift coefficient C_L:

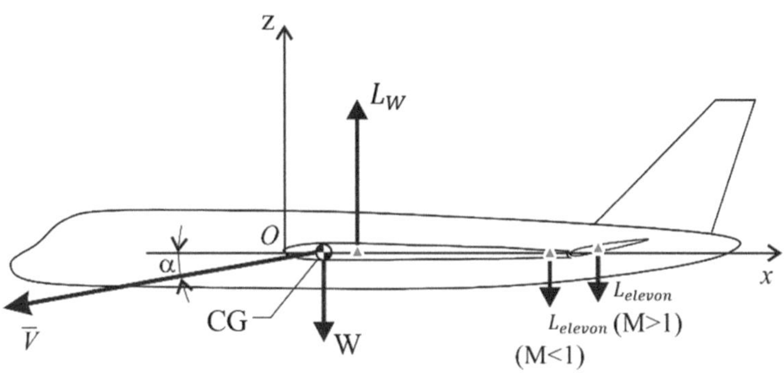

Fig. 17 Force diagram for a tailless aircraft

Table 3 Longitudinal balance and stability of the tailless layout

№	CG location, wing AC, aircraft AC	Trim	Stability
1	The center of gravity is ahead of the aircraft AC: $\frac{dC_m}{dC_L} < 0$	There are trim penalties	The aircraft is longitudinally statically stable with respect to angle of attack An (automatic) control system may be required for small stability margins
2	The center of gravity is behind the aircraft AC: $\frac{dC_m}{dC_L} > 0$	The lift due to deflected elevons increases the lift of the wing	The aircraft is statically longitudinally unstable with respect to the angle of attack The use of an (automatic) control system is necessary to ensure the stability for the aircraft

$$\delta_{elevon} = \frac{\overline{x}_{CG} - \overline{x}_{AC}}{C_{L_{\delta elevon}}\left(\overline{x}_{CG} - \overline{x}_{AC,elevon}\right)} C_L,$$
$$\frac{\Delta C_{L,elevon}(\delta_{elevon})}{C_L} = -\frac{\overline{X}_{CG} - \overline{X}_{AC}}{\left(\overline{X}_{CG} - \overline{X}_{AC,elevon}\right)} C_L. \tag{26}$$

Figure 18 shows typical variations of $\frac{\Delta C_{L,elevon(\delta_{elevon})}}{C_L}$ with the Mach number and the CG position $\overline{x}_{CG}$, and Fig. 19 shows the corresponding static stability margin of a tailless aircraft with a delta wing. It can be seen that at supersonic speeds the trim penalties increase significantly. At the same time, the supersonic trim penalties for aircraft with little stability or unstable aircraft at subsonic speeds are at a level of ≤30%. To reduce supersonic losses, a special wing design is used—a negative twist at tips or S-shaped airfoils. By selecting the value of $\left(\frac{C_{m,0}}{C_L}\right)_{cruise}$ for cruise, it is possible to significantly reduce the trim penalties at M >1, Fig. 20.

Due to the fact that the elevon deflection angles required for trim are determined only by the CG location (with the selected airfoils), the high-lift capabilities of the tailless configuration are limited and are enhanced at the rear location of the center of gravity as much as possible. Another way to increase the available maximum lift is the use of special aerodynamic devices on takeoff and landing; for example, the aircraft "Milan" (Dassault) and the Tu-144 aircraft used retractable canards equipped with powerful high-lift devices. This canard will not shift the AC with respect to the angle of attack (Fig. 1, sketch 20). Trimming the canard pitching moment requires the elevons to be deflected downward and creates an increase in lift. A similar method was used on the T-4 aircraft (Sukhoi), where a small movable canard was installed to facilitate balancing. The deck version of the light fighter aircraft LCA (India) uses an upwardly deflected part of the wing leading edge, which creates a pitching-up moment, for the same purpose. Split flaps that are deflected on the lower surface of the fuselage or wing and create virtually only pitching-up have also been known for a long time.

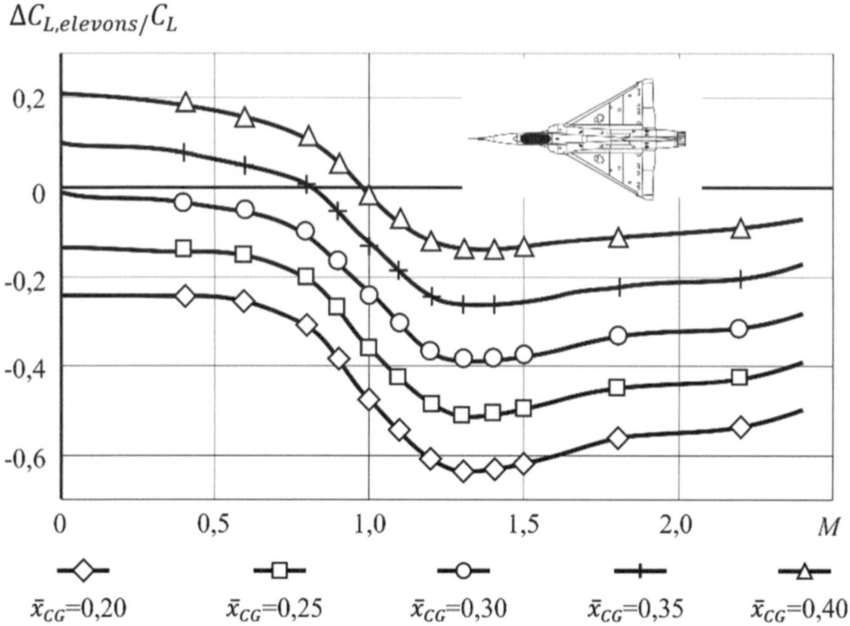

Fig. 18 Variation of $\Delta C_{L,elevons}/C_L$ as a function of the Mach number and CG location $\bar{x}_{CG}$

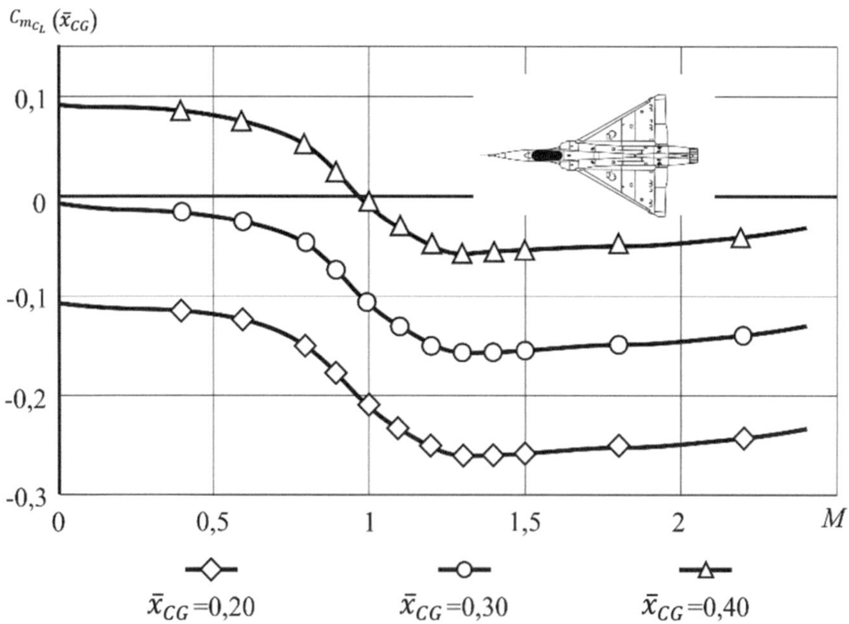

Fig. 19 Stability margin of a tailless aircraft

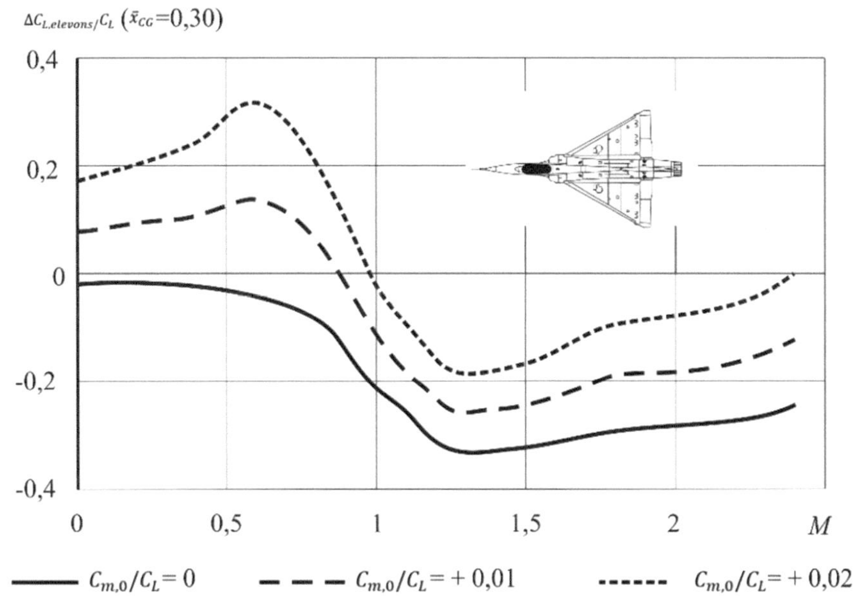

Fig. 20 Reduction of trim penalties at M > 1 as a result of $\left(C_{m,0}/C_L\right)_{cruise}$ optimization

When the elevons are deflected downward, the polar $C_L = f(C_D)$ of a tailless aircraft shifts up and to the right, as shown in Fig. 21a. Therefore, as stability decreases (instability increases), the trim resistance of a tailless configuration will improve, as illustrated by Fig. 21b. The optimal degree of instability of a tailless aircraft is usually much greater than that of a wing-tail configuration.

For a maneuvering tailless fighter aircraft, the nose-down pitching moment is important. Elevons at large angles of attack usually retain some efficiency $\Delta C_{m,elevon}(\alpha, \delta_{elevon})$. In Fig. 22, the aft center-of-gravity limit will be determined by the condition of providing a given $C_{m,nose-down}$ with the elevons deflected to their maximum angle. As a rule, the maximum achievable level of instability for the tailless configuration in terms of nose-down pitching moment margin is $C_{m_{C_L}} \approx 0.05$.

The flying-wing aircraft are a variation of the tailless aircraft. This layout has the lowest drag per unit surface area and can provide the highest cruise lift-to-drag ratio at subsonic speeds, with the trim capability provided mainly by the wing shape. This layout was used for low-maneuvering single-mode aircraft, such as bombers that do not experiencc supersonic flight speeds, such as the XB-35, YB-49 and B-2 firm Northrop. Supersonic aircraft using the flying wing configuration were not built due to the significant rise of drag and trim penalties at supersonic speeds.

With an increasing Mach number M, elevon hinge moments increase simultaneously with the trim deflection angles. Let's assume that the lift due to elevon deflection is applied approximately at the point $\overline{x}_{AC,elevon} \approx 50\%$(in terms of the elevon chord), and the hinge axis is located at the leading edge $\overline{x}_{AC,elevon} \approx 0\%$

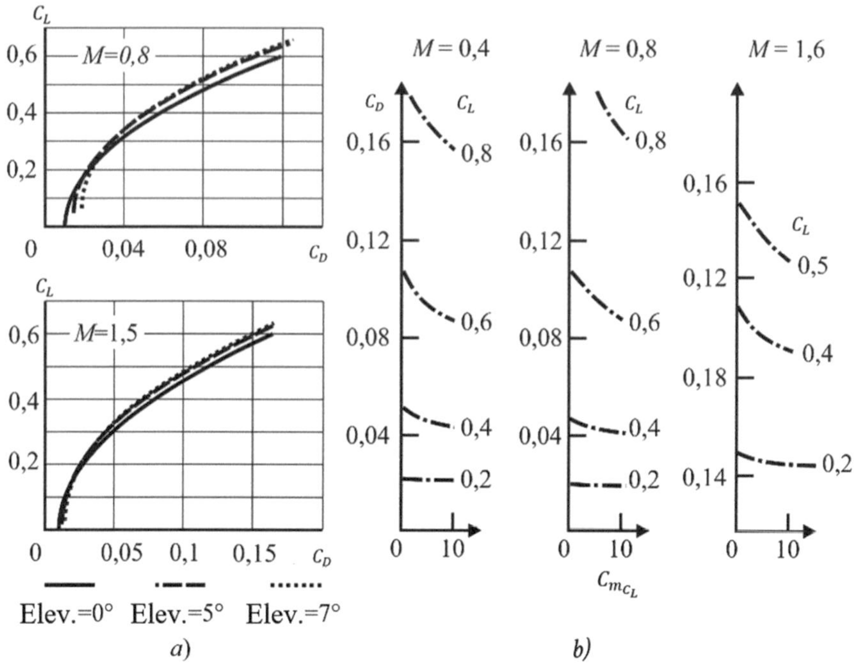

Fig. 21 A tailless aircraft polars affected by **a** elevon deflections and **b** instability margin

(for unbalanced controls). In this case, the parameter $M_{h,elevon}/(Wn_z c_{elevon})$ can be estimated as follows:

$$\frac{M_{h,elevon}}{Wn_z c_{elevon}} = \frac{\left[\overline{x}_h - \overline{x}_{AC,elevon}(M)\right]\left(\frac{C_{m,0}}{C_L} - C_{mC_L}(M)\right)}{\overline{x}_{AC}(M) - \overline{x}_{AC,elevon}(M)}. \tag{27}$$

A typical variation of this parameter, as well as the influence of the parameter $\frac{C_{m,0}}{C_L}$, characterizing the wing shape, can be seen from Fig. 23 (the corresponding values of C_{mC_L} are shown in Fig. 19).

2.5 *Features of Configurations with Variable Geometry of Lifting Surfaces*

Variable airframe geometry can be used as a means to combine good field and cruise characteristics within a single aircraft. Historically, various ways of changing the geometry of the load-bearing surfaces have been considered, the main ones being the following:

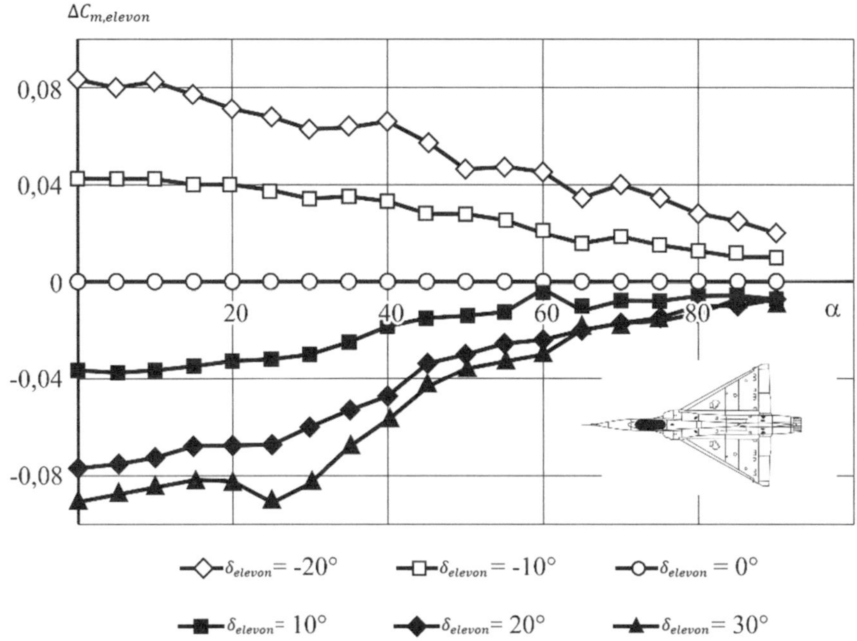

Fig. 22 Elevon authority of a tailless aircraft at high angles of attack

- sliding wings of various kinds (e.g., the LIG-7 project, Fig. 24a).
- wings with in-flight variable sweep angle, Fig. 24b.
- retractable and deflectable lifting surfaces of various types, Fig. 24c.

Due to technical problems, the first method has not found application. The third method is used, for example, on the aircraft XB-70 "Valkyrie", which used deflected downward wingtips to ensure directional stability at speeds with M = 3, Fig. 24c, as well as on the Starship 1 aircraft, which used a variable sweep canard. Most aircraft with variable geometry are wing-tail and implement the second method—the sweep angle is changed through rotating the wings about hinges at their roots. This approach is used by F-14 (Grumman), F-111 (General Dynamics), "Tornado" (Panavia), B-1 (Rockwell), Su-17, Su-24 (Sukhoi), MiG-23/27 (Mikoyan), Tu-22M3, Tu-160 (Tupolev). This could also include the experimental AD-1 (NASA) with a sliding asymmetric wing.

To ensure high values of lift in takeoff and landing modes, high-lift devices on the leading (slats) and trailing (flaps) edges are used. The high-lift devices are deployed in take-off and landing when the wing sweep angle is small. For roll control, ailerons, spoilers, and HT, differential deflection is used. At a sweep angle of $\Lambda_w = 70°$, ailerons become ineffective and for roll control differential deflection of the HT is used.

The feature of the wing-tail layout with a variable-sweep wing is that the aerodynamic center shifts in a wide range with the changing sweep angle. This range can

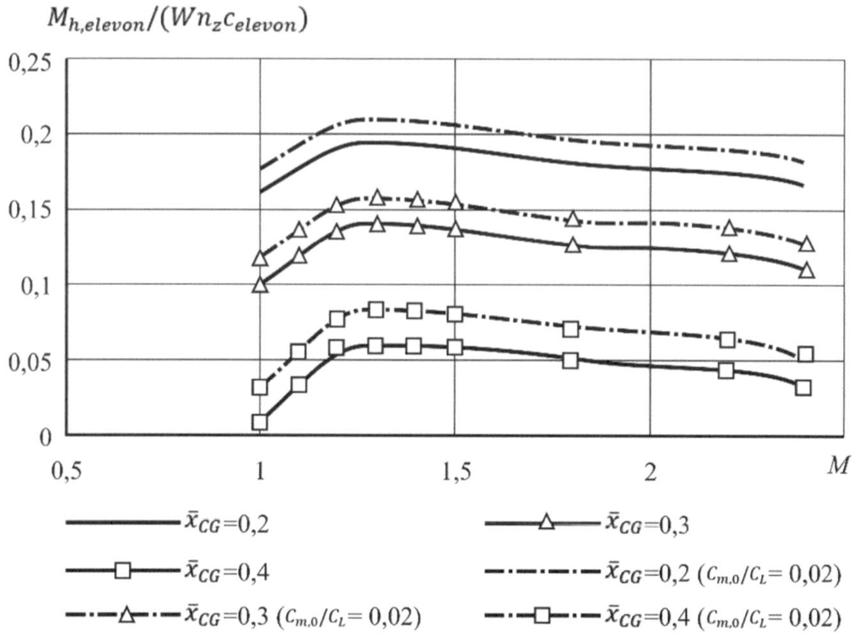

Fig. 23 Influence of the Mach number and the wing shape on the hinge moments of a tailless aircraft

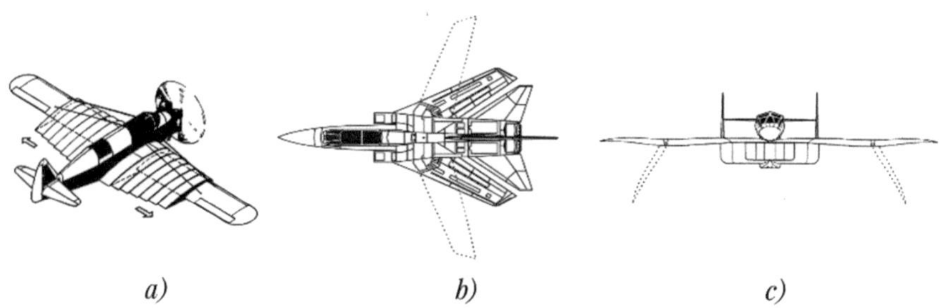

Fig. 24 Different ways of changing the wing planform geometry

reach up to $\Delta\bar{x}_{AC} = 45\%\ MAC\ at\ M < 1$, $\Delta\bar{x}_{AC} = 35\ldots 40\%\ MAC\ at\ M > 1$, and $\Delta\bar{x}_{AC} = 57\%\ MAC$ taking into account the transition from subsonic to supersonic speeds (in design and analysis of aircraft with variable geometry it is common to use the reference area and the MAC of the wing with the maximum sweep angle), which is illustrated in Fig. 25. The AC position for a wing-body combination will also shift within a similar range.

This AC location variation is significantly greater than that in a fixed-wing airplane, $\Delta\bar{x}_{AC} = 25\%MAC$ when transitioning from subsonic to supersonic speeds. The position of the aircraft center of gravity is selected near the AC at M < 1 at the

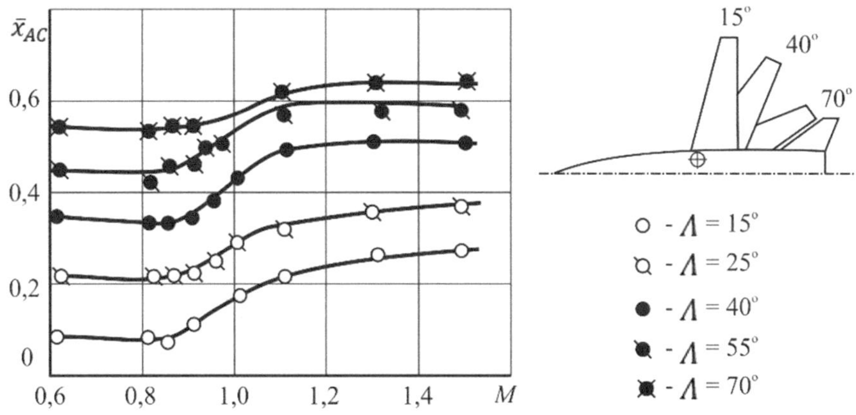

Fig. 25 Variation of aircraft aerodynamic center location as a function of Mach number M

minimum value of Λ, so the aircraft is excessively stable at M > 1, which makes it difficult to balance. It follows from (6) that to balance an airplane with variable wing sweep angle, it is necessary to increase the HT area, which means that when increasing the parameter $\left(C_{m_{C_L}}\right)_{W-B}$ and limited value of $C_{L,HTmax}$ it is necessary to increase the value of $\overline{S}_{HT}$. Taking into account the above-mentioned AC range, the required increase in the HT area ratio will be approximately 1.5…2.0 times the HT area ratio of an aircraft with fixed wing geometry. To give an example: on the Tornado aircraft (Fig. 24b) the HT area ratio is ~0.27 in relation to the wing with the maximum sweep and ~0.31 in relation to the wing with the minimum sweep, while the typical value of $\overline{S}_{HT}$ for aircraft with fixed geometry is 0.2.

When selecting the HT area for a variable wing sweep aircraft, essential limitations are, as a rule, the stability margin (degree of instability) at subsonic speeds at the minimum Λ, and the ability of the aircraft to be trimmed at the given n_z(i.e., C_L) at M > 1 and the maximum Λ, Fig. 26. The classical variable sweep design, therefore, allows for a good cruise performance at subsonic speeds at a small Λ, and a good performance when flying at high indicated speeds at low altitudes at the maximum Λ, because in these modes of flight, the magnitude of C_L is small and a large margin of longitudinal stability $C_{m_{C_L}}$ is not important.

Another way to expand the aircraft trimability at M > 1 and the maximum Λ is to use forward lifting surfaces that can be extended with the increasing sweep angle of the wings, thereby reducing the AC travel backwards. Such a solution is used on the Grumman F-14 aircraft (see Fig. 1, sketch 17).

For a canard aircraft with variable wing sweep angle, the AC travel will be significantly different from the AC behavior of a wing-tail aircraft. The AC displacement due to canard from $(\overline{x}_{AC})_{W-B}$ is determined by the change in the parameter $\overline{c}_{AC,y}$ and the ratio $C_{L,canard_\alpha}/C_{L,w_\alpha}$, which increases significantly with an increasing sweep angle due to the decreasing C_{L,w_α}. As a result, if the canard area is large enough, the

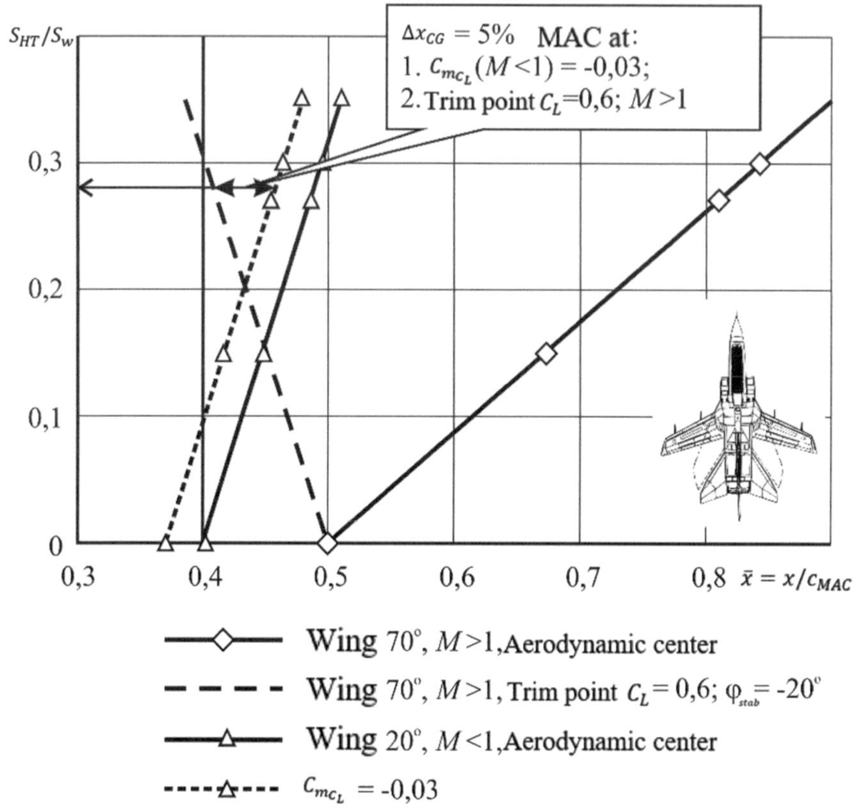

Fig. 26 Choice of HT area for a variable sweep aircraft

canard aircraft AC, unlike that of the wing-tail layout, may shift forward with the transition to M > 1.

The variable sweep canard was not implemented in practice, although such designs were considered—for example, in Fig. 27a shows one of the designs.

The reasons for this can be understood by analyzing the pitch balance of such a configuration. Figure 27b shows a diagram in terms of $\overline{S}_{canard}$ and $\overline{x}_{AC}$ for the aircraft presented in Fig. 27a. With the maximum sweep angle, the following design cases are critical for selection of the canard area:

- trimming the aircraft at the given C_L and M > 1;
- the allowable degree of instability at M <1.

At M < 1 and with the smallest sweep angle Λ(subsonic cruise configuration) the aircraft is statically stable with $C_{m_{C_L}} \approx -(0.20\ldots0.25)$. This degree of stability is far from optimal, which will lead to a deterioration in performance in subsonic cruise flight modes. Probably this is one of the reasons for the lack of variable-sweep canard aircraft.

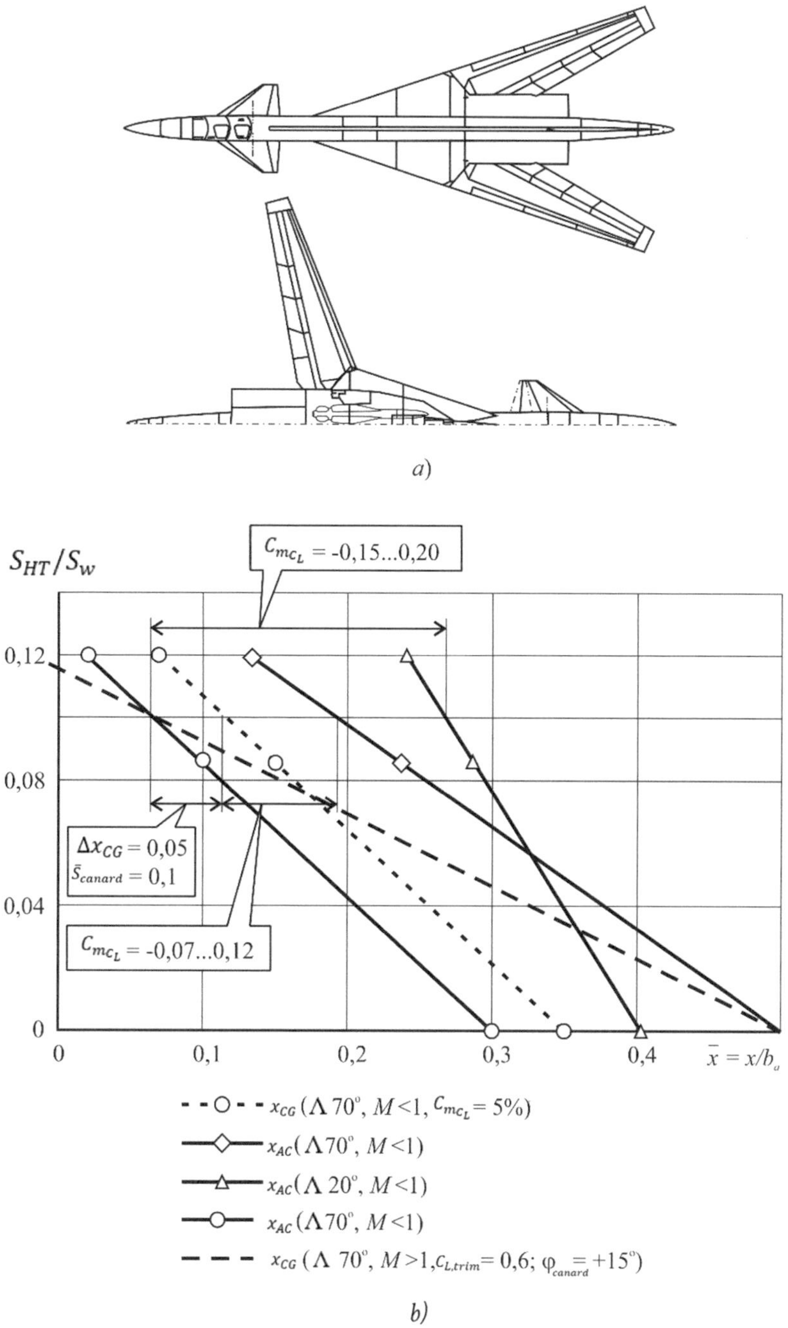

Fig. 27 Diagram for the determination of the foreplane area of a canard aircraft: **a** aircraft layout, **b** determination of the foreplane area

A variable-sweep canard is likely to have good trim characteristics in supersonic cruise flight due to a small stability margin, which in this case is $C_{m_{C_L}} \approx -(0.07\ldots0.12)$, Fig. 27b. Therefore, this layout is applicable for an aircraft with supersonic cruising flight. This conclusion is partially confirmed by the XB-70 "Valkyrie" aircraft designed to fly at M $\sim$ 3, since the downward deflected leading-edges, in terms of longitudinal characteristics, are equivalent to an increase in the wing sweep.

It should be taken into account that the mechanism for changing the wing sweep or another mechanism for changing the geometry of an aircraft will increase the weight of the structure, and also requires a solution to the problem of ensuring safety in case of possible failures. For combat aircraft, these issues have been solved, while the possibility of using variable sweep on passenger aircraft seems to be unlikely at this time.

2.6 Comparison of Different Layouts with Regard to Trimability. Performance Improvement Possibilities

Let us compare the considered aerodynamic layouts on the basis of the lift that can be generated to trim the aircraft.

At low flight speeds, the trimming lift affects the field characteristics of an aircraft. From the previous discussion, it can be seen that the highest lift coefficient $C_{L,max}$ during takeoff and landing is one of the defining ones. In the wing-tail configuration, due to multi-slotted high-lift devices, the maximum lift coefficient $C_{L,max}$ in trim may be as high as 2.5…3.0.

For a canard of tailless aircraft, the use of high-lift devices for landing is limited and the main means to achieve low landing speeds is to increase the wing area and use additional devices, partially discussed in Sects. 2.3 and 2.4.

In cruise, in order to obtain the highest possible lift coefficient C_L, high-lift devices are also used. On fighter aircraft, these are drooped leading-edges, flaps, flaperons, ailerons, which are deflected depending on the flight parameters. On passenger, transport and non-maneuverable combat aircraft, drooped leading-edges and flaps are used. A typical variation of the maximum lift coefficient with the longitudinal stability (instability) for the wing-tail layout and a canard aircraft can be seen in Fig. 28.

A comparison of the wing-tail and canard layouts in terms of induced drag in trim condition at subsonic velocities is shown in Fig. 29. The polars for trim-off condition of both aircraft are almost identical, Fig. 30a. At $C_{m_{C_L}} = 0,$ the canard configuration is inferior to the wing-tail layout, Fig. 30b. With the optimal value of $C_{m_{C_L}} = 0.15,$ the polar of a wing-tail aircraft in trim condition is slightly better than that of the canard configuration (Fig. 30c), while with the optimal value of $C_{m_{C_L}} = 0.30$ the polar of a canard aircraft is better than that of the wing-tail layout, Fig. 30c.

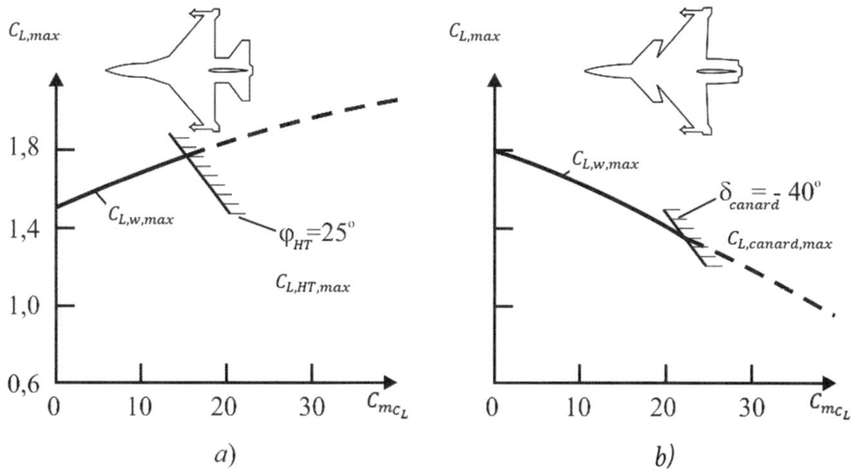

Fig. 28 Variation of $C_{L,max}$ as a function of C_{mC_L} for different aircraft layouts

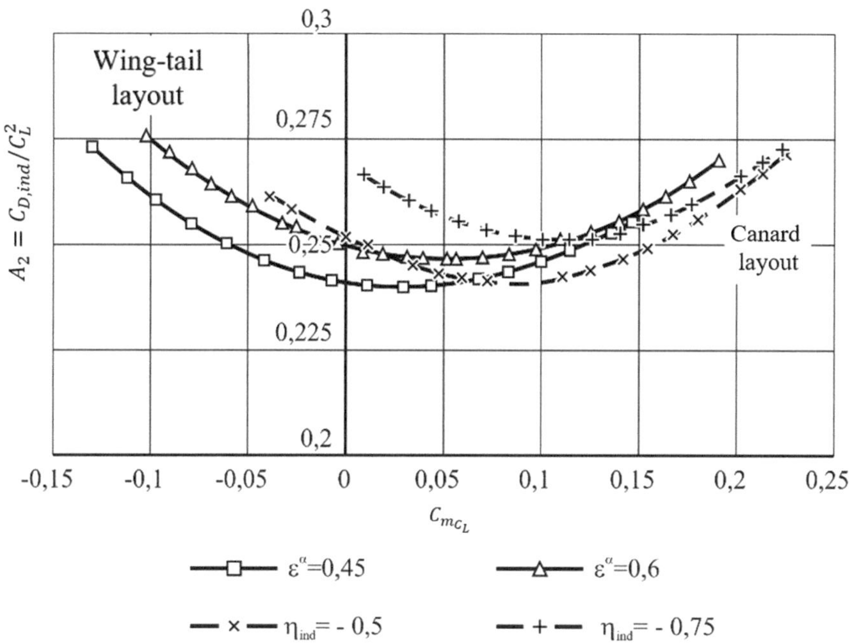

Fig. 29 Polars for different aircraft layouts

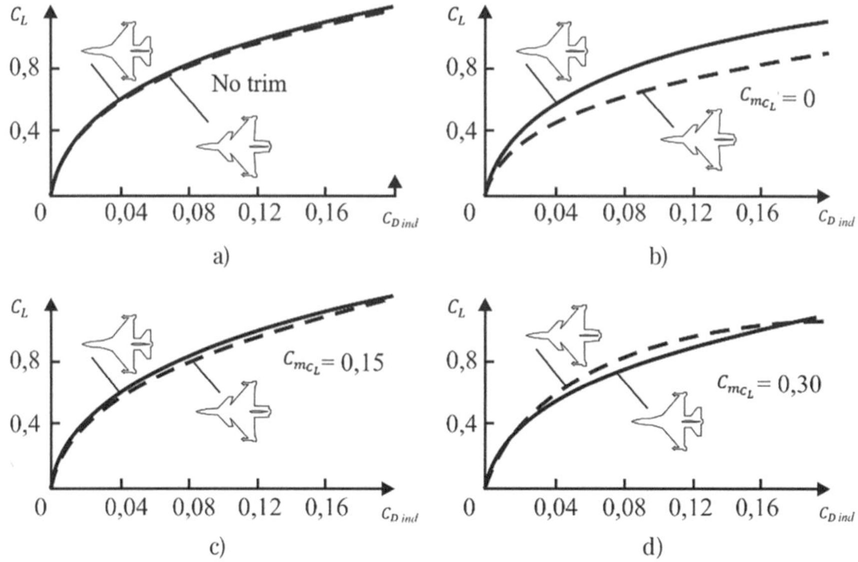

Fig. 30 $C_L = f(C_{D,ind})$ curves for the wing-tail and canard configurations

It should be borne in mind that the high-lift devices on the leading edge of a wing, located behind the foreplane, are not efficient because of the flow downwash due to the foreplane. Therefore, the inner portion of the wing of a canard aircraft doesn't have slats or drooped leading edge. Foreplane can be used as a high-lift device for this portion of the wing, but its deflection angle is not optimal, since it is determined by the requirement to trim the aircraft. Therefore, high-lift devices are somewhat more effective in the wing-tail configuration, where they can be used in the root portion of the wing with the optimized deflection angles.

In the transition to supersonic flight speeds, the AC travel of a canard aircraft is slightly smaller than that of a wing-tail aircraft, as shown in Fig. 31a. This is due to the fact that due to the reduced wing-tail interference (the parameter ε^{α} for the wing-tail layout and $|\eta_{ind}|$ for a canard aircraft) at M > 1 the AC travel per unit tail volume coefficient increases. Therefore, the HT shifts the AC farther backward, increasing the AC travel due wing, and the canard shifts it forward, reducing the wing AC shift. As a result, the canard layout will have a small degree of stability at supersonic speeds, Fig. 31b. Consequently, the wing-tail configuration compared to the canard configuration will suffer less trim drag penalties at subsonic speeds, but more at supersonic speeds.

The tailless layout has the lowest minimum drag C_{D0} at all Mach numbers among all configurations. As already noted, it is practically impossible to build in an instability of more than 5% of MAC on a tailless configuration aircraft. At the same time, despite the fact that the $C_L = f(C_{D,ind})$ curve of a tailless aircraft is worse than

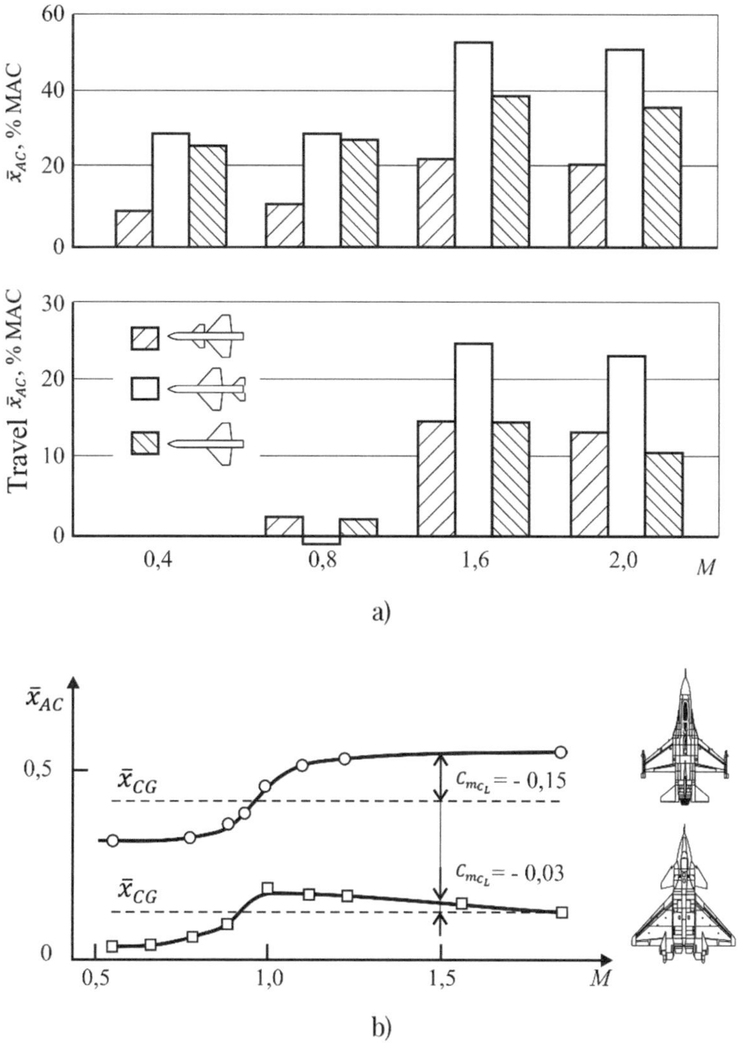

Fig. 31 AC travel as a function of the Mach number for aircraft of different layouts

that of the wing-tail layout and a canard aircraft, it can have an advantage of a low minimum drag. These features make the tailless layout the most advantageous for a small maneuverable aircraft, especially supersonic (Tu-144, "Concorde", SR-71, B-58). Also, many maneuverable fighter aircraft use this layout (F-106, "Mirage-3", "Mirage-2000").

The drag in the wing-tail and canard configurations is reduced by reducing the aerodynamic interference between the wing and the tail. The interference can be reduced, for example, in the following ways:

- wing-tail airplanes (Fig. 32a):

  the placement of the HT below the wing plane or the use of an anhedral angle for the HT;
  the placement of the HT above the wing plane by using the T- or V- tail or a dihedral angle for the HT (as a rule, on maneuverable or non-maneuverable aircraft, the pitching moment curves may be non-linear);

- canard airplanes:

  the placement of the canard above the wing plane, the use of dihedral for the foreplane (also, this improves the pitching moment curve at large angles of attack), Fig. 32b;
  the placement of the foreplane below the wing plane (which yields significant nonlinearities in the pitching moment curve) at medium angles of attack, so it is used only on low-maneuverability aircraft, for example, B. Rutan "Quickie", Fig. 1, sketch 9).

Important parameters in selecting aircraft layouts are the CG limits. The design CG travel is determined by the payload, fuel system configuration, and other factors. On the other hand, there are requirements for acceptable CG travel, trimability, control surface parameters, and control system authority. The above considerations related to trimability of aircraft of different configurations show that the required CG travel can be most easily ensured by using a wing-tail layout. For the canard configuration, the requirement for a wide CG range yields the use of a foreplane and elevons, as well as longitudinal static instability. In the tailless configuration, additional balancing surfaces may be required.

For non-maneuverable aircraft using canard and tailless configurations, in-flight fuel transfer can be used to change the position of the center of mass and maintain small margins of longitudinal stability at supersonic speeds. This technical solution is most effective with a long fuselage that can accommodate the necessary number of fuel tanks. Examples of fuel transfer aircraft are the Tu-144 (Tupolev) and T-4 (Sukhoi). On maneuverable fighter aircraft, fuel-shifting has not been used so far due to the design difficulties of its implementation on planes of such size and frequent and rapid changes of flight modes.

For maneuverable aircraft, it is important to ensure the design aft CG limit $x_{CG,aft}$ for the use of the entire range of possible payloads. From these positions, the most aft CG limit $\overline{x}_{CG,aft}$ (in terms of MAC) can be increased using the wing-tail layout. This, as a rule, is also true for the CG position measured in terms of the fuselage length. However, it should be taken into account that the translation of the CG measured in units of MAC $\overline{x}_{CG,aft}$ to the units of fuselage length $x_{CG,aft}$ will differ significantly for different aircraft configurations and the decision on the acceptability of one or another layout with respect to this indicator can be only be made based on the results of the detailed design.

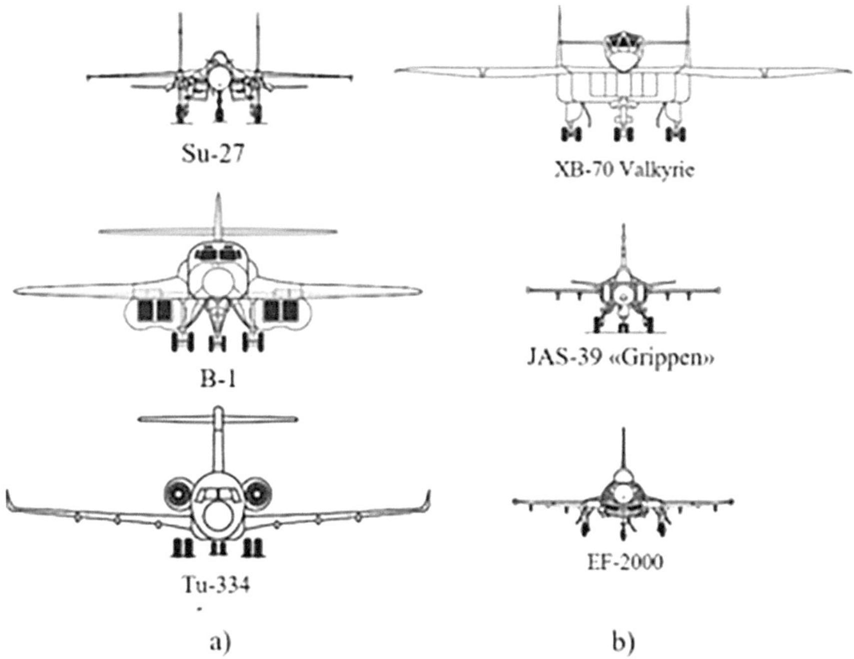

Fig. 32 Typical examples of various aircraft configurations: **a** wing-tail layout, **b** canard layout

When selecting the aerodynamic configuration for an aircraft with a booster or remote-control system, control system parameters, which include the control surfaces hinge moments (determining the structural weight and the actuator weight) and control cost in terms of energy spend—the total power of actuators (determining the power and weight of the aircraft hydraulic system and related units, such as coolers) should be taken into account.

The estimation of typical control surfaces' hinge moments in the form of the generalized dimensionless parameter $\frac{M_{h,controlsurf}}{Gn_z c_{MAC,controlsurf}}$ for different layouts was presented in Figs. 12, 16 and 24. As for all configurations the maximum values are reached at M > 1, Fig. 33 shows this parameter at M = 1,5 as a function of the longitudinal instability of an aircraft $C_{m_{C_L}}$.

The figure shows that the wing-tail layout has the smallest hinge moment, provided that the instability magnitude is optimized for each layout. At the same time, however, it is necessary to take into account the purpose of the aircraft (in this case, it is the design load factor), as well as the use of other control surfaces on the canard (foreplane) and tailless aircraft (elevons), as it allows to reduce the control hinge moments and make these layouts competitive with the wing-tail layout with respect to this parameter.

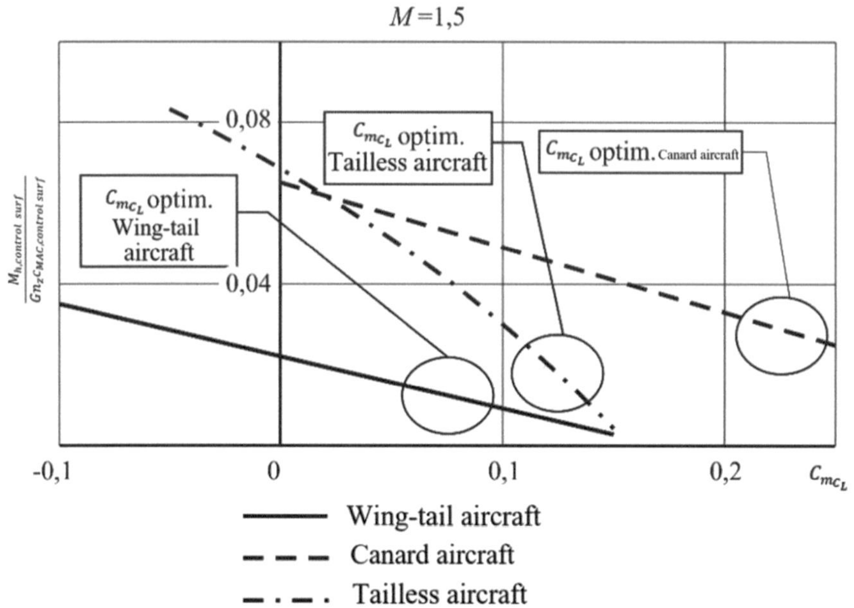

Fig. 33 Comparison of the hinge moments of aircraft of different layouts

The energy available for control is determined by the power of the actuators, which is proportional to the product of the maximum hinge moment and the maximum deflection rate of the control surface; when two control surfaces (e.g., foreplane and elevon) are used, their powers are summed. The maximum hinge moments for all configurations are achieved at supersonic speeds and the maximum deflection speed is achieved at subsonic speeds, where the aircraft may be longitudinally unstable. The required surface deflection rate increases rapidly as the degree of longitudinal instability increases.

For aircraft of the same class and similar size, the minimum required control power will be achieved, as a rule, by the wing-tail configuration. This is due to the lower required magnitude of longitudinal instability of the wing-tail configuration and the use of uncompensated control surfaces of the elevon type on the tailless and possibly canard aircraft, which will always have larger hinge moments than the compensated surfaces of a HT or a foreplane.

To conclude this section, let us briefly list the main features of different aerodynamic designs, which it is advisable to consider when selecting the aerodynamic design of an aircraft.

Wing-tail layout

- can provide the best flight performance, field, and maneuvering characteristics at subsonic speeds, and a high level of these characteristics at supersonic speeds,

- is associated with the lowest technical risk in terms of trim, stability and control system, even with longitudinal static instability,
- provides the widest possible operating CG range,
- generally, has the lowest energy costs for control,
- allows the use of variable wing geometry to ensure high supersonic cruise and field performance on a single aircraft.

Canard layout (using a foreplane and elevons)

- allows to achieve the best maneuvering characteristics and good supersonic flight performance,
- is associated with an increased technical risk in case of longitudinal static instability, requires a thorough development of the aerodynamic layout and control system at the design stage,
- will generally have a higher level of energy cost associated with the control system,
- a large wing area may be required to ensure good field performance,
- allows the use of variable wing geometry to improve the supersonic flight performance.

Tailless layout

- provides the best performance at supersonic speeds,
- the flying wing type provides the best performance at subsonic speeds,
- allows for a high level of maneuverability,
- is associated with an increased technical risk in case of longitudinal static instability, requires a thorough development of the aerodynamic layout and control system at the design stage,
- may require the use of a CG position control system via fuel transfer, especially for a multi-mode aircraft,
- will generally have a higher level of energy cost associated with the control system,
- to improve the field performance, an increase in wing area or the use of special aerodynamic devices can be required.

Additional considerations for the selection of aerodynamic layout

When choosing one of the three aerodynamic layouts, it should be borne in mind that due to the difficulty in obtaining large values of $\Delta C_{L,h-l}$ (especially for the tailless configuration), the tailless and canard layouts aircraft have to reach a high angle of attack during take-off and landing. This makes it impossible (or difficult) to use on such aircraft swept wings of large and medium aspect ratio, as this is associated with a very long landing gear.

As a consequence, only low-aspect ratio delta, blended-wing-body, ogival or sickle-shaped wings can be used for high-speed aircraft with the canard and tailless configurations. Because of the small aspect ratio such wings have a low lift-to-drag ratio in subsonic flight modes.

Therefore, the canard and tailless layouts should be used for aircraft whose primary mode of flight is supersonic flight. And it is natural that the wing-tail configuration is most appropriate for subsonic aircraft or aircraft in which the subsonic speed flight mode prevails over supersonic flight modes.

3 Selection of Aircraft Layout by Its Individual Features

Interference drag is caused by the mutual influence of aircraft parts. The mid-wing airplane causes the least interference drag. For this reason, most military aircraft feature a mid-wing unless the middle part of the wing, the wing carry-through structure, interferes with either the cargo compartment in the fuselage or the air ducts running from the nose of the aircraft to the engines located in the tail of the fuselage.

Typically, passenger and transport aircraft do not use this layout since the wing carry-through, passing in the middle of the fuselage, prevents the installation of a single passenger or cargo cabin. In wide-body passenger aircraft with a fuselage diameter of more than 5 m, it is possible to shift the wing up so that the distance between the upper surface of the wing carry-through and the upper fuselage outline is at least 3.5 m.

Almost the same interference as a mid-wing airplane is produced by a high-wing airplane (Fig. 34). The high-wing design has the following disadvantages:

- the landing gear cannot be placed on the wing, or (on small aircraft) the main landing gear supports are high and heavy. In this case, the landing gear is usually placed on the fuselage, loading it with large point forces. These forces, together with the loads arising from the fuselage sealing, significantly reduce the survivability of its structure, reducing its service life.

Fig. 34 Polars of aircraft with different wing positions with respect to the fuselage

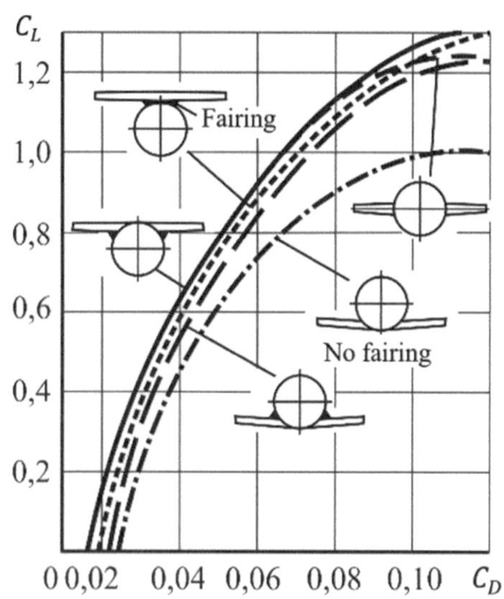

- in case of an emergency landing, the wings (especially if they carry the engines) tend to destroy the fuselage and the cargo or passenger cabin inside it. To eliminate this possibility, the fuselage structure in the wing area must be strengthened, which increases its weight.
- in the event of an emergency landing on water, the fuselage goes under the water's surface, making emergency evacuation of passengers difficult.

At the same time, aircraft with a high wing have the great advantage of a low distance from the bottom of the fuselage to the ground. This feature makes all existing military transport aircraft use a high wing. This makes it possible to easily and quickly load and unload military equipment, cargo, and people from the aircraft by air stairs or a ramp built into the lower part of the fuselage from the back or front.

The low-wing causes the greatest interference drag. Figure 34 shows that the installation of special fairings at the junction of the wing with the fuselage can significantly reduce this drag.

The advantage of a low wing is the possibility of placing the landing gear on it, ensuring greater safety for the aircraft in case of an emergency landing. Disadvantages of the low-wing are difficulties (because of the proximity of the wing to the ground) in placing the turbojet engine under the wing and the turboprop engine on the wing and the need to use a dihedral angle for the wing, which yields the need for automated lateral control for the aircraft.

Other aircraft configuration features may be analyzed in a similar way. For example, the attributes and layouts of a fuselage (cross-sectional shape, the length of the nose and tail, etc.) are discussed in Chap. 17.

4 Selection of the Type, Number and Layout of Engines for an Aircraft

4.1 Engine Types

The following types of engines are used for the propulsion system of modern aircraft: jet engines (JE), piston engines (PE) and liquid rocket engines (LRE). Jet engines, in their turn, are divided into gas turbine (GTE) and ramjet (RE).

At present, the most widespread in aviation are gas-turbine engines. This class of aircraft engines includes:

- turbojet engines;
- augmented turbojet engines;
- bypass turbojet engines;
- bypass augmented turbojet engines;
- turboprop engines;
- turbofan engines.

Piston engines are now installed only on very light aircraft.
Liquid propellant jet engines are used on experimental aircraft.
Ramjet engines will find application in hypersonic aircraft.

Simplicity of gas turbine engines, based on the use of only rotary motion, provides it with high reliability and long life, which caused the jet engines to become the main engines in aviation.

4.2 Engines for Subsonic Aircraft

The conventional type of engine for subsonic aircraft is a piston engine. Currently, only low-power piston engines are produced, and they are installed on very light aircraft: recreational, aerobatic, agricultural, etc. The low-power piston engine is still superior to low-power jet engines in specific fuel consumption, but this advantage is gradually decreasing, and the piston engine has begun to be replaced by other types of engines (turboprop and bypass turbojet) and on light aircraft.

The turboprop engine, until recently, has been less and less used in aviation since, at high flight speeds, it has insufficient power density and creates an increased noise level and vibration in the aircraft. However, considerably lower specific fuel consumption than that of turbofans (approximately by 1.5 times) in conditions of high prices of hydrocarbon fuel (kerosene) forces us to reevaluate the possibility of using the turbofans on subsonic aircraft designed for $M = 0.75–0.80$. At the same time, various variants of turbofans with multi-blade propellers of the fan type are investigated.

The most promising power plant for a subsonic aircraft is the bypass turbojet engine. The thrust produced by a modern turbojet engine per 1 kg of engine mass is 3–5 times that of a piston engine. This advantage becomes especially important if we take into account the fact that the requirement for power has increased by more than 20 times in comparison with the aircraft of the late 1940s of the last century, when the maximum power of a combustion engine was reached.

The desire to preserve the advantages of turbofan and bypass turbojet engines led to the development of propfan engines.

4.3 Engines for Supersonic Aircraft

Supersonic aircraft can be divided into two types: aircraft that perform a significant portion of the cruise flight at high subsonic speed (multirole fighters, etc.) and aircraft whose cruising flight is performed mainly at supersonic speed (supersonic transport, etc.).

For the first type of supersonic aircraft, the bypass augmented turbojet engine has the greatest application at present.

As bypass turbojet engines, bypass augmented turbojet engine is reasonable to use at high subsonic speeds. To achieve supersonic flight speeds (up to $M = 2.5$), a significant increase in engine thrust is required, which is achieved by turning on the afterburner. This increases thrust by 30…50%, however, specific fuel consumption increases with a factor of 2.0…2.5. In other words, it is difficult to obtain a greater range at supersonic speed.

In contrast to modern subsonic turbofans, supersonic turbofans have a relatively low by-pass ratio at $M = 0.5...1, 3$. For a better cost efficiency (reduction of specific fuel consumption), at subsonic flight and for a bypass augmented turbojet engine a greater bypass ratio is desirable. However, a decrease in the engine specific thrust with an increase in the bypass ratio leads to an increase in the engine diameter, which causes a significant increase in drag of the power plant at transonic and supersonic flight speeds (see Chap. 19).

For a long-range supersonic flight, there are two speed ranges: a speed range at $M \approx 2$, which has the advantage that the airframe can be constructed of conventional aluminum alloys, and a speed range at $M \approx 3$, which although posing some problems allows for a greater flight efficiency.

For the first speed range, both augmented turbojet engines and bypass augmented turbojet engines are used. The absence of a significant difference in the characteristics of non-bypass and bypass engines at supersonic speeds is explained by the fact that, when the bypass ratio is increased from 0 to 0.5, the specific fuel consumption decreases insignificantly (by about 1%). A further increase in the by-pass ratio worsens the characteristics of the power plant.

For a range of cruising flight speeds corresponding to $M \approx 3$, the best engine is an augmented turbojet engine with a low degree of pressure increase in the compressor $\pi_c^* = 3...4$ (at M > 3.5 the compressor becomes almost unnecessary).

4.4 Engines for Hypersonic Aircraft

The use of gas turbine engines for aircraft propulsion systems is limited by $M \approx 3.5$. For a high-speed flight, a subsonic combustion ramjet (M < 6) and a supersonic combustion ramjet (M < 6) would be more efficient.

The cost efficiency of an engine can be estimated in the first approximation by the value of its specific impulse J_f. The variation of J_f with Mach number for aircraft engines is shown in Fig. 35.

Liquid-propellant jet engines operating on chemical fuel have now reached a level of development where further increases in specific impulse are becoming increasingly slow and costly. In addition, the specific impulse of the liquid propellant rocket engine remains insufficient to install it on a hypersonic aircraft with a long endurance.

It should be noted that, despite the increase in specific fuel consumption at supersonic speeds, the full efficiency of an air-jet engine increases significantly. Total engine efficiency, as it is known, takes into account all losses in the process of

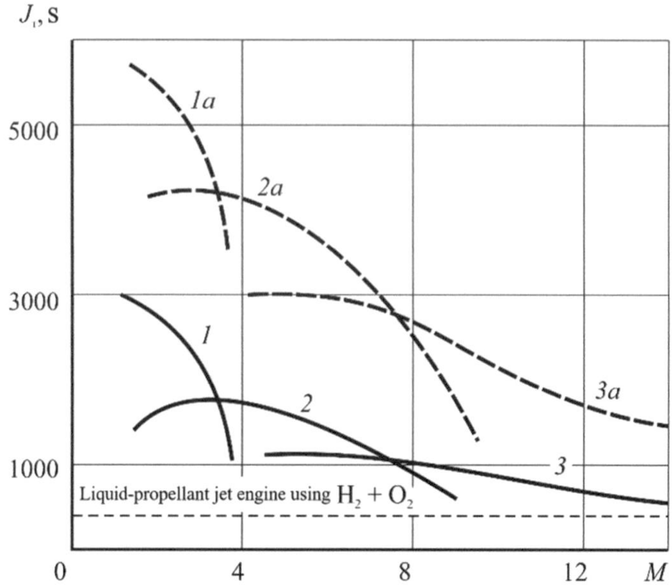

Fig. 35 The jet engine specific impulse J_f as a function of the Mach number for different types of kerosene-fueled engines (*1*; *2*; *3*) and hydrogen-fueled engines (*1a*; *2a*; *Za*): *1*, *1a*—jet engine; *2*, *2a*—ramjet with subsonic combustion; *3*, *3a*—ramjet with supersonic combustion

conversion of fuel energy into useful thrust work. For hydrocarbon fuel (kerosene), the total efficiency of a jet engine can be written as $\eta_T \approx 0.00082\frac{V}{C_p}$.

The total engine efficiency of a subsonic airplane (M = 0.85) is approximately 24%. At M = 2, it increases to 38% (which exceeds the thermal efficiency of the best modern power plants), and at M = 3 $\eta_T = 46\%$. The efficiency growth continues for hypersonic speeds achieved by a ramjet.

The most promising for hypersonic aircraft seem to be combined engines—turbo-ramjet engines.

4.5 Selecting the Number of Engines

The number of engines for an airplane power plant depends on a number of factors, determined both by the purpose of the aircraft and its basic parameters and performance characteristics. The contradictory influence of the number of engines on safety, efficiency and regularity of flights leads to the fact that the choice of the number of engines still remains a rather complex problem of aircraft design.

In the early period of aviation development, when the first multi-engine airplanes appeared, it turned out, for example, that regularity of flights on such airplanes was significantly lower than on light single-engine airplanes. The reason for this (although it sounds paradoxical nowadays) was a relatively large number of engines on these aircraft, because the frequency of engine failures is directly proportional to their number.

At present, the situation has changed significantly. The power plant of modern aircraft is designed with consideration of independent operation of engines. The life of jet engines reaches several thousand hours. Therefore, the probability of modern engine failure in flight has significantly decreased.

If safe takeoff and flight termination are ensured in case of failure of one engine, a passenger airplane with two engines is considered to be fully operational (for example, Tu-204, Tu-134, A-320, DC-9, B-737, B-757, etc.).

What is the basis for selecting the number of engines for the aircraft to be designed? In general terms, the requirements for all aircraft when selecting the number of engines can be formulated as follows:

- the aircraft must have the necessary take-off thrust-to-weight ratio;
- the aircraft must be sufficiently reliable and cost-efficient;
- the effective thrust of the propulsion system should be as large as possible;
- the relative cost of the motor should be as low as possible.

By determining the take-off thrust-to-weight ratio of the designed aircraft, it is possible to find the number of engines using the following ratio:

$$n_{eng} = \left(T_{take-off} / W_{take-off}\right) W_{take-off} / T_{take-off,i},$$

where $T_{take-off,i}$ is the take-off thrust of one engine.

The formal approach can provide the required value of $T_{take-off} / W_{take-off}$ with any number of engines (depending on the value of $T_{take-off,i}$). Therefore, when addressing this issue, it is also necessary to take into account the aircraft's reliability and efficiency requirements, the aircraft's specific purpose and the requirements imposed on its layout and propulsion system. The degree of engine readiness at the time of aircraft design completion also plays an important role in this matter.

Obviously, it is impossible to impose exactly the same requirements for reliability and efficiency on all aircraft without taking into account their purpose, type, and size. Therefore, it is difficult to give a single criterion for selecting the number of engines that would more or less equally meet the requirements for all aircraft. As a first approximation, the number of engines can only be related to the take-off weight of the aircraft.

On light military and civil aircraft, the number of engines is often determined by the availability of a production-ready engine, considerations of aircraft survivability (attack aircraft, fighter-bombers), and the characteristics of the aircraft layout. Heavy fighters and attack aircraft are usually equipped with two engines, not only to ensure their damage survivability but also to keep the expensive aircraft in peacetime in case one of the engines fails. Modern light aircraft, as a rule, are equipped with 1 or 2 engines (although there are examples when 3 engines are installed on light aircraft).

One of the decisive requirements for choosing the number of engines for heavy aircraft is to ensure a safe takeoff in the event of a single engine failure. The requirements for ensuring proper safety depending on the number of engines are outlined in Chap. 5. It also shows the algorithm for selecting the thrust-to-weight ratio of modern aircraft at the initial design stage.

The modern design methodology requires a normalized climb gradient at the third section of an airplane's takeoff, taking into account the failure of one of the engines. By assuming, as a first approximation, the same values of this gradient for two-, three-, and four-engined aircraft (and they are indeed close), we obtain the values of real thrust-to-weight ratios with all engines operating $\left(T_{take-off}/W_{take-off}\right)_n$. The graph in Fig. 36 shows that in this case, which is typical for day-to-day operation, the required thrust-to-weight ratio of a twin-engined aircraft is significantly greater than that of a three-engined and even more so than that of a four-engined aircraft.

The probability of engine failure on takeoff is extremely low (approximately one failure per 100,000 takeoffs) and in this case all three aircraft under consideration will be provided with an equal level of safety. In day-to-day operation, however, cases may arise (with a probability of about one per eight flights) requiring the realization of excess engine thrust for normal takeoff and departure from the ground. These include:

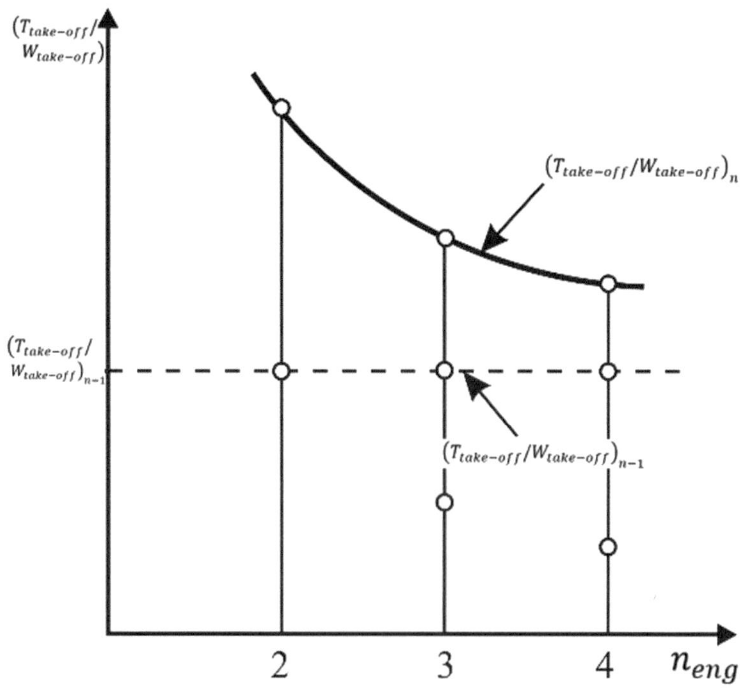

Fig. 36 Thrust-to-weight ratio as a function of the number of engines installed on an aircraft

- adverse weather conditions (cross wind; atmospheric turbulence, especially resulting in downward gusts; lower atmospheric pressure and higher air temperature);
- technical circumstances (wear of equipment, increased aircraft drag due to door covers, hatches, etc.)
- human factor, i.e. piloting errors.

The ability of an aircraft to counteract these circumstances is determined only by the excess thrust, and, therefore, a twin-engine aircraft on takeoff—the most difficult part of the flight envelope—is the most rcliable and safe, as it is less sensitive to the effects of the external environment. In addition, the probability of engine failure on takeoff (other conditions being equal) is smaller with a smaller number of engines on the aircraft, which is confirmed by statistics. As for the economic and operational sides of the issue, it will always be preferable to design and build aircraft with a smaller number of engines.

This does not at all exclude the expediency of building heavy long-range aircraft with three or four turbojet engines because they have better aerodynamic characteristics and a lower weight ratio due to their lower thrust-to-weight ratio. These properties enable multi-engine aircraft to have advantages when flying over 6–8 thousand kilometers. Airbus's assessment of rational areas of application for twin- and quad-engine aircraft confirms this conclusion.

4.6 Engine Placement on Aircraft

There are different options for placement of a turbojet engine on aircraft: in the fuselage, in the root of the wing, on pylons under the wing, under the wing (with the direct attachment of the engine nacelle to the wing), on pylons over the wing, over the wing (with the direct attachment of the nacelle to the wing), on the wing tips, on the tail of the fuselage, etc. All these configurations are used in jet aircraft (subsonic and supersonic, civil and military). Each configuration has certain advantages and disadvantages.

Let's take a look at the most common engine mounting options.

The placement of engines in the fuselage is used in almost all light military aircraft (multi-purpose fighters, etc.). The low aspect-ratio wing does not allow the engines to be mounted on or in the wing without compromising the high-lift devices on the wing. The engine nacelles would take up too much space, not only significantly reducing the span of the flaps, but also eliminating the possibility of suspending a combat load from the wing.

In addition, if the aircraft has one large jet engine (the thrust-to-weight ratio of fighters is the highest), then the fuselage is the only place to mount such an engine.

Placing engines in the root of the wing has been widely used in heavy subsonic jets for military and civilian applications (Tu-16, Tu-104, Tu-124, etc.; British aircraft

Vulcan, Victor, Comet, etc.). This option, while having certain advantages (the failure of one or two engines, placed on one side, does not cause large yawing and rolling moments; the air intake clearance prevents foreign objects from being sucked in; a relatively low aerodynamic drag of the propulsion system; etc.), has some significant shortcomings (especially for passenger aircraft). The list of disadvantages includes:

- proximity of the jet to the fuselage skin; intense noise in the passenger cabin;
- long air intakes reduce engine thrust on takeoff by 5–6% and by 2–3% in cruise flight;
- the fire originating in the engines may spread to the passenger cabin and fuel tanks (enhanced fire protection is required);
- if compressor or turbine blades are destroyed, the passenger cabin and fuel tanks may be hit (special armoring is required);
- the intakes on the leading edge of the wing and exhaust pipes on the trailing edge reduce the capability of the wing's high-lift devices;
- the installation of thrust reversal devices is hampered. A jet of gas directed downward and forward and reflected from the airfield surface can be sucked into the air intakes of the engines, causing a surge;
- poor engine operating conditions due to the maintenance difficulty;
- the wing volume available for fuel storage is significantly reduced;
- an increase in the structural weight due to a heavier wing (due to cutouts in the spars and removable panels) and due to long air intakes or exhaust pipes.

The placement of engines on pylons under the wing is widely used in modern heavy aircraft (subsonic and supersonic).

Such a configuration has the following advantages:

- the engines unload the structure in flight, reducing the bending moment, which leads to a reduction in the weight of the wing structure;
- the engines dampen the wing oscillations during flight in turbulent atmospheres and serve as anti-flutter balancers;
- the engine is easy to replace with another one (with larger dimensions);
- good engine maintenance conditions are ensured;
- the engine can be reliably isolated from the wing by means of firewalls in the pylon;
- less noise from the engines in the passenger cabin is ensured;
- the airplane structure is better insulated from the sound and pulsating pressures of the jet engines than the engines installed in the root of the wing;
- the engine can be easily fitted with thrust reversers and silencers;
- this configuration provides protection to passengers by means of the wing box in case of a blowout of the compressor or engine turbine blades.

In addition to the above advantages, the placement of engines on pylons under the wing has the following disadvantages:

- in the event of an engine stop, especially an outboard engine, a large yawing moment is generated;

- to ensure ground clearance during landing with the roll angle up to 4°, it is required to use a wing with a large dihedral, which worsens the stability and control characteristics of the aircraft and requires an automated control system;
- when engines are positioned low relative to the airfield surface, foreign objects may get into the air intakes;
- the use of pylons makes it difficult to install flaps over the entire wingspan, as the engine jet can destroy them during takeoff;
- the placement of engines at the aft portion of the fuselage is widespread in passenger aircraft.

On heavy aircraft with four engines, two engines are installed on each side of the fuselage on pylons (for example, the IL-62).

If the required thrust-to-weight ratio is provided by an odd number of engines (three engines), one of them must be installed in the plane of symmetry of the aircraft. In this case, the engine is located inside the tail portion of the fuselage, and its air intake is at the root of the vertical tail. The engines used in the Boeing 727, Lockheed L-1011, De Havilland Trident, and Soviet Yak-40, Yak-42, and Tu-154 are based on this design.

Positioning the engines at the tail of the fuselage when the engine nacelle is attached to the fuselage by means of a small pylon allows to:

- provide an aerodynamically "clean" wing, which increases its lift-to-drag ratio;
- maximize the wingspan to accommodate high-lift devices (flaps, slats, etc.), which improve the aircraft's takeoff and landing characteristics;
- determine the wing dihedral to ensure optimal yawing and rolling characteristics;
- improve the longitudinal and lateral stability of the aircraft:
- engine nacelles and pylons act as additional stabilizers;
- the HT is moved out of the wing wake;
- the yawing moment due to the failure of one engine is small;
- improve passenger comfort (compared to the configuration of the engines mounted at the root of the wings) by reducing noise, as the nacelles of the engines in this case are installed behind the pressurized cabin;
- improve (as compared to the configuration of the engines mounted at the root of the wings) the performance of the power plant and the aircraft as a whole by:

  creating sufficiently good conditions for engine maintenance;
  providing the possibility of replacing the engine nacelle when a larger thrust and size engine is required;

- the engine jet flows out at a high altitude relative to the airfield surface (less wear on runways and taxiways);
- prevent foreign objects from entering the engines during takeoff and landing due to the high position of the air intakes;
- improve the thrust reverser performance (as compared to engines placed in the root of the wing);
- create better emergency landing conditions for the aircraft.

However, along with these advantages, the configuration with the engines mounted on the aft fuselage has the following significant drawbacks:

- the weight of the aircraft structure is increased, which is due to:

  the additional weight of the fuselage tail structure due to the additional weight and inertial loads from the engines (the fuselage structure weight increases by about 7–11%);
  the increasing wing weight (by approximately 8–12%) due to the lack of the wing bending relief;
  the increasing vertical tail weight due to the installation of the horizontal tail on it;

- the empty aircraft center of gravity is shifted aft and the center of gravity of a fully loaded aircraft is shifted forward, thus creating difficulties in balancing the aircraft in flight (see Chap. 7);
- the nose of the fuselage is pushed forward, which negatively affects the aircraft's yawing and longitudinal stability;
- in the event of aircraft icing, there is a possibility of ice being dropped by wing de-icers into the engines.
- it is somewhat difficult to maintain engines high above the airfield surface.

The placement of engines under the wing (with the direct attachment of engine nacelles to the wing) is found on heavy supersonic aircraft. In this case, the engines can be installed either in a single nacelle located in the aircraft symmetry plane (aircraft "North American" XB-70) or in pairs in two nacelles (passenger aircraft Tu-144 and "Concorde").

Placement of engines in nacelles on the bottom of the wing is explained by the desire to use or even amplify the fields of increased pressures under the wing. When a system of shock waves is formed on an engine nacelle with specially profiled outer contours, an increased pressure—an additional lifting force (hence the term "compression lifting force") acts on the lower surface of the wing. As a result of the interaction of shock waves with the surface of the wing, the aerodynamic performance of the aircraft at supersonic cruise speeds improves significantly (for the XB-70 aircraft approximately by 20%). In addition, the angle of attack of the wing in cruise flight is reduced (as $\left({}^{L}/_{D}\right)_{max}$ is shifted to smaller angles), which also reduces drag.

Other advantages and disadvantages are similar to the configuration with the engines mounted on pylons under the wing.

Installation of engines over the wing on pylons, despite some increase in the flow circulation on the wing in the nacelle area and a good ground clearance of the air intakes, has not been widely used in aircraft for the following reasons:

- the aerodynamic drag of nacelles slightly increases due to their location in the zone of higher velocities;
- the mass and midsection of the nacelles increase due to the complex engine mounting system;
- increased noise level in the passenger compartment;

- difficulties with the installation of the thrust reversal;
- finally, and most importantly, the wing-box doesn't protect passengers from compressor or turbine blades in the event of blowout.

Installation of engines over the wing with the direct placement of the nacelle on the wing can be seen on aircraft AN-72 and AN-74, as well as on experimental short takeoff and landing aircraft. This configuration was used to exploit the Coanda effect, when the engine jet flowing along the upper surface of the wing increases lift. (see Chap. 12).

The choice of the type, number, and location of engines on an aircraft is inextricably linked to the problem of selecting the type and location of air intakes and nozzles. Efficient operation of these elements ensures, above all, the full realization of the engine's potential.

5 Aircraft Structural Design

In the general aircraft development cycle, the structural design starts at the preliminary design stage when the aircraft concept is being considered. However, it is possible to return to this task at subsequent design stages after the verification calculations if any design errors are identified. This leads to new iterative cycles of issuing design documentation and sometimes to making major changes to already-built aircraft.

The structural design process can be conventionally divided into two stages. The first stage is directly related to load transmission (the selection of the method for perception and transfer of forces) and the formation of a generalized structure of the main parts of the aircraft. The structure's configuration is determined by two groups of parameters: geometric (external linear dimensions, shape, and position in space) and topological (characterizing the arrangement of its elements relative to each other). The type of structure determines the choice of structural solutions associated with the way of perception and transformation of external and internal loads in the structural elements. For example, a structural element (e.g., a spar or a bulkhead) may be a truss, beam, or frame. The same structural members may have a shell structure, a monocoque structure, or a semi-monocoque structure.

6 Approach to the Selection of Aircraft Layout

As noted above, the selection of an aircraft configuration begins with the determination of the aircraft trimming technique. Then the specific configuration is refined according to its individual attributes, the variety of which is only partially shown in Fig. 1. The preceding sections of this chapter have described an approach to this process by evaluating each of the design attributes using examples of design selection in terms of wing-to-fuselage position, number of engines, and placement of engines on the aircraft. Layout options for other attributes should also be considered and evaluated. In doing so:

- it is required to ensure that the requirements to the aircraft are fulfilled. For example, for a passenger aircraft, this means ensuring the transportation of a given number of passengers, demonstrating the specified performance;
- the aerodynamic evaluation of the configuration is made in terms of ensuring the maximum lift-to-drag ratio and possibly higher values of $C_{L,max}$ during takeoff and landing (and maneuvering value of $C_{L,allow}$);
- the weight options are evaluated, taking into account the stiffness and strength requirements to the structure and the service life;
- the performance of the aircraft is considered in terms of the ease of aircraft maintenance.

Ultimately, the ability of a configuration to most fully meet the mission requirements and technical specification is evaluated.

Figure 37 shows a matrix of layout attributes for a passenger aircraft and a selection of configuration for a Tu-154 class and a McDonnell-Douglas DC-10 class aircraft.

In the context of computer-aided design system application, the matrix of layout attributes allows synthesizing the aircraft layout by selecting its geometric and technical features, i.e., it is a classifier of layout solutions. The matrix can serve as the basis for building programs to recognize aircraft prototypes by their geometric attributes. In this case, each cell of the matrix should correspond to a certain array of numeric and descriptive data stored in the engineering information bank and containing the characteristics of a particular design solution (availability of a prototype, repeatability, recommended range of application, aerodynamics, weight, and performance characteristics).

In the absence of a recognition program, the matrix is a reference for selecting the concept of a new aircraft.

In the process of creating the layout of a new aircraft, more than in any other stage of aircraft design, the designer's intuition and imagination are involved. The designer considers many functional relationships, the formalization and algorithmization of which are just being developed. This complicates the development of programs for the automated synthesis of aircraft configurations but by no means eliminates their appearance. Completely ignoring the use of computers at the stage of synthesis of aircraft layout at the present time, when the problem of the creation of artificial intelligence is getting closer to its solution, at least, is irrational. Without trusting the computer to make the final decision on the aircraft design, the designer can use the intellectual support of modern expert systems to point him in the direction of finding the best option.

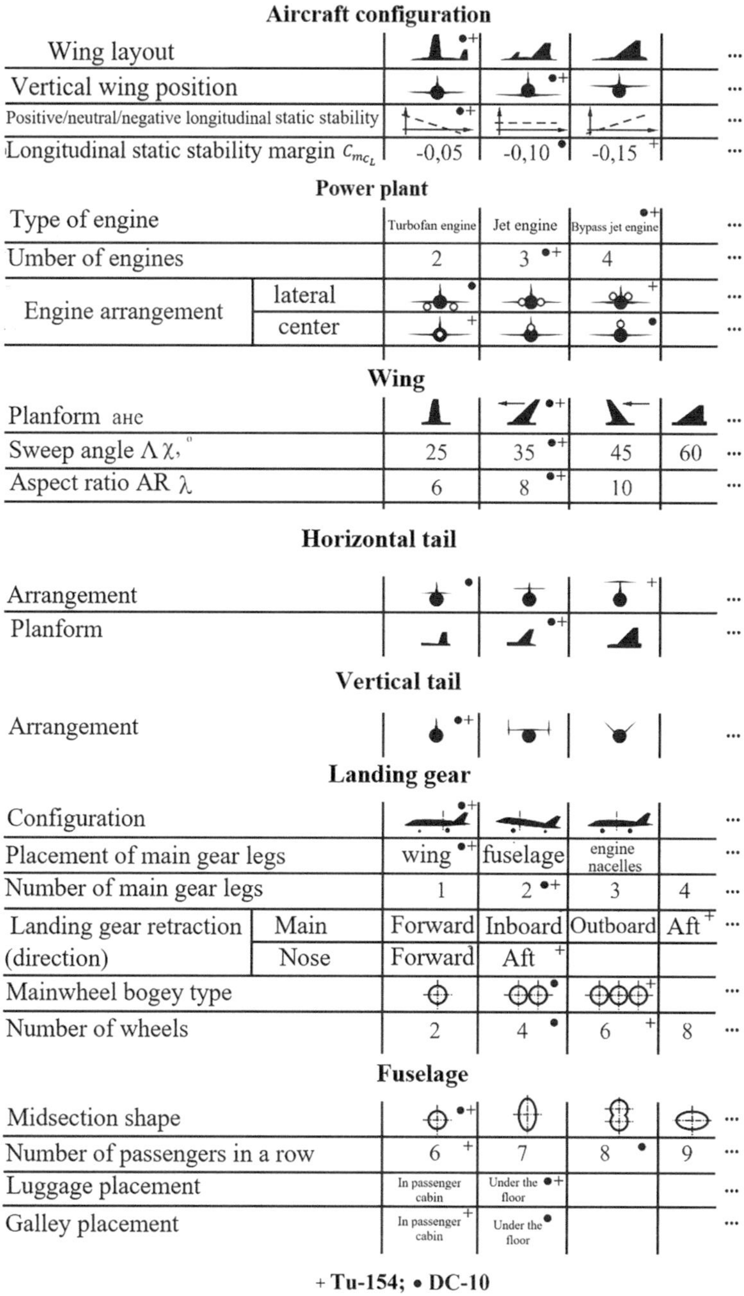

Fig. 37 Matrix of layout attributes under the floor

Variables		
Symbol	Description	Units
L_{HT}	Horizontal tail lift	N
A_W	Wing axial force	N
D_W	Wing drag	N
N_W	Wing normal force	N
R_W	Wing resultant aerodynamic force	N
α	Angle of attack	°
C_L	Lift coefficient	–
C_m	Pitching moment coefficient	–
$\bar{t}$	Thickness ratio	–
c_{MAC}	Wing mean aerodynamic chord	m
$c_{MAC,HT}$	Horizontal tail mean aerodynamic chord	m
W	Aircraft weight	N
n_z	Load factor component in z-direction	–
M	Pitching moment	N m
x_{CG}	Center of gravity MAC position	m
$x_{AC,W}$	Wing aerodynamic center MAC position	m
$x_{AC,HT}$	Horizontal tail aerodynamic center MAC position	m
l_{HT}	Horizontal tail moment arm	m
ε	Downwash angle	°
φ_{HT}	HT incidence angle relative to aircraft longitudinal axis	°
δ	Deflection angle of high-lift devices	°
k_{HT}	Horizontal tail efficiency factor	–
q_{HT}	Tail dynamic pressure	N/m^2
ε^{α}	Downwash angle derivative with respect to angle of attack	1/°
$C_{m_{C_L}}$	Longitudinal static stability margin	–
C_{L,W_α}	Wing lift coefficient derivative with respect to angle of attack	1/°
C_{L,HT_α}	Horizontal tail lift coefficient derivative with respect to angle of attack	1/°
$\bar{S}_{HT} = \frac{S_{HT}}{S_w}$	Horizontal tail area ratio	–
S_w	Wing area	m^2
S_{HT}	Horizontal tail area	m^2
$\bar{x}_{CG}$	Center of gravity location normalized by MAC	–
$\bar{x}_{AC,W}$	Wing aerodynamic center location normalized by MAC	–
$\bar{x}_{AC,HT}$	Horizontal tail aerodynamic center location normalized by MAC	–
$C_{m_{C_L}}$	Pitching moment coefficient derivative with respect to lift coefficient	–
C_{L,HT_α}	Horizontal tail lift coefficient derivative with respect to angle of attack	1/°
C_{L,W_α}	Wing lift coefficient derivative with respect to angle of attack	1/°

(continued)

(continued)

Variables		
Symbol	Description	Units
$\left(C_{m_{C_L}}\right)_{W-B}$	Wing-body pitching moment coefficient derivative with respect to lift coefficient	–
$\Delta C_{L,HL}$	Lift coefficient increase due to high-lift devices	–
$\overline{S}_{HT}$	Ratio of horizontal tail area to wing area	–
$\Delta X_{AC,HT}$	Aerodynamic center displacement due to horizontal tail	m
l_{HT}	Horizontal tail moment arm	m
C_{D0}	Zero lift drag coefficient	–
$C_{Di,W}$	Wing induced drag coefficient	–
$C_{Di,HT}$	Horizontal tail induced drag coefficient	–
AR	Aspect ratio	–
C_{mh-l}	Pitching moment coefficient due to high-lift devices	–
$x_{CGforward}$	Forward center of gravity limit	m
x_{CGaft}	Aft center of gravity limit	m
M_h	Hinge moment	N m
$\overline{X}_h$	HT hinge axis x-position normalized by HT MAC	–
W	Aircraft weight	N
L_{ind}	Canard-induced lift	N
x_{ind}	X-coordinate of point of application of canard-induced lift	m
l_{canard}	Canard moment arm	m
$C_{L,canard_\alpha}$	Canard lift curve slope	1/°
$\overline{S}_{canard}$	Canard-to-wing area ratio	–
$\overline{c}_{AC,y}$	Distance between canard and wing AC	m
$\overline{c}_{ind}$	Distance from the point of application of the induced force to the wing AC	–
V_{canard}	Canard volume coefficient	–
$\Delta C_{m,HL}$	Moment increase due to high-lift devices	–
α_{canard}	Canard angle of attack	°, rad
α_w	Wing angle of attack	°, rad
$\left(C_{L_\alpha}\right)_{canardfeath}$	Feathering canard lift slope	1/°
$C_{L,w}$	Wing lift coefficient	–
$M_{h,canard}$	Canard hinge moment	N m
φ_{canard}	Canard deflection angle	°
$x_{AC,elevon}$	Elevon aerodynamic center	m
L_{elevon}	Elevon lift	N
δ_{elevon}	Elevon deflection angle	°
$C_{L_{\delta elevon}}$	Tailless aircraft lift coefficient derivative with respect to elevon deflection angle	1/°

(continued)

(continued)

Variables		
Symbol	Description	Units
$C_{m,0}$	Zero lift pitching moment coefficient	–
$M_{h,elevon}$	Elevon hinge moment	N m
c_{elevon}	Elevon chord	m
Λ_w	Wing sweep angle	°
η_{T}	Total jet engine efficiency	–
$T_{take-off,i}$	Take-off thrust of one engine	N
n_{eng}	Number of engines	–
$T_{take-off}$	Aircraft take-off thrust	N

Reference

1. Kuchemann D (1979) The aerodynamic design of aircraft. AIAA, USA

Determination of Aircraft Design Parameters

Nikolai K. Liseitsev

Abstract The chapter lists out and explains several categories of dimensional and unitless aircraft design parameters. The chapter also presents a methodology for estimation of these parameters. In particular, the reader will learn how to determine the design wing loading and thrust loading, the aircraft gross mass using the unity equation (in two iterations), as well as the aircraft mass breakdown. The chapter presents some historical data on mass ratios of different components of various classes of aircraft. The chapter explains the Breguet equation for calculating aircraft range and shows how it can be used to estimate the required fuel capacity for a thermal aircraft. Finally, the chapter exemplifies the above calculations for a mid-range commercial airliner.

Keywords Unitless design parameters · Wing loading · Thrust loading · Aircraft gross mass · Unity equation · Breguet equation

1 Design Parameter Nomenclature

The number of parameters characterizing the shape and size of an aircraft in sufficient detail is quite large. It depends on the type of aircraft, the nature of the input information, and the formulation of the design problem. Thus, when designing an airplane for an off-the-shelf engine, the engine parameters are excluded from the search. When designing an aircraft that is a modification of an existing baseline variant, only the parameters of the modified aircraft parts are included in the list of unknowns.

The list of parameters should be determined *using the following rule:* the number of parameters should be the smallest required for the identification of the newly designed aircraft, the correct analysis of the project and the harmonization of its design parameters.

N. K. Liseitsev (✉)
Moscow Aviation Institute (National Research University), Moscow, Russia
e-mail: nikolai.liseitsev@yandex.ru

M. A. Pogosyan (ed.), *Aircraft Design*, Springer Aerospace Technology,
https://doi.org/10.1007/978-981-96-4599-2_5

At the conceptual design stage, within the framework of the selected layout, the designer applies the following design parameters.

1.1 Basic Dimensional Parameters of the Aircraft

These include:

- design take-off mass m_0, [kg];
- baseline wing area S, [m^2];
- takeoff thrust T_0, [daN].

When determining the design parameters of an aircraft and calculating its performance, it is convenient to use reduced parameters, called configuration parameters:

- take-off wing loading $\frac{m_0}{S}$, [daPa];
- take-off thrust-to-weight ratio $\frac{T_0}{W_0}$, [-].

These parameters represent the ratios between the aircraft mass and its wing area, as well as the thrust of engines and the aircraft weight. It is these parameters, along with the parameters characterizing the aerodynamic and weight perfection of the aircraft, as well as the gas-dynamic perfection of its propulsion system, that largely determine the level of its performance.

1.2 Shape and Size Parameters of the Main Parts of the Aircraft

Lift-Generating Surfaces (Wing, Tail):

- span b, b_{HT}, b_{VT}, [m];
- root and tip chord c_r, c_t, [m].

At the stage when the size of the aircraft is not yet determined, we use reduced geometric parameters:

- aspect ratio AR, AR_{HT}, AR_{VT};
- taper ratio λ, λ_{HT}, λ_{VT};
- root and tip airfoil thickness ratio $\frac{t}{c}$, [%];
- sweep angle of the quarter chord line or the mid-chord line $\Lambda_{c/4}$, $\Lambda_{c/2}$ [deg];
- wing angle-of-incidence i_w, [deg];

- wing dihedral angle Γ, [deg];
- areas of HT and VT, divided by the wing area $\overline{S}_{HT} = \frac{S_{HT}}{S}$, $\overline{S}_{VT} = \frac{S_{VT}}{S}$;
- properly non-dimensionalized spans and chords of rudders, ailerons, elevons and wing leading- and trailing-edge high-lift devices.

Fuselage:

- length of the fuselage, its forebody and afterbody l_f, $l_{f,fore}$, $l_{f,aft}$, [m];
- maximum midsection area S_{mid}, $[m^2]$;
- equivalent diameter of midsection $D_{mid} = \sqrt{\frac{4S_{mid}}{\pi}}$, [m];
- fuselage aspect ratio, forebody, afterbody aspect ratios $AR_f = \frac{l_f}{D_{mid}}$, $AR_{f,fore} = \frac{l_{f,fore}}{D_{mid}}$, $AR_{f,aft} = \frac{l_{f,aft}}{D_{mid}}$.

A specific fuselage configuration is also defined by the shape of the cross-sections, the planform forebody and afterbody angles, as well as the corresponding angles in the lateral projection (related to the design cockpit cutoff angle (forebody) and the design takeoff and landing clearance angle (afterbody)). For transport aircraft the shape of the afterbody is associated with the requirements for cargo loading and unloading. In addition to aerodynamic requirements, the fuselage dimensions are set by the requirements for accommodating a given target load and equipment, as well as the power plant (if it is placed in the fuselage).

Engines:

- dimensions of the engine (when placed inside the aircraft);
- nacelle dimensions (when placed on pylons).

Systems and Equipment:

- dimensions of the equipment. At the initial design stage, the dimensions of the equipment are often specified in terms of the required volumes to accommodate it.

Landing Gear:

- wheel base b, [m];
- distance between CG and main wheels e, [m];
- wheel track B, [m];
- height of the landing gear h, [m].

The specific configuration of the landing gear is determined by its layout, type of gear, number of legs; for a wheel gear, the additional parameters are the type and size of pneumatics, their number and placement on the gear.

A separate group of parameters are parameters that determine the relative position of the wing, the tail, and the engines. These include, for example:

- the properly reduced arms of HT and VT $\overline{l}_{HT}$, $\overline{l}_{VT}$;
- the ratio of the coordinate of the engine to the wing span $\frac{l_{eng}}{b}$, etc.

2 Methodology and Structure of the Algorithm for Determining Design Parameters

As it follows from the diagram in Fig. 2, the determination of design parameters is a part of the iterative process of aircraft design. The large dimensionality of the vector of design parameters and the variety of conditions and relationships that affect their selection (see Sect. 3.3), as well as the fact that many of them cannot be quantified at the early stages of design, makes it difficult to solve the problem of simultaneous determination of all the parameters describing the geometry of the aircraft. To solve it, the process of determining parameters is broken down into a number of consecutive steps, within which decisions are made to select the values of a particular group of parameters, given the values of other parameters, with subsequent refinement and coordination of them in an iterative process. The second method used to simplify the problem of determining the design parameters is to reduce the dimensionality of the vector of required design parameters by replacing the "zero" level parameters with generalized parameters.

Such parameters include, for example, $\frac{m_0}{S}$, $\frac{T_0}{W_0}$, volume coefficients of HT and HT, $V_{HT} = \frac{S_{HT} l_{HT}}{S c_{MAC}}$, $V_{VT} = \frac{S_{VT} l_{VT}}{Sb}$, etc.

The analysis of relations between parameters and performance, which are obtained from approximate solution of the first three equations of "aircraft feasibility", shows that the aircraft characteristics are functions of $\frac{m_0}{S}$, $\frac{T_0}{W_0}$; lift coefficient C_L and drag coefficient C_D for various aircraft configurations (takeoff, landing and cruise) and various segments of the flight profile; the specific fuel consumption of engines C_p at various flight modes; and the fuel mass ratio $\frac{m_f}{m_0}$.

Lift, drag coefficients and lift-to-drag ratio characterize the level of aerodynamic perfection of the aircraft. These parameters are not independent, the relationship between them is demonstrated by the aircraft polar. They are a function of geometric parameters of the aircraft, its surface quality, as well as the flight speed and altitude.

The specific fuel consumption C_p is a generalized parameter characterizing the gas-dynamic perfection of the power plant. It is a function of the gas-dynamic parameters of the engines, the power plant layout, as well as the speed, flight altitude, and engine operation mode.

The available fuel mass ratio $\frac{m_f}{m_0}$ is a generalized parameter characterizing the mass perfection of an aircraft and, together with the aerodynamic characteristics of the aircraft and gas-dynamic characteristics of the engine, determines the aircraft flight range.

Thus, setting the levels of aerodynamic and weight perfection of the aircraft, as well as gas-dynamic perfection of the power plant allows to reduce the task of determining the altogether design parameters of the aircraft to three main parameters: the required fuel mass ratio $\frac{m_f}{m_0}$, take-off wing loading $\frac{m_0}{S}$, and take-off thrust-to-weight ratio $\frac{T_0}{W_0}$.

By using the values of $\frac{m_0}{S}$, $\frac{T_0}{W_0}$ and $\frac{m_f}{m_0}$ iteratively, it is possible to determine all other parameters of the aircraft.

The analysis shows that the generalized parameters are very conservative and with the current level of scientific and technological advances, with the chosen layout and type of power plant for the aircraft of a specific mission, they vary within a very narrow range. This allows us predicting their values for certain design alternatives quite reliably. Therefore, it is possible to construct a project feasibility area in terms of $\frac{m_0}{S}$ and $\frac{T_0}{W_0}$. The boundaries of this area are defined by the performance requirements specified in the requirement specification and generalized parameters of technical perfection. Figure 1 presents a graphical interpretation of this area—a shaded area, bounded by the adopted ranges of the technical perfection parameters:

- maximum takeoff lift coefficient $C_{L,\max,take-off} = 2,6\ldots2,8$;
- maximum landing lift coefficient $C_{L,\max,landing} = 2,8\ldots3,0$;
- cruise lift coefficient $C_{L,cruise} = 0,55\ldots0,6$;
- climb lift-to-drag ratio $L/D_{climb} = 9\ldots10$.

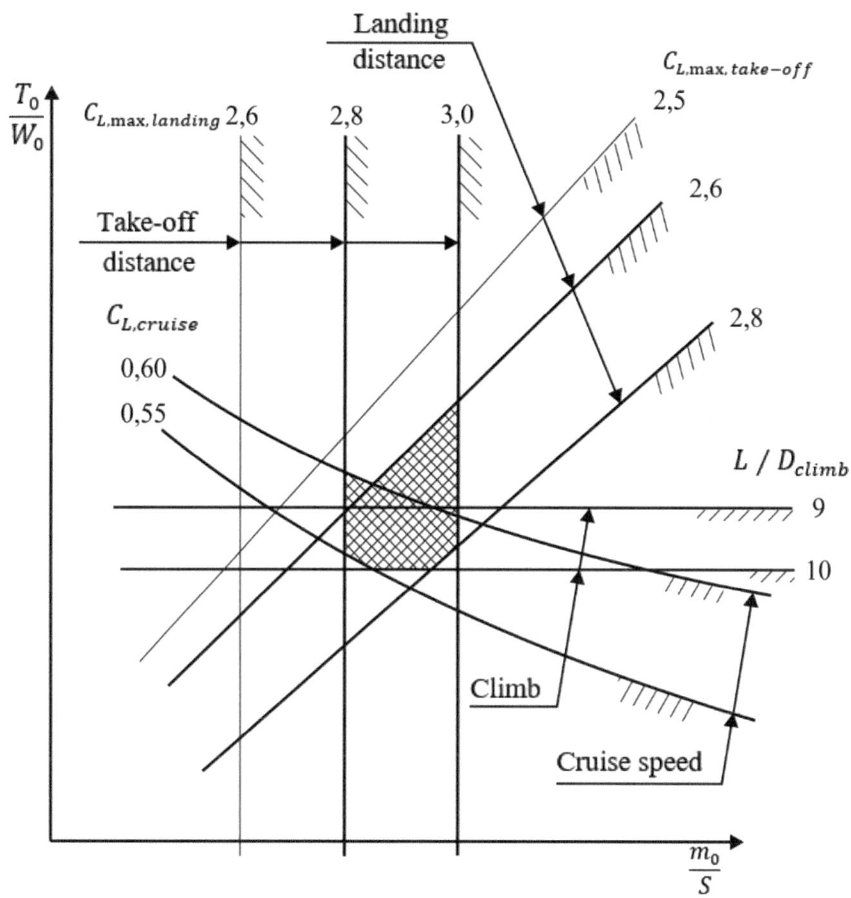

Fig. 1 Feasibility area in terms of $\frac{m_0}{S}$ and $\frac{T_0}{W_0}$

The boundaries of this area depend on the required flight performance and the technical solutions available to the designer, which provide a certain range of technical perfection parameters. For example, the range of $C_{L,max}$ during takeoff and landing is determined by the type of wing high-lift devices. It is clear that in case of unreasonably high level of required characteristics it may turn out that with the current technologies, the project is unfeasible.

With the present science and design practice, the success of a new aircraft is mainly determined by the new technical solutions delivering improvement in technical perfection parameters and, as a consequence, enhancement of technical and economic performance of the aircraft.

If there is a feasible area for an aircraft, one can specify the design point, i.e., a particular combination of $\frac{m_0}{S}$ and $\frac{T_0}{W_0}$.

To specify the design point in the feasible area, we will use the following reasoning. It is clear that of all the values of $\frac{m_0}{S}$ and $\frac{T_0}{W_0}$, lying in the feasible area, the most advantageous will be the values minimizing the takeoff weight of the aircraft.

In the first approximation, other things being equal, this condition will be met by the design with the smallest wing area and the lowest required engine thrust. Of course, these lowest values should not be less than those necessary to meet the most stringent constraints associated with the flight profile. Based on this, as a design wing loading one should select the smallest of the required values, $\frac{m_0}{S} = min\left\{\frac{m_0'}{S}, \frac{m_0}{S}'' \ldots \frac{m_0}{S}^i\right\}$, and as a design thrust loading—the largest value: $\frac{T_0}{W_0} = max\left\{\frac{T_0'}{W_0}\left(\frac{m_0}{S}\right), \frac{T_0}{W_0}''\left(\frac{m_0}{S}\right) \ldots \frac{T_0}{W_0}^i\left(\frac{m_0}{S}\right)\right\}$.

In curly brackets, we indicate the values of wing loading and thrust-to-weight ratio meeting the requirements dictated by airworthiness standards and the given flight profile.

Thus, for long-range passenger aircraft the key criterion for the choice of $\frac{m_0}{S}$, as a rule, is to meet the requirements for landing approach speed, cruising flight at a given altitude and speed, etc.

The required take-off thrust-to-weight ratio $\frac{T_0}{W_0}$ should be calculated for the following conditions: take-off from an airfield of a given class, the prescribed climb gradient with one engine out, cruise flight (see Sect. 4).

An even greater range of conditions and flight profile limitations is associated with a multi-mode combat aircraft flying at subsonic and supersonic speeds, at high and low altitudes, and conducting aerial combat.

The estimated range of configurational parameters required to achieve a given flight profile may be too large. This may mean, for example, that the wing loading of a passenger aircraft, selected for takeoff and landing on an airfield of a given class, may be too small to achieve the required cruise performance. Likewise, the high thrust-to-weight ratio of a fighter selected for takeoff from the deck of a ship makes it necessary to throttle its engine during subsonic cruise, which has a negative effect on fuel consumption.

In general, the following *rule* should guide the design.

If the values of $\frac{m_0}{S}$ and $\frac{T_0}{W_0}$, determined for a variety of design cases, are very different, one should try to find solutions to reduce the difference. For example, to increase the lowest tolerable value of $\frac{m_0}{S}$ by using more efficient high-lift devices or using a variable geometry wing. In order to reduce the required take-off thrust-to-weight ratio of a deck fighter $\frac{T_0}{W_0}$ it is possible to use launch boosters, catapult take-off or a ramp.

It should be noted that design practice has accumulated a significant arsenal of technical solutions to ensure the coordination of aircraft parameters, taking into account the specific requirements dictated by different phases of the flight profile. Let us list some of them.

Solutions related to the wing: high-lift devices on the leading and trailing edges of the wing, variable wing geometry, adaptive wing (perspective).

Solutions related to the tail: adjustable stabilizers, floating stabilizers, retractable stabilizers.

Solutions related to the fuselage: the forebody which can be deflected on takeoff and landing.

Solutions related to landing gear: landing gear drop-off after takeoff, additional (parking) undercarriage support, brake hook.

Solutions related to the engine: adjustable air intake, afterburner, thrust vector reversal and deviation, removable devices for thrust enhancement (horizontal and vertical), engines with variable gas-dynamic cycle (perspective).

In real design, other factors may be taken into account when determining the take-off wing loading and thrust loading. For example, today's widely used design practice based on the concept of aircraft family dictates the need to reduce the wing loading (to increase the wing area) of the baseline aircraft to enable the use of the same wing for aircraft modifications with a higher take-off weight designed for a longer range or a higher passenger capacity.

Having selected the design configuration parameters, one should proceed to determining the basic dimensional parameters of the aircraft. In general, the solution of this problem begins with determination of the first approximation takeoff mass m_0^I using historical data (see Sect. 3). Then, the wing area $S = \frac{m_0^I}{m_0/S}$ and the total take-off thrust of the engines $T_0^\Sigma = \frac{T_0}{W_0} W_0^I$ are calculated. Dividing T_0^Σ by the number of engines n_{eng}, we obtain the required take-off thrust of one engine $T_0 = \frac{T_0^\Sigma}{n_{eng}}$.

In real design, there may be other approaches to determining the dimensional parameters of an airplane. Thus, if the design is carried out for an existing engine or engine under development, the initial value of take-off mass can be determined by using the formula $m_0^I = \frac{T_0^\Sigma \cdot W_0}{T_0 g}$. It should be noted, however, that the value determined in this way may be very far from the design values of m_0 determined at later design stages, which would necessitate substantial redesign.

A similar approach to determining the first approximation takeoff mass can be applied when designing a modification with the wing of the baseline aircraft. Then, the first iteration takeoff weight can be determined using the equation $m_0^I = \frac{m_0}{S} S$.

After determining the basic dimensional parameters of the aircraft, using historical data on the dimensionless geometric parameters of the prototype aircraft (AR, Λ, $\frac{t}{c}$, λ, AR_f, $\bar{l}_{HT}$, $\bar{l}_{VT}$, $\bar{S}_{HT}$, $\bar{S}_{VT}$, V_{HT}, V_{VT} etc.), it is possible to calculate the dimensional geometric parameters of the main parts of the aircraft and to carry out preliminary scaling of the selected layout. By using general view drawing of the aircraft, one can refine the mass and aerodynamic characteristics of the aircraft and adjust the aircraft parameters such that the final geometry satisfies the three equations of aircraft feasibility, as well as the specified requirements and constraints.

First, use the preliminary general view drawing of the aircraft to predict its aerodynamic characteristics and replace the initially assumed historical values. The refined aerodynamic characteristics will be used to improve the design fuel mass ratio.

The geometric characteristics of the aircraft's parts can be used to refine the mass ratios of the aircraft assemblies and systems.

At the same time, most of the predicted mass ratios included in the equation for the second approximation takeoff mass depend on the take-off mass. Therefore, the second approximation takeoff mass calculation is performed iteratively (see Sect. 3.3).

The condition of exit from iteration is to satisfy the inequality.

$$\left|\frac{m_0^{i-1} - m_0^i}{m_0^i}\right| \leq \delta,$$

where:

m_0^i is the value of takeoff mass of the aircraft determined at the i-th step of iteration;

δ is a predetermined value (e.g., 0.05) characterizing the accuracy of solution the unity equation.

The next cycle of calculations is associated with the aircraft arrangement and mass analysis and leads to the correction of aircraft parameters satisfying the fourth and fifth equations of aircraft feasibility. Here, the problem of determining the rational mutual arrangement of the wing, fuselage, engines, stabilizers and landing gear, as well as the placement of cargo in the aircraft is solved. At the same time, the problem of aircraft general structural design satisfying the requirements of production and operational manufacturability (see Chap. 7) is solved.

Analysis of aircraft's stability and controllability characteristics makes it possible to correct such aircraft design parameters as wing dihedral angle, engine and horizontal stabilizer mounting angles, trim angles of control surfaces, etc. If necessary, the aerodynamic and mass-inertia properties of the aircraft are then refined and the aircraft parameters are matched once again.

In order to ensure that the flight performance and economic characteristics of the aircraft fully satisfy the design specification, it is necessary to perform verification calculations using generally accepted methods. Evaluating the results of these calculations, the designer, if necessary, modifies the inputs, primarily those associated with the selected design alternatives, to ensure that the calculated characteristics satisfy the design specification.

The block diagram of the above-described algorithm for determining the design parameters of the aircraft is presented in Fig. 2. Its structure makes it possible, in the presence of alternative options for fulfilling the requirements of the design specification, to solve the problem of optimizing the design parameters of the aircraft, either by direct-search method or by using formalized methods.

3 Determining the Take-off Mass of the Aircraft

3.1 Aircraft Mass Classification. Mass Balance Equation

Determining the takeoff mass of an aircraft and its components is one of the central tasks of the design process.

This is because, as we have seen, mass is one of the most important dimensional parameters of an aircraft and its determination practically begins the process of selecting and matching the design parameters of an aircraft that meet the specified requirements and constraints. At the same time, mass is also an integrated property of an aircraft, often determining the degree of design success. In other words, the mass (or its absolute and dimensionless components) simultaneously serves as a non-dimensional, and in some formulations of the problem, as an absolute criterion for assessing the efficiency of the aircraft at hand.

The problem of determining the takeoff mass of an aircraft is reduced to solving the following equation:

$$m_0 = \sum_{i=1}^{n} m_i,$$

where m_i is the mass of the aircraft components, with n depending on the type of aircraft and the stage of design.

Since most of the terms in the right-hand side of the above equation are functions of the takeoff mass of the aircraft ($m_i = f(m_0)$), and the value of the takeoff mass determines the basic dimensions of the aircraft and, in turn, depends on these dimensions, the process of determining the takeoff mass is iterative. Given the selected configuration of the airplane, its equipment and payload, as well as the known geometric parameters and materials of its parts, the determination of the airplane mass is not particularly difficult. The problem lies in the extent to which this mass (as a characteristic) matches the previously determined design mass (as a parameter). The difference between these two values determines, first and foremost, the number of iterations required to agree on and finally select the aircraft parameters and, consequently, the design time and cost. In addition, as shown in Chap. 3, a difference between the predicted mass and the design mass usually leads to a degradation of the aircraft performance and a failure to meet the requirements of the design specification.

This determines the relevance of the task of finding the design takeoff mass of an airplane with the highest possible accuracy. In this case, it has to be solved on the basis of very limited information available to the designer at the initial design stage.

A typical classification of aircraft masses is presented in Fig. 3.

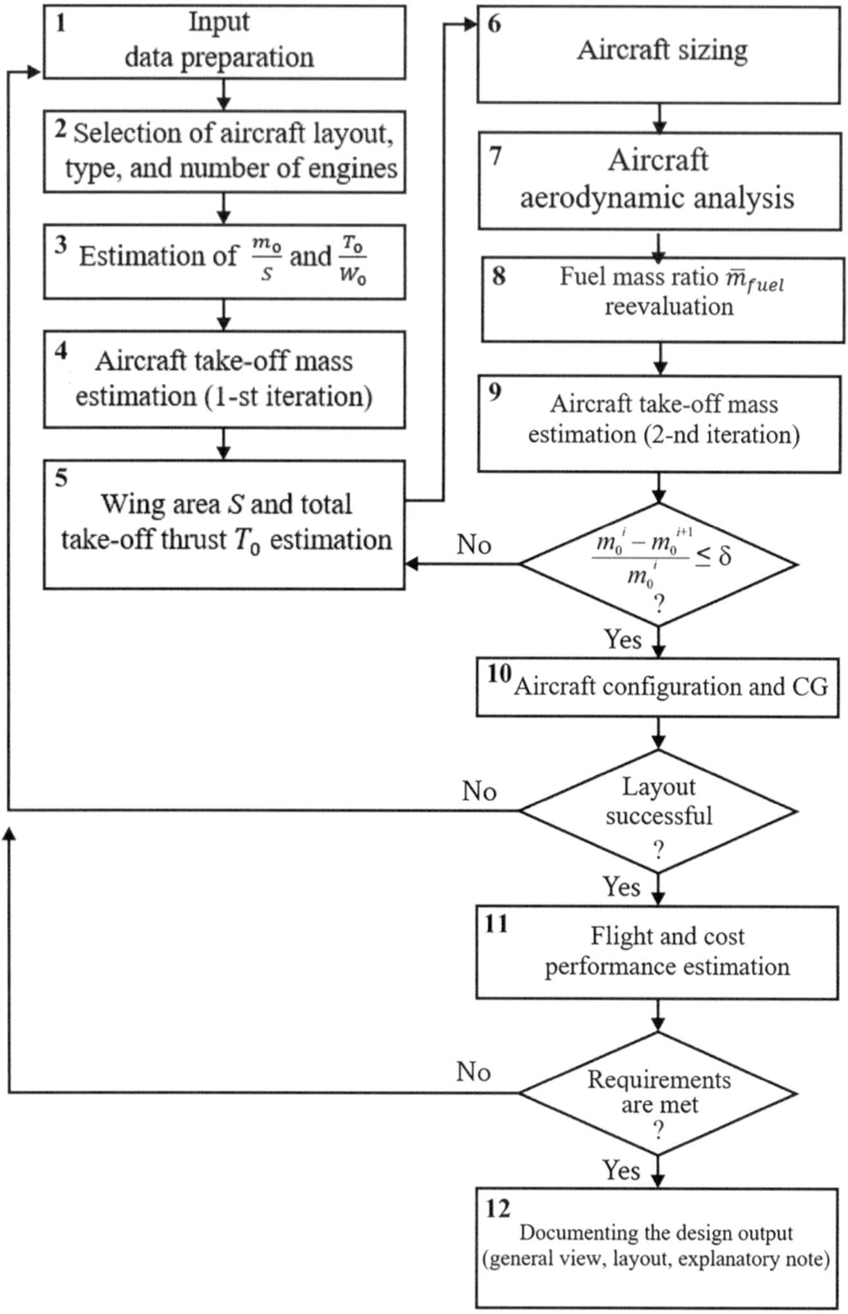

Fig. 2 Block diagram of the algorithm for determining aircraft parameters

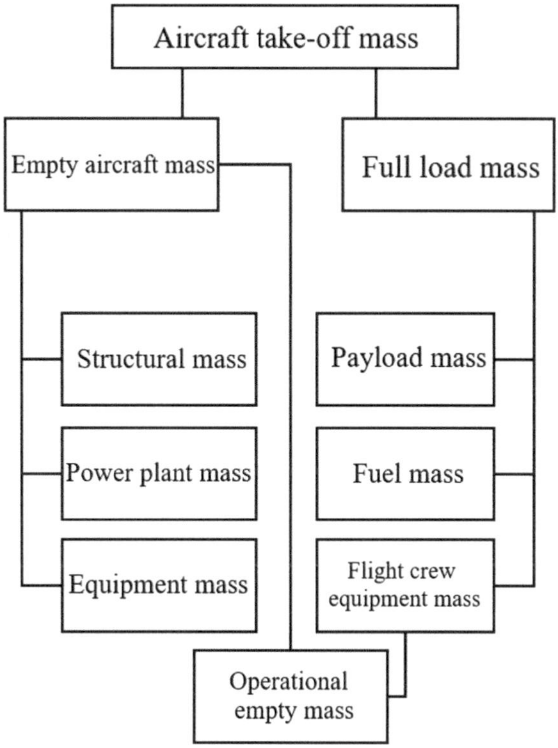

Fig. 3 Aircraft mass components

According to this diagram, the takeoff mass m_0 can be calculated using the following equation:

$$m_0 = m_{empty} + m_{FL} = m_{struct} + m_{PP} + m_{fuel} + m_{equip} + m_{FCEq} + m_{PL} \quad (1)$$

Equation 1 is called mass balance equation. Dividing its left and right parts by m_0, we obtain the same equation in a dimensionless form:

$1 = \overline{m}_{struct} + \overline{m}_{PP} + \overline{m}_{fuel} + \overline{m}_{equip} + \overline{m}_{FCEq} + \overline{m}_{PL}$,

where:

$\overline{m}_{struct} = \frac{m_{struct}}{m_0}$ = the mass ratio of the aircraft structure;
$\overline{m}_{PP} = \frac{m_{PP}}{m}$ = the mass ratio of the power plant;
$\overline{m}_{fuel} = \frac{m_{fuel}}{m_0}$ = the mass ratio of the fuel;
$\overline{m}_{equip} = \frac{m_{equip}}{m_0}$ = the mass ratio of the equipment;
$\overline{m}_{FCEq} = \frac{m_{FCEq}}{m_0}$ = the mass ratio of the flight crew equipment;
$\overline{m}_{PL} = \frac{m_{PL}}{m_0}$ = the mass ratio of the payload.

In this form, thanks to the works of Bolkhovitinov, Fomin [1, 2] and others, this equation is widely used in aircraft design. A typical nomenclature of the components included in the mass balance equation, which ensures uniformity of calculations and comparability of characteristics of different aircraft, is presented in Appendix I.

3.2 Determination of Aircraft Take-off Mass in the First Approximation

Let us rewrite the mass balance equation in the following manner:

$$1 = \overline{m}_{struct} + \overline{m}_{PP} + \overline{m}_{fuel} + \overline{m}_{equip} + \frac{m_{PL+}m_{FCEq}}{m_0},$$

$$m_0 = \frac{m_{PL+}m_{FCEq}}{1 - \left(\overline{m}_{struct} + \overline{m}_{PP} + \overline{m}_{fuel} + \overline{m}_{equip}\right)} \tag{2}$$

This equation is used to determine the aircraft mass in the first and second approximation.

The payload mass is usually specified in or determined from the technical design specification. For example, for a commercial aircraft with a given passenger capacity n_{pass}, if the amount of mail and cargo is not specified, the payload mass can be determined using the following equation:

$$m_{PL} = 90 \cdot n_{pass} + (10 \ldots 30) \cdot n_{pass}.$$

The larger is the plane, the smaller is the coefficient at n_{pass}.

The flight crew equipment mass, which includes the mass of flight and auxiliary crew with personal items, is also easy to determine based on the purpose of the aircraft and the relevant regulatory documents for its design. An example definition of m_{FCEq} for a long-range commercial aircraft is given in Sect. 4.

The mass of equipment (water and other liquids consumed in flight, unusable fuel, oil, containerized products, covers, rescue equipment, spare parts and tools), is specified in the technical design specification and other regulatory documents, or is derived from historical data.

It should be borne in mind that historical data are valid if they are processed for aircraft of similar design, size and purpose. Deviation from this rule, extrapolating from a small number of points may lead to unacceptable errors.

The historical data on aircraft of various classes are presented in Table 1.

Let's assume that the fuel mass ratio $\overline{m}_{fuel}$ does not depend on the takeoff mass of the aircraft. With this assumption, we can improve the accuracy of the take-off mass predicted in the first approximation by solving the equation of motion for the aircraft range. If the aircraft flies with constant values of lift-to-drag ratio $\left({}^{L}/_{D}\right)_{cr}$ and speed V_{cr}, and the aircraft mass reduction is compensated by the altitude gain, then we can use the following equation:

$$R_{cr} = \frac{V_{cr}\left({}^{L}/_{D}\right)_{cr}}{C_{p,cr}} ln \frac{m_{b,cr}}{m_{e,cr}} \tag{3}$$

where:

Table 1 Mass ratios of structure, power plant, equipment, and fuel for conventional take-off and landing aircraft

Purpose of the aircraft		$\overline{m}_{struct}$	$\overline{m}_{PP}$	$\overline{m}_{equip}$	$\overline{m}_{fuel}$
Subsonic commercial	Light	0,30…0,32	0,12…0,14	0,12…0,14	0,18…0,22
	Medium	0,28…0,30	0,10…0,12	0,10…0,12	0,26…0,30
	Heavy	0,25…0,27	0,08…0,10	0,09…0,11	0,35…0,40
Supersonic commercial		0,20… 0,24	0,08…0,10	0,07…0,09	0,45…0,52
Commuter aircraft		0,29… 0,31	0,14…0,16	0,12…0,14	0,12…0,18
Aerobatic		0,32…0,34	0,26…0,30	0,06…0,07	0,10…0,15
Agricultural		0,24…0,30	0,12…0,15	0,12…0,15	0,08…0,12
Light seaplanes		0,34…0,38	0,12…0,15	0,12…0,15	0,10…0,20
Motor gliders		0,48…0,52	0,08…0,10	0,06…0,08	0,08…0,12
Fighters		0,28…0,32	0,18…0,22	0,12…0,14	0,25…0,30
Bombers	Light	0,26…0,28	0,10…0,12	0,10…0,12	0,35…0,40
	Medium	0,22…0,24	0,08…0,10	0,07…0,10	0,45…0,50
	Heavy	0,18…0,20	0,06…0,08	0,06…0,08	0,55…0,60
Tactical transport	Light	0,30…0,32	0,12…0,14	0,16…0,18	0,20…0,25
	Medium	0,26…0,28	0,10…0,12	0,12…0,14	0,25…0,30
	Heavy	0,28…0,32	0,08…0,10	0,06…0,08	0,30…0,35

$m_{b,cr}$ = mass of the aircraft at the beginning of cruise, i.e., takeoff mass minus the mass of fuel consumed during takeoff and climbing to cruise altitude, and $m_{e,cr}$ = mass of the aircraft at the end of the cruise, i.e., its mass minus the mass of fuel consumed by the time of landing.

Let's rearrange the simplified Eq. 3 for the fuel ratio:

$$\overline{m}_{fuel} = \left[1 - \frac{1 - \overline{m}_{fuel,climb}}{exp\left(\frac{RC_p}{L/_{D} V} \right)_{cr}} \right] \frac{1}{1 - k_t} \tag{4}$$

Here:

k_t = factor associated with the fuel consumed by engines during start-up and warm-up, taxi to take-off point, runway approach, descent from cruise altitude, runway and taxi after landing, as well as air navigation fuel reserve; it depends on flight range and varies between 0.18…0.21 for short-range commercial aircraft and 0.13…0.17 for long-range planes.

$\overline{m}_{fuel,climb}$ = the fuel ratio required to climb to a cruising altitude, which is determined by the following formula:

$$\overline{m}_{fuel,climb} = \frac{\left(H_{cr} + \frac{V_{cr}^2}{2g}\right)C_p}{1300V_{cr}} \cdot \frac{\frac{T_0}{W_0}\left(L/D\right)_{cr}}{\frac{T_0}{W_0}\left(L/D\right)_{cr} - 1} \tag{5}$$

Here,

H_{cr} = cruising altitude, m;

V_{cr} = cruising speed, m/s;

$\left(L/D\right)_{cr}$ = cruising lift-to-drag ratio of the aircraft.

The following approximate equation can be used to estimate the fuel ratio $\overline{m}_{fuel}$ of a mid-range aircraft:

$$R_{cr} \approx 1020\frac{M_{cr}\left(L/D\right)_{cr}}{C_{p,cr}}\frac{\overline{m}_{fuel}}{\sqrt{1-\overline{m}_{fuel}}} \tag{6}$$

Let us also consider equations for cruising range for other simplified missions encountered in practice. If the aircraft flies with constant values of H_{cr} and $\left(L/D\right)_{cr}$, and the decrease in its mass is compensated by the change in flight speed, then

$$R_{cr} \approx 2\frac{V_{cr}\left(L/D\right)_{cr}}{C_{p,cr}}\left(1-\sqrt{\frac{m_{e,cr}}{m_{b,cr}}}\right) \tag{7}$$

If the plane flies with constant values of H_{cr} and $\left(L/D\right)_{cr}$, and its lift coefficient C_L changes according to the law $C_L = C_{L,b.cr}\frac{m_0}{m_{b,cr}}$, then

$$R_{cr} \approx \frac{V_{cr}}{C_{p,cr}\sqrt{\frac{C_{D0}}{\pi AR_{eff}}}} \cdot arctg\frac{1}{qS} \cdot \frac{m_{b,cr} - m_{e,cr}\frac{1}{\sqrt{C_{D0}\pi AR_{eff}}}}{1 + \frac{m_{b,cr}m_{e,cr}}{(qS)^2}\frac{1}{\sqrt{C_{D0}\pi AR_{eff}}}} \tag{8}$$

In the above equations, the indices at m mean:

b, cr = beginning of cruise mission,

e, cr = end of cruise mission.

Note that the ceiling flight mode $\left(\left(L/D\right)_{cr}, V_{cr}\right) = const$ is the most economical.

3.3 Determination of Take-off Mass in the Second Approximation

To determine the takeoff mass in the second approximation, the information derived from historical data and technical design specification is no longer sufficient. It is necessary to specify aircraft's geometric features, structural layout and structural materials, type of power plant, geometric parameters of airframe parts.

In accordance with the algorithm for selecting the design parameters of an aircraft, discussed in Sect. 2, to determine the mass of an aircraft in the second approximation, it is necessary:

- to set the geometric parameters, including AR, Λ, $\frac{t}{c}$, λ of the wing and tail, AR_f, V_{HT}, V_{VT}, C_{mC_L};
- to construct the polars of the aircraft, or determine the values of C_{D0} and $A = \frac{1}{\pi AR_{eff}}$;
- to determine the thrust-to-weight ratio required for the given flight profile.

Then, the algorithm for calculating the aircraft mass can be as follows. First, the required take-off thrust $\frac{T_0}{W_0}$ is specified for the aircraft. Based on the value of this thrust, either a specific engine (with a known mass), or a hypothetical engine is selected. Then its mass is determined by the formula:

$$m_{eng} = \frac{T_0}{W_0}\frac{\gamma_{eng}}{n_{eng}},$$

where:

γ_{eng} = specific engine weight;
n_{eng} = number of engines.

The relative mass of the power plant is determined using semi-empirical formulas like:

$$\overline{m}_{PP} = k\gamma_{eng}\frac{T_0}{W_0} + a,$$

where

$k = 1.25$ for fighters;
$k = 1.6$ for heavy subsonic aircraft;
$k = 1.75$ for heavy supersonic aircraft;
a = a coefficient associated with the location of the power plant on the aircraft:
$a = 0.08 \ldots 0.09$ if the engines are mounted at the root of the wing;
$a = 0.09 \ldots 0.11$ if the engines are installed on the fuselage afterbody;
$a = 0.11 \ldots 0.12$ if the engines are mounted on the wing.

Based on the results of the aerodynamic calculation, the amount of fuel required for the given flight profile and its mass in the second approximation are determined.

The most difficult part of the second approximation calculations is to determine the mass of the structure.

Given the complex dependence of structural mass on geometric and mass parameters of aircraft, it is defined as the sum of the masses of the wing, fuselage, tail and landing gear, i.e.

$$\overline{m}_{str} = \overline{m}_w + \overline{m}_{fus} + \overline{m}_{tail} + \overline{m}_{lg},$$

where

$\overline{m}_w = \frac{m_w}{m_0}$ = the mass ratio of the wing structure;
$\overline{m}_{fus} = \frac{m_{fus}}{m_0}$ = the mass ratio of the fuselage structure;
$\overline{m}_{tail} = \frac{m_{tail}}{m_0}$ = the mass ratio of the tail structure;
$\overline{m}_{lg} = \frac{m_{lg}}{Wm_0}$ = the mass ratio of the landing gear structure.

Determination of structural mass components is performed using theoretical and empirical equations derived for various types of aircraft. These formulas include a theoretical part reflecting the functional relationship between the mass of airframe parts and its shape and size parameters, the materials used, the applied loads. This makes it possible to conduct parametric studies and select the optimum parameters for the airframe parts, as well as take into account the specific features of the aircraft at hand. The correct selection of historical coefficients ensures the calculation accuracy required at this stage (~5%).

Similar formulas exist for other airframe parts as well as for equipment and controls. Summaries of such formulas for various aircraft are given, for example, in [3–5].

After determining the mass ratios included in Eq. 1, using the first approximation take-off mass value, solve Eq. 2.

Since only in rare cases this equation is solved at once (the sum of components equals 1), the following method is used to solve it. Taking as the starting point m_0, the sum of the components is calculated for different values of m_0. The value of m_0 at which the sum of components equals 1 will be the take-off mass of the aircraft in the second approximation. The graphical interpretation of this method is shown in Fig. 4.

The intersection of the curve (connecting at least three points) with the horizontal axis determines the value of m_0^{II}.

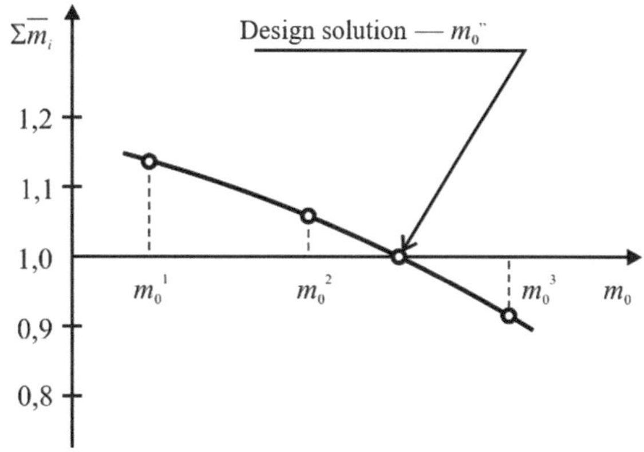

Fig. 4 Determination of the aircraft's take-off mass in the second approximation

In the graduation project, the condition for exiting the iterative process is 5% difference between two consecutive iterations, i.e.:

$$\frac{m_0^{II} - m_0^{I}}{m_0^{II}} \leq 0.05.$$

4 Determination of Design Parameters for a Mid-range Commercial Airliner

This section considers an example of determining design parameters for a commercial aircraft. This example is intended for use in the training process; the associated design requirements are presented in the Appendix.

4.1 Input Data

Based on the analysis of the requirements for the aircraft at hand, Tu-204-200 and Boeing-757-200 were selected as its main prototype aircraft. This made it possible to decide on the aircraft layout, type, number and placement of engines and a priori assign some of the parameters of the wing and tail, fuselage, as well as the values of the expected aerodynamic characteristics of the aircraft. In real design, decision-making on the above problems is based on the research and development carried out by the Design Bureau in close cooperation with the leading research institutes of the industry.

Wing Parameters

Based on the study of the wings of the prototypes, for the aircraft at hand we choose a swept wing, supercritical airfoil and special wingtips with the following parameters:

- aspect ratio $AR = 9.67$;
- taper ratio $\lambda = 3.93$;
- thickness ratio at the root $\left(\frac{t}{c}\right)_r = 0.145$;
- thickness ratio at the tip $\left(\frac{t}{c}\right)_t = 0.095$;
- quarter chord sweep $\Lambda_{c/4} = 29°$.

Average wing thickness ratio can be determined using the following equation: $\left(\frac{t}{c}\right)_{av} = \frac{\left(\frac{t}{c}\right)_r \lambda + \left(\frac{t}{c}\right)_t}{\lambda + 1}$, and for the aircraft at hand it equals 0.135.

Aileron Parameters

- chord ratio $\frac{c_{ail}}{c} = 0.25 \ldots 0.30$;
- span ratio $\frac{b_{ail}}{b} = 0.2 \ldots 0.4$;
- upward deflection angle $\delta_{ail}^{up} = 25°$;
- downward deflection angle $\delta_{ail}^{up} = 15 \ldots 25°$.

Flap Parameters

- chord ratio $\frac{c_{flap}}{c} = 0.3 \ldots 0.4$;
- maximum deflection angle $\delta_{flap} = 50 \ldots 60°$;
- area ratio $\frac{S_{flap}}{S} = 0.15 \ldots 0.25$.

Slat Parameters

- chord ratio $\frac{c_{slat}}{c} = 0.12 \ldots 0.18$;
- maximum deflection angle $\delta_{flap} = 20 \ldots 30°$.

Parameters of Horizontal and Vertical Stabilizers

- horizontal tail area ratio $\frac{S_{HT}}{S} = 0.25$;
- horizontal tail aspect ratio $AR_{HT} = 4.5$;
- horizontal tail taper ratio $\lambda_{HT} = 3.03$;
- horizontal tail sweep $\Lambda_{HT} = 33°$;
- vertical tail area ratio $\frac{S_{VT}}{S} = 0.201$;
- vertical tail aspect ratio $AR_{VT} = 1.68$;
- vertical tail taper ratio $\lambda_{VT} = 2.77$;
- vertical tail sweep $\Lambda_{VT} = 35°$;
- horizontal tail volume coefficient $V_{HT} = 0.96$;
- vertical tail volume coefficient $V_{VT} = 0.086$.

Fuselage Parameters

- fuselage aspect ratio $AR_f = 11$;
- equivalent diameter of midsection $D_{eq} = 3.95\,m$.

The parameters of a typical fuselage cross-section for a six-seat-row layout with one aisle (economy class) are shown in the table below.

Number of seats in cross section	n_{seat}	6
Aisle width	w_{aisle}, m	0,47
Passenger cabin width	w_{cabin}, m	3,5
Passenger cabin height	h_{cabin}, m	2,16
Luggage compartment width	w_{LC}, m	2,67
Luggage compartment height	h_{LC}, m	1,25
Mid-section width	w_{mid}, m	3,8
Mid-section height	h_{mid}, m	4,1

Expected Aerodynamic Performance

According to the prototype data, we set the following values of aerodynamic characteristics of the aircraft at different flight modes:

Aerodynamic characteristics	Takeoff	Landing	Cruising flight
Lift coefficient	1,9	2,9	0,6
Lift-to-drag ratio	14	10	17,5

4.2 Determining the Take-off Mass of the Aircraft in the First Approximation

To determine take-off mass in the first approximation, we use Eq. 2.

Payload mass m_{PL} depends on the number of passengers specified in the technical design specification and is determined by the formula:

$$m_{PL} = 75n_{pass} + q_l n_{pass} + 290\left(\theta_l - \frac{q_l n_{pass}}{120}\right),$$

where

- n_{pass} is the number of passengers, $n_{pass} = 204$ (from TRS);
- q_l is the mass of luggage free of charge for one passenger, $q_l = 20$ kg for a mid-range aircraft;
- θ_l is luggage space volume in m^3 determined according to the formula:

$$\theta_l = 0.254\, n_{pass} = 0.254 \cdot 204 = 51.816\,\text{m}^3.$$

Now,

$$m_{PL} = 9.8\left(75 \cdot 204 + 20 \cdot 204 + 290\left(51.816 - \frac{20 \cdot 204}{120}\right)\right) = 24546\,\text{kg}.$$

The mass of the service load m_{FCEq} is determined by the formula:

$$m_{FCEq} = 90n_p + 70n_{FA},$$

where

- n_p is the number of pilots according to technical requirements, $n_p = 3$;
- n_{FA} is the number of flight attendants according to TRS, $n_{FA} = 5$.

Therefore,

$$m_{FCEq} = 90 \cdot 3 + 50 \cdot 5 = 620\,\text{kg}.$$

The fuel mass ratio $\overline{m}_f$ is determined by formula (4).

Based on the TRS and the accepted input data, let's estimate the components included in this formula.

Cruising range: $R_{cr} = 0.95R = 0.95 \cdot 6200 = 5890\,\text{km}$.

Cruising specific fuel consumption: $C_{p,cr} = 0.595\,\text{kg}/(daN \times h)$.

Cruising lift-to-drag ratio: $\left(L/D\right)_{cr} = 17.5$.

Cruising speed: $V_{cr} = 850\,\text{km/h}$.

Historical coefficient for navigational fuel reserve and fuel reserve for descent from cruising altitude: $k_t = 0.17$ for $R = 5000\ldots7500\,\text{km}$.

The fuel mass ratio required to reach cruise altitude is determined by formula 5.

In this formula, $H_{cr} = 11000\,\text{m}$ (for the prototype aircraft, $\frac{T_0}{W_0} = 0.291$), $V_{cr} = 236\,\text{m/s}$; the values of $C_{p,cr}$ and $\left(L/D\right)_{cr}$ are indicated above.

The results of calculations are given below:

$$\overline{m}_{fuel,climb} = \frac{\left(11000 + \frac{236^2}{2 \cdot 9.81}\right)0.595}{1300 \cdot 236} \cdot \frac{0.291 \cdot 17.5}{0.291 \cdot 17.5 - 1} \approx 0.0334,$$

$$\overline{m}_{fuel} = \left[1 - \frac{1 - 0.0334}{exp\left(\frac{5890 \cdot 0.595}{17.5 \cdot 850}\right)}\right]\frac{1}{1 - 0.17} \approx 0.285,$$

The components $\overline{m}_{struct}$, $\overline{m}_{PP}$, $\overline{m}_{equip}$ are derived from historical data. For the aircraft at hand, we assume the following values:

$$\overline{m}_{struct} = 0.27, \overline{m}_{PP} = 0.09, \overline{m}_{equip} = 0.11.$$

Finally,

$$m_0 = \frac{24546 + 620}{1 - 0.27 - 0.09 - 0.11 - 0.285} \approx 102720.8\,\text{kg}.$$

4.3 Determination of the Take-off Wing Loading

As a rule, for the class of aircraft in question, the design conditions for determining the take-off wing loading are specified in Aviation rules [6]. This document prescribes under which conditions safe landing and cruising should be performed.

From the landing requirement, given that $V_{appr} = 1.3V_{min,land}$:

$$\left(\frac{m}{S}\right)_0^I = \frac{C_{L,\max land} V_{appr}^2}{30.2\left(1 - \overline{m}_{fuel}\right)},$$

where

- $C_{L,\max land}$ is the maximum value of C_L of the aircraft in landing configuration;
- V_{appr} is the approach speed in m/s.

We replace the variables with the corresponding values and obtain the following result:

$$\left(\frac{m}{S}\right)_0^I = \frac{2.9 \cdot 67.78^2}{30.2 \cdot (1 - 0.285)} \approx 6170Pa(617\,daPa).$$

In the graduation project, it is allowed to use in this equation the value of the fuel mass ratio from the mass balance equation. In real design, this value is determined by the ratio between the takeoff and landing masses of the aircraft, established in the technical requirements specification for the design of a particular aircraft, taking into account the fuel reserve and ensuring safe landing in emergency situations.

For cruise,

$$\left(\frac{m}{S}\right)_0^{II} = \frac{1}{1 - 0.6\overline{m}_{fuel}} C_{L,cr} q_{cr},$$

where

- $C_{L,cr}$ and q_{cr} are the lift coefficient and dynamic pressure at the cruise flight mode, respectively.

This formula is an approximation as it was derived under the assumption of some average aircraft mass at the cruising mode of flight.

By inserting the numerical values into the formula, we obtain:

$$\left(\frac{m}{S}\right)_0^{II} = \frac{1}{1 - 0.6 \cdot 0.285} \cdot 0.6 \cdot \frac{0.365 \cdot 236^2}{2 \cdot 10} \approx 736\, daPa.$$

Next, the value of the take-off wing loading is selected:

$$\left(\frac{m}{S}\right)_0 = min\left\{\left(\frac{m}{S}\right)_0^I; \left(\frac{m}{S}\right)_0^{II}\right\} = min\{617; 736\} = 617\, daPa.$$

This value is very close to the values of the take-off wing loading of the prototype aircraft (Tu-204-200: $\left(\frac{m}{S}\right)_0 = 595.6\, daPa$; Boeing-757-200: $\left(\frac{m}{S}\right)_0 = 613.7\, daPa$).

4.4 Determining the Take-off Thrust-To-Weight Ratio

The following conditions are used to determine the take-off thrust-to-weight ratio:

(a) *Climb with one engine out*

This condition satisfies the requirement of the Aviation Rules [6], to provide a standardized climb gradient on takeoff with one engine out:

$$\left(\frac{T}{W}\right)_0^I = 1.5\frac{n_{eng}}{n_{eng} - 1}\left(\frac{1}{\left(L/D\right)_{climb}} + tg\theta\right),$$

where

- n_{eng} stands for the number of engines on the aircraft, $n_{eng} = 2$ for the prototype aircraft;
- $\left(L/D\right)_{climb}$ stands for the climb lift-to-drag ratio.

For the aircraft at hand,

$$\left(\frac{T}{W}\right)_0^I = 1.5\frac{2}{2 - 1}\left(\frac{1}{14} + 0.024\right) \approx 0.286.$$

(b) *Horizontal flight*

$$\left(\frac{T}{W}\right)_0^{II} = \frac{1 - 0.6\overline{m}_{fuel}}{\left(L/D\right)_{cr}\xi_{HV}\varphi_t},$$

where:

- ξ_{HV} is the engine thrust coefficient dependent on the altitude and speed, and based on the prototype engine data; we assume $\xi_{HV} = 0.27$;
- φ_t is the cruise throttling factor corresponding to the unlimited engine operation time or to the optimum fuel consumption mode; we assume $\varphi_t = 0.8$.

Inserting the above values, we obtain

$$\left(\frac{T}{W}\right)_0^{II} = \frac{1 - 0.6 \cdot 0.285}{17.5 \cdot 0.27 \cdot 0.8} \approx 0.219.$$

(c) *Take-off performance (the given distance of take-off ground roll)*

$$\left(\frac{T}{W}\right)_0^{III} = 1.05\left(\frac{1.2\left(\frac{m}{S}\right)_0}{C_{Lmax,take-off}\, l_{TOGR} + \frac{1}{2}\left(3f_{TOGR} + \frac{1}{\left(L/D\right)_{TOGR}}\right)}\right),$$

where

- $\left(\frac{m}{S}\right)_0$ is the previously determined wing loading;
- $C_{Lmax,take-off}$ is the maximum value of aircraft lift coefficient in take-off configuration; historical data suggest that $C_{Lmax,take-off} = 2.4$ for the aircraft at hand;
- l_{TOGR} is the take-off ground roll distance, defined as $l_{TOGR} = 0.75L_{RWY}$. In accordance with the technical requirements specification, $L_{RWY} = 2600\,m$. Hence, $l_{TOGR} = 0.75 \cdot 2600 = 1950$ m;
- f_{TOGR} is the take-off friction coefficient; $f_{TOGR} = 0.02$ for concrete runways;
- $\left(L/D\right)_{TOGR}$ is the take-off ground roll lift-to-drag ratio; as per historical data, $\left(L/D\right)_{TOGR} = 10$ for sub sonic aircraft;

$$\left(\frac{T}{W}\right)_0^{III} = 1.05\left(\frac{1.2 \cdot 617}{2.4 \cdot 1950} + \frac{1}{2}\left(3 \cdot 0.02 + \frac{1}{10}\right)\right) \approx 0.25.$$

Let's determine the design value of thrust-to-weight ratio:

$$\left(\frac{T}{W}\right)_0 = max\left\{\left(\frac{T}{W}\right)_0^I; \left(\frac{T}{W}\right)_0^{II}; \left(\frac{T}{W}\right)_0^{III}\right\} = max\{0.286; 0.219; 0.25\} = 0.286.$$

The above value of take-off thrust-to-weight ratio is very close to that of the prototype aircraft (for Tu-204-200, $\left(\frac{T}{W}\right)_0 = 0.285$; for Boeing-757-200 $\left(\frac{T}{W}\right)_0 = 0.290$). By using $\left(\frac{\mathrm{m}}{\mathrm{S}}\right)_0$ and $\left(\frac{T}{W}\right)_0$ one can derive the values of the wing area S and the take-off thrust T_0, corresponding to the take-off mass of the first approximation:

$$S_w = \frac{m_0^I g}{10\left(\frac{\mathrm{m}}{\mathrm{S}}\right)_0},\ T_0 = \frac{\left(\frac{T}{W}\right)_0 m_0^I g}{10} \left(the\ unit\ of\ \left(\frac{\mathrm{m}}{\mathrm{S}}\right)_0 is\ daPa\right)$$

If the aircraft is powered by n_{eng}, then the required thrust of a single engine is

$$T_0^I = \frac{T_0}{n_{eng}}.$$

By using the value of T_0^I, the engine for the designed aircraft is selected.

The information obtained at this stage makes it possible to begin a more detailed analysis of the design based on the second approximation calculations aimed at refining the aerodynamic characteristics, propulsion system characteristics and mass characteristics of the aircraft.

Below an example of the second approximation calculations is presented.

4.5 Determining the Aircraft Take-off Mass in the Second Approximation

In the second approximation calculations, the take-off mass of the aircraft and its components are refined by replacing the historical values with predicted ones. The latter are calculated using semiempirical equations incorporating the information available at the given design stage.

(1) The mass ratio of the power plant is determined by the formula

$$\overline{m}_{PP} = \left(0.124 + \gamma_{eng}\right)\left(\frac{T}{W}\right)_0.$$

Here, γ_{eng} is the specific weight of the motor.

Use the calculated value of $\left(\frac{T}{W}\right)_0$ and the average historical value $\gamma_{eng} = 0.18$ and obtain the result as follows:

$$\overline{m}_{PP} = (0.124 + 0.18) \cdot 0.286 \approx 0.0869.$$

(B) The equipment mass ratio is determined by the formula:

$$\overline{m}_{equip} = \frac{45 n_{pass}}{m_0} + 0.04.$$

Therefore, for the aircraft at hand,

$$\overline{m}_{equip} = \frac{45 \cdot 204}{102720.82} + 0.04 \approx 0.1294.$$

(C) The structure mass ratio is defined as the sum of the mass ratios of its components:

$$\overline{m}_{str} = \overline{m}_w + \overline{m}_{fus} + \overline{m}_{tail} + \overline{m}_{LG},$$

where

$\overline{m}_w$ = mass ratio of the wing structure;
$\overline{m}_{fus}$ = mass ratio of the fuselage structure;
$\overline{m}_{tail}$ = mass ratio of the tail structure;
$\overline{m}_{LG}$ = mass ratio of the landing gear structure.

(a) The mass ratio of the wing structure is determined by the formula

$$\overline{m}_w = \frac{n_d AR \varphi 0.022 \sqrt[4]{m_0}}{\left(\frac{\mathrm{m}}{\mathrm{S}}\right)_0 \left(\frac{t}{c}\right)_{root}^{0.75} \left(cos\Lambda_{c/4}\right)^{1.5}} \left(0.85 + \frac{\frac{\left(\frac{t}{c}\right)_{root}}{\left(\frac{t}{c}\right)_{tip}} - 1}{\lambda + 3}\right) k_1 + \frac{4.5 k_2 k_3}{\left(\frac{m}{S}\right)_0} + 0.01,$$

n_d = the design load factor, calculated according to the formula:

$$n_d = 1.5 + \frac{1685}{\left(\frac{m}{S}\right)_0 \left(\frac{1}{cos\Lambda_{c/4}} + \frac{2}{AR}\right)} = 1.5 + \frac{1685}{\left(\frac{m}{S}\right)_0 \left(\frac{1}{0.874} + \frac{2}{9.67}\right)} \approx 3.521;$$

φ = wing bending relief coefficient associated with fuel and wing weights:

$$\varphi = 0.92 - 0.5\,\overline{m}_{fuel} - 0.1\,k_{eng},$$

where k_{eng} is the engine location coefficient. In our project, $k_{eng} = 1$, therefore,

$$\varphi = 0.92 - 0.5 \cdot 0.285 - 0.1 \cdot 1 = 0.6775;$$

k_1 = wing service life coefficient; $k_1 = 1.1$;

k_2 = high lift devices coefficient; for the aircraft at hand using slats, flaps, and spoilers, we assume $k_2 = 1.5$;

k_3 = wing fuel tanks sealing coefficient; let us set $k_3 = 1.1$.

Therefore,

$$\overline{m}_w = \frac{3.521 \cdot 9.67 \cdot 0.6775 \cdot 0.022\sqrt[4]{102720.82}}{617 \cdot 0.145^{0.75}(0.874)^{1.5}}\left(0.85 + \frac{\frac{0.145}{0.095} - 1}{3.93 + 3}\right) \cdot 1.1$$
$$+ \frac{4.5 \cdot 1.5 \cdot 1.1}{617} + 0.01,$$
$$\overline{m}_w = 0.1002.$$

(b) The mass ratio of the fuselage structure is determined by the formula

$$\overline{m}_{fus} = \frac{AR_f d_{fus}^2}{\left(m_0^I\right)^{k_1}} k_2 + k_3 + k_4 + k_5,$$

where

d_{fus} = fuselage diameter. Since the cross-section of our fuselage is not round, we use the equivalent diameter instead,

$d_{fus} = d_{fus}^{equiv} = \frac{H_{fus}+W_{fus}}{2} = 3.95m$, with the current fuselage midsection dimensions;

AR_f = fuselage aspect ratio,$AR_f = 11$;

k_1 = fuselage diameter coefficient, for a narrowbody aircraft with $d_{fus} = 4m$ $k_1 = 0.74$;

k_2 = engine location coefficient; $k_2 = 3.63 - 0.33d_{fus}$ if the engines are located on the wing. Therefore,

$k_2 = 3.63 - 0.33d_{fus} = 3.63 - 0.33 \cdot 3.95 = 2.3265$;$k_3$ = coefficient associated with the method of transportation of luggage and cargo; for the luggage and cargo stored in containers, $k_3 = 0.003$;

k_4 = coefficient associated with the place of attachment of the main landing gear; if the main gear is attached to the wing, $k_4 = 0$;

k_5 = coefficient associated with the place of landing gear retraction; if the main gear is retracted into the fuselage, $k_5 = 0.004$.

For the aircraft at hand,

$$\overline{m}_{fus} = \frac{11 \cdot 3.95^2}{102720.82^{0.74}} 2.3265 + 0.003 + 0 + 0.004 \approx 0.0851.$$

(c) To determine the tail mass ratio, we will use the formula

$$\overline{m}_{tail} = \frac{\overline{S}_{tail}}{\sqrt{\left(\frac{m}{S}\right)_0}} k_1 k_2 k_3 k_4,$$

where:

$\overline{S}_{tail}$ = sum of the ratio of HT and VT areas to the wing area:
$\overline{S}_{tail} = \frac{S_{HT}}{S} + \frac{S_{VT}}{S} = 0.25 + 0.201 = 0.451$;
k_1 = HT location coefficient; for the HT located on the fuselage, $k_1 = 1$;
k_2 = coefficient associated with the use of composite materials in the tail structure; if no composites are used in the tail structure, $k_2 = 1$;
k_3 = aircraft wing-tail layout coefficient; for a conventional layout,$k_3 = 1$;
k_4 = HT layout coefficient, for a HT with elevators, $k_4 = 1$.

Now,

$$\overline{m}_{tail} = \frac{0.468}{\sqrt{617}} 1 \cdot 1 \cdot 1 \cdot 1 \approx 0.0182.$$

(d) The landing gear mass ratio is determined by the formula

$$\overline{m}_{LG} = \frac{0.1h^2 \overline{m}_{land}}{m_0} k_1 k_2 k_3 k_4 + \frac{k_5}{\sqrt{p_{\mathrm{pn}}}},$$

where

h = height of the main landing gear; for the prototype aircraft, $h = 2.7$m;
$\overline{m}_{land}$ = aircraft landing mass ratio; $\overline{m}_{land} = 0.85$, since $0.2 < \overline{m}_{fuel} < 0.3$;
k_1 = landing gear service life coefficient; for subsonic commercial aircraft, $k_1 = 1.8$;
k_2 = coefficient associated with the main landing gear structural layout, for telescopic main gear, $k_2 = 1.2$;
k_3 = aircraft wing-tail layout coefficient; for a conventional layout, $k_3 = 1.4$
k_4 = coefficient associated with the number of main gears; there are two main gears on the aircraft, hence, $k_4 = 1$;
k_5 = runway type coefficient; for a concrete runway, $k_5 = 0.06$;
p_{pn} = pressure in the main gears' pneumatics; for an aircraft operating from concrete runways, we use high pressure pneumatics with $p_{pn} = 9\ daN/cm^2$;
m_0 = take-off mass in tonnes.

Therefore,

$$\overline{m}_{LG} = \frac{0.1 \cdot 2.7^2 \cdot 0.85}{10.72082} 1.8 \cdot 1.2 \cdot 1.4 \cdot 1 + \frac{0.06}{\sqrt{9}} \approx 0.0382.$$

Summing up the obtained values, we find,

$$\overline{m}_{str} = 0.1002 + 0.0851 + 0.0182 + 0.0382 = 0.2417.$$

By inserting the refined values of $\overline{m}_{str}$, and $\overline{m}_{PP}$, $\overline{m}_{equip}$ into the mass balance equation we obtain:

$$m_0^{II} = \frac{24546 + 620}{1 - 0.2417 - 0.0869 - 0.1294 - 0.285} \approx 97922.18\,\text{kg}.$$

Let us estimate the error associated with the takeoff mass of the aircraft:

$$\delta = \left|\frac{m_0^{II} - m_0^{I}}{m_0^{II}}\right| 100\%,$$

$$\delta = \left|\frac{97922.18 - 1027200.87}{97922.18}\right| \cdot 100 \approx 4.9\%,$$

Since the error is less than 5%, further iterations in determining the takeoff mass are not required and the current values of aircraft main parameters are used in the subsequent design stages.

4.6 Determination of Absolute Mass Values for an Aircraft

For a given aircraft take-off mass, the aircraft mass statement is formed using the formula $m_i = \overline{m}_i m_0^{II}$. For the aircraft at hand, this statement is presented below.

Take-off mass components	m_i, kg
Airframe	
– Wing	9812
– Fuselage	8333
– Tail	1782
– Landing gear	3741
Power plant	8509
Equipment	12,671
Payload	24,546
Flight crew equipment	620
Fuel	27,908
	97,922

Then, using the value of m_0^{II} we specify the wing area and the required total take-off thrust of the engines:

$$S_w = \frac{97922.18 \cdot 9.81}{10 \cdot 617} \approx 156\, m^2,$$
$$T_0 = \frac{0.286 \cdot 97922.18 \cdot 9.81}{10} \approx 27474 daN.$$

Then the required thrust of one engine:

$$T_0^1 = \frac{27474}{2} 13737\, daN.$$

4.7 Determination of Aircraft Geometric Parameters

In this section, we will summarize the previously defined geometry parameters of the main parts of the aircraft, as well as determine other geometry parameters.

Wing Parameters:

- area $S = 156\, m^2$;
- aspect ratio $AR = 9.67$;
- taper ratio $\lambda = 3.93$;
- quarter chord sweep angle $\Lambda_{c/4} = 29°$;
- wing span

$$b = \sqrt{S \cdot AR} = \sqrt{156 \cdot 9.67} \approx 38.84\, m;$$

- wing root chord

$$c_r = \frac{2S\lambda}{(\lambda + 1)b} = \frac{2 \cdot 156 \cdot 3.93}{(3.93 + 1) \cdot 38.84} \approx 6.404\, m;$$

- wing tip chord

$$c_t = \frac{c_r}{\lambda} = \frac{6.404}{3.93} = 1.629\, m;$$

– mean aerodynamic chord

$$c_{MAC} = \frac{2}{3} c_r \frac{1 + \lambda + \lambda^2}{\lambda(\lambda + 1)} = \frac{2}{3} \cdot 6.404 \cdot \frac{1 + 3.93 + 3.93^2}{3.93(3.93 + 1)} \approx 4.49\, m;$$

Parameters of the Horizontal Tail:

– HT area

$$S_{HT} = \frac{S_{HT}}{S_w} S_w = 0.25 \cdot 156 = 39\, m^2;$$

– HT span

$$b_{HT} = \sqrt{S_{HT} AR_{HT}} = \sqrt{41.652 \cdot 4.5} \approx 13.5\, m;$$

– HT root chord

$$c_r^{HT} = \frac{2 S_{HT} \lambda_{HT}}{(\lambda_{HT} + 1) b_{HT}} = \frac{2 \cdot 41.652 \cdot 3.03}{(3.03 + 1) \cdot 16.691} \approx 4.43\, m;$$

– HT tip chord

$$c_t^{HT} = \frac{c_r^{HT}}{\lambda_{HT}} = \frac{4.43}{3.03} \approx 1.46\, m;$$

– HT arm

$$l_{HT} = \frac{V_{HT} c_{MAC}}{\frac{S_{HT}}{S_w}} = \frac{0.96 \cdot 4.49}{0.25} \approx 17.24\, m;$$

Parameters of the Vertical Tail:

– VT area

$$S_{VT} = \frac{S_{VT}}{S_w} S_w = 0.201 \cdot 156 = 31.356\, m^2;$$

- VT span

$$b_{VT} = \sqrt{S_{VT} AR_{VT}} = \sqrt{31.356 \cdot 1.68} \approx 7.26\,m;$$

- VT root chord

$$c_r^{VT} = \frac{2 S_{VT} \lambda_{VT}}{(\lambda_{VT} + 1) b_{VT}} = \frac{2 \cdot 31.356 \cdot 2.77}{(2.77 + 1) \cdot 7.258} \approx 6.35\,m;$$

- VT tip chord

$$c_t^{VT} = \frac{c_r^{VT}}{\lambda_{VT}} = \frac{6.35}{2.77} \approx 2.29\,m;$$

- VT arm

$$l_{VT} = \frac{V_{VT} b}{\frac{S_{VT}}{S_w}} = \frac{0.086 \cdot 38.84}{0.201} \approx 16.62\,m;$$

Fuselage Parameters

- the fuselage length is determined by an empirical formula:

$$l_{fus} = 0.267 \frac{AR_{fus} b}{\sqrt{AR}} = 0.267 \frac{11 \cdot 38.84}{\sqrt{9.67}} \approx 36.683\,m.$$

Using the available data, it is possible to prepare a general view drawing of the aircraft in the first approximation (Fig. 5).

As for the wing and tail positioning with respect to the fuselage, we use the prototype aircraft data and set the distance from the fuselage nose to c_{MAC}, and the center of gravity location of $(0.27 \ldots 0.3) c_{MAC}$. We set the distance between the engines of $\sim \frac{1}{3} b$. Refinement of the general view is performed at the subsequent design stages.

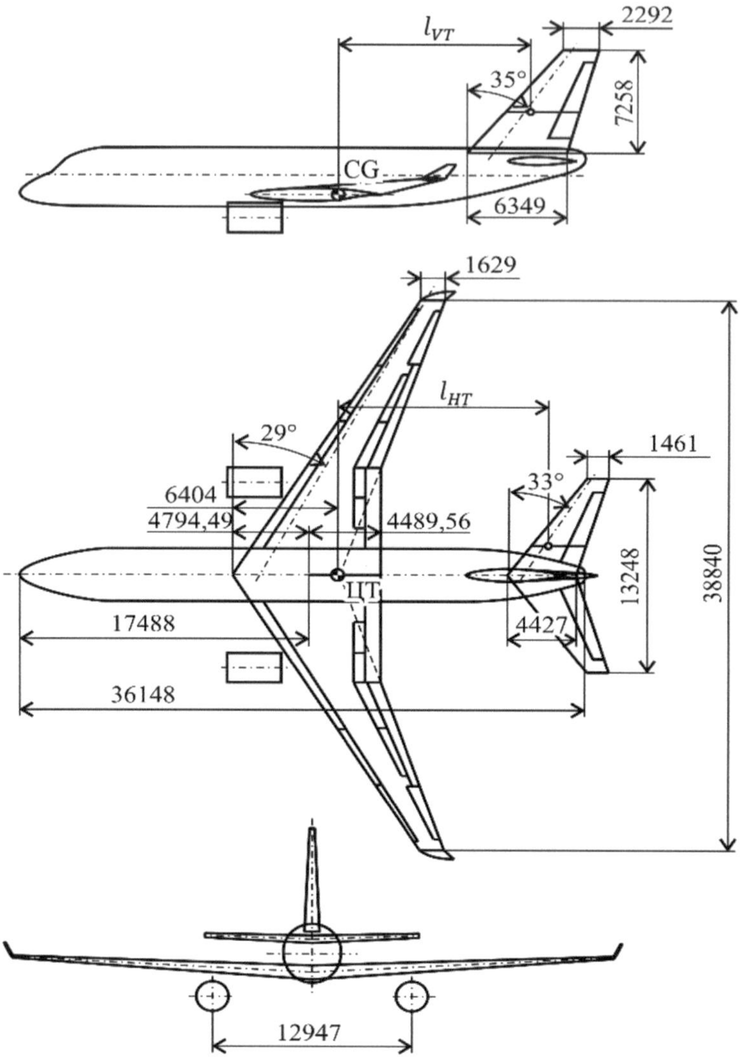

Fig. 5 General view of an aircraft

Variables		
Symbol	Description	Units
m_0	Design aircraft take-off mass	kg
S	Baseline wing area	m^2
S_w	Wing area	m^2
T_0	Engine takeoff thrust	daN

(continued)

(continued)

Variables		
Symbol	Description	Units
$\frac{m_0}{S}$	Take-off wing loading	daPa
b	Span	m
$\frac{T_0}{W_0}$	Take-off thrust-to-weight ratio	–
c_r	Root chord	m-
c_t	Tip chord	m
AR	Aspect ratio	–
λ	Taper ratio	–
$\frac{t}{c}$	Thickness ratio	–
$\Lambda_{c/4}$	Quarter chord line sweep angle	degrees
$\Lambda_{c/2}$	Mid-chord line sweep angle	degrees
i_w	Wing angle-of-incidence	degrees
$\overline{S}_{HT} = \frac{S_{HT}}{S}$	Horizontal tail area ratio	–
$\overline{S}_{VT} = \frac{S_{VT}}{S}$	Vertical tail area ratio	–
l_f	Fuselage length	m
S_{mid}	Midsection area	m^2
$l_{f,fore}$	Forebody length	m
$l_{f,aft}$	Afterbody length	m
D_{mid}	Midsection equivalent diameter	m
AR_f	Fuselage aspect ratio	–
$AR_{f,fore}$	Forebody aspect ratio	–
$AR_{f,aft}$	Afterbody aspect ratio	–
b	Wheel base	m
e	X-distance of CG to main wheels	m
B	Wheel track	m
h	Landing gear height	m
$\overline{S}_{HT} = \frac{S_{HT}}{S_w}$	Horizontal tail area ratio	–
$\overline{l}_{HT}$	Horizontal tail arm ratio	–
$\overline{l}_{VT}$	Vertical tail arm ratio	–
V_{HT}	Horizontal tail volume coefficient	–
V_{VT}	Vertical tail volume coefficient	–
C_L	Lift coefficient	–
C_D	Drag coefficient	–
$C_{L,\max,take-off}$	Maximum takeoff lift coefficient	–
$C_{L,\max,landing}$	Maximum landing lift coefficient	–
$C_{L,cruise}$	Cruise lift coefficient	–

(continued)

(continued)

Variables		
Symbol	Description	Units
L/D_{climb}	Climb lift-to-drag ratio	–
n_{eng}	Number if engines	–
m_0^i	i-th iteration aircraft takeoff mass	kg
δ	Unity equation accuracy	– (%)
m_{empty}	Empty aircraft amass	kg
m_{FL}	Full load mass	kg
m_{struct}	Structural mass	kg
m_{PP}	Power plant mass	kg
m_{fuel}	Fuel mass	kg
m_{equip}	Equipment mass	kg
m_{FCEq}	Flight crew equipment mass	kg
m_{PL}	Payload mass	kg
R_{cr}	Cruise range	km
V_{cr}	Cruise speed	km/h
H_{cr}	Cruise altitude	km
$\left(L/D\right)_{cr}$	Cruise lift-to-drag ratio	–
C_{D0}	Zero lift drag coefficient	–
$\left(\frac{t}{c}\right)_{av}$	Average wing thickness ratio	–
δ	Deflection angle	degrees
w	Width	m
h	Height	m
c	Chord length	m
V_{appr}	Approach speed	m/s
$V_{min,land}$	Minimum landing speed	m/s
q_{cr}	Cruise dynamic pressure	Pa
ξ_{HV}	Engine thrust coefficient	–
φ_t	The cruise throttling factor	–
l_{TOGR}	Take-off ground roll distance	m
f_{TOGR}	Take-off friction coefficient	–
$\left(L/D\right)_{TOGR}$	Take-off ground roll lift-to-drag ratio	–
γ_{eng}	Engine specific weight	–
n_d	Design load factor	–
c_{MAC}	Mean aerodynamic chord length	m

References

1. Bolkhovitinov VF (1962) Puti razvitiya letatel'nyh apparatov (Ways of development of aircraft). Oborongiz, Moscow
2. Fomin NA (1961) Proektirovanie samolyotov (Aircraft Design). Oborongiz, Moscow
3. Eger SM, Mishin VF, Liseitsev NK et al (2005) Proektirovanie samolyotov (Aircraft Design). Logos, Moscow
4. Sheinin VM, Kozlovsky VI (1984) Vesovoe proektirovanie i effektivnost' passazhirskih samoletov (Weight design and efficiency of passenger aircraft). Mashinostroenie, Moscow
5. Roskam J (1985) Airplane design. Part V: component weight estimation. Roskam Aviation and Engineering Corporation, Kansas
6. Aviation Rules. Part 25—Transport civil category airplanes

About Aircraft Weight Analysis (Weight Design)

Nikolai K. Liseitsev

Abstract The chapter explains the essence of the 'weight design' concept and applies this concept to estimate an airplane's gross weight in a third iteration, including the fuel, power plant, and airframe weights. The chapter also shows how an aircraft's gross mass may be recalculated if some changes to the baseline design are introduced. The chapter discusses the concept of mass budgeting and control and how these approaches can be realized in engineering practice. Finally, the chapter explains the moment of inertia of aircraft and its prediction.

Keywords Aircraft weight analysis · Weight recalculation · Mass budgeting · Moment of inertia

The "weight design" concept was suggested by V.M. Sheinin [1]. Weight design incorporates:

- methods for calculating and monitoring the mass of an aircraft and its components at all design stages;
- methods for mass reevaluation when the parameters and characteristics of the aircraft are changed;
- weight budgeting and weight control;
- determination of the moments of inertia of the aircraft.

Let us briefly consider the content of these components of weight analysis.

N. K. Liseitsev (✉)
Moscow Aviation Institute (National Research University), Moscow, Russia
e-mail: nikolai.liseitsev@yandex.ru

M. A. Pogosyan (ed.), *Aircraft Design*, Springer Aerospace Technology,
https://doi.org/10.1007/978-981-96-4599-2_6

1 Determination of Aircraft Mass in the Third Approximation

Once the layout of an aircraft has been established and its general view has been drawn, it is possible to calculate the mass of the aircraft in the third approximation. The purpose of this approximation is to specify the components of the aircraft's takeoff mass on the basis of the information obtained at this stage of design.

Fuel mass ratio refinement $\overline{m}_{fuel}^{III}$. Based on the takeoff mass m_0^{II}, refined aerodynamic characteristics and the characteristics of the selected engine, the fuel required for takeoff, climb and acceleration, horizontal flight and descent are determined.

1. The first step is to determine the navigational fuel reserve m_{NFR}:
   - for a commercial airliner, it shall ensure the safe flight from the airfield of intended landing to the alternate airfield at a distance of 450 km and the loiter at the alternate airfield of 30 min at an altitude of 400 m;
   - for a military aircraft, it is usually set as a range of 8…10% for fighters and about 5% for a long-range attack aircraft. The unusable fuel mass ratio is assumed to be $\overline{m}_{Ufuel} = (0.003\ldots0.005)$.
2. *Refinement of the engine and air intake masses with regard to* the power plant layout.
3. *Refinement of the mass of the airframe units* on the basis of element-by-element calculations.
4. *Refinement of the wing area* S_w to ensure the specified flight performance.
5. *Preparation of a detailed equipment specification and determination of the mass of individual systems or individual pieces of equipment.*

Upon completion of the third approximation calculations, the weight "appearance" of the aircraft is established.

After that, the limit weights are assigned and issued to the teams producing technical and analysis documentation (drawings, specifications, strength analysis reports, etc.).

Note that in aircraft design, we have to deal with forces, so along with the concept of mass, the concept of weight is widely used, as well as the terms of weight design, weight calculations, weight culture, etc. When using these concepts, it should be remembered that the relationship between weight W and mass m is determined by expression $W = mg$.

The weight budgeting reflects a certain technical policy. For example, it may be as follows: structural designers and strength engineers assign tolerances of 5…10% to the selected sizes. Since they work sequentially, there should theoretically be formed a safety margin:$f\prime = (1.05\ldots1.1)^2 = 1.1\ldots1.2$; therefore, the structure is always 10…20% heavier than the optimum one. Taking this into account, the predicted weights of the third approximation are reduced by approximately 10%.

The weight budget for design teams is issued with a factor of 0.9–0.95, given that all approximations are made on the basis of historical data and incorporate theoretical and practical design overweighting. There may be other ideas for weight policy.

In the process of producing working drawings, design teams predict the weight of each part, assembly, and unit. Special weight teams perform weight control, random checks, collect weight data on units, and calculate the real, practical weight of the aircraft. Prior to the commencement of flight tests, the aircraft is weighed in various versions: empty, with equipment, and with fuel.

Note that in practice, each design bureau involved in aircraft design uses its own approaches and methods for determining aircraft mass.

In the last decade, due to the introduction of computers and numerical methods for analyzing the stress–strain state of a structure, codes for predicting the mass of the load-bearing structure have been actively developed. This undoubtedly leads to an improvement in the quality of mass prediction, especially for nonconventional structures.

2 Recalculation of Aircraft Mass

When one or two parameters of the aircraft or its parts are changed, the new value of takeoff mass or mass of units (wing, fuselage, etc.) is easier and faster to find by recalculation from the original to the new value, rather than by iteration. Such recalculation is expedient to perform during mass analysis of the aircraft, when evaluating the influence of individual parameters, when modified units are used, when there is a prototype. The basic formula for takeoff mass recalculation, based on the finite increment method, proposed by A.A. Badiagin, is as follows

$$m_{0,new} = m_{0,in} \pm \sum_{k=1}^{n} \frac{\delta m_0}{\delta i_k} \Delta i_k, \tag{1}$$

where i is the variable,

k is the number of variables.

In case of additional mass Δm_{add} we have

$$m_{0,new} = m_{0,in} + \chi_m \Delta m_{add}, \tag{2}$$

where χ_m is the take-off mass growth factor associated with the additional mass ($\chi_m = const$, see Sect. 3)

For m_0 recalculation, specific values g_i can also be used:

$$m_{0,new} = m_{0,in} \pm \chi_m (g_i \Delta i), \tag{3}$$

where $g_i = \frac{m_{unit}^{in}}{i \approx const}$.

For example, if the fuselage length changes ($\Delta i = \Delta l_{fus}$), then

$$m_{0,new} = m_{0,in} \pm \chi_m \left(g_{i,fus} \Delta l_{fus} \right), \tag{4}$$

where $g_{i,fus} = \frac{m_{fus}^{in}}{l_{fus}^{init} \approx const}$.

If the wing area increases ($\Delta i = \Delta S$), then

$$m_{0,new} = m_{0,in} \pm \chi_m (g_S \Delta S), \tag{5}$$

where

$g_S = \frac{m_w^{in}}{S \approx const}$ = the mass of 1 m^2 of the wing;

χ_m = growth factor of m_0 under the given constraints.

When the takeoff mass changes, the mass ratio of the aircraft structure can be recalculated as follows:

$$\overline{m}_{str} = n_{limit} \left(0.005 \sqrt[3]{m_0} + \frac{0.082}{\sqrt[3]{m_0}} \right) + 0.061. \tag{6}$$

Formula (3) is applicable to wing-tail non-maneuvering aircraft and has the following limitations:$n_{limit} = 3.5 \ldots 5.5$; $m_0 = 10 \ldots 10^3 t$. Before using expression (6), its coefficients should be corrected for the initial mass ratio of the structure; then, the calculation for $\overline{m}_{str}$ is performed with the new value of m_0.

If the wing area and wing loading are changed, the new value of wing mass can be found as follows:

$$m_w^{new} = m_{w,HLoff}^{init} \frac{S_{new}}{2S_{init}} \left(1 + \frac{\left(\frac{m}{S}\right)_0^{new}}{\left(\frac{m}{S}\right)_0^{init}} \sqrt{\frac{S_{new}}{S_{init}}} \right) + m_{HL}^{init} \frac{S_{HL}^{new}}{S_{HL}^{init}}, \tag{7}$$

where

$m_{w,HLoff}^{init}$ = initial mass of the wing minus the high-lift devices mass;

S_{HL}^{init}, S_{HL}^{new} = initial and new area of high-lift devices.

If only the take-off mass changes, it is convenient to predict the wing mass ratio using the equation which for a heavy nonmaneuverable aircraft is as follows:

$$\overline{m}_w \approx 0.0147 \sqrt[3]{m_0} + 0.0275. \tag{8}$$

Here, the unit of m_0 is tonne.

Equation (8) is used in the same way as Eq. (6).

Also, it is convenient to recalculate the wing mass assuming that the mass of 1 m^2 of wing structure is constant (provided that the change in its area is relatively small ($\le 15\%$)). In this case

$$m_w^{new} = m_w^{init} \pm (g_w \Delta S), \tag{9}$$

where $g_w = \frac{m_w^{init}}{S_{init}} \approx const.$

The mass of the tail or fuselage can be recalculated in the same way. For example, if the fuselage length is increased:

$$m_{fus}^{new} = m_{fus}^{init} \pm \left(g_{l,fus} \Delta l_{fus}\right), \tag{10}$$

where $m_{fus}^{new} = \frac{m_{fus}^{init}}{l_{fus}^{init}} \approx const.$

Fuselage mass ratio of a heavy non-maneuvering aircraft as a function of take-off mass is as follows:

$$\overline{m}_{fus} = 0.005\sqrt[3]{m_0} + 0.2m_0^{-(1/3)} + 0.03, \tag{11}$$

where the unit of m_0 is tonne.

Expression (11) can be used in the same way as expression (6).

In order to transition from theoretical mass m^* to real mass, coefficient k_{unit} is used:

$$m_{unit} = k_{unit} m^*.$$

For example, for the wing of a non-maneuvering aircraft, according to historical data,

$$k_{unit} = k_w \approx 4.5 \frac{2000 - m_0}{2000 + m_0},$$

where the unit of m_0 is tonne.

For the fuselage of a non-maneuvering aircraft,

$$k_{unit} = k_{fus} \approx 6 \frac{2000 - m_0}{2000 + m_0}.$$

With an increasing m_0, the load experienced by the structure increases, the number of secondary structural elements decreases. Therefore, with an increasing m_0, the value of k_{unit} decreases.

The numerical value of k_{unit} can be found not only empirically (using statistics), but also theoretically by determining the mass of the secondary structural elements of the unit, the mass associated with design traditions, operation and manufacturing costs, as well as the mass spent on ensuring the service life, rigidity, reliability and to meet other requirements.

3 Mass Budgeting and Control

These tasks include:

- development of methods to minimize the structural mass and take-off mass of an aircraft;
- development of the baseline data, constraints, and a budgeting system for mass predictions;
- drawing up action plans to reduce aircraft mass;
- monitoring the implementation of the above plans;
- mass control;
- development of a system for tracking mass changes during aircraft design;
- development of a mass control system in prototype and serial production of aircraft;
- development of a system of financial and moral incentives to reduce the mass of aircraft details, parts, and units.

A complex system of weight budgeting and control has to be created in order to control the increase in mass and keep it within the specified (designed) limits. The actual mass of an empty aircraft, as a rule, exceeds its theoretical design value. This fact is an objective law.

Some objective reasons for the growth of empty airplane mass during its design are:

- insufficient depth of design elaboration due to limited or inaccurate information at the initial design stage, when the mass of the aircraft is to be determined;
- additional customer requirements;
- refinement of design data, deviations from the design to improve overall aircraft performance (most often accompanied by an increase in mass);
- difficulties in implementation of new technologies (new engines, materials, equipment, etc.).
- refinement after static and dynamic structural tests; refinement after flight.

The weight analyst should take into account that:

- coincidence of actual and design values of empty aircraft mass is possible within some tolerance.
- insufficient accuracy in the weight prediction can lead to overweighting.

To reduce aircraft weight growth, it is required to separate overweighting, which is based on subjective factors such as lack of experience or time, from objective reasons for weight growth.

The experience shows that the takeoff mass of an aircraft increases by 6–8% on average when a prototype is designed and built and by 1.5–2% when it is put into series production, all in all, by 7.5–10% (as compared to the initial theoretical mass).

Various methods are used to restrain aircraft mass growth, such as:

- applying the concept of weight reserve when an empty aircraft mass increase is planned in advance. The reserve in this case is used to adjust the weight budgets.
- planning for aircraft modifications at early design stages;

- using strength reserves, which in most cases are revealed by tests. According to this method, the structural analysis is conducted for somewhat underestimated loads.
- the method of redesigning, when engineering drawings of units, parts, and details are redesigned without significant time and cost and without fundamental changes (it is known that any design can be improved and lightened by specifying the loads, more careful design, introduction of new design layouts, new materials, technology, etc.).

In the process of weight control at all design stages, the weight budget for each aircraft unit is adjusted as the design progresses.

4 On the Take-Off Weight Growth Coefficient Associated with Added Weights of Aircraft Parts

When designing an aircraft, one often encounters the problem of changing the mass of a unit, equipment group or payload. The question arises: how much should the takeoff mass be adjusted so that the basic performance and strength of the aircraft remain the same?

Such tasks are typical for the conceptual design stage, when, based on variations in parameters, the basic dimensions are selected and the mass of the aircraft is predicted. However, at the subsequent design stages (even for modifications of existing aircraft), these tasks are not excluded, since the detail design of the aircraft is usually accompanied by some deviations of the mass of the units from their values established during conceptual design.

So, let us assume that at some stage of aircraft design it became necessary to introduce some additional mass m_{add}. If, when changing the mass of any unit by $(m_{add})_1$, some additional mass $(m_{add})_2$ is needed to correct possible changes in the aircraft center-of-gravity, to increase the structural rigidity, etc., then we will assume that the additional mass m_{add} is the sum of all these components, i.e. $m_{add} = (m_{add})_1 + (m_{add})_2 + \ldots$.

Let's assume that we know the dependences of the masses of all components on the takeoff mass of the aircraft, as well as the values that the designer considers necessary to leave unchanged, for example, $\left(\left(\frac{m}{S}\right)_0, \left(\frac{T}{W}\right)_0\right) = const$, or $(S, T_0) = const$, or $\overline{m}_{fuel} = const$, etc.

It is required to determine the increase in take-off mass associated with meeting the performance and strength requirements.

Let us consider the aircraft mass balance equation with the additional mass:

$$m_0 = m_{empty} + m_{fuel} + m_{PL} + m_{FCEq} + m_{add}, \tag{12}$$

where

m_{empty}= mass of empty aircraft, $m_{empty} = f_1(m_0)$;

m_{fuel} = mass of fuel, $m_{fuel} = f_2(m_0)$;

$m_{PL,}m_{FCEq}$= payload and flight crew equipment mass, $m_{PL,}m_{FCEq} = const$. It follows from (12) that $m_{add} = m_0 - m_{empty} - m_{fuel} - m_{PL} - m_{FCEq}$.

Let us take the partial derivative of m_0:

$$\frac{\delta m_{add}}{\delta m_0} = 1 - \frac{\delta}{\delta m_0}\left(m_{empty} + m_{fuel}\right).$$

Since it is required to determine the derivative of $\frac{\delta m_0}{\delta m_{add}}$, from the aforementioned equation follows

$$\frac{\delta m_0}{\delta m_{add}} = \frac{1}{1 - \frac{\delta}{\delta m_0}\left(m_{empty} + m_{fuel}\right)}. \tag{13}$$

This is the takeoff mass growth coefficient due to the additional mass. Let us denote it for the sake of brevity as follows:.

$$\frac{\delta m_0}{\delta m_{add}} = \chi_m.$$

The numerical values of χ_m can vary from one to infinity. The latter is possible when $\frac{\delta}{\delta m_0}\left(m_{empty} + m_{fuel}\right) = 1$.

If $\frac{\delta}{\delta m_0}\left(m_{empty} + m_{fuel}\right) = 0$, then $\chi_m = 1$. Thus, $\left(m_{empty}, m_{fuel}\right) = const$ and $(m_0)_{new} = (m_0)_{init} \pm m_{add}$, which corresponds to the case when all flight characteristics of the aircraft and the strength of its units change (this case is possible when modifying an existing aircraft). It is important to note that the value of χ_m may be greater than unity only if, with the introduction of additional mass, the designer seeks to maintain some of the aircraft properties unchanged—the flight performance, strength, etc.

Let us consider the physics of the derivative χ_m when $\chi_m > 1$.

Let's assume that on the aircraft at hand it was necessary to install some additional equipment with a mass m_{add}, which does not deteriorate aerodynamics and does not require an increase in the volume of the aircraft. It is necessary to modify the aircraft, preserving the basic flight characteristics, payload, as well as the structural strength. To keep the performance unchanged, the wing loading and take-off thrust loading should be the same.

As the mass of the aircraft increases by m_{add}, the wing, fuselage, landing gear, and tail loading also increases. Consequently, in order to maintain the structural strength, it is necessary to increase the size and mass of structural members. However, in order to fulfill the condition of $\left(\frac{m}{S}\right)_0 = const$, it is necessary to change the wing dimensions, and for $\left(\frac{T}{W}\right)_0 = const$—the engine thrust, which in itself would entail a change in the wing and power unit mass. Further, with an increase in the wing size, it is necessary to increase the size and mass of the tail to keep the aircraft stability characteristics unchanged. Finally, in order to maintain the range, it will obviously be necessary to increase the initial fuel mass:$\overline{m}_{fuel} = const$. The given increase in the mass of individual parts of the aircraft and the initial fuel mass will result in an increase in the takeoff mass of the aircraft, which, in turn, will entail an increase in the load experienced by aircraft units and increase their mass. As a result, a new, secondary increase in structural and fuel mass will be required.

The process ends when the takeoff mass reaches the value obtained from solving the mass balance equation.

Thus, the aircraft added mass m_{add} increases the takeoff mass by a much larger amount $m_{add}\chi_m$.

The derivative $\frac{\delta m_{empty}}{\delta m_0}$, which is included in (13), can be determined by the formula

$$\frac{\delta m_{empty}}{\delta m_0} = \frac{\delta m_w}{\delta m_0} + \frac{\delta m_{fus}}{\delta m_0} + \frac{\delta m_{tail}}{\delta m_0} + \frac{\delta m_{LG}}{\delta m_0} + \frac{\delta m_{PP}}{\delta m_0} + \frac{\delta m_{equip}}{\delta m_0}.$$

Here, $m_w, m_{fus}, m_{tail}, m_{LG}, m_{PP}, m_{equip}$ are the weight of the wing, fuselage, tail, landing gear, power plant, equipment. Weight formulas of the wing and other units depending on m_0 are usually known.

When determining the derivative $\frac{\delta m_{fuel}}{\delta m_0}$, also included in (13), it should be taken into account that the fuel mass is linearly related to the takeoff mass, and therefore $\frac{\delta m_{fuel}}{\delta m_0} = \overline{m}_{fuel}$. For approximate calculations, the following dependence derived from the Breguet formula can be used:

$$\frac{\delta m_{fuel}}{\delta m_0} = \overline{m}_{fuel} = 1 - e^{\frac{RC_p}{\frac{V_{flight}L}{D}}} + \overline{m}_{TF} \tag{14}$$

If $\overline{m}_{fuel} \leq 0.2$, then formula (14) can be simplified. By expanding the second term into a series and limiting the first two terms of the expansion, we obtain an error of $\pm$ 2%.

$$\frac{\delta m_{fuel}}{\delta m_0} = \overline{m}_{fuel} \approx \frac{1.1RC_p}{\frac{V_{flight}L}{D}} + \overline{m}_{TF}, \tag{15}$$

where

1.1 = correction factor;

R = design range, km;

C_p, $\frac{L}{D}$ = average values of specific fuel consumption and lift-to-drag ratio;

V_{flight} = flight speed, km/h;

$\overline{m}_{TF}$ = trapped fuel ratio.

As an approximation, it can be assumed that $\frac{\delta m_{empty}}{\delta m_0} = \overline{m}_{empty}$. Then expression (13), taking into account (14), will take a simple form:

$$\chi_m = \frac{\delta m_0}{\delta m_{add}} = \frac{1}{1 - \overline{m}_{empty} - \overline{m}_{fuel}} \tag{16}$$

This formula can be applied in the early stages of aircraft design.

The growth coefficient χ_m can also be determined based on the principle of linear similarity:

$$\chi_m = \frac{1}{1 - 0.5k_{sim}\overline{m}_{empty} - \overline{m}_{fuel}}, \quad (17)$$

where $k_{sim} = 2.2 \ldots 2.6$ is the linear similarity coefficient.

Formula (17), like (16), can be applied in the early stages of design.

Table 1 shows the values of takeoff weight mass factor χ_m for commercial jets. With $(S, T_0 = const)$ the value of χ_m is significantly lower than at $\left(\frac{m_0}{S}, \left(\frac{T}{W}\right)_0 = const\right)$. This is explained by the fact that the additional mass at $\left(\frac{m_0}{S}, \left(\frac{T}{W}\right)_0 = const\right)$ entails an increase in the mass of all the take-off mass components except the commercial and service loads, since usually $\left(m_{pay}, m_{FCEq} = const\right)$. At $(S, T_0 = const)$, the power plant mass and the part of the wing area-dependent mass remain unchanged.

Take-off mass growth factor values for commercial airliner

The above calculation method for χ_m does not consider the origin of the additional mass m_{add}. Whichever airplane unit (wing, fuselage, etc.) produces additional mass, the change in take-off mass will be the same and equal to χ_m. A more accurate calculation of χ_m should distingush the location of additional mass formation. The calculations show that the farther away from the airplane's center of mass the additional mass is generated, the greater the value of χ_m. Compared to the above formulas, the value of χ_m may increase by 15…20% if the additional mass occurs on the wing or afterbody.

The derivative χ_m not only simplifies the determination of new takeoff mass at any design stage, but also makes it possible to estimate the mass perfection of the aircraft at hand. Obviously, the greater the value of χ_m, the more time and cost is required to reduce the mass of each unit, component and part.

The method of take-off mass gradients simplifies evaluation of the design and engineering solutions related to structural design, aerodynamics, life and cost of airframe structures. For more details on takeoff mass growth gradients (coefficients), see. [3]. However, note that the limited accuracy, as well as the increasing use of computer-aided design methods, reduce the scope of the gradient method.

Table 1 Values of take-off mass growth coefficient for commercial airliners

Aircraft type	Practical flight range with maximum payload, km	$\frac{m_0}{S}, \left(\frac{T}{W}\right)_0 = const$	$S, T_0 = const$
Commuters	600 … 1200	2,1 … 2,2	1,5 … 1,6
Mid-range	1600 … 2200	2,3 … 2,6	1,7 … 1,8
Long-range	4500 … 5500	3,2 … 3,8	2,2 … 2,6
Supersonic commercial	6000 … 6500	8 … 10	3,5 … 4,0

5 Determination of Aircraft Moments of Inertia

An important task in weight analysis is to determine the inertial properties of the aircraft. Knowledge of these properties allows for solving a number of important design problems.

They include the airframe's dynamic characteristics, whose natural frequencies of oscillation depend on the moments of inertia. Without inertial properties, it is impossible to analyze aircraft stability and controllability to design its control system. Determining inertial loads during aircraft rotation and designing mechanisms for landing gear retraction, engine rotation, etc. are also impossible without knowledge of the moments of inertia.

The problem of correctly determining the moments of inertia has become particularly acute due to the increase in aircraft size and takeoff weight. Since the moments of inertia increase in proportion to the fourth or fifth power of linear size, the inertia of aircraft and, accordingly, the required control forces, increase rapidly as aircraft size increases. The introduction of high-speed aircraft with swept wings and low aspect ratio wings led to a dramatic change in the traditional ratio of the aircraft moment of inertia around the X and Y axes. This had a negative effect on the stability characteristics of the aircraft.

The main measure of inertia of any body is its *axial moments of inertia* (I_x, I_y, I_z in a rectangular coordinate system). Their values are determined from the expression (for the X -axis):

$$I_x = \iiint 2x^2 dm,$$

where $2x$ is the distance of the elementary mass to the X-axis; dm is the elementary mass of the body.

Integration is performed over the volume occupied by the body V.

For the Y and Z axes we have, respectively,

$$I_y = \iiint 2y^2 dm, I_z = \iiint 2z^2 dm.$$

The concept *of radius of inertia* is also used in design practice. Its values for the axes X, Y, Z are defined, respectively, as

$$i_x = \sqrt{I_x/m}, i_y = \sqrt{I_y/m}, i_z = \sqrt{I_z/m}.$$

To analyze the motion of bodies with many degrees of freedom, such as aircraft, we use the concept of product of inertia. Its values with respect to the corresponding axes are defined by expressions:

$$I_{XY} = \iiint xydm; I_{XZ} = \iiint xzdm; I_{YZ} = \iiint yzdm.$$

These moments are important when dynamically balancing rotating bodies.

A distinction is made between intrinsic and transverse moments of inertia.

Intrinsic moment of inertia I^i is the moment of inertia of an element of a system, considered with respect to its own central axes parallel to the axes of the system in question.

Transferred moment of inertia I^t is the moment of inertia of an element of a system, considered as a material point with a mass concentrated in the center of its mass, relative to the axes of the given system. These moments are calculated as follows:

$$I_x^t = x^2 m, I_{xy}^t = xy^2 m,$$

where *x, y* are the distances from the center of mass of the body (element of the system) to the axes of the system.

Both experimental and computational methods are used to determine the moments of inertia. Experimental methods, due to their cumbersome and expensive nature, are mainly used for verification and are applicable primarily to individual aircraft units (rudders, ailerons). The analytical methods based on generalization of integration results for bodies and planar figures of various shapes are widely spread. The comparative simplicity of aircraft shapes and the presence of a plane of symmetry, as well as the use of previously derived expressions ensure an acceptable calculation effort with the error not exceeding 3…5%.

In the first approximation, the axial moments of inertia of the aircraft can be found as follows ($kg \bullet m^2$):

$$I_x = \frac{m_0 b^2}{12} \varphi_x;$$

$$I_z = \frac{m_0}{12}\left(b^2 + l_f^2\right)\varphi_z;$$

$$I_y = \frac{m_0 b^2}{12} \varphi_y,$$

where m_0 is the aircraft take-off mass (kg); b, l_f are the wing span and fuselage length (m); φ_x, φ_y, φ_z are coefficients (see Fig. 1).

In recent years, numerical methods for calculating moments of inertia based on finite-element models have been increasingly used. Their application ensures a reduction in labor intensity and an improvement in calculation accuracy.

The reader can find more information about weight analysis methods in [1] and [2]. These books also provide various methods for determining the takeoff weight and its components necessary for the course and diploma projects. It should be remembered that the widely used theoretical and statistical equations for determining the weight of aircraft components require constant adjustment and should be aligned with the progress in aircraft design methods.

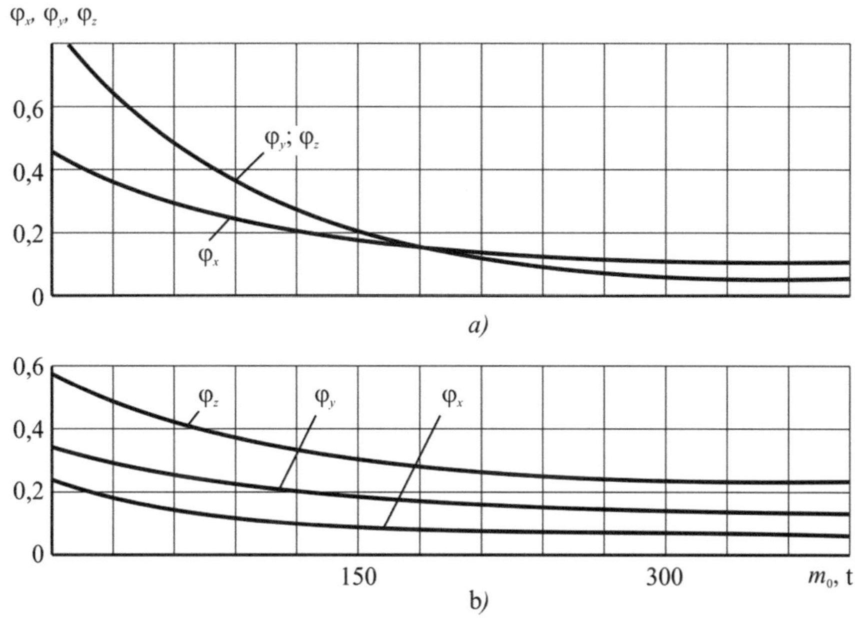

Fig. 1 Aircraft moment of inertia coefficients: **a**—Conventional aircraft; **b**—Vertical take-off and landing aircraft

Variables		
Symbol	Description	Units
m_0^i	Take-off mass, calculated in i-th iteration	*kg*
m_{NFR}	Navigational fuel reserve mass	*kg*
$\overline{m}_{Ufuel}$	The unusable fuel mass ratio, $\frac{\delta m_{fuel}}{\delta m_0} = \overline{m}_{fuel}$	–
S_w	Wing area	m^2
$f\prime$	Safety margin	–
Δm_{add}	Additional mass	*kg*
$m_{0,new}$	Mass recalculated using the take-off mass growth factor	*kg*
χ_m	Take-off mass growth factor, $\chi_m = \frac{\delta m_0}{\delta m_{add}}$	–
n_{limit}	Limit load factor	–
$\overline{m}_{str}$	Fuel—to-take-off mass ratio, $\overline{m}_{fuel} = \frac{m_{fuel}}{m_0}$	–
S	Wing area	m^2
$\overline{m}_w$	Wing mass—to-take-off mass ratio, $\overline{m}_w = \frac{m_w}{m_0}$	–
$\overline{m}_{fus}$	Fuselage mass—to-take-off mass ratio, $\overline{m}_{fus} = \frac{m_{fus}}{m_0}$	–
m^*	Theoretical mass	*kg*
k_{unit}	Transitional coefficient from theoretical to real mass of a unit	–

(continued)

(continued)

Variables		
Symbol	Description	Units
$\left(\frac{m}{S}\right)_0$	Take-off wing loading	kg/m^2
$\left(\frac{T}{W}\right)_0$	Thrust-to weight ratio	–
T_0	Take-off thrust	N
m_{empty}	Empty aircraft mass	kg
m_{fuel}	Fuel mass	kg
m_{PL}	Payload mass	kg
m_{FCEq}	Flight crew equipment mass	kg
m_w	Wing mass	kg
m_{fus}	Fuselage mass	kg
m_{tail}	Tail mass	kg
m_{LG}	Landing gear mass	kg
m_{PP}	Power plant mass	kg
m_{equip}	Equipment mass	kg
R	Range	km
C_p	Specific fuel consumption	$g/(N \bullet s)$
V_{flight}	Flight speed	km/h
$\overline{m}_{TF}$	Trapped fuel ratio	–
$\frac{L}{D}$	Lift-to-drag ratio	–
k_{sim}	$k_{sim} = 2.2 \ldots 2.6$ is the linear similarity coefficient	–
I_x, I_y, I_z	Axial moments of inertia	$kg \bullet m^2$
dm	Elementary mass of the body	kg
V	Volume occupied by the body	m^3
i_x, i_x, i_x	Radius of inertia	m
I_{XY}, I_{XY}, I_{XY}	Product of inertia	$kg \bullet m^2$
I^i	Intrinsic moment of inertia	$kg \bullet m^2$
I^t	Transferred moment of inertia	$kg \bullet m^2$
b	Wing span	m
l_f	Fuselage length	m
$\varphi_x, \varphi_y, \varphi_z$	Aircraft moment of inertia coefficients:	–

References

1. Sheinin VM, Kozlovsky VI (1984) Vesovoe proektirovanie i effektivnost' passazhirskih samoletov (Weight design and efficiency of passenger aircraft). Mashinostroenie, Moscow

2. Eger SM, Mishin VF, Liseitsev NK et al (2005) Proektirovanie samolyotov (Aircraft Design). Logos, Moscow
3. Novozhilov GV (ed) (1976) Teoriya i praktika proektirovaniya passazhirskih samolyotov (Theory and practice of designing passenger aircraft). Nauka, Moscow

Aircraft Layout

Nikolai K. Liseitsev

Abstract The chapter explores in detail the three aspects of aircraft configuration and sizing: aerodynamic, internal, and structural configuration and sizing. From the chapter, the reader will learn how to reduce the trim penalties, what the idea behind area rule is, as well as how the local aerodynamics of an aircraft can be enhanced. The chapter explains what parameters are used to estimate the lateral static stability of an aircraft and how the aircraft geometry can be adjucted to ensure the required level of static lateral stability. This chapter describes some methods for laying out the components of an aircraft and estimating its center of gravity, as well as the center of gravity envelope. The concept of the mean aerodynamic chord is introduced and explained. The chapter offers an algorithm for calculating the center of gravity for an aircraft and provides a sample weight report. The chapter also presents some approaches to ensuring the design center of gravity envelope by design methods. In addition, the chapter discusses special design layouts, such as aft-podded engines, as well as the design of cockpits and passenger cabins. Finally, some guidelines for aircraft structural design are presented.

Keywords Aerodynamic design · Internal design · Structural design · Trim penalties · Area rule · Static stability

1 Layout Definition and Objectives

In Russian, the word "layout" has two meanings. As a noun, it is a document (graphic or digital) that unambiguously defines the three-dimensional positioning of the parts of the object at hand. As a verb, it means the process of positioning of these parts (sizing).

Sizing is the process of aircraft synthesis at both preliminary and conceptual design stages. The most common objectives of aircraft sizing are to:

N. K. Liseitsev (✉)
Moscow Aviation Institute (National Research University), Moscow, Russia
e-mail: nikolai.liseitsev@yandex.ru

M. A. Pogosyan (ed.), *Aircraft Design*, Springer Aerospace Technology,
https://doi.org/10.1007/978-981-96-4599-2_7

- position the functional elements (payload, crew, equipment, and fuel) within the given aerodynamic configuration, corresponding to the basic design parameters and weight characteristics of the aircraft; this operation, among other things, makes it possible to assess the feasibility of the aircraft;
- achieve the densest possible arrangement, which leads to a reduction in its mass and wetted area;
- refine the geometric parameters and the position of aircraft units with respect to the arrangement density, aerodynamic properties, stability and controllability characteristics;
- conduct aircraft weight analysis and determine its center of mass location;
- revise the aircraft arrangement if the resulting CG envelope does not meet the requirements for the forward and aft CG limits;
- conduct structural design and positioning of the aircraft components.

Aircraft layout is a result of three interrelated processes: aerodynamic, internal and structural design, which are performed by the designer in parallel.

The problem associated with the layout is that the requirements for the aircraft parameters are often conflicting and imply incompatible actions. Thus, to reduce drag and achieve a higher lift-to-drag ratio, preference is given to small fuselage cross-sections, thinner airfoils. This can cause problems with positioning payloads, fuel, etc. At the same time, the structural height of various units also decreases, which leads to strength and stiffness problems. Also, the requirement to place a large payload inside the fuselage, apart from increasing the cross-sectional areas of the unit, can lead to significant curvature of load paths due to the impossibility of placing the structural members in the most suitable locations, which also entails problems with strength and rigidity of the structure and unjustified increase of its mass. In turn, overdesigning the structural members or straightening the load paths negatively affects the aerodynamic characteristics of the aircraft, reduces the density of the layout, increases the aircraft size, and, consequently, its mass.

2 Aerodynamic Layout

2.1 Aerodynamic Layout Objectives

Let's list the most important objectives associated with the aerodynamic layout.

- The aircraft with the given economic and flight performance *must have the smallest possible dimensions*. This generally leads to a mass and wetted-area reduction.
- *At* a given *speed, the aircraft must have the maximum lift-to-drag ratio to* ensure the minimum fuel consumption. In this regard, the aircraft should have the lowest possible drag and, in particular, trim penalties.
- In takeoff and landing, *the aircraft should have the greatest possible* $C_{L,max}$ while providing the design safety margin, allowing it to fly with a smaller wing area.

– In all flight modes, *the aircraft must demonstrate the required stability and controllability margins.*
– The *most favorable conditions for propulsion system operation*, determined by the lowest possible losses at the entrance to the engines and at the exit from the nozzles, must be *ensured* on the aircraft.
– *A favorable aerodynamic interference* of the airframe components should be achieved.
– *Critical flight modes* (e.g. high speeds or large angles of attack) *must not have dangerous consequences* (flutter, buffeting, deep stall, stall, etc.); measures must be taken to prevent entry into such modes and to allow exit from such modes to normal ones.

2.2 Reducing Trim Penalties

One of the primary problems of aerodynamic design is, as shown earlier, to reduce trim penalties. These penalties occur as soon as a trim force, whether it is directed upward or downward, is generated on the aircraft's horizontal tail or wing elevons (in the case of a tailless aircraft).

In Chap. 4, in considering the approaches to trimming an aircraft, the following conclusions were drawn:

1. Trimming the aircraft leads to an increase in $A = 1/\pi eAR$ and thus a decrease in lift-to-drag ratio.
2. The magnitude of trim penalties is determined by the longitudinal static stability $C_{m_{C_L}}$ or $\left(C_{m_{C_L}}\right)_{tail-off}$ and the parameters of the horizontal tail (S_{HT}, l_{HT}, $C_{L_\alpha,HT}$).
3. As a rule, in the canard layout, the value of $C_{L,w}$ is reduced due to the foreplane, which reduces the wing induced drag and trim penalties with the same static stability margin, see Fig. 1. The same figure shows that $C_{L,w}$ of the wing-tail layout is increased, as the wing should compensate the negative vertical force on the horizontal tail.

$$TrP = \frac{(L/D)_{max,trimmed}}{(L/D)_{max}\left(C_{m_{C_L}} = 0\right)}$$

Figure 2 demonstrates that with a typical static stability margin for wing-tail long-range commercial airliner $\left(C_{m_{C_L}} = -0.15\right)$, aerodynamic efficiency penalties reach up to 7 … 10%, and with the forward CG limit and CG envelope of 15 … 20%, these penalties can increase by 30% and more. Accordingly, the flight range decreases and the hourly and kilometer fuel consumption increases.

With the introduction of automatic control systems, the designers of modern aircraft reduce the aircraft design static stability $\left(C_{m_{C_L}} = -0.02\ldots0.05\right)$, develop aircraft with neutral stability $\left(C_{m_{C_L}} = 0\right)$ and even statically unstable aircraft $\left(C_{m_{C_L}} = 0.07\right)$.

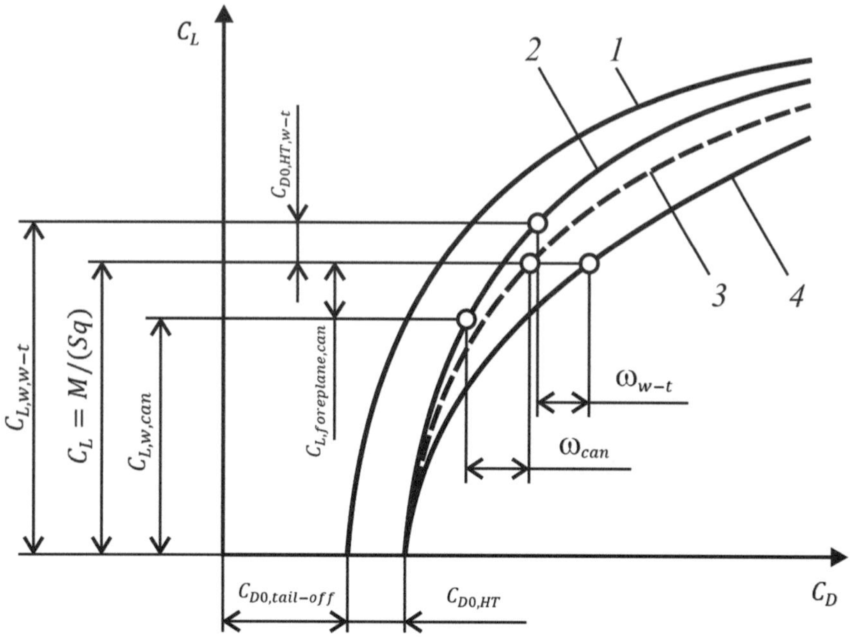

Fig. 1 Trimmed polars for a wing-tail and a canard aircraft: 1—aircraft without horizontal tail; 2—aircraft with horizontal tail at $C_{mC_L} = 0$; 3—canard trimmed polar; 4—wing-tail trimmed polar (both polars at $C_{mC_L} < 0$); ω = trim penalties

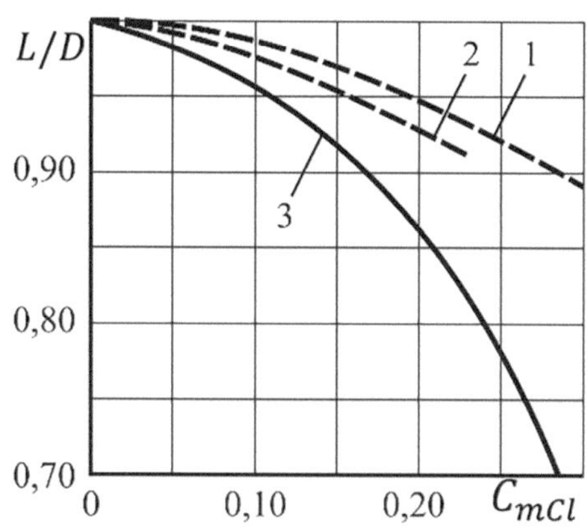

Fig. 2 Aerodynamic penalties of various layouts as a function of C_{mC_L}: wing-tail layout; canard layout; 1—$\overline{S}_{HT} = \frac{S_{HT}}{S} = 0.12, \overline{l}_{HT} = \frac{b_{HT}}{c_{MAC}} = 1.5$; 2—$\overline{S}_{HT} = 0.08, \overline{l}_{HT} = 1.5$; 3—$\overline{S}_{HT} = \frac{S_{HT}}{S} = 0.12, \overline{l}_{HT} = \frac{b_{HT}}{c_{MAC}} = 1$

This also explains why designers of maneuverable and business aircraft are turning more and more often to the canard layout, which demonstrates lower trim penalties for an acceptable level of static stability (instability). At the same time, due to its reliability and available historical and research data, the wing-tail layout with reduced static stability margins continues to be the mainstream for subsonic long-range commercial airliners.

In more detail, the issue of trim penalties reduction through optimization of static stability (instability) was discussed in Sect. 4.2.6.

2.3 The Area Rule

During aerodynamic design, it is necessary to keep in mind that a system consisting of optimal parts is not, in general, optimal. This means that an airplane composed of individually optimized parts (wing, fuselage, tail, power units and landing gear nacelles) may not be optimal due to the negative aerodynamic influence of these parts on each other.

On the other side, this interference can reduce the overall aerodynamic drag of the aircraft and increase its lift-to-drag ratio. An example of this is the use of the area rule. This rule can be formulated as follows: *to ensure the minimum drag of the aircraft at transonic and subsonic speeds, the total cross-sectional area distribution of all elements of the aircraft should correspond to the area distribution of the equivalent streamlined body* [1, 4].

The meaning of the area rule is illustrated in Fig. 3, where the cross-section distribution of a non-area-ruled and an area-ruled aircraft are compared.

The results of the application of the area rule are shown in Figs. 4 and 5, where the curves, each point of which corresponds to an optimum aircraft designed to fly at a given M, are shown.

As can be seen from these figures, the application of the area rule leads to a 25 … 30% decrease in the $C_{D,wave}$ growth in the transonic velocity zone. With further growth of velocity, the influence of the area rule decreases, and at M = 1.8 … 2.0 it practically does not work any more. All of this applies to a flight at $C_L = 0.05 \ldots 0.5$. At higher values, due to a significant increase in the induced drag, the effect of the area rule decreases.

The practical use of the area rule is illustrated by the supersonic bomber B-58 "Hustler" (USA) (Fig. 6).

All previously stated about the area rule refers to the possibility of reducing the wave drag of the aircraft in the transonic flight region (for M = 0.95 … 1.3). For an aircraft flying at a high M, it is advisable to apply a modified, so-called "supersonic" area rule. This rule involves plotting areas of cross-sections of the aerodynamic configuration obtained not in planes perpendicular to its longitudinal axis, but instead in the sections set at the Mach cone angle (Fig. 7). Sometimes, to simplify the geometric constructions, two vertical planes are used, each of which is rotated relative to the plane of symmetry of the aircraft at the Mach cone angle.

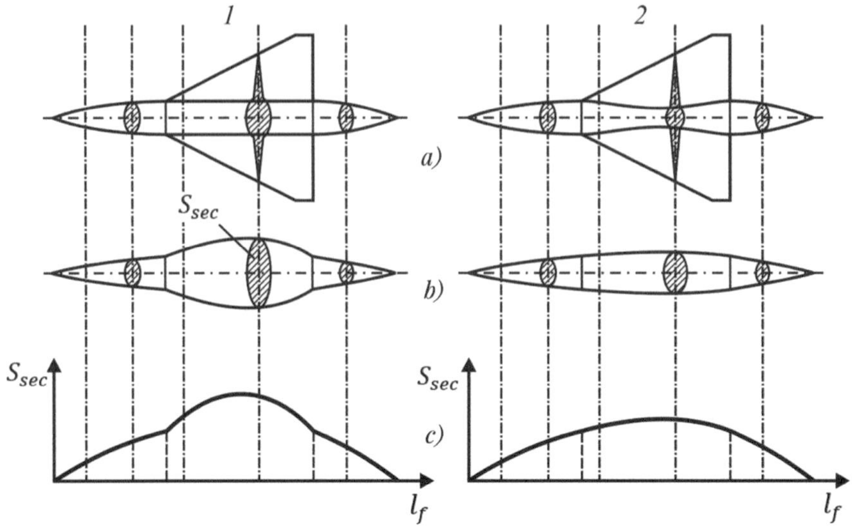

Fig. 3 Non-area-ruled wing-body combination (1) and area-ruled wing-body combination (2): **a**—wing-body combination; **b**—equivalent body of rotation; **c**—distribution of cross-sectional areas

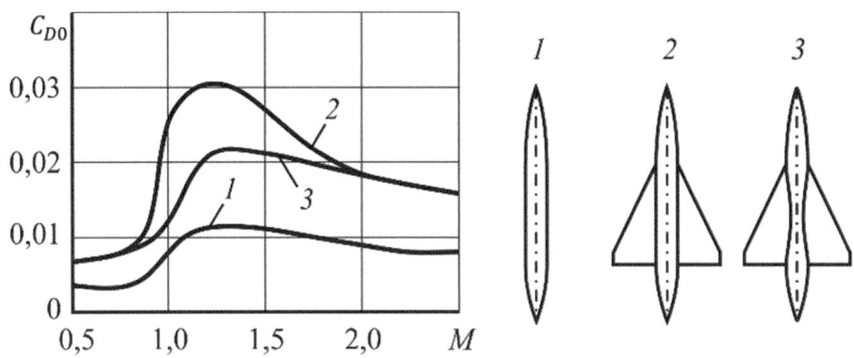

Fig. 4 Body waisting effect on the drag coefficient of a body-wing combination

Therefore, the optimal for transonic and low supersonic velocities symmetric body should be transformed for a supersonic flight: its largest midsection shifts from the middle to its end. Examples of supersonic area plots of some aircraft are shown in Fig. 8. There, the magnitude of the midsection shift in fractions of the body length as a function of M is also shown.

Along with the general (for the entire aircraft) area rule, the so-called local area rule for some difficult to streamline areas is also used. The shape of these areas should be refined by experiment, with the final goal of reducing the aerodynamic interference. Examples of the local area rule applications are shown in Fig. 9.

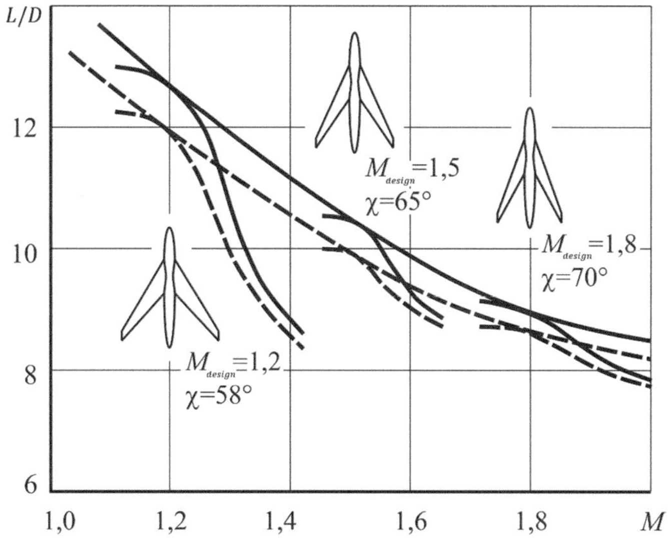

Fig. 5 Effect of body waisting on the lift-to-drag ratio of an aircraft: The curves correspond to the optimum aircraft for the design M: area-ruled aircraft; non-area-ruled aircraft

2.4 Aircraft Local Aerodynamics

Aircraft local aerodynamics is improved by:

- using favorable vortex effects (special devices for vortex management, etc. see Fig. 10);
- installation of fairings, special wingtips and Whitcomb winglets (Fig. 11);
- application of the "area rule" and the "stagnated flow rule";
- *Note: The "stagnated flow rule" recommends that units which may cause additional drag should be placed downstream other units.*
- considering the local direction of streamlines for the orientation of individual units (planform view rotation of nacelles located under the wing; rotation "towards the flow" of air intakes, placed on the lateral surfaces of the fuselage, etc.);
- using feathering surfaces and units in areas with significant changes in the local direction of streamlines when changing the flight mode (tilt missile pylons on the wing of variable sweep Su-24 aircraft).

2.5 Horizontal Tail Positioning

To ensure the longitudinal static stability of an aircraft, its center of mass should be ahead of its aerodynamic center; the distance between these points, divided by the wing mean aerodynamic chord, determines the longitudinal stability margin:

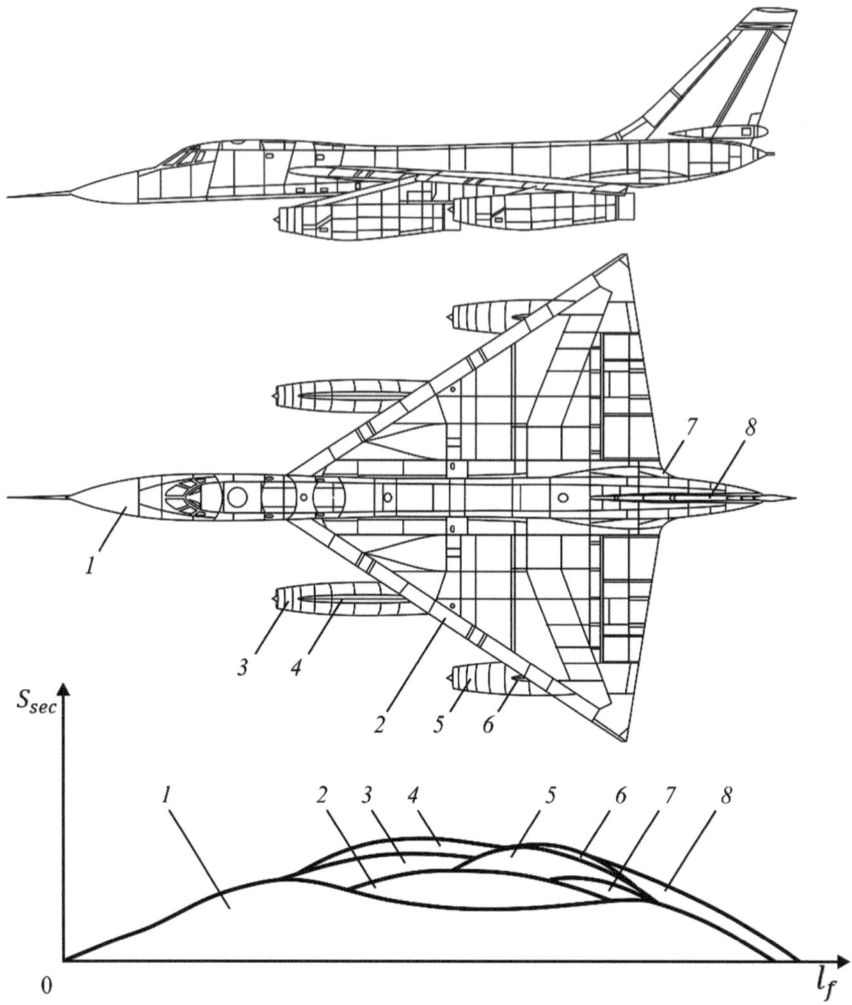

Fig. 6 Cross-sectional area distribution along the length of the B-58 "Hustler" aircraft (USA): 1—fuselage; 2—wing; 3—inner engine nacelles; 4, 6—pylons; 5—external engine nacelles; 7—fairing; 8—tail

$$C_{m_{C_L}} = \overline{X}_{CG} - \overline{X}_{AC} < 0$$

In the wing-tail layout, the horizontal tail moves the aerodynamic center backward; in the canard layout, the aerodynamic center is shifted forward. The magnitude of this displacement for the wing-tail can be estimated as follows:

$$\Delta\bar{x}_{F,HT} = k_{HT} V_{HT} \frac{C_{L_{\alpha,HT}}}{C_{L_\alpha}} (1 - \varepsilon^\alpha), \tag{1}$$

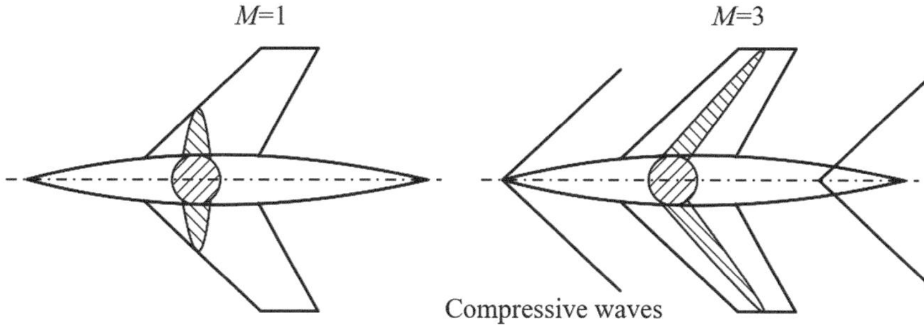

Fig. 7 Cross sections for transonic and supersonic area plots

where

ε^{α} is the derivative of the downwash angle at the horizontal tail with respect to the angle of attack of the wing;

$k_{HT} = \frac{q_{HT}}{q}$ is the flow deceleration coefficient in the area of the horizontal tail. For the canard layout, in most cases, $k_{HT} = 1 \; and \; \varepsilon^{\alpha} \approx 0$.

For some design longitudinal static stability margin $C_{m_{C_L,design}}$ and with a known CG position $\overline{X}_{CG}$ the required displacement of the aerodynamic center of the wing-body combination due to the horizontal tail is determined by the equation

$$\pm\Delta\overline{x}_{F,HT} = C_{m_{C_L,design}} - \overline{X}_{CG} + \overline{X}_{AC,tail-off}. \tag{2}$$

The minus sign at $\Delta\overline{x}_{F,HT}$ refers to the wing-tail configuration and the plus sign refers to the canard layout.

Comparing Eqs. (1) and (2) and using as inputs $C_{m_{C_L,design}}$, $\overline{X}_{CG}$ and $\overline{X}_{AC,tail-off}$ (determined from the parameters of the wing, fuselage, engine nacelles, etc.), one can predict the required value of V_{HT}. k_{HT}, C_{L_α},$C_{L_{\alpha,HT}}$, and ε^{α} are derived from historical data, including wind-tunnel tests of similar models, or analytically predicted (see Chap. 19).

Since V_{HT} is the product of $\overline{l}_{HT}$ and $\overline{S}_{HT}$, the required volume coefficient can be obtained either by a larger horizontal tail area and a smaller arm or, on the contrary, by a larger arm and a smaller area.

Note that the horizontal tail efficiency and the longitudinal controllability of the aircraft increase in proportion to HT arm and that the aircraft damping is proportional to l_{HT}^2. However, increasing the horizontal tail arm yields a longer fuselage, which is always associated with a greater weight than increasing the tail area. As a consequence, the process of aircraft design with respect to V_{HT} is iterative and subject to optimization.

For the preliminary aircraft design, one may refer to the values of volume coefficient and tail arm given in Table 1 for different aircraft classes.

After preliminary selection of the geometric parameters of the tail (AR_{HT}, λ_{HT},$\left(\frac{t}{c}\right)_{HT}$, and Λ_{HT}—see Chap. 19) the value of V_{HT} should be refined. For this purpose, the "boundary lines" method can be used. This method allows the designer

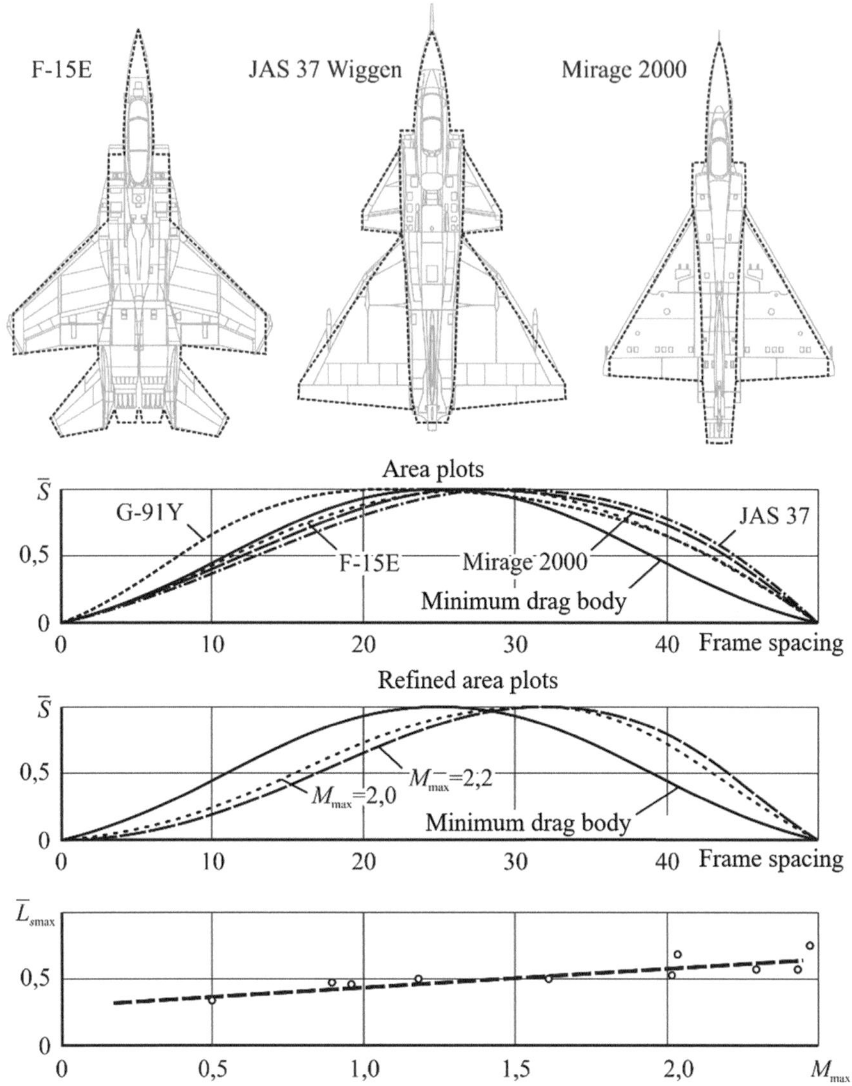

Fig. 8 Area plots of various aircraft and their comparison with the minimum drag body plot. Dependence of the distance of the largest midsection on M $\bar{l}_{S,max} = \frac{l_{S,max}}{l_{body}}$

to immediately answer the question what value can be selected for V_{HT} to meet the requirements for stability and controllability characteristics and to ensure the design CG envelope.

For longitudinal stability and controllability, the most important of these requirements are:

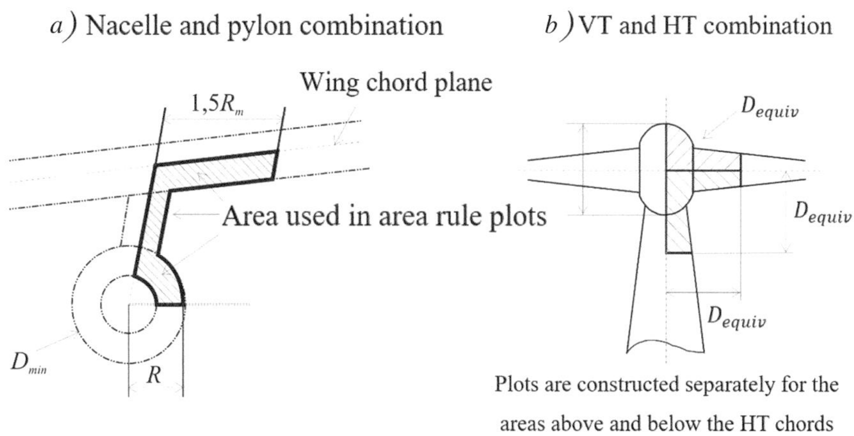

Fig. 9 Cross sections of high-drag aircraft components, subject to the local area rule

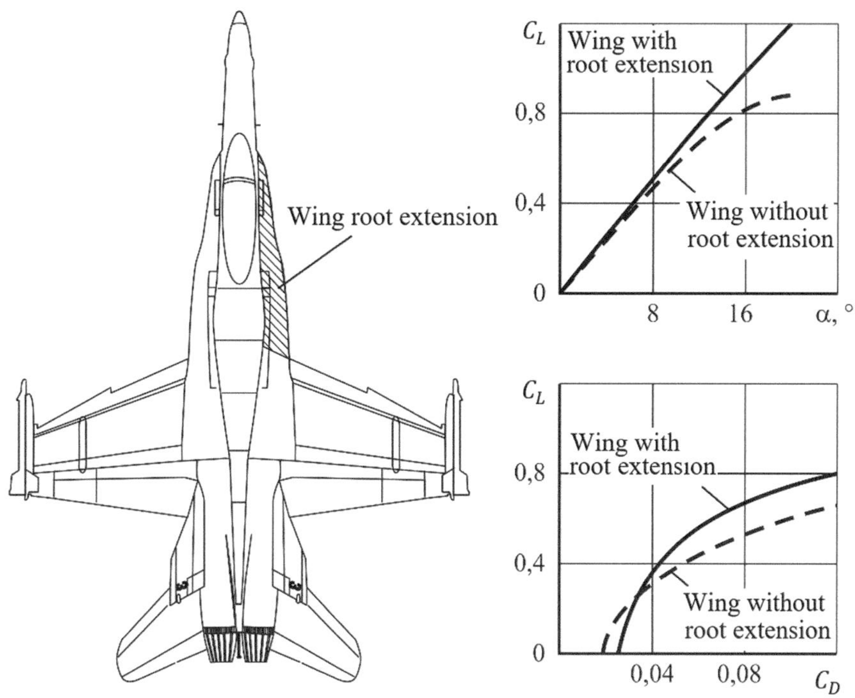

Fig. 10 Wing extensions used to improve aircraft aerodynamic performance

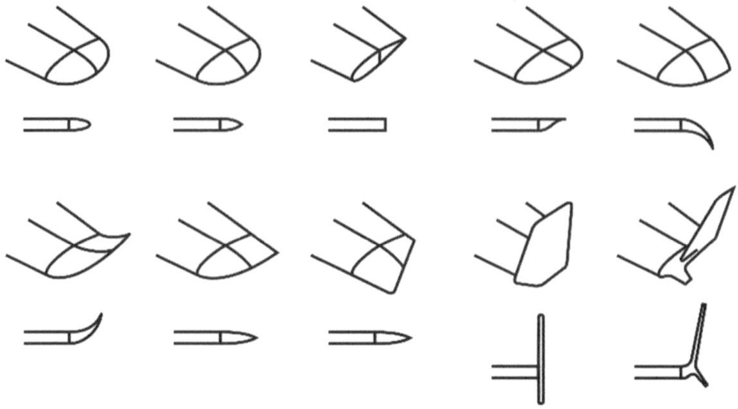

Fig. 11 Wing tips reducing induced drag

- to ensure the minimum allowable longitudinal static stability margin (with the aft CG limit $\overline{X}_{CG,aft}$);
- to ensure the aircraft trim in takeoff and landing with the high-lift devices extended and forward CG location.

Taking into account expression (1), the condition for the first requirement can be written as

$$\overline{X}_{CG,aft} - \overline{X}_{AC,tail-off} - kV_{HT}\frac{a_{HT}}{a_w}(1 - \varepsilon^{\alpha}) = C_{m_{C_{L,min}}}. \tag{3}$$

Here $a_{HT} = C_{L_{\alpha,HT}}$, $a_w = C_{L_\alpha}$. When formulating the inequality, it should be kept in mind that for a stable airplane, $C_{m_{C_{L,min}}} < 0$.

The required efficiency of the pitch control surfaces for aircraft trim in takeoff and landing with the extended high-lift devices and forward CG location $\overline{X}_{CG,f}$ can be derived from the equilibrium state equation at these modes. For example, for the case of takeoff,

$$\begin{aligned} C_m &= C_{m0,tail-off} + \Delta C_{m,HL} + \Delta C_{m,eng} \\ &+ C_{L_\alpha}\left(\overline{X}_{CG,f} - \overline{X}_{AC,tail-off}\right)\alpha_{unstick} \\ &- k_{HT}a_{HT}V_{HT}\left(\alpha_{unstick} - \varepsilon_{0,HL-off} - \varepsilon^{\alpha}_{HL-off}\alpha_{unstick} - \Delta\varepsilon_{HL} - \Delta\varepsilon_{gr} + \overline{S}_{st}\varphi_{st} + n_{elev}\delta_{elev}\right) = 0 \end{aligned} \tag{4}$$

In this equation, index "*HL*" denotes extended high-lift devices; indices "*tail-off*" and "*HL-off*" stand for the configuration without horizontal tail and with the high-lift devices retracted; the index "*gr*" denotes ground effect;

$\overline{S}_{st} = \frac{S_{st}}{S_{HT}}$ is the ratio of the stabilizer area to the total area of horizontal tail;

$n_{elev} = \sqrt{\overline{S}_{elev}} = \sqrt{\frac{S_{elev}}{S_{HT}}}$ is the elevator efficiency factor (in case of all-movable stabilizer, the term $n_{elev}\delta_{elev}$ should be omitted, and for the stabilizer deflection angle

Table 1 Volume coefficient and tail moment arm values for different aircraft classes

	V_{HT}	V_{VT}	$\frac{l_{HT}}{c_{MAC}}$	$\frac{l_{VT}}{b}$	$\frac{S_{HT}}{S}$	$\frac{S_{VT}}{S}$
General aviation aircraft A						
Light aircraft with a single piston engine	0,52…0,80	0,028…0,45	2,5…3,6	0,35…0,45	0,18…0,24	0,08…0,14
Light aircraft with two piston engines	0,6…1,05	0,046…0,08	2,5…3,9	0,37…0,45	0,2…0,3	0,11…0,19
Aerobatic	0,52…0,54	0,056…0,073	2,4…2,6	0,47…0,51	0,21…0,22	0,12…0,15
Agricultural	0,38…0,5	0,023…0,035	2,3…2,8	0,3…0,36	0,16…0,21	0,06…0,1
Commercial airliner						
Commuter with a turboprop	0,90…1,32	0,07…1,00	3,8…5,4	0,36…0,46	0,20…0,28	0,16…0,24
Airliners with bypass turbojets	0,70…0,96	0,060…0,090	3,0…4,0	0,39…0,51	0,20…0,27	0,14…0,20
Executive aircraft with bypass turbojets	0,6…0,85	0,061…0,085	2,4…3,6	0,34…0,44	0,21…0,27	0,17…0,21
Cargo aircraft	0,65…1,0	0,06…0,09	3,5…4,5	0,34…0,44	0,18…0,25	0,15…0,21
Trainer aircraft						
Trainer with a piston or turboprop engine	0,6…0,72	0,045…0,072	2,9…3,1	0,42…0,47	0,19…0,24	0,10…0,16
Subsonic advanced trainer with a bypass turbojet engine	0,35…0,6	0,061…0,081	1,6…2,42	0,35…0,48	0,19…0,24	0,14…0,19
Supersonic trainer with a bypass turbojet engine / Light combat aircraft	0,2…0,33	0,07…0,111	1,12…1,51	0,42…0,47	0,18…0,23	0,17…0,23
Combat aircraft						
Fighter with all-movable HT	0,19…0,25	0,053…0,082	0,9…1,4	0,27…0.0,47	0,17…0,26	0,17…0,24
Fighter with all-movable HT and VT	0,19…0,25	0,030…0,048	0,9…1,4	0.3…0.4	0,17…0,26	0,1…0,15
Stormtroopers	0,4…0,5	0,05…0,08	2,0…2,5	0,3…0,4	0,18…0,21	0,16…0,21

(continued)

Table 1 (continued)

	V_{HT}	V_{VT}	$\frac{l_{HT}}{c_{MAC}}$	$\frac{l_{VT}}{b}$	$\frac{S_{HT}}{S}$	$\frac{S_{VT}}{S}$
Subsonic long-range aircraft						
Long-range stratosphere manned and unmanned aircraft	0,28…0,52	0,017…0,025	2,3…3,3	0,15…0,24	0,1…0,21	0,08…0,12
Unmanned medium-altitude, long-range aircraft	0,6…0,78	0,026…0,032	2,6…3,7	0,17…0,3	0,2…0,23	0,15…0,21

assume $\varphi_{st} = 0.8\varphi_{st,max}$; if the aircraft is supposed to use adjustable stabilizer, then for takeoff and landing modes assume $\varphi_{st} = \varphi_{st,TO}$ or $\varphi_{st} = \varphi_{st,land}$ respectively);

$\Delta C_{m,eng} = \frac{T_{eng} z_{eng}}{q_{unstick} S c_{MAC}}$ is the engine thrust moment coefficient.

From inequality (3), we obtain the following constraint for the horizontal tail volume coefficient:

$$V_{HT} \geq \frac{-C_{m_{C_{L,min}}} + \overline{X}_{CG,aft} - \overline{X}_{AC,tail-off}}{1 - \varepsilon^{\alpha}} \frac{a_w}{k_{HT} a_{HT}}, \tag{5}$$

and from Eq. (4),

$$V_{HT} \geq \frac{C_{m0,tail-off} + \Delta C_{m,HL} + \Delta C_{m,eng} + C_{L_\alpha}(\overline{X}_{CG,f} - \overline{X}_{AC,tail-off})\alpha_{unstick}}{k_{HT} a_{HT}\left(\alpha_{unstick} - \varepsilon_{0,HL-off} - \varepsilon^{\alpha}_{HL-off}\alpha_{unstick} - \Delta\varepsilon_{HL} - \Delta\varepsilon_{gr} + \overline{S}_{st}\varphi_{st} + n_{elev}\delta_{elev}\right)}. \tag{6}$$

Plotting the expressions (5) and (6) in terms of $\overline{X}_{CG}$ and V_{HT}, we obtain the so-called "cross" that limits the feasible area for V_{HT} for a particular value of CG envelope $\overline{X}_{CG,f} - \overline{X}_{CG,aft}$ (Fig. 12).

If $\overline{X}_{CG,f}$ and $\overline{X}_{CG,aft}$ have already been determined (see Sect. 3.3), the required values of V_{HT} for each CG position can be obtained using the appropriate boundary lines and selecting the highest value. If the CG envelope have not yet been determined, by setting some value of $V_{HT} > V_{HT,min}$ the permissible CG envelope can be obtained and then achieved by an appropriate configuration.

The allowable CG travel can be expanded for the selected value of V_{HT} by increasing a_{HT} (e.g., by increasing the tail aspect ratio, as shown in Fig. 12 with a dotted line). However, keep in mind that the ultimate design goal is to select such tail parameters that would result in the lowest tail weight (and as the aspect ratio increases, the weight increases). Therefore, the horizontal tail geometric parameters should be optimized with respect to its mass.

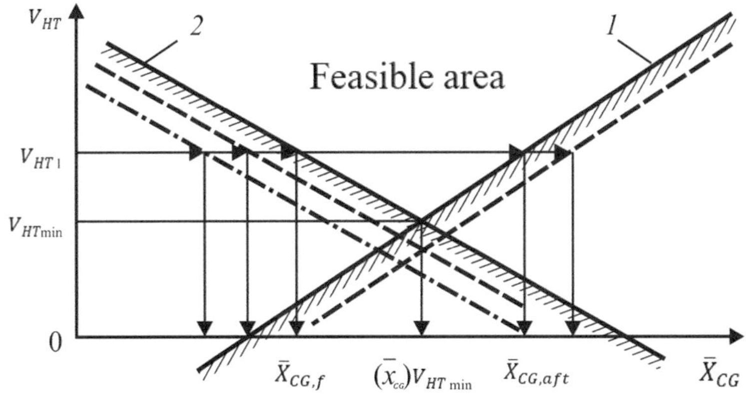

Fig. 12 The boundary lines ("cross") limiting the feasible area of V_{HT} and X_{CG} for the minimum allowable static stability $C_{m_{C_L,min}}$ (1) and aircraft trim at large angles of attack (2): expansion of the feasible area when the derivative $a_{HT} = C_{L_{\alpha,HT}}$ increases; expansion of the feasible area due to variable incidence angle stabiliser

For a heavy high-speed aircraft, V_{HT} in the case of a forward CG location may be selected for ensuring pitch control to achieve $\alpha_{unstick}$ at $V = 0.85V_{unstick}$ within the limited runway length.

For a maneuvering supersonic aircraft, the critical condition for the selection of V_{HT} in the case of a forward CG location is usually to ensure a sufficient pitch control efficiency for balancing the aircraft at maximum load factor, especially at supersonic speeds and high altitudes.

For a low-speed aircraft with a manual control system, the V_{HT} range is additionally limited by the following constraints:

- the aircraft longitudinal damping ratio;
- the control stick force variation;
- the control force derivative with respect to the load factor, etc.

On high-speed aircraft with power-controlled systems, the required controllability is provided by the control system (see Chap. 20).

Trimming peculiarities of aircraft of different aerodynamic layouts are given in Chap. 4.

In a wing-tail layout, the vertical position of the horizontal tail should be selected.

In flight, a zone of turbulent wake is formed behind the wing. The dynamic pressure and angles of attack of the horizontal tail are affected by the parameters of this flow: the tail efficiency factor $k_{HT} = \frac{q_{local}}{q}$ and the downwash angle ε, which depend on the relative position of the wing and the tail, as well as on the flight mode (Fig. 13).

When the horizontal tail gets blanketed by the wing, the horizontal tail lift curve slope may change significantly and the aircraft may become unstable (the tendency

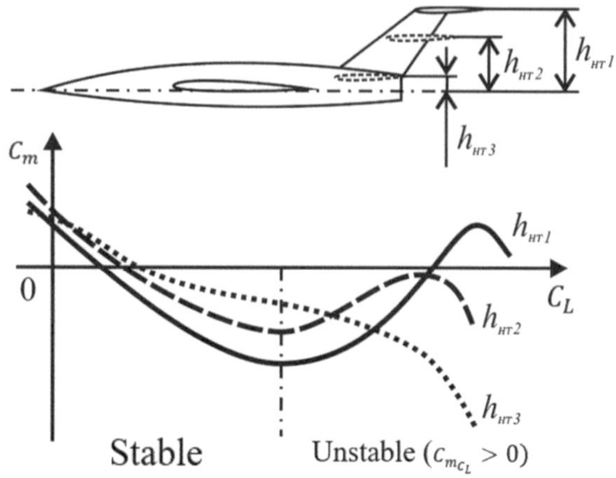

Fig. 13 Influence of horizontal tail height on moment curve $C_m = f(C_L)$

to increase the angle of attack further with the aircraft pitching-up). This is clearly illustrated in Fig. 13.

In case of an aircraft with small aspect ratio wings and a sufficiently long afterbody (supersonic aircraft) the zone of the maximum turbulent wake usually shifts upwards. Therefore, in such aircraft the horizontal tail is located on the fuselage (Fig. 14).

On supersonic aircraft of all types, horizontal tail can be placed below the wing plane, at the lower afterbody.

For subsonic aircraft with relatively low flight speeds ($V < 800$ km/h) and relatively long afterbody ($\bar{l}_{HT} = 2.7 \ldots 3.0$) the horizontal tail can be mounted with an upward offset $h_{HT} = (0.15 \ldots 0.20)c_{inb}$, where c_{inb} is the length of the wing chord at the side of the fuselage.

There are the airplanes which, due to their design features (for example, aft podded engines), have a T-tail. When the critical angle of attack is exceeded, such aircraft can enter a deep stall, which results in the aircraft sort of hanging at a large angle of attack, not being able to reduce it. Then it loses speed and stalls. The substance of the phenomenon is that due to an accidental excess of the critical angle of attack (for example, a strong vertical gust), the wing wake rises and blankets the tail. As a consequence, an additional pitching-up moment is generated, causing a further increase in the angle of attack. Since the horizontal tail is blanketed by a highly turbulent wake, the rudder authority, even if it is fully deflected downward, may be insufficient to counteract the pitching-up tendency.

The moment curve $C_m = f(\alpha)$ for an airplane with a T-tail is shown in Fig. 15.

In aircraft with a low horizontal tail (dashed in Fig. 15), after wing stall and the upward departure of the turbulent wake, the downwash in the horizontal tail area decreases. This causes the pitching-down moment, which reduces the angle of attack of the aircraft, since the rudders are not blanketed by the wake and retain their effectiveness.

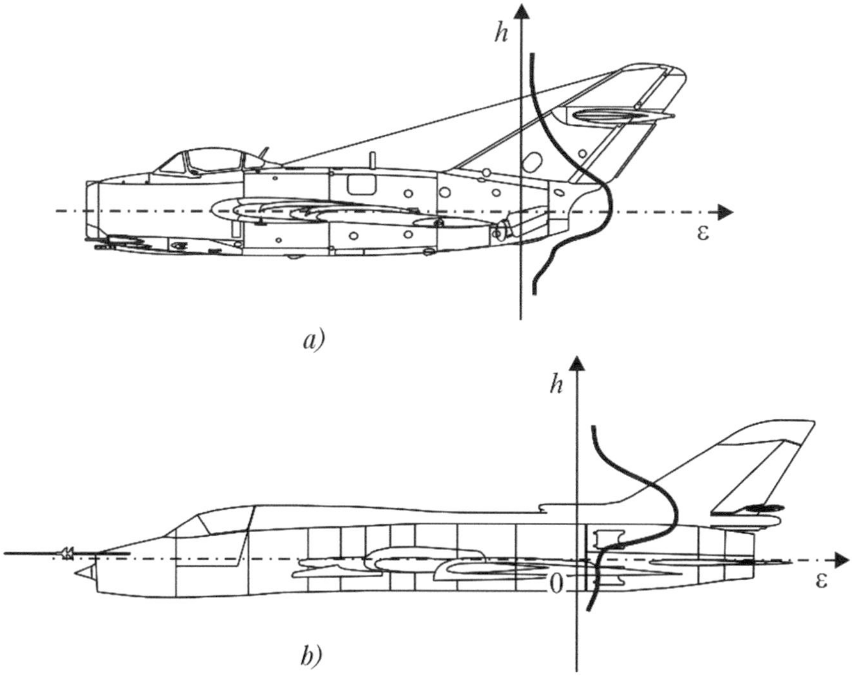

Fig. 14 Vertical downwash angle variation and placement of horizontal tail: **a**—an aircraft with a medium-aspect ratio swept wing and a short fuselage (high horizontal tail); **b**—an aircraft with a low-aspect ratio wing with a relatively long fuselage (low horizontal tail)

To prevent a T-tail airplane stalling, pitch control systems usually include special automatic devices alerting to the approaching dangerous mode (e.g., shaking of a steering wheel).

Deep stall can be avoided by increasing the aspect ratio of the horizontal tail so that its tips do not enter the wing wake.

Recall that in a canard aircraft, the foreplane operates almost in an undisturbed flow, and in this case, $k_{HT} = 1$ and $\varepsilon^{\alpha} \approx 0$.

2.6 Vertical Tail Parameters and Wing Dihedral Angle

To ensure the required natural lateral stability and controllability for an aircraft with the given wing geometry, the designer should select an appropriate vertical tail volume coefficient $V_{VT} = \frac{S_{VT} l_{VT}}{Sb}$ and wing dihedral angle Γ, as these parameters have the most significant impact on the derivatives C_{L_β}, C_{N_β} and $C_{L_{\omega_z}}$, to the greatest extent determining the characteristics of lateral stability and controllability.

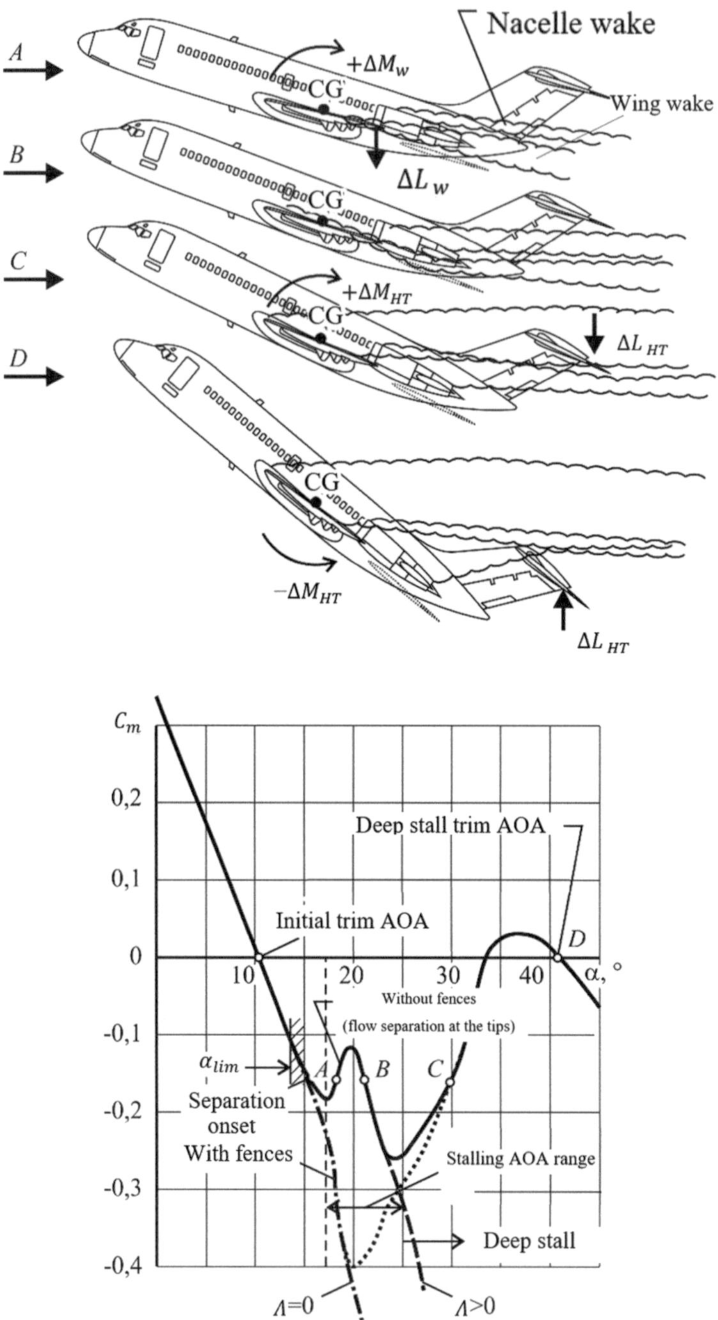

Fig. 15 Stall development on a swept wing airplane with a T-tail and its moment curve$C_m = f(\alpha)$: The moment curve represents the wind-tunnel test results for the Douglas DC-9 aircraft model. For comparison, the moment curves for a straight wing aircraft and a swept wing aircraft with a low horizontal tail are shown

For a low-speed aircraft with a manual control system, the choice of vertical tail, the wing dihedral angle Γ, yawing and rolling control efficiency is defined by the following basic requirements:

1. Factor $\chi = \frac{|\overline{\omega}_{xmax}|}{|\overline{\omega}_{zmax}|}$ representing the ratio of rolling to yawing angular velocities should not exceed certain limits. At high values of χ, the aircraft responds to disturbances in flight by rocking from wing to wing; at low values of χ, it demonstrates a tendency for spiral instability (supersonic aircraft without automated control would have values $\chi \approx 10$ and more).
2. The lateral short-period damping must be sufficient.
3. The lag in aircraft response to control deflection shall not exceed the design limits.
4. Trim capability and sufficient lateral controllability of the aircraft at landing with a given crosswind and at takeoff with one engine out (i.e. with incomplete and asymmetrical thrust in a multi-engine aircraft) must be ensured.
5. The authority of the roll and yaw controls must be sufficient to perform the specified maneuvers in all possible flight modes and to reliably recover the aircraft from stall.
6. The spiral motion of the aircraft should not have strongly pronounced instability (the permissible time of parameter deviation doubling from the initial values in the spiral motion is limited).

For high-speed aircraft with booster control systems, requirements 1, 2 and 6 are usually met by the use of special automatic systems that act independently of the pilot (yaw dampers and automatic sliding controls).

The parameters of the vertical tail and the rudder are selected by successive approximations using the boundary line method. These lines, after completing the preliminary layout and determining the aerodynamic characteristics (e.g., the derivative $a_{VT} = C_{Y_\beta}$) and the vertical tail arm l_{VT}, can be plotted in terms of $S_{VT} - \Gamma$ or $n_{elev} - V_{VT}$ (at subsonic speeds, the rudder effectiveness $n_{rud} \approx \sqrt{S_{rud}/S_{VT}}$).

The application of the boundary line method can be demonstrated through the following example: determine n_{rud} and V_{VT} for the aircraft trim under one engine out condition in takeoff and cross wind landing, as well as meeting the requirement of an acceptable value of χ.

The crosswind landing is one of the main design cases when setting the aircraft effective dihedral C_{l_β}. The sideslip angle under these conditions is determined by the maximum crosswind speed specified in the aircraft requirements W_{ymax}, at which safe landing should be performed $\left(\beta_{max} \approx W_{ymax}/V_{land}\right)$.

From the equation of equilibrium of lateral forces and moments in a straight flight with a sideslip angle, we can determine the feasible area in terms of n_{rud} and V_{VT}:

$$V_{VT} > \frac{C_{n_{\beta,VT-off}}}{a_{VT}(1 - n_{rud}\delta_{rud}/\beta_{max})}, \tag{7}$$

where $\delta_{rud} = 0.8\delta_{rud,max}$.

In case of one engine out at takeoff (on an airplane with several engines moved away from the symmetry axis), an unbalanced thrust yields a yawing moment N_{thrust}. This moment is greatest when the engines are operating in full-thrust mode. This condition is also critical for the choice of vertical tail geometry and rudder effectiveness, since they must counteract the unbalanced thrust moment at a sufficiently low speed (usually assumed to be $= V_{unstick} = \sqrt{\frac{2m_0 g}{C_{L,unstick}\rho S}}$).

From aircraft equilibrium in a straight flight with asymmetric thrust follows that

$$V_{VT} > \frac{m_{z,thrust}}{n_{rud} a_{VT} \delta_{rud}}, \tag{8}$$

where $m_{z,thrust} = \frac{\sum_{i=1}^{n_{eng}-1} T_{0,i} y_i}{q_{unstick} Sb}$, y_i is the distance from the engine axis to the symmetry axis.

The ratio of maximum rolling to yawing angular velocities at lateral oscillations of the aircraft is determined by

$$\chi = \frac{|\overline{\omega}_{xmax}|}{|\overline{\omega}_{zmax}|} \approx \frac{\overline{C}_{l_\beta}}{\overline{C}_{n_\beta}} \frac{1}{\sqrt{1 - \frac{1}{\mu} \frac{\left(\overline{C}_l^{\omega_x}\right)^2}{\overline{C}_{n_\beta}}}}, \tag{9}$$

where $\overline{C}_{l_\beta} = \frac{C_{l_\beta}}{\overline{r}_x^2}$, $\overline{C}_l^{\omega_x} = \frac{C_l^{\omega_x}}{\overline{r}_x^2}$, $\overline{C}_{n_\beta} = \frac{C_{n_\beta}}{\overline{r}_z^2}$.

For initial calculations, the above expression can be simplified:

$$\chi \approx \frac{\overline{C}_{l_\beta}}{\overline{C}_{n_\beta}} = \frac{C_{l_\beta} \overline{r}_z^2}{C_{n_\beta} \overline{r}_x^2} = \frac{C_{l_\beta} I_z}{C_{n_\beta} I_x}, \tag{10}$$

where $\overline{r}_x^2 = \frac{4I_x}{mb^2}$, $\overline{r}_z^2 = \frac{4I_z}{mb^2}$.

Now, it is possible to estimate the static lateral stability margin meeting the requirement of $\chi < \chi_{allow}$ (recall that $C_{n_\beta} < 0$):

$$C_{n_\beta} < \frac{C_{l_\beta} I_z}{\chi_{allow} I_x} \tag{11}$$

The ratio of moments of inertia I_z / I_x for each class of aircraft can be derived from historical data or calculated.

The value of C_{l_β}, which characterizes rolling stability $\left(C_{l_\beta} < 0\right)$, depends on the shape and size of the wing and vertical tail, as well as on the wing dihedral angle Γ. This can be clearly seen from the approximate expression determining the value of C_{l_β}:

$$C_{l_\beta} = C_{l_{\beta,w}} + C_{l_{\beta,VT}} = C_{l_{\beta,0}} - \frac{1}{2} C_{L_\alpha} \overline{y}_{CG,w} \Gamma cos^2 \Lambda - \frac{1}{2} C_L \overline{y}_{CG,w} sin\Lambda - k_{VT} a_{VT} \frac{S_{VT}}{S} \frac{z_{VT}}{b}, \quad (12)$$

where Λ is the wing sweep angle; Γ is the wing dihedral angle; z_{VT} is the distance from the VT lateral force to the aircraft centerline (Fig. 16); $\overline{y}_{CG,w}$ is the distance from the aircraft centerline to the center of gravity of the wing, divided by the wing half span; for a tapered wing $\overline{y}_{CG,w} = \frac{1}{3} \frac{\lambda+2}{\lambda-1}$ (λ is wing taper ratio); b is the wingspan.

As it follows from expression (12), on an airplane with the given parameters $S, \Lambda, \overline{S}_{VT}, C_{L_\alpha}$ the rolling stability C_{l_β} can be reduced by using an anhedral angle. This improves lateral dynamics at high angles of attack, as it was done, for example, on the Tu-104 and Tu-114. In this case, the second term in expression (12) becomes positive and the negative value of the derivative C_{l_β} decreases. However, as can be seen from Fig. 17, a, at small angles of attack (flying at high speeds), this measure may lead to an excessive reduction in the absolute value of C_{l_β}, which may result in a reverse roll response to rudder deflection.

Additionally, due to the lowered wingtips it is necessary to increase the landing gear length. As a consequence, at present, the problem of ensuring the design lateral dynamic characteristics with a dihedral angle is solved by using various automatic means (sideslip and yaw dampers).

The derivative C_{n_β} mainly depends on the ratio of the areas of the forebody and vertical tail projected on the aircraft symmetry plane (see Fig. 16):

$$C_{n_\beta} \approx -0.06 C_L^2 tg\Lambda - \alpha C_{l_{\beta,w}} - k_{VT} a_{VT} V_{VT} + k_\beta \frac{h_f l_f^2}{Sb} \quad (13)$$

where $k_\beta \approx 0.32\left(\overline{x}_f - 0.25\right) + \frac{1.5}{AR_f^{1.5}}$; $\overline{x}_f = \frac{x_f}{l_f}$; $AR_f = \frac{h_f}{l_f}$.

The increase in the length of fuselages l_f and their forebody x_f in modern high-speed aircraft leads to an increase in the "destabilizing" last term in expression (13) and, therefore, to a decrease in the absolute value of C_{n_β}. At the same time, expression

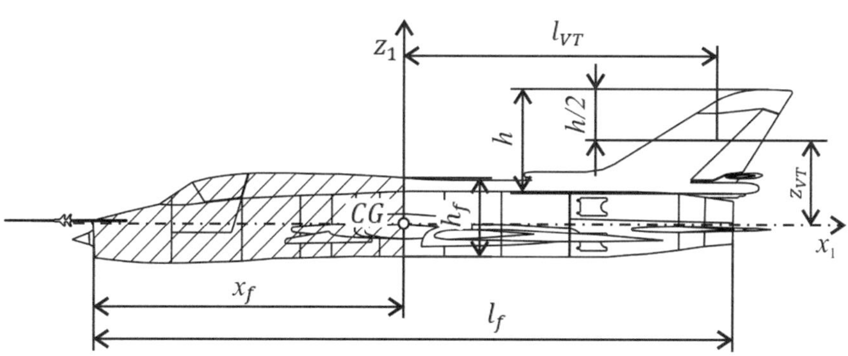

Fig. 16 Aircraft geometric parameters included in the expression for the lateral static stability derivatives C_{n_β} and C_{l_β}

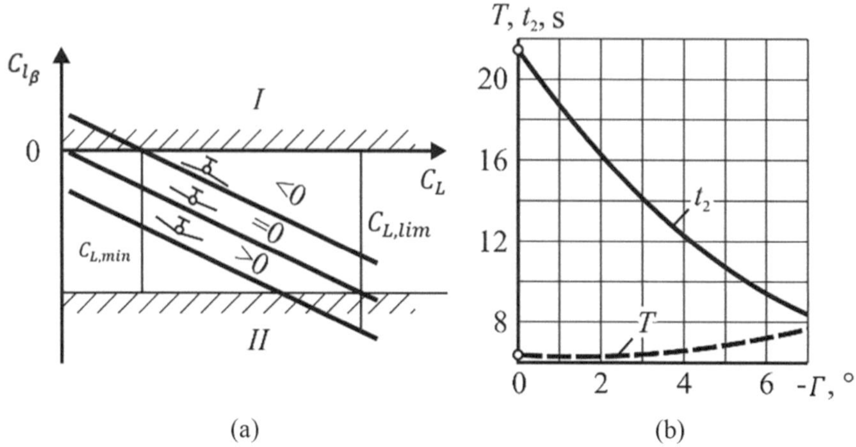

Fig. 17 Influence of dihedral angle ($\Gamma < 0, \Gamma = 0, \Gamma > 0$) on aircraft lateral static and dynamic characteristics **a**—$C_{l_\beta} = f(C_L)$ for aircraft with swept wings; **b**—lateral dynamic stability of the Boeing B-47 (USA) aircraft during approach to landing ($H = 0$, $V = 240$km/h, flaps deflected); T-period of lateral short-period oscillations; t_2—time of halving the oscillation amplitude); *I*—zone of lateral instability; *II*—zone of excessive lateral instability

(13) shows that with the increasing angle of attack (landing approach or flight at high altitudes), stability decreases due to the adverse effect of $C_{l_{\beta,w}} < 0$ and increasing blanketing of the vertical tail by the fuselage, which results in a decreasing VT efficiency coefficient $k_{VT} = \frac{V_{VT}^2}{V^2} < 1$.

At higher M, the derivative C_{n_β} decreases due to a decrease in the efficiency of the vertical tail ($a_{VT} = \frac{\delta C_{Y,VT}}{\delta \beta}$ decreases).

All this leads to the need to appropriately increase the vertical tail area to achieve yawing stability, especially in supersonic flight conditions at high altitudes.

It is also impossible to solve the problems of general design of high-speed aircraft without ensuring the design lateral oscillations damping ratio, dependent mainly on $C_n^{\omega_z}$.

From an approximate expression for this derivative

$$C_n^{\omega_z} \approx \left(C_n^{\omega_z}\right)_{VT+fus} = -2.4 k_{VT} a_{VT} \frac{S_{VT}}{S} \frac{l_{VT}^2}{b^2} \tag{14}$$

follows that it decreases rapidly at high supersonic speeds (due to reduction of a_{VT}) and at large angles of attack (due to flow deceleration and reduction of k_{VT}). The most effective way to increase it with the given wing parameters is to increase the vertical tail arm and its area. Since both are associated with an increase in aircraft weight, the problem of ensuring the required values of $C_n^{\omega_z}$ on modern aircraft is solved by installing a yaw damper.

Returning to the boundary line method used to determine aircraft feasible area in terms of V_{VT} and n_{rudder}, let us transform expressions (12) and (13):

$$C_{l_\beta} = C_{l_{\beta,w}} - k_{VT} a_{VT} V_{VT} \frac{z_{VT}}{b},$$

$$C_{n_\beta} = C_{n_{\beta,VT-off}} + C_{n_{\beta,VT}} = C_{n_{\beta,VT-off}} - k_{VT} a_{VT} V_{VT}.$$

If we put the expressions for C_{n_β} and C_{l_β} in (11), it will take the following form:

$$C_{n_{\beta,VT-off}} - k_{VT} a_{VT} V_{VT} < \left[C_{l_{\beta,w}} - k_{VT} a_{VT} V_{VT} \frac{z_{VT}}{b} \right] \frac{I_z}{I_x \chi_{allow}}.$$

Finally,

$$V_{VT} > \frac{C_{n_{\beta,VT-off}} - \frac{I_z}{I_x \chi_{allow}} C_{l_{\beta,w}}}{k_{VT} a_{VT} \left(1 - \frac{z_{VT}}{b} \frac{I_z}{I_x \chi_{allow}}\right)}. \tag{15}$$

The constraint curves (8), (9) and (15) are shown in Fig. 18a. These curves represent the boundaries of the aircraft feasible area in terms of V_{VT} and n_{rudder}.

In computer-aided design, C_{n_β} and C_{l_β} meeting the given requirements can be estimated as early as at the stage of preliminary design. After that, the vertical tail parameters and the wing dihedral angle can be determined using expressions (12) and (13). For this purpose, it is convenient, assuming some appropriate values for the required geometric parameters, to plot C_{n_β} and C_{l_β} variations with $\bar{l}_{VT}$, as shown in Fig. 18b.

At the conclusion of this section, let's note another important task of aerodynamic layout of the aircraft. This is to integrate the airframe and powerplant, in particular by specifying the position and dimensions of air intakes and nozzles, taking into account their properties described in Chap. 18.

The result of aerodynamic layout is a draft of general view of the aircraft with the wing area derived from the previously obtained values of the take-off weight and the take-off wing loading, as well as the given values of the tail volume coefficients. The dimensions (volume) of the fuselage in this case are determined based on the average statistical density of the layout of aircraft of this class. This configuration is the starting point for the internal aircraft layout (Fig. 19) and is essentially a shell to be filled with functional elements.

3 Internal Layout and Aircraft Center of Gravity

3.1 *Objectives and Constraints*

The purpose of aircraft internal layout is to place within the previously defined aerodynamic configuration functional elements (payload, crew, equipment, fuel, etc.) required to perform the given mission.

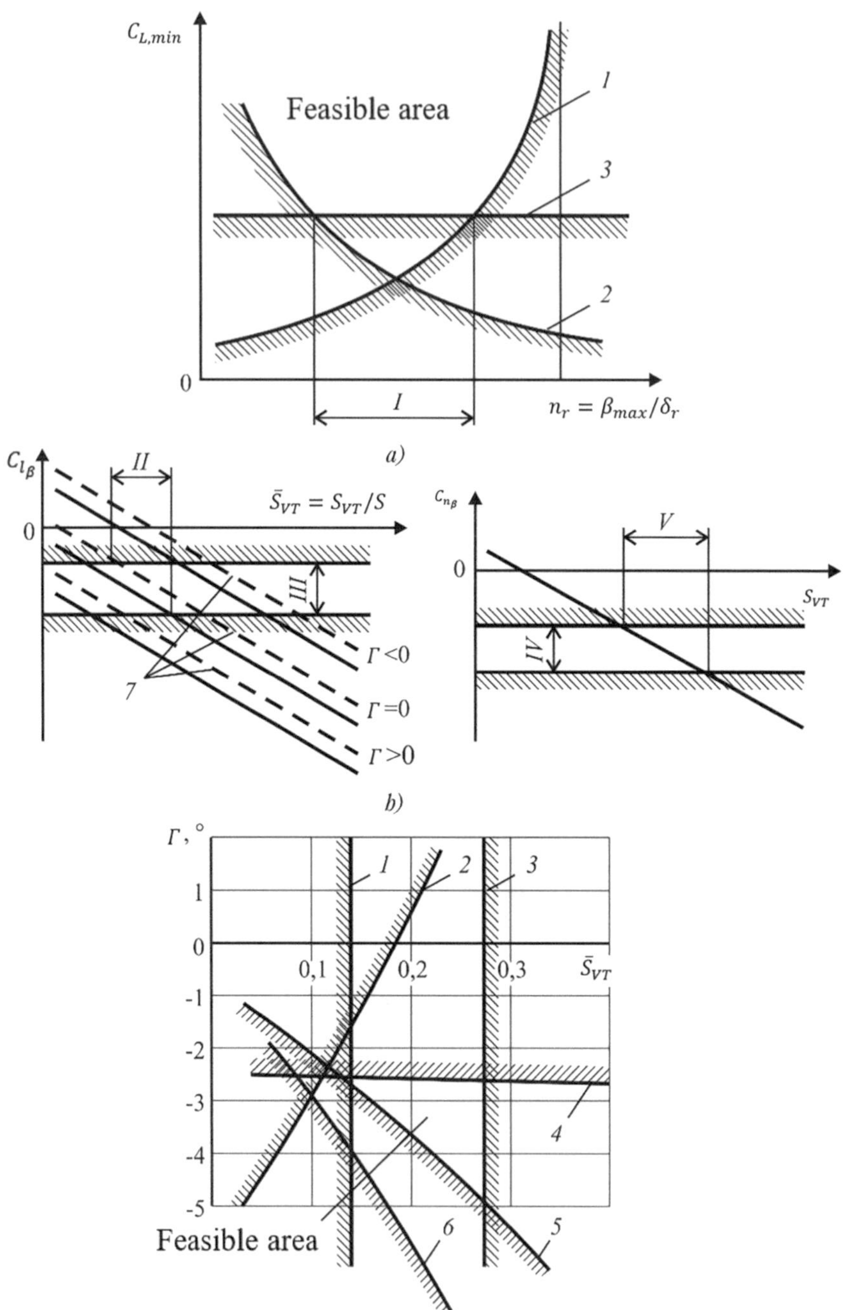
$C_{L,min}$
Feasible area
1
3
2
0
I
$n_r = \beta_{max}/\delta_r$
a)
C_{l_β}
II
$\bar{S}_{VT} = S_{VT}/S$
0
III
Γ <0
Γ =0
Γ >0
7
C_{n_β}
V
0
S_{VT}
IV
b)
Γ, °
1
0
-1
-2
-3
-4
-5
0,1
0,2
0,3
$\bar{S}_{VT}$
1
2
3
4
5
6
Feasible area
c)

◂**Fig. 18** Determination of vertical tail parameters and wing dihedral Γ: **a**—aircraft feasible area plotted in terms of VT volume coefficient and rudder efficiency with the following boundaries: 1—crosswind landing at sideslip angle β_{max}; 2—takeoff with asymmetric thrust; 3—rolling to yawing oscillations ratio ($\chi \leq \chi_{allow}$); **b**—VT area ratio corresponding to the design ranges of C_{l_β} and C_{n_β}; **c**—sample feasible area constructed for a light subsonic aircraft with swept wings with the following boundaries: 1—maximum allowable damping period, during which the amplitude of lateral short-period oscillations reduces by half; 2—χ upper limit; 3—maximum deflection of rudder at crosswind landing; 4—maximum deflection of ailerons at crosswind landing; 5—χ lower limit; 6—maximum allowable degree of spiral instability (minimum time of initial spiral amplitude increase by a factor of two); 7—high-lift devices deflected; I—the minimum range at which the rudder is efficient; II—$\overline{S}_{VT}$ at $\Gamma = 0$; III—the recommended range of C_{l_β}; IV—the recommended range ofC_{n_β}; V—$\overline{S}_{VT}$ range

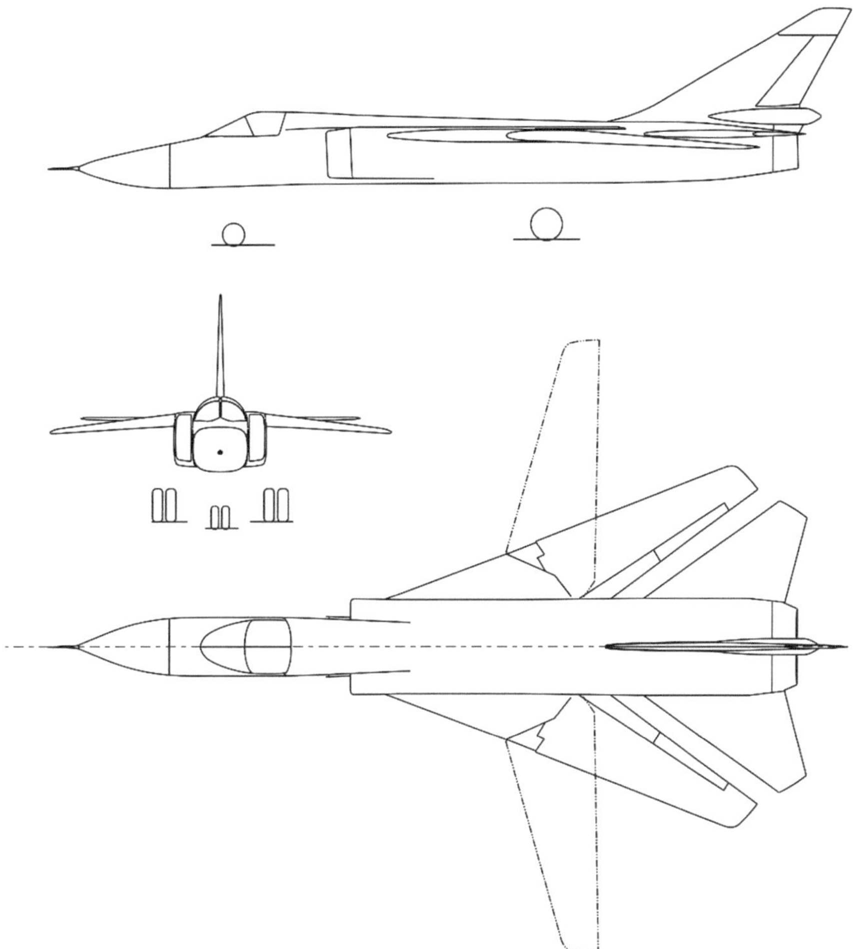

Fig. 19 Preliminary aircraft configuration resulting from aerodynamic layout

The placement of aircraft components and systems is subject to rather stringent constraints. Let us list them.

- The airframe units feature complex geometric forms.
- There are a number of components whose placement in the aircraft is strictly defined (control panels, antennas, photo equipment, etc.). When positioning some components, it is required to take into account the direction of flight.
- For some components, their influence on each other must be taken into account when arranging them. For example, the closest possible placement of the components to each other is desirable if this allows to reduce the weight of the cables connecting them or increase their anti-interference capacity; in some cases, the components may be incompatible with each other (due to their thermal, electromagnetic, etc. influence on each other).
- It is necessary to provide service areas for the on-board items (i.e. to enable their installation, replacement and maintenance). Requirements for access to these facilities can be divided into 2 categories—"For Flight" and "For Ground Maintenance".

It should be noted that currently there is a tendency to arrange all or most of the elements of a system in one particular compartment of the aircraft, called a service compartment. This approach reduces the length and weight of interconnecting lines and facilitates the placement, set-up, operation conditions and technical operation of these systems.

The following requirements must be met when laying out the aircraft:

- The specified aircraft center of gravity location with respect to the body reference frame. This requirement is essential for ensuring the stability and controllability of the aircraft. In some cases, additional constraints are set on the moment of inertia of different items, if it affects the dynamic characteristics of the aircraft.
- Reducing the mass of the aircraft. The main layout-related means of mass reduction is through the reduction of the length and, as a consequence, the mass of the communication lines (in modern aircraft communication network has a mass commensurate with the mass of equipment). The careful layout also allows reducing the weight of the structural elements required to install the equipment on aircraft.
- The mass reduction requirement is related to the CG requirement. This is due to the fact that when the position of the CG goes beyond the established limits, to move it, additional balance weights must be installed, which is the weight penalty associated with the given layout option.
- An increased utilization of the aircraft's compartment volumes. Increased compartment volumes caused by an incorrect layout result in both increased airplane drag and increased compartment and overall airplane mass.
- Ensuring the given reliability of aircraft elements. Reliability is largely dependent on the operating conditions of the equipment at hand (levels of vibration, changes in pressure, temperature conditions, etc.), which in turn are determined by the location of this equipment in the aircraft compartments.

- A reduction of installation and maintenance man-hours. This is achieved by ensuring access to the equipment corresponding to the required maintenance ratio (or reliability), the labor intensity of its installation and dismantling.
- The ergonomic requirements associated with the reduction of crew physical and psychological stress which improves flight safety. This mainly refers to the arrangement of the cockpit instruments and controls.

Obviously, these requirements are contradictory. For example, the labor costs reduction associated with installation and maintenance will require a less dense layout, which will worsen the use of the volume of the compartment and lead to an increase in its weight. Therefore, aircraft laying out involves finding a reasonable compromise between these requirements.

Aircraft layout can be internal (placement of onboard equipment in compartments, etc.), or external (placement of antennas, solar panels, emitters, etc.). It is not always possible to draw a clear distinction between internal and external layout, since some items (especially this is typical for antennas), although placed inside the aircraft, must provide a directional pattern outside the compartment in which they are placed.

Configuration-related problems arise at all stages of aircraft design. At the conceptual design stage, project brief and historical data are used to generate layout options. At the preliminary design stage, the layout data are determined with sufficient accuracy to select the baseline variant. This variant is used for further design studies, strength analysis, and specification of control system requirements.

3.2 Development of Internal Layout Methods

At the dawn of aviation, the creation of new, more perfect aircraft was carried out by rebuilding and modifying existing models. Without a scientific foundation, such reconstruction, carried out by trial and error, did not even require making appropriate drawings.

The introduction of the blueprint method of aircraft design allowed the search for acceptable design solutions on the blueprint and separated the process of aircraft design from its production. With regard to layout analysis, graphic modeling is carried out by *the drawing method*. It consists in making drawings of several layout variants. Since the labor intensity of this method is very significant, the layout objects are drawn in a simplified manner, and the number of options elaborated by the designer is very limited. This method evolved into *the cutout method* according to which in all variants of the layout the composition of the elements remains unchanged. This makes it possible to avoid drawing the objects anew each time during the design process and use their cutouts (cut out orthogonal views of the elements to be assembled on a reduced scale). The cutouts made in this way are moved along the aircraft drawing until an acceptable (from the designer's point of view) arrangement is obtained, which is then transferred to the drawing. The cutout method significantly increases labor efficiency and features a high degree of clarity, although the design process is separated from the mass and balance analysis of the aircraft. However, in the case of a high density arrangement, typical for modern aircraft, even the most careful implementation of the drawing method does not guarantee the exclusion of intersection of assembled objects.

The Mock-up method implies full-scale modeling of aircraft compartments and elements in reasonable detail. The models are made from easily machinable materials (wood, plywood, foam plastic, light alloys), although even with these materials labor intensity remains high. The advantage of the method, besides the total exclusion of intersection of the objects to be integrated, is the highest visual clarity allowing preliminary evaluation of the aircraft operating characteristics. This method can also use real aircraft items instead of physical models. Despite good detailing, the use of this method is limited by its technical complexity.

The fundamental disadvantage of the physical modeling method is its low productivity.

Manual layout time contributes significantly to the total design time of an aircraft, bringing its obsolescence time closer. Also, the quality of the layout is largely determined by the individual abilities of the designer.

The main way to automate the layout process and bring it into compliance with modern requirements of aircraft design is the use of CAD. CAD makes possible the transition from physical to mathematical and geometric modeling, which allows to reduce design time (due to the speed of modern computers), improve the quality of design solutions due to the possibility of considering a significant number of options for these solutions, and obtain any relevant information about the selected design solutions.

Solid, parametric, etc. modeling allows feasibility analysis of the layout-related solutions in the virtual space.

The three-dimensional electronic model not only replaces the full-scale mock-up of the aircraft, but also serves as the basis for the development of modern aircraft production processes.

3.3 Algorithms for Internal Layout

The internal layout of an aircraft is a very complex process, which is difficult to formalize. This is due to the variety of factors affecting the layout results and the ambiguity of their influence on these results. For this reason, despite significant efforts, little progress has been made in automating the layout process. Design bureaus are staffed by highly qualified specialists with extensive practical experience, who have accumulated a significant database and a set of rules, techniques for solving layout problems for aircraft of a particular type. It should be noted that combining this experience with the capabilities of modern 3D modeling tools made it possible to improve the quality and reduce the time required for internal layout.

The internal layout algorithm, called the linear algorithm, is intended for use in the training process. This algorithm can be successfully used for an aircraft with an established position of functional elements relative to each other. Such aircraft include, for example, conventional subsonic passenger aircraft as well as modifications of combat aircraft based on the baseline design. According to this method, components are placed in a certain order along the longitudinal axis of the aircraft.

Before conducting the longitudinal layout of the aircraft a number of alternative versions of the fuselage cross-sections, differing in the number of passenger seats arranged in one row (Fig. 20) are usually formed. In this case the radius of the upper "passenger" part of the fuselage is determined as half the sum of the width of the seat units (taking into account the dimensions of armrests), the width of aisles between seats and gaps between the seat structure and internal fuselage panels. The radius of the fuselage lower "cargo" part is defined from the condition of accommodation (including clearances) of the biggest cargo container. The upper and lower fuselage section radii obtained in this way are internal and must be increased by the thickness of the thermal and acoustic insulation as well as the thickness of the fuselage structure to obtain external dimensions of the section.

As can be seen from the figure, the circular cross-section of the fuselage, which is optimal in terms of aerodynamics, strength, production, is often modified into a surface described by various radii. This is done to better use the cross-sectional area of the fuselage to accommodate the given items of passenger and cargo compartments and thereby reduce its midsection.

Some dimensions of passenger seats, aisles, overhead stowage bins, depending on the class of the cabin, as well as dimensions of standard cargo and luggage containers are given in Chap. 10.

Once several versions of the fuselage cross-sections are formed, for each of them a fuselage planform layout is worked out.

For subsonic passenger aircraft, the order of laying out the functional elements is usually as follows: large radar antenna—service compartment with electronic equipment—cockpit—service crew compartment—auxiliary areas of the passenger cabin including galleys, dressing modules, lavatory modules, etc. —I-th passenger compartment—passenger cabin auxiliary areas-2—II-nd passenger compartment—passenger cabin auxiliary areas-3, etc. Further, a fuselage tail compartment is formed, which may contain passenger cabin auxiliary areas, luggage compartment (for light aircraft), as well as, outside the sealed area of the fuselage, a compartment for placement of aircraft equipment, auxiliary power plant, etc. Using this method, the size of the aircraft fuselage is determined virtually "in one go".

When arranging the passenger compartment, recommendations set out in specific regulations should be taken into account, e.g. the inadmissibility of passenger seats in the plane of the propeller rotation, the undesirability of adjacent galley and toilet facilities, etc. (see Chap. 10).

Then, the fuselage is streamlined and the wing, landing gear, tail and power plant are placed with respect to the fuselage.

Examples of the linear layout of airliner fuselages are presented in Fig. 21.

Currently, a number of institutes and design organizations have developed layout codes based on the principle of "linear placement" of functional elements, which allow for the automated internal layout of subsonic passenger aircraft [2–4].

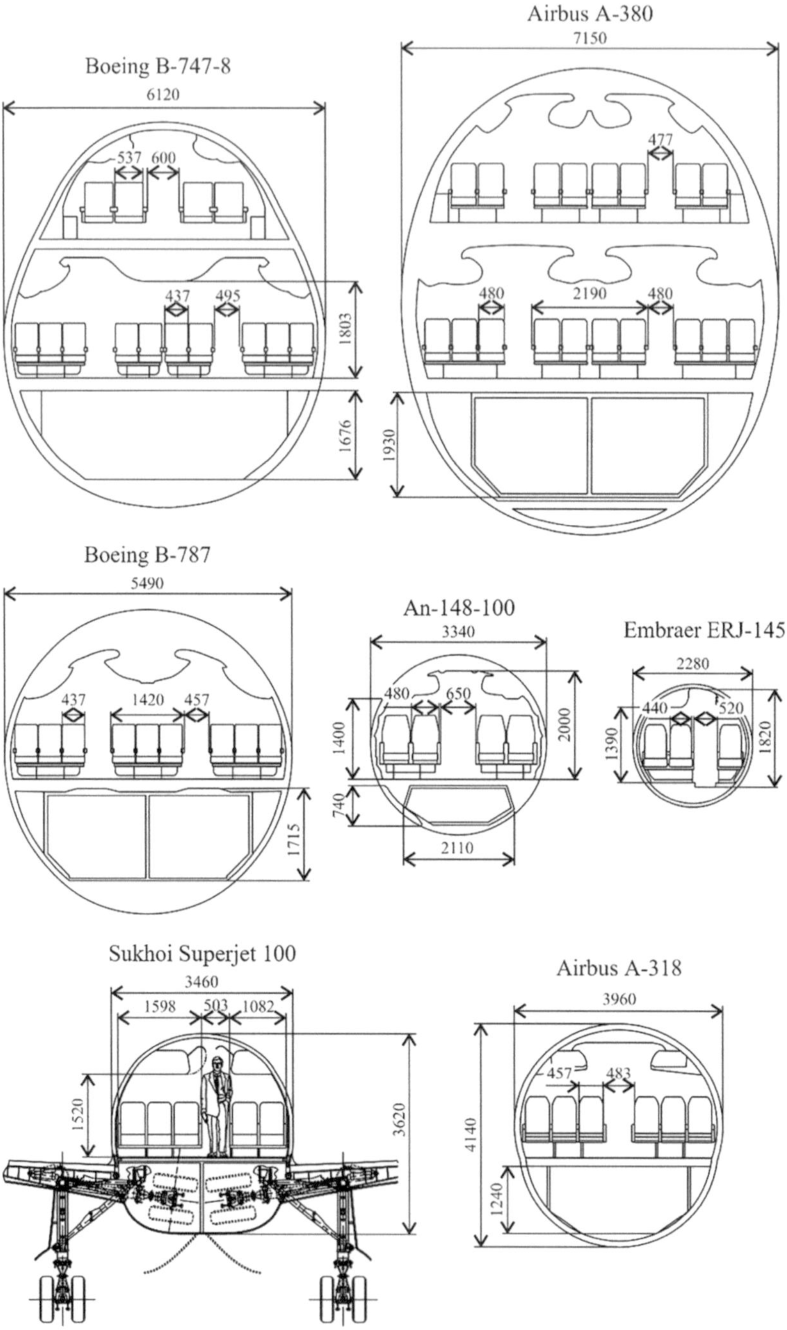

Fig. 20 Fuselage cross sections of airliners

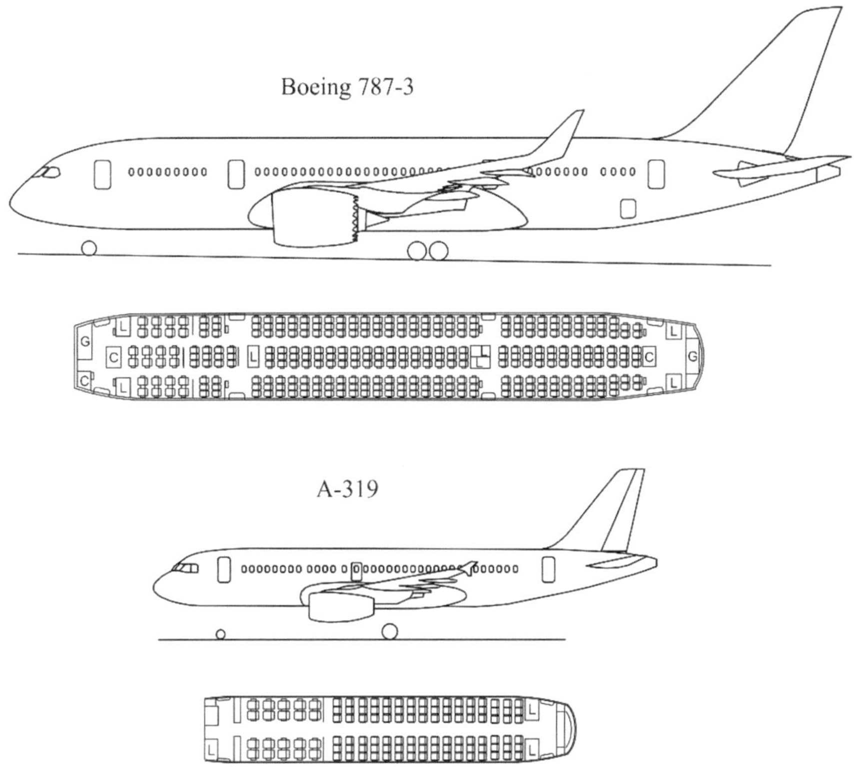

Fig. 21 Linear fuselage layout of wide-body and narrow-body airliners

While having such obvious advantages as simplicity and speed, the described layout technique cannot be used for other types of aircraft, as well as for unconventional aircraft. In that case, an algorithm based on the analysis of designers' actions can be used. Below one possible algorithm is presented (for more detail, see [4]). It includes the following steps.

1. Formation of an appropriate aerodynamic configuration of the aircraft for the most typical mode of flight. Determination of the aircraft's overall dimensions in the first approximation, using the previously obtained values of take-off weight, wing loading and thrust-to-weight ratio, as well as the average layout density typical for the aircraft of the given class. Calculation of absolute parameters and placement of wings and tail. Thus, the starting point for internal configuration is the layout drawing of the aircraft (Fig. 19).
2. Allocating areas in the cross sections, as well as on the side and planform views of the layout drawing for the airframe structure and wiring (controls, fuel lines, electrical harnesses, etc.). This can be done by using historical data on previously built planes. Typically, 7–10% of the cross-sectional area is reserved (Fig. 22a).

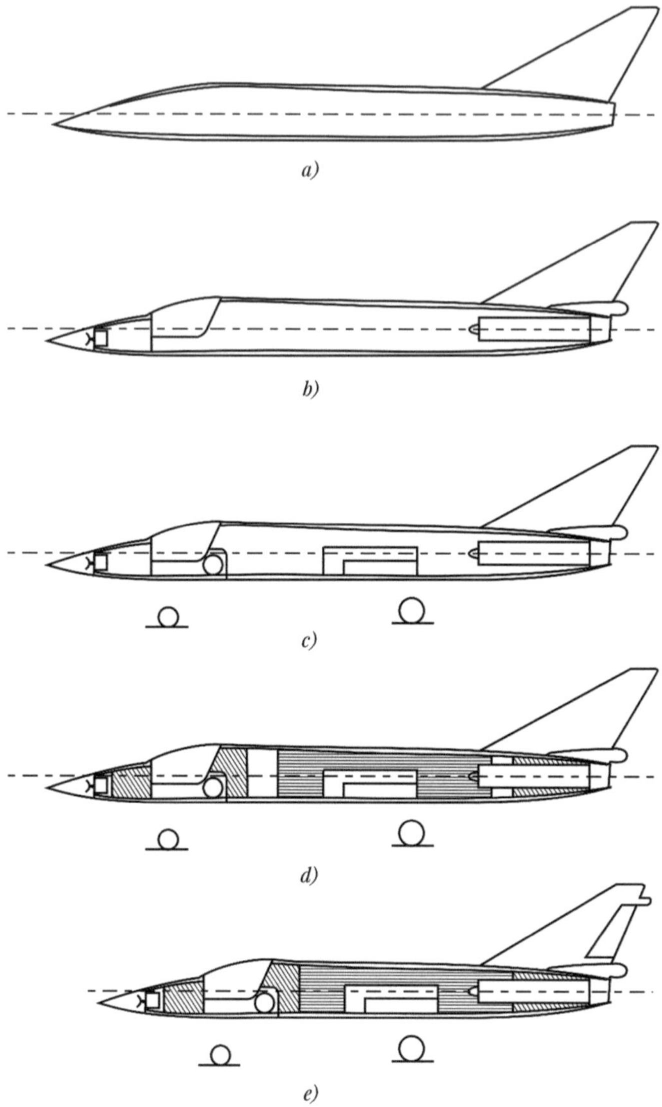

Fig. 22 Sequence of operations in the free (random) layout method

3. Placement of functional elements using the side view of the aircraft. All operational constraints must be taken into account. Since this stage of the laying-out process is the most critical, let us dwell on it in more detail:

   3.1. The first step is the placement of elements with forward or aft "open areas", i.e., those that require certain "forward or aft view" areas for normal operation. These include large radar antennas, cockpit, aft gun mount, brake parachute, etc. (Fig. 22b). The dimensions of these functional elements can be taken from special catalogues or from prototype aircraft data.

3.2. At the next stage, the elements whose position with respect to the center of mass of the aircraft (drop cargo, landing gear, etc.) is critical are placed (Fig. 22c).

3.3. Then, the elements which can be arranged relatively freely along the length of the aircraft (large avionics units, auxiliary power unit, aerial camera equipment, etc.) are placed.

4. The fuselage layout is completed by the placement of small equipment and fuel. Equipment, as a lower weight element, is placed starting at the ends and moving to the middle of the fuselage. Before placing, the total volume of equipment is divided into parts, and each of them is placed until the empty volume of frame space (the space between the frames) is completely filled so as to bring the center of gravity of empty aircraft to a predetermined location. The fuel is distributed in the same manner, but staring at the center of the fuselage and moving to its ends.
5. After all the elements have been placed and the empty volumes remaining inside the plane are identified (see Fig. 22d), the layout is generated again. The procedure of iterative reduction of the configuration dimensionality should be conducted in such a way that the change of the aircraft volume goes with decreasing steps until the existing voids are eliminated (Fig. 22d).

In conclusion, it should be noted that increasing layout density is subject to constraints related to the placement of individual layout elements (listed in Sect. 3.1). This narrows the possibilities of formalization and hence full automation of the layout process.

Final evaluation of CG envelope is performed using the aircraft CG diagram (Fig. 24) and weight report (Table 1). This process is described in more detail in Sect. 3.4.

3.4 Aircraft Center of Gravity. Calculations of First and Subsequent Approximations

Center of gravity is the distance expressed in fractions of the mean aerodynamic chord from the leading edge of the mean aerodynamic chord of the wing to the center of mass of the aircraft ($\overline{X}_{CG} = \frac{X_{CG}}{c_{MAC}}$ Fig. 23).

Since the aircraft's CG location depends on its loading and the phase of flight (takeoff, landing, etc.), we distinguish between the forward CG limit $\left(\overline{X}_{CG,f}\right)$ and the aft CG limit $\left(\overline{X}_{CG,aft}\right)$. The difference between the forward and aft limits constitutes *the operational CG envelope* $\Delta\overline{X}_{CG}^{op} = \overline{X}_{CG,f} - \overline{X}_{CG,aft}$ (see Fig. 12).

The smaller the CG envelope is, the smaller tail volume coefficient is required to ensure the given longitudinal static stability margin, the smaller are the deflection

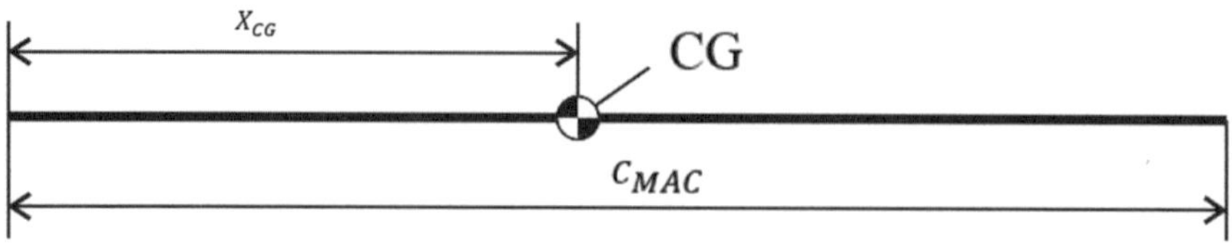

Fig. 23 Determination of aircraft center of gravity

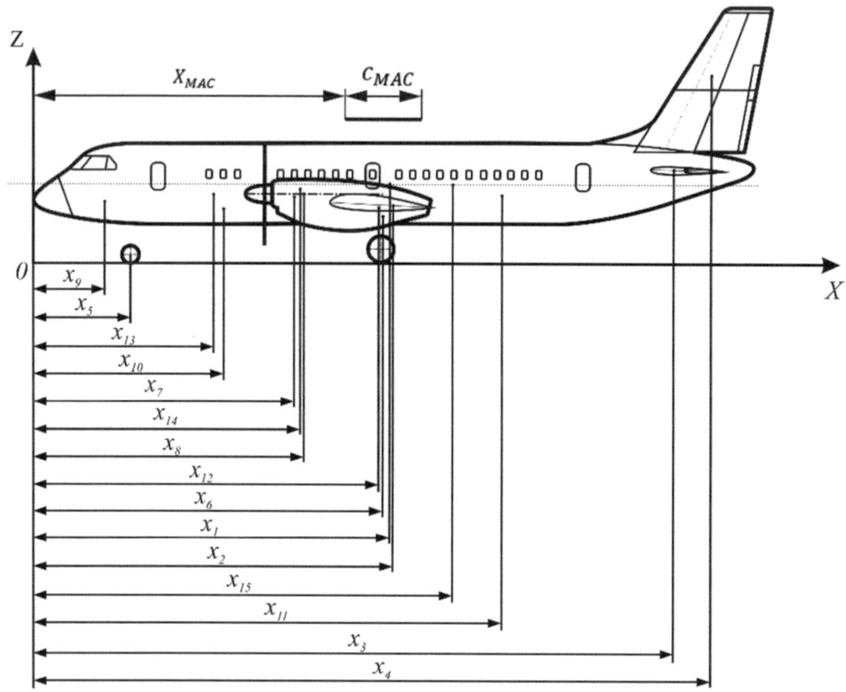

Fig. 24 Aircraft layout diagram

angles of control surfaces required to trim the aircraft on takeoff and landing with deployed high-lift devices. Consequently, the fewer trim penalties it will suffer.

The CG operational range is set in the aircraft design specification as a compromise between the requirements of the customer (payload positioning) and the aircraft designer (reduction of trim penalties).

Let us consider the CG-determination algorithm intended for use in the training process.

The input data for the determination of the CG are:

- the aeroplane layout diagram, which is an elaborated version of its inner layout (a side view of the aeroplane with the skin removed) made to a certain scale;
- the second approximation weight statement;

– the airplane's MAC length and location.

The CG-determination process begins with the drawing of the aircraft layout diagram. An example of the layout of the IL-114 regional passenger aircraft is shown in Fig. 24. In order to make the drawing easier to read, the components are represented on it by their center coordinates.

The CG location is determined with respect to two axes: horizontal *OX* and vertical *OZ* (Fig. 24). When calculating the CG location, the origin of the reference frame may be placed at:

- fuselage nose, so that all load coordinates are positive, thus avoiding errors in static coefficient summation (used in the training process). However, if the fuselage nose has to be shifted forward or backward in the analysis process, then the selected origin point will be detached from any physical point of the airframe;
- at the leading edge of the root chord of the wing or (better) of some structural element of the wing (the axis of the front or rear center wing spar). In this case, the loads located in front of the origin of the reference frame will have a negative value of abscissa.

The *X*-axis is usually aligned with the fuselage reference line or with the runway level (training process).

The next step is to determine the length of the mean aerodynamic chord c_{MAC} and its coordinates X_{MAC} and Z_{MAC}. Recall that c_{MAC} it is a chord of a rectangular wing equivalent to the wing of the designed aircraft with respect to the pitching moment coefficient. The MAC calculation procedure for a tapered wing is shown in Fig. 25

In general, the MAC length and its position (leading-edge) relative to the Y-axis can be estimated as follows:

$$c_{MAC} = \frac{2}{S}\int_{0}^{b/2} c^2 dy,$$

$$X_{MAC} = \frac{2}{S}\int_{0}^{b/2} xcdy,$$

where S is the wing area, m^2; $b/2$ is half of the wing span, m; c and x are the local length of the wing chord and the distance from the leading edge of this chord to the axis.

If the wing has a large dihedral angle Γ, it is necessary to determine the MAC vertical station with respect to the *XOZ* plane; Z_{MAC} is determined as follows:

$$Z_{MAC} = \frac{2}{S}\int_{0}^{b/2} zcdy.$$

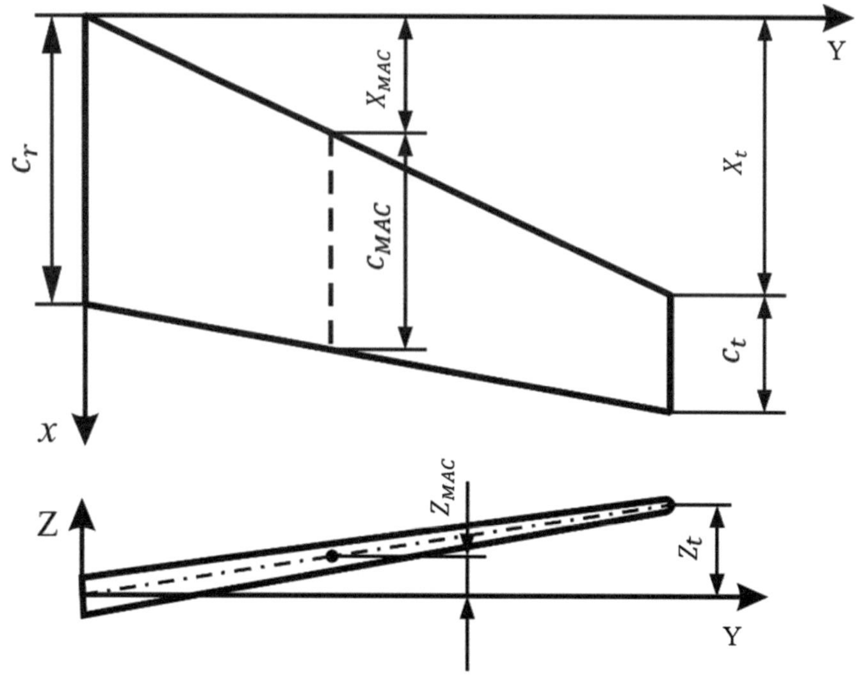

Fig. 25 MAC calculation procedure for a tapered wing

If the wing has a geometric twist, then the angular position of the MAC relative to the chord of the wing root section is defined as

$$\varphi_{MAC} = \frac{2}{S}\int_{0}^{b/2} \varphi c dy.$$

Values z and φ in the above equations are the local values of the MAC vertical station and twist angle.

For a tapered wing (Fig. 25), these equations take the following form:

$$c_{MAC} = \frac{2}{3}\frac{c_r^2 + c_r c_t + c_t^2}{c_r + c_t} = \frac{2}{3}c_r\frac{1+\lambda+\lambda^2}{\lambda(1+\lambda)},$$

$$X_{MAC} = \frac{X_t}{3}\frac{c_r + 2c_t}{c_r + c_t} = \frac{X_t}{3}\frac{\lambda+2}{\lambda+1},$$

$$Z_{MAC} = \frac{Z_t}{3}\frac{c_r + 2c_t}{c_r + c_t} = \frac{Z_t}{3}\frac{\lambda+2}{\lambda+1},$$

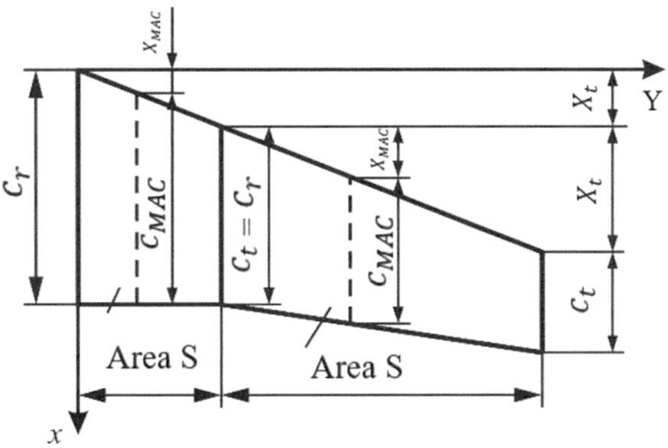

Fig. 26 Diagram for determination of the MAC of a compound wing Z_{MACi} and φ_{MACi} can be determined in the same way (Fig. 27)

$$\varphi_{MAC} = \frac{\varphi_t}{3}\frac{c_r + 2c_t}{c_r + c_t} = \frac{\varphi_t}{3}\frac{\lambda + 2}{\lambda + 1},$$

where c_r is the tapered wing root chord; c_t is the tapered wing tip chord;$\lambda = c_r/c_t$ is the wing taper ratio; X_t, Z_t, and φ_t identify the position of the tip chord.

The Y-station of the MAC can be determined as follows:

$$Y_{MAC} = \frac{b}{3}\frac{c_r + 2c_t}{c_r + c_t} = \frac{b}{3}\frac{\lambda + 2}{\lambda + 1}.$$

If a wing is composed of two or more trapezoids (Fig. 26), first determine $c_{MACi}, X_{MACi}, Z_{MACi}, \varphi_{MACi}$ for each trapezoid, and then use the following expressions to find the MAC parameters of the entire wing:

$$c_{MAC} = \frac{c'_{MAC}S' + c''_{MAC}S'' + \ldots + c^i_{MAC}S^i}{S\prime + S'' + \ldots + S^i},$$

$$X_{MAC} = \frac{X'_{MAC}S' + (X'_t + X''_{MAC})S'' + \ldots + (X'_t + X''_t + \ldots X^i_{MAC})S^i}{S' + S'' + \ldots + S^i}.$$

The vertical position of the aircraft's center of gravity Z_{CG} is of great importance in calculating its dynamics; it is used in placing the main landing gear wheels in the down position. It must also be taken into account when it is far from the wing, for example, in a high-wing configuration (Fig. 27):

$$X'_{CG} = \frac{X_{CG} - X_{MAC}}{cos\varphi} + [(Z_{MAC} - Z_{CG}) - (X_{MAC} - X_{CG})]sin\varphi.$$

All of the above equations are also used to determine the length of the mean aerodynamic chord of the horizontal and vertical tail. In this case, the horizontal tail area includes the intra-fuselage part formed by the extension of the leading and trailing

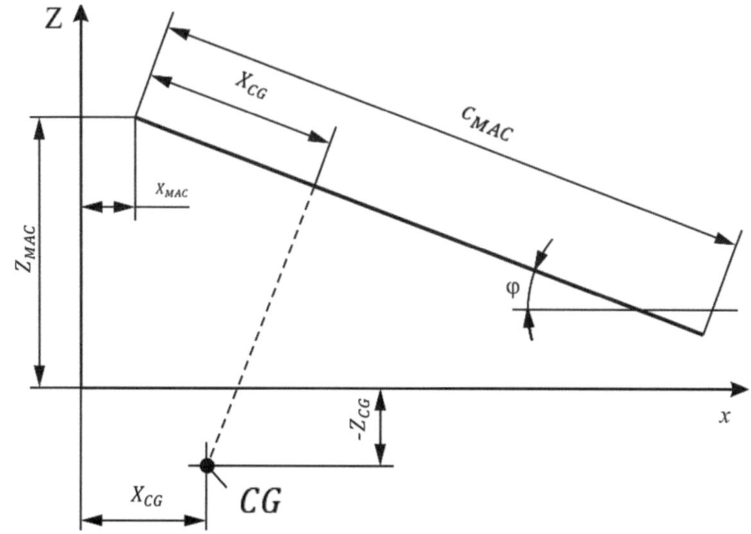

Fig. 27 Determination of CG when it is far from the MAC plane: φ is the angle between MAC and horizontal plane *XOY*

edges of the tail to the aircraft centerline; the intersection of these lines with the aircraft centerline defines the root chord.

The root chord of the vertical tail is taken from the upper fuselage outline and goes normally parallel to the fuselage axis (*X-axis*) through the intersection of the trailing edge of the vertical tail with the upper fuselage outline.

Let us also give a graphical method of determining the length and position of the mean aerodynamic chord of lifting surfaces. On the planform view of the surface (Fig. 28), draw a line connecting the center points of the side-of-body chord and tip chord. Then, lay off the length of the tip chord on the continuation of the side-of-body chord, and the length of the side-of-body chord on the continuation of the tip chord (in the opposite direction).

Fig. 28 Graphical determination of MAC

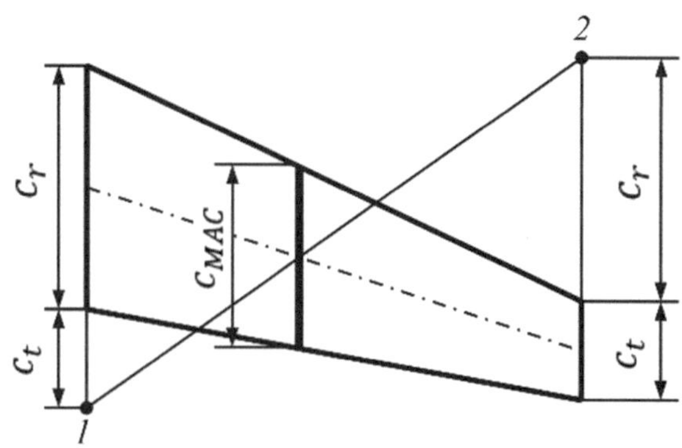

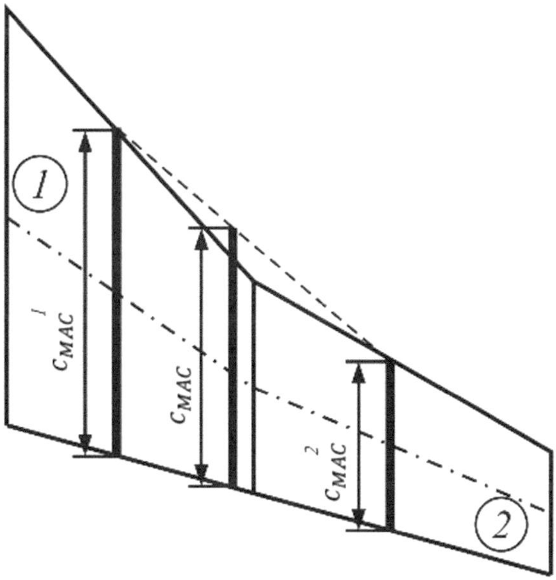

Fig. 29 Graphical determination of the MAC of a wing consisting of two trapezoids

Draw a line through points *1* and *2*. The intersection of this line with the centerline of the wing determines the position and length of the MAC.

If the wing has a more complex planform shape, for example, consists of two trapezoids (wing with a leading-edge extension) (Fig. 29), then the following algorithm can be used.

Using the method described above, determine the MAC for each part of the wing. Then, using them as root and tip chords, determine the MAC for the entire wing. It can be seen that in this case, the MAC can go beyond the leading edge of the wing. This circumstance is used to reduce the trim losses associated with the movement of the aerodynamic center of the wing during the transition to supersonic flight. A similar method can be used to determine the MAC in more complex cases (crescent wing).

After determining the MAC length and position, it is plotted on the layout diagram and its station with respect to the given reference frame is determined. Figure 24 shows an example of the MAC positioning with respect to *X-axis*, since for the aircraft shown in the figure vertical station of the MAC is not relevant. In order not to obscure the drawing, the MAC is shown out of its limits.

The next step is the preparation of the weight report. For each item in the list, its weight W_i, center of gravity and X-static moment W_iX_i are specified. In general, Z_i—coordinate and value of Z-static moment are also given.

An example of the weight report for the aircraft shown in Fig. 24 is presented in Table 2.

CG locating can be determined as follows:

Table 2 Weight report

No. n/a	Unit, system, load	W_i, daN	X_i м	W_iX_i
	Design			
1	Wing	2677	13.06	34,961.62
2	Fuselage	2456	13.44	2456
3	Horizontal tail	362	23.9	362
4	Vertical tail	230	25.3	230
5	Front landing gear	171	3.38	171
6	Main landing gear	767	12.95	767
	Propulsion system			
7	Engines (with nacelles)	1383	9.77	13,512
8	Other components of PS	1076	10.11	10,878
	Equipment			
9	Equipment in the forward compartment	1097	3.06	3356.82
10	Equipment in the middle compartment	1548	7.177	11,110
11	Equipment in the rear compartment	1935	17.48	33,823,8
	Empty aircraft	13,702	12.06	165,196.1
	Fuel			
12	In the wing	2471	12.9	31,875.9
	Ferry flight case	16,173	12.09	197,062
	Payload			
13	Payload in the front compartment	1200	6,79	8148
14	Payload in the middle compartment	2000	9.97	19,940
15	Payload in the rear compartment	2800	15.68	43,904
	Critical landing case	19,702	12.04	237,188.1
	Total	23,210	11.86	269,064

$$X_{CG} = \frac{\sum(W_iX_i)}{W_i} \text{ and } Z_{CG} = \frac{\sum(W_iZ_i)}{W_i}.$$

$$\overline{X}_{CG} = \frac{X_{CG} - X_{MAC}}{c_{MAC}}.$$

$\overline{X}_{CG}$ is usually expressed as a percentage (multiplied by 100%).

In the *first approximation balance analysis,* it is common to assume that:

(a) the wing center of mass is located at 40 … 42% c_{MAC};
(b) the tail center of mass is located at 45 … 50% $c_{MAC,HT}$ or $c_{MAC,VT}$;
(c) the fuselage center of mass is located at approximately 50% of the fuselage length for a straight wing configuration or 60% for a swept wing configuration;

(d) the center of mass of fuel and fuel system are located in the center of mass of the wing fuel compartments; for fuselage fuel compartments (tanks), the volume and center of mass of each tank should be calculated.

The maximum volume of fuel placed in the wing can be estimated as follows:

$$V_{f,w} \approx 0.7 \frac{(t/c)_{aver} S^{3/2}}{AR^{1/2}}.$$

By multiplying it by the fuel density (for a bypass turbojet it is equal to 0.78 … 0.83 g/cm^3) the fuel weight in the wing can be calculated;

(e) The center of mass of equipment is located in the center of mass of the fuselage, i.e. at 50 … 60% of the fuselage length;
(f) The typical value of pilot's mass is 100 kg (on military aircraft) and 80 kg (on civil aircraft); flight attendant's mass is 70 kg; passenger's mass is 75 kg and luggage mass is 15 kg.

The coordinates of loads are taken from the layout drawing (Fig. 30), which is a side view of the aircraft or a longitudinal section of the aircraft on as large a scale as possible. On this drawing, all loads and units of the aircraft are shown. In the process of laying out, the relative positions of units and loads change, so the process of creating a layout drawing is as iterative as the laying out process itself.

Acceptable CG locations are the following:

- straight wing aircraft: 0.20–0.25;
- swept wing aircraft ($\Lambda = 35...40°$): 0.26–0.30;
- swept wing aircraft ($\Lambda = 50...55°$): 0.30–0.34;
- small aspect ratio delta wing aircraft: 0.32–0.36.

If the current CG location doesn't fall within the above ranges, it is necessary to move the center of mass of the aircraft. This is possible in two ways:

(a) by moving individual (most heavy) loads. Then the shift of the center of mass of the aircraft will be $\Delta\overline{X}_{CG}$,

$$\Delta\overline{X}_{CG} = \frac{M_i}{M} \frac{X_{i0} - X'_i}{c_{MAC}},$$

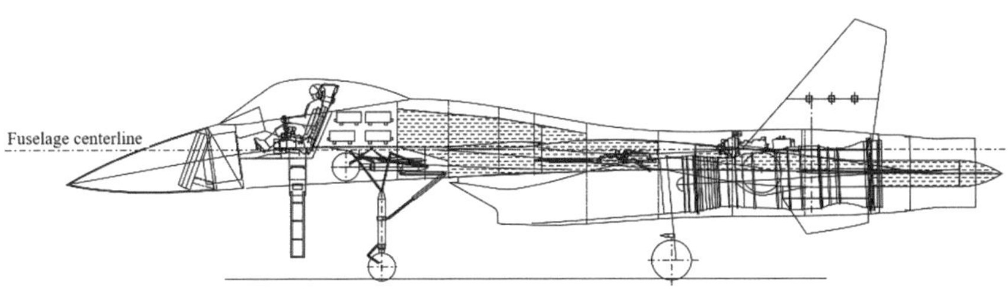

Fig. 30 Light aircraft layout (hypothetical aircraft)

where M_i stands for the mass of the shifted load; M is the aircraft mass; X_{i0} is the initial load coordinate; X'_i is the new load coordinate;

(b) by moving the entire fuselage with the loads placed on it relative to the wing.

For this purpose, the mass of the aircraft is divided into two groups:

- the mass of the wing group, which includes the mass of the wing, the mass of all loads on the wing (engines, fuel, etc.), and the mass of the main landing gear, regardless of where it is mounted (on the fuselage or on the wing);
- the mass of the fuselage group, including the rest of the aircraft mass.

Then the necessary shift of the fuselage group relative to the wing group will be

$$\Delta\overline{X}_{FG} = \frac{\Delta X^*_{CG} M c_{MAC}}{M_{FG}},$$

where M_{FG} is the total mass of the fuselage group; ΔX^*_{CG} is the value of the required shift of the aircraft center of mass.

During the second approximation balance analysis:

- the weight report of the aircraft is refined by dividing the weights into smaller groups (a typical second approximation weight report is given in annex I);
- the aerodynamic center of the aircraft is calculated;
- the center-of-gravity is adjusted to ensure the given $C_{m_{C_{L,min}}}$.

Balance analysis is performed for the following basic cases:

(a) aircraft maximum allowable mass with a full load (commercial or combat); tanks filled with the appropriate amount of fuel;
(b) aircraft maximum allowable mass with maximum fuel and correspondingly reduced load;
(c) aircraft with maximum fuel and no load (ferry flight);
(d) aircraft with a full load without fuel (limit landing case);
(e) empty aircraft with no load or fuel.
  - with equipment (with crew)
  - with no equipment.

The latter option is necessary to assess the aircraft's parking behavior. All other cases are considered with retracted and extended landing gear.

As a result of these calculations, the CG envelope is created, which determines the dimensions of the tail and, consequently, the airplane mass.

If the CG envelope doesn't match the given volume coefficient V_{HT}, it is possible to reduce the value of V_{HT} by reducing the horizontal tail area and thereby reduce the mass of the aircraft.

If the resulting CG envelope is greater than the allowable CG envelope, it is necessary to:

- expand the allowable CG envelope by shifting backward the aft CG limit by increasing S_{HT} or decreasing the stability margin $C_{m_{C_{L,min}}}$, e.g., by introducing appropriate automation into the control system;
- expand the allowable CG envelope by shifting forward the forward CG limit and increasing the efficiency of the horizontal tail (increasing the angles of the rudders or stabilizer);
- reduce the resulting CG envelope.

In order to reduce the CG envelope, one should ensure that:

- the fuel center of gravity is close to the aircraft center of gravity so that the aircraft center of gravity is displaced by no more than 5% c_{MAC} when the fuel is burned;
- the in-flight drop loads are located so that when they are dropped, the aircraft CG moves within 3% c_{MAC};
- the variable loads (e.g., passengers) are as close to the aircraft's CG as possible, so that they cause minimal changes in CG in any loading scenarios.

Taking these measures into account, the CG envelope should be no more than 10% c_{MAC} for military aircraft and 20% c_{MAC} for passenger and transport aircraft.

Usually, a CG diagram (see Figs. 31 and 32) is constructed as a result of the above analysis.

3.5 *Special Layout Options*

Let's consider the effect of engine placement on aircraft CG location.

Aft-podded engines generate a large static moment relative to the center of gravity of the aircraft, and in order to balance this moment, i.e. to ensure the aft CG limit, the forebody has to be moved far forward. Consequently, the passenger cabin CG is ahead of an empty aircraft CG. When passengers are boarding, the aircraft CG shifts forward—there are "scissors" between the aft CG limit of an empty aircraft and the forward CG limit of a fully loaded aircraft (Fig. 31).

As a result of this:

(a) To prevent the tail strike of an empty aircraft, the wheels of the main landing gear should be shifted backward with respect to their typical position. With such a rear main wheels position, takeoff of a fully loaded airplane requires either an increase in the horizontal tail area or a deflection of the stabilizer and elevator to very large angles, which increases the negative load on the tail, the design wing lift and the takeoff distance;

(b) A flight of a fully loaded aircraft with a forward CG location is associated with high trim penalties (due to a large value of $C_{m_{C_{L,min}}}$) and, consequently, increased fuel consumption;

(c) In a ferry flight, in order to ensure the minimum allowable stability margin, a balance weight should be used or the forward fuel tanks should be filled with unusable fuel.

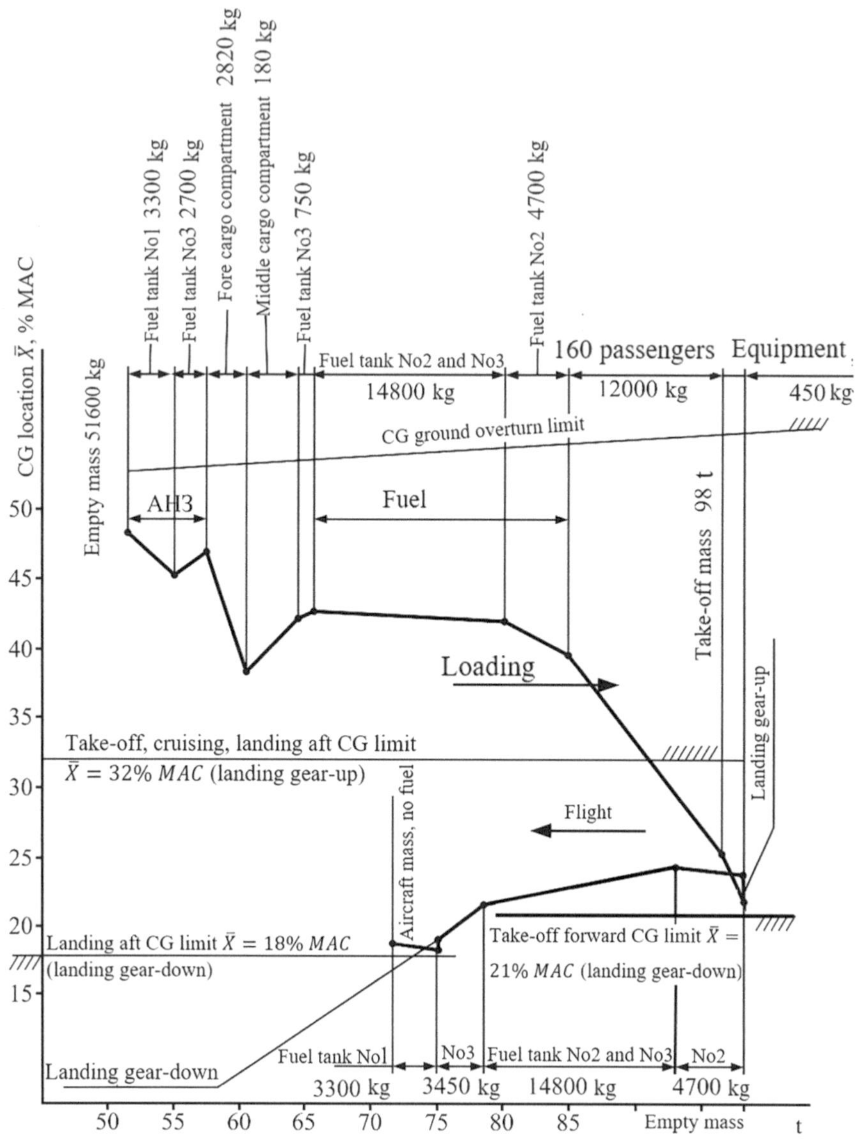

Fig. 31 CG diagram of an aft-podded aircraft

Installation of engines under the wing with a large offset forward from the leading edge requires a large balancing moment arm. Therefore, the passenger cabin is shifted backward, its center of gravity is located behind the center of gravity of an empty aircraft, and when passengers are boarding, the center of gravity goes back. Thus, in this case there is a "scissors" between the aft CG limit of a fully loaded aircraft and the forward CG limit of an empty aircraft (Fig. 32).

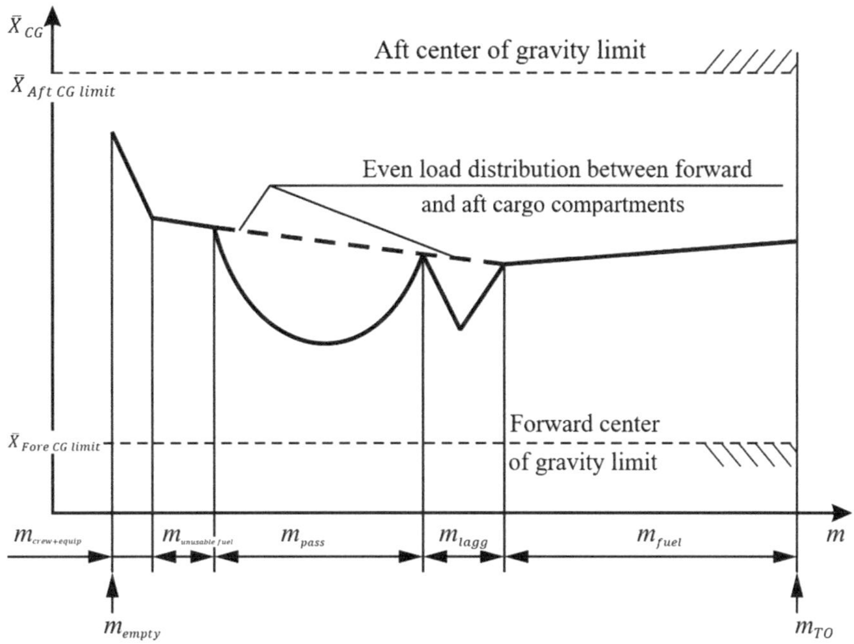

Fig. 32 CG diagram of an aircraft with engines under the wing

However, in doing so:

(a) a fully loaded aircraft takes off with a small CG arm relative to the wheels of the main landing gear, which reduces the horizontal tail loading;
(b) a fully loaded aircraft features a small value of $C_{m_{C_L}}$, and the trim penalties in this case are minimal;
(c) take-off in the ferry case very rarely requires the installation of balance weights.

The third approximation of balance analysis is carried out after the aircraft structural design and systems are worked out and it is possible to specify the masses and location of individual structural and systems elements.

This is usually associated with the completion of the conceptual design stage. However, balance analysis continues through all subsequent stages of aircraft development.

3.6 Aircraft Crew Accommodation

An important objective of the internal layout is crew accommodation. The crew of modern passenger aircraft usually consists of two people: the aircraft commander—first pilot and co-pilot, who is responsible for the duties of navigator and radio operator. On aircrafts performing long-range or special flights, it is possible to increase the number of simultaneously operating flight crew members up to 4 people by installing two additional seats (like on A-380) to accommodate navigator, pilot-inspector and others.

Figure 33 shows the layout of the forebody and cockpit of a modern long-range commercial airliner. Figure 34 describes some geometric dimensions of the cockpit, and Fig. 35 gives an idea of the required cockpit visibility.

It is important to provide the crew with the particular cockpit visibility. The requirements for the crew visibility are as follows:

1. Unobstructed view area for the left-hand pilot:
   - to the left 20°;
   - to the right 30°;
   - downwards (along the pilot's seat axis) at least 16°;
   - upwards (along the axis of the pilot's seat) 20°.
2. In the range of angles of 20 … 45° to the left in azimuth, one support strut is permitted in the cockpit structure.
3. The left-aft view should ensure the visibility of half of the wingtip (on the IL-86, this necessitated a backward azimuth visibility of up to 145° instead of the standard 135°).
4. The width of the view obstructions associated with the cockpit structural elements shall not be greater than 70 mm.
5. The distance from the pilot's eye to the inner surface of the glass in the plane of the pilot's seat (dimension L_c in Fig. 34) must be at least 500 mm.

   Cockpit floor line.

6. When plotting view diagrams from the point C_x it is allowed to move the head distance of $P = 100 \dots 120$ mm (within the length of the harness).

The required left-hand view diagram for a commercial airliner is shown in Fig. 35.

When arranging the cockpit, it is recommended to take into account the pilots' wishes:

- when wheels are touching the runway, it is desirable to see the landing strip at a distance not exceeding 50 … 70 m;
- on final approach at night, it is desirable to be able to see at least 5 … 6 runway lead-in lights at the decision altitude (at 30 m for ICAO Category II landings).

Fulfilment of these conditions requires a downward view directly in front of the pilot up to 19 ... 22°. On the Boeing 747 aircraft this angle is 18.5°, and on the IL-86 aircraft—20°.

The basic dimensions of a fighter aircraft's cockpit are shown in Fig. 36 and Table 3.

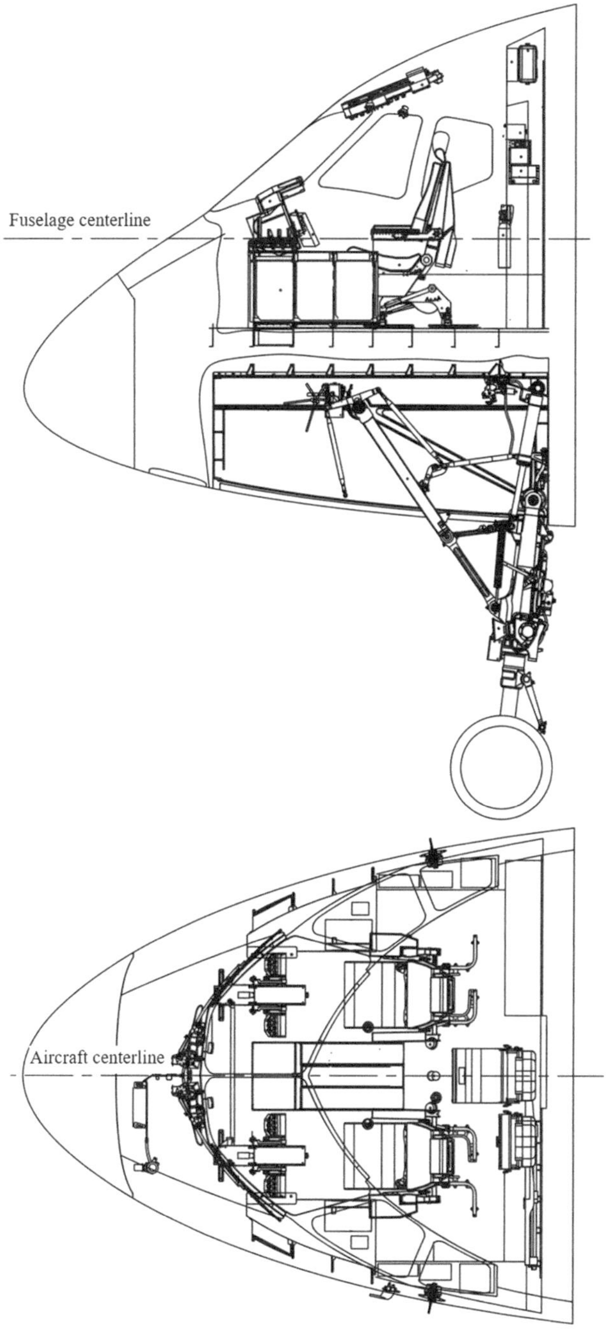

Fig. 33 Forebody and cockpit layout of a modern long-range commercial airliner

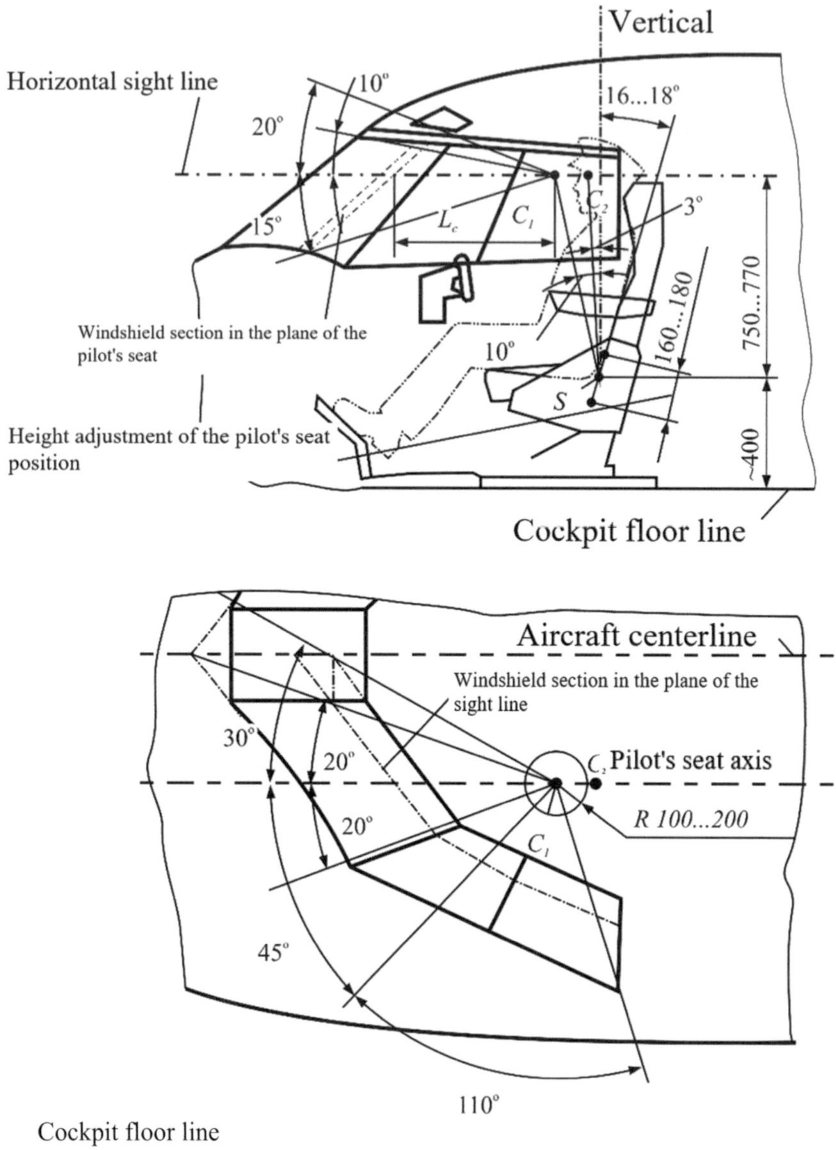

Fig. 34 View diagram: C_1-pilot's eye position during aircraft take-off and landing; C_2-pilot's eye position in cruise

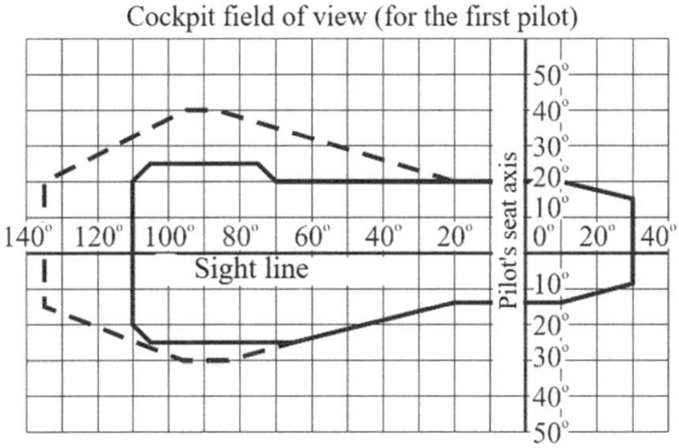

Fig. 35 Cockpit field of view, required field of view, desirable field of view

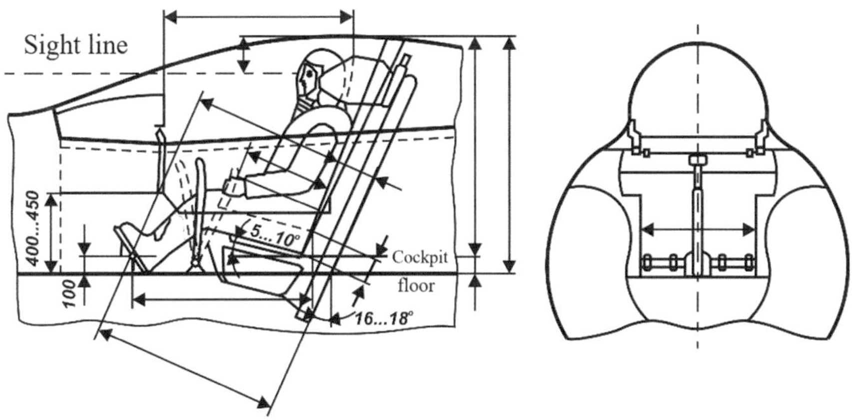

Fig. 36 Basic dimensions of the fighter aircraft cockpit

4 Structural Layout

4.1 Structural Layout Objectives

The main objectives of the structural layout are to develop an aircraft structure that would provide:

(a) a minimum structural mass;
(b) a combination of the structural elements and the volumes used to accommodate the payload, crew, equipment, propulsion system;
(c) the compliance with the maintenance and production requirements;
d) static and dynamic stability of the structure exposed to the free stream flow;
(e) the design service life and fail-safe characteristics.

Table 3 The basic dimensions of a fighter aircraft's cockpit

Dimension, mm	Description
А = 950…975	Length of emergency exit
Б = 375…400	Distance between control stick and soft backrest at medium bucket adjustment
В = 185…200	Size of the seat structure perpendicular to the tilt frame of the cockpit
Г = 800…	Length of the cockpit emergency exit without seat
Д = 850…950	Cockpit length in relation to centre backrest (distance between backrest and dashboard)
Е = 750	Distance from headrest to equipment panel
Ж = 210	Distance from sight line to the canopy outline
З = I + К	Cockpit height from floor to canopy outline
И = 1020…1120	Distance from the upper point of the canopy to the rear edge of the rigid member of the seat plane
К = 100…200	Distance from the floor line to the rear edge of the seat with the bucket in the lower position
Л = 180	Height adjustment of the seat
М = 750	Cockpit width with dropped canopy at crew member's shoulders level
Н = 700	Cockpit working width in the area of crew member shoulders
О = 900	Cockpit width at crew member's elbows
П = 600	Distance between vertical consoles at crew member's shin level
Р = 880	Distance from the rear edge of the seat in top position to the footboard
Angle 16…18°	Installation angle of the seat
Angle 5…10°	Seat plane inclination

In order to meet these requirements, the following recommendations should be followed (Fig. 37):

- make maximum use of in-flight unloading of units (e.g. by using structural fuel tanks in the wing to reduce the bending moment acting on it in flight);
- use straight and short load paths (for example, to use center wing, carry-through or, in a triangular wing, let the wing spars go perpendicular to the aircraft centerline);
- take the acting bending and torsional moments at the maximum arm (for which purpose, for example, the carrythrough passing through the "integral" fuselage can be designed with an increased structural depth);
- combine the functions of the structural members and ensure that they are fully functional in different load cases (for example, just two reinforced members in the afterbody of a subsonic jet transport can be used as the base for the VT spar mount, the adjustable HT cutout coaming, the base for the HT rotation unit mount, the base for the HT drive mount);

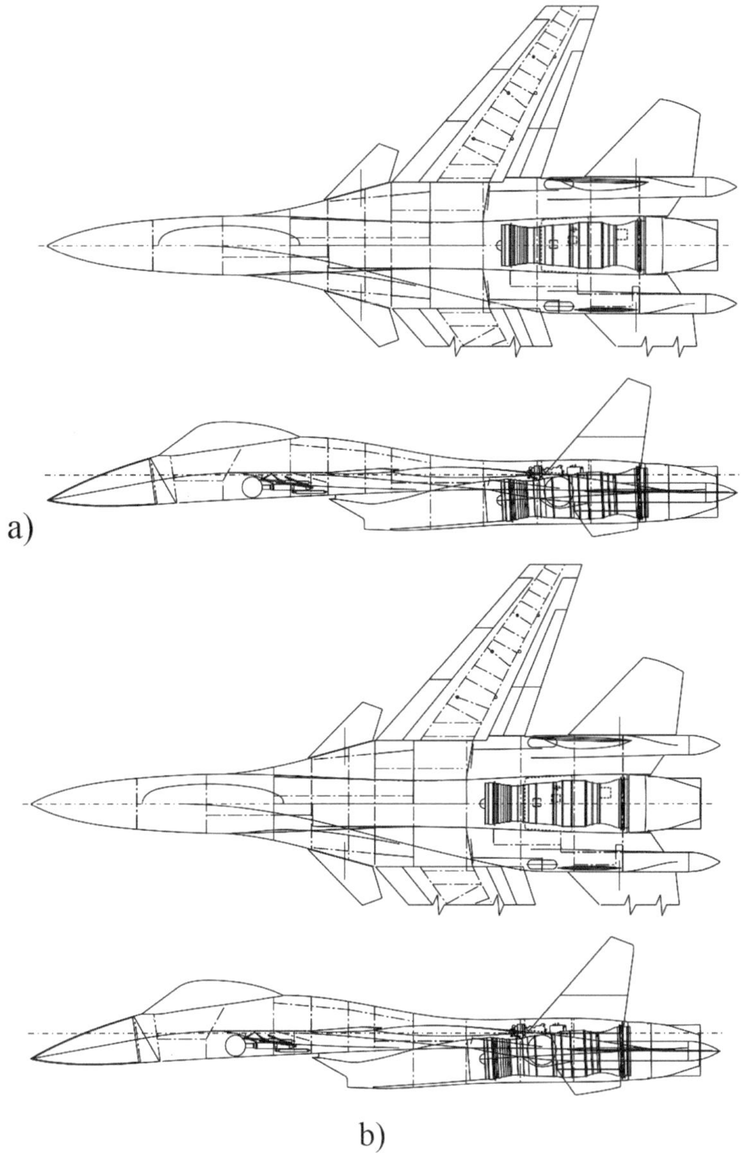

Fig. 37 Possible structural layouts of a fighter aircraft **a** and business class aircraft **b** (only the main structural elements are shown)

- combine production and maintenance breaks, as well as structural and maintenance compartments and hatches (the production break of a fighter afterbody serves a maintenance beak, since it is used for engine replacement, and in the landing gear well there are hydraulic system units that require frequent inspection and maintenance);
- hatches and cutouts in units should be made in less loaded areas, with their main axes aligned with the load path; avoid placing several cutouts in the same unit (for example, in forebody of a long-range airliner, avoid placing in the same section the landing gear well, entrance and galley doors);
- structural layout diagrams of individual units must be linked (for example, if a wing uses a torsion box, the fuselage must feature a center section).

In the structural layout of an airplane, partial relocation of components is possible to combine functions and reduce the number of structural members. As a consequence, the CG location may need to be determined once again for the new structural layout.

4.2 Square-Cube Law and Aircraft Structural Design Optimization

The development of aviation technology, the complication of tasks to be solved, the requirement to increase the economic efficiency of civil aircraft operation entail steady growth of the aircraft size and increase of its take-off and structural weight.

If the size of the aircraft increases, the area (wing, tail) and surface (fuselage and other parts) increase in proportion to the square of the linear growth factor. If the structural layout does not change, i.e. all its elements increase in the same ratio in width, height and length, then the mass of the structural elements grows in proportion to the cube of the linear growth factor. In this case, naturally, it is assumed that the stress value and distribution do not change.

If we assume that the wing loading$p$, thrust-to-weight ratio $(T/W)_0$, and the flight range determined by the fuel ratio $\overline{m}_{fuel}$, remain the same, the "square-cube" law can be formulated in the following form. *As the linear dimensions of the aircraft increase, the takeoff mass increases in proportion to the square of the linear growth factor, the mass of the structure increases in proportion to the cube of the linear growth factor, and the payload mass ratio decreases linearly.*

Using this law, some scientists predicted a weight barrier in aircraft design, i.e. some limit weight at which the empty mass ratio approaches unity with the payload mass ratio approaching zero. In fact, the structural mass ratio also increases, but rather slowly. Approaches preventing the mass increase:

- structural layout optimization;
- new materials with a lower specific weight and a higher level of allowable stress (if this level is allowed by fatigue strength requirements);

– refinement of methods for estimating aircraft static, dynamic and fatigue strength;
– reduction of aerodynamic loads.

Despite all these measures, the structural weight ratio still increases as the size of the aircraft increases. This makes the problem of optimizing structural layout especially important at the earliest stages of design.

This problem could be solved using finite element methods for analyzing the stress–strain state of structures and numerical methods for their optimization.

Speaking about the weight limits of an aircraft, one should not forget about its size limits associated with the operational constraints. An illustration of this is the experience of the introduction into service of the heaviest commercial airliner A-380 by Airbus.

4.3 Manufacturability and Maintainability in Aircraft Structural Design

One of the main objectives of structural design is to establish the maintenance and production breaks in the aircraft structure. On the one hand, the high-volume production require partitioning of the aircraft structure into individual assemblies in order to reduce the aircraft assembly cycles. On the other hand, each maintenance and production break is an additional mass and sometimes an additional stress concentrator leading to fatigue failure. For example, on one of the medium-range airplanes, the maintenance break between the inner wing and outer wing increases the wing weight by 350… 400 kg. And it may turn out that the increase in economical efficiency of aircraft operation due to the reduction of the number of breaks will be such that it is economically more feasible to build larger shops or plants than to make additional breaks in the structure.

In parallel with structural design, production documentation is being produced, since the production processes may also affect the structural design features. For example, the size of the wing panels often determines the distance between the wing spars and the spacing of production breaks. The production documentation reflects new trends in aircraft production and specific features of the given production plant (its production capabilities in terms of space, workshop size and equipment), the approximate number of the aircraft ordered, and anticipated aircraft operation conditions (for example, operation in tropical climates, etc.). The production documentation also sets limitations on the range of materials used in the design of the aircraft, the range of sections and sheets, fasteners (bolts, rivets, etc.).

To enhance maintainability, free access must be provided to areas where units and wiring of aircraft equipment and control systems are located, as well as to the structure to be inspected in aircraft operation. This requires the use of appropriate hatches or removable panels in the aircraft structure in addition to hatches providing accommodation for the target load (for passenger aircraft - in addition to entrance and service doors, baggage and emergency hatches and window openings).

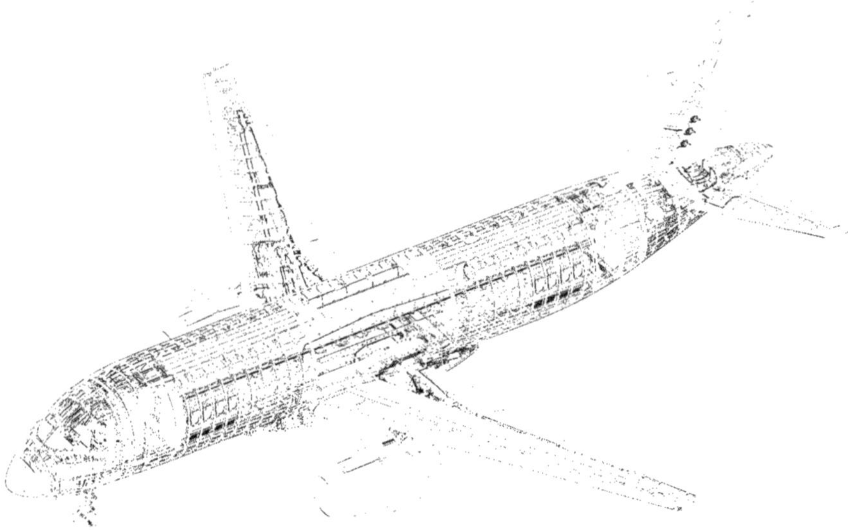

Fig. 38 Passenger aircraft layout

An example of a commercial airliner's layout is shown in Fig. 38.

4.4 Automation of the Aircraft Structural Design

Although in aircraft design, the structural design is one of the most complex creative processes, with the development of aviation technology, methods and software tools based on a high level of formalization of the design process have developed and are successfully being used.

The central place in structural design is occupied by a model representing elastic interaction of individual structural elements, which is based on the laws of solid deformable body mechanics. Along with numerous analytical methods, modern practice gives preference to the use of discrete mathematical models based on the finite element method (FEM) to analyze the stress–strain state (SSS). It makes it possible to achieve a high degree of adequacy in representing the properties of the investigated thin-walled structural systems with highly irregular properties (Fig. 39).

Modern software packages cover the whole structural layout process, from general configuration to detailed sizing. These packages may include modeling and executing tools, and the software for all kinds of multidisciplinary optimization (Fig. 40).

In the software, numerical and empirical analysis methods can be used. Among the first group the most widespread are methods based on the design of theoretically optimal structures and, in particular, the method of finding a structural layout delivering the minimum weight to the aircraft. In formal-heuristic design concepts the

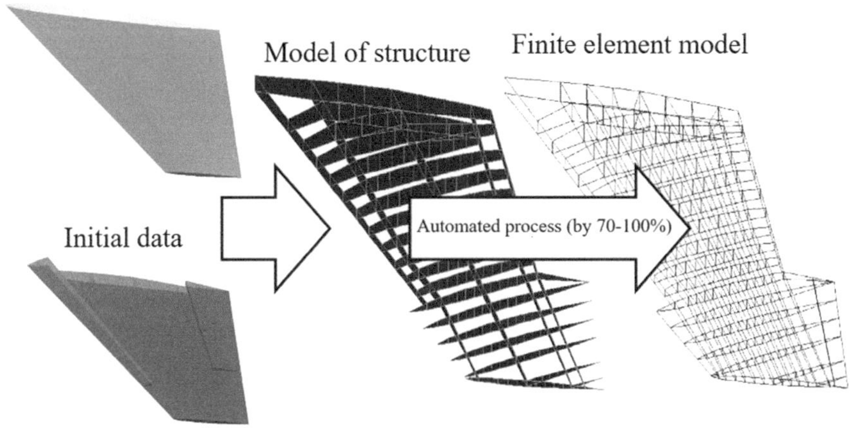

Fig. 39 Finite element models in automated structural layout systems [6]

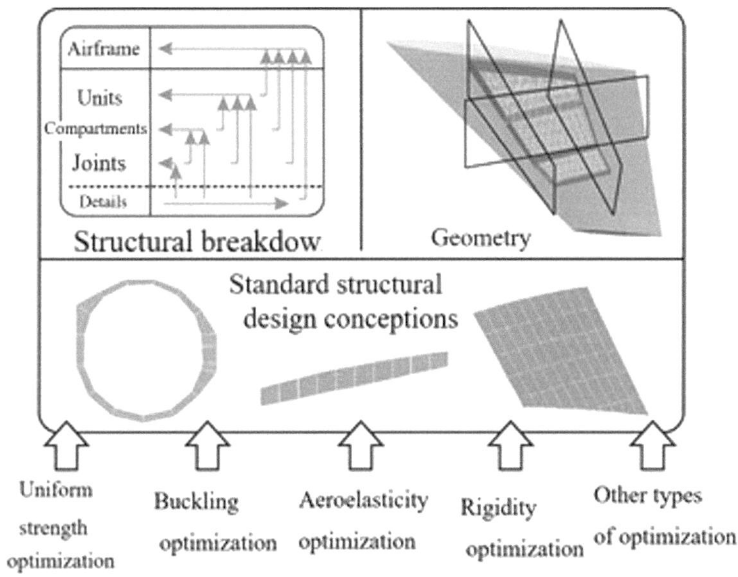

Fig. 40 Structural design software diagram

central place is given to the synthesis and analysis of a structural model to improve the selected measure of merit.

The existing codes are provided with powerful visualization tools that support synthesis of structural layout at any design stage, help investigate fundamentally new designs, and improve existing ones. As an example, Fig. 41 shows the stress–strain state of an arbitrary fragment of an aircraft structure.

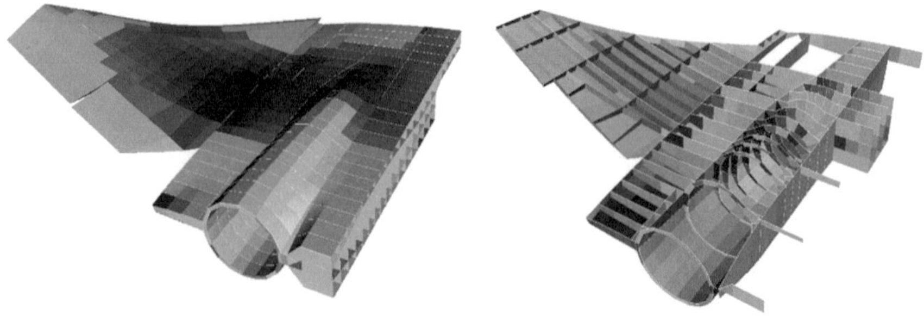

Fig. 41 Stress–strain state of an aircraft's structure. Shading shows stresses experienced by the structure

5 Shaping and Laying Out an Aircraft

5.1 Shaping an Aircraft

All three components of the sizing process eventually come together, culminating in aircraft's final shape.

Formation of the external contours of the aircraft is usually performed in two stages. At the first stage, the preliminary shaping is performed and the position of the main structural members is determined. At the second stage, the external shape of the aircraft is refined.

As a rule, nowadays, both steps are performed by means of a computer. Therefore, the described below shaping method using batt lines and reference lines, as well as the graphoanalytical method of determining the intermediate wing sections are of more historical than practical interest.

By **laying out the shape** we mean the graphic or analytical determination of the coordinates of the unit outline in a number of planes.

Preliminary shaping of a unit can be carried out using the butt and reference lines. First, draw the outline of the unit in side and plan views. Then, proceed to shaping using the diagram of combined sections, the side view of batt lines and the planform view of horizontal lines (Fig. 42). On the diagram of combined sections, the points defined by the outlines in the lateral and horizontal views (zero batt lines and half-breadths) should be plotted along the vertical and horizontal axes, and then these points should be connected by curves, defining the shape of the frames.

Shaping of wings, tail, pylons and other conical-shape units can be made both graphically and analytically.

The following input data are required: wing planform, spacing of ribs, collection of wing airfoils, and lengths of transition zones (from one airfoil to another).

For each airfoil, the family, chord and relative thickness must be known.

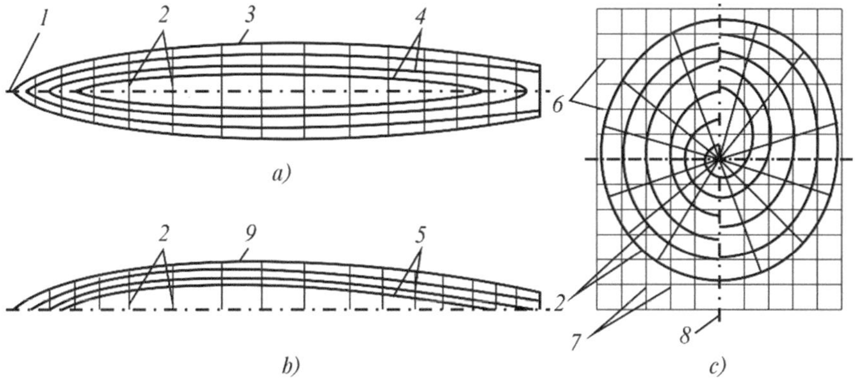

Fig. 42 Fuselage shaping using butt lines and reference lines: **a**—side view; **b**—planform view; **c**—diagram of combined sections; 1—fuselage reference line; 2—frames; 3—zero batt line; 4—batt lines; 5—half-breadth plan; 6—traces of the planes of the reference horizontal lines; 7—traces of the planes of the batt lines; 8—fuselage centerline and zero batt line plane; 9—half-breadth plan

The *graphical method* is based on the construction of a diagram of combined sections (Fig. 43). The root airfoil (in the simplest case) is drawn on the diagram in such a way that the rib planes are parallel to the plane of the diagram. After that, the lines of equal chord percentages, i.e. forming the conical surface of the wing, are drawn.

All lines of equal percentages of the chord are divided in proportion to the distance between the ribs, and through the corresponding points, smooth curves of the intermediate airfoils are drawn. The graphical method is of little use due to its insufficient accuracy.

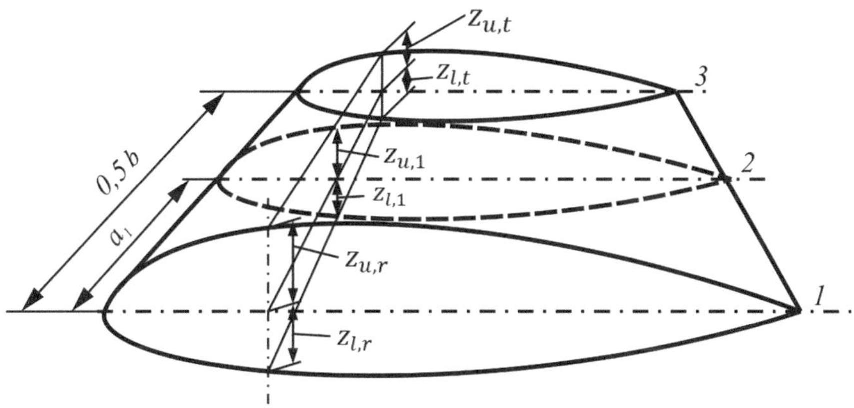

Fig. 43 Determination of the ordinates of an intermediate section of wing or tail: 1—root chord; 2—intermediate chord; 3—tip chord

The *analytical method* of shaping units with conical surfaces uses geometric properties of trapezoids composed by generator lines and ordinates of root and tip sections, with all required dimensions determined by calculation. Usually the upper (z_u) and lower (z_l) ordinates are determined separately.

The results should be entered in tables. The ordinates can be determined as follows:

$$z_{u,1} = z_{u,t}\frac{2a_1}{b} + z_{u,r}\left(1 - \frac{2a_1}{b}\right), \tag{16}$$

where $z_{u,t}$ is the ordinate of the tip section;$z_{u,r}$ is the ordinate of the root section;a_1 is the distance from the root section to the rib.

In the most advanced automated aircraft design systems, preliminary shaping is performed by the designer in an interactive dialogue mode with the computer. The designer can correct the outline of the aircraft surface directly on the screen, and the computer will take this correction into account when approximating the surface by one of the mathematical methods. In this case, the designer's experience and intuition are supplemented by a large memory capacity and the huge computational capabilities of modern computers.

After completing preliminary shaping of the aircraft surface, it is necessary to describe its shape mathematically. Let us consider the basic requirements for the corresponding methods.

Aerodynamic smoothness can be ensured by continuity of the first derivative. For critical units that significantly affect the flow characteristics of the aircraft as a whole (wings, forebody, engine nacelles, etc.) it is necessary to ensure the continuity of the second derivative and limited values of the third derivative in the longitudinal direction. The requirement for smoothness of aircraft cross sections is usually by an order of magnitude lower than the requirement for smoothness of longitudinal sections.

With respect to operational, structural, and production requirements, mathematical methods for formulating the outlines should provide:

- smooth inclusion of cylindrical, conical, parabolic and other surfaces such as cylindrical windshields of pilot cabins, wing and tail conical surfaces, special antenna fairings with specified shape, etc. into the aircraft surface;
- smooth inclusion of flat areas into the aircraft surface representing hatches for cameras and astronautical instruments, glazing of cabins, bases of special sensors;
- straightness of the given portions of some structural members, for example, the upper cap of a wing spar.

Along with the above requirements, mathematical methods for shaping aircraft contours should fully and unambiguously describe the surface of the aircraft. This requirement became obligatory recently in connection with wide introduction of CNC machines in aircraft production.

The variety of requirements for mathematical methods of defining the surface of aircraft units, as well as the diversity of types of units themselves, gave rise to many methods of outline description, of which the most commonly used ones at the present time are the following:

- second-order curve method;
- arc-of-circle method;
- power equations;
- spline functions;
- polynomial equations [1].

5.2 Automated Aircraft Shaping

The top-level design systems were initially created in the largest aerospace corporations (Lockheed (CADAM), Dassault (CATIA), McDonnell-Douglas (Unigraphics) and Matra (EUCLID)) [5].

Among domestic means of design work automation, CREDO, BPIO SIC ASK and others can be mentioned.

In contrast, the lower level systems (AutoCAD Mechanical Desktop and T-Flex by Top Systemes, etc.) have undergone twenty years of evolutionary development. During development they were focused on a wide range of industrial and other areas (geodesy, architecture, etc.).

Mid-level CAD-systems, as a fundamentally new class of software products, appeared in the early 90's. It was the era of Windows and UNIX competition. A breakthrough was made by RTS Corporation, which created the program Pro/Engineer such that the system was relatively independent of the hardware platform, so RTS was able to adapt it for Windows NT already in 1994. These systems significantly displaced both their "lighter" colleagues, working on PC (AutoCAD-like systems), and "heavier" complexes, originally oriented on workstations (Unigraphics, CATIA, etc.). CAD-systems of medium level have taken over good capabilities of solid-state and surface modeling from "heavy" systems, and from "light" ones—openness of interface and accessible price. The spheres of application of different graphic systems are shown in Fig. 44, and examples of their use are shown in Figs. 45, 46.

The open interface of midlevel systems has become a sufficient condition for development firms to actively start creating their own codes that use the 3D geometry developed in midlevel systems. Such a scheme provides the user with a very flexible solution for design, production and other tasks.

At the moment, the market of mid-level CAD-systems is represented by a number of software products, which work on PC in Windows environment, have an open interface and through parameterization. The main representatives here are packages with the geometric core Parasolid. SolidWorks, developed by the firm of the same name. Another well-known package—Solid Edge by Unigraphics Solutions—is offered as a lower level in the Solid Edge and Unigraphics suite. Systems that have also risen to this level include AutoCAD Mechanical Desktop and T-Flex from Top Systemes.

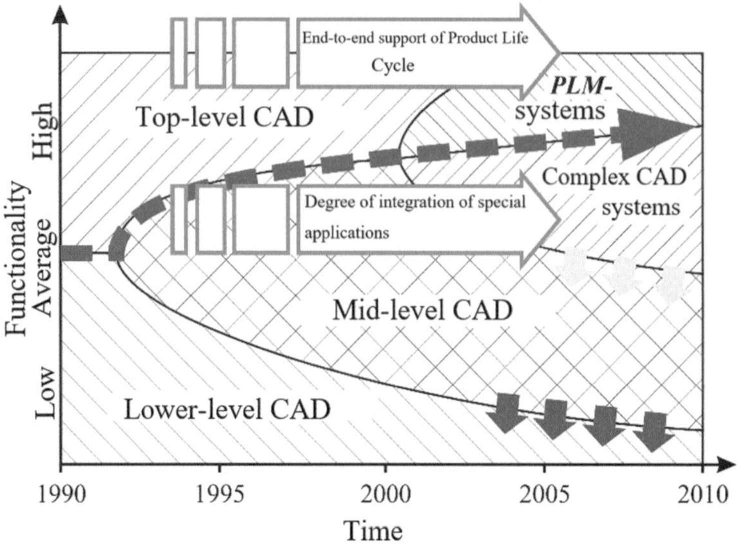

Fig. 44 Scope of various design systems

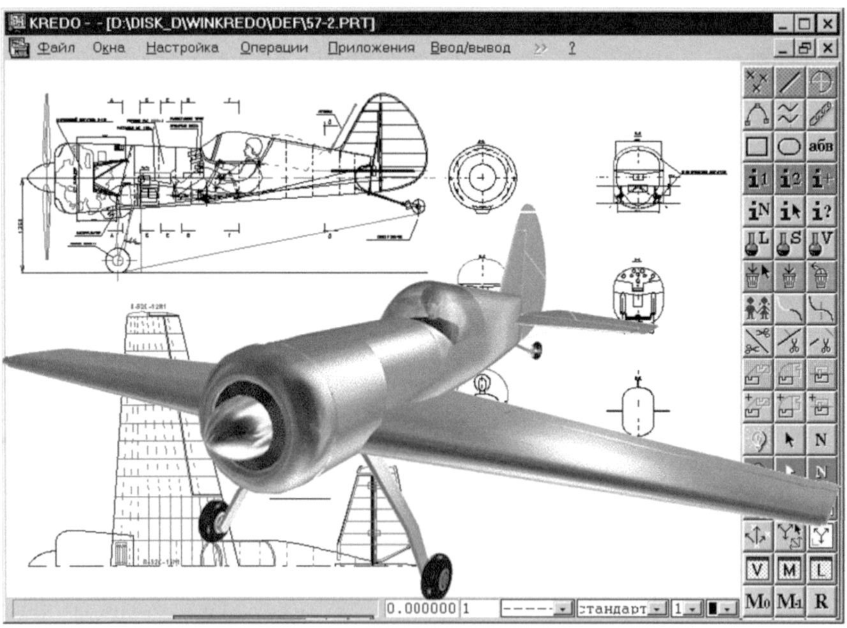

Fig. 45 Using a mid-level CAD system for sizing a light aircraft

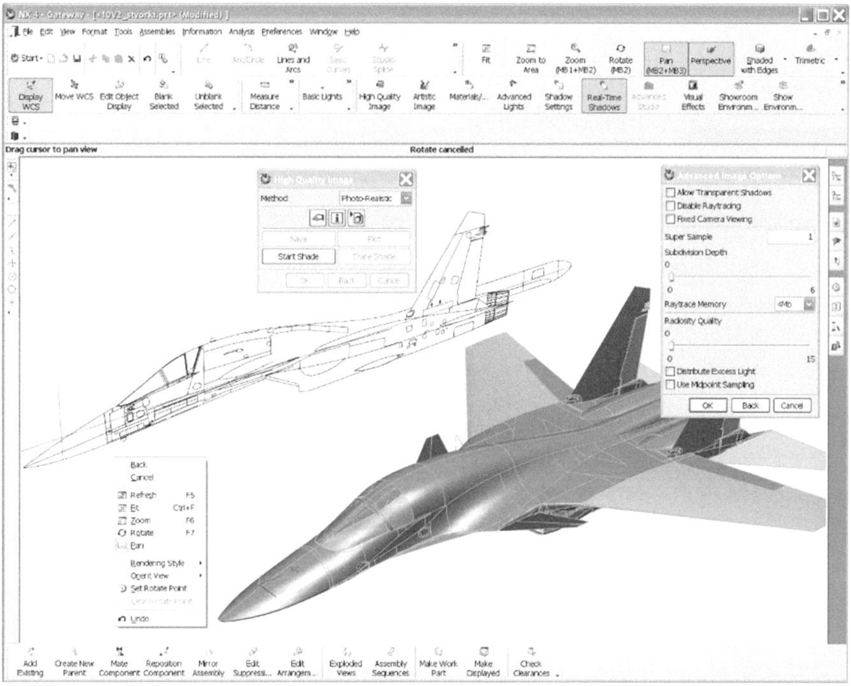

Fig. 46 Sizing and 3-D modeling a maneuvering aircraft

Their peculiarity is that they themselves are the pinnacle in the development of the "light" segment of CAD-systems.

5.3 *Construction of the Lines Drawing*

After the outline of the airplane unit has been defined and modeled on the computer, one may proceed to making the lines drawing (Fig. 47).

A lines drawing usually presents the structural layout of the unit and its large functional elements (in a dashed line).

The following requirements apply to the lines drawings:

- the drawing must present the unit layout, shown by a sufficient number of sections and views;
- the surface of the unit must be defined mathematically with unambiguous definition of any point on the surface;
- when specifying the unit in the local reference frame, the relationship of the local system to the main coordinate system of the aircraft must be specified;
- for a wing and tail, airfoil data shall be given in a dimensionless form;

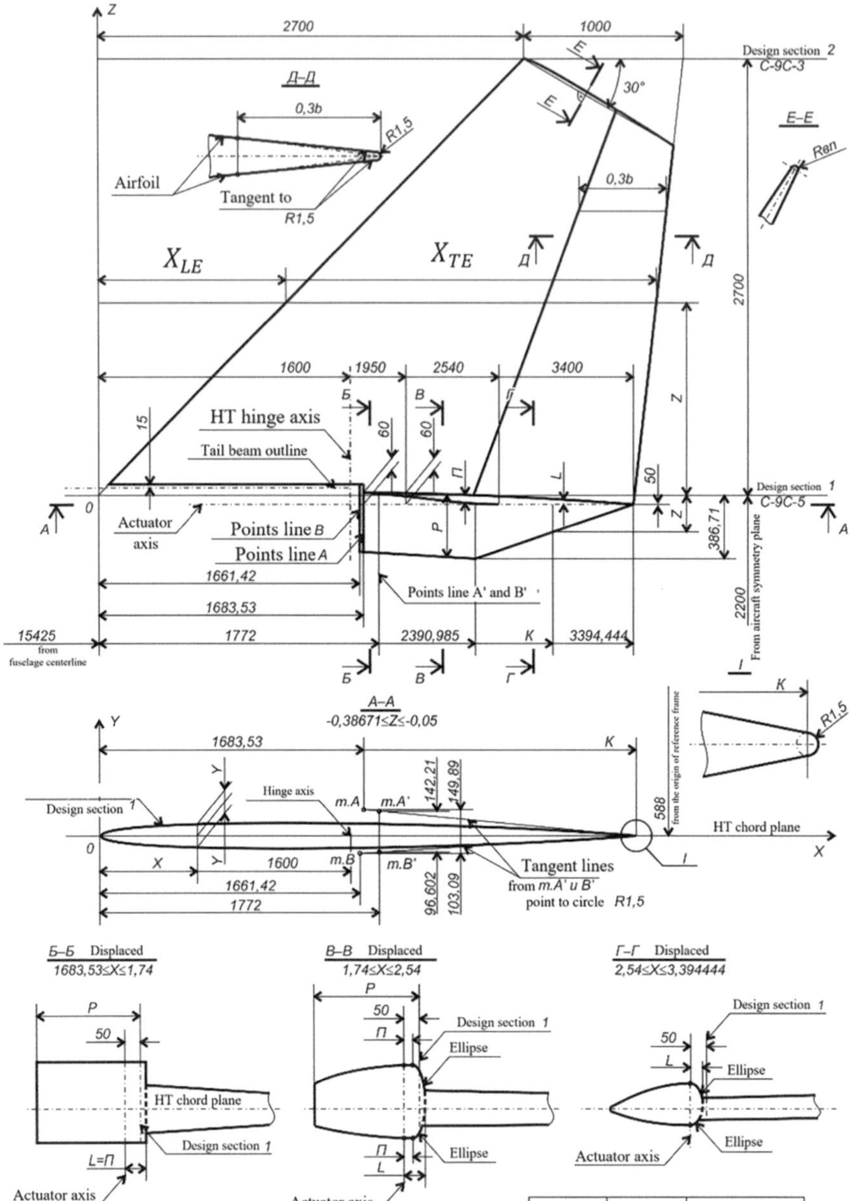

P	...	... ≤ X ≤ ...
L	...	... ≤ X ≤ ...
П	...	... ≤ X ≤ ...
b	...	... ≤ Z ≤ ...
K	...	... ≤ Z ≤ ...

Fig. 47 Lines drawing

– as additional information should be given geometric parameters of the unit required for aerodynamic calculations (wing aspect ratio and taper ratio, MAC, the nacelle intake diameter ratio, aspect ratio of the forebody and afterbody, etc.);

6 Sizing Documentation

6.1 Aircraft Layout Drawing

In the process of preliminary design and in the course of graduation thesis preparation, a layout drawing (diagram) of the aircraft should be developed. As a rule, it presents two views with a sufficient number of cross-sections reflecting the main layout and design solutions. The scale of the layout drawing should be as large as possible.

The layout drawing is primarily required to size the aircraft, for which purpose it depicts the position of as many functional elements as possible (crew, payload, equipment, etc.) and the structural members of the main units (landing gear, tail, fuel tanks, canopy, etc.).

In the process of aircraft sizing, necessary changes and additions are made to the aircraft layout drawing and then it is finalized. An example of the layout drawing of a short-range commercial airliner based on the Tu-334 aircraft as part of the graduation project is shown in Fig. 48.

The resulting layout drawing is one of the most important parts of the preliminary and conceptual design (or diploma project) and serves as source material for the development of the general view of the aircraft.

6.2 General View of the Aircraft

Based on the layout drawing, a general view of the aircraft is developed (not the other way around!). Usually, the general view drawing presents three views of the aircraft, specifying the overall dimensions and other dimensions of the aircraft, allowing the positioning of its units relative to each other. For example, for the wing and horizontal tail, span (for VT—its height), sweep, incidence angle should be indicated. The general view drawing also includes some data on aircraft performance (maximum landing angle, distance from the runway surface to the lower edge of the air intake, etc.). Figure 49 shows the general view drawing of the aircraft presented in Fig. 48.

During the detailed design process, the general view drawing is further refined and data required for aerodynamic and other aircraft analysis (surface areas of individual aircraft parts, their arms, etc.) are added to it.

Other aircraft top-level drawings are: overall view, landing diagram, configuration of hatches and antennas, breakdown diagram, etc.

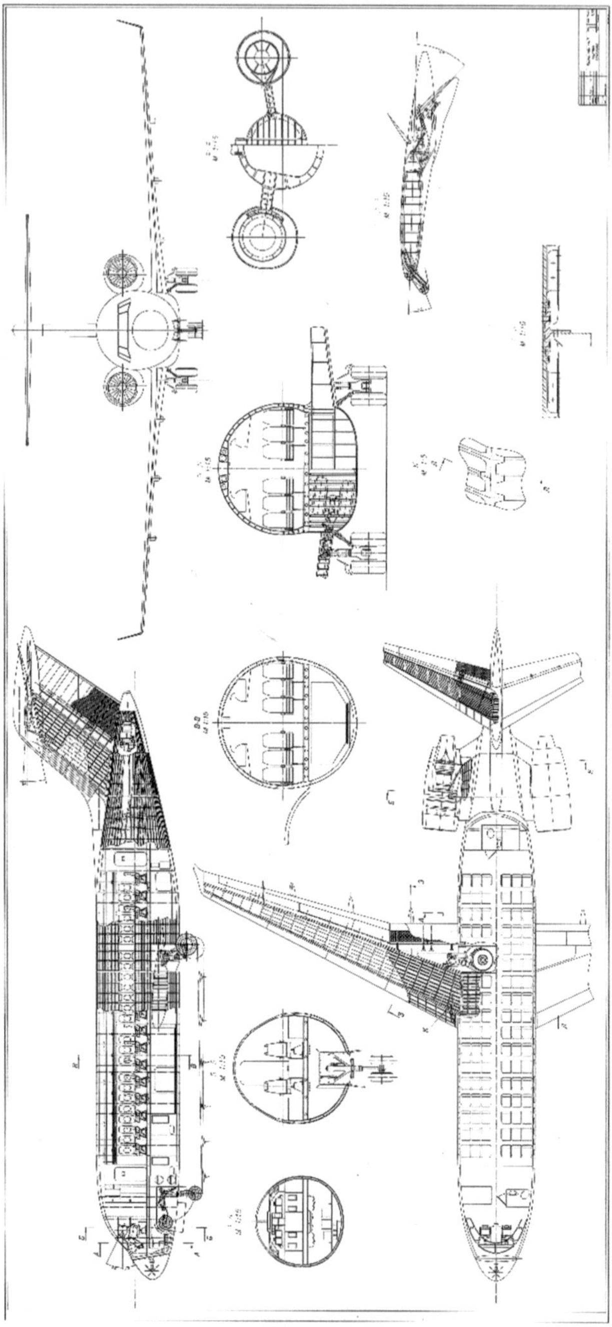

Fig. 48 Layout drawing

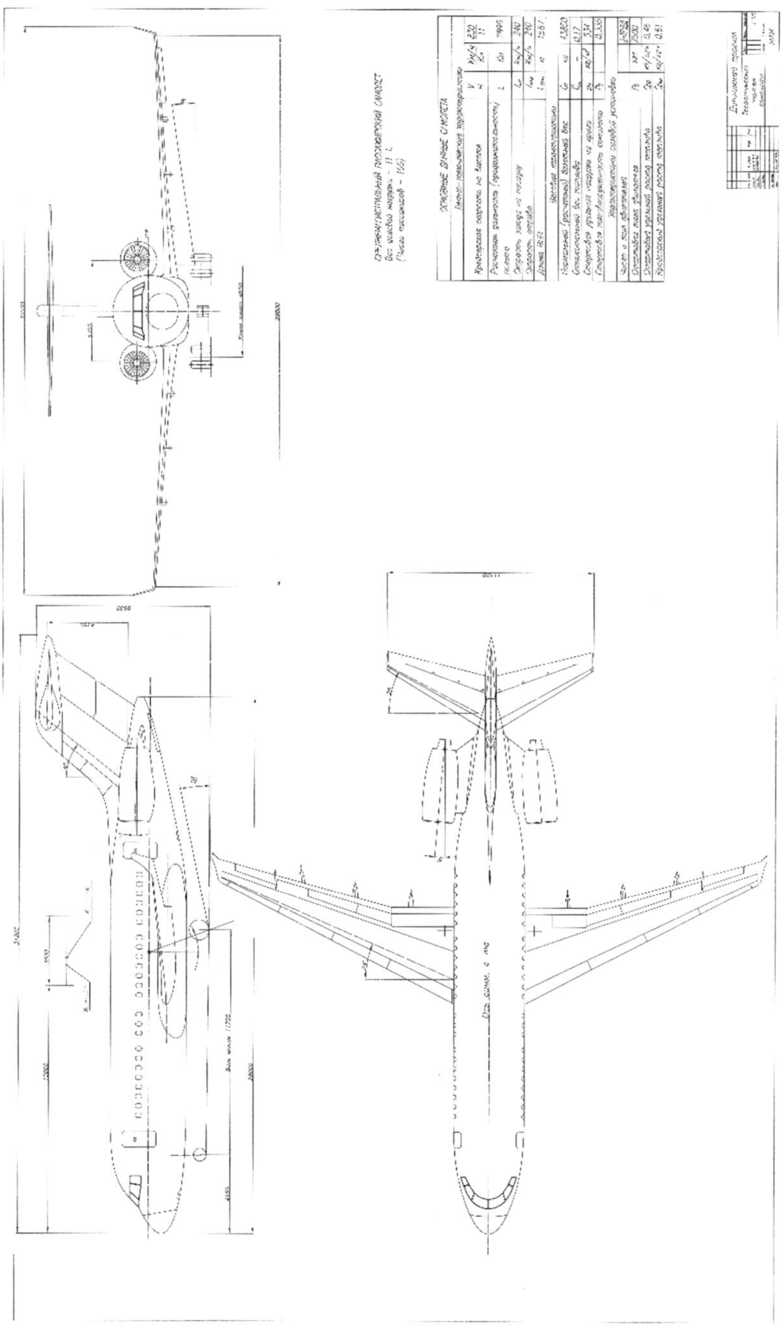

Fig. 49 Example of general view drawing

Variables

Symbol	Description	Units
$C_{L,max}$	Maximum lift coefficient	–
$A = 1/\pi eAR$	Polar slope angle	rad
e	Oswald efficiency factor	–
C_{mC_L}	Aircraft longitudinal static stability margin	–
$\left(C_{mC_L}\right)_{tail-off}$	Aircraft longitudinal static stability margin without a tail	–
S_{HT}	Horizontal tail area	m^2
l_{HT}	Horizontal tail arm	m
$C_{L_\alpha},_{HT}$	Horizontal tail lift curve slope	1/grad
$C_{L,w}$	Wing lift coefficient	
$\overline{S}_{HT} = \frac{S_{HT}}{S}$	Horizontal tail-to-wing area ratio	–
$\bar{l}_{HT} = \frac{b_{HT}}{c_{MAC}}$	Dimensionless HT moment arm	–
S_{HT}	Horizontal tail area	m^2
S	Wing area	m^2
b_{HT}	Horizontal tail arm	m
c_{MAC}	Wing mean aerodynamic chord	m
$TrP = \frac{(L/D)_{max,trimmed}}{(L/D)_{max}\left(C_{mC_L}=0\right)}$	Aerodynamic trim penalties	–
$C_{D,wave}$	Aircraft wave drag coefficient	–
C_L	Aircraft lift coefficient	–
$\bar{l}_{S,max} = \frac{l_{S,max}}{l_{body}}$	The distance from the body nose to the largest midsection in terms of length of the body	–
$\overline{X}_{CG} = \frac{X_{CG}}{c_{MAC}}$	Center-of-gravity location in terms of the Mean Aerodynamic Chord length	–
$\overline{X}_{AC} = \frac{X_{AC}}{c_{MAC}}$	Aerodynamic Center location in terms of the Mean Aerodynamic Chord length	–
$k_{HT} = \frac{q_{HT}}{q}$	The flow deceleration coefficient in the area of the horizontal tail	–
$V_{HT} = \frac{S_{HT} l_{HT}}{S c_{MAC}}$	Horizontal tail volume coefficient	–
$C_{L_{\alpha,HT}}$	Horizontal tail lift curve slope	1/grad
C_{L_α}	Aircraft lift curve slope	1/grad
ε^α	The derivative of the downwash angle at the horizontal tail with respect to the angle of attack of the wing	–
$C_{mC_{L,design}}$	Design static stability margin	–
$\overline{X}_{AC,tail-off}$	The wing-body combination aerodynamic center in terms of the MAC length	–
$\bar{l}_{HT} = \frac{l_{HT}}{c_{MAC}}$	Horizontal tail arm ratio	–

(continued)

(continued)

Variables		
Symbol	Description	Units
$\overline{S}_{HT} = \frac{S_{HT}}{S}$	Horizontal tail area ratio	–
$V_{VT} = \frac{S_{VT} l_{VT}}{Sb}$	Vertical tail volume coefficient	–
b	Wing span	m
$AR_{HT} = \left(\frac{b^2}{S}\right)_{HT}$	Horizontal tail aspect ratio	–
$\lambda_{\text{HT}} = \left(\frac{c_{tip}}{c_{root}}\right)_{HT}$	Horizontal tail taper ratio	–
$\left(\frac{t}{c}\right)_{HT}$	Horizontal tail sweep angle	–
Λ_{HT}	Horizontal tail sweep angle	degrees
$\overline{X}_{CG,aft}$	Aft CG limit in terms of the MAC length	–
$a_{HT} = C_{L_{\alpha,HT}}$	Horizontal tail lift curve slope	1/deg
$a_w = C_{L_\alpha}$	Wing lift curve slope	1/deg
$\overline{X}_{CG,f}$	Forward CG limit in terms of the MAC length	–
C_m	Pitching moment coefficient	–
$C_{m0,tail-off}$	Zero-lift pitching moment coefficient for a wing-body combination	–
$\Delta C_{m,HL}$	Pitching moment coefficient change due to high-lift devices extension	–
$\Delta C_{m,eng} = \frac{T_{eng} z_{eng}}{q_{unstick} S c_{MAC}}$	Pitching moment coefficient change due to engine thrust	–
$\alpha_{unstick}$	Unstick angle of attack	degrees
$\varepsilon_{0,HL-off}$	Downwash angle with the high-lift devices retracted	degrees
$\varepsilon^{\alpha}_{HL-off}$	Derivative of the downwash angle with respect to angle of attack (angle with the high-lift devices retracted)	–
$\Delta\varepsilon_{HL}$	Downwash angle change due to the high-lift devices extended	degrees
T_{eng}	Engine thrust	N
z_{eng}	Engine Z-coordinate	m
$q_{unstick}$	Unstick dynamic pressure	N/m^2
$\Delta\varepsilon_{gr}$	Downwash angle change due to the ground effect	degrees
$\overline{S}_{st} = \frac{S_{st}}{S_{HT}}$	Horizontal stabilizer area ratio	–
φ_{st}	Horizontal stabilizer angle	degrees
δ_{elev}	Elevator deflection angle	degrees
$n_{elev} = \sqrt{\overline{S}_{elev}}$	Elevator efficiency factor	–
$\overline{S}_{elev} = \frac{S_{elev}}{S_{HT}}$	Elevator area ratio	–
S_{HT}	Horizontal tail area	m^2

(continued)

(continued)

Variables		
Symbol	Description	Units
$k_{HT} = \frac{q_{local}}{q}$	Horizontal tail efficiency factor	–
ε	Downwash angle	degrees
C_m	Pitching moment coefficient	–
$\bar{l}_{HT}$	Horizontal tail arm in terms of the Mean Aerodynamic Chord length	–
c_{inb}	Inboard chord length	m
h_{HT}	Horizontal tail vertical offset	m
$V_{VT} = \frac{S_{VT} l_{VT}}{Sb}$	Horizontal tail volume coefficient	–
Γ	Wing dihedral angle	degrees
C_{L_β}	Effective dihedral	1/deg
C_{N_β}	Directional stability derivative	1/deg
$C_l^{\omega_z}$	Derivative of rolling moment coefficient with respect to angular velocity around z-axis	sec/deg
$a_{VT} = C_{Y_\beta}$	Derivative of side force with respect to sideslip angle	1/deg
l_{VT}	Vertical tail arm	m
S_{VT}	Vertical tail area	m^2
$n_{elev} \approx \sqrt{S_{elev}/S_{VT}}$	Rudder effectiveness	–
C_{l_β}	Effective dihedral	1/deg
W_{ymax}	Maximum crosswind speed	m/s
$C_{n_{\beta,VT-off}}$	Directional stability derivative for aircraft without VT	1/deg
$\chi = \frac{\lvert\overline{\omega}_{xmax}\rvert}{\lvert\overline{\omega}_{zmax}\rvert}$	Factor representing the ratio of rolling to yawing angular velocities	–
δ_{rud}	Rudder deflection angle	deg
N_{thrust}	Yawing moment due to unbalanced thrust	$N \cdot m$
I_x, I_z	Moments of inertia	
b	Wing span	m
$AR_f = \frac{h_f}{l_f}$	Fuselage aspect ratio	–
l_f	Fuselage length	m
h_f	Fuselage height	m
$k_{VT} = \frac{V_{VT}^2}{V^2}$	Vertical tail efficiency factor	–
C_{Y_β}	Side force derivative	1/deg
$a_{VT} = \frac{\delta C_{Y,VT}}{\delta\beta}$	Side force due to VT derivative with respect to sideslip angle	1/deg
$C_n^{\omega_z}$	Yawing moment coefficient derivative with respect to angular velocity about z-axis	sec/deg
$\overline{X}_{CG} = \frac{X_{CG}}{c_{MAC}}$	Scaled CG location	–

(continued)

(continued)

Variables		
Symbol	Description	Units
c_{MAC}	Mean aerodynamic chord length	m
$\overline{X}_{CG,f}$	Forward center of gravity limit	–, %
$\overline{X}_{CG,aft}$	Aft center of gravity limit	–, %
$\Delta\overline{X}_{CG}^{op} = \overline{X}_{CG,f} - \overline{X}_{CG,aft}$	Operational center of gravity envelope	–
X_{MAC} and Z_{MAC}	MAC X- and Z- coordinates	m
φ	Angle between MAC and horizontal plane XOY	degrees
$c_{MAC,HT}$	Horizontal tail mean aerodynamic chord	m
$c_{MAC,VT}$	Vertical tail mean aerodynamic chord	m
Λ	Sweep angle	degrees
$\Delta\overline{X}_{CG}$	Center-of-gravity shift parallel to X-axis	m
M_i	Mass of the shifted load	kg
M	Aircraft mass	kg
X_{i0}	Initial X-coordinate of a load	m
$X\prime_i$	New X-coordinate of a load	m
$\Delta\overline{X}_{FG}$	The necessary shift of the fuselage group relative to the wing group	m
$C_{m_{C_{L,min}}}$	Longitudinal static stability margin	–
S_{HT}	Horizontal tail area	m^2

References

1. Eger SM, Mishin VF, Liseitsev NK et al (2005) Proektirovanie samolyotov (Aircraft Design). Logos, Moscow
2. Eger SM, Liseitsev NK, Samoilovich OS (1986) Osnovy avtomatizirovannogo proektirovaniya samoletov (Fundamentals of Computer Aided Aircraft Design). Mashinostroenie, Moscow
3. Kuprikov MYu (eds) (2012) Komponovka samoletov (Aircraft layout), MAI, Moscow
4. Mal'chevskii VV (2011) Matrichno-topologicheskij metod sinteza skhemy i komponovki samoleta (opyt avtomatizacii tvorcheskoj deyatel'nosti konstruktora) (Matrix-topological method of synthesis of the scheme and layout of the aircraft (experience of automation of the creative activity of the designer), MAI, Moscow
5. Kuprikov MYu, Markin LV (2013) Tekhnologiya avtomatizacii proektirovaniya i podderzhki zhiznennogo cikla samoleta (Technology of automation of design and support of the aircraft life cycle), MAI, Moscow
6. Pen'kov EA (2007) Automation of the formation of complex structural layout. Polyot 1: 40–48

Relationships Between Aircraft Parameters and Performance. Ways to Improve Performance

Nikolai K. Liseitsev and Viktor A. Kliagin

Abstract This chapter introduces the concept of flight profile and shows what parameters describe each segment of a flight profile. The chapter offers equations for estimating the field performance, cruise, climb, and ceiling performance, as well as the maneuvering characteristics of an airplane.

Keywords Flight profile · Flight performance · Maneuvering characteristics

As shown, in aircraft design, the parameters ensuring the given flight performance and technical and economic characteristics are to be determined.

The desired performance is set with regard to the competitiveness in the global aviation market, or the need to solve newly emerging national economic or defense objectives.

It is clear that the solution of this problem requires the designer to have a full understanding of the relationships between the aircraft parameters and characteristics, the ability to analyze the consequences of various design decisions for all segments of the flight profile. A typical flight profile of a single-mode (transport) aircraft is shown in Fig. 1.

Each of the segments has its own specific requirements for aircraft parameters. Thus, for takeoff, the critical requirements are: takeoff from a runway of a given length, flotation on airfields of a given class, with the required takeoff speed V_2 and climb gradient in case of failure of one engine.

The cruising segment of the flight determines the desired thrust of the engine at a given altitude and flight speed, the amount of fuel required for the flight at a given range, and the wing area which ensures the flight at a C_L close to the C_L, corresponding to $(L/D)_{max}$.

Some aircraft parameters are affected by the landing requirements: landing gradient, maximum allowable approach speed and runway length limit. Critical are

N. K. Liseitsev (✉) · V. A. Kliagin
Moscow Aviation Institute (National Research University), Moscow, Russia
e-mail: nikolai.liseitsev@yandex.ru

M. A. Pogosyan (ed.), *Aircraft Design*, Springer Aerospace Technology,
https://doi.org/10.1007/978-981-96-4599-2_8

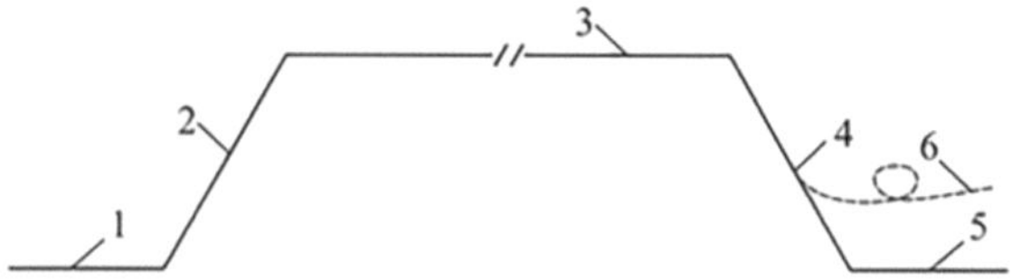

Fig. 1 Typical flight profile of a transport aircraft: 1—takeoff; 2—climb to H_{cruise}; 3—cruising; 4—descent; 5—landing; 6—loiter in the airfield area, leaving for an alternate aerodrome

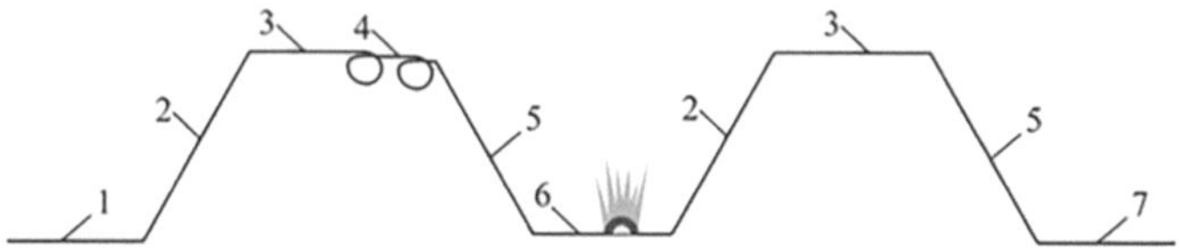

Fig. 2 Flight profile of a frontline attack aircraft: 1—takeoff; 2—climb; 3—cruising flight; 4—patrol; 5—descent; 6—overcoming air defense zone, combat; 7—landing

the cases of a crosswind landing, as well as a single-engine go-around from a height of 20 m at a speed of 1.2 V_{min} and a vertical climb speed of 2 m/s.

Even more challenging is the problem of selecting and matching the parameters of a multi-mode aircraft designed to conduct combat operations, such as a frontline attack aircraft. A possible flight profile for such an aircraft is shown in Fig. 2.

The need to fly at subsonic and supersonic speeds at high and low altitudes, to overcome the low-altitude air defense zone makes it extremely difficult to choose the parameters of such an aircraft.

In view of the above, let us consider the main relationships between the parameters and flight performance of the aircraft.

1 Field Performance

It is known that the field performance of an airplane primarily depends on its minimum flight speed. Its allowable value can be derived from the equality of aircraft lift to its weight:

$$L = m_0 g \rightarrow C_L \rho \frac{V^2}{2} S = m_0 g, \text{ hence } V_{\min,allow} \sqrt{\frac{2p}{C_{L,allow}\rho}}, p = \frac{m_0}{S}.$$

Note that the above equations do not reflect the influence of the vertical component of the thrust of the engines on V_{min}, which is noticeable when the thrust vector is deflected.

As follows from this formula, with the given wing loading and the air density, $V_{\min,allow}$ is the function of the lift coefficient, which depends on the wing configuration, including twist, as well as the type of the high-lift devices.

The above formula corresponds to the wing stalling speed. According to the Aviation Rules (AR), an aircraft should take off at a safety speed V_2, which is 20% higher than V_{min} with 2 or 3 engines and 15% higher than $V_{\min,allow}$ with 4 engines. In practice, this means that the maximum available C_L meeting the safety requirements is $C_{L,V_2} = C_{L,max,TO}/1.44$ for aircraft with 2 or 3 engines and $C_{L,V_2} = C_{L,max,TO}/1.3225$ for aircraft with 4 engines.

Even more stringent minimum airspeed constraint is associated with landing. Thus, for civil aircraft the landing approach speed (V_{app} speed at the runway edge at 15 m height) must be 30% higher than V_{min} in landing configuration, i.e. $C_{L,app} = C_{Lmax,land}/1.69$.

A typical lift curve for a modern subsonic passenger aircraft (Tu-204) with a supercritical wing in the main flight modes is shown in Fig. 3.

For military aircraft, all flight modes, including takeoff and landing, should be performed at C_L not exceeding $C_{L,allow}$, i.e., within the linear range of lift and moment curves (Fig. 4).

Let's take a closer look at the field characteristics.

(a) Takeoff

A full takeoff includes acceleration from a stationary position on the runway to the unstick speed (ground roll) and accelerated flight to an altitude of $H = 400$ m.

Full takeoff is divided into two parts: takeoff and initial climb (Fig. 5).

The following speeds have been established in accordance with the airworthiness standards for aircraft:

- the speed at which failure of the critical engine occurs, V_1;
- the safe take-off speed, i.e. the lowest speed at which
- the initial climb is allowed, V_2;
- the speed corresponding to the start of high-lift devices retraction, V_3.

During takeoff, the aircraft accelerates to V_{LOF}, detaches from the runway and climbs until it reaches V_2 and $H = 10.7m$ (obstacle clearance height). The distance covered during this period is called the take-off distance L_{TO}. It is equal to the sum of the takeoff ground run distance L_{GR} and the length of the first climb segment L_1, $L_{TO} = L_{GR} + L_1$. As a rule, landing gear retraction begins on the first climb segment, usually at an altitude of 3–5 m. During the second climb segment, the aircraft accelerates to the speed of safe climb with the landing gear retracted. It ends at the end of landing gear retraction.

During the third takeoff segment, the acceleration continues with the aircraft reaching the altitude of the retraction of high-lift devices $H_{HL} = 120$ m and the speed of the retraction of high-lift devices V_3. During the fourth segment, the retraction of high-lift devices (transition from the takeoff configuration to the cruising configuration) is carried out with reaching an altitude $H = 400$ m above runway level and a speed $V \geq 1.25V_{st}$ that should be equal to or greater than the control

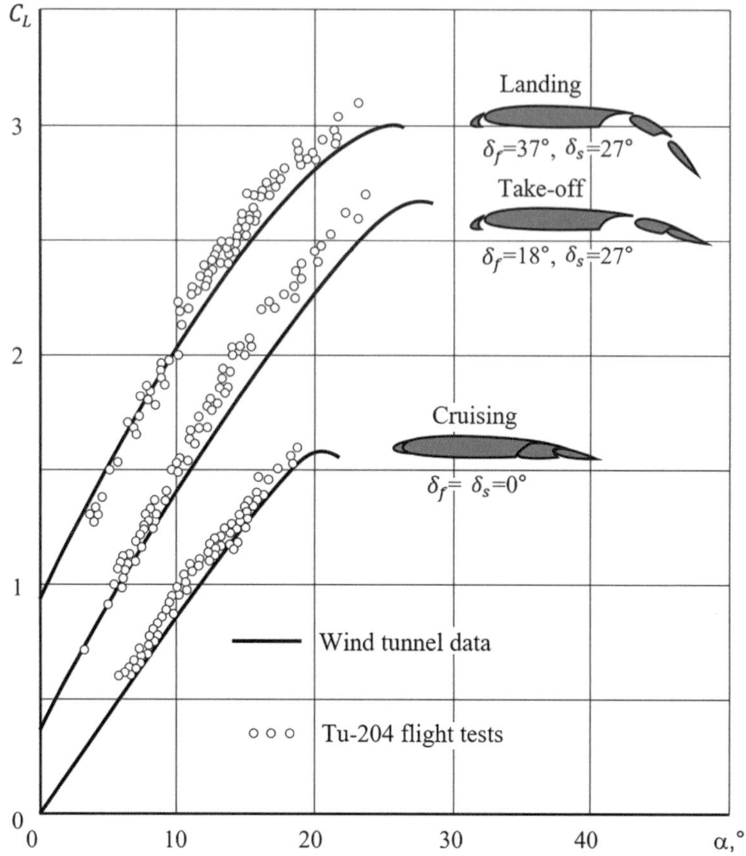

Fig. 3 Lift curve $C_L = f(\alpha)$ for an airliner with a supercritical wing ($\Lambda = 28°$, $AR = 10$) in cruising, take-off and landing configurations

speed V_c (V_{st} is the stalling speed in the cruising configuration). Control speed is the speed at which the basic maneuvers of the aircraft in question can be performed. For example, for heavy aircraft, it is determined as follows: when turning at this speed with a roll angle of 15°, the required C_L must not exceed the highest value allowed in normal flight operation of the aircraft $C_{L,allow}$ under the given flight conditions.

The value V_c can be determined using the following relation:

$n_z W = C_{L,allow} \frac{\rho_H V_c^2}{2}$,

where $V_c = 3.6\sqrt{\frac{2n_z G}{\rho_H C_{L,allow} S}}$.

Assuming that $n_z = \frac{1}{cos 15°} = 1.04$, we obtain $V_c = 5.18\sqrt{\frac{p}{\rho_H C_{L,allow}}}, p = \frac{W_{aircr}}{S}$.

Therefore, for $H = 400m$ $V_c = 14.9\sqrt{\frac{p}{C_{L,allow}}}$.

Let us explain the meaning of the speed V_1. The takeoff of a two- or multi-engine aircraft, in the course of which failure of one engine occurred, is called an engine

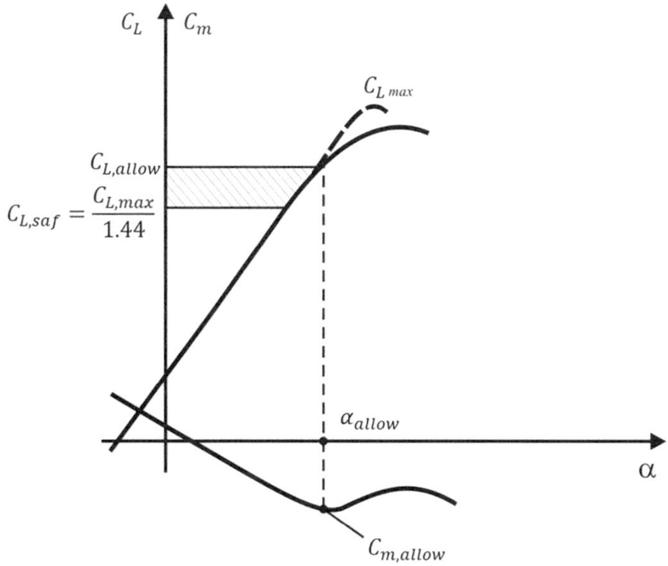

Fig. 4 Allowable and safety C_L

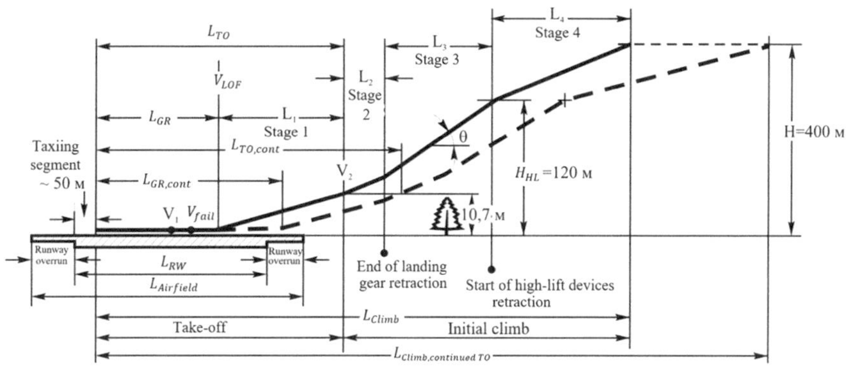

Fig. 5 Nomenclature for T-O segments

failure takeoff. And the most dangerous is the failure of the critical engine, i.e., the one whose failure causes the most adverse consequences in the behavior of the aircraft (usually, this engine is the farthest from the symmetry axis of the aircraft). If the critical engine fails on takeoff, depending on the speed at which it occurs, the takeoff may be aborted or continued. An *aborted takeoff* is a takeoff that begins as a normal takeoff and ends with the aircraft braking to a complete stop. The distance the aircraft travels from the beginning of the takeoff to its full stop is called the *distance of the aborted takeoff*.

A *continued* takeoff is a takeoff that begins as a normal takeoff and ends with an inoperative critical engine, the failure of which occurred at a speed of$V_{fail} < V_1$. The distance covered by the aircraft from the beginning of the ground roll to $= 10.7m$, is called the continued take-off distance$L_{TO,cont}$. There is some speed at which the distances of aborted and continued takeoff are equal. This speed is called balanced takeoff speed$V_{TO,bal}$. For each aircraft type, it depends onW_0, atmospheric conditions, runway surface condition. The takeoff distance corresponding to the continued takeoff with critical engine failure at a balanced takeoff speed is called *the balanced takeoff distance*$L_{TO,bal}$.

The runaway length for a continued takeoff with critical engine failure at the balanced speed is called a balanced ground roll length $L_{GR,bal}$. The notion of balanced take-off length is usually used for passenger aircraft, while balanced ground roll length is used for other aircraft.

As noted, the airworthiness standards require the speed at the end of takeoff segment 1 to be $V_2 > 1.2V_{st}$ for aircraft with 2 and 3 engines and $V_2 > 1.15V_{st}$ for aircraft with 4 or more engines.

The rotation speed is $V_{rot} = 1.05V_{st}$.

The aircraft getaway speed at $V_{GA,min} \cong 1.1V_{st}$ if $n_{eng} = 2; 3$ $V_{GA,min} \cong 1.07V_{st}$ if $n_{eng} \geq 4$.

Let us analyze these requirements. Since$= C_L S \frac{\rho V^2}{2}$, for the standard conditions at $H = 0m$ $W = C_L S \frac{V^2}{16}$ and, therefore, $V_{st}^2 = 16 \frac{W_0/S}{C_{L,max,TO}} = 16 \frac{p_0}{C_{L,max,TO}}$.

Then if $n_{eng} = 2; 3$ $V_2^2 = (1.2V_{st})^2 = 23 \frac{p_0}{C_{L,max,TO}}$.

If $n_{eng} \geq 4$ $V_2^2 = (1.15V_{st})^2 = 21.1 \frac{p_0}{C_{L,max,TO}}$.

This means that the allowable C_L at the first segment of takeoff which is $C_{L,V_2} = C_{L,max,TO}/1.44$, is significantly less than the maximum possible with the chosen type of high-lift devices (Fig. 6).

The analysis of the above relations shows that with the same aerodynamic characteristics, the higher is the wing loading, the higher is the takeoff speed, and consequently, the higher is L_{TO} which requires an increase in the runway length and complicates the takeoff. On the other hand, an increase in $C_{L,max,TO}$ makes it possible to provide the specified takeoff and landing characteristics with a smaller wing area S or a larger wing loading (reduction of W_{str}). With the $C_{L,max}$ increasing from 1.8 to 2.0 the wing area of an airliner with the take-off weight of ~ 100 tons reduces from 200 m^2 to 178 m^2. Assuming that $\overline{m}_w \cong 0.1$ and the mass of 1 m^2 of the wing structure is 45 kg, this will reduce the take-off weight of the airplane by 990 kg (assuming that all other units and systems of the airplane remain unchanged).

This shows the importance of wing high-lift devices and especially of the slats.

The ground roll distance can be determined as follows:

$$L_{GR} = \frac{V_{GA}^2}{2g} \frac{1}{\left(\frac{T_{av}}{W_0} - \frac{2f_{GR}}{3} - 1/(L/D)_{GR}\right)}.$$

If we assume $V_{GA} = 1.15V_{st}$ and $V_{GA}^2 = 23.6 \frac{p_0}{C_{L,max,TO}}$,

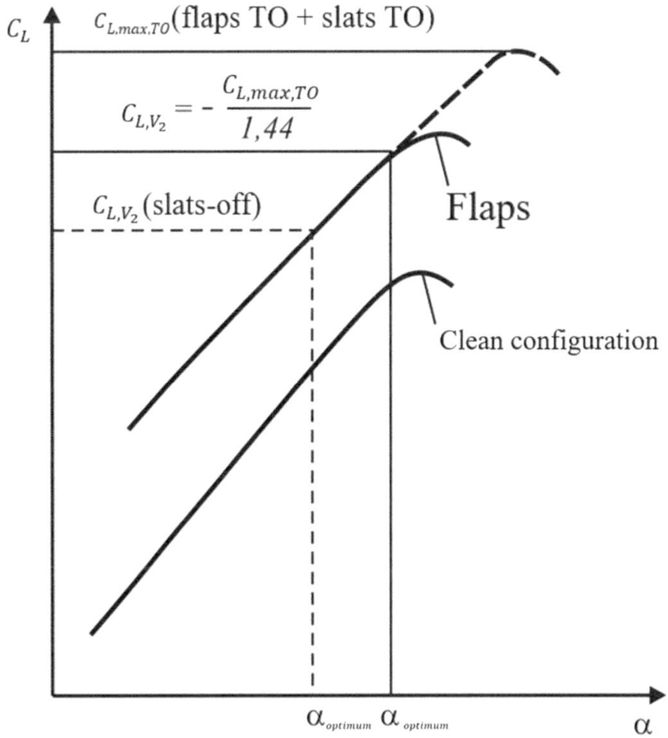

Fig. 6 Effect of wing high-lift devices on $C_{L,max,TO}$

then $L_{GR} \cong \frac{1.2}{C_{L,max,TO}} \frac{p_0}{\overline{T}_{av} - 1/3\left(2f_{GR} + 1/(L/D)_{GR}\right)}$, where $\overline{T}_{av} = \frac{T_{av}}{W_0}$.

For preliminary calculations it can be assumed $L = 5 \ldots 6$ for supersonic aircraft and 8 … 10 for subsonic aircraft. The values are f_{GR} given in Table 1.

Let us represent the impact of $\overline{T}_{av}$ on L_{GR} graphically (Fig. 7).

The figure shows the change of the ground roll distance with increasing $\overline{T}_{av}$ and $(L/D)_{GR}$ by 20% as a function of thrust loading.

Table 1 Friction coefficient at ground roll

Runway	f_{GR}
Dry concrete	0,02
Wet concrete	0,03
Hard soil	0,04
Wet grass	0,06
Grass	0,08
Rolled snow or ice	0,02

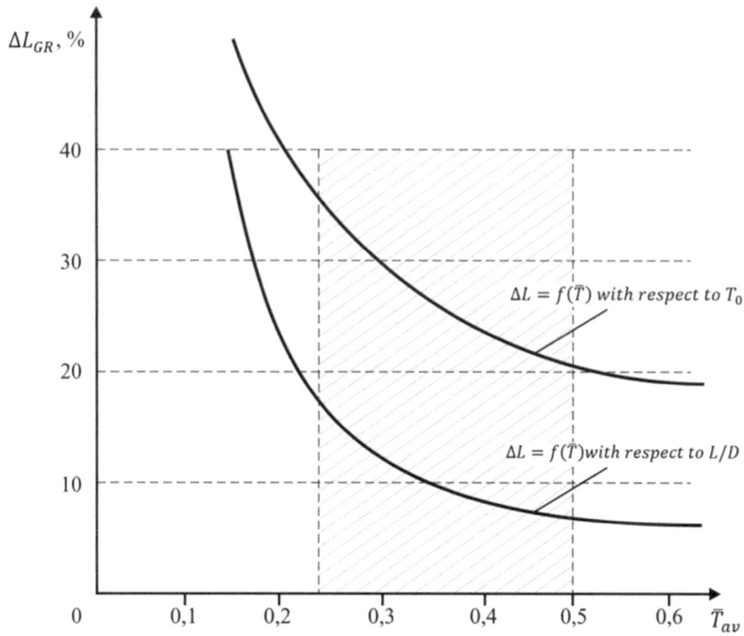

Fig. 7 Effect of thrust-to-weight ratio on ground roll distance

As can be seen from the graph, the effect of thrust loading and lift-to-drag ratio increases with decreasing thrust loading. With the typical values of thrust loading $\overline{T}_0$ for a commercial jet (0.25 … 0.35, M < 1), a 20% increase in thrust loading decreases the ground roll distance by 36 … 23%; a 20% increase in lift-to-drag ratio decreases the ground roll distance by 18 … 6%. That is why the main factor determining the aircraft take-off ground roll distance is the thrust-to-weight ratio.

The typical values of $C_{L,max}$ for take-off and landing configurations for subsonic and transonic aircraft with wings of various sweep, are shown in Fig. 8.

(b) Climb

The climb requirements are also governed by the airworthiness standards, in particular the climb gradient. The total climb gradient with an inoperative critical engine must be $\eta_{t,climb} \geq tg\theta \cdot 100\%$. With all engines running, $\eta_{t,climb} = 5.0\%$ during segment 3 and $\eta_{t,climb} = 3.0\%$ during segment 4.

The total climb gradient at different TO segments as a function of the number of engines is presented in Table 2.

From the diagram of forces shown in Fig. 9,

$$T_{climb} - W_0 sin\theta = C_D q S,$$

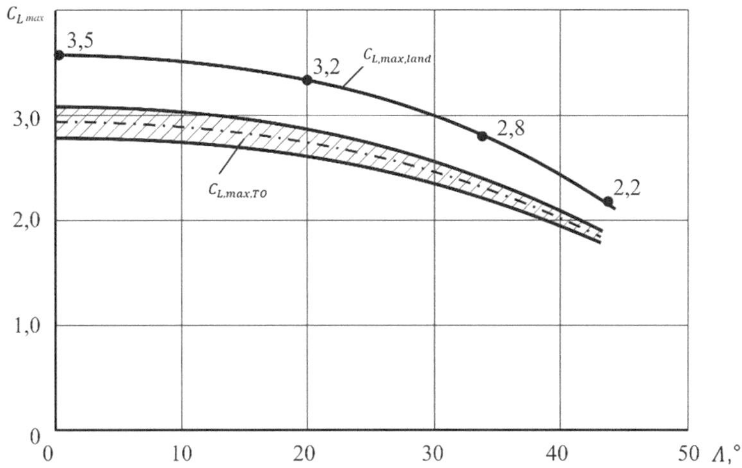

Fig. 8 Typical values of $C_{L,max}$ at take-off and landing

Table 2 Total climb gradient at different TO segments as a function of the number of engines

Number of engines on the aircraft	Total climb gradient $\eta_{t,climb}$, %			
	Segment 1	Segment 2	Segment 3	Segment 4
2	0,0	0,5	2,4	1,2
3	0,5	1,1	2,7	1,5
4	0,5	1,3	3,0	1,7

Fig. 9 Diagram of forces acting on a non-maneuvering aircraft during takeoff

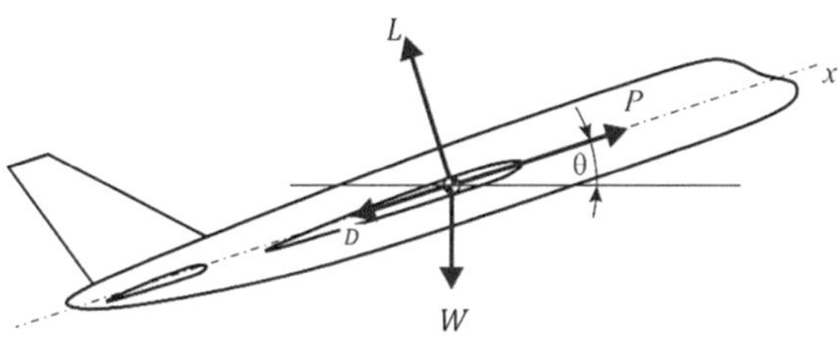

$$W_0 cos\theta = C_L q S.$$

As follows from the table, θ_{climb} is small, so we can assume that $cos\theta \approx 1$, and $sin\theta \approx tg\theta$.

Dividing one expression by the other, we obtain.

$T_{climb} = W_0\left(\frac{1}{(L/D)_{climb}} + tg\theta\right)$ or $\overline{T}_{climb} = \frac{1}{(L/D)_{climb}} + tg\theta$.

According to the standards:

(a) the high-lift devices are in takeoff position and this determines $(L/D)_{climb}$;
(b) the thrust of the engines corresponds to the design case of$t^{\circ}_{climb} = \pm 30^{\circ}$, $p_{atm} = 750 mmHg$, $M = 0.3 \ldots 0.4$.

The thrust-to-weight ratio is selected to ensure the required climb gradients for the case of a continued takeoff with one failed engine.

Taking this into account, the climb thrust-to-weight ratio should be or $\overline{T}_{climb} = \frac{n_{eng}}{n_{eng}-1}\left(\frac{1}{(L/D)_{climb}} + tg\theta\right)$, where n_{eng} is the number of engines, and the take-off thrust-to-weight ratio can be determined as follows:

$$\overline{T}_0 = k_V \frac{n_{eng}}{n_{eng}-1}\left(\frac{1}{(L/D)_{climb}} + tg\theta\right).$$

k_V is the factor associated with the change in the take-off thrust due to the particular takeoff conditions. Its value can be determined from the corresponding altitude-speed performance of the given engine. For approximate calculations, assume $k_V = 1.3 \ldots 1.5$. Then, for the configuration with the high-lift devices deflected to the takeoff position, the landing gear retracted, we obtain the following equation:

for a twin-engined aircraft, $\overline{T}_0 = k_V 2\left(\frac{1}{(L/D)_{climb}} + 0.025\right)$;

for an aircraft with three engines, $\overline{T}_0 = k_V 1.5\left(\frac{1}{(L/D)_{climb}} + 0.027\right)$;

for an aircraft with four engines, $\overline{T}_0 = k_V 1.33\left(\frac{1}{(L/D)_{climb}} + 0.03\right)$.

This means that, with the same $(L/D)_{climb}$, for example,$(L/D)_{climb} = 10$, the required take-off thrust-to-weight ratio of a twin-engined aircraft will be 23.8% higher than that of a three-engine aircraft, and 30.7% higher than that of a four-engine aircraft. Since the weight ratio of propulsion system $\overline{m}_{PP}$ is directly proportional to the take-off thrust-to-weight ratio, with equal engine weight ratios, the aircraft with the greater number of engines will have an advantage. In addition, an airliner with more engines have a smaller climb lift-to-drag ratio, therefore, requires more efficient high-lift devices and a larger $\Delta C_{L,max}$. Nevertheless, the majority of modern airliners, including intercontinental airliners, are equipped with two engines. An explanation of this fact is given in Chap. 10.

(c) Landing approach and landing

Landing is a segment of flight profile, which has a significant impact on the choice of the design parameters of an aircraft, since this choice affects not only the runway length required for landing and, consequently, the possibility of operating from airfields of a certain class, but also the flight safety. It is known that at least 70% of all accidents are caused by landings.

The landing approach and landing conditions of civil aircraft are also regulated by the airworthiness standards. The normalized landing pattern is shown in Fig. 10.

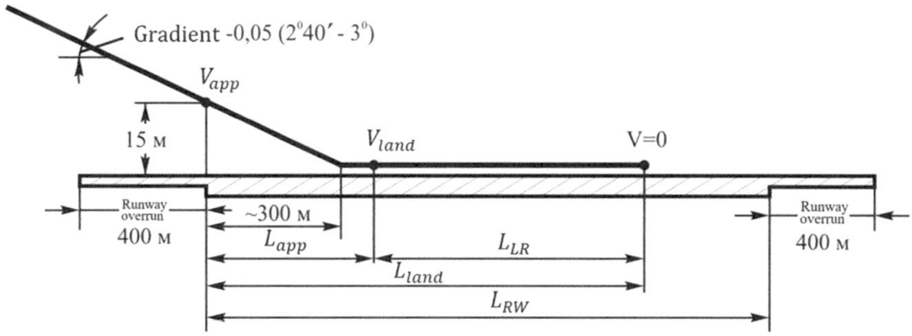

Fig. 10 Aircraft landing

The aircraft approaches at a gradient of approximately 0.05(24°0′ ... 3°), reaching the approach speed V_{app}. This speed is usually specified in the aircraft design specification. Its value $V_{app} \cong 210 \ldots 230 km/h$ is determined for the automatic approach according to ICAO category II and III *(ICAO—International Commercial Aviation Organization)*: $V_{app} \geq 1.3V_{land,min}$. This is a very stringent requirement, which often determines the wing area (wing loading) and the requirements for high-lift devices. Indeed, since $V_{min,land} = \sqrt{\frac{2W_{land}/S}{C_{L,max,land}\rho}}, V_{app} = 1.3\sqrt{\frac{2W_{land}/S}{C_{L,max,land}\rho}}$, the required area will be $S = \frac{2W_{land}}{\rho V_{app}^2 C_{L,max,land}/1.69}$.

The value $C_{L,max,land}/1.69$ determines the required value of $C_{L,safe}$ at landing. The equation shows that the required wing area is determined by the value W_{land} and $C_{L,max}$. The larger is $C_{L,max}$, the smaller is the landing mass, and consequently, the less is the mass of empty airplane. The effect of high-lift devices and especially slats on $C_{L,max,land}$ is illustrated by the graph in Fig. 11.

This necessitates their use on most modern aircraft. The graph also illustrates the effect of the flaps on ΔC_L. The more effective they are, the smaller is the approach angle of attack, and the better is the view from the cockpit.

An important aspect in determining the aircraft parameters meeting the landing constraints is the choice of its landing mass m_{land}. The assumption that $m_{land} = m_0 - m_{fuel} = m_0(1 - \overline{m}_{fuel})$ is true for a landing with fully burned fuel and does not reflect the reality due to the need to have the ICAO regulated aeronautical fuel reserve to await landing in the airfield area or reach an alternate airfield in case of unforeseen circumstances. On the other hand, the landing mass determines the design loading at landing, and thereby affects the mass of the structure. It is no coincidence that civil and often combat aircraft, in case of impossibility of the emergency fuel draining during emergency landing, have to fly in the airfield area to burn fuel.

The value m_{land} is usually set by the designer after consultation with the customer. For preliminary design, historical values of m_{land}/m_0 for various types of aircraft are given in Table 3. $C_{L,max}$ on takeoff and landing is also given here.

A modern runway is an expensive concrete structure with an underlying soil drainage system. In these circumstances, the task of reducing the required runway length is relevant to aircraft design. In this connection, let's consider the ways of

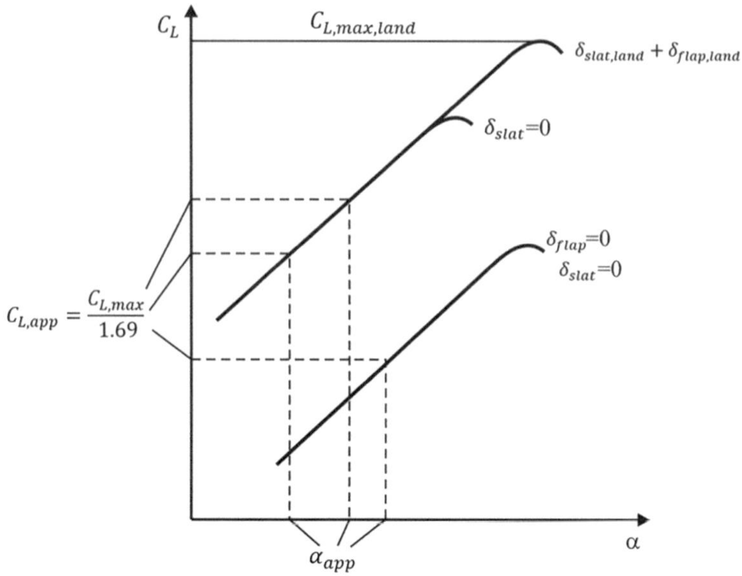

Fig. 11 Effect of high-lift devices on $C_{L,app}$ and α_{app}

Table 3 Historical data on m_{land}/m_0 for various types of aircraft

Type of aircraft	m_{land}/m_0			$C_{L,max,TO}$	$C_{L,max,land}$
	min	average	max		
Single-engine propeller-driven	0,95	0,997	1,0	1,3…1,9	1,6…2,3
Twin-engine propeller-driven	0,88	0,99	1,0	1,4…2,0	1,6…2,5
Agricultural	0,7	0,94	1,0	1,3…1,9	1,3…1,9
Commuter with a turboprop	0,92	0,98	1,0	1,7…2,1	1,9…3,3
Executive aircraft with bypass turbojets	0,69	0,88	0,96	1,6…2,2	1,6…2,6
Passenger and transport vehicles with bypass turbojet engines	0,65	0,84	1,0	1,6…2,2	1,8…2,8
Training aircraft	0,87	0,99	1,0	1,4…2,0	1,6…2,2
Fuel tankers, bombers	0,68	0,76	0,83	1,6…2,2	1,8…3,0
Seaplanes, amphibious aircraft, deck-based planes	0,79	–	0,95	1,6…2,2	1,8…3,4
Supersonic passenger aircraft	0,63	0,75	0,88	1,6…2,0	1,8…2,2
Fighters	0,78	–	1,0	1,4…2,0	1,6…2,1

reducing the landing distance $L_{land} = L_{app} + L_{LR}$, which determines the runway length required by the airworthiness standards. According to the standards, at landing in dry weather $L_{RW,dry} = 1.67L_{land}$, and in wet weather $L_{RW,wet} = 1.15L_{RW,dry} = 1.92L_{land}$. Coefficients 1.67 and 1.92 are used to allow for the possibility of either icing of the runway surface, or formation of such a layer of water, at which the wheels of the landing gear enter the aquaplaning mode during run.

The minimum value of L_{app} when the aircraft approaches the runway without flare is $L_{app} = 15/tg25°0\prime \cong 300m$. At $V_{app} = 250km/h(70m/s)$, $V_z = 3.5m/s$. However, pilots avoid such a "rough", although permissible, landing. Before landing they level the aircraft to nearly horizontal flight and decrease speed to $V_{land} \geq V_{min,land}$ and $V_z \approx 0.1 \ldots 0.3m/s$. In this case the length of the approach section increases to 500…700 m depending on the flare height.

The greatest effect on the landing distance is associated with L_{LR}. The landing run distance can be determined as follows:

$$L_{LR} = \frac{0.94}{C_{L,max,land}} \frac{p_{land}}{1/3\left(\frac{1}{\left(\frac{L}{D}\right)_{LR}} + 2f_{GR}\right) + 10(T_{rev} - T_{LT})/W_{land}}$$

where T_{rev} is the reverse thrust of the engines,

T_{LT} is the low throttle thrust of non-reversible engines.

This equation defines the following ways to reduce L_{LR}:

- to increase $C_{L,max,land}$ by using powerful high-lift devices;
- to increase the landing friction coefficient f_{GR} by applying more efficient braking systems;
- to reduce $\left(\frac{L}{D}\right)_{LR}$ through rapid opening of the spoilers, which allows effective braking to begin immediately, since the friction force $F_{fr} = f(W_{land} - C_L qS)$, and the spoilers reduce C_L by the factor of 2… 3;
- to use thrust reversal;
- to reduce the wing loading p_{land}.

For more details on how to shorten the runway length, see Chap. 12.

2 Range

To analyze the influence of aircraft parameters on the cruise range, we can use the Breguet equation as a first approximation:

$$R = 3.6\frac{(L/D)V}{C_p} ln\frac{W_0}{W_{end}} = \frac{(L/D)Ma}{C_p} ln\frac{W_0}{W_{end}}.$$

It follows from the equation that range is directly proportional to lift-to-drag ratio L/D, flight speed V (m/s) and the logarithm of the aircraft weight (mass) ratio at the beginning and end of flight. It is inversely proportional to the specific fuel consumption C_p. The terms in front of the logarithm are called the range factor. Its value is determined by the current level of aerodynamic perfection of the aircraft and gasdynamic perfection of its propulsion system. Let us first consider the effect of the range coefficient on R.

Let us denote the range coefficient $r = \frac{Ma}{C_p}(L/D)$, a is the speed of sound. Figure 12 shows the variation of r and its components with M for the altitude range of $11...0km$, $a = const$.

The above plots are based on the specific fuel consumption of the most advantageous engines for the speed range under consideration. Thus, for $M = 0.5$, it is the specific fuel consumption of a turbofan, for $M = 2$, it is a ramjet, for $M = 4$, it is a ramjet. The lift-to-drag ratio strongly depends on the aircraft shape and M_{cr}. The figure shows the average values of L/D for long-range aircraft.The first maximum of r is reached at M, corresponding to the subsonic region, and the second one at $M \approx 3$. Let us consider in more detail the transonic velocity range and determine the effect of sweep on L/D and $(L/D)M$. This can be done by using the graph shown in Fig. 13.

This graph can be used for the selection of wing sweep angle if the range is the main criterion (swept wings are checked for take-off and landing requirements).

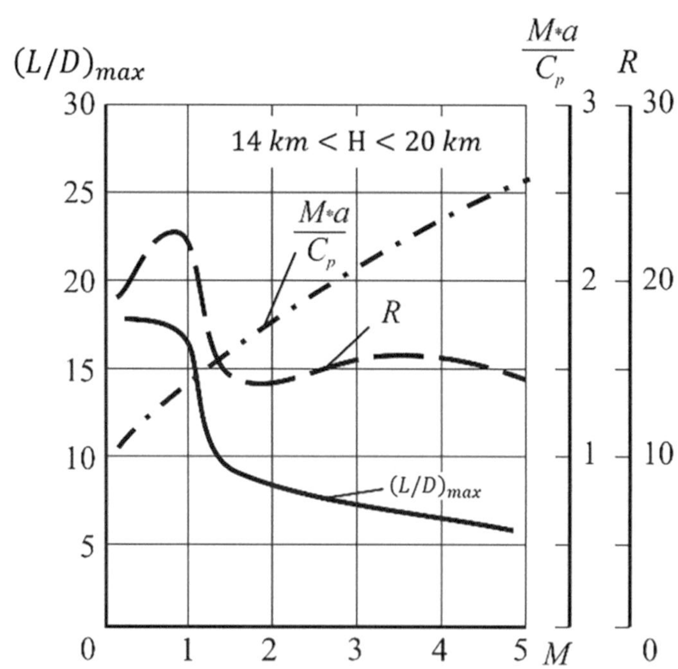

Fig. 12 Range coefficient and its components as a function of Mach number

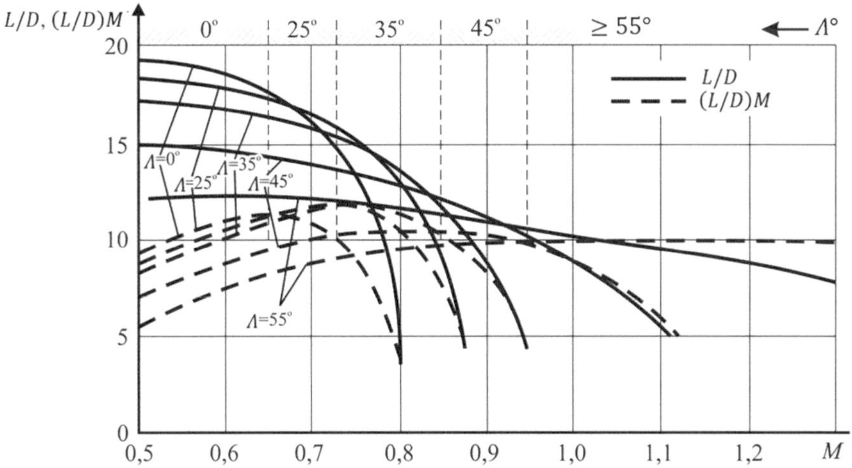

Fig. 13 L/D, $(L/D)M$ in transonic range for different sweep angles

Cruising usually takes place at speeds up to M_{cr}, i.e. before the sharp increase in drag due to the local supersonic flow zones and the beginning of the formation of shock waves (Fig. 14).

Now consider the effect of the second term of the Breguet equation. For a transport (passenger) airplane, with the weight changing only due to fuel burn, i.e. $W_{end} = W_0 - W_{fuel}$, the Breguet equation will take the following form: $R = r \cdot ln\frac{1}{1-\overline{W}_{fuel}}$. This leads to an obvious conclusion: the increase of the fuel ratio is a direct way to increase a flight range. It is clear that keeping the takeoff weight of the airplane constant could only be achieved by weight improvement of the airplane, and first of all by reducing

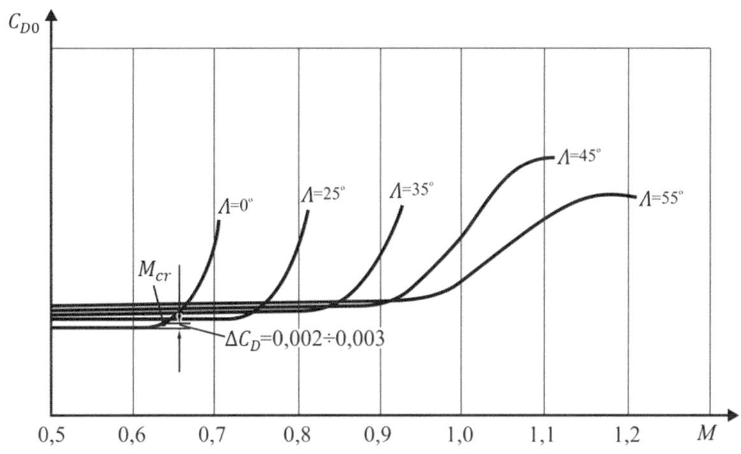

Fig. 14 Critical Mach number for wings of different sweep

Fig. 15 Change in relative range due to reduced mass of the structure

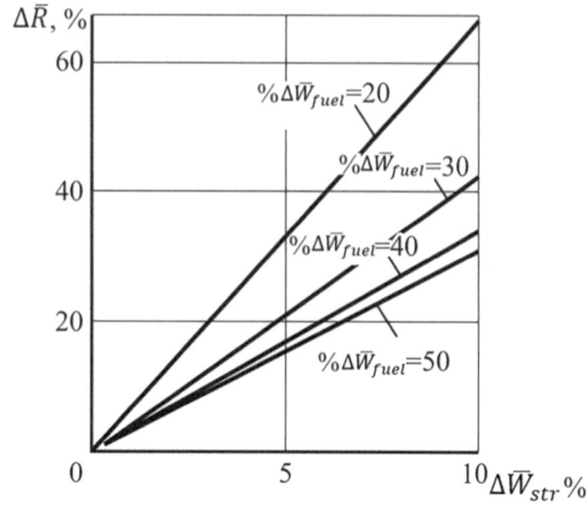

the structural weight by ΔW_{str}. Then, its flight range will be $R\prime = r \cdot ln\frac{1}{1-\overline{W}_{fuel}-\Delta\overline{W}_{str}}$. Let us estimate the relative range changes due to weight reduction:

$$\overline{R} = \frac{R\prime}{R} = \frac{\frac{ln1}{1} - \overline{W}_{fuel} - \Delta\overline{W}_{str}}{\frac{ln1}{1} - \overline{W}_{fuel}} \text{and } \Delta\overline{R} = \overline{R} - 1.$$

For example:
$\overline{W}_{fuel} = 0.2 \rightarrow \Delta\overline{R} = 6.7\Delta\overline{W}_{str}$,
$\overline{W}_{fuel} = 0.3 \rightarrow \Delta\overline{R} = 4.35\Delta\overline{W}_{str}$,
$\overline{W}_{fuel} = 0.4 \rightarrow \Delta\overline{R} = 3.5\Delta\overline{W}_{str}$.

It can be seen that the change in relative range is 6.7 … 3.5 times greater than the change in structural weight ratio. Moreover, the lower the fuel weight (fighters, attack aircraft, short- and medium-range passenger aircraft), the greater the reduction of empty aircraft weight (Fig. 15).

Thus, reducing the weight of an empty aircraft is the most effective way to increase range.

Sometimes, in order to increase the flight range, it is possible to carry additional fuel. However, this way is significantly less effective, as it leads to an increase in the wing loading, and, in turn, to an increase in C_L and decrease in the flight speed and range.

At the same time, for some types of military aircraft, increasing the range (endurance) of flight by using external fuel tanks and aircraft refueling in flight is a widely used method.

In conclusion, we remind you that an important reserve for increasing the range is the optimization of the trajectory and speed of flight, provided by the coordination of the layout parameters with the power plant characteristics.

3 Maximum Flight Speed

The maximum flight speed can be assessed by comparing the required T_r and available thrust T_a, and since $T_r = C_D\frac{\rho V^2}{2}S$, we can write down $T_a = C_D\frac{\rho V_{max}^2}{2}$. Therefore, $V_{max} = \sqrt{\frac{2T}{c_D\rho S}}$. By multiplying the numerator and denominator of the sub-root expression by the weight of the aircraft W, we obtain $V_{max} = \sqrt{\frac{2p\overline{T}}{c_D\rho}}$.

As follows from this formula, V_{max} at a given altitude depends on the wing loading, thrust-to-weight ratio $\overline{T}$ and drag coefficientC_D. Let us consider in more detail the influence of these factors onV_{max}.

The thrust-to-weight ratio $\overline{T}$ depends on the take-off thrust $\overline{T}_0 = \frac{T_0}{m_0 g}$, flight speed and altitude, as well as the engine operating mode (see Chap. 18). Typical thrust and specific fuel consumption relationships for turbojets of a moderate bypass ratio m designed for subsonic flight speeds are shown in Figs. 16. and 17.

The above curves can be approximated as follows:

$$\xi(\mathrm{H, M}) = \frac{\mathrm{T}_{HV}}{T_0} = \left[1.707 - \left(0.803 + 0.085m + 0.005m^2\right)sin\left(\frac{157M + 108}{1 + 0.18m}\right)\right]\left(1 - \frac{0.055H}{1000}\right)$$

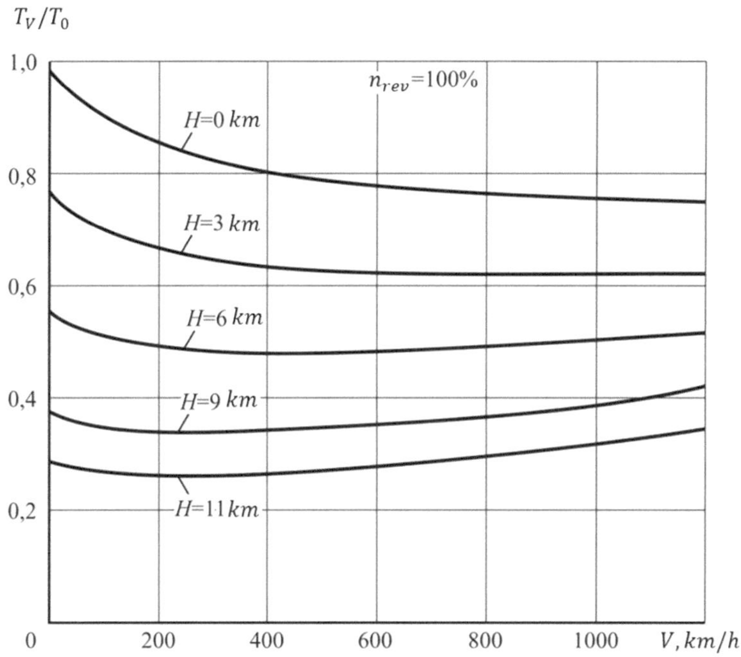

Fig. 16 Altitude and speed performance of bypass turbojets ($m = 1.5$)

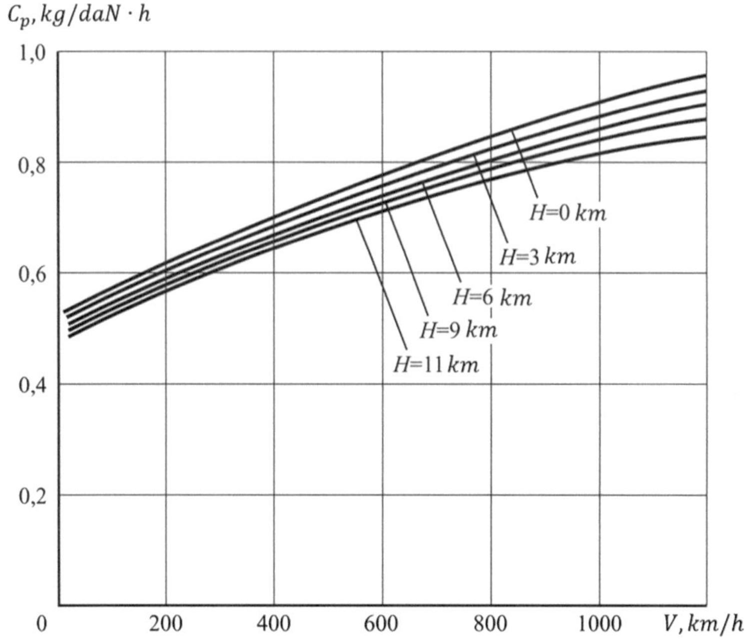

Fig. 17 Bypass turbojets ($m = 1.5$) fuel consumption

$$C_p(H, M) = 0.9\left[\frac{0.82}{1 + 0.525\sqrt[3]{m}} + M\left(0.494 - \frac{0.0145H}{1010}\right)\right]K_F$$

In these equations: M is the Mach number; m is the engine bypass ratio, H is the flight altitude, [m]; K_F is the coefficient related to the specific fuel consumption of the engine under the given conditions. For further analysis, we will use simplified dependencies, reflecting the physical substance of the process:

$$T_{HV} = T_0 \xi_V \Delta^{0.85} \text{ for } H \leq 11\ km \text{ and } T_{HV} = 1.2T_0 \xi_V \Delta \text{ for } H > 11 km.$$

In these equations, ξ_V is the factor related to thrust change due to velocity. Relative air density $\Delta = (1 - \alpha_1 H)^{4.225}$, where according to ISA, $\alpha_1 = 0.02261/km$, H is flight altitude in km.

Using the above relations we obtain the following equations for V_{max}:

$$V_{max} \cong 14.4\sqrt{\frac{\xi_V p \overline{T}_0}{C_D \Delta^{0.15}}} \text{ for } H \leq 11km \text{ and } V_{max} \cong 15.7\sqrt{\frac{\xi_V p \overline{T}_0}{C_D}} \text{ for } H > 11km.$$

It follows from these equations that increasing the take-off thrust is a direct way to increase the maximum speed. However, it should be emphasized that this goal cannot be achieved simply by increasing the engine thrust (power). Recall the unsuccessful attempts to significantly increase the flight speed of fighter aircraft in the 1940s. This is because an increase in thrust leads to an increase in the weight of the power plant as well as the aircraft structure weight, i.e., an increase in takeoff weight. So what are the ways to increase the takeoff thrust? The answer to this question is obvious from the following simple relation:

$$\overline{T}_0 = \frac{T_0}{W_0}\frac{W_{PP}}{W_{PP}} = \frac{\overline{m}_{PP}}{\gamma}$$

$\overline{m}_{PP}$ is the mass ratio of the power plant;
γ is the specific weight of the power plant.
On this basis, the take-off thrust can be increased in the following ways:

- by reducing the specific weight of the power plant (improvement of the engine, transition to a new type of power plant);
- by increasing the value of $\overline{m}_{PP} = 1 - \overline{m}_{str} - \overline{m}_{fuel} - \overline{m}_{equip} - \overline{m}_{pay}$ by reducing C_p of the engine (reduction of $\overline{m}_{fuel}$); a similar effect is produced by the weight improvement of the aircraft (reduction of $\overline{m}_{str}$).

Next, let's consider the ways to increase the maximum speed due to the friction coefficient.

For any aircraft configuration at a given M, the corresponding dependency of drag coefficient on lift coefficient (polar) can be plotted (Fig. 18):$C_D = C_{D0} + AC_L^2$. Note that for simplicity, the polar presented in Fig. 18 does not take into account the wing twist typical of modern supersonic maneuvering airplanes.

For aerodynamic analysis, it is more convenient to use the curves $C_{D0} = f(M)$ and $A = f(M)$. For a fighter aircraft, the example plots are shown in Fig. 19.

The value of the maximum lift-to-drag ratio can be determined as follows:

$$(L/D)_{max} = \frac{1}{2\sqrt{C_{D0}A}}.$$

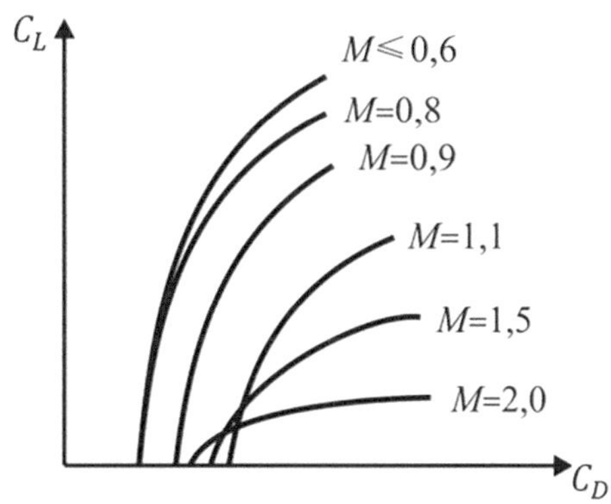

Fig. 18 Supersonic aircraft polars

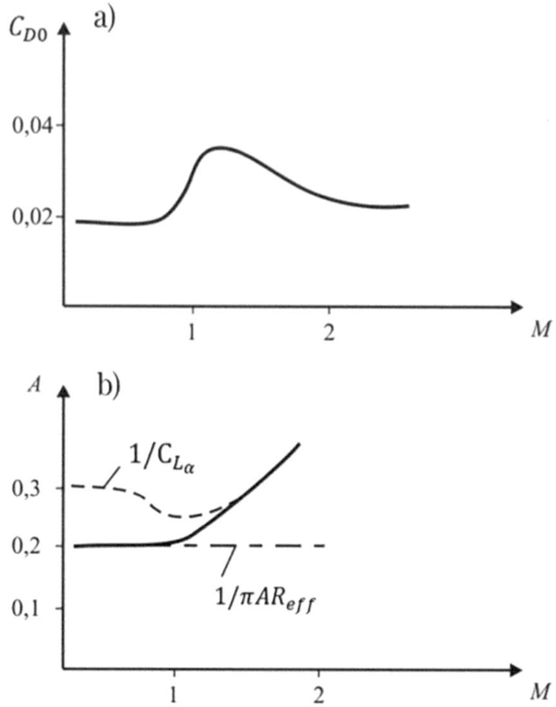

Fig. 19 $C_{D0} = f(M)$ and $A = f(M)$ curves for a fighter aircraft [2]

The coefficient A mainly depends on the load-carrying capacity (C_{L_α}) and on the leading edge suction capability of the wing.

The value C_{L_α} of the wing of the given planform shape depends on the flight Mach number (Fig. 20).

The smaller is Λ and the larger is R, the greater are the changes to C_{L_α} of the wing and aircraft.

Particularly strong changes to C_{L_α} due to Λ are observed for wings with small sweep angle ($\Lambda = 0 \ldots 30°$) and sufficiently large aspect ratio, for which both an

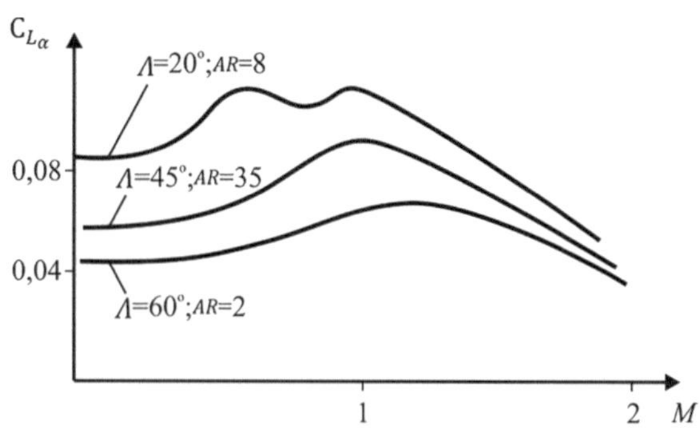

Fig. 20 $C_{L_\alpha} = f(M)$ curves for wings of various planform shapes [2]

increase and a decrease of $C_L(C_{L_\alpha})$ are possible with the increasing M. It should be noted that at $M \geq 2$ the influence of Λ and AR on C_{L_α} is weak.

Coefficient A also depends on the leading edge suction capability of the wing. With the airfoil immersed in the flow, the leading edge critical point (the point where the flow divides into upper and lower parts) is located at some distance from the leading edge. When the leading edge is rounded, the pressure in the region of the leading-edge is decreased, which causes some forward suction force, which leads to a decrease in drag compared to the case when the leading edge is sharp and the pressure drop near it does not lead to a decrease in drag or when there is no pressure drop at all.

The suction near the leading edge, even if it is rounded, is not observed at the supersonic flight speeds when the normal velocity component V_n exceeds the local speed of sound, i.e. when the leading edge of the wing is located ahead of the Mach cone, the top of which coincides with the top of the wing (Fig. 21).

In this case, the leading edge is called *supersonic.*

The larger is C_{L_α} and the higher is the leading-edge suction, the smaller is A, i.e., the less is C_{Di}. At subsonic speeds, $M < M_{cr}$, with the suction area due the rounded leading edge, A is minimal and approximately constant: $A = \frac{1}{\pi AR_{eff}}$, where $AR_{eff} = f(\Lambda, AR)$.

At supersonic speeds and with a supersonic leading edge, $A = \frac{1}{C_{L\alpha}}$, and it grows with the increasing Mach number approximately in proportion to $\sqrt{M^2 - 1}$, so that $A \approx \frac{\sqrt{M^2-1}}{4}$. When the Mach number decreases and with a subsonic leading edge, A decreases gradually approaching the value of $A = \frac{1}{\pi AR_{eff}}$.

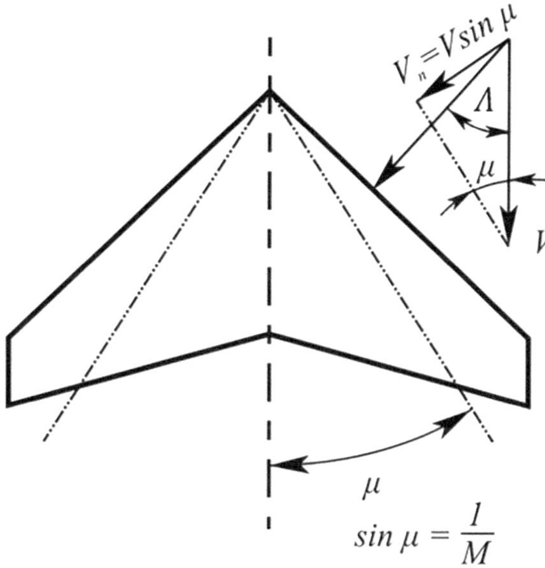

Fig. 21 Supersonic leading edge

In the case of a wing with a sharp leading edge (if the wingtip is not deflected) for all $M, A = \frac{1}{C_{L\alpha}}$ (dashed line in Fig. 19).

These considerations determine the choice of Λ, AR, as well as the shape of the leading-edge (reduction of C_{Di}).

In most aircraft whose primary flight mode is subsonic, the leading edge must be rounded to reduce C_{Di}.

For supersonic speeds, the leading edge should be sharpened to provide a larger $(L/D)_{max}$, since such a wing has less C_{D0}, and the shape of the leading edge has almost no effect on C_{L_α}. The reduction of $\frac{t}{c}$ leads to a reduction in the wing wave drag at transonic speeds as well.

The wing sweep Λ has the strongest effect on $C_{D,wave}$ at transonic speeds, because in this case M_{cr} increases. However, at the Mach numbers, at which the leading edge of the wing is supersonic, this effect is virtually absent. Therefore, with a supersonic leading edge, a further increase in Λ doesn't reduce the wave drag. At the same time, in order to reduce the induced drag, it is advantageous, on the contrary, to slightly reduce the sweep (C_{L_α} grows). Reduction of AR also contributes to the reduction of the wave drag. However, it is known that this leads to a deterioration of the load-carrying properties of the wing (C_{L_α} drops) and an increase in the induced drag (Fig. 22).

At relatively high supersonic speeds ($M = 1.8 \ldots 2.3$), the variation in AR and Λ do not have such a significant effect on the value of $(L/D)_{max}$ and they can be selected to meet other requirements, for example, maneuvering at transonic speeds. This fact is clearly illustrated by the graph $(L/D)_{max} = f(M)$, Fig. 23.

It can be seen that a sharp leading edge leads to a significant reduction in $(L/D)_{max}$ at subsonic speeds due to the growth of C_{Di}. It should also be borne in mind that the normal acceleration during flight in turbulent atmosphere at low altitude and high speed is proportional to VC_{L_α} and, in order to meet the requirements for comfort and strength, a wing of small aspect ratio and large sweep should have a smaller C_{L_α}.

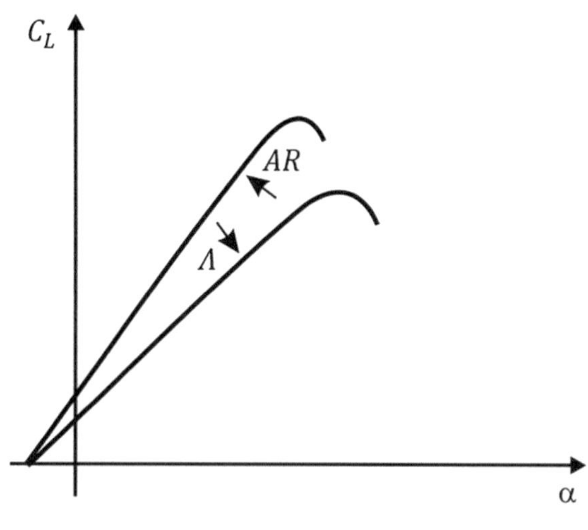

Fig. 22 Impact of AR and Λ on C_{L_α} of a wing

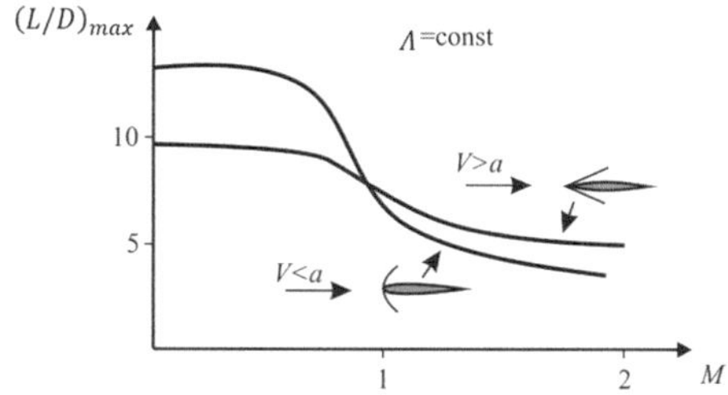

Fig. 23 $(L/D)_{max}$ variation with M for wings with rounded and sharp leading-edges

C_{Di} is strongly affected by fuselage aspect ratio $AR_f = \frac{l_f}{d_{mid}}$. With increasing aspect ratio, friction drag increases and pressure drag (wave drag) decreases. Since at subsonic speeds the main share of drag is friction drag, it is common to use $AR = 4\ldots5$. At supersonic speeds, the wave drag can account for a significant portion (sometimes 50–60%) of the total drag. Therefore, the increase in aspect ratio is beneficial as long as the increase in friction drag due to the increase of the wetted surface is not so significant. At $M = 2\ldots2.5$, the optimum values of fuselage aspect ratio are 8…10.

Let's consider the effects of V_{max} on aircraft drag. The total aircraft drag can be estimated as follows:

$$D = D_{wing} + D_{tail} + D_f + D_n + D_{int} + D_i, \text{ or}$$

$$C_D qS = C_{D,wing} qS + C_{D,t} qS_t + C_{D,f} qS_{mf} + C_{D,n} qS_{mn} + C_{D,int} qS + C_{D,i} qS,$$

$$C_D = C_{D,wing} + C_{D,t}\frac{S_t}{S} + C_{D,f}\frac{S_{mf}}{S} + C_{D,n}\frac{S_{mn}}{S} + C_{D,int} + C_{D,i} = C_{D,0} + C_{D,i}, \text{ since.}$$

$$C_{D,0} = C_{D,wing} + C_{D,t}\frac{S_t}{S} + C_{D,f}\frac{S_{mf}}{S} + C_{D,n}\frac{S_{mn}}{S}.$$

Let's represent C_D in terms of wing loading:
$C_{D,wing} + C_{D,t}\frac{S_t}{S} = K_t C_{D,wing}$ where $K_t = 1 + \frac{C_{D,t}}{C_{D,wing}}\frac{S_t}{S} \approx 1.2\ldots1.4$.
$C_{D,f}\frac{S_{mf}}{S} + C_{D,n}\frac{S_{mn}}{S} = C_{D,f}\frac{S_{mf,mn}}{S}$, where $S_{mf,mn} = S_{mf} + S_{mn}$, $C_{D,i} = AC_L^2 = A\frac{p^2}{q^2}$
where $A = \frac{1+\delta}{\pi AR_{eff}}$ for $M < M_{cr}$ and $A \approx \frac{\sqrt{M^2-1}}{4} = \frac{1}{C_{L\alpha}}$ for $M \geq 1.1\ldots1.5$.
Then,
$C_D = C_{D,wing}K_t + C_{D,int} + C_{D,f}\frac{S_{mf,mn}G}{SG} + A\frac{p^2}{q^2}$, $AR_{eff} = \frac{AR_{wing}}{1+\frac{S_{mf,mn}}{S}}$.
Let's introduce the following designations:
$F_1 = C_{D,wing}K_t + C_{D,int}$;
$F_2 = C_{D,f}\frac{S_{mf,mn}}{G}$;
$F_3 = \frac{A}{q^2}$;
$\frac{S_{f,n}}{G}$ Is the fuselage midsection and nacelle loading.
Then, $C_D = F_1 + F_2 p + F_3 p^3$.

On the other hand, $C_{D,0} = C_{D,p} + C_{D,wave}$.

The above considerations can be presented on the graph $C_D = f(M)$ (Fig. 24), where the total drag coefficient is resented in terms of its components.

The above equations and graph show the ways in which aircraft C_D can be reduced in order to increase V_{max}:

(a) by increasing the wing aspect ratio for subsonic aircraft;
(b) by reducing the drag associated with the “steps” in the joints, skin waviness;
(c) by reducing the pressure (form) drag associated with protruding elements (antennas, control system elements, fairings), using fillets, streamlining;
(d) by reducing the wave drag, that is, by increasing M_{cr} which can be achieved through increasing Λ, decreasing $\frac{t}{c}$, the use of supercritical airfoils and adequate shaping of aircraft surfaces (sharpening of the forebody, etc.).

Note that at supersonic speeds, $C_{D,wave} = \frac{k\left(\frac{t}{c}\right)^2}{\sqrt{M^2-1}}$, that is, it depends on the thickness ratio squared.

The effect of wing loading variation on V_{max} cannot be considered in isolation from C_D, since we have seen that $C_D = f(p)$.

It can be shown that $V_{max} = \sqrt{\frac{2\overline{T}}{\rho}\frac{1}{F_1 p^{-1} + F_2 + F_3 p}}$.

Obviously, there is an optimum value of wing loading at which the value p/C_D reaches a maximum. This value depends on the value of thrust-to-weight ratio as shown on the graph (Fig. 25).

With increasing thrust-to-weight ratio the optimum wing loading shifts to the right. Graphical interpretation of this phenomenon is shown in Fig. 26, where $F = F_1 p^{-1} + F_2 + F_3 p$.

Note: With increasing velocity, the fraction of $C_{D,i}$ decreases, because the required C_L is small: $C_L = \frac{p}{\rho V^2/2}$, $C_{D,i} = A\left(\frac{p}{\rho V^2/2}\right)^2$. Therefore, for a supersonic flight, *AR* can be small, since low AR is associated with low $C_{D,wave}$.

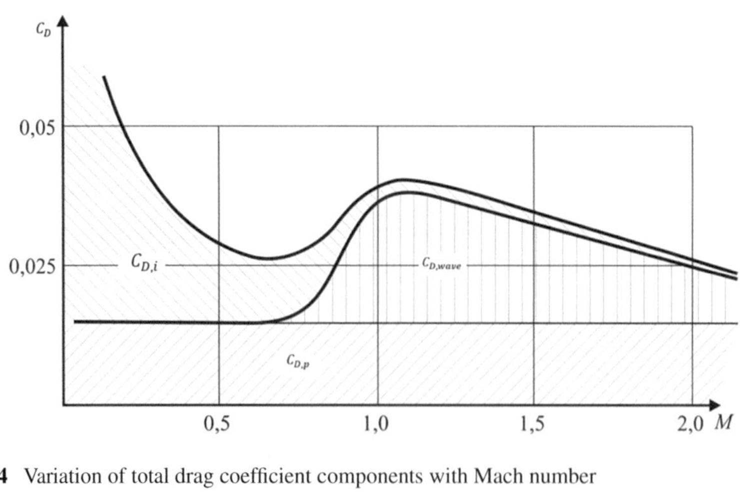

Fig. 24 Variation of total drag coefficient components with Mach number

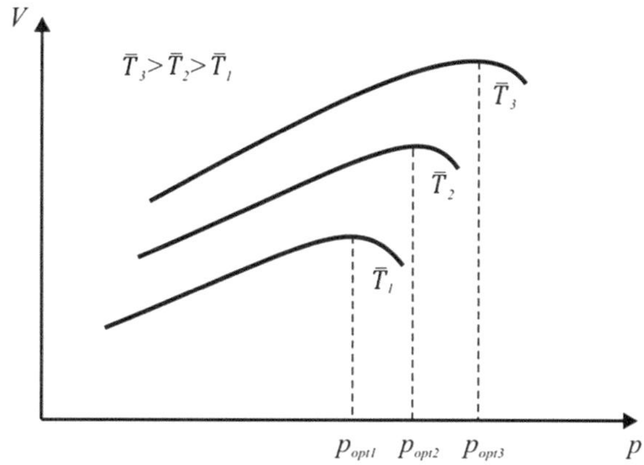

Fig. 25 Relationship between thrust-to-weight ratio $\overline{T}$ and optimum wing loading p_{opt}

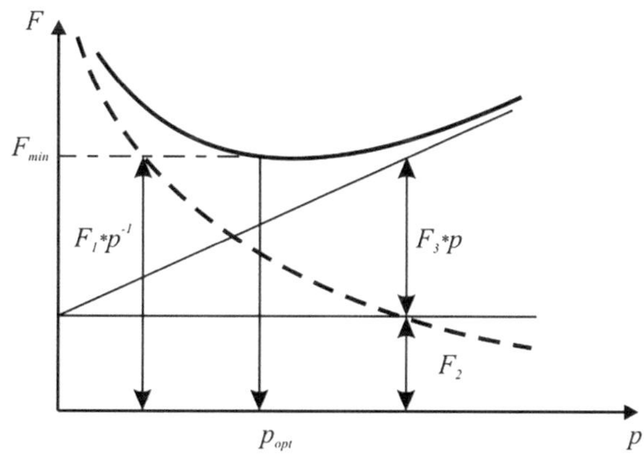

Fig. 26 Optimal wing loading

4 Aircraft Rate of Climb and Ceiling

4.1 Rate of Climb

Rate of climb can be derived as follows:

$WV_z = (T - D)V_{climb}$, therefore, $V_z = \left(\overline{T} - q\frac{C_D}{p}\right)V_{climb}$.

An approximate relation for determining the maximum ground speed is as follows:

$V_{max} = 15.3\sqrt{\frac{p\left(\xi\overline{T}_0\right)^3}{C_{D,0}}}$.

From this equation it can be seen that:

- the main parameters determining the value of V_z, are take-off thrust T_0 or excess thrust $\Delta T = T - D$;
- with decreasing $C_{D,0}$, V_z and ΔT increase;
- with increasing wing loading p up to some limit, V_z slightly increases, and then decreases due to the growth of C_D, and in particular $C_{D,0}$.

The optimum value of the specific wing loading corresponding to $V_{\max}$, can be determined as follows:$p_{V_{z,max}} = q\sqrt{\frac{(L/D)_{opt} C_D}{A}}$.

4.2 Aircraft Ceiling

Airplane ceiling is the maximum altitude it can achieve.

At the ceiling altitude, the aircraft is capable of flying in a single mode corresponding to a certain angle of attack. In this mode, the aircraft performs a horizontal steady flight at a speed that is maximum (because there is no thrust reserve to increase speed), minimum (because an increase in angle of attack leads to deceleration with descent), and the most advantageous: $\frac{L}{D} = \left(\frac{L}{D}\right)_{max}$.

At the ceiling altitude, $T_{avail} = T_{req} - D$, in any steady horizontal flight $T_{req} = W/\left(\frac{L}{D}\right)$, consequently, at the ceiling, $\overline{T} = 1/\left(\frac{L}{D}\right)_{max}$, and since $\left(\frac{L}{D}\right)_{max} = \frac{1}{2\sqrt{AC_{D,0}}}$, then $\overline{T} = 2\sqrt{AC_{D,0}}$.

For H > 11*km*, $\overline{T} = 1.2\Delta\xi\overline{\mathrm{T}}_0$ $H > 11$, therefore, $\Delta = \frac{2\sqrt{AC_{D,0}}}{1.2\xi\overline{\mathrm{T}}_0} = \frac{1.67\sqrt{AC_{D,0}}}{\xi\overline{\mathrm{T}}_0}$.

By applying the values of A, we obtain:

for subsonic aircraft, $\Delta_{req} = 1.67\sqrt{\frac{C_{D,0}/\pi AR_{eff}}{\xi\overline{\mathrm{T}}}}$;

for supersonic aircraft, $\Delta_{req} = 0.83\sqrt{\frac{C_{D,0}\sqrt{M^2-1}}{\xi\overline{\mathrm{T}}}}$.

Thus, with decreasing $C_{D,0}$ and increasing AR and $\overline{T}$ the aircraft ceiling increases (this conclusion is valid for subsonic aircraft).

Thus, for example, the american reconnaissance airplane TR-1 with a ceiling of 27,730 m has the following characteristics: $AR = 10.6, \overline{\mathrm{T}}_0 = 0.42, p_0 = 195 daN/m^2$.

The flight speed at the ceiling can be determined as follows:

$$q_H = \frac{p}{C_L\left(\frac{L}{D}\right)_{max}};\ \left(\frac{L}{D}\right)_{max} = \sqrt{\frac{C_{D,0}}{A}};$$

$$q_H = \sqrt{\frac{Ap^2}{C_{D,0}}};\ C_{L,\left(\frac{L}{D}\right)_{max}} = 2C_{D,0};$$

$$q_H = \frac{T}{C_{L,\left(\frac{L}{D}\right)_{max}} S} = \frac{p\overline{T}}{2C_{D,0}}$$

The ceiling flight speed as affected by p and $\overline{T}$. The greater is p when flying at the ceiling, the greater the thrust-to-weight ratio should be (Fig. 27).

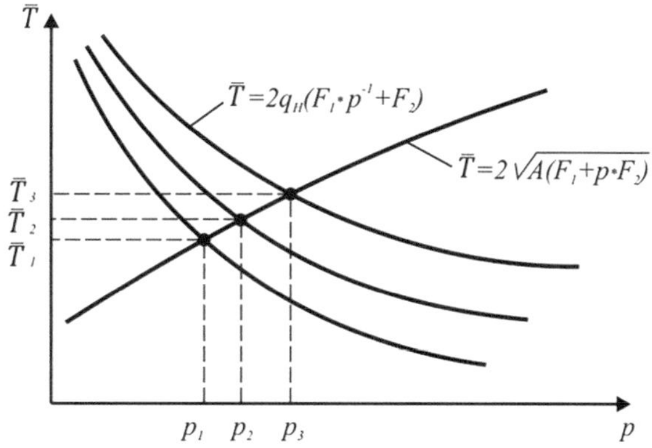

Fig. 27 Wing loading and thrust-to-weight ratio at ceiling

5 Aircraft Maneuvering

An airplane maneuver is in general any change in its position in space, flight configuration, position of controls, and engine operation modes.

Maneuvering is referred to the coordinates of the center of mass of the aircraft (X, Y, Z) and the magnitude and direction of the velocity vector applied at the center of mass (V, θ, ψ). Based on this, maneuverability is usually understood as the ability of aircraft to change the position of its center of mass in space, i.e. the speed, altitude and direction of flight.

As known from the flight mechanics, the rate of change of the velocity vector at the aircraft center of mass can be determined as follows:

$$\dot{V} = g(n_x - sin\theta); \dot{\theta} = \frac{g}{V}(n_z cos\gamma - cos\theta); \dot{\psi} = \frac{g}{Vcos\theta} n_z sin\gamma.$$

I.e., they depend on the magnitude of load factor n, which is determined by the magnitude and direction of the thrust force of the engines and the components of aerodynamic forces.

Thus, in a flight with zero sideslip angle,

$$n_z = \frac{T - D}{mg}; n_x = \frac{D + Tsin(\alpha + \varphi_T)}{mg}; n_{y=0}.$$

In the general case, aircraft maneuver occurs with a change in energy level $H_{en} = Vn_x$.

Available ranges of speed, altitude and load factor are called *general* maneuvering *indicators*, and the characteristics of individual maneuvers (climb, acceleration, turn, etc.) are called *particular indicators*. The available normal load factor n_z is a function of lift which can be produced at the given altitude and speed:

$$n_z \le n_{z,avail} = \frac{C_{L,allow} qS + T(\alpha_{allow} + \varphi_T)}{mg},$$

φ_T = the thrust vector deflection angle.
At small $(\alpha + \varphi_T)$,
$n_{z,avail} = C_{L,allow}\frac{q}{p} + \overline{T}(\alpha_{allow} + \varphi_T)$.
Since the second term is significantly smaller, we can write down:

$$n_{z,avail} \cong C_{L,allow}\frac{q}{p}.$$

Therefore, a greater $C_{L,allow}$(more efficient high-lift devices), a lower wing loading, a greater thrust-to-weight ratio (to a lesser extent - for conventional aircraft without thrust vector rotation) increase *the available normal load factor*.

As it is known, at high dynamic pressure, the available load factor is limited by the applied load factor related to the strength requirements and other limitations: $n_z \le n_z^{appl}$. Applied load factors for maneuvering aircraft are of the order of $8 \div 9$, and for non-maneuvering—$2 \div 3$.

The load factor is $n_x = \frac{T - C_D q S}{G} = \overline{T} - C_D\frac{q}{p}$, i.e., it increases with increasing available thrust-to-weight ratio and decreasing $\frac{C_D}{p} = F_1\rho^{-1} + F_2 + F_3 p$. For a constant altitude, the relationship between n_x and n_z can be determined using the equation $n_x \cong \overline{T} - n_z\frac{D}{L}$.

A representative plot for the longitudinal load factor n_x at $H = const$ with the constraints of $C_{L,allow}$, n_z^{appl}, and M_{limit} is shown in Fig. 28.

Let us also mention such important characteristics of aircraft maneuvering in the horizontal plane as turn radius and turn time.

When *a turn is coordinated* (a steady turn without sideslip), the turn radius is $R_{turn} = \frac{V^2}{g n_z sin\gamma}$ or $R_{turn} = \frac{V^2}{g\sqrt{n_z^2 - 1}}$, and $T_{turn} = \frac{2\pi V}{g\sqrt{n_z^2 - 1}}$, i.e., the greater is the allowable normal load factor, the smaller is the turn radius and the shorter is the turn execution time. Recall that a turn is a full 360° turn of an aircraft.

Let us consider aircraft energy characteristics. As noted in [2], the main energy characteristics of subsonic and especially low-speed aircraft were altitude, vertical speed, and climb time. They characterize only the potential energy, which is clearly insufficient for modern supersonic maneuvering airplanes.

Fig. 28 Longitudinal load factor of a maneuvering aircraft at $H = const$

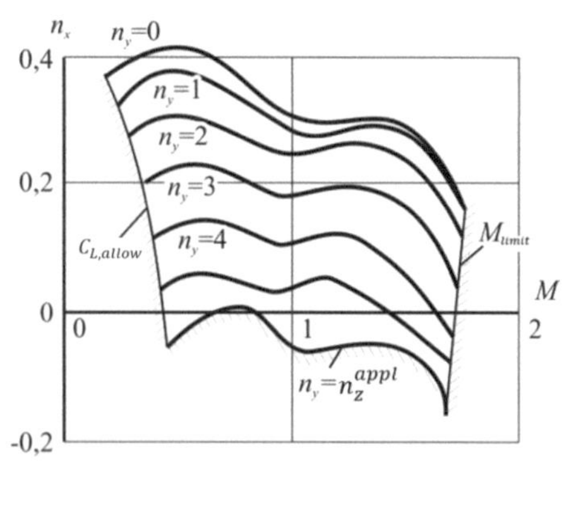

Indeed, a subsonic aircraft flying at a speed of 900 km/h (250 m/sec) at an altitude of 15 km has the following total specific mechanical energy (energy level) ($kg \bullet m/kg$) :

$$H_{en} = \frac{V^2}{2g} + H = \frac{250}{2g} + 15000 = 1820,$$

and the kinetic energy is 17.6% of the total mechanical energy. For a modern supersonic aircraft having a speed of 900 m/s ($M \approx 3$) at an altitude of 25 km, the energy level is $H_{en} = \frac{900^2}{2g} + 25000 = 63300$. The kinetic energy is already 62.2%.

The total mechanical energy reserve of an aircraft H_{en} is the most important characteristic of its combat capability and advantage. However, if two aircraft in combat have the same energy level, and one of them can store energy faster than the other, after some time the advantage will be on its side. Thus, the rate of total mechanical energy storage, referred to as *the energy rate of climb* (specific excess power), $\frac{dH_{en}}{dt} = V_z^*$, is the most important characteristic of an aircraft's combat capability, since it determines its ability to change its energy by changing speed and (or) altitude:

$$V_z^* = Vn_x = Vsin\theta + \frac{V}{g}\frac{dV}{dt}, \; n_x = \frac{T - D}{W} = \overline{T} - C_D\frac{q}{p}.$$

The accumulated mechanical energy depending on θ can be used to increase the height (potential energy) and to increase the speed (kinetic energy). If $\theta = 0$, then $V_z = 0$, therefore, all of the energy is used for acceleration. If $sin\theta = n_x$, then $\frac{dV}{dt} = 0$, and the aircraft flies at a constant velocity and all the energy goes to gain altitude.

A typical dependence of the energy rate of climb on altitude and Mach number, which is usually given in aircraft technical specifications, is shown in Fig. 29.

With increasing velocity, the specific excess power increases up to the values corresponding to the beginning of the drag growth. After this it decreases as the specific excess thrust decreases faster than the velocity increases. V_z^* is maximal at sea level and equal to zero at the boundary of the steady flight region. The magnitude and shape of the change of V_z^* depends mainly on $\overline{T}$ and C_D.

Naturally, if the tactical situation allows, it is necessary to perform climb in such a mode that at each altitude the value of V_z^* is maximal, which provides the fastest accumulation of energy and, therefore, the shortest time to reach the given altitude and speed.

Progress in aerodynamic design, aircraft automated control systems, and propulsion systems (increased thrust-to-weight ratio, thrust vector control) made it possible for the latest modifications of the domestic Su-27 and MiG-29 fighters to employ supra-maneuverability modes, which enhance their combat capabilities (see Chap. 11).

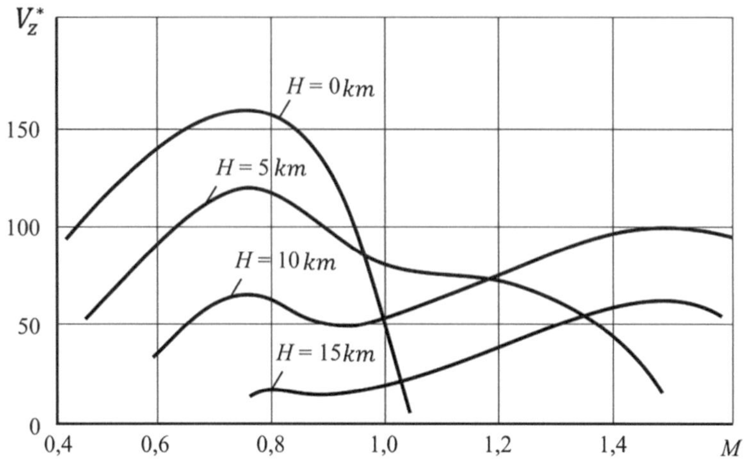

Fig. 29 Dependence of energy rate of climb on altitude and Mach number

The analysis of relationships between aircraft parameters and characteristics demonstrates a complex and often contradictory influence of individual aircraft parameters on its performance. The substance of such contradictions is demonstrated by the example of the influence of wing parameters of multi-mode combat aircraft on its characteristics, presented in Table 4.

In this table, the + sign means improvement of the characteristic when the parameter increases,—the opposite, 0 - the effect is insignificant, (+)—the trend is valid within certain limits.

This circumstance makes it extremely difficult to select the parameters of an aircraft for a particular purpose, which requires the designer not only to have a profound knowledge in the basic disciplines, but also to be able to solve complex interdisciplinary problems of aircraft design optimization as an element of a combat

Table 4 Influence of wing parameters of a multi-mode combat aircraft on its performance

Parameters Characteristics	Λ	AR	$\frac{t}{c}$	$S\prime$	f
V_z^*	+	0	–	–	0
Prolonged overloading, $M < 1$	–	+	(+)	+	+
Prolonged overloading, $M > 1$	+	–	–	(+)	–
M_{max}, ceiling	+	–	–	–	–
Subsonic cruising and loiter	–	+	+	+	+
V_{max}, $H = 0$	+	0	–	–	–
Gust load factor	+	–	0	–	0
Field performance	–	+	+	+	+

or other aviation system (see Chap. 9). The approach to the selection of criteria for optimizing parameters of such aircraft is discussed below.

6 On Criteria for Evaluating the Effectiveness of Combat Maneuvering Aircraft

In determining parameters of combat maneuvering aircraft an important role is played by the criteria selected to compare options for technical solutions. These criteria are related both to the aircraft and its mission. Such criteria are called performance indicators or efficiency criteria. The main principle of their choice is the tight alignment of the criteria with the objectives to be achieved. Two types of performance indicators can be adopted to assess the expected effectiveness of combat operations:

- the probability of an event if the task is solved by causing a certain damage (destruction of a ground target, preservation of a defended object, destruction of at least a given number of targets, etc.),
- the mathematical expectation (average value) of the damage inflicted on the enemy, if the goal is to inflict as much damage as possible or to prevent damage from enemy actions against our targets (for example, the mathematical expectation of the number of destroyed aerial targets).

To assess the effectiveness of elementary actions (target detection, issuance of target designation, etc.) a single indicator is enough. To assess large in volume, complex in physical nature phenomena, it is necessary to involve several performance indicators, one of which should be the main, corresponding to the main goal of action, the rest—additional, characterizing the state of forces and resources, spatial, temporal and other conditions and constraints.

If a complex mission has been broken down into steps, each of which is evaluated independently, then the main and specific criteria should be used. The specific criteria evaluate the effectiveness of solving subtasks. The final result of the mission is evaluated by the main criterion. To estimate it, as a rule, it is necessary to calculate the specific criteria. Specific criteria like the main criterion can also be used to understand the influence of some factors on the success of the mission.

Since combat aircraft are primarily designed to destroy air and ground targets, the following criteria can be used as combat effectiveness criteria.

When Destroying Aerial Targets

- the mathematical expectation of the number of air targets hit per operation;
- the probability of a timely flight to destroy air targets;
- the probability of striking enemy aircraft in aerial combat;

When Destroying Ground Targets

- the expectation (average value) of the number of hit homogeneous targets per operation for a given duration;
- probability of engaging a single small ground target in a single sortie;
- the mathematical expectation of the relative affected area of the target, where;
- the probability of timely departure when striking a mobile ground target. The condition for a timely strike is that the time of impact should not exceed the waiting time of the moving target, that is, the time during which the moving object (target) stays in a certain location;
- the probability of overcoming an air defence system;
- The methods for calculating these indicators are given in the special literature [3].

The combat capabilities of an aircraft are determined by its performance characteristics, as well as the technical characteristics of its special equipment and armament. The performance characteristics determining the combat capabilities include:

- field characteristics;
- operational range of speeds and altitudes of straight horizontal flight;
- range and endurance depending on the mass and type of target load;
- maneuvering characteristics, which in turn depend on the mass of the target load, the layout parameters of the aircraft, and the characteristics of its power plant.

Operational speed range for a straight horizontal flight determines: the remoteness of the enemy air interception line from the guarded object; time of the timely air strike against the enemy troops; the probability of successfully overcoming the enemy air defense; the operability and amount of the enemy reconnaissance information and the available number of combat sorties.

To quantify the combat capabilities of aircraft combat systems, the concept of combat capability has been introduced [4, 5]:

$$B_P = k\frac{P_0^4 P_i^3 \overline{T} n_{xb} J U \frac{L}{D}}{p^2 S_{wet} \sigma^{1/2}},$$

P_0 = firepower, which depends on the number and characteristics of weapons;

P_i = information capability, depending on the characteristics of the sighting and targeting system and the airborne defence system;

$\overline{T}$ = thrust-to-weight ratio of the aircraft in flight configuration and fuel weight corresponding to aerial combat;

n_{xb} = the maximum absolute value of tangential load factor in deceleration with the breaking flaps deployed.

J = aircraft survivability;

U = aircraft stability and controllability factor at an allowable angle of attack;

$\frac{L}{D}$ = aircraft lift-to-drag ratio;

p = wing loading at configuration and weight corresponding to aerial combat;

S_{wet} = aircraft wetted area;

$\sigma^{1/2}$ = radar cross section;

k = dimensional conversion factor.

In the above equation, five factors are related to the flight capabilities of the aircraft, three factors serve to assess the aircraft as an object of attack, and two factors characterize the capabilities of special equipment and weaponry.

When calculating combat capability using this formula, it should be taken into account that the value obtained is not absolute, since it is determined by the subjective opinion of experts and, in addition, the formula itself is not a strict mathematical expression. Thus, for example, a two-fold decrease in the thrust-to-weight ratio of the aircraft leads to a considerable decrease in the combat capability of the aircraft, which cannot be compensated even by a many-fold increase in other factors, for example, such as $P_0, P_i, J, U, L/D$.

In addition, most of the factors listed in the equation are interrelated and do not have an unambiguous influence on combat capability. For example, increasing the number of guided missiles in accordance with the expression significantly increases the combat capability of a fighter aircraft. In reality, there is an optimal number and type of weapons required for successful aerial combat. Increasing the number of weapons beyond the optimal number leads to a significant increase in the drag and mass of the aircraft. As a result the maneuverability and other flight characteristics of the aircraft can deteriorate to an extent that making the employment of weapons impossible at all, and their presence on board the aircraft useless.

A summary of the relationships between the parameters and performance of a combat aircraft is given in Table 5.

By changing the aircraft design parameters given in Table 5, the composition and characteristics of the avionics, the visibility characteristics, the designers develop a design that meets the customer requirements in terms of efficiency.

Assessment of the effectiveness of various design solutions is based on the simulation of the typical combat operations. In such modeling, design solutions are determined to ensure the maximum efficiency at a given cost or the minimum cost at a given efficiency.

Variables		
Symbol	Description	Units
H_{cruise}	Cruising altitude	m
$(L/D)_{max}$	Maximum lift-to-drag ratio	–
C_L	Lift coefficient	–
V_{min}	Minimum speed	m/s
L	Lift	N
m_0	Take-off mass	kg
g	Gravity acceleration	m/s^2
ρ	Density	kg/m^3
S	Wing area	m^2
$C_{L,allow}$	Allowable lift coefficient	–
$V_{\min,allow}$	Minimum allowable velocity	m/s
V_2	Safety speed	m/s
V_{app}	Approach speed	m/s

(continued)

(continued)

Variables		
Symbol	Description	Units
δ_f	Flap deflection angle	degrees
δ_s	Slat deflection angle	degrees
Λ	Sweep angle	degrees
AR	Aspect ratio	–
C_m	Pitching moment coefficient	–
$C_{L,saf}$	Safety lift coefficient	–
L_{RW}	Runway length	m
L_{CGR}	Continued ground roll distance	m
V_f	Failure speed	m/s
L_1	Length of the first climb segment	m
ρ_H	Density at altitude H	kg/m^3
L_{RW}	Runway length	m
$L_{GR,cont}$	The ground roll distance of continued take-off	m
L_{GR}	Ground roll distance	m
L_{TO}	Take-off distance	m
$L_{TO,cont}$	Continued take-off distance	m
V_1, V_{fail}	The speed at which failure of the critical engine occurs	m/s
V_{LOF}	Lift-off velocity	m/s
H_{HL}	The altitude of retraction of high-lift devices	m
V_2	Safe take-off speed	m/s
V_3	The speed corresponding to the start of retraction of high-lift devices	m/s
L_1	First climb segment length	m
V_{st}	Stalling speed	m/s
V_c	Control speed	m/s
n_z	Z-component of the load factor	–
$p = \frac{W_{aircr}}{S}$	Wing loading	N/m^2
$V_{TO,bal}$	Balanced takeoff speed	m/s
W_0	Take-off weight	N
$L_{TO,bal}$	Balanced takeoff distance	m
V_{GA}	Getaway speed	m/s
n_{eng}	Number of engines	–
C_{L,V_2}	The allowable C_L at the first segment of takeoff	–
$C_{L,max,TO}$	Take-off maximum lift coefficient	–
T_{av}	Average thrust	N
f_{GR}	Ground roll friction coefficient	–
$\overline{T}_{av} = \frac{T_{av}}{W_0}$	Average thrust loading	–
$\eta_{t,climb}$	Total climb gradient	%
θ	Climb angle	
V_{land}	Landing speed	m/s

(continued)

(continued)

Variables		
Symbol	Description	Units
L_{land}	Landing distance	m
L_{app}	Approach distance	m
L_{LR}	Landing roll distance	m
V_{app}	Approach speed	m/s
$C_{L,max,land}$	Maximum landing lift coefficient	–
$C_{L,app} = \frac{C_{L,max}}{1.69}$	Approach lift coefficient	–
δ_{slat}	Slat deflection angle	degrees
δ_{flap}	Flap deflection angle	degrees
$L_{RW,dry}$	Runway length in dry weather	m
$L_{RW,wet}$	Runway length in wet weather	m
p_{land}	Wing loading during landing	N/m^2
W_{land}	Landing weight	N
T_{rev}	Reverse thrust of engines	N
T_{LT}	The low throttle thrust of non-reversible engines	N
K_F	The coefficient related to the specific fuel consumption of the engine under the given conditions	–
m	Engine bypass ratio	–
C_p	Specific fuel consumption	kg/daN·h
n_{rev}	Number of revolutions	–
T_r	Required thrust	N
T_a	Available thrust	N
F_{fr}	Friction force	N
p_{land}	Landing wing loading	N/m^2
W_{end}	The aircraft weight at the end of cruise	N
R	Range	km
$r = \frac{Ma}{C_p}(L/D)$	Range coefficient	–
a	Speed of sound	m/s
Λ	Wing sweep angle	degrees
M_{cr}	Critical Mach number	–
W_{str}	Structure weight	N
T_{HV}	Thrust set for the given velocity and altitude	N
ξ_V	Thrust change due to velocity	–
Δ	Relative air density	–
$\overline{m}_{PP}$	Mass ratio of the power plant	–
γ	Specific weight of the power plant	
C_{Di}	Induced drag coefficient	–
$\frac{t}{c}$	Thickness ratio	–

(continued)

(continued)

Variables		
Symbol	Description	Units
$C_{D,wave}$	Wave drag coefficient	–
$AR_f = \frac{l_f}{d_{mid}}$	Fuselage aspect ratio	–
d_{mid}	Fuselage midsection diameter	m
l_f	Fuselage length	m
$C_{D,wing}$	Wing drag coefficient	–
q	Dynamic pressure	N/m^2
S	Wing area	m^2
$C_{D,t}$	Tail drag coefficient	–
$C_{D,f}$	Fuselage drag coefficient	–
S_t	Tail area	m^2
S_{mf}	Fuselage midsection area	m^2
$C_{D,n}$	Nacelle drag coefficient	–
S_{mn}	Nacelle midsection area	m^2
$C_{D,int}$	Interference drag coefficient	–
$C_{D,i}$	Induced drag coefficient	–
$C_{D,wave}$	Wave drag coefficient	–
$C_{D,p}$	Pressure drag coefficient	–
$C_{D,0}$	Zero lift drag coefficient	–
$AR_{eff} = ARe$	Effective aspect ratio	–
e	Oswald factor	–
μ	Semiapex of the Mach cone	degrees
V_{climb}	Climb speed	m/s
p	Wing loading	kg/m^2 (H/m^2)
q_H	Dynamic pressure at altitude H	H/m^2
θ	Pitch angle	degrees
γ	Roll angle	degrees
ψ	Yaw angle	degrees
φ_T	Thrust vector deflection angle	degrees
H_{en}	Specific mechanical energy (energy level)	m

Table 5 Summary of the relationships between the parameters and performance of a combat aircraft

No.	Parameter	Designation	Effect on performance
1	Thrust-to-weight ratio	$\overline{T}$	With increasing $\overline{T}$: – the maximum speed increases; – the static ceiling increases; – the thrust limit of the power plant normal load factor increases; – the angular velocity of sustained turns increases; – the tangential load factor increases; – the energy climb rate increases; – the acceleration time to the set speed decreases; – the take-off roll distance and time decrease; – the space required to perform sustained maneuvers decreases
2	Wing loading	p	With decreasing p: – the minimum allowable speed decreases; – the landing and unstick speeds increase; – the take-off and landing roll distances decrease; – the space required to perform forced maneuvers decreases; – the available normal load factor at subsonic speeds increases; – the angular velocity of forced turns increases; – the thrust limit of the power plant normal load factor increases; – the maximum range increases; – the oscillations when flying in a turbulent atmosphere increase
3	Lift-to-drag ratio	L/D	With increasing L/D: – the range and endurance increase; – the static ceiling increases; – the thrust limit of the power plant normal load factor increases; – the angular velocity of the sustained turn increases; – the tangential load factor increases; – the energy climb rate increases; – the glide range when the propulsion system is inoperative increases; – the acceleration time to the set speed decreases; – the space required to perform sustained maneuvers decreases
4	Available lift coefficient	$C_{L,\ avail}$	With increasing $C_{L,\ avail}$: – the available normal load factor at subsonic speeds increases; – the angular velocity of the wind-up turn maneuver increases; – the minimum allowable speed decreases; – the space required to execute wind-up turn maneuver decreases; – the safety of flight in turbulence improves
5	Available normal load factor	$n_{z,\ avail}$	With increasing $n_{z,\ avail}$: – the angular velocity of the wind-up turn maneuver increases; – the safety of flight in turbulence improves; – the space required to execute wind-up turn maneuver decreases

References

1. Fomin NA (1961) Proektirovanie samoletov (Aircraft design). Oborongiz, Moscow
2. Lysenko NM (ed) (1984) Aerodinamika i dinamika polyota manevrennyh-samoletov (Aerodynamics and flight dynamics of maneuverable aircraft). Voenizdat, Moscow
3. Myshkin LV (2006) Prognozirovanie razvitiya aviacionnoj tekhniki (Forecasting the development of aviation technology). Fizmatlit, Moscow
4. Babich VK (1983) Istrebiteli menyayut taktiku (Fighters change tactics). Voenizdat, Moscow
5. Levitskiy SV, Sviridov NA (2008) Dinamika polyota (Flight dynamics), VVIA named after Prof. Zhukovsky NE, Moscow
6. Abchuk VA (1979) Spravochnik po issledovaniyu operacij (Handbook of Operations Research). Voenizdat, Moscow

Optimization in Design. Methods of Solving Optimization Problems

Nikolai K. Liseitsev

Abstract The chapter discusses the role of optimization in design, the types of optimization problems, and the difference between acceptable and optimal design solutions. The chapter shows how an optimal design problem can be set (including the explanation of simple and compound measures of merit, variables, and limitations). In particular, the chapter discusses Lagrange multipliers method, linear and integer programming, descent methods (including the gradient descent method and the method of the fastest descent), the Newton–Raphson method, and random search methods.

Keywords Mathematical optimization · Optimal design solution · Measure of merit

1 Role and Place of Optimization in Design. Types of Optimization Problems

As was shown in Chap. 2, many design problems can be reduced to the problems of parametric analysis and synthesis. However, prior to analysis, it is necessary to formulate a design alternative in one way or another, while in the parametric synthesis, apart from this, it is required to select the best alternative on the basis of decision-making procedures. It is here that optimization procedures play an important role. On the one hand, they ensure the search for the reasonable parameters within the selected feasible area, and on the other hand, they are a tool for making the reasonable decisions in the presence of alternatives.

The inclusion of optimization procedures in the design process transforms it into an optimal design. It should be noted that the search for extrema is not the only task of optimal design. An equally important task is to study the behavior of the objective function near the extremum, which allows making balanced decisions in conditions

---

N. K. Liseitsev (✉)
Moscow Aviation Institute (National Research University), Moscow, Russia
e-mail: nikolai.liseitsev@yandex.ru

M. A. Pogosyan (ed.), *Aircraft Design*, Springer Aerospace Technology,
https://doi.org/10.1007/978-981-96-4599-2_9

where not all the constraints can be formalized. Finally, an important task is to study the sensitivity of the project to changes in the initial design data.

The term "optimum" was introduced in the eighteenth century by H. Leibniz, taking as a basis the Latin word "*optimus*", which means "the best". The concept of optimum in technical design has some difference from the similar concept in mathematics. By "optimal design solution" or "optimal design", is meant such a solution or design that is the best of the feasible solutions or designs. "Acceptable design solution" means that the solution satisfies all the requirements for the projected object. These requirements narrow down the range of possible solutions, playing the role of constraints in solving the problem.

In practice, under tight constraints, finding at least one acceptable solution is the most important design problem. In the general case, however, the number of acceptable solutions, especially if the constraints are not too tight, can be quite large. This is where the problem of choosing the best (optimal) design solution arises. This can be done only if the acceptable design solution is quantifiable, allowing comparison of options in terms of effectiveness or utility.

There are two approaches to the implementation of the optimization problem-solving structure presented in Fig. 2.6. In one approach, the optimal variant is found by enumerating the parameters according to the logic defined by the designer, who plays the role of feedback in the model. Grid search, scanning, "blind" search methods use this approach [1].

Another approach involves the inclusion of a special formalized management strategy into the model feedback. The logical process of forming the sequence of projects to be improved is carried out automatically on the basis of information on the results of previous iterations. This approach is implemented by means of ordered search methods for extremum of functions and functionals.

Each approach has its own advantages and disadvantages. For example, grid search methods do not require prior specification of the objective function and constraints. The sensitivity of the project to changes in the variables is obvious from the analysis results. Besides, when solving the task, the designer is always in the center of events, controlling the computational process. The mentioned advantages explain the wide use of this approach in real design practice. However, this approach is ineffective when investigating multi-parameter problems. Due to the natural limitations on the number of parameters to be varied, it does not provide truly optimal solutions.

The advantage of formalized methods for finding the optimal set of parameters is that they are potentially more efficient when the dimensionality of the problem is large. They are applicable to a wide class of problems, including those in which human intuition in choosing an improvement strategy is underdeveloped. Moreover, they minimize human biases. The main requirement of such methods is a rigorous quantification of the research area, which is very difficult and often requires simplification of the real problem. In addition, it diminishes the role of experience and intuition of the designer.

There are the following types of optimization problems associated with aircraft design.

1. *The parametric type of problems* represents cases where the optimization objective can be expressed as a function F of a certain number of variables (design parameters):

$$F = f(x_i), i = 1, \overline{n} \tag{1}$$

In this case, F is called the objective function and its arguments are called the parameters. As the vector x changes, the objective function changes, forming a surface in a $n + 1$-dimensional space. In practice, in most cases, a number of constraints are imposed a priori on individual parameters and (or) their combination. These constraints determine the permissible area of variation of the design parameters x.

Under certain assumptions, it is possible to reduce the problems of optimization of both basic aircraft parameters, and structural parameters. An example problem is the minimization of aircraft takeoff weight or the cost of transportation of a ton-kilometer of payload by selecting optimal design parameters ($\frac{W}{S}$, AR, $\frac{t}{c}$, Λ, m_0^*, etc.)

2. *The dynamic type of problems is* associated with processes that are described by a system of ordinary differential equations:

$$\frac{dy_i}{dt} = f\left(y_i u_j\right), i = \overline{1, n}, j = \overline{1, r} \tag{2}$$

where: y_i are the state variables or phase variables; u_j are the control variables or control parameters affecting the course of the process.

The optimization goal is to find a change in control $u_j(t)$ with time t, which provides an extremum of the functional that expresses some objective function F related to the transition of the object from a given initial phase state $y_i(t_0) = y_{i0}$ to the prescribed final phase state $y_i(t_1) = y_{i1}$:

$$F = \int_{t_0}^{t_1} f_0(y_1, \ldots, y_n; u_1, \ldots, u_r)dt = min \tag{3}$$

An example of tasks of the second type is finding the optimal trajectory (e.g., for takeoff and landing) of an aircraft, which provides minimum time of climb (or descent) or minimum fuel consumption.

3. *Tasks of joint optimization of design parameters of an aircraft and its flight trajectory.* The general formulation of such a problem, which combines problems of the first and second types, is considered below. Let us denote the aircraft and engine parameters by the vector x. Let us denote the trajectory parameters characterized by phase coordinates y_i *by* vector Y. The components of this vector can be velocity V, altitude H, range R, flight path angle Θ, and fuel ratio $\overline{m}_f$, i.e.:

$$Y = \left\{V, H, L, \Theta, \overline{m}_f, \right\}.$$

Denote the control variables by the vector U. In our case:

$$U = \{U_1, U_2\},$$

where U_1 is the control variable determining the law of change of aerodynamic forces, it can beα, C_L, etc.; U_2 is the control variable determining the engine mode, it is either throttle setting, or turbine shaft rotational rate n_{rev}, or thrust factor:

$$R_{eng} = \frac{T}{T_{max}}\left(\frac{n_{rev}}{n_{rev,max}}\right)$$

Parameters X take values from the set X_{allow}.

Variables Y and U at each t and x can take values from some set $V(t, x)$, which is determined by the allowable load factor, the minimum flight speed, the maximum Mach number, etc.

The problem is set for the pairs of vector-functions $Y(t)$, $U(t)$ and parameters X satisfying the above constraints to find the element $\left(\overline{Y}\left(\overline{t}\right), \overline{U}\left(\overline{t}\right), \overline{X}\right)$, which delivers the minimum value to the functional J, i.e.:

$$\overline{J} = \left(\overline{Y}\left(\overline{t}\right), \overline{U}\left(\overline{t}\right), \overline{X}\right) = \min J(Y, U, X) \tag{4}$$

Any generalized efficiency criterion can be used as a functional. There are two types of minimization problems:

- minimization of the control functional with fixed parameters (optimal control without parameters);
- minimization of the functional by varying the input parameters obtained
- for the optimum control (non-linear mathematical programming):

The joint solution of these two problems is carried out by the method of successive approximations.

The problem of the described type can be solved on the basis of an integrated model, the structure of which is presented in Fig. 1.

In practice, at the early design stages, most practical problems can be reduced to problems of the first type, so we will focus on them.

2 Setting the Optimal Design Problem

In order to set an optimization problem it is necessary to:

(1) define optimization objective (select the objective function F);
(2) choose the parameters that most significantly affect the objective function (x_i, $i = 1, \ldots, n$);

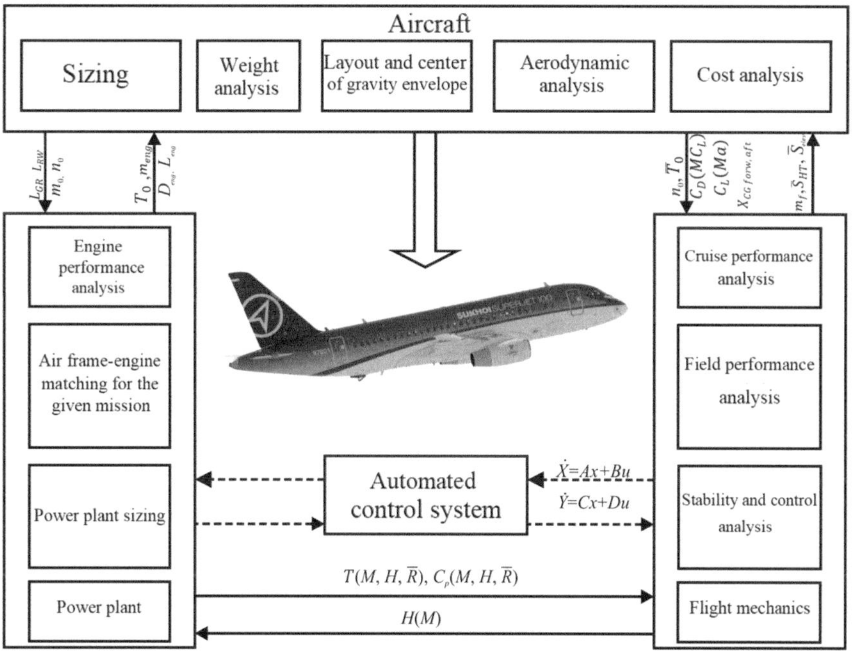

Fig. 1 Structure of joint optimization of airframe, power plant and control system parameters

(3) present the objective function in terms of optimization parameters $F = f(x_i)$ in such a mathematical form which, on the one hand, would reflect most closely the true impact of each parameter, on the other hand, would provide the greatest simplicity of the forthcoming optimization process;
(4) provide a mathematical description of all constraints on individual parameters and (or) their combination.

Such constraints can be expressed in the form of equations:

$$\psi_j(x_i) = B; \quad j = 1, \ 2, \ldots, m \tag{5}$$

or in the form of inequalities, when permissible upper and lower limits of parameter variation are specified:

$$x_i^{Lower} \leq x_i \leq x_i^{Upper} \tag{6}$$

In addition, constraints of the inequality type can also be imposed on the combination (function) parameters $E_k(x_i)$:

$$C_k^{Lower} \leq E_k(x_i) \leq C_k^{Upper}, \ \mathrm{k} = 1{,}2, \ldots, \ \mathrm{p} \tag{7}$$

The constraints listed above define the allowable range of variation of the design parameters X. This area in abbreviated notation is defined as follows:

$$X_{allow} = \left\{ \frac{X}{g_j(X)} \leq 0, j = 1,2, \ldots, n \right\} \quad (8)$$

Here, $g_j(X)$ are the limitations.

Any vector X belonging to the allowable domain ($X \in X_{allow}$) defines an allowable version of the project, and the set of such vectors forms the allowable domain of the project parameters.

In the allowable domain of the project parameters (project feasibility area), a project can exist whose parameters deliver extremum to the objective function. Suppose that we are looking for the minimum of the objective function. Then the optimal design problem is formulated as follows: find a vector of design parameters $X^* = (x_1, x_2, \ldots, x_m)$, which delivers a minimum value to the optimality criterion when the system of inequalities (8) is satisfied. A brief notation of this problem is:

$$\min F(x),$$

$$X \in X_{allow},$$

$$X_{allow} = \left\{ \frac{X}{g_j(X)} \leq 0, j = 1,2, \ldots, n \right\} \quad (9)$$

This notation means that the subset of design parameters X_{allow} of the n–dimensional Euclidean space R^n consists of all those vectors $X = (x_1, x_2, \ldots, x_m)$ for which the given constraints are satisfied. The notation in the form (9) corresponds to the most complicated problem of finding a constrained extremum of a multi-parameter function.

A particular kind of the above problem is the problem of finding an extremum of a real function of n real variables containing no constraints:

$$\min F(x),$$

$$X ! R^n \quad (9')$$

Finally, the simplest of the optimization problems is the problem of finding an extremum of a real function of a real variable:

$$\min F(x),$$

$$a \leq x \leq b \quad (9'')$$

where a and b are the given boundaries of the variable.

Depending on the formulation of the optimization problem, type of the objective function, presence and type of constraints, certain optimization methods are used. They are considered in the next section.

3 Optimization Methods

Optimization methods can be divided into two large groups—analytical methods and numerical methods. Analytical methods include *differential calculus, variational calculus, Lagrange multipliers method*. Application of these methods imposes a significant constraint on the optimization problem being solved which is that they require an analytical expression for the objective function (functional) which, moreover, should be at least once differentiable and have a finite number of discontinuities. To use the classical methods of identification of function and functional extrema–differential and variational calculus—in the general case, the absence of constraints on the required parameters is also mandatory. The method of Lagrange multipliers is used in the presence of constraints in the form of equalities. Lagrange multipliers are also used in variational calculus and some numerical optimization methods.

In the problems of classical analysis of the unconditional extremum of a function of one or many variables the necessary condition for the optimum is equality to zero of the gradient of the function $F(X)$:

$$''F\left(X^*\right) = \left(\frac{dF}{dx_1}, \ldots, \frac{dF}{dx_n}\right) = 0. \tag{10}$$

If there are constraints in the form of equations, the optimal solution for X^* will be determined by the ratio:

$$''L\left(X^*, \lambda^*\right) = 0, \tag{11}$$

where $L(X) = F(X) + \lambda^T g(X)$ is the Lagrange function, and $\lambda^T = (\lambda_1, \ldots, \lambda_m)$ is the vector of uncertain factors. Sufficient optimality conditions are related to the notion of Hesse matrix (Hessian) H, which is a matrix of second partial derivatives of the function $F(X)$:

$$H = \left\| \frac{d^2F(X)}{dx_i dx_j} \right\|, \; i, j = 1, \bar{n} \tag{12}$$

To be X^* a minima point, it is sufficient to fulfill the following condition:

$$\Delta X^T H\left(X^*\right)\Delta X \geq 0 \text{ for any } \Delta X = 0.$$

The notion of gradient and hessian is used in many algorithms of extremum search.

The methods of differential calculus find application in solving particular optimization problems when studying relatively simple models with one or three variables.

If the objective function is represented in the form of functionals, methods of calculus of variations are applied. These methods find application in optimizing the parameters of the flight path and the engine thrust change over time. In this case the optimization problem is reduced to integration of the system of differential Euler equations. Other examples of the use of variational calculus methods in solving aircraft design optimization problems, are the optimum spanwise thickness ratio distribution for a wing and the optimum twist of a supersonic aircraft wing to maximize the lift-to-drag ratio. Attempts to solve more complex aircraft optimization problems using variational calculus encounter significant difficulties associated with the need to solve the cumbersome boundary problem.

As a whole, the mathematical apparatus of optimization theory briefly described above, which had a noticeable influence on the development of mathematics in the XIX century, met the practical needs of designers only to a small extent. The emergence in the second half of the last century of high-speed computers and the related development of numerical optimization methods opened up new possibilities for the practical implementation of optimal design methods.

This is the period in which a new applied discipline called operational research began to emerge. The initial stage in the development of this discipline involved attempts to apply scientific principles to the design of combat operations in a manner that takes into account the careful use of scarce resources. This discipline gradually combined the previously existing and newly created methods of optimizing "operations", meaning by this term any purposeful action, contributing to the balanced choice of the method of action, a variant of the plan, i.e. decision-making. Decision-making tasks have much in common for diverse areas of human activity. This commonality served as the basis and impetus for the development of operations research and its wide application in planning, management and design. A set of numerical optimization methods developed mainly within the framework of this discipline made it possible in the late 1960s to form a branch of applied mathematics that was called *mathematical programming* or optimization theory.

The first, most significant, results of using mathematical programming were achieved when solving linear programming problems, i.e. problem (9), in which both objective function and constraints were linear functions of parameters. A large range of practical problems encountered in such spheres of human activity as organization and planning of production activities, distribution and use of various resources, transportation, military affairs, etc. can be reduced to this class.

At present, methods of solving *linear programming* problems are rather well developed. These methods can be divided into general and special ones depending on the range of problems to be solved. Among general methods, the *simplex method* and its variants are the most widespread. This includes the modified method of solving multipliers of L.V. Kantorovich.

Special methods include linear bifurcation method, distributive method, etc.

A special class of mathematical programming problems are the problems in which variables are indivisible into parts, i.e. they are integers. In aircraft design, such variables may be, for example, the number of engines, the number of crew members, the number of windows, the number of hydraulic systems, the number of wing structural members, etc. Such problems are called *integer (discrete) programming* problems. In these problems the area of admissible solutions, in which it is necessary to select discrete points, is non-convex and disconnected. To solve such problems it was necessary to develop special methods, among which the most famous are cutoff methods, e.g. Gomori's method.

Let us also mention combinatorial methods for solving integer programming problems. Among them, the most famous is *the branch and bound method.*

It allows to solve both linear and nonlinear discrete programming problems.

Successful application of linear programming methods for solving optimization problems in a wide variety of fields, as well as their extension to nonlinear problems admitting acceptable linear approximation, stimulated work of specialists in the field of operational research on other problems, which could be reduced to solving optimization problems. Thus, on the basis of the optimality principle, R. Bellman developed a rather general computational method of solving mathematical programming problems, known as *dynamic programming.*

Dynamic programming is focused primarily on solving optimization problems related to the study of systems whose characteristics depend on time. These include, for example, problems of automatic control of an aircraft, economic planning, etc.

L.S. Pontryagin's works on further development of the variational calculus led to the formulation of *the maximum principle.* This allowed us to put the theory of optimal control on a rigorous mathematical basis and opened up wide possibilities for its practical application.

By the same period, a number of computer-oriented algorithms were developed to find local optima of multi-parameter functions in the presence and absence of parameter constraints. However, in the general case, problems (9) in which both the target function and the constraints are non-linear dependences on X, called *nonlinear mathematical programming* problems, do not offer a single algorithm or a class of algorithms for their solution having universal applicability. The difficulty of solving such problems is explained by the fact that under nonlinear constraints the admissible region of parameter variation X_{allow} may not be a convex set or even consist of a number of non-convex regions. If for a convex function $F(X)$ and a convex set X_{allow}, as we have already mentioned, the local minimum is at the same time a global minimum, then for a nonconvex region X_{allow}, even for a convex function, the nonlinear programming problem $F(X)$ can be multi-extreme, which significantly complicates its solution.

Nevertheless, for a number of practically interesting partial problems of nonlinear programming very effective numerical methods of solution have been developed. Thus, if $F(X)$ is the quadratic function and the constraints are linear functions, then the problem (9) is called *a quadratic programming* problem. For the majority of such problems, universal algorithms have been developed that provide an optimal solution for a finite number of iterative process steps.

An optimization technique for nonlinear programming problems in which the objective function and constraints are polynomials with respect to non-negative variables $x_1, x_2, \ldots, x_n$ with positive coefficients was developed:

$$F(X) = \sum_{i=1}^{k} p_i(X),$$

where $p_i(X) = c_i x_1^{a_{i1}} x_2^{a_{i2}} \ldots x_{1n}^{a_{in}}$, $c_i > 0, x_j > 0, j = 1, \ldots, n$, and the power coefficients $a_{i1}, a_{i2}, \ldots, a_{im}$ are arbitrary real numbers.

Such problems, called *geometric programming* problems, are often encountered in design practice.

Before discussing methods for solving problem (9) of a general form, let us consider methods for solving particular problems of the form (9′), i.e. unconstrained optimization problems.

Mastery of methods for solving such problems is necessary because algorithms of minimization with constraints are constructed on the basis of algorithms of minimization without constraints. Besides, an optimization problem with constraints is often solved by transforming it into an optimization problem without constraints.

The solution of problem (9′) is complicated in general form. Only in the case of convexity of the objective function, the optimal solution X^* can be found by fulfilling the necessary optimality conditions–equality to zero of the gradient of the function, i.e. $''F(X^*) = 0$. However, for the majority of practical problems the convexity either does not exist or cannot be verified. In this case to solve the problem it is recommended to choose several initial search points and then take the lowest of all endpoints.

Let us consider numerical methods for solving problem (9′), which are based on procedures for systematically obtaining a sequence of vectors, i.e. points $X^0, X^1, \ldots, X^k$ in R^n such that

$$F\left(X^0\right) > F\left(X^1\right) > \ldots > F\left(X^k\right) > \ldots,$$

and estimate their efficiency in locating the minimum points. Such methods are called *descent methods.*

Solution of the problem begins with the choice of the starting point $\left(X^0\right)$. One tries to use any available information about the behavior of the function $F(X)$ in order to choose X^0 not too far from the minimum point. After choosing the starting point, it is necessary to decide in which direction the next points are supposed to be located and then what is the step size in the chosen direction. In any descent method the sequence $\{X\}$ obeys the condition:

$$X^{k+1} = X^k + t^k d^k, k = 0,1 \ldots,$$

where d^k is the direction, $\|t^k d^k\|$ is the step value.

If where d^k is normalized $\left(\|d^k\| = 1\right)$, the step value will be equal to $\left|t^k\right|$.

By changing the selection procedure for d^k and t^k, it is possible to vary the descent methods. Depending on the method of selecting the direction of the next step, the descent methods are divided into three groups, namely:

- methods that only use the information about function values to select direction; they are called zero-order or *direct methods*;
- methods using in addition the first derivatives of functions, called *first-order methods*;
- methods that require computing the second derivatives of functions, called *second-order methods*.

The choice of the step value largely determines the efficiency of extremum search, including the required number of steps (iterations), on which the computer-intensity depends. Among many methods of step selection the most effective ones are based on minimization of function $F(X)$ in chosen direction, i.e. solving problem (9″) of one-parameter optimization.

Graphical interpretation of direct methods (a) and methods using derivative values (b) is presented in Fig. 2a and b. It can be seen that they differ in the choice of search direction.

In first-order methods, the direction of motion at each step coincides with the antigradient of the function. In these methods, called gradient methods, the point X^{k+1} is chosen with respect to X^k in the direction r^k:

$$X^{k+1} = X^k - tr^k,$$

where $r^k = {''}F\left(X^k\right)$ is gradient of function $F(X)$ in arbitrary point X^k, t is some positive scalar value called search step.

By choosing the magnitude of t, it is possible to satisfy the condition $F\left(X^{k+1}\right) < F\left(X^k\right)$. Depending on the method of step t selection, there are varieties of gradient methods, among which the most famous are *the gradient descent method and the method of the fastest descent*.

In the first method, we set a step of small enough value and calculate the gradient at each step (Fig. 2b). This circumstance can lead to excessively high computer-intensity. In the method of the least-cost descent the step value is calculated optimally by solving the one-dimensional minimization problem meeting the following constraint:

$$F\left(X^k - t''F\left(X^k\right)\right) = \min_{t \geq 0} F\left(X^k - t''F\left(X^k\right)\right).$$

In this method it is not required to compute the gradient at each step. The new direction is chosen only if the objective function no longer improves in the former one. This can lead to a gain in the number of computer operations. A number of other methods belong to first-order methods, in particular, a very effective method of conjugate directions and its special case, the conjugate gradient method.

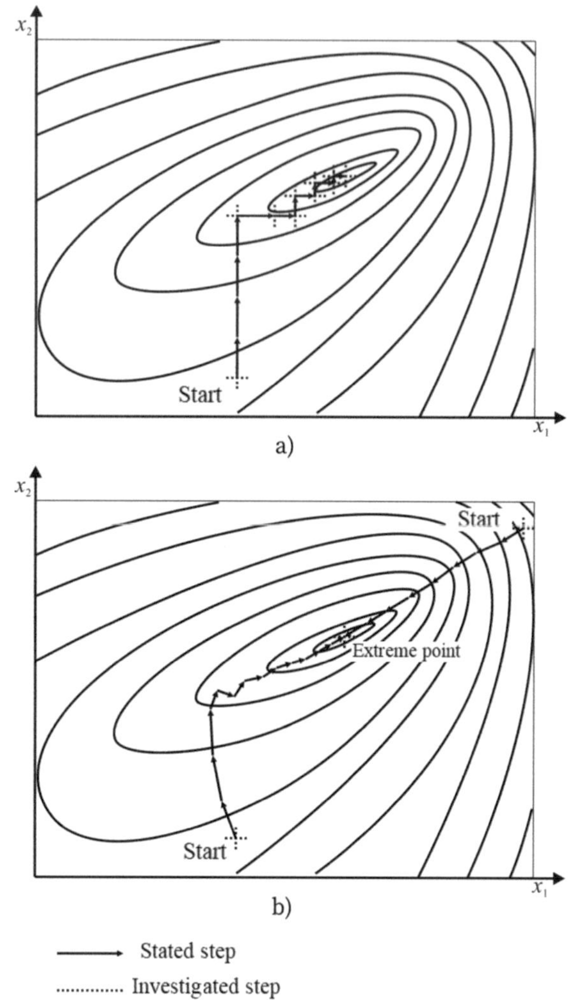

Fig. 2 Graphical interpretation of extremum search by descent methods

In gradient methods, the process of extremum search stops if the following conditions are fulfilled for all components of the gradient vector:

$$\left|\frac{dF\left(X^k\right)}{dx_j}\right| \le \delta, j = -\overline{1, n},$$

where δ is the analysis accuracy.

Essential disadvantage of gradient methods is their weak convergence in case the optimum lies in a “ravine” or on a long narrow “ridge”. In this case trajectory of gradient methods is represents a rather slow zigzag motion along the bottom of “ravine” or “ridge” to the minimum point.

There are a number of techniques to avoid 'ravine' effects, e.g. by changing the scale of independent variables, which changes the topography of the objective function response surface (equal level line trajectories). However, the most effective way to overcome such effects is to switch to *second-order* methods as the second derivatives of the minimizing function respond to the curvature of level lines. These methods are a generalization of *Newton's method*, which can be interpreted as a method for sequentially finding points of minimum quadratic approximations of the function $F(X)$. The algorithm of Newton's method is as follows:

$$X^{k+1} = X^k - H^{-1}\left(X^k\right)''F\left(X^k\right),$$

where $H(X^k)$ is the Hesse matrix of the objective function calculated at X^k. It is clear that in the case of a strictly quadratic function $F(X)$ Newton's method gives the solution of the problem in a single step. However, in the general case of the non-square function $F(X)$, this method is potentially more effective in finding the minimum than the gradient methods, since the quadratic function locally approximates the function $F(X)$, more exactly than the linear one, which is the basis of the gradient methods.

The disadvantages of Newton's method are mainly related to significant computational difficulties caused by the necessity to calculate the second derivatives and the Hesse matrix circulation. In addition, the method does not provide convergence if the objective function is not strongly convex and the initial approximation is chosen poorly.

There are also a number of methods representing modifications of Newton's method aimed at eliminating its drawbacks. Thus, to reduce the difficulties associated with the calculation of the second derivatives, in *the variable metric method* (the Fletcher-Powell method) the minimum is calculated as follows:

$$X^{k+1} = X^k - tH^{-1}\left(X^0\right)''F\left(X^k\right).$$

The Hesse matrix is evaluated using approximate formula without calculating the second partial derivatives. The step size is determined by one-dimensional minimization of the objective function on the half line $H^{-1}{}''F\left(X^k\right)$.

The Newton–Raphson method, in addition to determining the search direction, introduces a procedure for selecting the step length along it:

$$X^{k+1} = X^k - tH^{-1}\left(X^k\right)''F\left(X^k\right).$$

Note that methods of variable metrics and conjugate directions have good convergence, so their application is preferable for solving parametric optimization problems.

What is common for first- and second-order methods, as we have seen, is that all of them involve search for local minima on a sequence of directions (half lines). This search is a part of the general minimization procedure. Finding a local minimum of the objective function X^k along a given direction is a one-parameter minimization problem (9″). A sufficient number of algorithms for solving this problem have been developed. Let us name three of them which are the most frequently used. These are *the golden section method, the Fibonacci method*, and *the chord method*.

The first two methods do not require calculating the derivative of the function. Their main idea is to reduce the number of iterations and to enclose the minimum in sequentially decreasing nested intervals $[a_i, b_i]$. To this end, the function is calculated for two interior points l_i (left) and r_i (right) of the current interval $[a_i, b_i]$. Either $[a_i, r_i]$, or $[l_i, b_i]$ is used as a new smaller interval $\left[a_{i+1}, b_{i+1}\right]$. Both methods are based on that the remaining interior point l_i or r_i of the interval $\left[a_{i+1}, b_{i+1}\right]$ is used in the next step as one of the interior points of that interval. Both methods are applicable for optimizing unimodal (having one extremum) functions. In this case the Fibonacci method provides for the maximum reduction of the interval for a given number of iterations. However the Golden Ratio method is somewhat easier to implement and therefore, generally speaking, is preferable.

The chord method is based on the use of a discrete version of Newton's method for finding the zero of the derivative $F'(x)$ in the interval $[a_i, b_i]$, if it exists. Newton's method involves linearization of $F'(x)$ in the neighborhood of the current point and choosing the zero of this linearized function as the starting point of the next iteration.

Let us now consider *direct methods* of unconditional optimization. In a number of cases of practical interest, these methods prove to be the only possible ones. This refers to the cases when calculation of gradients and hessians of a complex function by numerical methods is either impossible or may require more computer time than calculation of function values necessary for finding the minimum. Such situation may occur if the objective function has discontinuities or is defined not in an implicit form, but by a system of equations related to different subsystems of some system (which is often the case when building models of complex systems or processes). Another case where zero-order methods can compete with first- and second-order methods is the case where the objective function has several local extrema.

A characteristic feature of direct methods is their heuristic nature, the lack of strict justification. This circumstance requires experimental verification of convergence and efficiency of methods by solving specific problems.

Let us name the most famous of the methods of this group: *the coordinate descent method* (Gauss–Seidel method), *the method of configurations* (Hook and Geaves method), *the deformable polyhedron method* (Nelder and Mead method), *Rosenbrock method, Powell method*.

The simplest of them is the coordinate descent method, which involves the sequential minimization of the objective function with respect to separate variables. In general case the method does not provide exploration of the function during one cycle, which requires many iterations. Besides, the method is inefficient in minimization of ravine functions. Application of this method is expedient in combination with other methods.

The method of configurations in its essence is a fairly general method of direct search. According to this method, for a selected starting point X^0 by changing one or more values of the vector components its vicinity is examined (Fig. 2a). After finding an acceptable direction, the function is computed in this direction at a gradually increasing step. This continues as long as the search in this direction continues to lead to points X with smaller values of the function $F(X)$. If no point with smaller function values can be found in the adopted direction, the step size is reduced. After several successive decreases of the step size, the adopted configuration (direction) is abandoned and a new survey of the neighborhood is undertaken.

The method allows accounting for the constraints imposed both on individual variables and on the search area, such as irregular boundaries and isolated forbidden regions. The disadvantage of the configuration method is that it can "get stuck" near the local minimum.

The deformable polyhedron method is based on the configuration method. Here the function of n variables is minimized using $n + 1$ vertices of some deformable polyhedron in the space of these variables R^n. A vertex (point) in R^n, in which the value $F(X)$ is maximal, is projected through the "center of gravity" of the remaining vertices. The improved (lower) values of the objective function are found by consecutive replacement of the point with the maximum value of $F(X)$ by "better" points until the minimum is found.

Rosenbrock's method, which is also an evolution of the coordinate descent method, uses a procedure to rotate the orthogonal coordinate axes so that one of them coincides with the direction of the ravine to overcome the difficulties of finding the minimum in the presence of narrow ravines.

In a more complex Powell's method, the extremum is searched along conjugate rather than orthogonal directions. It also improves the linear search procedure, allowing finding the global optimum instead of the local one.

Random search methods occupy a special group among direct methods. In these methods, in contrast to the previously considered deterministic methods, the choice of points in the objective function definition area is carried out randomly.

When using these methods, the success largely depends on the a priori knowledge of the possible location of optimal points. These methods can be especially effective for optimization problems with discrete variables (i.e. integer programming problems) as well as mixed problems with discrete and continuous variables. Among the random search methods the best known are *random point search and random ray search.*

Let us now consider ways of solving the nonlinear programming problem with constraints, i.e. problem (9), which is of most interest for the designer. There are two different approaches to solving such problems. The first one is based on direct consideration of constraints when solving the problem. There are a number of methods based on this approach. These include, for example, *the method of possible directions* (Zeutendijk's method), *the projective gradient method*, and *approximating linear programming methods.*

The second approach is based on attempts to transform problems with constraints to simpler problems without constraints. This goal can be achieved by replacing variables or modifying the objective function. The most widespread is *the penalty functions method,* the main idea of which is to reduce the constrained extremum problem to a sequence of unconstrained optimization problems by using penalty functions.

Let us briefly describe the essence of some of the above mentioned methods of constrained optimization.

The method of possible directions implements the following idea. The objective function is minimized as a function without constraints until a constraint is encountered. Then a search direction is found that allows reducing the target function without violating the constraints. Such direction is called admissible (possible). Denote it by S^k, then the new point will be determined by the relation

$$X^{k+1} = X^k + h^k S^k.$$

The point X^{k+1}, and consequently, the direction S^k will be admissible if the following conditions are satisfied:

$$\frac{dg^T(X^k)}{dX} S^k \leq 0 \tag{13}$$

Solving the problem

$$\min{''}F^T(X^k)S^k \tag{14}$$

under condition (13), we define a vector of admissible directions S^k along which the objective function has the largest decreasing rate. The step size h^k along the chosen direction can be determined by solving the following single-parameter minimization problem:

$$F(X^k + h^k S^k) = minF(X^k hS^k),$$

$$g(X^k + hS^k) \leq 0, h \geq 0.$$

The advantage of the method of possible directions is its universality; the disadvantages of the method are the impossibility to take into account the constraints in the form of equations and the large amount of calculations at each iteration.

The projective gradient method is a modification of the method of possible directions. In contrast to the latter, in this method when a point X^k gets into the boundary region the admissible direction of search S^k is determined not by solving problem (14), but by projecting the antigradient $''F(X^k)$ on a polyhedron which is a linear approximation of the admissible set near the point X^k. This makes it possible to take into account both inequality-type and equality-type constraints.

The essence of approximating linear programming methods is to reduce the nonlinear programming problem to a linear programming problem by replacing the nonlinear objective function and constraint function with a sequence of approximating linear functions. In some algorithms this goal is achieved by using linear interpolation of nonlinear functions, in others—by Taylor series expansion in the vicinity of point X^k.

In the penalty functions method the constrained optimization problem is reduced to the equivalent problem of unconstrained optimization by transforming the objective function. The new objective function $F'(X)$ is formed by adding to the objective function $F(X)$ a penalty function composed of constraining conditions in such a way that approaching the boundary of the feasibility area leads to a sharp increase in the new objective function, i.e. the violation of constraints is penalized by worsening $F(X)$.

Depending on whether the unconstrained extremum solution is inside or outside the initial feasible area, there are two types of algorithms for solving problems by the penalty function method: the internal penalty function algorithm and the external penalty function algorithm.

In the first case, the search for the optimum must start from the feasible area and its path will lie entirely within this area.

This is achieved by forming a new objective function, e.g. of the form:

$$F'(X) = F(X) + R_k \sum_{i=1}^{m} \frac{1}{g_i(X)},$$

where $R_k > 0$ is the weighting factor.

When approaching the boundary of the feasible area from within this area, any element of the constraint vector tends to zero and hence the penalty function approaches infinity. The disadvantage of this algorithm is the necessity to choose an initial point inside the feasible area and its inapplicability in the case of constraints in the form of equalities.

In the second case, the search can start from any point, including those outside the feasible area. In this case, the penalty function is chosen in such a way that the values of the new objective function in the feasible area are exactly or approximately equal to the values of the original objective function, and outside this area they significantly exceed the values of $F(X)$. A possible form of such new objective function is as follows:

$$F'(X) = F(X) + R_k \sum_{i=1}^{m} \delta g_i(X),$$

$$\delta_i = 0 \text{ if } g(X) \leq 0$$

$$\delta_i = 0 \text{ if } g(X) > 0$$

The optimization methods considered above refer to the study of deterministic (non-random) functions and processes, but in design practice it is necessary to solve optimization problems involving random factors. Such tasks are solved by stochastic programming methods. Usually while solving such tasks mathematical expectations of values are considered, i.e. stochastic problem is reduced to deterministic one, which is not strict, or linear programming is involved to solve stochastic problems. These methods are applied in the analysis of models describing conflict situations.

In aircraft design practice, there may be simultaneous optimization tasks for several measures of merit. Suppose a designer is tasked with obtaining the best values for several aircraft characteristics, for example, to maximize the flight range, minimize the required runway length and take-off weight of the aircraft. As a rule, these characteristics are contradictory, and optimization with respect to each of them would lead to different values of design parameters X. In cases where it is not possible to find a generalized measure of merit that includes these partial criteria, the task of multi-criteria (vector) optimization arises. For a multi-criteria problem in general case the solution is not optimal for any particular case. At the same time it is a compromise for the vector criterion:

$$F(X) = [F_1(X), \ldots, F_m(X)].$$

Such a solution is called a trade-off region or a Pareto-optimal solution region. Such optimality means that it is impossible to improve the value of one of the partial criteria without worsening the value of at least one of the other criteria. To define a Pareto minimum it is necessary to pass from vector optimization problem to a nonlinear optimization problem with a specially designed scalar objective function by solving previously the convolution problem of vector optimality criterion.

The way of convolution of the vector optimality criterion depends on the information about the degree of comparability of private optimality criteria.

4 Concluding Remarks

In conclusion, let us consider some recommendations on the choice and practical implementation of optimization methods in solving design problems.

To solve an optimization problem, an engineer needs to select a method that unambiguously determines the order of operations leading to the solution. Of course, the very definition of the task allows you to discard some methods as unacceptable. As we have shown, solution methods are different for problems with weak nonlinearity and a small number of variables and for problems with a pronounced nonlinearity with a small and a large number of variables; for problems with a single extremum and with many local extrema. Nevertheless, the choice is still big enough and in order to make it the designer should be qualified enough. They must have a good understanding both of the physical essence of the problem and peculiarities of the optimization algorithm. The concept of applying optimization programs according to the "black box" principle can lead to solution results quite far from optimal ones.

Experience shows that there is no universal method that can be successfully applied to a wide range of practical problems of optimizing the appearance and structures of an aircraft. In this connection, when developing aircraft design optimization programs, we followed the path of applying a combination of different methods in such programs, for example, several direct methods and several first- and second-order methods. This allows different methods to be used for the initial and final phases of problem solving. Developed computer-aided design systems have a special optimization subsystem, which is a complex software package. It includes a combination of different methods of extremum search united by a special control program, which performs transition to different methods depending on the behavior of the objective function at different iterations.

Such a system should feature both automatic and interactive mode, allowing the user to make changes both to the model and to the initial data, as well as to use various empirical techniques allowing accelerating convergence. A possible structure of such a system is presented in Fig. 3.

The experience of using optimization software with aircraft analysis models has shown that this does not require significant modification of the latter. The refinements are mainly related to the need for scaling of the parameters, constraints and objective functions. Scaling facilitates finding an extremum, since it is necessary to estimate distances when choosing the direction and magnitude of the step. This operation requires that all parameters have the same dimensionality or be dimensionless at all.

In addition to the above benefit, scaling leads to similarity of different optimization problems, facilitates analysis of optimization results by allowing comparison of the relative "contribution" to the change in criterion of each variable.

In the scaled form all variables have the order of one. The following scale algorithm can be proposed.

1. Define the scaled variables $\overline{x}_i$:

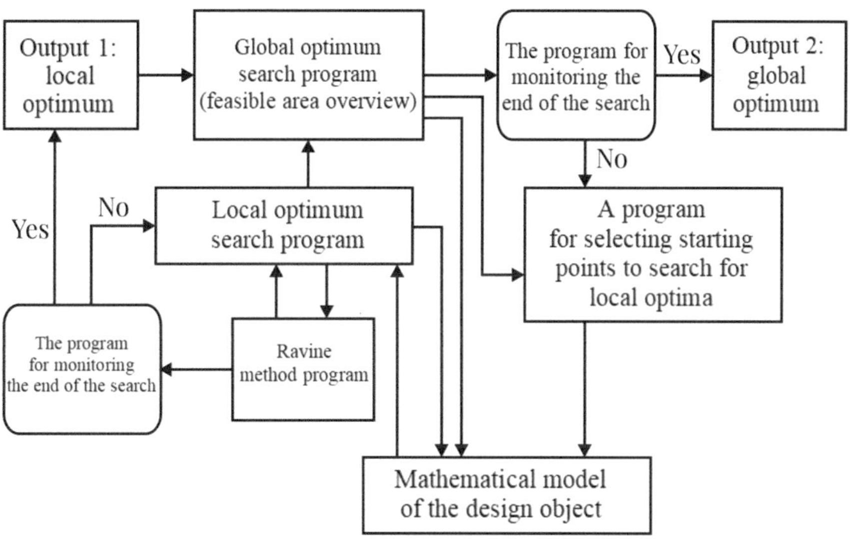

Fig. 3 Optimization system structure

$$\begin{cases} \overline{x}_i \underline{\Delta} \frac{x_i - x_i^l}{x_i^u - x_i^l} & i = 1,2,\ldots,n \\ 0 \leq \overline{x}_i \leq 1 \end{cases}$$

2. Define scaled constraints in the form of equations:

$$\overline{\psi}_j(x_i) \underline{\Delta} \frac{\psi_j(x_i)}{B} = 1, \quad j = 1,\ 2,\ \ldots,\ m$$

3. Define scaled constraints in the form of inequalities:

$$\overline{E}_k(x_i) \underline{\Delta} \frac{E_k(x_i)}{C_k^{\hat{a}}} \leq 1 \left(\text{or } \overline{E}_k \geq 1\right), k = 1, 2, \ldots, p.$$

In these equations, the symbol $\underline{\Delta}$ means equality by definition, and the indices “u” and “l” are the upper and lower limits of parameter variation, respectively.

The objective function can also be scaled using its value at the initial calculation point given by the vector of input parameters $X_{input,i}$ or, equivalently, $-\overline{X}_{input,i}$. The value of the scaled objective function will be equal to:

$$\underline{F}(X) \underline{\Delta} \frac{F(X)}{F\left(X_{input,\ i}\right)}.$$

It can be argued that with current advances in mathematical programming almost any optimization problem of nonlinear systems with several hundreds of variables and constraints can be solved. Nevertheless, there are certain restrictions on the dimensionality of the problem to be solved. In this case the number of variables rather than the number of constraints is more critical. The matter is that the growing dimension of the problem not only complicates the process of analyzing the results but, more importantly, reduces the convergence of the optimization process. This is a consequence of computational accuracy when dealing with complex non-linear functions.

The practice of application of formal optimization methods confirms their high efficiency in solving aircraft optimization problems. Thus, the optimization of hypersonic passenger aircraft with the help of Boeing’s AESOP (Automated Engineering and Scientistic Optimization Program) developed by specialists for five parameters only (p, $\frac{T}{W}$, AR_{wing}, AR_{fus}, cabin pressure differential) allowed to increase the objective function (number of passengers) by 15% compared to the nominal value obtained in the traditional design.

Nevertheless, due to a number of reasons, the use of modern methods of optimal design has not yet become an everyday practice in the activities of aircraft engineers [4]. Progress in this direction should be considered as a very effective way to improve the quality of engineering solutions.

References

1. Aoki M (1971) Introduction to optimization techniques: Fundamentals and applications of nonlinear programming. Macmillan, New York
2. Shoup T (1979) A practical guide to computer methods for engineers. Prentice-Hall, New Jersey
3. Haug EJ, Arora JA (1979) Applied optimal design. John Wiley&Sons Ltd., Chichester
4. Ashley H (1982) On making things the best—aeronautical uses of optimization. J Aircraft 19:1

Design Features of Commercial Airliners and Cargo Aircraft

Valentin A. Kiselev

Abstract The chapter introduces efficiency indicators for a commercial airliner, discusses and compares the features of airliner's aerodynamic configurations, the potential of increasing its size and weight, and promising innovative layouts, such as multi-deck fuselages, "tailless," and "flying wing" aircraft. The chapter also provides details of the design of passenger cabins, emergency and safety equipment, baggage and cargo compartments, galley, and lavatory. The chapter introduces a method for defining the optimal cruise speed and thrust capacity for a typical commercial airliner. The last part of the chapter is devoted to the features of supersonic passenger aircraft, including the issues of aerodynamic heating and sonic boom.

Keywords Efficiency indicator · Tailless aircraft · Flying wing aircraft · Passenger cabin · Optimal cruise speed · Supersonic passenger aircraft

1 Main Trends and History of Commercial Airliners Development

The criterion for evaluating the efficiency of civil aircraft is the total cost, which is based on the transportation cost per tonne-kilometre. The cost can be estimated as follows:

$$a = \frac{100A}{k_{pay} m_{pay} V_{service}},$$

where A is the cost of aircraft operation, rubles/hour;

m_{pay} is payload mass, t;

$V_{service}$ is cruising speed, km/h;

k_{pay} is payload factor.

V. A. Kiselev (✉)
Moscow Aviation Institute (National Research University), Moscow, Russia
e-mail: valentinkiselev36@mail.ru

M. A. Pogosyan (ed.), *Aircraft Design*, Springer Aerospace Technology,
https://doi.org/10.1007/978-981-96-4599-2_10

Analysis of this equation shows that the efficiency of a commercial airliner can be increased if:

- the payload increases;
- the service speed increases;
- the aircraft operation cost decreases.

Aircraft operation cost is tightly interconnected with its fuel consumption. Fuel consumption can be estimated as follows:

$$q_{fuel} = \frac{mg}{L/D} C_p, \tag{1}$$

where m, L/D and C_p—respectively, the average values of aircraft mass, lift-to-drag ratio, and engine specific fuel consumption.

It follows from (1) that the reduction in the aircraft fuel consumption is associated with a reduction in the average mass of the aircraft (mainly due to the reduction of empty aircraft mass), with an increase in the lift-to-drag ratio and with a reduction in the engine specific fuel consumption. These conclusions also determine the ways for further development and improvement of commercial airliners and cargo aircraft.

The British were the pioneers in civil jet aviation. In 1952, the Comet jet began to perform regular flights. However, the Comets were unlucky. Fatigue failure of the pressurized passenger cabin led to a disaster. The causes for the failure were found only after lengthy research. Until they were identified and eliminated, the Comet flights were suspended. (Earlier and more successful was the operation of the passenger turboprop Wycount aircraft.) In 1956, a first-born of Russian civil jets TU-104 made its way to Moscow-Irkutsk line. First generation passenger jets, of course, were created based on developments and experience accumulated during the design and operation of military jet aircraft. It is known that TU-104 was a modification of bomber TU-16 (fundamental difference consisted in new fuselage with passenger pressurized cabin). The designers of "Comet", undoubtedly, used the experience of designing British bombers. "Comet" had four and Tu-104—two engines, which were located at the root of the wing in the same way as the prototype—bomber. But such a successful layout solution was poorly consistent with the requirements of comfort in the passenger cabin and led to a significant noise and vibration in it. The engines, located near the edge of the passenger cabin could lead to its depressurization from the probable destruction of the turbine or compressor blades, to cause fire. The use of this layout for bombers was explained by better aerodynamics associated with the reduced drag. However, a reduction in comfort and safety seemed unacceptable for an airliner. Therefore, the layout solutions implemented by the French in the passenger jet Caravelle, were perceived as revolutionary and very relevant. The highlight was that the French placed two turbojet engines in the afterbody. As a result, the wing was "aerodynamically clean" and the passenger cabin was moved out of the zone of maximum sound pressure of the engines (the plane of rotation of the turbine, compressor and the area of the gas jet from the nozzle). Noise and vibration reduction has solved the passenger comfort problems.

As a result, the entire second generation of passenger jets was created under the influence of the Caravelle—with aft podded engines. In England, the aft podded aircraft were: BAC. 111 (2 engines), DH.121 (3 engines), "Vickers" VC.10 (4 engines), in USSR—Tu-134 (2 engines), Tu-154 (3 engines), IL-62 (4 engines), Yak-40 and Yak-42 (3 engines). Even American designers, after a very successful for its time passenger Boeing B-707 and Douglas DC. 8 with four engines located on the pylons on the wings (life has shown great benefit of such a layout), could not resist and paid tribute to the aft podded configuration—Boeing B-727 (3 engines), DC.9 (2 engines). Not all firms appreciated properly the inherent disadvantages of the aft podded configuration. Therefore, many of them became apparent too late. BAC.111 and Tu-134 crashed. Complex investigations revealed the reason: the airplane entered the so-called prolonged, "closed" stall when the airplane reached the angles of attack of 25–30° and above. The aircraft was sort of "locked" in this position with its nose up, losing speed and stalling. The critical angles were reached when the aircraft was caught in a powerful ascending flow, a gust of air. Such powerful gusts at high altitudes are very rare, but every aircraft, as a rule, gets into them. However, as it turned out, only aircraft with aft podded engines proved to be unstable in this mode. At critical angles of attack the wake from nacelles blankets the horizontal tail of the T-tail configuration, making it ineffective. The T-tail is also heavier than conventional. In addition, the fuselage portion, which the engines are attached to, is relatively heavy. As it turned out, for an aircraft with aft-podded engines, the benefit of "aerodynamically clean" wing tends to be outweighed by the increased aerodynamic drag due to the mutual influence of nacelles and afterbody.

Another disadvantage is related to the large center of gravity travel common for the aircraft with aft-podded engines. The aft podded engines cause the center of gravity of the aircraft to be shifted backward. The wing also shifts backwards. Consequently, the fuselage and passenger cabin turn out to be divided by the wing into unequal parts—a long forebody and a short afterbody. The payload (passengers, luggage, cargo) moves the center of gravity forward. As a result, the center of gravity envelope has exceeded all previously known limits. How can this problem be solved? The designers of the Caravelle and the IL-62, decided to go the usual way. Let the true CG envelope is enormous, but the aircraft should fly only when its CG location is moderate, which is characteristic of the former aircraft with engines on the wing. Therefore, the wing and the main landing gear were placed relative to the forward center of gravity limit (full loading). What will happen when the passengers get off and the center of gravity travels backwards? Will the airplane flip to the tail? To avoid it, on IL-62, they applied additional tail landing gear. On Caravelle, the role of tail support was played by passenger boarding ramp in the afterbody (after passengers unloading, the aircraft leans on it until the fuel truck fills the wing fuel tanks). This is on the ground, but how to fly, if the CG is shifted back and the airplane is unstable in flight? The IL-62 has a balancing tank in the forebody, in which, if there is no payload, it is filled with … water. After all, the fuel should not be placed in the fuselage next to the passenger cabin—it's fire hazardous. On the Caravelle, ballast is loaded into the forward cargo compartment during the ferry flight. In cruise, the aircraft flies with small CG travel, which requires less trim loads and a smaller tail.

Probably, the designers of the VC.10, DH.121, BAC.111 did not like these restrictions. They wanted to solve the problem radically—to ensure the ability to fly with the existing huge CG limits. Therefore, it was necessary to place the wing and main landing gear struts relative to the aft limit of the center of gravity (aircraft without payload). In this case, the airplane would never flip to the tail and will always be stable in flight. But the problem arises when the aircraft is fully loaded. It consists in that the huge arm of main landing gear relative to the center of gravity makes it difficult for the nosewheel to lift when the aircraft takes off. It is also difficult to trim the aircraft in flight: greater forces on the horizontal tail are required, which increases trim drag. These problems are solved by substantially increasing the area (and mass) of horizontal tail.

The difficulties and disadvantages listed above significantly undermined the attractiveness of the aft-podded layout, and the subsequent generation of passenger jets featured engines mounted on wings. "The "firestarter" here was Boeing with the Boeing B-737. The firm deliberately went for the danger of having the engine air intakes low over the runway surface. And all subsequent aircrafts: the giant Boeing B-747, the long-haul Boeing B-757, -767, -777 and -787 had the engines on the wing. For many designers, the rejection of the aft podded layout was not so unambiguous and was associated with many hesitations. Thus, the IL-86 was initially designed with four engines on the back of the fuselage, as the previous IL-62 (a mock-up was built). And only a deeper analysis of the weight and performance characteristics, as well as the example of foreign companies forced the designers to make the right decision. There were also hesitations and mistakes abroad. Thus, the British initially decided not to participate in the creation of the European 300-seat Airbus A.300, and built their own BAC.311 with the same dimensions and characteristics, but using the aft-podded layout, which had been adopted for the previous 100-seat aircraft of the second generation BAC. 111. Worse weight characteristics contributed to the decision to abandon the design of the BAC.311.

Thus, if you compare the two layouts, the engines-on-wing configuration is lighter, but the "aerodynamically clean" wing provides the aft-podded layout with somewhat higher lift-to-drag ratio. But it should be borne in mind that the ratio of advantages and disadvantages of these layouts depends on the aircraft size and range. The weight advantage of the engine-on-wing layout increases with the size of the aircraft being compared. On the other hand, the weight penalty associated with aft-podded designs decreases with decreasing aircraft size. The impact of aerodynamic advantages on the economic efficiency of aircraft operation increases with increasing flight range. An aircraft with a higher lift-to-drag ratio requires less engine thrust, other things being equal, and therefore less fuel. Therefore, there are areas of more advantageous application of a particular layout. For example, for a short—mid- range airliner with large passenger capacity, the layout with engines on the wing will be more advantageous, as it provides lighter structures, while the worse aerodynamics will not significantly reduce economics due to a moderate range of flight. These are the European Airbus A.300B and our IL-86. On the other hand, small long range aircraft will derive benefit from the aft podded layout, as at long range aerodynamic advantages will play the key role and the aircraft structural weight will be small because of the small size of the aircraft. These are executive aircraft, most of which feature the aft podded layout.

On the other hand, it is impossible to give preference to one or another layout for small, short-range aircraft. In this case, the weight penalty associated with the aft-podded layout is low (size is limited), and the aerodynamic advantage at short range is not significant. Examples of engines-on-wing aeroplanes are Boeing B-737, VFW.614, BAe.146, Emb.145; Tu-134, Yak-40, DS-9, BAC.111, F.28 are aft-podded engines aircraft. It should also be noted that advances in aerodynamic matching of pylons to wing led to a significant improvement of the aerodynamic performance of aircraft and the overwhelming use of engines-on-wing layout for long- and medium-range passenger aircraft (Tu-204; Boeing B-757, -767, -777, -787; A.310, 320, 330, 340, 350). At the same time, for small aircraft, the aft-podded layout is the only way to raise the engines to a sufficient height above the ground.

Now, let us consider three-engine aircraft with two engines attached to the wings and the third one attached to the aftrebody. These are the Lockheed L.1011, DC.10, MD.11 (USA). This mixed configuration uses the advantages of each of the two considered above and gets rid of their disadvantages. Thus, the weight penalty is much less than that of the aft-podded configuration, since only one of the three engines loads the fuselage. At the same time, the aerodynamics is better, since the interference drag of the afterbody is reduced. The mixed configuration doesn't require a heavy T-tail. Furthermore, nothing prevents having large passenger doors in the tail section of the fuselage, which improves aircraft safety and performance.

2 Increasing the Size of Passenger and Cargo Aircraft

Passenger and cargo traffic in the world is steadily increasing. The size of aircraft increases accordingly, otherwise airports cannot cope with aircraft queues on the ground and in the air caused by the growing number of aircraft. The IL-62 and Boeing B-707 seeming giant not so long ago can now be described only as middle-size... However, among reasons that restrain aircraft size growth we should first of all turn our attention to the problems of ensuring weight perfection of giant airplanes. In air transport, the most economically efficient aircraft is the one which, with a constant flight range and other conditions being equal, has the maximum payload ratio—the ratio of the payload weight to the take-off weight of the aircraft [3]. The increase or decrease of payload with variation in aircraft size is mainly determined by the change in the structure weight ratio, thus affecting the economic efficiency of the aircraft.

The struggle to reduce the weight of the structure—wing, fuselage, tail, landing gear—began since the inception of aviation. At the same time, designers began to investigate the relationship between the size and weight of aircraft parts. The square-cube law was discovered. In accordance with it the weight of the structure and the whole plane grows in proportion to the cube of increase in linear dimensions. An increase in aircraft weight outstripping an increase in lift must inevitably limit the increase in size. In order to keep the performance and field characteristics with increasing aircraft size, it is necessary to ensure that the wing loading (the ratio of takeoff weight to wing area) remains constant. But this is only possible if the weight of the aircraft grows in proportion to the square of the linear dimensions, not the cube. For the weight of the structure—the main component of the aircraft weight—this is practically impossible under the condition of geometric similarity and for the same structural layout. Therefore, the weight ratio of the structure will inevitably increase with increasing aircraft size, while the share of payload weight will decrease. At the same time, the value of wing loading will increase in proportion to the increase in size. The statistics confirms this tendency. For example, Tu-134 has ${}^{W_0}\!/_S = 370\ daN/m^2$, Tu-154—470 daN/m^2, IL-62—580 daN/m^2, Boeing 747−740 daN/m^2 [1]. This, of course, increases the landing and takeoff speeds and the required runway length, but otherwise it would not be possible to create large aircraft, since with an increase in the wing loading the weight of the wing structure decreases due to a decrease in its area (at constant W_0). Therefore, sometimes relatively small airliners operating from long runways have high wing loading and powerful high-lift devices. For example, the modern Tu-334 has a significant wing loading for its size, ${}^{W_0}\!/_S = 550\ daN/m^2$. With the growth of dimensions, the excess of structural strength due to various technological factors decreases, the stability of the compressed elements increases. The structure is used more fully, more efficiently. But this circumstance leads to a more intensive growth of the weight of the structure with increasing dimensions. Figure 1 shows the change of payload weight with increasing dimensions of a medium-range airliner. We first calculate the weight of some notional aircraft that would not carry any payload ($W_{pay} = 0$). Then, we take a small commercial load and obtain a new, increased weight of the aircraft. If we continue this calculation by continuously increasing the passenger capacity, we can see that the payload weight increases rapidly but reaches a maximum and decreases as the weight ratio of the aircraft increases with the size of the aircraft. This maximum corresponds to the aircraft dimensions up to which their increase is accompanied by the increase of effective stresses and decrease of excessive strength, and further increase of stress is impossible, and the increase of dimensions and loading leads to intensive growth of the structure weight. With the number of seats over 600, the double-deck fuselage becomes advantageous, having in comparison with single-deck fuselage the shorter length, higher loading density, lesser structural weight. When the number of seats is over 1600, the triple-deck fuselage is advantageous. Transition to multi-deck configurations is one of the ways of rational "breaking" of geometric similarity with increasing number of seats. It is known, that with the increase of flight range the influence of aerodynamic perfection of the aircraft increases in comparison with the influence of weight, and vice versa. Therefore, the optimum value of the fuselage midsection area decreases as the design range increases. A vivid example to this rule can be the change of profitability of single- and double-deck fuselages with the variation of passenger capacity and flight range. Single-deck fuselages provide greater lift-to-drag ratio, but they are heavier than double-deck fuselages. Therefore, single-deck fuselages are preferred for aircraft with less passenger capacity and greater range, while increased passenger capacity and reduced range make a

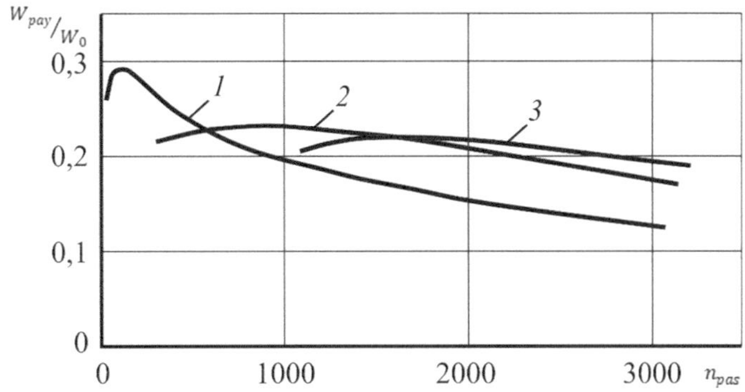

Fig. 1 Payload weight as a function of aircraft passenger capacity: 1—single-deck fuselage; 2—double-deck fuselage; 3—triple-deck fuselage

double-deck layout more advantageous. Aircraft with maximum wingspan up to 80 m, fuselage length up to 90 m, fuselage diameter up to 8 m, main wheel track up to 16 m, number of main landing gear legs up to 4, and number of wheels on each leg up to 6 are under consideration.

Another important direction in "fighting" the limitations imposed by the "square-cube" law is the use of the fact that this law does not take into account the possibility of changes in the loads acting on the wing structure as the wing size increases. In reality, wing structures take the load which is less than the lift produced by the wing. It is less by the weight of the wing structure and all loads placed inside the wing (fuel, cargo, equipment) or installed on it (engines, landing gear, etc.). This unloading of the wing allows for a significant reduction in the weight of its structure. Therefore, with increasing aircraft size, it becomes possible and advisable to locate more and more of aircraft weight components in the wing. This idea is expressed in the aerodynamic layout "flying wing", which was very popular in the first half of the last century. But at that time, this layout could not provide a gain in efficiency as compared with the conventional one. To accommodate passengers and cargo, large volumes and areas of such a wing were required. With the small weights and sizes of airplanes typical for that time, the required considerable wing area resulted in a low wing loading and low cruise lift coefficient. Therefore, the potentially high lift-to-drag ratio inherent in the "flying wing" in cruise was not achieved. The wing turned out to be oversized. Increasing flight speed only exacerbated this defect: with increasing dynamic pressure lift coefficient and lift-to-drag ratio decreased even more. It was more advantageous to use a smaller wing area to achieve the maximum lift-to-drag in cruise, and to place the payload in the elongated fuselage with a small mid-section area. The use of the conventional layout allowed to reduce the total aerodynamic drag.

Calculations show that the layout of flying wing is beneficial for aircraft with the passenger capacity of 1000 and more passengers, when the aircraft weight (and

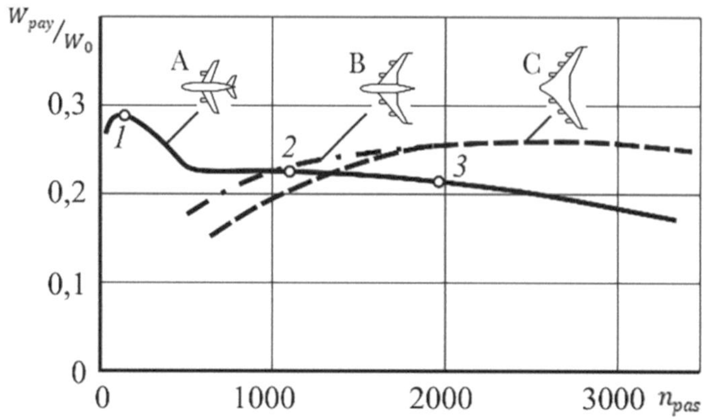

Fig. 2 Variation of payload weight with passenger capacity for various aerodynamic laouts: **a**—wing-tail aircraft (1—single-deck fuselage; 2—double-deck fuselage; 3—triple-deck fuselage); **b**—tailless aircraft, passenger cabins are located in the wings and nacelles; **c**—flying wing aircraft

wing loading) is so increased, that the size of the wing, selected from the condition of achieving the maximum lift-to-drag ratio in cruise, becomes sufficient to accommodate passengers, luggage, fuel, and equipment [2].

To date, design methods have already been created and weight equations have been obtained to determine the change in weight of aircraft components as the function of aircraft size. Figure 2 shows a comparison of payload weight ratios of medium-range airliners of different aerodynamic configurations.

With a passenger capacity exceeding 1000 seats, conventional aircraft with a double-deck fuselage (type A.380) begin to yield in payload weight ratio to a flying wing. Partial placement of passengers in the fuselage-nacelle does not over-size the wing and ensure that in cruise, the lift-to-drag ratio is high.

Thus, the use of new types of layouts, including multi-deck fuselages, "tailless" and "flying wing" aircraft, allows the designers to increase the payload ratio.

The introduction of more perfect structures, materials, equipment, engines, will also contribute to the increase of the economic efficiency of thousand-seater aircrafts. Recently, various studies of prospective giant passenger and transport aircraft have been conducted abroad and in Russia. The Boeing firm announced its intention to begin flight tests of a model of prospective heavy "Blended wing body" aircraft—BWB concept, considering the design of an 800-seater aircraft with the wingspan of 88.1 m, the length of 48.8 m and the height of 12.2 m (Fig. 3).

The concept considered by different firms had passenger capacity of 750–800 seats (three classes), a wingspan of 85–100 m (some had the wingtips that could be folded upwards to reduce the span at parking and during taxiing). As a rule, all researchers compared a BWB layout to the conventional one. Comparison showed that the BWB layout provides some weight savings and has potentially higher maximum lift-to-drag ratio. But achievement of a high lift-to-drag ratio in cruise seems to be problematic, because the wing loading of BWB remains approximately 1,5 times less, than that

Fig. 3 Tailless BWB

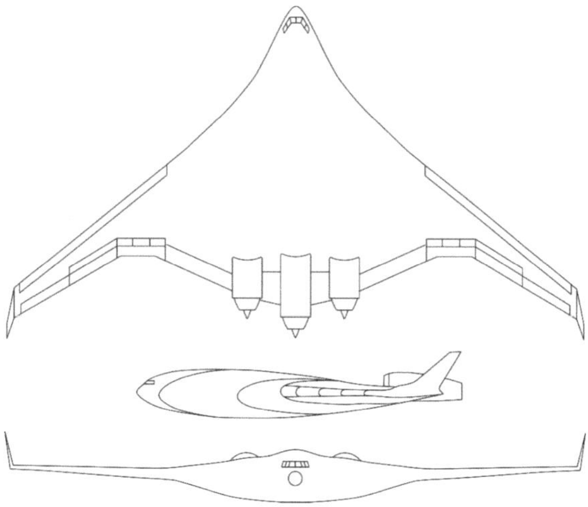

of the equivalent wing-tail configuration. It means that the BWB layout does not yet provide any decisive advantages for the considered dimensions. It is a pity that firms have not investigated layouts with passenger capacity of 1200–1500 seats or more, but that appears to be yet to come.

An even greater variety of design solutions is related to the projects of super-heavy cargo aircraft developed over the last decades to address the problems of increasing cargo traffic, especially in the context of the energy and other crises. At the same time, it is important to develop a classification of configurations for giant aircraft, which contributes to the systematization of the trends under consideration. The success of any classification depends on whether the most important feature or criterion is selected as its basis. The classification by aerodynamic configurations applied to super-heavy aircraft proved to be unable to assess the fundamental, qualitative differences in individual design solutions. In many projects (Fig. 4), the marked tendency to place more and more payload in wing can be clearly seen. However, in practice, this trend cannot be detected by means of classification by aerodynamic configurations, since the options of placing all loads not only in the fuselage, but also in the wing can be implemented in any of the three known aerodynamic configurations. For example, the Lockheed "Spenlouder" layout with the payload distributed along the wingspan has all the weight, structural and design features typical for the "flying wing" layout; however, the addition of horizontal stabilizers installed on the wing tips makes this aircraft attribute to the conventional aerodynamic layout with a conventional tail instead of the "flying wing". Therefore, it is advisable to classify super-heavy cargo aircraft by their main distinctive features associated with the location of their payload. It is this classification of these aircraft that makes it possible to identify the main differences in design and layout solutions affecting their weight, aerodynamic and performance characteristics and, consequently, their economic efficiency indicators.

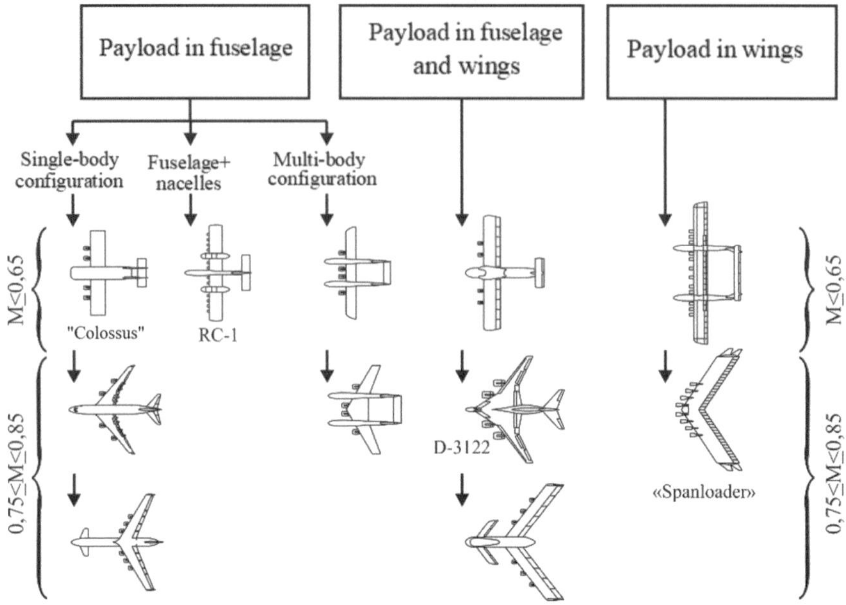

Fig. 4 Classification of super-heavy cargo aircraft

The classification diagram depicted in Fig. 4 includes three main classes of super-heavy cargo aircraft, differing in the location of payload. Cargo placement in the fuselage can be made both for a single fuselage layout (all modern cargo aircraft, projects of aircraft with a load-bearing fuselage "Shin", "Colossus", etc.), and for a two-fuselage layout (projects of McDonnell-Douglas and Lockheed), which provides greater wing unloading, cheaper production. Similar advantages can be achieved by placing payloads into nacelles-containers located on the wings of a giant aircraft (for example, the RC.I project by Boeing). It is the multi-fuselage layouts as well as the nacelle-container arrangement along the wingspan that are clear examples of the struggle to reduce structural weight while increasing aircraft size.

The tendency to place a significant part of payload in the wing (along with the fuselage) is expressed in mixed layout options. The use of thick supercritical wing airfoils (with the thickness ratio up to 21%) ("Spenlouder"), or larger wing root extensions (the draft aircraft D.3122), or delta wings will allow placing in the wing, as in the fuselage, the payload of considerable size, such as standard containers with a cross section of 2.44×2.44 m.

Finally, the heaviest aircraft designs involve placing the entire payload inside a giant wing. The use of supercritical airfoils with high structural depth facilitates the solution of this problem. Characteristically, in some designs, to ensure the longitudinal stability and controllability additional horizontal stabilizers are installed on the wingtips.

Figure 4 shows example layouts in two variants: with a straight wing ($M = 0.65$) and with a swept wing ($0.75 \le M \le 0.85$). Thus, the proposed classification also takes into account differences in aircraft flight speeds.

3 Passenger Cabin Layout

The placement of payload in a passenger aircraft is primarily related to the layout of the passenger cabin. When designing a passenger aircraft, it is important to accommodate the given number of passengers and cargo with the lowest possible empty weight an aerodynamic drag. However, the passenger cabin design should also allow aircraft scaling (changing the number of passengers).

Currently, three different classes of cabins, differing from each other in terms of passenger seating comfort and conditions of passenger service in flight, are used in air passenger transportation practice.

In the highest, I class, the width of seats and the distance between rows (seat pitch) are increased, passengers are offered more food (kitchens are enlarged), etc. In II, Business or Tourist class, and in III, or Economy class, narrower seats and smaller row spacing are used respectively.

3.1 Determining the Required Width of the Passenger Cabin

The main dimensions of passenger seats are shown in Fig. 5 and given in Table 1.

The basic requirements for the design and layout of passenger seats are as follows:

1. Passenger seats are usually made as blocks of two or three seats. For Class I cabins only two-seat blocks are used; for Class II and III both two-seat and three-seat blocks are possible.

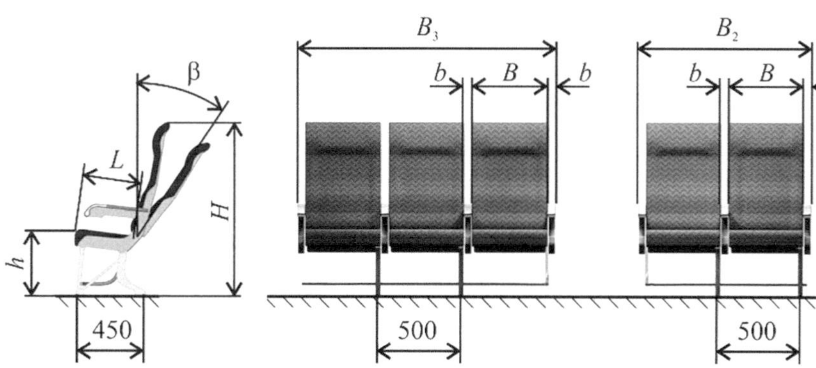

Fig. 5 Basic dimensions of passenger seats

Table 1 Main dimensions of passenger seats

Passenger seat classes	Distance between armrests *B*, mm	Width of armrests *b*, mm	Seat cushion to backrest length *L*, mm	Seat height above floor *h*, mm	Seat height with backrest *H*, mm	Backrest inclination, degrees	Width of seat unit, mm	
							B2	*B3*
I	500	70	500	445	1140	45	1260–1450	–
III	440	50	470	445	1120	25–36	1030–1070	1520–1570

2. Seats in Class I shall have a backrest which reclines to 45° from vertical and a removable footrest to provide a semi-reclining position for passengers; the backrests of Class II and III seats shall recline to 36° and 25° respectively.
3. The seats in classes II and III shall have a backrest which can be freely folded forwards so that in the event of an accident a rear-row passenger doesn't bump his head against the back of the front seat.
4. Seat cushions in Class III cabins should be reclining (like theatre seating) due to small seat spacing.
5. Each Class I seat shall have two armrests not less than 70 mm wide. In Class II and III seating units, the middle armrests shall be made one between the seats; their width shall not be less than 50 mm.
6. When several seating units are installed in the same row, the width of the aisle between these units at a height from the floor of 635 mm (zone of the armrests) or more shall be not less than 510 mm for passenger numbers 20 to 299 and not less than 650 mm for passenger numbers 300 and more.
7. On aircraft with only one main aisle in the passenger cabin, there should be no more than three seats on each side of the aisle. Consequently, there should be two aisles in the passenger cabin when the number of passengers in a row is greater than six, and three aisles when the number of passengers is greater than twelve.
8. The gap between the inner panelling of the passenger compartment and the seat structure (along the armrest or across the top of the back) shall be not less than 50 mm for seats of Class I and not less than 30 mm for seats of Class II and III.
9. The thickness of the passenger cabin wall, including load-bearing structure, heat and sound insulation and decorative finish, should be 100…130 mm.

The above conditions allow determining the fuselage width required for layout reasons at 635 … 650 mm above the floor of the passenger cabin. It is determined as follows:

$$B_f = B_2 n_2 + B_3 n_3 + c_{aisle} n_{aisle} + 2\delta_1 + 2\delta_2, \tag{2}$$

where B_2 and n_2, B_3 and n_3 are, respectively, the width and number of two-seat and three-seat blocks of seats; c_{aisle} and n_{aisle} are, respectively, the width and number of main aisles; δ_1 is clearance between the seat and the inner surface of the cabin wall; δ_1 is the thickness of the cabin wall.

The required width of the fuselage is equal to its diameter (with a circular fuselage cross-section) only if the armrests of the seats are in the area of the circle centre (the floor line is 550 … 650 mm below the maximum width or diameter of the fuselage).

The method for selecting the fuselage diameter based on the layout considerations only is shown above. However, the choice of diameter is also influenced by aerodynamics (ensuring a high lift-to-drag ratio of the aircraft as a whole) and strength (ensuring minimum structural weight). This optimization problem is discussed in Chap. 17.

3.2 Determining the Required Passenger Cabin Length

The length of passenger cabin is based on the following considerations:

1. Based on the given number of passengers and the selected number of seats in one row, the required number of rows of seats is determined. As a rule, at the end of the passenger cabin, the width of the cabin narrows due to the streamlined shape of the aft fuselage. Consequently, the last rows have to be made with a reduced number of seats, and to accommodate a given number of passengers the required number of rows of seats must be increased by 1–2 rows (in cabins with two aisles—by 3–4 rows);
2. The distance t between the rows of seats (pitch) is a multiple of 30 mm (in Western Europe and the USA—a multiple of one inch—25.4 mm):

Cabin class	I	II	III
t, mm	980 … 1080	840 … 870	780 … 810

3. The minimum distance l_1 from the cabin front bulkhead plane to the first row of seats:

Cabin class	I	II	III
l_1, mm	630	615	585

4. The minimum distance l_2 from the cabin rear bulkhead plane to the front mounting of the last row of seats:

Cabin class	I	II	III
l_2, mm	1000	800	750

The dimension l_2 is required to be such that, with the backrest fully reclined, the clearance between the seat structure and the bulkhead is not less than 20 mm; the requirements 2…4 are shown in Fig. 6.

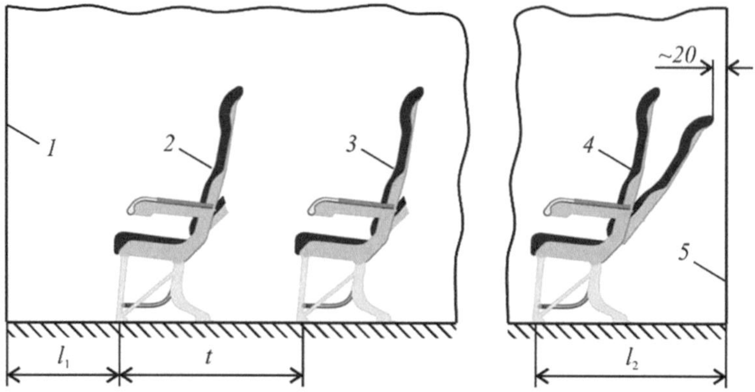

Fig. 6 Arrangement of passenger seats in relation to the front and rear partitions of the cabin: 1—front wall of the passenger cabin; 2—front row of seats; 3—second row of seats; 4—last row of seats; 5—rear wall of the passenger cabin

5. The ratio of the width of the passenger cabin to its length k_f (cabin shape factor) should be chosen so that it falls within the range of $k_f = 0.2 \ldots 0.5$. If $k_f < 0.2$, then the passenger cabin becomes similar to a "pipe", and if $k_f > 0.5$, the cabin looks "squeezed".
   Due to this condition in case of a large number of passengers it is necessary to divide the passenger cabin into two or more cabins.
6. The height of the passenger cabin should be no less than 1900 mm and no more than 2500 mm.

When the dimensions of the cabin are determined, the specific volume of the cabin (compartment) γ_f per passenger (Table 2) can be estimated.

Based on the above conditions, the required length of the passenger cabin (mm) is determined as follows:

$$L_{c,required} = l_1 + (i_{row} - 1)t + l_2, \tag{3}$$

where i_{row} is the number of rows of seats.

Example Determine the required length of the passenger cabin for 180 passengers assuming the layout of the tourist (II) class with 6 seats in a row.

Table 2 Minimum values of specific volume of passenger cabin γ_f, m^3/person

Fuselage diameter, mm	Flight duration, h					
	Up to 1	1 … 2	2 … 4	4 … 6	6 … 8	8 … 10
<4000 * > 5000	0,84 0,96	0,85 0,98	0,92 1,06	0,98 1,13	1,2 1,27	1,2 1,36

* The data are given for Class III cabins. Class I and II cabins have higher γ_f values

According to (3)

$$L_{c,required} = 615 + \left(\frac{180}{6} - 1\right)840 + 800 = 25775\ mm.$$

According to Eq. (2) with 6 seats in a row we will have an outside width of the fuselage of 3850 mm and an inside width (without taking into account the thickness of the walls, i.e. without $2\delta_2$) –3550 mm.

Then the shape factor k_f will be equal to

$$k_f = B_{cab,internal}/L_{cab} = 3550/25775 = 0.138.$$

Since the minimum permissible value of $k_f = 0.2$, we divide the cabin into 2 salons of 90 people each. Then the length of the front cabin

$$L'_{sal} = 615 + \left(\frac{90}{6} - 1\right)840 + 800 = 13175\ mm,$$

$$k'_f = \frac{3550}{13175} = 0.28,$$

i.e. the value of k_f now is within the permissible limits.

The required length of the rear cabin, given the narrowing of the aft fuselage, will be longer (one more row of seats should be placed, since the rear three rows will have twin rather than triple seat blocks).

$$L''_{sal} = 13175 + 840 = 14015\,mm.$$

Let's carry out a control test for compliance with the requirements for the specific volume γ_f. For the inner diameter (inner width) of the fuselage of 3550 mm and for the cabin height of 2100 mm, the internal cross-sectional area of the cabin will be 6.6 m^2. The specific volume of the forward cabin:

$$\gamma_f = \frac{13.175 \cdot 6.6}{90} \approx 0.97 m^3/person,$$

i.e., if the flight endurance do not exceeds 4 h, this arrangement can be considered acceptable.

4 Emergency and Safety Equipment

4.1 Location of Emergency Exits

The layout of a passenger cabin is very much influenced by the location of the safety equipment, which must provide:

- emergency evacuation of all passengers and crew on the ground in not more than 90 sec with the landing gear extended and retracted, and when the landing gear is retracted on one side only;
- emergency disembarkation of passengers and crew into onboard emergency floating facilities (inflatable rafts) in case of landing of the aircraft on the water surface.

These requirements are primarily addressed by emergency exits. The number of passenger aircraft emergency exits is provided by Aviation Rules, Chap. 25 and is given in Table 3.

If the number of passengers falls between 179 and 299, additional pairs of emergency exits are required based on Type III hatches for 35 additional passengers (over 179), Type II hatches for 40 additional passengers, Type I hatches for 45 additional passengers, and Type "A" hatches for 100 additional passengers.

When the number of passenger seats exceeds 299, each emergency exit in the side of the fuselage shall be an exit of type "A" or type I. Each pair of Type "A" exits shall have a maximum of 100 passenger seats and each pair of Type I exits shall have a maximum of 45 passenger seats.

If the tail exit is provided which permits a passenger evacuation rate at least equal to that of type III exit when the aircraft is in the most unfavourable position for opening the hatch, this exit may be treated as additional 12 … 15 passengers.

The exits shall be rectangular in shape with a corner radius equal to 1/6 of the exit width for type "A" and 1/3 of the exit width for other types. The dimensions of the exits are as follows:

Table 3 Number of emergency exits on each side of a passenger aircraft fuselage

Number of passenger seats (not including flight attendant seats)	Type I	Type II	Type III	Type IV
1 … 10	–	–	–	1
11 … 19	–	–	1	–
20 … 39		1	1	
40 … 79	1	–	1	
80 … 109	1	–	2	
110 … 139	2	–	1	
140 … 179	2	–	2	

Exit type	«A»	I	II	III	IV
Width, mm	1070	610	510	510	480
Height, mm	1830	1220	1120	915	660

The exits of type "A", I must be at floor level. The exits of other types shall also be at floor level if they are not located above the wing. If they are located above the wing, however, the thresholds must not exceed the following values.

Exit type	II	III	IV
Threshold height inside aircraft, mm	250	510	740
Threshold height outside, mm	430	690	910

When laying out emergency exits and passenger cabin (the placement of rows of seats along the length of the cabin), the emergency exits of type I, II must have a free passage width of at least 510 mm, and to the emergency exit of type "A"—a width of at least 910 mm.

In the immediate vicinity of the emergency exits of type "A", I and II, which are not in the wing area, there must be rubber escape slides that automatically inflate when ejected (Fig. 7). At the emergency exit of the "A" type the inflatable escape slides must allow two simultaneous parallel flows of evacuating passengers.

Modern aircraft with pressurized cabins allow for emergency landing on water surface and a sufficiently long stay afloat after such landing (not less than 15 … 20 min). In the process of aircraft development, a dynamically similar model of the aircraft is built and catapulted into the water, simulating an emergency landing of the aircraft on the water surface. The recommendations for such an emergency landing are practiced and the degree (grade) of sea disturbance safe for landing is established.

If the flight is over water for not more than one hour, each passenger is provided with an inflatable lifejacket placed next to the passenger seat. In case of a longer flight over water, inflatable rafts shall be installed on board near the emergency exits to accommodate all passengers and crew (the number of rafts shall be determined by their carrying capacity).

If the escape hatches are below the waterline, the aircraft shall have at least two easily accessible overhead hatches of at least Type III exit size.

4.2 *Passenger Cabin Doors and Windows*

Passenger aircraft with more than 80 passengers should have at least two entrance doors, preferably located at the front and rear ends of the passenger cabin. Entrance doors should be located on the left side of the aircraft. Aircraft with more than 250 seats may have entrance doors located on both sides.

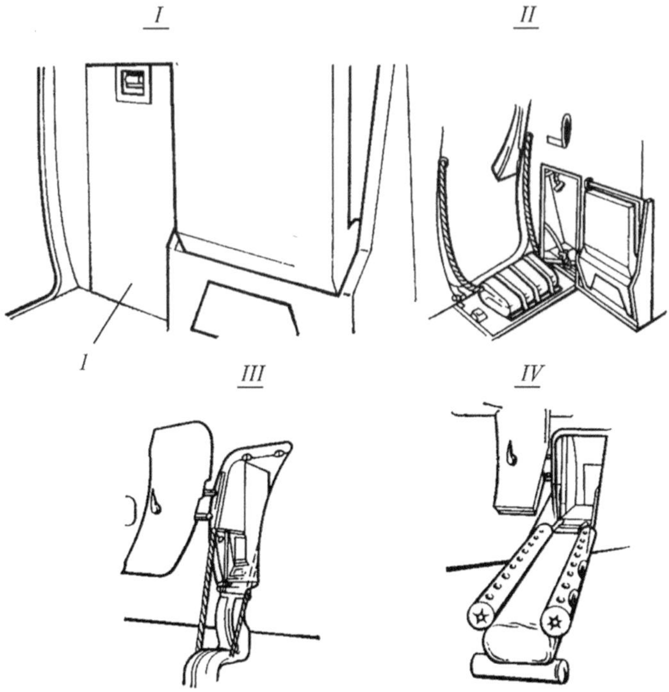

Fig. 7 Location of escape inflatable slides: I—flight position (slide bay cover closed); II—slide in "Ready to drop" position; III—door open, slide thrown out; IV—slide automatically inflated and ready for passenger descent; 1—gangway bay cover

Every passenger entrance door may be classified as an emergency exit if its dimensions meet or exceed the requirements for Type I and Type "A" emergency exits. The entrance door shall not have a threshold above the floor of the cabin.

Entrance doors must open outward. This requirement poses a very great challenge to door design, as this design must ensure that the door opening is sealed once the door is closed, the door cannot be opened in flight, there is little force in opening the door on the ground and little wind load in the open position when parked at the aerodrome.

It should be borne in mind that the door with type "A" dimensions is loaded with 120…150 kN due the cabin pressurization.

The windows in the passenger cabin must provide the possibility to use them conveniently from the seats located at the side of the cabin, when changing the pitch of the seats. Structurally, these are small windows (200…230 mm width, 320…350 mm height) located between the ribs of the cabin frame (window spacing of 500…510 mm).

5 Placement of Flight Attendants. Laying Out Domestic Equipment and Baggage Compartment

5.1 Placement of Flight Attendants

The crew consists of flight attendants to maintain order in the passenger area, prepare and distribute meals to passengers in flight, instruct them in the use of various passenger equipment and emergency systems, and provide medical assistance. Their number in the aircraft is determined as follows:

(a) two flight attendants should be available for the first class cabin with up to 30…32 passengers. If the number of I class passengers is more than 30… …32, one flight attendant per 15…16 passengers should be provided;
(b) in II and III classes–one flight attendant per 25… 30 passengers.

Each flight attendant should have a seat with a seat-belt; the seat may be reclining and should preferably be placed so that the flight attendant can see all passengers in the cabin.

5.2 Laying Out Baggage and Cargo Compartments

Airliners must have baggage and cargo compartments. For aircraft operating at altitudes above 3000 m, these spaces should be located in the pressurized part of the fuselage.

In case of containerless loading of baggage and cargo the required volume of baggage and cargo space (m^3) shall be determined by the number of passengers and mass of mail or cargo specified for the aircraft according to the following formula:

$$V_b = \frac{15n_{pas,max}}{120} + \frac{m_{pay} - 90n_{pas}}{290}. \tag{4}$$

Here $n_{pas,max}$ is the maximum number of passengers; m_{pay} is the given payload; 120 and 290—baggage and mail or cargo density respectively, kg/m^3.

When baggage and cargo are loaded on board in containers, the required volume (m^3) of containers is

$$V_c = \frac{15n_{pas,max}}{250} + \frac{m_{pay} - 90n_{pas}}{350}, \tag{5}$$

where 250 and 350 are, respectively, the density of baggage and mail or cargo in kg/m^3.

The increase in density compared to containerless loading is due to the possibility of denser stacking.

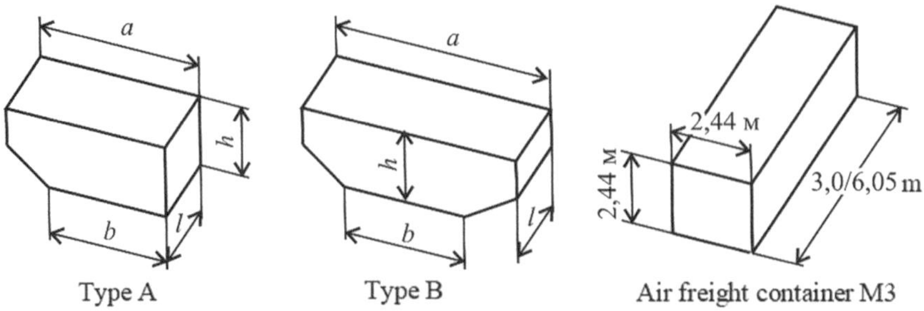

Fig. 8 Standard airline units

Table 4 Dimensions of standard airline units

	a, м	*h*, м	*l*, м	*b*, м	Type
LD-1	2.337	1.626	1.530	1.560	A
LD-2	1.560	1.626	1.530	1.194	A
LD-3	2.001	1.626	1.530	1.560	A
LD-6	4.064	1.626	1.530	3175	B
LD-8	3.175	1.626	1.530	2.440	B
LD-29	4.720	1.626	2.240	3.175	B

On aircraft with a fuselage diameter greater than 3000 mm, baggage and cargo compartments are normally located under the passenger compartment floor and are divided into compartments so that the loading of these compartments can adjust the position of the aircraft centre of mass. The height of the baggage and cargo compartments shall be not less than 1100 mm. Luggage hatches for each of the compartments shall be positioned so that they can be accessed by trucks. The dimensions of the luggage hatches are selected on the basis of loading containers either specially created for this aircraft or corresponding to those used on international lines (Fig. 8, Table 4).

5.3 Galley Layout

In order to provide in-flight meals to passengers, one or two galleys are installed in the passenger compartment, where containers with pre-packaged food and beverages are loaded at the aerodrome. During the flight flight attendants only heat up the required amount of food and deliver or distribute served trays to passengers on special trolleys.

Aircraft are equipped with a single galley if there is a single passenger cabin and the number of passengers does not exceed 100 people. If there are more than one passenger cabin, then regardless of the number of passengers in the aircraft there shall be two kitchens, the total volume of which (m^3)

$$V_g \approx (0.12 \ldots 0.14) n_{pass}, \tag{6}$$

and the floor area (m^2).

$$S_g \approx (0.1 \ldots 0.05) n_{pass}, \tag{7}$$

where a coefficient of 0.1 applies to Class I galleys and 0.05 to Class II and III galleys.

On wide-body aircraft, it is advisable to place the galley in the lower floor under the floor of the passenger compartment, feeding served trays in containers or trolleys upwards on special lifts for distribution.

The following should be kept in mind when developing the aircraft weight specification.

(a) the mass of the galley with equipment is 11 … 12 kg per person, including 4 … 5 kg/person of removable equipment;
(b) the mass of food is 2.5 kg per passenger of I class and 1.2 … 1.6 kg per passenger of II and III classes for a flight time of more than 6 … 8 h.

On aircraft with less than 250 passengers, the galley shall be loaded through the service doors placed on the right side of the aircraft. The dimensions of the service doors should be at least 1220 × 610 mm.

5.4 Laying Out Lavatory Modules

The number of lavatories on an aircraft depends on its endurance and number of passengers.

Endurance, h	Less than 2	2…4	2…4	Over 4	Over 4	Over 4
Number of passengers	Up to 120	Up to 120	Over 120	Less than 120	120…200	Over 200
Number of passengers per lavatory	50	40	45…50	30	40	45…50

The floor area of the lavatory module must be at least 1.5…1.6 m^2, the width must be at least 1 m.

Prior to flight, the aircraft is refilled with water for domestic needs (galleys) and for hygienic purposes (lavatories). Required amount of water, dm^3.

$$V_w = (0.3 + 0.15E)\left(n_{pass} + n_{cr} + n_{FA}\right), \tag{8}$$

where E is endurance, h; n_{cr}, n_{FA} are number of crew members and flight attendants, respectively. Water tanks should have 10% more volume.

Drain tanks shall hold the amount of water determined by the formula (8) and have additional volume of ~0.15 dm^3 per hour per person as well as hold chemical liquid for disinfection of 12% of the total volume, i.e. the required volume of drain tanks, dm^3,

$$V_{DT} \geq 1.12(0.3 + 0.3E)\left(n_{pass} + n_{cr} + n_{FA}\right). \tag{9}$$

5.5 *Coat Room Layout*

The most uncertain in passenger aircraft layout is the coat room problem.

The floor area of the coat room must be (m^2).

$$S_{CR} = (0.035 \ldots 0.050)n_{pass} \tag{10}$$

and the volume of this room (m^3) should be

$$V_{CR} = (0.05 \ldots 0.08)n_{pass} \tag{11}$$

A reasonable solution to this problem is to provide coat rooms with removable partitions so that additional passenger seats can be installed in the aircraft in the summer instead of a coat room. Coat rooms should be located either near the entrance doors or in the areas where noise inside the passenger cabin is greatest.

6 Features of Aerodynamic Layout of Passenger Aircraft. Selection of Thrust-To-Weight Ratio

6.1 *Selecting the Wing Area*

As shown in Chap. 5, the wing area of a passenger aircraft is selected to provide the given landing approach speed.

$$S = \frac{30.2m_{land}g}{V_{appr}^2 C_{L,max,land}} \tag{12}$$

where m_{land} is aircraft mass at landing (kg) with a given mass of payload and fuel reserve; V_{appr} is the given landing approach speed, m/s; $C_{L,max,land}$ is the maximum lift coefficient in landing configuration.

6.2 Towards Selection of Optimum Cruising Speed and Basic Wing Parameters

The most advantageous cruising speed for subsonic passenger aircraft and such wing parameters as aspect ratio, sweep angle, airfoil thickness can be obtained in finding the extremum of the function, which depends on several contradictory parameters.

On the one hand, the cruising speed V_{cr} determines the cost of transportation, and the higher V_{cr}, the lower the cost of transportation a and the more economical the passenger aircraft. For subsonic passenger aircraft, for example, in order to fly at $M = 0.8...0.9$, a wing with a sweep $\Lambda = 25\ldots35°$ and profile thickness $\frac{t}{c} = 0.1\ldots0.14$ are required.

The cost of operating the aircraft per flight hour A strongly depends on the fuel cost A_f, which, depending on the fuel price, varies from 35…40% to 70…75% of the total value of A. First of all, minimization of A requires minimization of A_f, which is a function of aircraft fuel consumption per flight hour (q_t, see formula 1).

Thus, the task of minimization of a and A in terms of aerodynamic layout of the passenger aircraft is reduced to ensuring the maximum value of the cruising lift-to-drag ratio of the aircraft. In particular, this means finding the minimum value of drag of the airplane, and since

$$C_D = C_{D0} + \frac{1+\omega}{\pi AR_{eff}} C_L^2 \tag{13}$$

(where ω is the aircraft trim penalties factor; AR_{eff} is the effective wing aspect ratio), then, C_{D0}, as well as the trim penalties (see Chap. 4) must be minimized, while the wing must have the greatest possible aspect ratio (by increasing the span).

The increase in span is associated with a significant increase in wing weight and aggravation of aeroelasticity problems (aileron reversal, flutter). As a consequence, it is necessary to reduce the sweep angle of the wing and, consequently, reduce the value of M_{crit}. With the growth of fuel price the optimum value of V_{cruise} and sweep angle decreases, and the optimum value of the wing aspect ratio increases. The widespread use of modern supercritical profiles makes it possible either to increase the cruising speed or to increase the thickness of the wing profile and thus reduce its weight, which also makes it possible to increase the wingspan.

6.3 Selection of Thrust-To-Weight Ratio and Number of Engines for a Subsonic Commercial Airliner

This task can be solved if the basic aircraft performance parameters are given in the technical requirements: payload W_{pay}, design range R, cruising altitude H and speed V, airfield class - runway length L_{RW}.

$$\left(W_{pay}, R, V, H, L_{RW}\right) = \text{const.}$$

Chapter 5 shows that for passenger aircraft the most important requirement affecting the thrust-to-weight ratio is to ensure the continued takeoff with one engine-off at the design temperature and pressure.

The required take-off thrust with all engines operating under the design conditions ($t = +30\,°C, P = 730\,mmHg$) is (see Sect. 8.1).

For a twin-engine aircraft

$$T_0 = 3m_0 g\left(\frac{1}{(L/D)_{climb}} + 0.025\right); \tag{14}$$

for an aircraft with three engines

$$T_0 = 2.25m_0 g\left(\frac{1}{(L/D)_{climb}} + 0.027\right); \tag{15}$$

for an aircraft with four engines

$$T_0 = 2m_0 g\left(\frac{1}{(L/D)_{climb}} + 0.030\right); \tag{16}$$

The minimum thrust-to-weight ratio means that the aircraft has a lower mass ratio of the propulsion system $\overline{m}_{PP}$ and the mass ratio of empty aircraft $\overline{m}_{empty}$. The maximum required thrust-to-weight ratio is achieved with twin-engine aircraft. This delivers better take-off characteristics (shorter ground roll and take-off distances), which makes it expedient to use such aircraft as short-range airliner or as a commuter with shorter runway lengths.

On the other hand, the choice of the number of engines on the aircraft is related to their by-pass ratio. This is due to the effect of the by-pass ratio m on the intensity of thrust drop with increasing flight altitude—Fig. 9.

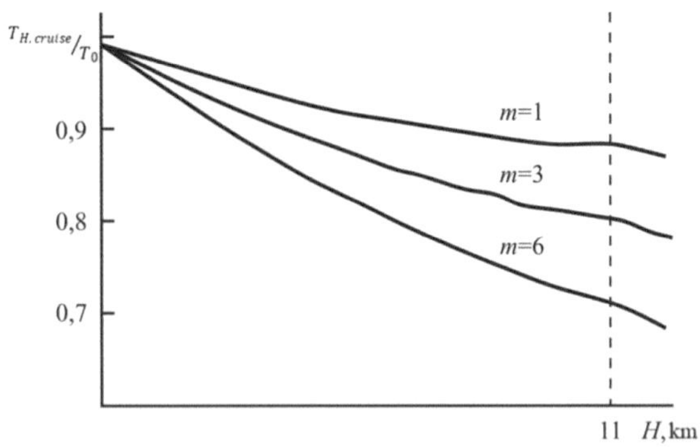

Fig. 9 $T_{H,cruise}/T_0$ for by-pass engines

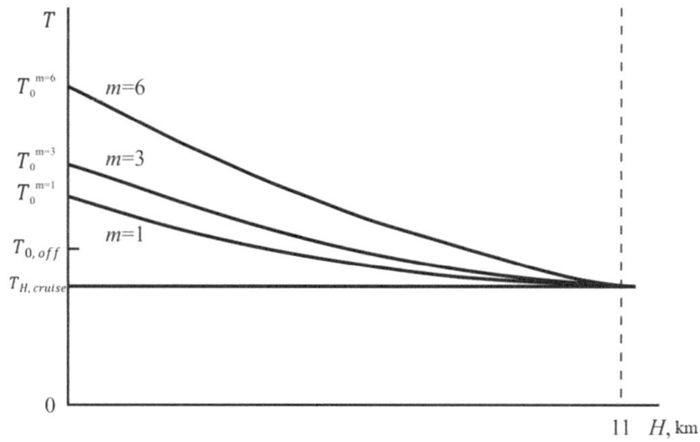

Fig. 10 Increase in takeoff thrust ($H = 0$) depending on by-pass ratio at constant thrust at cruising altitude ($m = 1...6$)

As the by-pass ratio increases, the drop in thrust in the cruise increases compared to the takeoff mode. With increasing by-pass ratio, the excess thrust at takeoff increases.

This is illustrated in Fig. 10, which shows, for constant thrust at cruising altitude, $H \approx 11\ km$, the increase in thrust at takeoff for aircraft with engines with different by-pass ratios, $m = 1...6$.

$T_{H,cruise}$ is the cruising thrust, $T_{0,off}$ is the thrust required for continued takeoff in case of engine failure, $T_0^{m=1}$, $T_0^{m=3}$, $T_0^{m=6}$ are takeoff thrust of aircraft having engines with, respectively. $m = 1$, $m = 3$, $m = 6$.

Thus, on takeoff, the thrust of aircraft with $m = 1$ will be about a quarter of the thrust required when one engine fails ($T_{0,off}$ is the same for all three aircraft operating from the same runway). And, therefore, four engines can be selected for this aircraft. The thrust of engines with $m = 3$ increases even more at takeoff and will be about a third greater than the required takeoff thrust if one engine fails, and the aircraft can be a three-engine aircraft. Accordingly, the thrust of the engines with $m = 6$ on takeoff will be even greater, for example, twice the required takeoff thrust when one engine fails, i.e. two engines are sufficient. Hence, it can be seen that all three compared aircraft have equal thrust in cruising and will be able to continue takeoff in case of failure of one engine. Consequently, increasing the degree of by-pass ratio gives the opportunity to reduce their number on the aircraft. And the aircraft with a smaller number of engines is cheaper to produce, easier to operate, but most importantly, the increase in m reduces fuel costs and increases efficiency.

Of course, the above example is only a scheme to illustrate the principle. In reality, the choice of the number of engines is often determined by the availability or unavailability of a suitable engine. For example, when creating the giant Boeing 747, there could be no other option with a smaller number of engines, since a more powerful engine did not exist. It is also true that the by-pass ratio of the first and second generation aircraft ($m = 0...0.8$) steadily increased, and in order to improve efficiency, reached for the latest generations of aircraft the value of 6, which made it possible to reduce fuel consumption by about half. As a result, long-range, multi-passenger aircraft with only two engines (A330, 350, Boeing 777, 787) appeared.

7 Features of Supersonic Airliner Design

As shown at the beginning of this chapter, one of the ways for further development of passenger aircraft is to increase cruising speeds. Sooner or later the systematic operation of supersonic airliners will begin on airlines. The failures of the first supersonic airliners of Concorde type are not connected with their performance, but with the fact that with a sharp rise in fuel prices their operational costs grew accordingly. It should be noted that the first generation of Concorde was designed during the second generation of subsonic airliners of VC.10, Il-62, Boeing 727 types, with which the aircraft could compete successfully. However, during the design, testing, beginning of operation of supersonic airliners the subsonic passenger aviation made a great leap: there appeared airliners of the third generation, wide-body aircraft (Boeing 747, IL-86, A.300B, etc.), the by-pass ratio increased several times (to $m = 4...6$), specific fuel consumption reduced approximately by half. As a result, supersonic airliners became unprofitable. Therefore, the leading design bureaus and research institutes continued the development of supersonic airliners of next generations. The most in-depth studies concerned the aerodynamic configuration of the wing and the entire aircraft. "Concorde" (100–144 seats) has a lift-to-drag ratio of about 7.5 at the Mach number of 2. It was the starting point of the ongoing studies. The aircraft shown in Fig. 11 (designed by McDonnell-Douglas) corresponds to the second generation of supersonic airliners, has a length of 94.5 m, a wingspan of 41.5 m, a wing area of 920 m^2, a takeoff weight of 340 tons, and it was designed to carry 273 passengers over a range of 8300 km at a cruising speed corresponding to $M = 2.2$. The wing is of small aspect ratio with the wing root extension sweep angle of 71°, the outer wing sweep angle of 57°. The airfoil thickness ratio is approximately 3%. The predicted lift-to-drag ratio, confirmed through testing in the supersonic wind-tunnel, is 9.1. It is believed that the maximum lift-to drag ratio of 9.6 may be achieved through further optimization of the layout.

Improvements concern all aspects of supersonic aircraft flight: improved materials, noise and acoustic shock, propulsion system, atmospheric pollution, improved field performance.

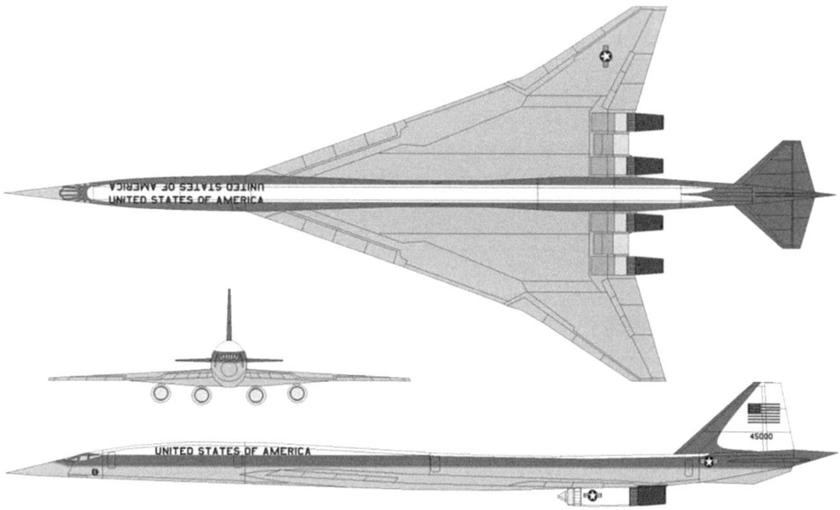

Fig. 11 General view of the second generation supersonic airliner

7.1 Selecting the Cruising Speed

The flight range of an aircraft can be determined by using the expression.

$$R = 1020\frac{{}^{L}\!/\!{}_{D}M}{C_p}ln\frac{m_0}{m_{final}}, \tag{17}$$

where ${}^{L}\!/\!{}_{D}$, M, C_p stand for the lift-to-drag ratio, flight Mach number, and cruise specific fuel consumption; m_0 and m_{final} are initial (take-off) and final (landing) values of the aircraft mass.

The variation of possible values of *L/D, LM/D*, and LM/DC_p depending on the flight Mach number for aircraft optimally designed for each respective Mach number, is shown in Fig. 12. The figure shows:

- in the zone of supersonic flight speeds, the lift-to-drag ratio of aircraft is sharply reduced;
- the values of $\frac{{}^{L}\!/\!{}_{D}M}{C_p}ln\frac{m_0}{m_{final}}$ are the same at flight Mach numbers equal to 1.1…1.2 and 2.0…2.5.

As will be shown below, the use of aluminum-based alloys in a supersonic airliner is possible only up to $M = 2, 0...2, 5$.

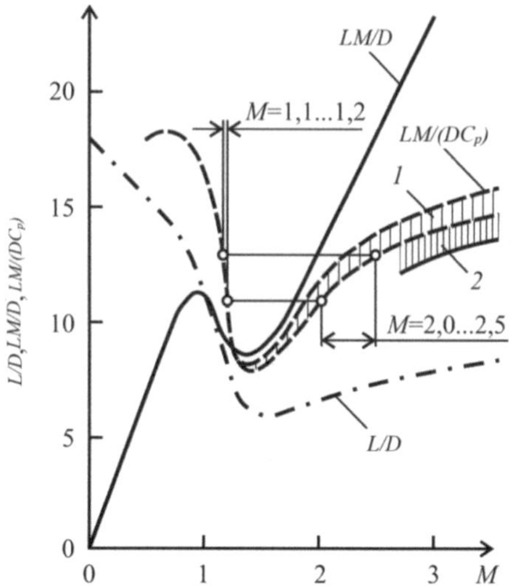

Fig. 12 Variation of *L/D*, *LM/D*, and *LM/(DC$_p$)* with the flight Mach number. The curves correspond to the optimally designed aircraft for specific flight Mach numbers: 1—accounting for the change in $ln\frac{m_0}{m_{final}}$ due to thermal insulation, cooling systems and due to pressure losses at engine inlet; 2—effect of transition to titanium alloys and steel

7.2 *Aerodynamic Heating*

A number of difficulties arise in the design and operation of supersonic airliners. One of them is the aerodynamic heating of the structure that occurs due to adiabatic compression of the air in front of the flying aircraft. The temperature of total air deceleration in front of the leading edge of the wing, tail, fuselage, etc. t_{total} (degrees Celsius) is equal to

$$t_{total} = T_H + T_H 0.2M^2 - 273.2, \tag{18}$$

where T_H is the absolute air temperature at the flight altitude, K.

For stratospheric flight (H = 11,000 km),

$$t_{total} = 216.7 + 43.3M^2 - 273.3 = 43.3M^2 - 56.5.$$

The values of t_{total} for different flight Mach numbers at an altitude of 11,000 m are as follows:

Flight mach number	1.0	1.2	1.3	1.5	1.7	2.0	2.2	2.5	2.7	3.0
t_{total}, °C	–13.3	6.0	16.7	41.0	68.5	116.5	159.0	214.0	259.0	336.0

The skin surface temperature T_w can be obtained from the heat balance equation:

$$\alpha(T_{BL} - T_W) - \sigma \cdot \varepsilon \left[\left(\frac{T_W}{100} \right)^4 - \left(\frac{T_H}{100} \right)^4 \right] = 0, \tag{19}$$

where α is the heat transfer coefficient; T_{BL} is the boundary layer temperature, K; T_w is the wall temperature, K; σ is the radiant coefficient; ε is the degree of blackness; T_H is the ambient air temperature, K.

Below, the values of T_w for an area 1 m away from the leading edge (where $T_w \approx T_{total}$) for an aircraft with a polished aluminium alloy skin ($\varepsilon = 0.8$, $\sigma\varepsilon = 3.92$) flying at an altitude of 12,000 m, are given.

Mach number	1.0	1.2	1.3	1.5	1.7	2.0	2.2	2.5	2.7	3.0
T_w,°C	–17.5	–1.0	8.0	30.0	52.0	94.0	124.0	174.0	210.0	271.0

In case of aerodynamic heating, there are problems of material use, durability of non-metallic structural elements, metal creep (elongation under load and heating), thermo-fatigue strength of the structure due to local stresses at uneven heating (during acceleration after takeoff) and uneven cooling (during braking before landing). In particular, the use of aluminum-based alloys as the main structural materials is possible only up to a temperature of 150… 180 °C, which, as noted above, corresponds to a flight with a Mach number of 2.0… 2.5.

Titanium alloys are acceptable for use up to temperatures of 280…300 °C, which determines their application up to $M = 3.0...3.2$. At higher flight speeds it is necessary to use steel alloys.

It should be borne in mind that in the Mach number range of 1.1…1.2 the aircraft temperature in flight is around 0 °C and there are no problems related to structural heating, no special systems or measures for cooling are required; the flight speed increases by approximately 30% as compared to flight at $M = 0.85...0.9$.

7.3 Sonic Boom

An airplane flying at supersonic speed produces shock waves (leading and trailing). When these waves reach the ground (at the Mach number exceeding 1.2), the change in pressure behind the leading edges of these waves is perceived as a thunderclap, which is called "sonic boom" (Fig. 13).

The intensity of this impact ΔP depends on the speed and altitude of flight, on the mass and shape of the aircraft. It can be estimated (daN/m^2) using the following expressions:

$$\Delta P = \sqrt{(\Delta P_v)^2 + (\Delta P_m)^2}, \tag{20}$$

$$\Delta P_v = 0.43 k_{ref} k_v \frac{\left(M^2 - 1\right)^{1/8}}{H^{3/4}} \cdot \frac{D_{equiv}}{l^{1/4}} \sqrt{p_0 p_H} \left[\frac{\cos(\mu + \theta)}{cos\mu}\right]^{3/4}, \tag{21}$$

$$\Delta P_m = 0.363 k_{ref} k_m \frac{\left(M^2 - 1\right)^{3/8}}{M} \cdot \frac{\sqrt{mg}}{H^{3/4}} \frac{\sqrt{p_0}}{c^{1/4}} \left[\frac{\cos(\mu + \theta)}{sin\mu} \right]^{3/4}, \quad (22)$$

where k_{ref} is the factor of reflection from the ground surface ($k_{ref} = 1.6 \ldots 2.0$); k_ν is the body shape factor ($k_\nu = 1.5 \ldots 1.6$ for aircraft with small wing aspect ratios); k_m is the lift distribution factor ($k_m = 1.4 \ldots 1.63$); H is the flight altitude; D_{equiv} is the aircraft equivalent diameter; l is the aircraft length; c is the wing root chord; p_0 is the air pressure at the ground level ($H = 0$); p_H is the air pressure at the flight altitude; $sin\mu = 1/M$; θ is inclination angle of flight path.

The predicted values of ΔP may vary within ± (30…50%) depending on θ, the aircraft maneuver in the horizontal plane and meteorological conditions due to the possibility of focusing the sonic boom.

The impact of sonic boom on humans, animals and structures is tentatively shown in Fig. 14. The impact is highly subjective.

There are other difficulties in supersonic aircraft operation, such as the danger of space radiation when solar activity intensifies, the impact of acoustic loads from pulsations in the boundary layer on the aircraft's skin and a number of others, but they still need a deeper investigation.

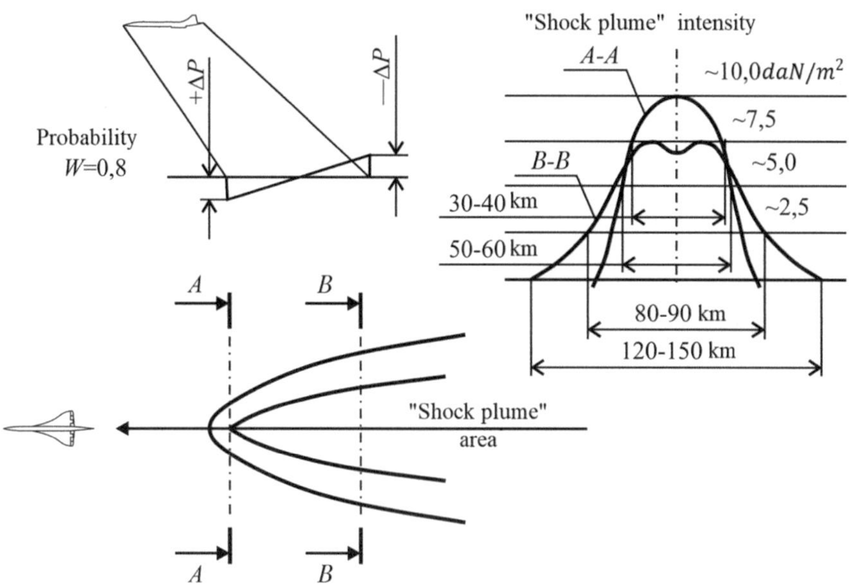

Fig. 13 Diagram of "sonic boom" and "shock plume". The cross-section distribution along the plume is given for an aircraft with a mass $m_0 = 180 \ldots 200t$ flying at a height of 18…20 km with a speed corresponding to the Mach number of 2.0…2.5

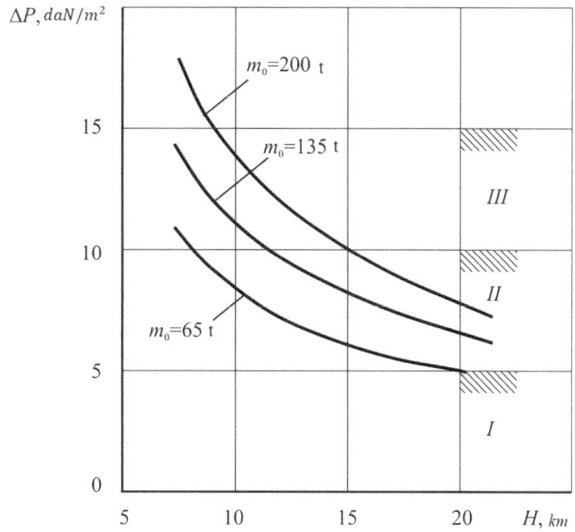

Fig. 14 Variation of the intensity of sound impact near the ground with the flight altitude and aircraft mass at the Mach number of 2.0…2.2: I—impression of distant explosion; II—sence of close impact; III—possible damage of large mirror glasses, ear pain

8 Features of Cargo Aircraft Design

8.1 Convertible Aircraft

According to the world statistics, the volume of air cargo transportation has been growing faster than the volume of passenger transportation for the last 20…25 years (average annual growth of cargo transportation is 12…15%, passenger transportation growth is 8…10%). In this connection modifications of passenger planes are used for air cargo transportation. It is possible through removing the passenger equipment, strengthening the floor of the passenger deck for the loading of 400 daN/m^2, and opening a big cargo door in the left side of the fuselage forebody at floor level. The cargo door is made load-bearing and included in the fuselage structure with locks.

Depending on the volume of cargo transportation, either the entire passenger cabin or its front part is modified, as it is shown in Fig. 15. Cargo placement in the front part of the fuselage is related to the safety of passengers carried in the rear compartment. On a number of convertible and cargo aircraft (Boeing 747F, An-124, C.5A) cargo hatches are located in the forward part of the fuselage and the cockpit is raised to the second floor of the aircraft, making room for the hatch on the first floor (see Fig. 16).

Such modified aircraft can also be used for cargo and passenger transportation. In this case passenger seats are attached to special platforms, which in turn are attached to nodes on the power cargo floor; some of the furnishings (toilets, kitchens, etc.) are retained inside the aircraft. In this case the aircraft can be quickly converted from cargo to passenger or serve as a cargo-passenger airplane.

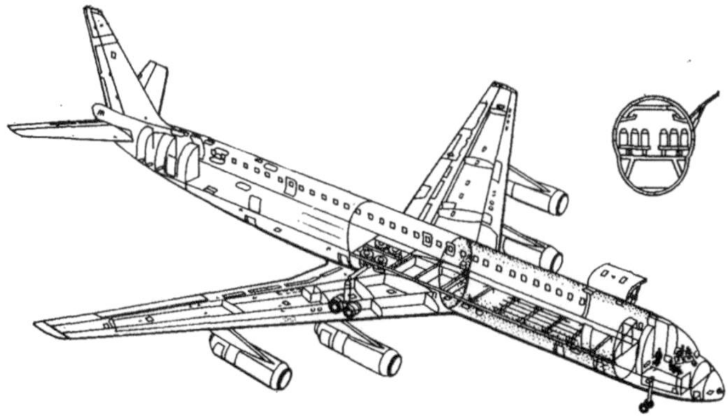

Fig. 15 A cargo-passenger aircraft (cargo compartment in the forebody)

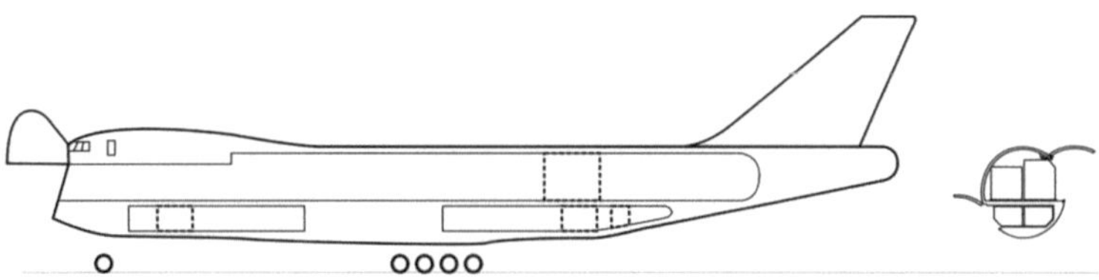

Fig. 16 Cargo version of the Boeing 747 F aircraft

8.2 Special Cargo Aircraft

Just as in other types of transport (automobile, railway and waterways) passenger and cargo vehicles (cars, ships) are made special, so in the development of air transportation it is natural that special cargo aircraft will appear.

A special feature of cargo planes is that they are made with high wings so that the fuselage can be as close to the airfield surface as possible to facilitate loading and unloading operations. It is desirable that the level of the cargo floor when the aircraft is parked on the ground is at the same level as the area of the truck body.

For loading, a hatch is made in the afterbody, the downwardly tilting part of which serves as a loading ramp. On some cargo planes, it was suggested to make the tail part of the fuselage tilting sideways (Canadair CL-44 aircraft) or the nose part of the fuselage to provide loading and unloading. The landing gear of a number of cargo aircraft (An-124) allows changing the parking clearance and fuselage tilt angle, which lowers the cargo floor and facilitates loading and unloading operations (Fig. 17).

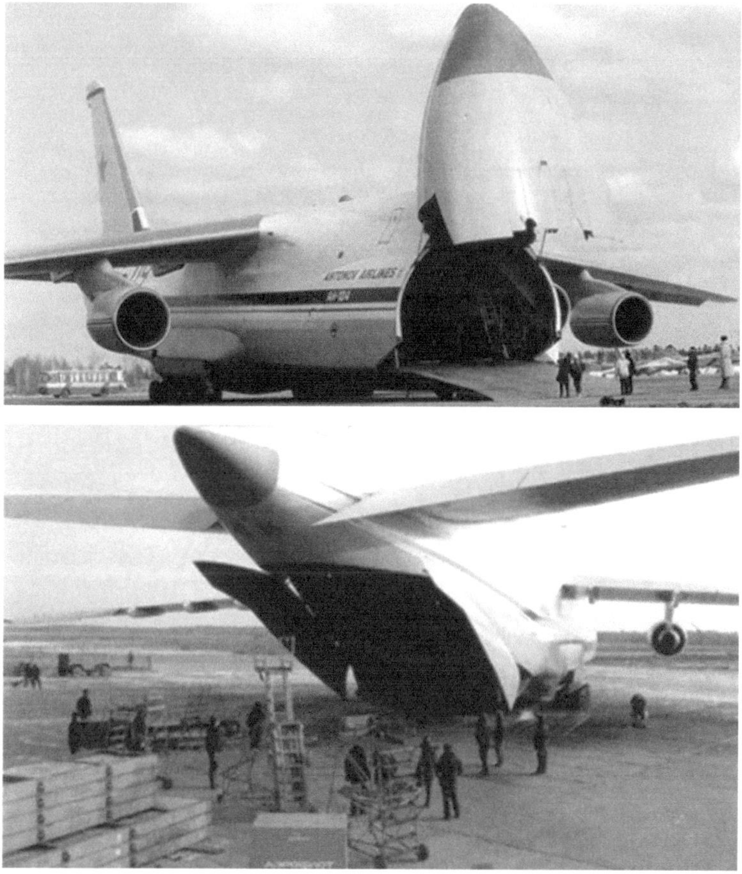

Fig. 17 An-124 "Ruslan" transport aircraft. The wingspan is 73,3 m, the length is 69,1 m, the wing area is 628,5 m^2. The maximum takeoff mass is 405000 kg; payload mass is 120000 kg. The cruising speed is 800–850 km/h. The flight range with a maximum payload is 4500 km

Variables		
Symbol	Description	Units
a	Civil aircraft total cost	*Roubles*
A	Aircraft operation cost	$roubles/hour$
m_{pay}	Payload mass	t
$V_{service}$	Service speed	km/h
k_{pay}	Payload factor	–
q_{fuel}	Fuel consumption	kg/km
C_p	Specific fuel consumption	$kg/(km \cdot h)$
n_{pas}	Number of passengers	–

(continued)

(continued)

Variables		
Symbol	Description	Units
k_f	Cabin shape factor	–
γ_f	Cabin (compartment) specific volume per passenger	$m^3/person$
V_b	Baggage and cargo volume	m^3
V_c	Cargo containers volume	m^3
V_g	Galley volume	m^3
S_g	Galley floor area	m^3
m_{land}	Aircraft mass at landing	kg
V_{appr}	Landing approach speed	m/s
$C_{L,max,land}$	Maximum lift coefficient in landing configuration	–
V_{cr}	Cruising speed	m/s
Λ	Sweep angle	Degrees
$\frac{t}{c}$	Airfoil thickness	–
C_D	Aircraft drag coefficient	–
C_{D0}	Aircraft zero-lift drag coefficient	–
ω	Trim penalties factor	–
AR_{eff}	Effective wing aspect ratio	–
C_L	Aircraft lift coefficient	–
M_{crit}	Critical Mach number	–
W_{pay}	Payload weight	N
R	Range	km
H	Flight altitude	km
L_{RW}	Runway length	m
T_0	Take-off thrust	N
$(L/D)_{climb}$	Climb lift-to-drag ratio	–
$\overline{m}_{PP}$	Power plant mass ratio	–
$\overline{m}_{empty}$	Empty aircraft mass ratio	–
m	By-pass ratio	–
$T_{H,cruise}/T_0$	Ratio of thrust at the cruising altitude to the take-off thrust	–
t_{total}	The temperature of total air deceleration in front of the leading edge	Degrees celsius
T_H	Absolute air temperature at the flight altitude	K
α	Heat transfer coefficient	–
ΔP	Intensity of sound impact at the ground level	daN/m^2
θ	Inclination angle of flight path	Degrees

References

1. Eger SM (1964) Proektirovanie passazhirskih reaktivnyh samolyotov (Design of passenger jet aircraft) Mashinostroenie, Moscow
2. Kiselev VA (1977) Voprosy komponovki passazhirskih samolyotov (Passenger Aircraft Layout Issues). MAI, Moscow
3. Sheinin VM, Kozlovsky VI (1984) Vesovoe proektirovanie i effektivnost' passazhirskih samoletov (Weight design and efficiency of passenger aircraft). Mashinostroenie, Moscow

Design Features of Combat Maneuvering Aircraft

Mikhail A. Pogosyan and Dmitry Yu. Strelets

Abstract The chapter expores in detail aircraft maneuverability, the relevant load factors and performance parameters, and provides historical data on maneuvering aircraft. The chapter explains the concept and purpose of wing extension and the use of high-lift devices for improving the manoeuvering characteristics of an airplane. The chapter offers some design approaches to ensure the stability and controllability requirements for combat maneuvering aircraft. Finally, some new technologies for advanced combat maneuvering aircraft are discussed.

Keywords Aircraft maneuverability · Wing extension · Static stability · Controllability

Aircraft maneuverability is its ability to change the magnitude and (or) the direction of its velocity vector. The faster an aircraft's velocity vector changes, the greater its maneuverability (see Chap. 8).

Both private and general maneuverability indicators are used to assess aircraft maneuverability characteristics. The former include angular velocities and radii of aircraft trajectory curvature when performing maneuvers (aerobatics), as well as maneuver execution time. The latter include normal (n_z) and tangential (n_x) load factors [6].

Aircraft are usually subdivided into classes according to the value of the maximum load factors (n_z).

- maneuvering ($7 < n_z < 9$);
- of limited maneuverability ($5 < n_z < 7$);
- non-maneuvering ($n_z \leq 3$).

Recently, a class of super maneuvering aircraft has emerged, which includes aircraft capable of performing controlled flight at supercritical angles of attack.

M. A. Pogosyan · D. Yu. Strelets (✉)
Moscow Aviation Institute (National Research University), Moscow, Russia
e-mail: maksmai33@gmail.com

M. A. Pogosyan (ed.), *Aircraft Design*, Springer Aerospace Technology,
https://doi.org/10.1007/978-981-96-4599-2_11

These maneuverability indicators directly depend on the design and aerodynamic characteristics of the aircraft, the parameters of its power plant and control system.

Given that the reader is already familiar with the scientific and methodological foundations of general aircraft design, this chapter deals only with the specific problems of combat maneuvering aircraft design, as well as new technologies that are used or will be used in their development. Of course, it is impossible to put an end to this process.

1 Combat Maneuvering Aircraft Mission Requirements

Combat maneuvering aircraft are mainly multifunctional fighters (MFF), which are part of the frontline aviation of the Russian Air Force, designed to solve tasks of gaining air superiority by destroying air and ground targets day and night, in simple and difficult meteorological conditions.

To accomplish this task, they must to:

- conduct all types of aerial combat and ground attack;
- perform all typical aerobatics, while ensuring the piloting by means of the automated control system.

The destruction of aerial targets by MFFs is carried out during aerial combat, which are subdivided into:

- long-range air-to-air combat using long- and medium-range air-to-air missiles;
- close air combat within visual range using short-range missiles or cannon weapons.

The most demanding maneuvering characteristics are required in close air combat, where maneuvering with the maximum permissible g-loads to reach weapons conditions in the shortest possible time is necessary. However, long-range air combat and attacks on ground targets may also require the use of extreme maneuvering capabilities, such as anti-missile maneuvers.

At the beginning of the design process, the general customer, the Russian Ministry of Defence, defines the mission requirements for a future combat manoeuvering aircraft. These requirements are formalized and issued to the lead contractor in the form of military operational requirements, which, as a rule, set the following parameters:

- M_{max} (at an altitude of 11 km);
- V_{max} (at ground level at sea level);
- V_z^* (maximum excess specific power);
- $n_{z,lim}$ (maximum operational load factor);
- $n_{z,st}$ (steady turn load factor);
- $R_{\max}$ (maximum tactical radius when flying at altitude and near the ground);
- acceleration time at altitudes of 1000 m, 5000 m and 10,000 m in specified speed ranges.

The maneuverability indicators are determined for a design payload and a fixed fuel mass, which is specified as a percentage of the fuel mass required to achieve a given tactical radius.

The maneuvering performance estimated in flight tests includes:

- the characteristics set in the technical specifications;
- the range of maneuver entry parametrs, without exceeding the maximum allowable flight speed, Mach number, load factor and angle of attack, the minimum allowable maneuver speed; maneuvering aircraft must be capable of performing maneuvers at speeds up to zero.

2 Relationship Between Manoeuvring Performance and Design Parameters

It is convenient to analyze the aircraft maneuvering performance using the g-load polar. In contrast to the Lilienthal polar, which presents the relationship between C_L and C_D, the g-load polar is plotted in terms of n_x and n_z (Fig. 1).

These parameters have an unambiguous relationship with the main maneuverability indicators:

- the angular acceleration $\dot{\psi} = \frac{g}{V}\sqrt{n_z^2 - 1}$;
- the excess specific power $V_z^* = n_x V$.

Specific power V_z^* can be estimated as follows:

$$V_z^* = n_x V = \frac{T - D}{mg} V.$$

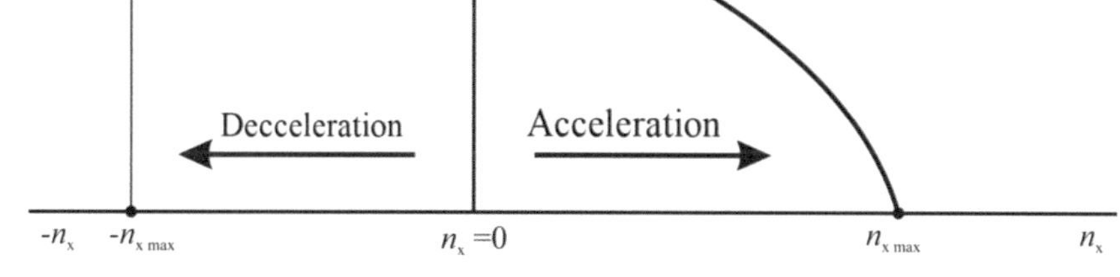

Fig. 1 G-load polar

The g-load polar can also display additional information showing the various overload limits:

- $n_{z,max}$ = strength constraint;
- $n_{z,allow}$ = a limit corresponding to the maximum allowable angles of attack at which the performance of the aircraft ensures safe and comfortable piloting;
- $n_{z,phys}$ = a constraint related to the pilot's physiological capabilities, determined by the pilot's cockpit ergonomics and his ability to tolerate high g-loads while retaining the ability to control the aircraft.

The point in the g-load polar at which $n_x = 0$ is called the steady-state g-load ($n_{z,steady}$). It indicates the level of overload and the corresponding angular velocity that a maneuvering aircraft can achieve while maintaining a constant airspeed.

To analyze the relationship between the maneuverability indicators and the main design parameters, we will use the analytical dependence, which allows the designer to choose the main directions of design. Such dependence can be obtained by using the following relations:

$$C_D = C_{D0} + AC_L^2,$$

$$A = \frac{1}{\pi eAR},$$

$$T = T_0\xi,$$

$$C_L = \frac{n_z mg}{qS},$$

$$\overline{m}_f = \frac{m_f}{m_0},$$

$$m = m_0 - m_f = m_0\left(1 - k\overline{m}_f\right),$$

where k is the consumed fuel factor (usually when analyzing maneuverability characteristics it is assumed to be 0.5), e is the Oswald coefficient, which takes into account the deterioration of the load-bearing properties of the aircraft as a result of violation of the ellipticity of the load distribution on the wingspan.

The assumption of the quadratic variation of the induced drag with the lift coefficient is reasonable for a comparison of the various design options.

Substituting the above expressions into the load factor equation yields.

$$n_x = \overline{T}_0\frac{\xi}{1 - \kappa\overline{m}_f} - \frac{D_0}{m_0 g\left(1 - \kappa\overline{m}_f\right)} - \frac{1}{\pi eAR}\frac{{m_0 g}/{S}\left(1 - \kappa\overline{m}_f\right)}{q}n_z^2.$$

Assuming $n_x = 0$ and neglecting the second term of the equation (which is acceptable for aircraft with a high thrust-to-weight ratio), we obtain a simplified formula for calculating steady-state g-load:

$$n_z \cong \sqrt{\frac{\overline{T}_0 \xi \pi e AR}{(1 - \kappa \overline{m}_f)^2 m_0 g / S}}.$$

Let us use it to analyze the effect of changes in the aircraft parameters on the steady-state g-load. Let us identify the ratio $\frac{\overline{T}_0 AR}{m_0 g / S}$, which includes the most important design parameters: wing aspect ratio, aircraft thrust-to-weight ratio, and specific wing loading. It can be called a generalized maneuverability indicator.

Consider how this varied in different maneuvering aircraft of generations III, IV and V (Fig. 2). The graph shows that for a number of aircraft this parameter increased linearly. However, some aircraft, such as the F-16A, the MiG-29, which are not inferior to the F-15A in their maneuvering performance, have a smaller parameter value. This is explained by the fact that the F-15A aircraft has a relatively simple aerodynamic configuration, which does not feature complex high-lift devices and wings of complex planform shape, while the aircraft F-16A, MiG-29, have a modern complex aerodynamic configuration. Due to this fact, the high level of steady-state g-loads is achieved.

Let us consider the influence of the parameters included in the above equation by dividing them into the following groups:

- power plant group;
- aerodynamic layout group;
- group of major n_z constraints.

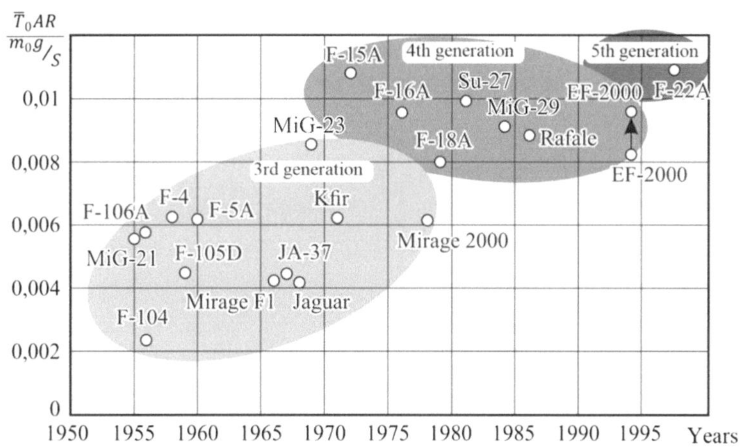

Fig. 2 The generalized maneuverability indicator for generation III, IV, and V maneuvering aircraft

Influence of main power plant parameters on aircraft maneuverability

The thrust-to-weight ratio of an aircraft in flight is determined by the ratio of the effective thrust of the propulsion system, i.e. taking into account thrust losses in the air intake and nozzle, to the weight of the aircraft. To increase it, it is important to limit the aircraft weight growth due to the engine weight and fuel weight. Advances in engine design have contributed to this goal.

The main characteristics of the engines of Generation III and IV maneuvering aircraft are presented in Table 1.

In the table, m is bypass ratio; T^*_{st} is stagnation temperature before the turbine; T_{sp} is specific thrust, $T_{sp} = \frac{T_0}{G}$; γ_{eng} is specific engine weight, $\gamma_{eng} = \frac{m_{eng}g}{T_0}$; $\overline{T}_{fr}$ is frontal thrust, $\overline{T}_{fr} = \frac{T_0}{S_{FrEng}}$; $\overline{T}_{unit}$ is thrust per unit length, $\overline{T}_{unit} = \frac{T_0}{l_{eng}}$.

Table 1 shows that with the change of generations of aircraft there was a change of engine type. Engines with some advanced parameters were created. It occurred due to the development of heat-resistant cooled turbine blades that allowed to sharply raise temperature T^*_{st}. Also, a new engine type with the by-pass ratio of 0.6 was introduced, which had a significant influence on aircraft weight, geometry, as well as specific fuel consumption.

This made it possible to significantly increase the thrust-to-weight ratio of aircraft without increasing the mass ratio of the propulsion system, since $\overline{m}_{PP} = k_{PP}\gamma_{eng}\overline{T}_0$. This can be seen in Fig. 3, which shows the relationship between take-off thrust-to-weight ratio for different generations of aircraft.

The performance of maneuvering aircraft is significantly affected by the integration of the power plant and airframe, in particular the parameters and layout of air intakes and nozzles (see Chap. 18).

At the same time, the aerodynamic characteristics of the aircraft were improved, both by reducing $C_{D,o}$ and $C_{D,i}$.

The aircraft drag at near-zero angles of attack is determined by:

- friction drag;
- drag component depending on the shape of the aircraft;
- wave drag at flight speeds exceeding the critical Mach number.

Table 1 Engine performance of Generation III and Generation IV maneuvering aircraft

Engine characteristics	Generation III with turbojet engines	Generation IV with by-pass turbojet engines	Ratio
m	0	0,6	–
T^*_{st}	1350°K	1650°K	1,22
T_{sp}	90	120	1,33
γ_{eng}	0,18	0,125	0,7
$\overline{T}_{fr}$	16	19,7	1,23
$\overline{T}_{unit}$	0,57	0,73	1,28

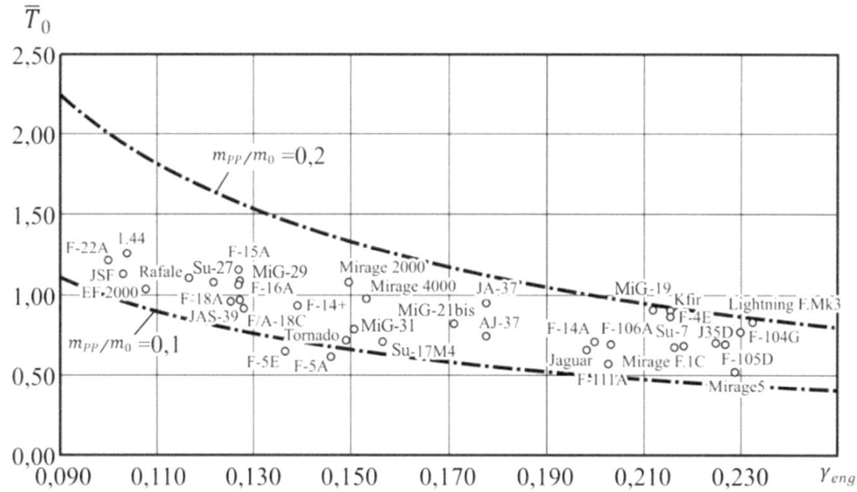

Fig. 3 Take-off thrust-to-weight ratio versus engine weight ratio

The friction drag depends on the wetted area and the state of the boundary layer. The state of the boundary layer, in turn, is determined by the local Mach number, surface roughness and irregularities.

The quality of the aircraft surface depends on the level of design and technological solutions. The greatest influence is exerted by:

- the elements of the aircraft control system (rockers, hinges, fairings)—8%;
- the information systems (antennas, temperature sensors, slots)—4.7%;
- possible surface deviations from the theoretical contour—14%.

It follows that to reduce the aerodynamic drag at near-zero angles of attack, it is necessary to strive to minimize the influence of the above factors.

The induced drag is represented by the last term in the g-load polar equation:

$$\frac{1}{\pi eAR}\frac{{m_0 g}/{S}\left(1-\kappa\overline{m}_f\right)}{q}n_z^2.$$

When analyzing this ratio, first of all, one should pay attention to the ratio ${m_0 g}/{SAR}$. Let us consider some historical data on this parameter for some maneuvering aircraft (Table 2).

It can be seen that this ratio has been minimized for the Generation III aircraft (MiG-21, Mirage IIIC) through the use of the tailless configuration.

The MiG-25 interceptor aircraft is given in this table as an example aircraft of limited maneuverability for which there was no need to minimize ${m_0 g}/{SAR}$.

Operating steady-state g-loads are significantly higher for Generation IV aircraft than those of Generation III aircraft. This is achieved by the right choice of the ratio of wing area, mass and aspect ratio. For V generation aircraft, the ratio ${m_0 g}/{SAR}$ can

Table 2 Historical data on m_0g/SAR for manoeuvring aircraft

Parameters	MiG-21	Mirage IIIC	MiG-25	F-15	F-16	MiG-29	Su-27
	Aerodynamic layout						
	Wing-tail	Tailless	Wing-tail	Wing-tail	Wing-tail	Wing-tail	Wing-tail
m_0g/S, daN/m^2	350	250	600	320	400	380	360
AR	2,2	1,8	3,0	3,0	3,2	3,4	3,4
m_0g/SAR, daN/m^2	159	139	200	107	125	112	105

be improved by increasing the wing aspect ratio and using composite materials for wing structures.

At the stage of preliminary design, it seems reasonable to select a moderate aspect ratio of (3.4…4.0) with a wing loading of $(300\ldots400\,daN/m^2)$. The choice of leading edge sweep angle is dictated by the need to reduce drag and increase the critical Mach number. However, a too large wing sweep is associated with the reduction in the load carrying characteristics (Fig. 4). Therefore, wing sweep is 40…45° for most typical for Generation IV aircraft. The exception is the F-18 aircraft designed for deck deployment, due to which it uses a wing with a sweep of 27°, which provides good field performance.

Consider some directions for improving g-polar by optimizing the aerodynamic layout.

Opportunities to improve the g-polar are mainly related to reducing the aircraft induced drag. The most effective ways to reduce it are:

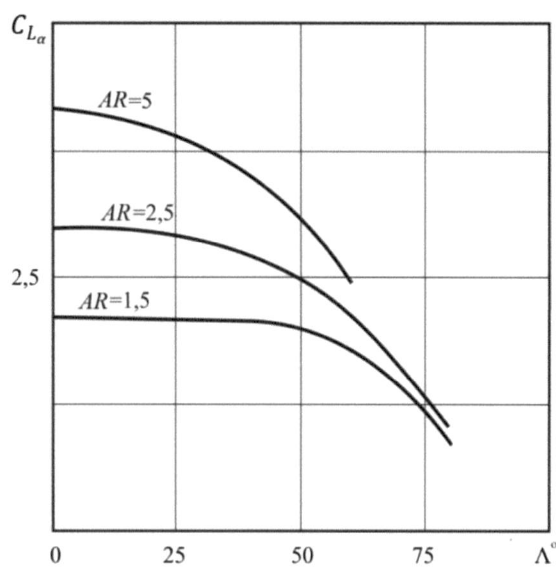

Fig. 4 Effect of sweep and aspect ratio on the lift capacity of wings

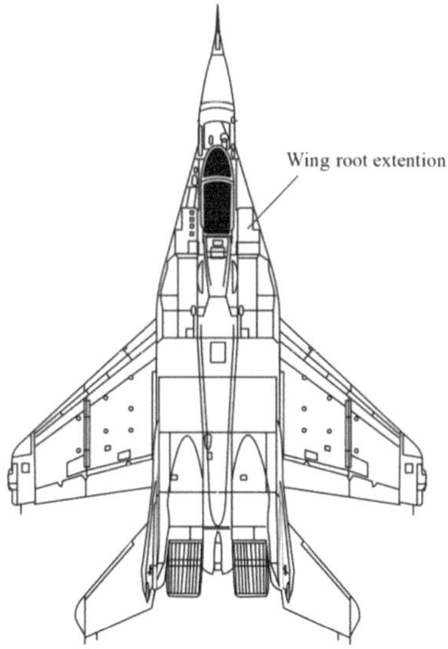

Fig. 5 Wing with wing root extention

- wing extention;
- wing twist;
- deflecting the high-lift devices during maneuvering.

Influence of wing extention. As noted, the optimal wing parameters for modern maneuvering aircraft are a sweep of 40...45°, an aspect ratio of 3.4... 4.0 (for V generation aircraft wing sweep increased due to the requirements of supersonic cruise flight, flight at transcritical angles of attack and small signature). However, at high angles of attack, flow separation is experienced with a consequence of decreased lift capability and lift-to-drag ratio. To avoid this, the wings of the most advanced modern fighters use wing root extentions with a leading-edge sweep of 75 . . . 80° and an area ratio of 0.12.

Figure 5 shows a typical shape of such a wing. With a large sweep wing root extention, at high angles of attack, a stable vortex system is formed that prevents the wing flow separation. At the same time, there is an improvement in the lift curve.

Wing root extention also has a significant effect on the longitudinal and lateral stability characteristics. The effect of wing root extention on lift and moment curves is demonstrated in Fig. 6.

Shaping the wing root extention is a complex process. Suffice it to say that about 30 different variants were investigated in the F-16 and about 10 variants in the MiG-29. The parameters to be varied were the planform shape, dimensions, leading edge sweep, airfoils, and incidence angle.

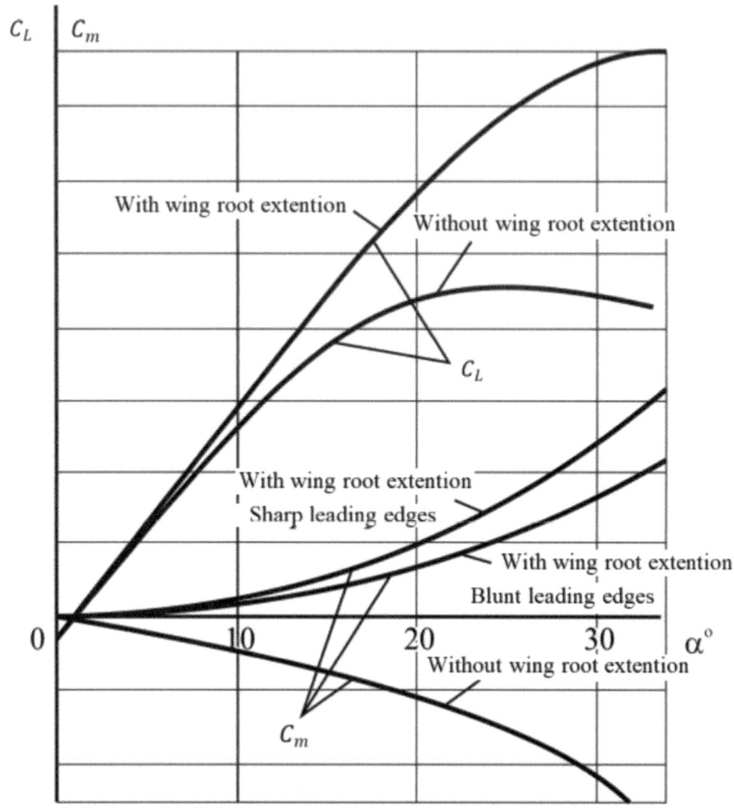

Fig. 6 Influence of surge on aerodynamic performance

Effect of wing twist. Wing shape is defined in terms of:

- the geometric twist spanwise variation: $\varphi_w = f(\overline{y})$;
- the airfoil camber spanwise variation $\overline{f} = f(\overline{y})$;
- the variation of the leading edge deflection angle along the wingspan $\overline{\delta}_{LE} = f(\overline{y})$.

Typical values of these parameters as functions of wing spanwise position for a maneuvering airplane are shown in Fig. 7a. They affect aircraft aerodynamic characteristics (Fig. 7b).

High-lift devices as a means to improve aircraft maneuvering characteristics On III generation aircraft, slats and flaps were used for takeoff and landing. However, the requirement to improve maneuverability led to the use of high-lift devices at large angles of attack during maneuvering. Typically, drooped leading edges and flaps are employed in maneuvering.

The main requirements for the high-lift devices are to enhance lift, lift-to-drag ratio, stability, controllability and to eliminate the buffet due to flow separation at high angles of attack.

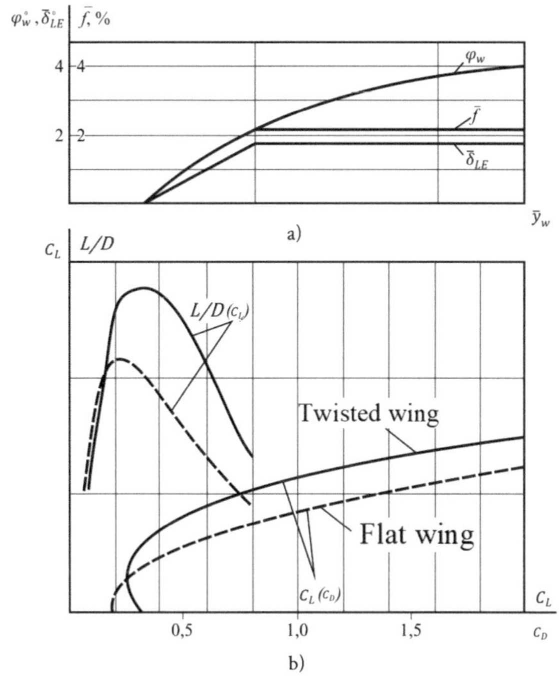

Fig. 7 Wing twist (**a**) and its effect **b** on aircraft aerodynamic performance

Since the use of drooped leading-edges significantly affects directional stability, these devices must be capable of deflecting quickly to avoid the loss of aircraft stability. Currently, two laws of leading-edge control are implemented:

- deflection law proportional to the change of angle of attack with feedback (Su-27, F-16 aircraft);
- discrete leading-edge deflection at the angles of attack exceeding 10° (Mig-29).

Figure 8 shows the dependency of lift-to-drag ratio on lift coefficient for the Mach number of 0.6. It can be seen that by varying the deflection angles of high-lift devices with angles of attack, it is possible to achieve the minimum drag for a given (design) value of the lift coefficient. On the right side of the figure typical deflection programs for high-lift devices are shown, with feedback at the top and a conventional discrete one at the bottom.

For each flight Mach number, graphs like those shown in Fig. 8 will be different, hence, the leading-edges and flaps (elevons) deflection program must depend on the Mach number. The high-lift devices control system that ensures smooth (within the required accuracy) change of the flap and the leading-edge angles depending on the angle of attack and the Mach number, is called adaptive. The use of adaptive control system significantly improves the performance of the wing, especially at high angles of attack. Adaptive control system is used, for example, on F-22A, Su-33, MiG-35 aircraft.

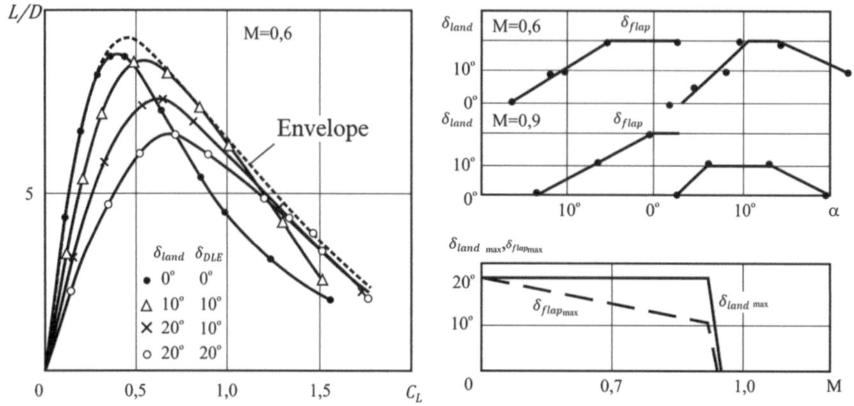

Fig. 8 Flap and drooped leading-edge deflection programs and associated lift-to-drag ratio

The choice of wing trailing edge deflection laws is usually simplified, as its deflection does not have a significant impact on the stability and controllability characteristics.

When the aircraft reaches large angles of attack, a flow separation begins, which gradually develops along the wingspan towards the fuselage. This results in aerodynamic buffet of the aircraft. It can be eliminated by ensuring a smooth mating of the leading-edge and wing surfaces in the deflected position.

There are the following ways of drooped leading-edge-wing mating:

- a simple drooped leading-edge (F-16, Su-27 aircraft) (Fig. 9a);
- a leading-edge with an additional controlled tab (MiG-29 aircraft) (Fig. 9b);
- multi-sectional leading-edges (Fig. 9c);
- adaptive leading-edges with elastic surface (Fig. 9d).

3 Main Ways to Meet Stability and Controllability Requirements for Combat Maneuvering Aircraft

As stated above, the limit value of one of the main parameters of a maneuvering fighter aircraft $n_{z,allow}$ or its corresponding allowable angle of attack α_{allow} depends on both the load-bearing characteristics and the stability and controllability characteristics. The stability and controllability characteristics should be the focus of the designer's attention from the earliest stages of design process. Let us consider the basic layout solutions that provide the required values of these characteristics for maneuvering aircraft.

For Generation IV maneuvering aircraft, the most powerful means of influencing the longitudinal stability are the size, shape and profiling of the wing root extention,

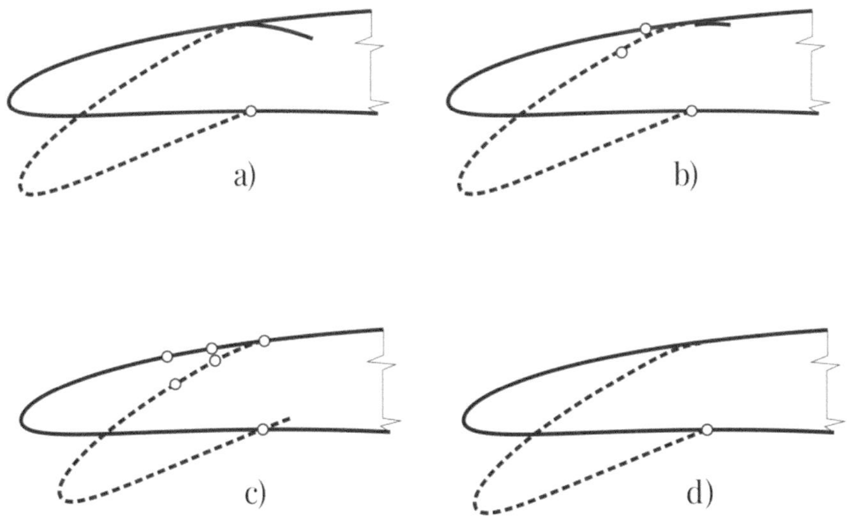

Fig. 9 Drooped leading-edge-wing mating

as well as the size and position of the horizontal tail. Let us illustrate this with some examples.

Figure 6 shows the change in longitudinal stability of an aircraft without horizontal tail, without a wing root extention and with a wing root extention with different leading-edge shapes. It can be seen that the wing root extention introduces large nose-up pitching moment due to the stability of the vortex system operating up to very large angles of attack. The blunt leading edges of the wing root extention reduce the efficiency of the vortex system and significantly change the pitching moment curve.

Installation of horizontal tail of different size also affects the pitching moment curve (Fig. 10). This is because, in the presence of a vortex system, the contribution of the horizontal tail to the pitching moment increases significantly as the angle of attack increases, which allows, by selecting its area, to correct the logitudinal stability.

For maneuvering aircraft of the IV generation, the typical wing root extention and horizontal tail parameters are as follows:

– for an airplane with a mechanical control system:

$\frac{S_{WRE}}{S_w} = 0.12$, blunt leading edge:

$$\Lambda_{WRE} = 70\ldots 80°;$$

$$AR_{HT} = 0.25\ldots 0.3.$$

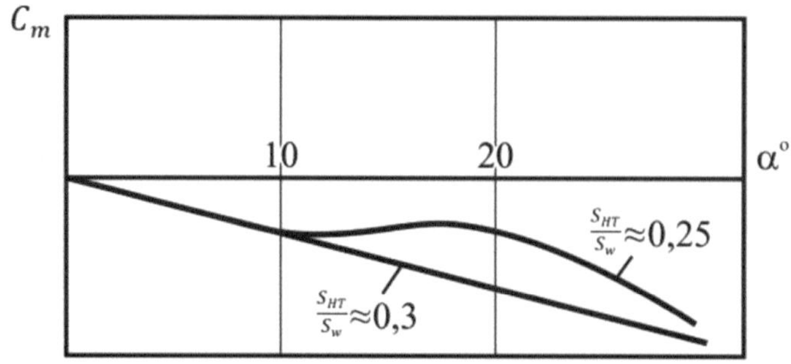

Fig. 10 Influence of the area ratio of the horizontal tail on the pithing moment curve

- for an aircraft with an automated control system:

$\frac{S_{WRE}}{S_w} = 0.12$, sharp leading edge:

$$\Lambda_{WRE} = 70 \ldots 80°;$$

$AR_{HT} = 0.2 \ldots 0.25$ (in some cases due to the peculiarities of aircraft behavior at high angles of attack $AR_{HT} = 0.3$).

These parameters can be used as inputs when laying out a maneuvering aircraft. They were implemented on the aircraft MiG-29, Su-27, F-16. In all cases, in order aviod blanketing the horizontal tail by wing wake at high angles of attack, it is recommended to lower it below the wing chord by $\frac{h_{HT}}{c_{MAC}} = 0.1 \ldots 0.25$.

Let us consider the basic design solutions that ensure lateral stability of the aircraft. Currently, there are two types of vertical tail: single- and twin vertical tail. Single vertical tail, as a rule, is used for single-engine aircraft, but this does not exclude its use on twin-engine layout. The advantage of single vertical tail over twin tail is better performance at high angles of attack due to its span. The disadvantages of single vertical tail are as follows:

- increased load;
- deterioration of signature;
- deteriorating ejection conditions;
- storage difficulty.

For twin-engine aircraft, a twin vertical tail layout is most commonly used. The greatest difficulty in its design is placing it in an optimal position with respect to the wakes coming off all parts of the aircraft. As an example, let's consider the interaction of the twin vertical tail with the aircraft's vortex system at large angles of attack (Fig. 11). It can be seen from the figure that a vortex system is formed consisting of 5 vortices—2 wing tip vortices, 1 canopy vortex, and 2 wing root extention vortices. Of this system, the most influence on the vertical tail is exerted by the vortices produced by the wing root extentions, as the most powerful and closely located ones. The vertical position of the wing root extention vortices depends on the

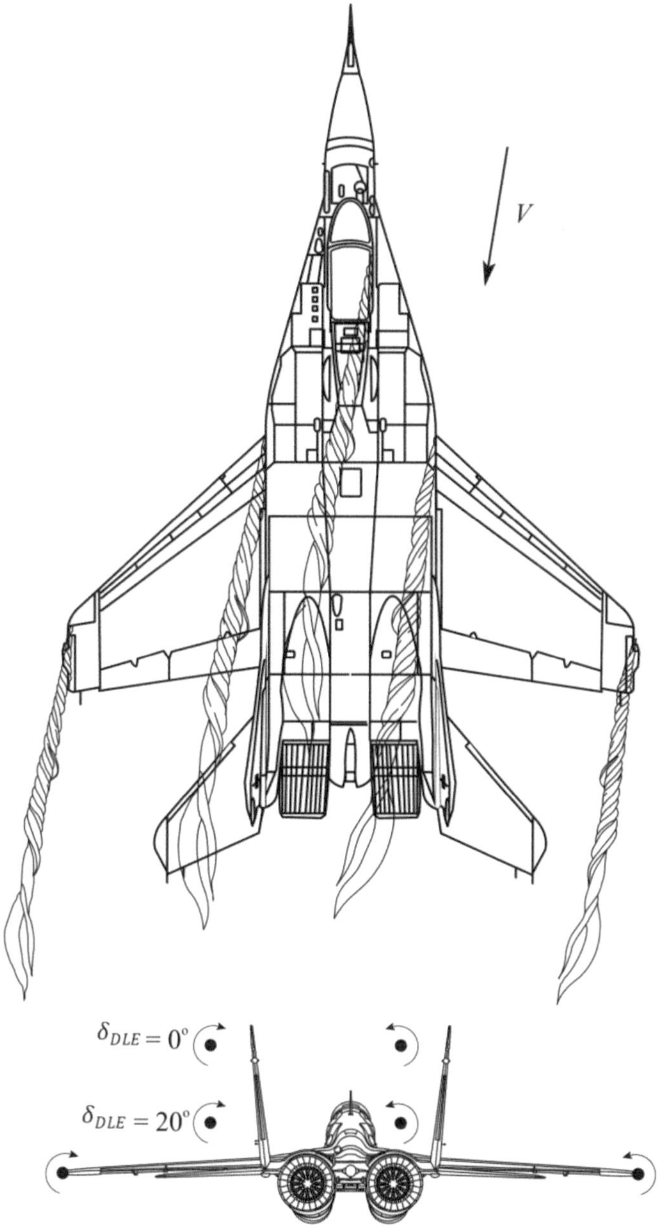

Fig. 11 Interaction of twin vertical tail design and aircraft vortex system at large angles of attack

deflection of the wing leading edges. When the leading edges are deflected, due to suction forces, the vortices are pressed against the upper surface of the aircraft and forces act on the outer vertical tail to reduce the sideslip angle. This physical pattern is consistent with wind tunnel tests.

Figure 12 shows the effect of vertical tail aspect ratio. It can be seen that aspect ratio increases C_{n_β}, but practically do not increase the angle of attack at which $C_{n_\beta} = 0$.

The effect of verical tail position on the stability is shown in Fig. 13. Shifting the tail does not increase the angle of attack where $C_{n_\beta} = 0$, but only reduces the stability due to a decrease in tail momet arm. Thus, to begin designing an airplane with a twin vertical tail, it is advisable to use an aspect ratio of 1.4 and place it at the greatest possible distance from the center of gravity. The interation of the tail with the wing vortices including the configuration with deflected leading-edges must be taken into account. The recommended tail volume coefficient range is 0.07… 0.09.

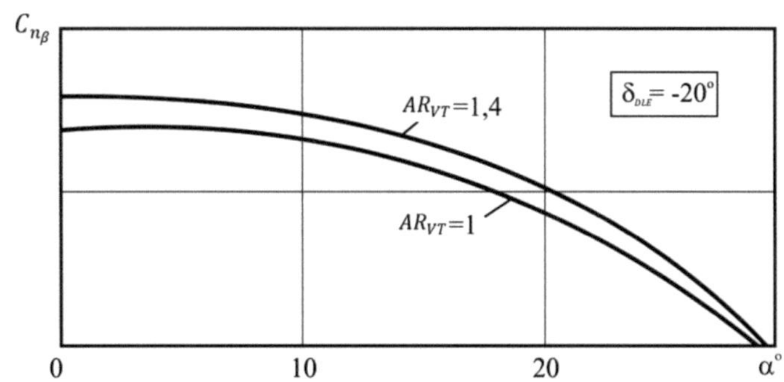

Fig. 12 Effect of vertical tail aspect ratio on aircraft directional stability

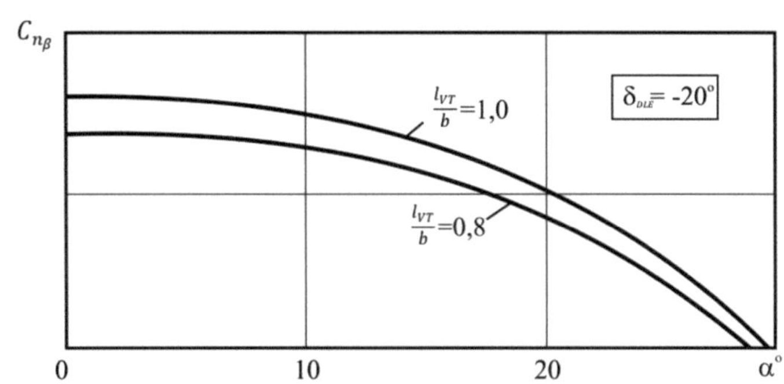

Fig. 13 Effect of vertical tail position on aircraft directional stability

4 Development Trends of Combat Maneuvering Aircraft

Development of combat mancuvcring aircraft is inseparably connected with transformation of the corresponding requirements. According to experts, the aircrafts of future generations of frontline aviation will have to operate under a high degree of uncertainty, in conditions of intensive omnidirectional irradiation and counteraction of technically highly equipped enemy in the air, in space and on the ground. In this connection, advanced aircraft must have: high survivability, which enables them to continue the mission for the whole period of military operations; high combat readiness both on the ground and in the air for timely detection of targets and countermeasures; low visibility for covert movement over the theater of military operations.

Progress in the development of next-generation precision-guided weapons, combined with advances in electronic operations planning systems, will lead to significant changes in air combat techniques, ranging from rear midcourse attacks to air combat beyond visual range at various altitudes and directions. Combat will predominantly be conducted at medium range, and when fully confident of winning over the enemy, it will move into close maneuvering combat.

Conducting air combat within medium range requires the ability to fly at supersonic speeds and the ability to perform at these speeds a number of intense dynamic maneuvers.

The range of combat missions that the aircraft has to perform requires a multichannel, multi-band, jammer-protected information system onboard, with the capability to reliably acquire and track an adversary beyond visual range at different angles of view, and in the case of multiple targets, select the most threatening targets and identify additional targets for further attack.

The main contribution to improving the performance of the next generation of combat aircraft is associated with:

- new avionics;
- low-visibility technology ("Stealth"), which significantly reduces radar, infrared and optical wavelengths;
- a new generation of engines with a deflected thrust vector in two planes, reduced fuel consumption and gas temperature at the nozzle outlet;
- a new generation of high-precision, highly intelligent weapons with the ability to launch in a dynamic aircraft flight mode;
- new aerodynamic layouts.

Advances in electronic equipment made it possible to create powerful onboard computing systems with speeds of several billion operations per second, as well as powerful radars, air-to-air missile guidance devices, air situational awareness calculators, navigation systems, etc. The creation of such airborne equipment led to the development of all-angle airborne weapon systems. Such systems make it possible to create a circular information zone around the aircraft and a corresponding zone for missile launches. Simplistically, this zone can be viewed as an ellipsoid with axes of several tens of kilometers. The major axis in this ellipsoid is about three times the minor axis. This detection and kill zone has led *to new requirements for air combat tactics.*

Maneuvering must be accomplished in such a way that one of the aircraft can "cover" the enemy aircraft with its launch area while being out of the enemy's range. This is possible when maneuvering at large radii along trajectories corresponding to the engagement zones of the two aircraft, followed by one of them entering the enemy's rear hemisphere, where the enemy zone is much smaller than that of the attacking aircraft in the forward hemisphere. This position can be achieved through prolonged maneuvering with a flight speed greater than that of the enemy. In this case, a new requirement is imposed on aerial combat tactics—conducting aerial combat at supersonic flight speeds. This does not mean that the next generations of aircraft have reduced requirements for engagement at subsonic speeds, as there are many targets for which the air-to-air combat mode at $M < 1$ is optimal. Therefore, in addition to the traditional requirement to conduct air combat at subsonic flight speeds for advanced aircraft, a new requirement for supersonic maneuvering appears.

The estimations of combat effectiveness have shown that the number of the target load delivered to the combat area depending on the flight speed corresponds to the ratio $\overline{m}_{load} \cong \frac{m_{load,V=1000}}{m_{load,V=2000}} = 0.6$, and the probability of the carrier hitting the enemy when reaching the area corresponds to the ratio $\overline{P} \cong \frac{P_{V=1000}}{P_{V=2000}} = 1.3$. Hence the requirement to increase the cruising speed up to $M = 1.5...2$ arises.

Based on the above, it can be concluded that the generation of advanced fighter aircraft should have two modes of cruise flight: subsonic and supersonic, i.e. should be a generation of multi-mode aircraft.

A characteristic feature of next-generation fighters will be new maneuvering modes. Such modes include:

- maneuver modes with direct control of the lifting force;
- maneuver modes at overcritical angles of attack.

The physical principles of maneuvers with direct control of lifting force can be explained by using the lift curve for different flap positions (Fig. 14).

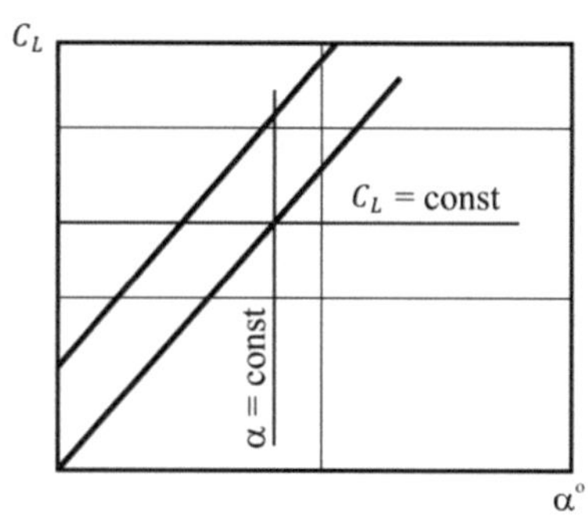

Fig. 14 Lift curve for various flap positions

This graph shows that the same lift coefficient may be achieved at different angles of attack and different values of the lift coefficient may correspond to one angle of attack. From this, one can conclude that it is possible to fly to maintain the same trajectory angle θ at a costant load factor $n_z \approx const$ at different angular positions of the fuselage.

The control parameter at $\theta = const$ is the flap deflection angle δ_3. By controlling the flap deflection angle at $\theta = const$ it is possible to ensure the flight along a constant trajectory when the position of the fuselage axis changes. This mode of the direct lift control system (DLCS) can be used in various maneuvers, for example, the aiming mode by changing the position of the fuselage axis by applying flaps.

The conducted researches show that the DLCS shortens the time of aiming several times as compared to the control through angle of attack, when the moments of inertia of the aircraft near the center of gravity are taken into account and a pilot has to control three parameters: α, ϑ, θ.

Maneuvering at high angles of attack is divided into two types:

- super maneuverability at the angles of attack where the aircraft responses to the contol inputs;
- hyper maneuverability, which is flying at the angles of attack of up to 80…120° with the aircrfaft being stable, but without metered control capabilities.

Super maneuverability allows the use of large thrust vector components, which significantly reduces the turn space and affects the turn time (Fig. 15). It can be seen that the trajectory change is not only due to aerodynamic forces, but also due to the thrust component of the engine thrust.

In order to eliminate the loss of speed during such maneuvers, an altitude gain is usually made in the first stage, and an altitude reduction is made in the final stage, during which the speed increases, reaching the initial value.

A comparison of the space at a steady-state turn at $\alpha = 25°$ and during super maneuvering shows its reduction by about 2–2.5 times (Fig. 16). The use of such maneuvers makes it possible to make an anticipatory launch of missiles in aerial combat due to a rapid turn in the detection zone, as well as vigorously exit from it due to abrupt deceleration.

The hyper maneuverability is an uncontrolled dynamic throwing on angles of attack $\alpha = 80\ldots120°$. This throwing is realized by creating large momentum at low angles of attack, achieved by deflecting all controls, followed by an increase in the angle of attack. The departure from large angles of attack is achieved due to large pitching moments produced at the angles of attack exceeding *70°* due to a shift of the center of pressure backward because of the flow separation from the aircraft surface (Fig. 17). Such a process lasts approximately 3 s and allows a preemptive launch of missiles.

The requirements for a super maneuvering aircraft are as follows:

- dynamic lateral stability over the entire range of angles of attack up to 120°;
- control authority sufficient to reduce large angles of attack;
- the capability of the control system to prevent stall.

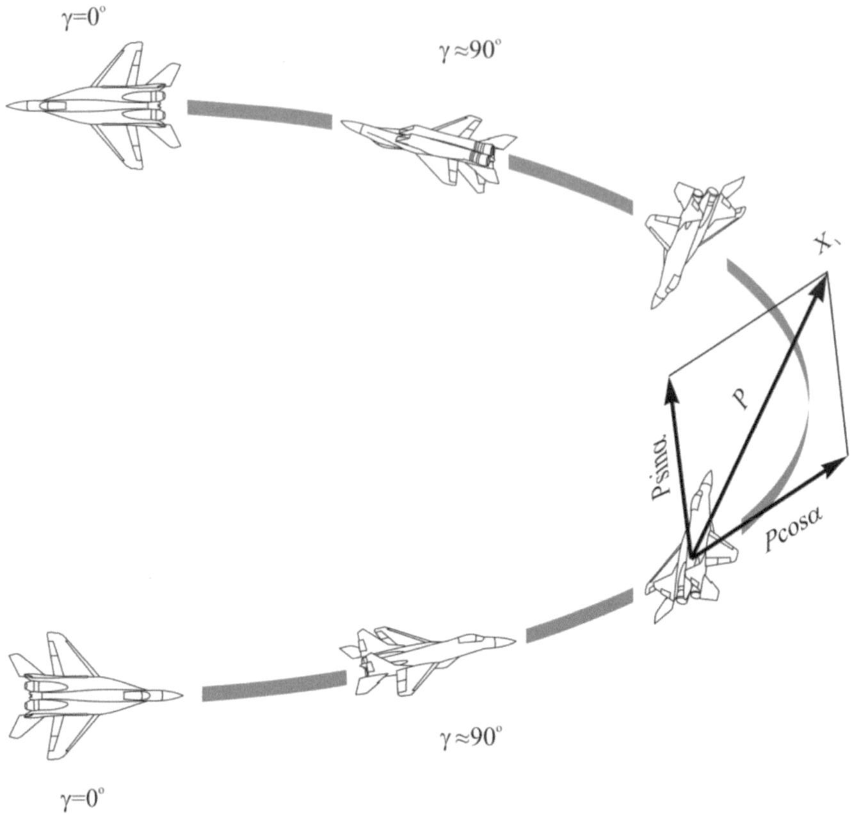

Fig. 15 Aircraft super maneuverability

Fig. 16 Comparison of the space required for a typical turn and supermaneuvering

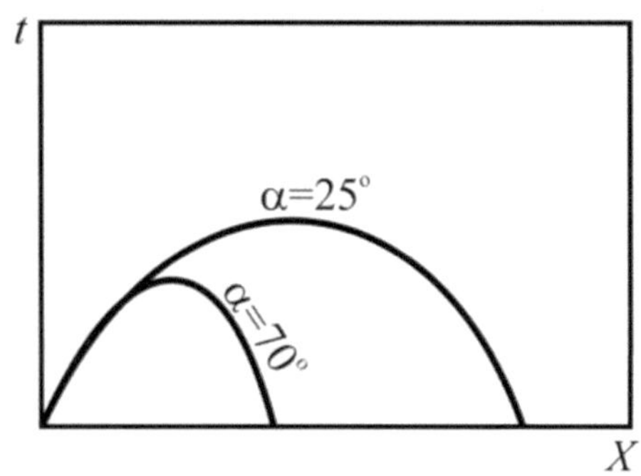

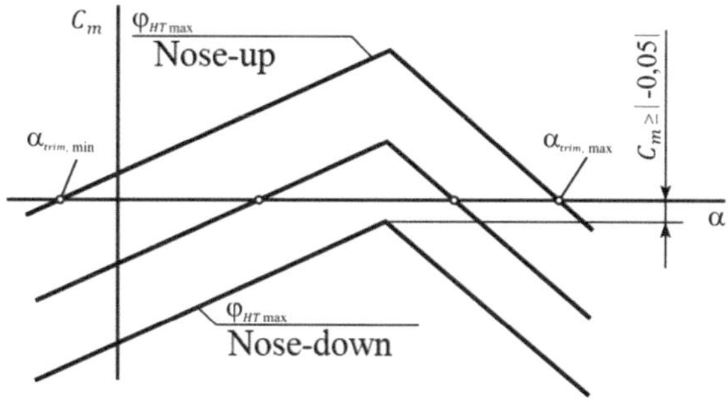

Fig. 17 Moment curve delivering super and hypermaneuverability

Gas-dynamic control, such as deflected thrust vector (DTV), is used to extend the types of maneuvers at non-critical angles of attack.

The purpose of DVT is to produce additional forces to control the aircraft at high angles of attack and low speed. The operating range of aircrafts without DVT is quite narrow and practically limited only to Hook and Cobra maneuvers, when the aircraft is practically uncontrollable and its stability is mainly due the short duration of the maneuver. It is possible to dramatically improve controllability by using a rotary engine nozzle. When the jet is deflected, two components of the engine thrust appear: one passes through the center of mass and is directed along the aircraft axis, the other is perpendicular to it. Depending on the orientation of the nozzle rotation axis, the control moments in longitudinal and lateral motion are generated when deflecting the nozzle (Fig. 18a and b). For twin-engine layout, deflection of nozzles in opposite directions allows creating rolling moments (Fig. 18c).

Rotary nozzle development and control is a complex technical task. The simplest one is the axis design, implemented on the Su-30MKI and F-22 aircraft. More complicated is the dual-axis design used on the MiG-29OVT, F-16 MATV "VISTA", F-15 "ACTIV", which provides independent pitch, yaw and roll control. Developed jointly by TsAGI and Sukhoi Design Bureau, the V-shaped position of the Su-30MKI's single-axis circular nozzles (Fig. 18g) provides control in all three axes of the twin-engine aircraft within the single-axis configuration. The use of DVT makes it possible to significantly extend the range of maneuvers.

5 New Technologies for Advanced Combat Maneuvering Aircraft

Aerodynamic layouts

The aircraft layout is significantly influenced by the requirements for the tactics of aircraft application in the future theater of war, as well as the achievements in the field of computer technology and software, the development of a new generation of aviation weapons. In this connection specialists of different countries for several

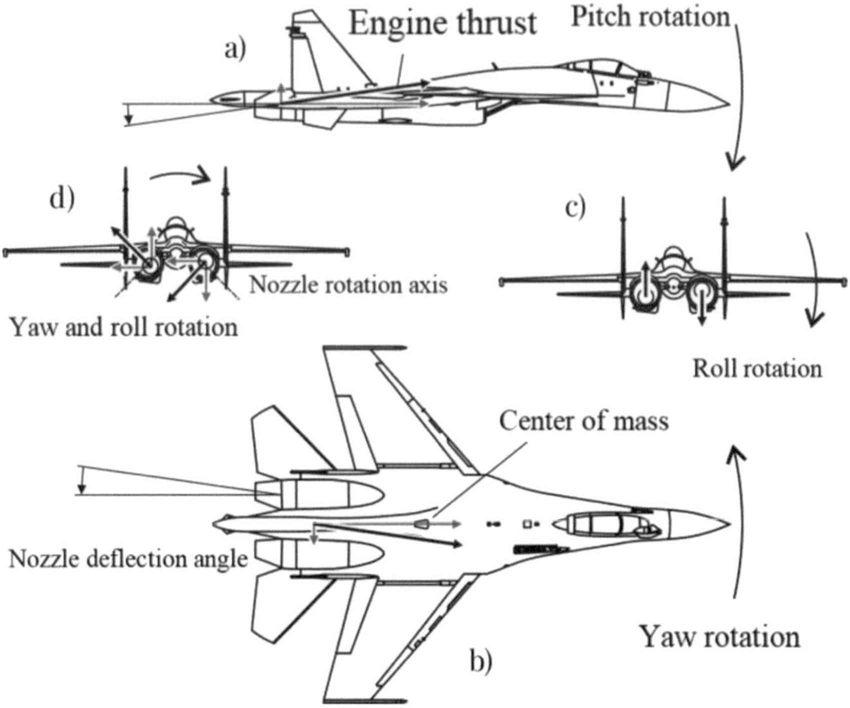

Fig. 18 Forces produced on an aircraft when the engine nozzle is deflected

years have been making attempts to create aircraft combining advanced performance and extremely low signature in one configuration, which could not be obtained on the aircraft of the previous generation.

Solving this problem began approximately in the mid-1980s (F-22A, 1.44, Su-47, etc.) and continues to this day (J-20, PAK FA). Note, that by the specified period, our industry, unlike the U.S., which has accumulated experience in comprehensive implementation of the Stealth technology for subsonic non-swept aircraft (B-2, F-117A), has not had such experience. Besides, there was an opinion among many specialists that the Stealth technology was deeply inconsistent with the image of an airplane that had high flight performances. For this reason, the only way to reduce the radar effective area of the aircraft was to use radio absorbing materials. This resulted in reducing the radar effective area of the MiG-29 and Su-27 aircraft by a factor of about 20. As a result, the aerodynamic shapes of the first Russian advanced aircraft turned out to be more traditional (see Fig. 19).

The appearance of the Russian aircraft 1.44 was significantly influenced by the work carried out in the direction of increasing the load capacity and efficiency by using the interaction of the engine jet and the flow field around the aircraft (see Chap. 12 and Fig. 20).

Fig. 19 Examples of advanced combat maneuvering aircraft

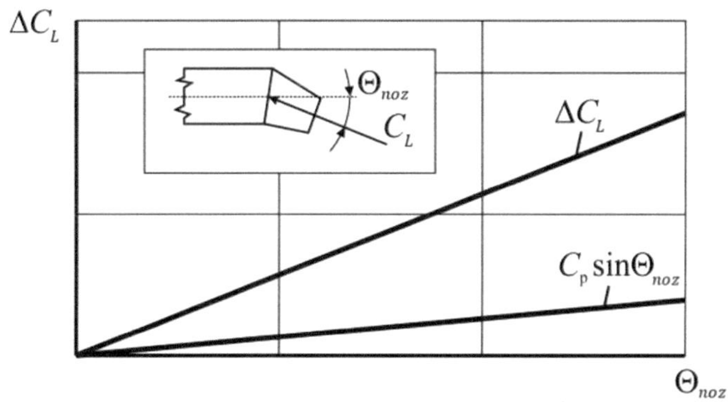

Fig. 20 Increment in C_L due to jet blowing

To increase these effects, it is necessary to bring the engine jet closer to the wing. However, in the traditional arrangement of engines on the aft fuselage, the deflection of the thrust vector leads to a nose-down pitching moment. For a wing-tail aircraft this usually yields high trim penalties.

The layout that meets the requirements for optimum interaction of the engine jet and circulation with minimal trim penalties is possible if the fins are located in front of the wing, i.e. for the canard layout. Interaction of its elements is shown in Fig. 21.

Flat nozzles are used with the following objectives:

- to increase the load capacity and lift-to-drag ratio during maneuve in conjunction with the aircraft's high-lift devices;
- to produce control and balancing moments at different flight modes;
- to increase drag through opening the nozzle flaps for abrupt braking;
- to reverse the thrust vector.

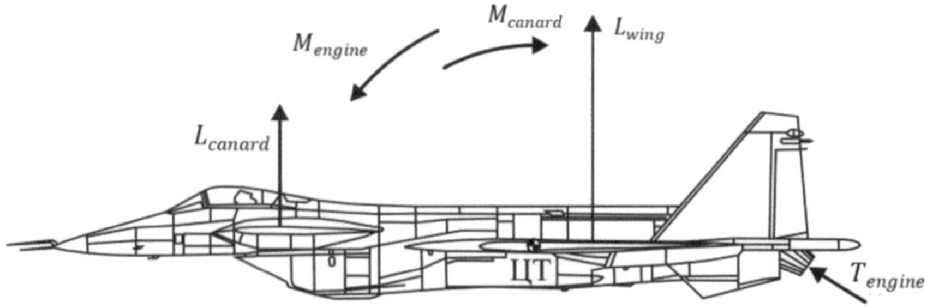

Fig. 21 The system of forces and moments for the 1.44 aircraft

The positive interaction between the engine jet and the airframe is best ensured at large ratios of L/h, where L stands for the nozzle width, h is the nozzle height. However, this leads to an increase in internal hydraulic losses due to an increase in the internal surface required for a smooth transition from the circular cross section of the afterburner to a flat nozzle, an increase in the nozzle weight and a deterioration in the steepness of the outer contours, so at present, for fighters, $\frac{L}{h} \approx 0.7 \ldots 1.0$.

The main feature of the aerodynamic configuration of another Russian prospective aircraft Su-47 is the forward-swept wing (FSW).

The main advantages of this layout are a significant improvement in maneuverability without the use of deflected thrust vector engines, a large lift compared to the aft-swept wings of the same area, a large range at subsonic speeds due to the lower trim penalties, a better controllability at low speeds, better operating conditions for the high-lift devices and an excellent stalling behaviour, an increased internal volume in the junction of the wing and fuselage.

For a supersonic highly maneuvering airplane with forward-swept wings, the main problems are its divergence (torsion followed by collapse), ensuring static and dynamic strength, efficiency of controls, high structural weight.

Divergence safety issues are most intrinsic to a forward-swept wing due to its geometric characteristics resulting in increased local angles of attack when the structure is deformed. The importance of static strength is related to the high values of maximum operational g-loads required to provide the design maneuverability. The high flight speeds and the significant wing aspect ratio cause problems related to the dynamic strength of the wing structure, as well as the effectiveness of the wing controls. The solution of these problems should fit within the specified weight limit to ensure high flight performance.

The above-mentioned problems are solved by increasing the strength and stiffness through the use of composite materials (CM) and advanced structural and technological methods that ensure the fabrication of structures with the specified parameters. To reduce the negative effects of the forward sweep, along with monofilament layers arranged symmetrically, an unbalanced layer was introduced into the design of the wing panels. This unbalanced (not mirrored) composite layer with different layup angles creates the so-called "open contour" effect, shifting, as shown in Fig. 22,

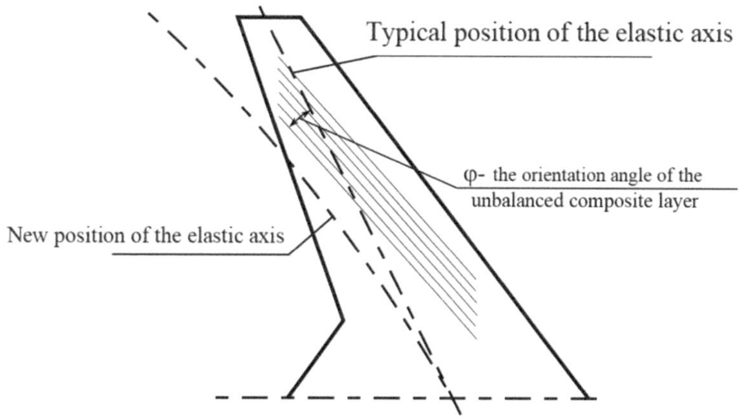

Fig. 22 Displacement of the elastic axis due to the introduction of an unbalanced composite layer with different layup angles

the stiffness axis outside the wing planform. Since the position of the elastic axis depends on the thickness of a given layer in the package and the angles of its orientation, a comprehensive study of the effect of this solution on the whole range of the above mentioned characteristics and problems was carried out. As a result, the selected structural design and the thickness of the CM package ensured the absence of control surface reversal and divergence over the entire range of operational flight speeds.

The following can be noted in the aerodynamic layout of the F-22A aircraft:

- this is the wing-tail configuration, statically longitudinally unstable with a low value of the horizontal tail volume coefficient. The peculiarity of this configuration is that the horizontal tail is blanketed by the low aspect ratio wing. In this arrangement, the horizontal tail being exposed to the wing wake doesn't move the aircraft aerodynamic center until it leaves the wake zone, remaining effective withing the full operating range of deflection angles;
- it features an integral aerodynamic configuration, which is characterized by a streamlined wing-fuselage juncture. It has high subsonic lift capabilities, and due to the fairly smooth distribution of cross-sectional areas over the airframe it experiences relatively small wave drag. Integral aerodynamic layout also allows for almost linear lift curve up to $\alpha = 30°$;
- it has an adaptive low aspect ratio wing.

The next generation aircraft feature the wings with the aspect ratio of 2…2.4. This is due to the fact that large supersonic flight ranges require not only a high lift-to-drag ratio at $M > 1$, which is possible with a significant reduction in the mid-section of the aircraft, but also to reduce the lift coefficient at which the maximum lift-to-drag ratio is achieved. And this is provided only by a wing of small aspect ratio. In addition, for the V-th generation of engines the ratio of afterburning thrust and thrust changed in favor of the latter, which increased the difference in $(L/D)M/C_p$ for subsonic and supersonic flight. At subsonic flight, relatively small lift capacity of low aspect ratio wings is compensated by an increased wing area, i.e. a lower wing loading.

The leading edge of the F-22A wing is equipped with drooped leading-edges, while ailerons and flaps are located on the trailing edge. Movable elements of the wing can deflect depending on the flight mode by signals of a special auxiliary system integrated with the aircraft control system, i.e. automatically adapt to it. This makes it possible to prevent wing flow separation up to large angles of attack.

The F-22A aircraft is equipped with a low-signature flat nozzle with a large value of *L/h* and a deflectable thrust vector, which ensures an increase of the lift coefficient value at high angles of attack due to supercirculation on the upper fuselage surface and generation of the required nose-down pitching moments at super maneuverability modes.

Small radar cross section of F-22A aircraft is also provided by using low-reflective aircraft shapes with a minimum number of joints and hatches on the outer surface; internal placement of weapons; use of input and output devices with reduced demasking properties (oblique air intakes with curved channels to prevent radar exposure to engine compressor blades, flat nozzles, etc.); use of radar absorbing materials and special technologies to maintain low signature during operation.

The PAK FA aircraft has a wing-tail layout with a large horizontal tail providing the necessary longitudinal control authority in all modes of flight, as well as the lateral moments at differential deflection at high angles of attack and high flight speeds. In order to reduce the tendency to pich up when the angle of attack increases, a small aspect ratio wing with a straight leading edge was selected.

The selected aspect ratio and sweep contribute to the production of a stable vortex system along the leading edge of the wing, which provides a large value of the critical angle of attack and a large value of $C_{L,max}$, and a highly developed spatial flow field around the wing reduces the probability of asymmetric flow separation. The selected wing shape also gives a minimum increase in C_m with increasing angle of attack.

In order to reduce radar signature, the forebody is made "faceted" with a side rib, passing into the front of the wing extention, which also contributes to the symmetry of the flow field near the aircraft at high angles of attack. To produce the nose-down pitching moment at large and ultra-large angles of attack, the front of the wing root extention is made rotary. Deflection of the leading-edge of the wing root extention at low angles also increases the lift-to-drag ratio during maneuvers.

The airplane is equipped with a small all-movable vertical stabilizer, which ensures the lateral stability and control authority, makes it possible to slightly reduce the airplane drag by reducing the wetted area and improve the polar by adaptive deflection of the leading-edges, flaperons, ailerons and vertical stabilizers.

The aircraft has a small degree of longitudinal static instability, which makes it possible to trim the aircraft at the aft center of gravity at the angles of attack up to $60\ldots70^\circ$ and generate the nose-down pitching moments at all angles of attack.

The trim penalties at $M > 1$ are reduced due to the positive value of $C_{m,0}$, provided primarily due to the profiling of the lifting fuselage, air intakes and flattened-faceted forebody.

PAK FA additional qualities, which were not achieved by the aircraft of previous generations:

- flight at M > 1 without afterburner with in-fuselage weapon placement;
- good acceleration at low angles of attack to minimize aircraft energy recovery time after a maneuver with loss of speed and altitude;
- the possibility of obtaining large steady-state g-loads;
- low signature when changing from air-to-air to air-to-surface weapons.

Further development of aerodynamic layout for next generation combat maneuvering aircraft is dictated, first of all, by the nature of future military conflicts. The war of “engines” becomes a war of information, in which the probability of victory will depend to a greater extent not on the characteristics in the battle itself, but on the ability to take a preferred position to attack, evade in time or successfully withdraw from the battle. These tactics are based on revolutionary development of some branches of science, in particular electronics and software, emergence of “intelligent”, highly energetic weapons with dynamic launch allowing to release aircraft from necessity of fixing its position relative to the target by means of aerodynamics. These factors already allow us to take a new look at the task of achieving situational superiority in air combat with a probable enemy, focusing not on increasing the characteristics of the aircraft itself, but on providing intellectual support and stealth of its movement over the theater of military operations.

The emergence of new qualities of aircraft that do not fit into the framework of old concepts of warfare allows, on the one hand, to solve a number of contradictions that will inevitably arise on the evolutionary path of aircraft technology (for example, human physiology and capabilities of aircraft, cost-effectiveness of new and deeply modernized aircraft of the previous generation, etc.), and, on the other hand, requires the development of new methods of analysis and appropriate models that adequately reflect the impact of new qualities of aircraft on its combat effectiveness.

At the present time in our country and abroad there is an active formation of scientific and technical groundwork (Fig. 23), the first results of which allow us to assert that the features of the aerodynamic configuration of the next generation aircraft will be:

- the high degree of integration and flattening of the layout, optimizing it to ensure low signature and reduce destabilizing yawing moment of the fuselage;
- the high lift-to-drag ratio of small aspect ratio wings due to the lower friction drag and better shaping of the fuselage cross sections;
- the transition to a mixed system of aerogasdynamic control and balancing;
- the weight reduction of aircraft systems by optimizing the required capacities of actuators, the number of control surfaces, reducing the degree of static longitudinal instability;
- the internal placement of the main weapon.

Fig. 23 Examples of aerodynamic layouts of advanced combat aircraft systems

Weight reduction of combat maneuvering aircraft

The following major factors can reduce the weight of combat maneuvering aircraft:

– optimization of the structural layout using the modern methods of engineering analysis;
– transition to integral layouts with a more favorable distribution of aerodynamic and mass forces over the airframe;

- application of new technological methods of manufacturing and connection of airframe parts;
- expansion of application of composite materials, light aluminum and titanium alloys;
- reduction of the weight of avionics and propulsion system.

The impact of these factors leads to a significant reduction in the ratio of the airframe weight to $W_p n_{z,max}$, showing the value of the total load acting on the airframe (Fig. 24). The graph shows that due to new structural materials and manufacturing solutions, the weight ratio of the 5th generation airframe has decreased significantly compared to the 4th generation.

One of the main factors that triggered these changes is the increased use of composite materials in airframe design. For the V-th generation aircraft the share of composite materials reaches 30% due to their use mainly in structural elements (Fig. 25).

In order to create airframes with reduced weight and longer service life, the following materials are used:

- a new generation of aluminum-lithium weldable alloys, with a 15–20% reduction of the weight of structures;
- layered composite materials for fuselage skin, in which fatigue crack propagation rate is 10 times lower than that of existing aluminium alloys;
- high-strength titanium alloys for load-bearing parts;
- fundamentally new polymer adhesive composites (PAC) with adjustable strength and heat resistance.

A significant factor in aircraft weight reduction is the reduction of avionics weight, which occurs due to transition to a new element base and integral structures of

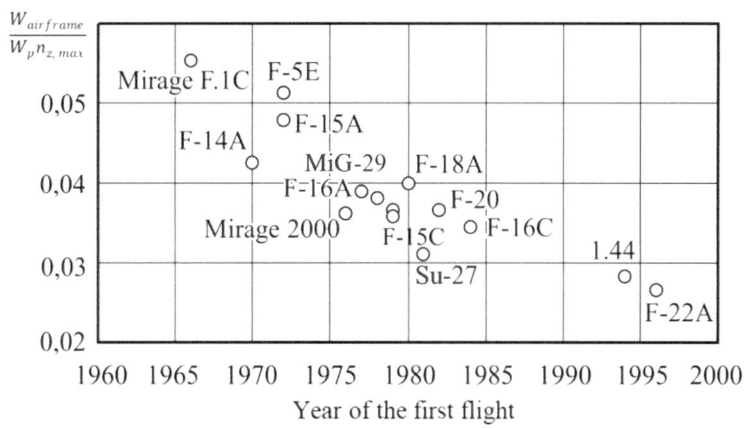

Fig. 24 Retrospective change in airframe weight ratio

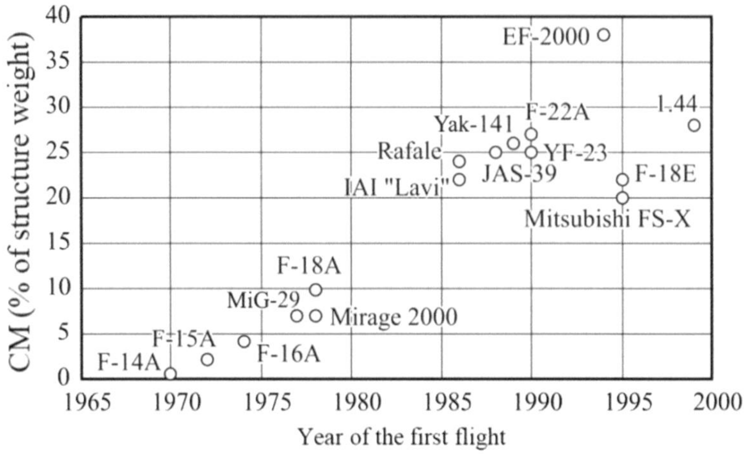

Fig. 25 Use of carbon fibre reinforced plastics in airframe design

onboard systems organization (Fig. 26). It can be seen that the mass ratio of the aircraft avionics was growing until about 1980, then got a steady downward trend.

The aircraft weight is significantly affected by the weight of the propulsion system. This issue is discussed in more detail below. The only thing to note here is that the specific engine weight for V-generation engines is $\gamma = 0.1$, and for next-generation engines, this parameter is expected to be reduced by about half.

In addition to the above factors leading to a reduction in the weight of maneuvering combat aircraft, there are a number of reasons leading to an increase in its weight. These reasons are caused mainly by the requirements of combat application. Chief among them are the requirements of increased combat payload and increased

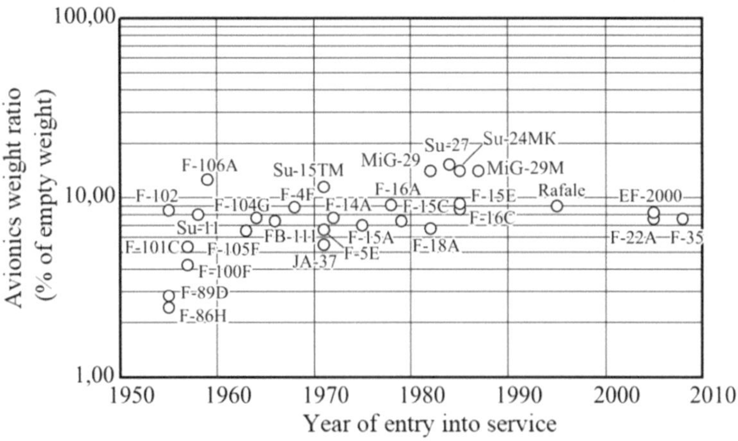

Fig. 26 Avionics weight as a percentage of the empty weight

range. Figure 27 shows an increase in the maximum combat load on fighters, but the combat load placed in the compartments does not tend to increase. This is due to the difficulties in placing large cargo compartments on this aircraft type.

Due to the need to improve performance there is a steady tendency of growth of the fuel weight ratio (Fig. 28).

The sum of the above factors increases the weight of fighter aircraft. However, this trend is limited by the rising cost of these aircraft, which makes them unaffordable for today's combat market.

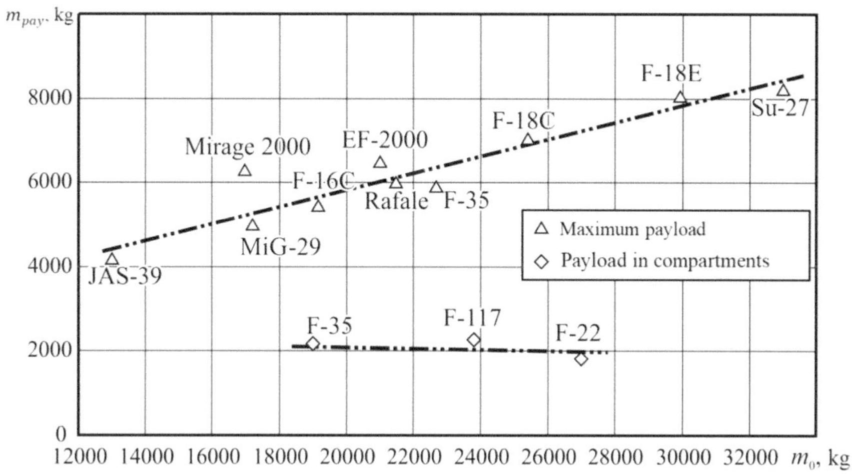

Fig. 27 Combat payload mass

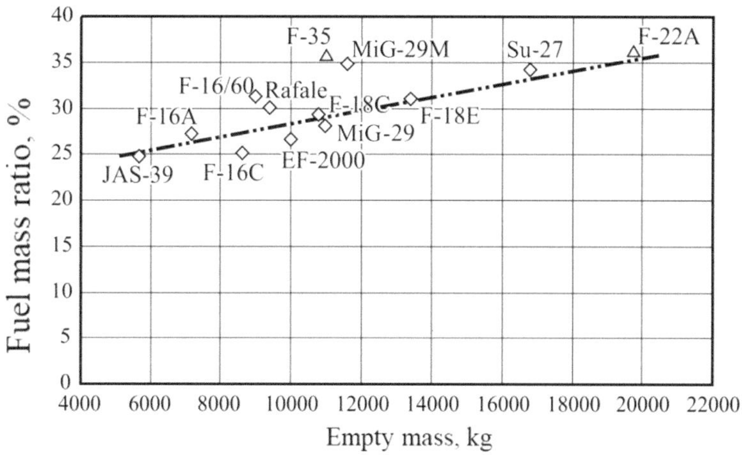

Fig. 28 Fuel mass ratio growth

Propulsion Systems

Priority areas in the development of next-generation fighter propulsion systems can include:

- extensive integration of the engine, input and output devices, power plant systems with the airframe and aircraft systems based on structural transformation and functional integration;
- further intensification of key cycle parameters;
- further improvement in terms of specific mass and volume;
- development of digital electronic engine control systems;
- development of new metallic and non-metallic materials and structures;
- signature reduction;
- reduction of the time and cost of engine development;
- increased operational reliability and safety.

Since 80's of the last century intensive researches have been conducted on the specified directions both in our country and abroad. As a result, the V-th generation engine appeared in the 90's of the last century. This engine, for example F119-PW-100 (USA), has high thrust-to-afterburner ratio (about 0,8 … 0,85) and small specific fuel consumption during supersonic flight, 20–25% smaller weight ratio, 15–20% higher efficiency during maximum and afterburner modes. Such a significant improvement of V-generation engine characteristics was achieved due to the use of:

- large-chord monocrystalline blades designed using three-dimensional flow analysis methods;
- heat-resistant materials;
- monolithic wheels in which discs and blades are made in one piece;
- digital electronic control systems and integrated diagnostic systems.

The modern engines are equipped with a flat, low-signature nozzle with a deflected thrust vector. The air intakes are located on the sides of the fuselage. This arrangement of air intakes at a distance of two calibers (widths) of air intakes between them allows to reduce the fuselage side projection area located in front of the center of gravity of the aircraft, eliminates mutual influence of engines and contributes to the formation of two vortices, rather than one combined, thereby reducing the possibility of foreign objects entering the engine inlet.

In order to reduce the radar signature from the side view, the air intakes have parallelogram cross-sections, the sides of which are parallel to the edges of the main elements of the airframe.

Studies have shown that the radar signature of the propulsion system in the forward hemisphere along the aircraft's longitudinal axis can account for up to 70% of the total signature of the aircraft. To reduce the signature of the propulsion system, in addition to the sweep of the leading edges of the air intakes, artificial curvature of the channels was introduced and radio absorbing coatings were applied.

According to specialists' opinion, the next generation of military aircraft engines should have 1,4…2 times reduced specific weight ($\gamma = 0.05 \ldots 0.06$), 15…30%

increased efficiency of combat application, 1,5…2 times improved dynamic characteristics, increased life to airframe service life, increased by 60…80% reliability, 2 times reduced labor input in maintenance and 1,5 times reduced life cycle cost. According to published data, these results will be achieved through the use of a set of advanced technologies, the main of which are presented in Fig. 29 [4].

Development of avionics

The functions performed by modern avionics systems have expanded in comparison to the forties of the last century from simple radio communication to flight control, weapons control, radio-electronic countermeasures.

The avionics of a combat aircraft shall perform the following tasks:

- determining the exact position of the aircraft in space;
- acquisition and processing of information on air and ground conditions;
- effective control of the aircraft and weapons by directive commands from the pilot and external systems;
- comprehensive countermeasures against the enemy;
- communication with external control systems and between aircraft of the group;
- monitoring and recording of flight information;
- providing the pilot with the information he needs to make a decision in a convenient form;
- optimization of maneuvering and aerial combat tasks;
- development of warning signals of dangerous flight conditions.

These tasks are well known and have not changed for decades. Only the requirements to the quality of their solution change, and, consequently, the transition

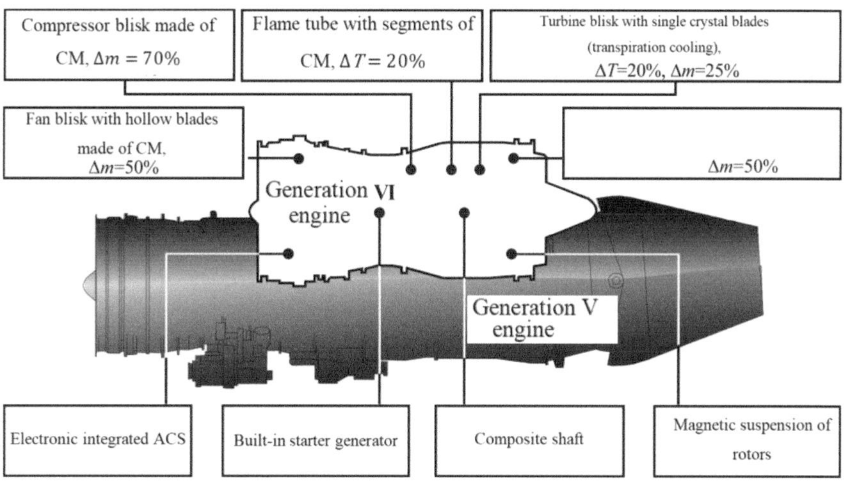

Fig. 29 Key technologies of the Generation VI engine

from one generation of aircraft to another changes the principles of construction (architecture), composition and requirements to the technical characteristics of avionics.

These requirements include:

- openness and reconfigurability of the architecture;
- modularity of design elements and software;
- building the intelligence of on-board systems;
- the secrecy of the work;
- hardware, information and functional integration;
- omnidirectional information field and increased accuracy of all information and targeting systems;
- increased reliability of operation;
- weight reduction.

For V-generation aircrafts, the architecture of on-board systems was changed, whereby instead of a distributed system of on-board computer systems with local computing devices (Fig. 30a), all tasks must be solved by the central computer, which constitutes the so-called "core of the on-board computer system" (Fig. 30b). This approach makes it possible to organize a single integrated computing environment, in which there is initially no rigid distribution of computing facilities over the information channels of the system and there are no hardware-implemented subsystems, which, in turn, significantly increases the fail-safety and noise immunity of the combat complex as a whole. However, such organization of onboard systems requires significant growth of intelligence of onboard systems.

This problem can be solved through the transition to onboard computers with performance about 50 milliards per second and memory about 100 Gb, that exceeds capacity of 4th generation aircraft computers by several orders of magnitude.

Transition to integration of onboard equipment systems and to a new element base will make it possible to reverse the tendency towards increasing the avionics weight (Fig. 26) with a significant increase in accuracy and speed of solving all tasks.

Improvement of avionics, on the one hand, resulted in the increase of onboard equipment reliability (Fig. 31) and the reduction of maintenance labor intensity related to flight hours (Fig. 32), and on the other hand, in the growth of relative cost of onboard equipment in total cost of the whole aircraft (Fig. 33).

The emergence of powerful on-board computers, the use of modern computer technologies to control the aircraft, avionics and armament, the development of radar with an active phased antenna array, allowed to reach a new level of information systems.

The next step in radar improvement should be considered an "integrated antenna unit level avionics system," developed under the Rafe Race program. It is supposed to be used on the F-35 aircraft.

The development of optronic targeting and target acquisition systems began with the Generation IV aircraft.

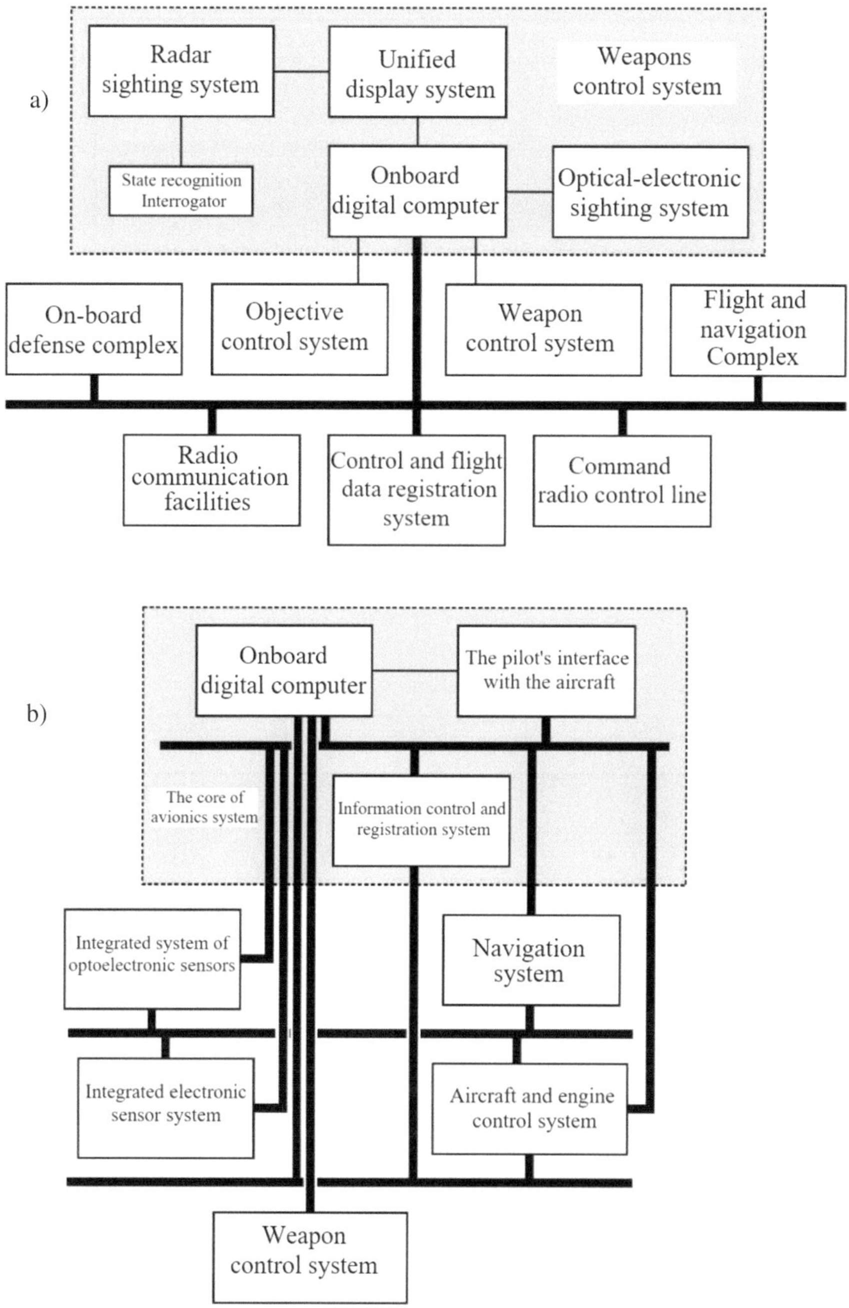

Fig. 30 Structural diagrams of aircraft avionics of IV (**a**) and V (**b**) generations

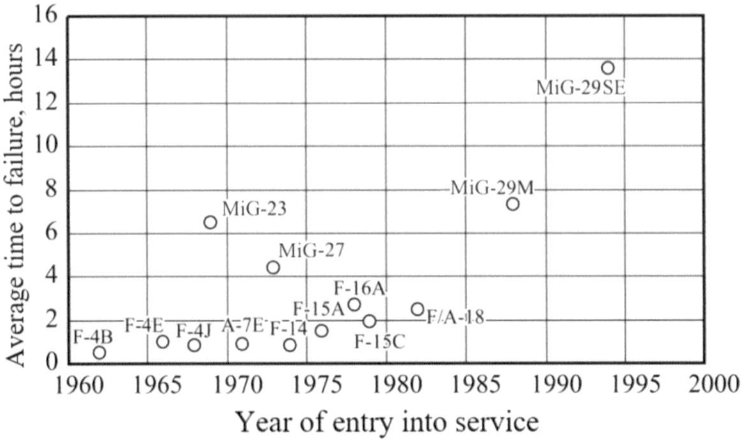

Fig. 31 Improving aircraft reliability

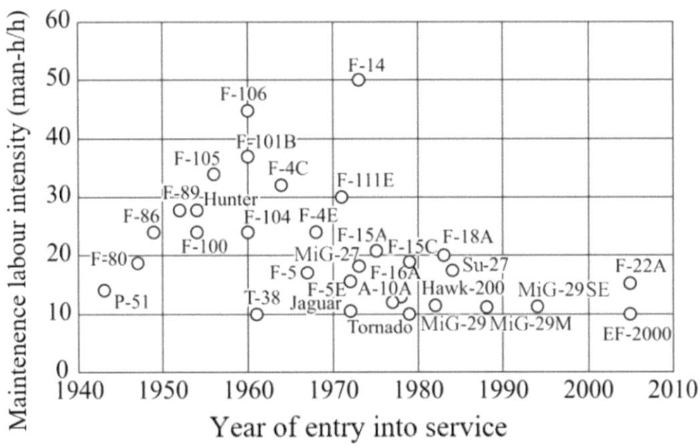

Fig. 32 Trends in maintenance and repair

Laser television sighting systems ensure the use of all types of aviation weapons with television homing warheads and guided missiles, as well as the use of precision weapons with laser beam guidance.

A complex of radio-electronic countermeasures is used to conduct radio-technical reconnaissance and create interference in the radar and infrared wavelength ranges. The greatest prospects in this area are associated with the development of a digital system of radio-electronic countermeasures, which at the hardware and functional level is integrated with all radio-electronic systems, including radars.

According to experts, the main direction of increasing the efficiency of advanced aviation combat systems is the development and improvement of the information and control field (ICF) of the pilot's cockpit. The concept of a multiscreen display

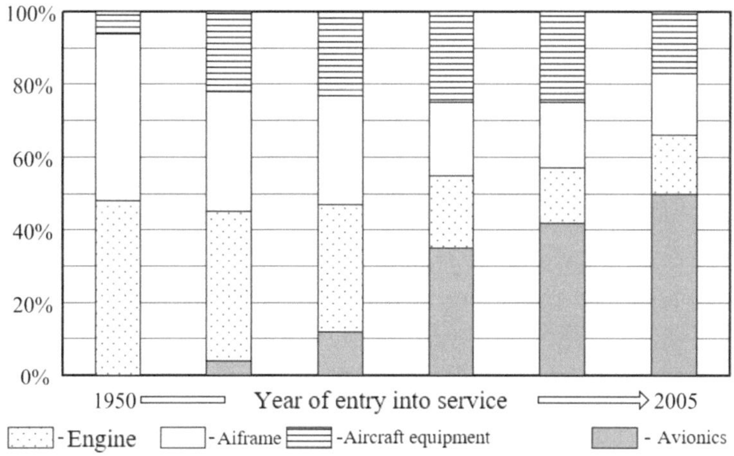

Fig. 33 Fighters cost components

system using wide-angle displays with holographic optical elements, color image indicators with large screen sizes and multifunctional button framing is a determining factor for building the 5th generation cockpit ICF. The multifunctional displays are complemented by a helmet-mounted indicator which can be used both as a targeting device for missile launches and as an information display device.

The accuracy of determining its coordinates on V-generation aircraft is ensured by the free inertial navigation systems based on laser gyroscopes and satellite system, which gives approximately twice the accuracy of similar characteristics of inertial platform systems.

Trends in the development of aircraft systems. Control systems

The development of control systems is based on the development of powerful onboard computers with a performance of several billion operations per second. Such onboard computers make it possible to move to a complex aircraft control system (ACS), which creates artificial stability of an unstable aircraft, providing control of the aircraft on the ground and in the air. Control on the ground is provided by the front strut wheel turning and wheel braking systems. Air control is provided by electro-hydraulic actuators of control surfaces and servo drives of turning engine nozzles. The control system ensures comfortable airplane control, limitation of critical modes (Fig. 34), including elimination of stall, automation of combat modes and automatic aircraft withdrawal from dangerous altitude (active safety), which virtually eliminates accidents due to pilot's errors and loss of orientation (Fig. 35). Short landing at high angles of attack (Fig. 36) is possible with the help of ACS. The development of integrated control system capabilities will significantly improve reliability, flight safety and combat effectiveness of aircraft systems.

Energy systems. Currently, three secondary power systems are used on aircraft: power supply system, hydraulic system, pneumatic system. Such configuration of the

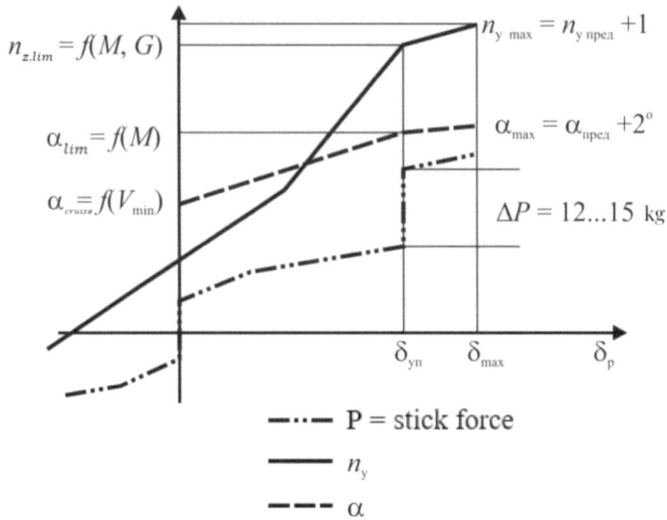

Fig. 34 Automatic limitation of α and n_z

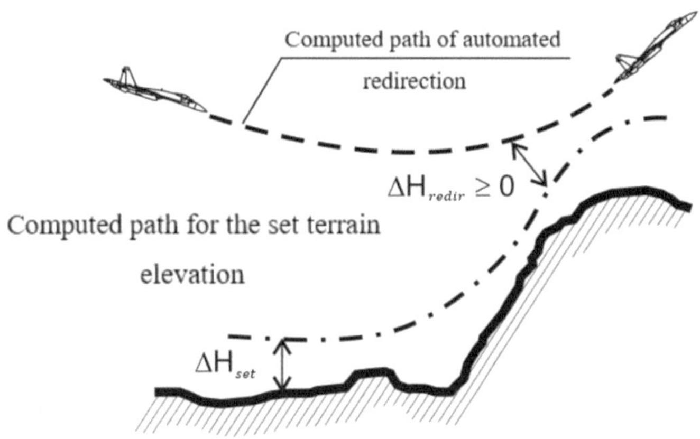

Fig. 35 Active safety system

onboard power supply system for advanced aircraft is not optimal, requires significant costs for its operation and causes significant difficulties in the integration of onboard equipment. One of the promising solutions to this issue is the concept of the aircraft with fully electrical equipment ("all-electric aircraft" or AEA). AEA is an aircraft with a single centralized power supply system that provides for all power needs of the aircraft.

At AEA, electric power will be used to power the most energy-intensive systems, which traditionally used hydraulic and pneumatic power. Implementation of the AEA concept implies a significant change in the appearance of the aircraft propulsion

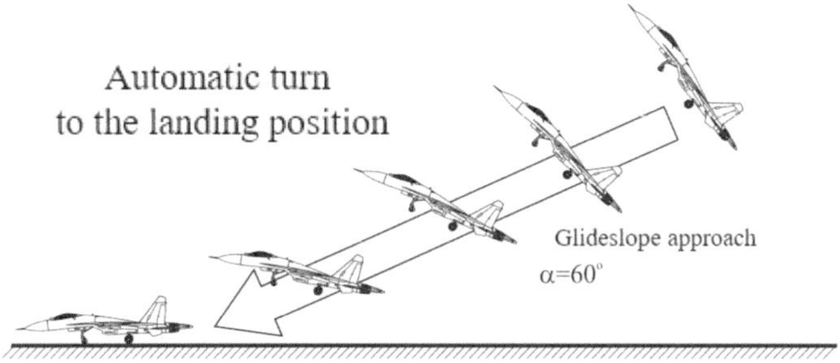

Fig. 36 Short landing at high angles of attack

system due to the elimination of the air conditioning system air extraction, elimination of the hydraulic system drive, integration of the electric power source with the aircraft engine, wide use of electric drives in the jet engine control systems, etc. "Electric" jet engine will not have a drive box. In the electric starting system of the aircraft engine it is supposed to use the built-in starter-generator. Auxiliary power unit in this case will be used only to generate electricity. This will make it possible to reduce the engine mass and midsection by 10–15%, reduce the fuel system heat load by 10–20 °C, reduce maintenance costs by 2–3 times and, as a consequence, improve the aircraft performance.

One of the most energy-intensive aircraft systems is the air conditioning system (ACS). The power taken from the power plant of modern aircraft to ensure the operation of this system significantly exceeds the power required for the air conditioning system, and the parameters of the air taken from the aircraft engines (temperature and pressure) are significantly higher than the values required for ACS. If stand-alone compressors driven by electric motors are used, the electric drive power in all modes will correspond exactly to the value required for ACS operation.

Studies have shown that the electrification of the ACS will provide:

- the reduction of power losses by up to 30%;
- the weight reduction of the pipeline system and units by 20–30%;
- the reduction of fuel consumption by 1–2%.

The key point in implementation of AEA concept is the creation of electromechanical drives of flight control systems. As is known, currently, flight control systems universally use hydraulic actuators, which receive energy for their operation from centralized hydraulic systems. On aircrafts with fully electrified equipment these actuators must be replaced by electric actuators supplied by electric power sources. Modern electromechanical drives based on contactless DC motors with permanent magnet excitation or valve inductor motors are already on a par with or even superior to hydraulic drives by a number of indicators.

Studies have shown that the weight of the power part of the flight control system can be reduced by 20% in the next 5–7 years and by 40% in 10–15 years due to implementation of the AEA concept. Nevertheless, introduction of purely electromechanical actuators in flight control systems is restrained for a number of reasons. It is believed that the use of purely electromechanical actuators for the main control surfaces of the aircraft (elevator, rudder, ailerons) is currently premature. This is primarily due to the relatively low reliability of the mechanical gearbox drive, which converts the rotational motion of the motor shaft in the progressive movement of the output link of the drive. In addition, the electromechanical drive with a planetary reduction gear and ball screw gear is still inferior to a conventional electrohydraulic drive in terms of dynamic characteristics.

Higher dynamic characteristics, sufficient for use in primary flight control system, have electrohydrostatic actuators (EHS-drives), which are essentially electric drives with a "hydraulic" gearbox.

The EHS drive is a combination of an adjustable, non-contact DC motor and a hydrostatic transmission based on a hydraulic cylinder and an unregulated reversible pump.

On the F-35 aircraft the flight control system uses only electrohydrostatic steering actuators powered by electric power sources. Due to this, the central hydraulic system of the aircraft is completely eliminated, which allowed to reduce the weight of the control system by ~300 kg (Fig. 37).

This type of drive, which requires the use of autonomous hydraulic systems, seems to be an intermediate solution to the problem. In the future, it should be replaced by an electromechanical drive.

Increase in the level of electrification of "all-electric aircraft", naturally, will be accompanied by an increase in capacity of both electric power sources and power supply system as a whole. Currently, the most promising for AEA are AC power

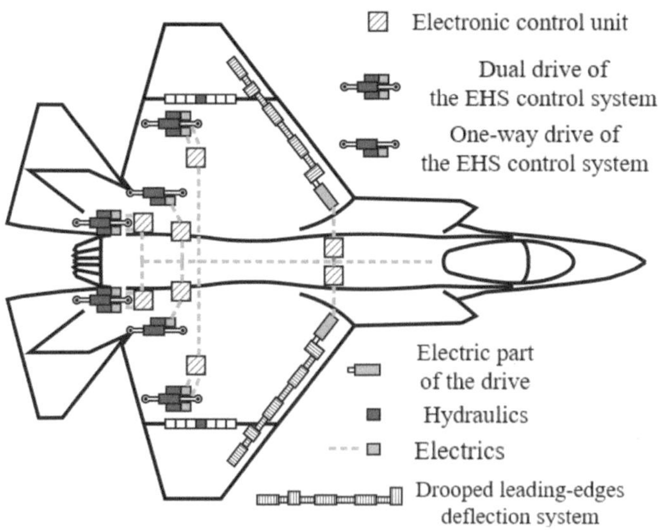

Fig. 37 F-35 EHS system

supply system of alternating frequency (in the range of 360–800 Hz) and DC power supply system of increased voltage ($U = 270B$).

It is proposed to raise the AC voltage level of the AEA by a factor of 2 and the DC voltage level by a factor of 10 with respect to the currently accepted levels. An even higher DC voltage level of up to 540 V is being considered.

Speaking about the total effect of applying the AEA concept, it should be noted that, according to the conducted studies, it provides the maximum benefit for large aircraft. For fighters the benefits from its implementation are not so great, but as the elements that can create the basis for the electrification of aircraft are improved, these benefits should become more and more tangible. For example, the use of new magnetic materials manufactured using nanotechnology, data transmission devices, microprocessor technology and power electronics allows the benefits of the AEA concept to spread to an increasing range of aircraft.

6 Ensuring Low Signature of Aircraft

One of the important tasks addressed in the process of designing combat aircraft is to develop a set of measures aimed at reducing its signature, since it largely determines the effectiveness of the aircraft in the face of enemy countermeasures. The main directions and ways of solving this problem are briefly discussed.

Ensuring low radar signature

The radar acquisition range of an aircraft depends not so much on the size of the aircraft as on the stealth technology applied during its construction (Fig. 38). Stealth technology refers to a set of various design and technological means that ensure a low signature of any aircraft in the radar and infrared wavelength ranges, which ultimately increases its combat effectiveness. Stealth technology provides, in particular, for the use of special forms of surfaces, radio absorbing materials, coatings, special modes of use of onboard equipment, etc.

The magnitude of signature depends on the level of the radar signal reflected by the target. The term radar cross-section refers to the equivalent area of a flat plate. Sometimes the radar cross-section is interpreted in terms of the cross section of an equivalent isotropic reflector. It is a quantitative measure of the ratio of the power density of the signal scattered in the direction of the receiver to the power density of the radar wave incident on the target, taking their vector properties into account.

An estimate of the contribution to the radar cross-section of various aircraft elements is presented in Fig. 39.

The air intake and exhaust systems can be categorized as the most challenging components in terms of reducing their radar signature. They are oriented in the most tactically significant directions: in the frontal area, where the enemy has time to see the aircraft approaching and counteract it, and on the tail side, the area of the classic "shot after".

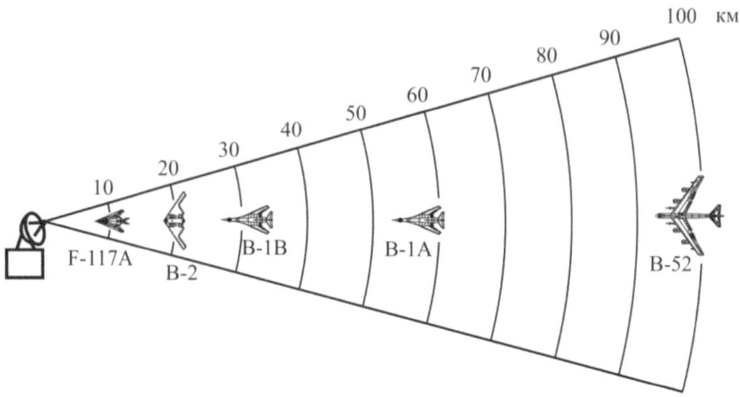

Fig. 38 Acquisition range of some aircraft

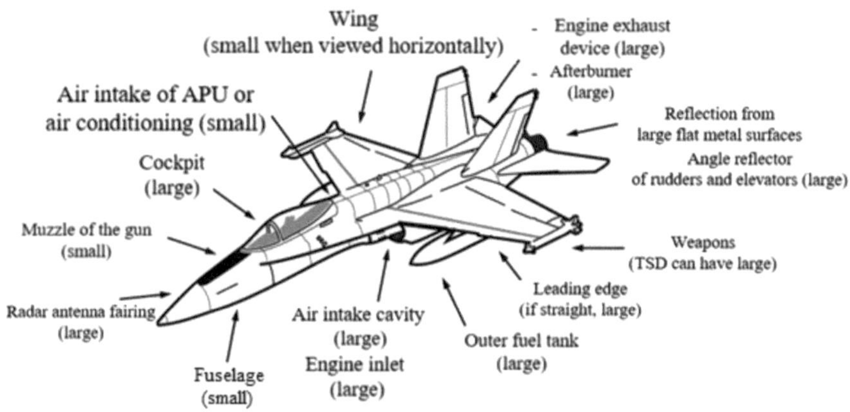

Fig. 39 Qualitative assessment of the contribution to the radar cross-section made by different parts of the aircraft

Untreated air intakes and exhaust are often the most reflective parts of an aircraft. Both of these components act as re-reflective cavities. In doing so, the air intake/ exhaust channel sends radar energy to the compressor/turbine that reflects it and ejects it back to the enemy's irradiating radar.

The key to reducing the radar cross section of an air intake is to understand how radar waves travel along air intake or exhaust system channels. This is highly dependent on the ratio of frequency or wavelength to channel width. When the wavelength is more than twice the channel width, the wave is too large to enter and be reflected from the air intake (Fig. 40). Energy transfer along the channel is essentially zero. A 90 cm wide channel would block waves longer than 1.8 m, or with a frequency below 150 MHz.

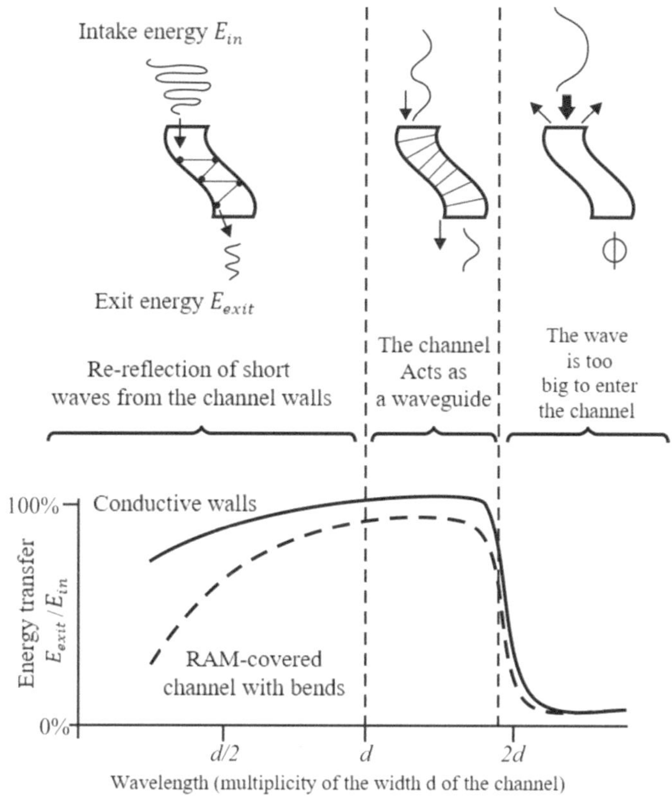

Fig. 40 Peculiarities of radio wave propagation in the air intake channel

When the wave is reduced to 1–2 widths of the channel, it becomes a waveguide and effectively carries radar energy along the bends with little loss. In this zone, the radio absorbing material (RAM) covering the channel can only attenuate the wave boundaries and therefore gives a moderate effect.

The wavelength, which is significantly smaller than the channel width, tends to travel as in free space, reflecting off the walls like a beam of light. In this case, if the wall is conductive, the energy loss is quite low, but if it is covered with RAM, the re-reflections cause a strong attenuation of radar energy.

The main activities of aircraft designers to reduce radar cross-section are presented in Fig. 41.

Providing small signature in the infrared range

Similar to radio waves, infrared radiation (IRR) is also part of the electromagnetic spectrum with a wavelength in the range of 0.77 to 1000 μm. IRR emanates from any body at a temperature above absolute zero (–273 °C). The total amount and distribution of thermal radiation as a function of wavelength depends on the material

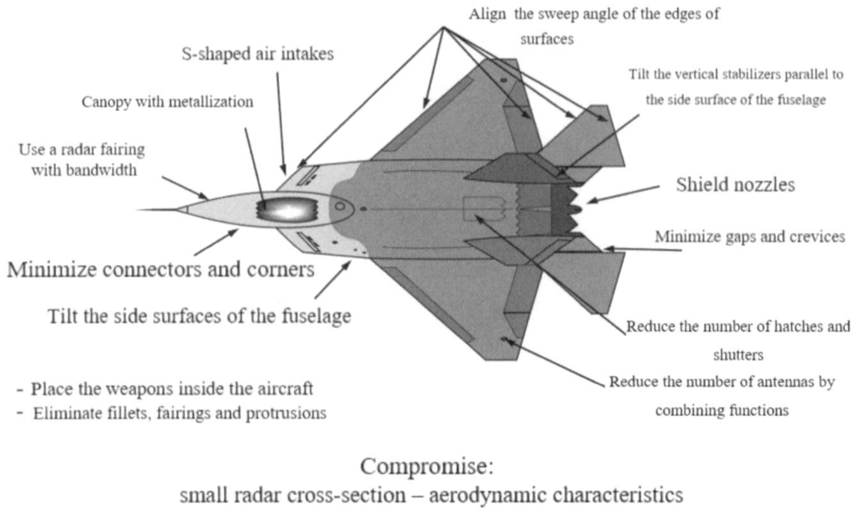

Fig. 41 Areas of focus for designers to ensure a low radar cross-section

and temperature of the body. For solids, such as the surface of an airplane, the thermal radiation is uniformly distributed over a wide range of wavelengths. For jet engine gases, the radiation emanates over a very narrow range of wavelengths.

Stephan-Boltzmann law approximates the radiated energy M (Watt/m^2) by the relation $= \varepsilon\sigma T^4$, where ε is the radiation coefficient (or degree of blackness, it is a function of wavelength), σ is the Stephan-Boltzmann constant, equal to $5.67 \cdot 10^{-8} watt/\left(m^2 \cdot K^4\right)$, T is the temperature in [K]. The emission coefficient is the ratio of the energy emitted by the surface to the energy emitted by an ideal radiation source at the same temperature. Thus, this coefficient is usually less than unity.

The contribution to the infrared signature of an aircraft comes from two components, the engine exhaust system and the aircraft skin heating due to friction in the atmosphere. At low speeds, the second component is negligible compared to the first. The engine radiates energy from the hot parts and from the exhaust jet itself (Fig. 42). The main variables available for reduction of IRR are temperature and emission coefficient, which are largely effective available for masking along the line of sight.

Thermal seekers of air-to-air missiles are typically most sensitive in the mid-infrared (3–5 μ), which is the natural emission frequency of hot carbon monoxide exhaust gases. Longwave IRR seekers (8–12 μ) detect heat from solar heating and aerodynamic friction of the airframe. Both frequency ranges are used in airborne IRR seekers.

A decrease in temperature can result in a dramatic decrease in the emitted IRR energy. The above equation is a very large simplification because it neglects the change in emission frequency with temperature. At typical hot metal temperatures of the engine exhaust system, the exponential relationship will be more likely to

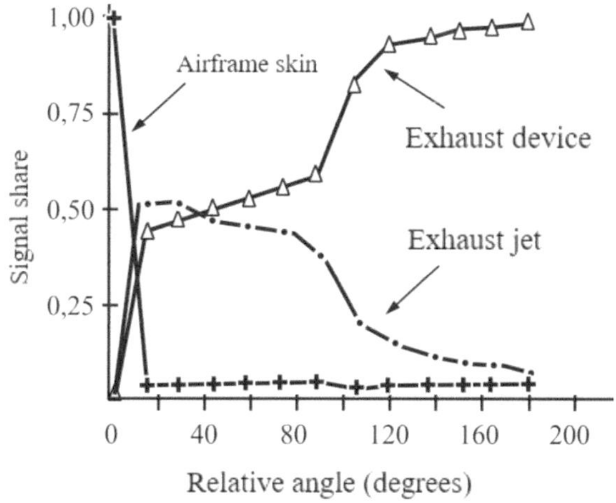

Fig. 42 Components of IRR emission of engine exhaust system

be close to eight rather than four. In any case, it is clear that a small reduction in temperature can have many more effects than the reduction in blackbody degree that can be achieved. For these reasons, the primary focus in the field of IRR reduction has been on temperature reduction and masking. Temperature reduction in the exhaust duct is achieved in various ways.

Eliminating the use of afterburner where supersonic cruising is concerned significantly lowers the temperature of the exhaust tract. The significantly more work that the turbine takes off in the case of turboprop or turboshaft engines results in a cooler exhaust. For a jet engine, where hot gas is the driving force, this option is less attractive. Lockheed F-117F, Northrop B-2 and YF-23 aircraft engines have exhaust systems at the top of the fuselage to prevent hot parts from being seen from below.

Rectangular nozzles also create an unstable exhaust stream that collapses more intensely, entrapping atmospheric air with more mixing and therefore cooling faster.

When designing the coating that can attenuate the emissivity, it is necessary to ensure that the material does not contribute to radar reflection. The use of multilayer materials to reduce both radar and IRR reflection requires films several angstroms thick. However, engines create carbon deposits on the surfaces of the exhaust system, which have a very high emissivity value. Therefore, engine exhaust must reduce the production of carbon particles as much as possible to zero. Otherwise the exhaust system will be covered by a carbon layer with a high emission factor after only a few hours of operation. For the above reasons, work is carried out to reduce the emission factor on both the surfaces of the air intakes and the outer surfaces of the aircraft airframe.

To summarize, IRR emission reduction is achievable with the scrupulous application of design techniques to reduce temperature and mask components with hot parts. As with radar cross section reduction, the sources of the dominant IRR components must be identified first and then their radiation is minimized by an integrated approach to the overall system design.

Optical aircraft signature

In the visible wavelength range, the following means can be used on a fighter:

- binocular optical sighting devices of anti-aircraft missile systems and anti-aircraft small arms systems;
- monocular sightings of anti-aircraft missile systems;
- laser rangefinders of air defence and anti-aircraft artillery systems.

In order to reduce the fighter's visibility in the optical range, it can be recommended:

- to use special low-reflection camouflage paints or camouflage coatings;
- to use artificial illumination of shaded areas of the aircraft to eliminate shadows by special light sources with the use of light filters;
- to apply special design to the cockpit lantern and glazing of the aircraft to eliminate glare;
- to introduce additives to the chemical composition of glazing to reduce light reflection;
- to use low-smoke fuels that do not form an inversion trail in flight.

Variables		
Symbol	Description	Units
n_z	Normal load factor	–
n_x	Tangential load factor	–
M_{max}	Maximum Mach Number	–
V_z^*	Maximum excess specific power	m/s
$n_{z,st}$	Steady turn load factor	–
$n_{z,phys}$	Load factor constraint related to pilots' physiological capabilities	–
$n_{z,limt}$	Limit load factor	–
$\dot{\psi}$	Angular acceleration about Y-axis	m/s^2
T	Thrust	N
D	Drag	N
$n_{z,steady}$	Steady-state g-load	–
T_0	Take-off thrust	N
m_f	Fuel mass	kg
m_0	Take-off mass	kg
e	Oswald coefficient	–
$\overline{T}_0$	Take-off thrust-to-weight ratio	–
$\frac{\overline{T}_0 AR}{m_0 g/S}$	Generalized maneuverability indicator	–
m	Bypass ratio	–
T_{st}^*	Stagnation temperature before the turbine	K

(continued)

(continued)

Variables		
Symbol	Description	Units
T_{sp}	Specific thrust	–
γ_{eng}	Specific engine weight	–
$\overline{T}_{fr}$	Frontal thrust	N/m^2
$\overline{T}_{unit}$	Thrust per unit length	N/m
$C_{D,o}$	Zero lift drag coefficient	–
$C_{D,i}$	Induced drag coefficient	–
δ_{flap}	Flap deflection angle	Degrees
δ_{DLE}	Drooped leading-edge deflection angle	Degrees
α_{allow}	Allowable angle of attack	Degrees
C_m	Pitching moment coefficient	–
$\frac{S_{HT}}{S_w}$	Horizontal tail area ratio	–
$\frac{S_{WRE}}{S_w}$	Wing root extention area ratio	–
Λ_{WRE}	Wing root extention sweep angle	Degrees
AR_{HT}	Horizontal tail aspect ratio	–
c_{MAC}	Mean aerodynamic chord length	m
$C_{n\beta}$	Yawing moment coefficient derivative with respect to angle of sideslip	rad^{-1}
AR_{VT}	Vertical tail aspect ratio	–
l_{VT}	Vertical tail moment arm	m
b	Wing span	m
$M = \varepsilon\sigma T^4$	Radiated energy	$Watt/m^2$
ε	Radiation coefficient (or degree of blackness)	–
σ	Stephan-Boltzmann constant	$Watt/(m^2 \cdot K^4)$
T	Temperature	K

References

1. Matveenko AM, Akimov AI, Akopov MG et al (2004) Proektirovanie, konstrukcii i sistemy samoletov i vertoletov (Design, construction and systems of aircraft and helicopters). In: Matveenko AM (ed) Mechanical engineering. Encyclopedia, Masinostroenie, Moscow
2. Voronovich S, Kargopoltsev V, Kutakhov V (2009) All-electric aircraft. Aviapanorama 2:23–27
3. Hamlyn D (1999) Avionics. The core avionics system for future fighters. In: Abstracts of the international conference "Fighters-99", London
4. Kablov EN (2000) VIAM: materialy i tekhnologii dlya novogo pokoleniya aviacionnoj tekhniki (VIAM: materials and technologies for a new generation of aviation equipment), Aerospace Courier, 5

5. Blaze M (1999) Combat aircraft of the future—manned and unmanned variants. In: Abstracts of the international conference "Fighters-99", London
6. Samoylovich OS (1992) Kratkij kurs proektirovaniya samoletov (Aircraft Design Brief Course). MAI, Moscow
7. Bratukhin AG (2000) (eds) Aviastroenie: letatel'nye apparaty, dvigateli, sistemy, tekhnologii (Aircraft industry: aircraft, engines, systems, technologies), Mashinostroenie, Moscow
8. Whitford R (1999) Designing a more modern fighter: Designing a fighter as a series of advanced solutions and concessions. In: Abstracts of the international conference "Fighters-99", London
9. Simonov MP, Pogosyan MA et al (2000) Forward-swept wing aircraft. Russia patent No 48183:23
10. Andreev YuV (1999) Osobennosti proektirovaniya i perspektivy razvitiya manevrennyh samoletov (Design features and development prospects of maneuverable aircraft). MAI, Moscow

Design Features of Short and Vertical Take-Off and Landing Aircraft

Nikolai K. Liseitsev

Abstract The chapter starts by justifying the short-and vertical take-off and landing aircraft (S/VTOL). The chapter offers equations for the analysis of the performance of S/VTOL. The chapter also provides a categorization and brief description of powered high-lift devices intended to facilitate short take-off and landing. In addition, the reader will find some information on passive, active, and combined boundary layer control methods. The chapter discusses some problems of VTOL aircraft design and ways to solve them, the features of VTOL layouts, as well as how to determine the basic design parameters of a VTOL aircraft, including the thrust-to-weight ratio, wing loading, and gross weight.

Keywords S/VTOL flight performance · Powered high-lift devices · Boundary layer control methods · Design problems

1 Introductory Remarks

The desire to increase the aircraft efficiency constantly stimulated an increase in the payload m_{pay}, range R and, consequently, the aircraft take-off mass m_0, as well as the flight speed V. It was achieved mainly through an increase in wing loading p_0, the use of thin swept wings, improvement of aircraft aerodynamics and power plants. Thus, the typical wing loading increased from $p_0 = 50\ldots100$ daN/m^2 (1925–1930), up to $p_0 = 450\ldots750$ daN/m^2 for modern passenger and transport aircraft. The consequence of this was a sharp increase in the size and cost of airfields. An increase in aircraft mass led to an increase in load on landing gear wheels, which, along with an increase in pneumatic pressure, reduced the flotation of the landing gear on the airfield and necessitated the creation of concreted runways. The aerodromes turned into complex technical facilities that required huge investments in construction and maintenance. Considerable dimensions of a modern airfield (the required area is of about 900 ha), as well as restrictions concerning permissible noise level on the

N. K. Liseitsev (✉)
Moscow Aviation Institute (National Research University), Moscow, Russia
e-mail: nikolai.liseitsev@yandex.ru

M. A. Pogosyan (ed.), *Aircraft Design*, Springer Aerospace Technology,
https://doi.org/10.1007/978-981-96-4599-2_12

ground, made it necessary to build airfields at a considerable distance from the cities. With the growth of air traffic, airports became congested both in the air and on the ground, which led to a significant increase in unproductive waiting time for takeoff and landing permission. Significant time spent delivering passengers and cargo to the airport, waiting for permission for takeoff and landing leads to a reduction in door-to-door delivery speed and reduces the attractiveness of aviation as a fast mode of transport, especially when traveling relatively short distances, where aircraft are inferior to road and rail means of transport.

In the second half of the last century, many aviation specialists associated overcoming the above difficulties and contradictions with the creation of short- or vertical take-off and landing aircraft. This stimulated a large amount of research and development in this area of aircraft engineering.

Specific properties of this class of aircraft are: short runway, steep takeoff and landing trajectories, low takeoff and landing speeds and accelerations, the possibility to design wing based on cruise flight conditions. These aircraft will address the following issues:

- developing the airline network and increasing the volume of air transportation in hard-to-reach areas, especially Siberia and the Far East;
- improving the efficiency of existing airfields due to increased flight regularity, reduction of noise on the ground and more rational use of airspace;
- increasing the speed of passenger and cargo delivery between major administrative and industrial centres by bringing airfields closer to the city centre.

The advantages of using S/VTOL for military purposes, in terms of operational and tactical capabilities, are quite evident.

The above-mentioned properties of S/VTOL determine the specific features of design of both short take-off and landing aircraft, which nowadays usually include aircraft with runway length of about 600 m, and vertical take-off and landing aircraft, which require runway dimensions commensurate with the size of the aircraft.

2 Short Take-Off and Landing Aircraft

2.1 Features of Short Take-Off and Landing

Let's consider on which parameters the takeoff and landing characteristics of an aircraft depend.

The take-off ground roll distance is determined by the following equation:

$$L_{TOGR} = \frac{V_{unstick}^2}{2a_x}. \tag{1}$$

Here, a_x is the average acceleration during take-off ground roll. It is determined by the ratio of acceleration and deceleration forces. In first approximation, neglecting the influence of changes in aerodynamic forces during acceleration, we can write down:

$$a_x = g\left(\overline{T}_{av} - \frac{3}{2}f_1 - \frac{1}{2(L/D)_{GR}}\right).$$

Here, $\overline{T}_{av} = \frac{T_{av}}{m_0 g}$, which is the average thrust-to-weight ratio at take-off ground roll; f_1 is the wheel rolling friction coefficient.

Similarly, for the landing ground roll distance it is possible to write:

$$L_{LGR} = \frac{V_{land}^2}{2|a_x|}, \tag{2}$$

where $|a_x|$ is the average deceleration during the landing ground roll. The following approximation is valid for it:

$$a_x = \left(\frac{T_{rev} - T_{idl}}{m_{land}\, g} - \frac{1}{3(L/D)_{LGR}} - \frac{2}{3}f_2\right),$$

where T_{rev}, T_{idl} are the reverse thrust and idling thrust; $(L/D)_{LGR}$ is the lift-to-drag ratio during landing ground roll; f_2 is the rolling friction coefficient.

Velocities $V_{unstick}$ and V_{land} are obtained from the minimum steady-state flight speed or stall speed V_{st}.

For an aircraft with a horizontal thrust vector,

$$V_{st} = \sqrt{\frac{2mg/S}{\rho C_{L,max,TO(land)}}}. \tag{3}$$

According to Aviation Rules $V_{unstick} \geq 1.1V_{st}$ at $n_{eng} = 2; 3$ and $V_{unstick} \geq 1.07V_{st}$ at $n_{eng} = 4$. The approach spead is $V_{app} = 1.3V_{st}$.

Analysis of Eqs. (1) and (2) shows that there are two ways to reduce the aicraft ground roll. The first way is to increase the acceleration at takeoff and deceleration at landing. On takeoff, this is achieved by increasing the thrust of engines or by installation of augmentors. During landing, engine thrust reversal, air brakes, undercarriage wheel braking, as well as various arresting devices on the runway are used. Note that aircraft parameters affect a_x in different ways. Thus, during take-off ground roll it is desirable for the aircraft to produce minimum drag at high lift, and during landing, on the contrary, —as high drag as possible with low lift. The ground roll distances are differently influenced by the thrust-to-weight ratio and the friction coefficients f_1 and f_2.

The second way is to reduce the minimum speed of the aircraft. By substituting in Eq. (1) the value of $V_{unstick}$, we obtain an approximate expression for determining the takeoff ground roll distance:

$$L_{TOGR} \approx \frac{p_0}{\rho g C_{L,unstick}(T_{av} - f_1)}. \tag{4}$$

It is true when the angle of attack of the wing is constant, there is no wind and no runway inclination. The unstick lift coefficient can be estimated as follows:

$$C_{L,unstick} = 0.827 C_{L,max,TO} \left(n_{eng} = 2; 3\right).$$

The nomogram (Fig. 1) based on Eq. (4) allows us to analyze the influence of aircraft parameters on L_{TOGR}. It can be seen from Fig. 1 that the short takeoff requirement imposes very strict limitations on the permissible value of $q_0 = \frac{p_0}{C_{L,unstick}}$. Thus, in order to provide $L_{TOGR} = 200m$, q_0 should fall within the range of 45–235 daN/m^2, depending on the value of starting thrust-to-wcight ratio.

The short landing requirement imposes no less stringent restrictions on p_0 and $C_{L,unstick}$. Dependence of the required landing distance on the wing loading and approach lift coefficient is shown in Fig. 2. When using conventional high-lift devices, taking into account engine failure and wind gusts, the value of approach lift coefficient usually does not exceed 1.6–1.9. This value corresponds to the approach speed, approximately 1.3 times higher than stall speed in case of engine failure. This establishes the values of the wing loading required for short takeoff in the range of 90–175 daN/m^2.

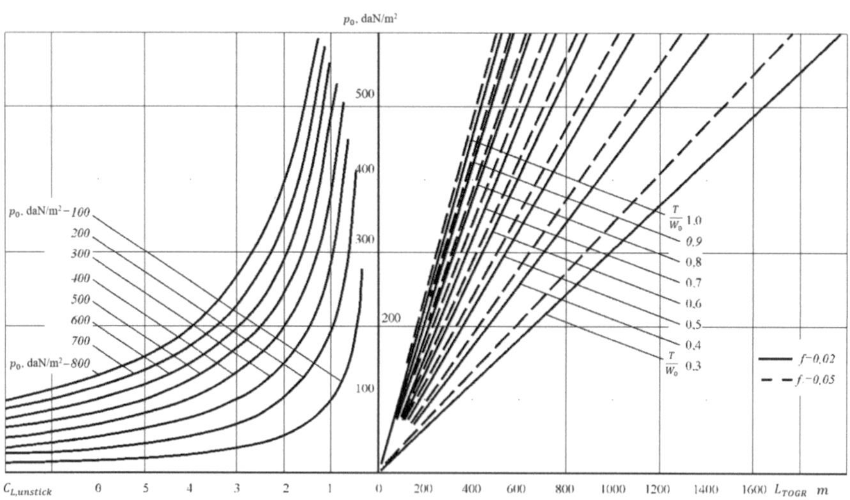

Fig. 1 Dependence of take-off geound roll distance on main aircraft parameters $q_0 = \frac{m_0 g}{C_{L,unstick} S}$

High values of $C_{L,max}$ can be achieved though the use of high-lift devices over the entire wingspan. For this reason, a number of aircraft have flaps over the entire wingspan, with the outer sections of the flaps simultaneously acting as ailerons. Low wing loading of aircraft leads to reduced comfort for passengers and crew when flying in turbulent atmosphere, complicates landing in crosswinds and makes the aircraft heavier. Figure 2 shows that for short takeoff and landing with acceptable values of wing loading (360–490 daN/m^2) and conventional control system (shaded area on Fig. 2) $C_{L,appr}$ must be 4–5. The further increase in $C_{L,appr}$ apparently demands jet control similar to the control system of VTOL aircraft. Also, more effective powered high-lift devices are required.

Let us consider the main features of STOL aircraft with powered high-lift devices. Classification of these devices is presented in Fig. 3. The methods for increasing lift at low speeds can be divided into three groups. The first group includes methods based on increasing the wing load capacity by controlling the boundary layer.

The second group includes combined methods that use supercirculation and engine jet vectoring by means of high-lift devices. The methods of the third group are based on the principle of direct lift generation using different thrust sources. This way makes it possible to perform short and vertical takeoff and landing.

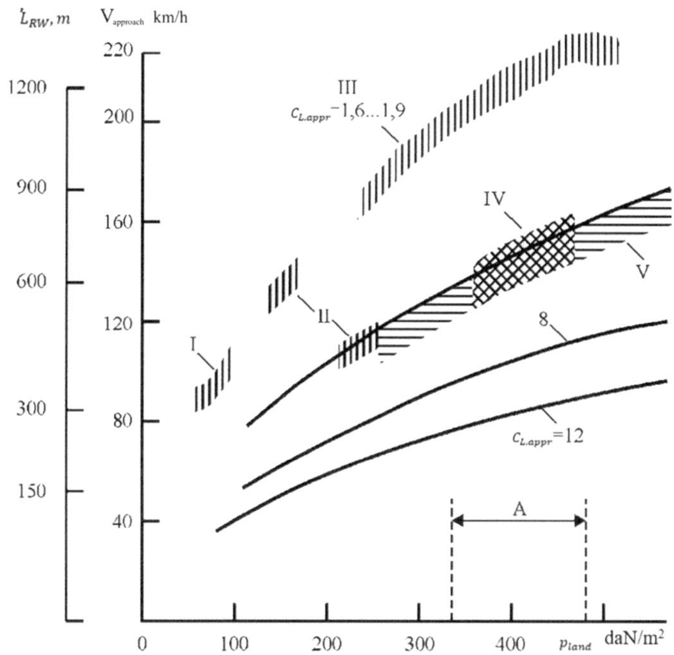

Fig. 2 Impact of wing loading and approach lift coefficient on landing distance A—acceptable range of p_{land}; I—light aircraft; II—rotary wing aircraft; III—conventional take-off and landing jets; IV—short and vertical take-off and landing jets; V—approximate limit of conventional control

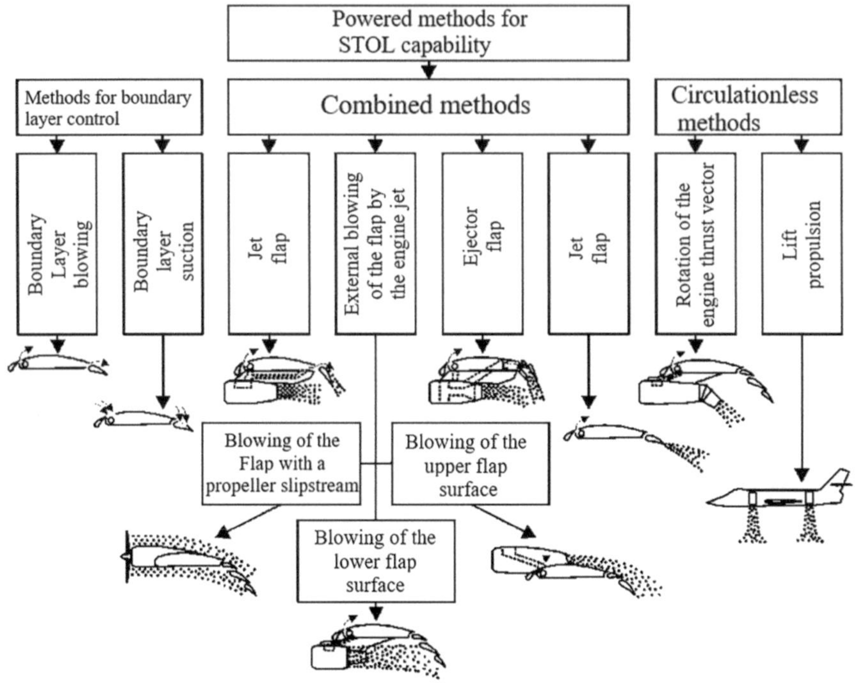

Fig. 3 Classification of powered methods for STOL capability

2.2 Boundary Layer Control

The essence of this method is to ensure a noncontinuous wing flow in a sufficiently large range of angles of attack by increasing the energy of the boundary layer. In this sense, the boundary layer control (BLC) does not differ from the principle of operation of slats and flaps. The only difference is in the source and magnitude of the energy transferred to the boundary layer.

The idea of introducing air with high energy into the boundary layer to prevent stall was put forward long ago. However, its practical implementation became possible only with the advent of turbojet engines.

Of the various methods of BLC, the most common is the method based on blowing out through a slot or a system of slots a thin jet of air flowing with high velocity tangentially to the surface of the wing. Blowing out is placed in the zone of the most probable onset of flow separation, i.e. in the leading edge and on the "knee" of the flap. The BLC by means of boundary layer suction is used less frequently due to the difficulties of its implementation. Boundary layer blowing is characterized by the impulse coefficient of the blown out jet:

$$C_\mu = \frac{m_{air} V_s}{qS}. \tag{5}$$

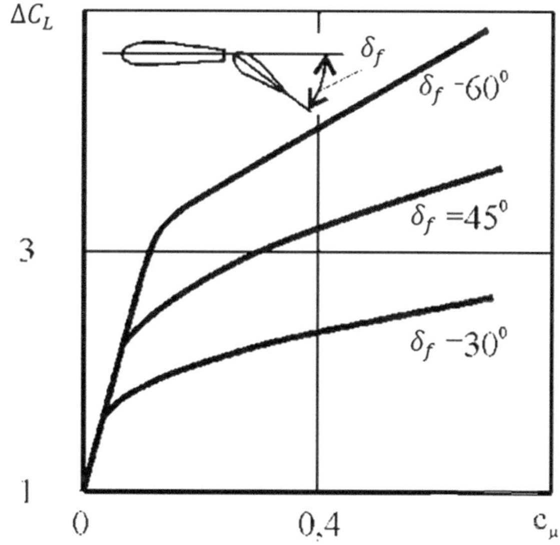

Fig. 4 ΔC_L as a function of C_μ and δ_f

Here, $m_{air} = \frac{G_{air}}{g}$ is the mass flow rate of the blowing system; V_s is the jet velocity; $q = \frac{\rho_\infty V_\infty^2}{2}$ is the freestream dynamic pressure; S is the wing area affected by the blowing system.

The increment of lift coefficient due to BLC is mainly determined by two parameters—the value of the impulse coefficient of the blown out jet and the deflection angle of the flap.

The curves $\Delta C_L = f(C_\mu)$ (Fig. 4) have a characteristic kink, beyond which the rate of growth of ΔC_L sharply decreases. The kink location determines the value of the coefficient $C_{\mu,A}$, at which attached flow on the foil is still observed.

It is of interest to determine the value of the impulse coefficient $C_{\mu,A}$. In the general case, it depends not only on the deflection angle of the flap, but also on the slot width, angle of attack and velocity ratio V_s/V_∞. The following equation can be used for approximate calculations:

$$C_{\mu,A} = (1.5 \ldots 2.0)\delta_f^2 10^{-5}. \quad (6)$$

Here, δ_f is the flap deflection angle in degrees. Higher values of $C_{\mu,A}$ correspond to larger values of the slot width.

With the value of $C_{\mu,A}$, it is possible to determine the value of impulse $m_{air}V_s$, which provides the required lift coefficient for the selected wing loading:

$$C_L = \frac{mg/S}{q} \text{ and } C_\mu = \frac{m_{air}V_S}{qS} \text{ yield}$$

$$m_{air}V_S = \frac{mgC_\mu}{C_L}. \quad (7)$$

The relationship between the magnitude of the impulse and the parameters of the air blown through the slot, under the condition of isentropic expansion, is determined by the following relation:

$$m_{air}V_s = \frac{2\nu}{\nu - 1}RT_a\left(1 - \frac{p_\infty}{p_a}\right)^{\frac{\nu-1}{\nu}}\gamma_{air}S_{slot}, \tag{8}$$

where $\nu = \frac{C_p}{C_V}$ is the specific heat ratio (for air, $\nu = 1.4$); R is the gas constant (for air, $R = 29.27$ m/k); T_a is the absolute temperature of air in the channel, bringing air to nozzle, K; p_∞ is the freestream air pressure; p_a is the total pressure of air in the channel before nozzle; γ_{air} is the specific weight of air (at standard conditions, $\gamma_{air} = 1,224$ daN/m^3); S_{slot} is the slot area, m^2.

By varying the value of pressure in the duct, as well as the area of the slot, it is possible to obtain the required impulse of air taken from the engine compressor.

The engine thrust loss due to air extraction can be estimated using the following approximate equation:

$$\overline{P} = 1 - k\overline{G}_{air}, \tag{9}$$

where $\overline{P} = \frac{P_{bleed}}{P}$ is the ratio of thrust with air bleeding to thrust without air bleeding; $\overline{G}_{air} = \frac{G_{air}}{G_{air}\Sigma}$ is the ratio of the bleeded air flow to the total air flow; $k = 1.5\ldots2.0$ is the coefficient, depending on the engine by-pass ratio; larger values of k correspond to greater engine by-pass ratios.

It should be noted that the use of high velocity compressed air in the channel is the most advantageous way to achieve a given value of the jet momentum, as it allows to manage less air (less thrust loss), facilitates the problem of piping layout in the wing, and also leads to increased bleeding efficiency due to increased ratio V_s/V_∞. The application of the BLC makes it possible to obtain $C_{L,max} = 4\ldots5$. However, the realization of such high values of $C_{L,max}$ of the wing requires a corresponding increase in the efficiency of the fins, in particular, through the application of BLC to it. Note also that in order to use BLC, it is important to ensure uniformity of blowing on the left and right wings and fins, both in normal operation and in case of engine failure. This significantly complicates the design of the aircraft.

An example of practical application of BLC is its use on the thin small aspect ratio wing ($AR = 2.4$) of the Lockheed's F-104 Starfighter aircraft. BLC made it possible to reduce the very high landing speed of the aircraft by 10%.

2.3 Combined Methods

These methods can include deflection of the propeller slipstream by means of flaps, external and internal blowing of flaps by the jet from turbojet engines, as well as increasing the wing loading by means of jet flaps.

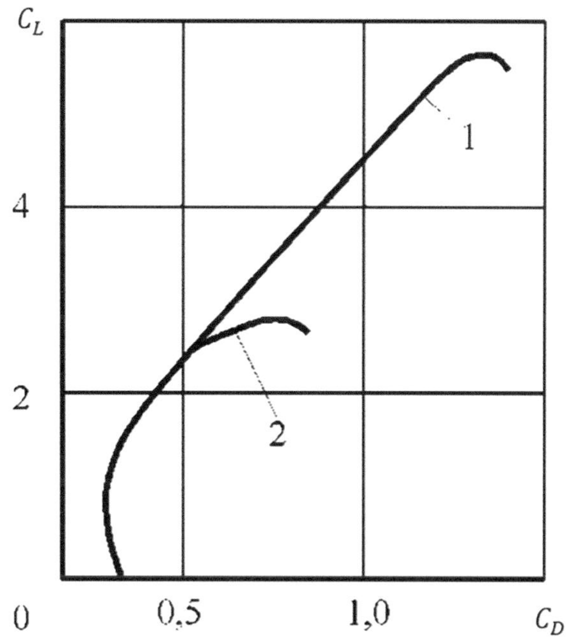

Fig. 5 Wing polar for $\delta_{flap} = 45°$: 1—with wing blowing; 2—without wing blowing by propellers

The first method is an effective means of reducing takeoff and landing distances for propeller aircraft. Its essence is to prevent stalling at very large flap deflection angles by means of propeller jet energy, to increase effective wing blowing speed.

As an example, a Breguet 941 airplane, the wing of which has a triple-slotted flap blown by the flow from four propellers driven by four turbofans, can be mentioned. By deflecting the flap to 45° it is possible to obtain $C_{L,max} \geq 5.5$ (Fig. 5). The application of such high-lift system on the Breguet 941 aircraft with a takeoff weight of 21 t provides the takeoff distance of 285 m and the landing distance of 255 m.

Another example of successful implementation of this method is the AN-70 transport aircraft with four 5th generation propfan engines (Fig. 6). This airplane has unique field performance that allow its operation from unpaved runways of 800 m length with 20 t of payload and the range of 3000 km. Such characteristics are achieved due to the record high value of $C_{L,max} \geq 6$.

If the amount of air energy introduced into the boundary layer exceeds that required to maintain attached flow at moderate angles of attack, a phenomenon known as *supercirculation* occurs, resulting in an increase in the lift coefficient to a very large value depending on the amount of air movement diverted from the propulsion system into the wing. The engine exhaust is directed into the wing and ejected over the flaps, allowing for the aircraft to produce $C_{L,max} \geq 7$. This approach was implemented in England on the experimental HS.126 aircraft. It provides the greatest aerodynamic effect, but its practical application presents a difficult engineering problem associated with the design of ducts, insulation and the use of the internal volume of the wing.

Fig. 6 An-70 aircraft (Ukraine, Russia)

The flap with the blown bottom side eliminates the need to complicate the design of the wing by internal ducts, but introduces some new problems. During takeoff and landing, the engine jet suspended under the wing is directed to the deflected flaps. In this case, the lift is produced not only due to the change of thrust vector direction, but also by supercirculation due to the spreading of the flap blowing effect to the entire surface. The efficiency of the flap is increased by the wake descending from the flap. Flow separation is prevented as part of the jet passes through the slots of the flaps and introduces energy into the boundary layer.

The lift coefficient of a wing with such a flap can be represented as a sum:

$$C_L = C_{L,C_{\mu=0}} + \eta C_\mu \sin\left(\delta_{jet} + \alpha\right) + \Delta C_{L,\Gamma}, \tag{10}$$

where $C_{L,C_{\mu=0}}$ is the wing lift coefficient without jet blowing; η is the static thrust recovery coefficient; δ_{jet} is the jet flap deflectison angle; α is the angle of attack; $\Delta C_{L,\Gamma}$ is the lift increment due to supercirculation. A typical curve $C_L = f\left(C_\mu\right)$ is shown in Fig. 7.

With this type of high-lift devices, the axial force also changes. The axial force coefficient due to flap blowing can be predicted as follows:

$$C_D = \eta C_\mu \cos\left(\delta_{jet} + \alpha\right) - \frac{C_{L,C_{\mu=0}} + \Delta C_{L,\Gamma}}{\pi (AR)_{eff}} - C_{D,0}. \tag{11}$$

Studies have shown that with rationally chosen parameters of the flap and engine nozzle position, the static thrust recovery factor η can reach values up to 0.85–0.90 at the maximum jet deflection angles. Practical implementation of this concept, which

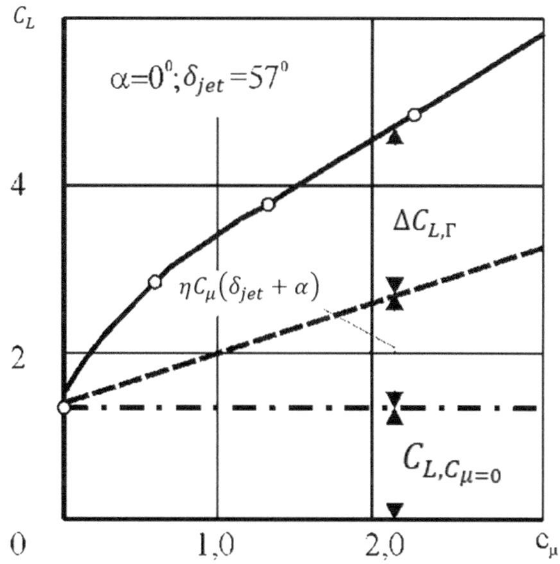

Fig. 7 C_L as a function of C_μ for external blowing of the lower flap surface by engine jet

emerged in the mid-50 s, became possible only after the emergence of engines with a large by-pass ratio, providing an acceptable take-off and cruise thrust ratio and having a relatively low jet temperature (aircraft YC-15).

Blowing of the upper flap and wing surface to increase lift is based on the use of the Coanda effect— "sticking" of the jet which is directed tangentially to a surface. This phenomenon was first systematically investigated in the late thirties. Although Coand's efforts were mainly aimed at improvement of ejectors, this effect was used in all BLC systems. Later, it was shown that powerful exhaust jets of large bypass ratio jet engines could be deflected by this method, at large angles and without excessive losses.

The exhaust jet of the jet engine flows along the upper surface of the wing, blowing the flap, the deflection of which changes the direction of the thrust vector of the engine. The lift increases due to supercirculation, as well as due to deflecting the thrust vector. In this case, a sufficiently high thrust recovery factor is achieved. Figure 8 shows the aerodynamic characteristics of the YC-14 model at low speeds obtained in a wind tunnel. Note that the application of boundary layer blowing in the area of the wing leading-edge significantly increases the critical angles of attack and the maximum lift coefficient (by about 20%).

When estimating the characteristics of lifting and propulsive systems which are based on external blowing of the flap by engine jet, instead of the impulse coefficient C_μ it is convenient to use the equivalent thrust ratio C_t, which can be determined as follows:

$$C_t = \frac{P(V)}{qS}, \tag{12}$$

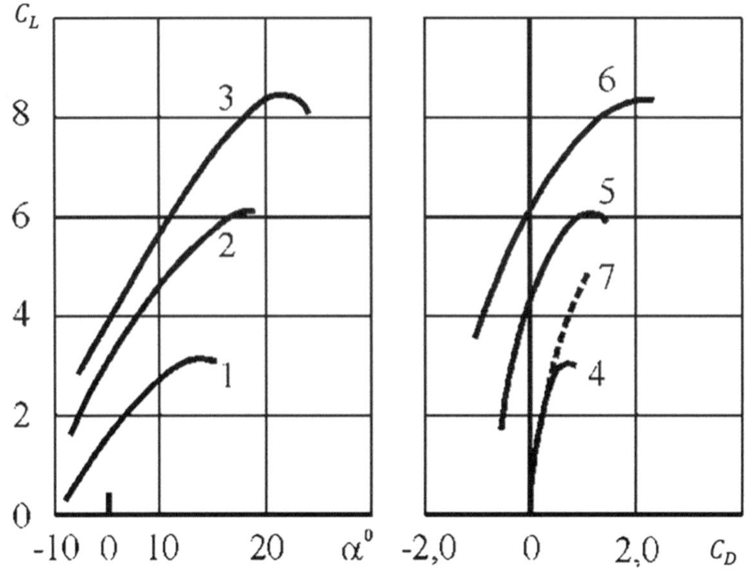

Fig. 8 Aerodynamic characteristics of the YC-14 model aircraft (landing mode, all engines running, model without horizontal tail, untrimmed, no leading-edge deflection): 1—($C_t = 0$, non-operating engines); 2—$C_\mu = 0,039$; 3—$C_t = 2.0$; $C_\mu = 0,067$; 4—$C_t = 0$; 5—$C_t = 1.0$; 6—$C_t = 2.0$; 7—$C_D = C_{D,0} + C_L^2/(\pi AR)$

where $P(V)$ is the total thrust of engines, which depends on the flight speed.

In the design of a STOL aircraft with externally blown flaps, a number of complex problems have to be solved. One of these problems is aircraft stability and trimming capability. The large lift applied to a point behind the center of gravity causes a large nose-down pitching moment and, therefore, leads to the need to have a large horizontal tail to trim the aircraft. In addition, according to experimental data, the tail of the wing-tail configuration can even have a destibilizing effect.

Studies have shown that the use of a T-tail with an area of order of $0.25S_w$ delivers the aircraft the required stability and trimming capability at $C_{L,max} \approx 6$.

It is also a significant challenge to provide the necessary performance when a single engine fails, as this results in asymmetric lift and, consequently, roll moments. This requires the development of aerodynamic control that provides sufficient roll authority to eliminate the asymmetry caused by engine failure without additional loss of lift. The roll moment can be balanced by reducing the deflection angle of the flap on the side with the engines running and increasing it on the side with the failed engine. A roll control method based on taking high pressure air from the engine compressor and diverting it to the ailerons to provide the BLC is also possible. When an engine fails, twice as much air must be supplied to the deflected aileron (on the side with the failed engine) as to the aileron on the side with the running engine.

Fig. 9 An-72 aircraft

An example of quite successful implementation of the concept of blowing the upper surface of the wing and flap is the domestic serial aircraft AN-72 and its modifications (Fig. 9). However, due to the difficulty of solving the above problems related to the failure of one engine, the actual $C_{L,max}$ of this aircraft is almost half as much as theoretically calculated. As a consequence, aircraft of this type, although they have good takeoff and landing characteristics, cannot be classified as STOL.

A wing with an internal flow flap (ejector flap) was developed on the basis of the jet flap. In such a wing, a thin jet of exhaust gases flows along the span into the channel formed by the upper and lower sections of the deflected flap (see Fig. 3). The channel along the wingspan protrudes over the upper surface, which allows a part of the flow from the upper surface of the wing to flow through the channel. The diffuser-shaped channel walls are formed by parts of the flap when it is deflected partially (at takeoff) or completely (at landing). The amount of air flowing through the channel depends on the efficiency of mixing the primary gas flow with the air flow induced from the upper surface of the wing. The secondary flow induced by the ejecting action of the primary gas flow increases its thrust and therefore creates lift. The effective coefficient of the system is equal to the coefficient of thrust of the primary jet minus losses in the channel and nozzle plus the thrust increment due to air suction from the flow. The thrust increase coefficient (ratio of thrust at the channel outlet to the primary flow thrust) achieved at present is 1.4. Studies have shown that the wing concept with this type of flaps can meet the basic requirements for a transport jet STOL aircraft.

Such a flap is more efficient than an externally blown flap because the flow from the engine can be more evenly distributed over the span. In addition, the concept of such a flap seems particularly promising for reducing induced drag as well as for

Fig. 10 Jet flap

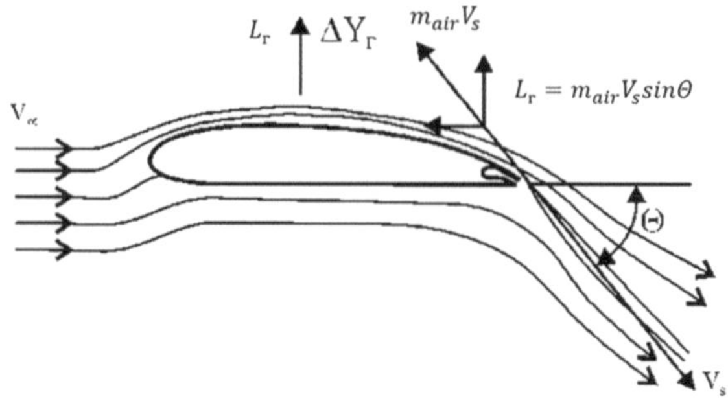

providing controllability during engine failure. The latter derives from the possibility of controlling the distribution of the increased lift over the wingspan by means of an internal flow flap. If the problem of engine failure is solved for a turbofan engine by mechanical connection of shafts, then for an airplane with a bypass turbojet engine, the connection of gas channels can be used.

The jet flap is a flat stream of air flowing out at high speed through the trailing edge of the wing at an angle Θ with respect to the lower surface of the wing (Fig. 10). The jet flap produces lift by changing the velocity field around the wing and redistributing the pressure on its upper and lower surfaces; in addition, due to momentum of the outflowing jet $m_{air}V_s$, it produces a vertical component of force $m_{air}V_s sin\Theta$, which, however, is significantly less than the aerodynamic increment. The effectiveness of the jet flap depends on the momentum coefficient of the blown out jet and the angle Θ. The following interpolation equation can be used for approximate calculations of the lift coefficient increment for the jet flap:

$$\Delta C_L = 3.9\sqrt{C_\mu}\sin\Theta. \tag{13}$$

The use of jet flaps makes it possible to obtain fairly large values of lift coefficient, but at the same time it requires C_μ significantly higher than for the BLC. The effectiveness of the jet flap significantly decreases with decreasing wing aspect ratio; in addition, near the ground surface, the jet flap does not provide the estimated values of ΔC_L, due to the ground effect, which makes it inefficient during take-off and landing.

2.4 Circulationless Methods for Lift Generation

The growth of aircraft thrust-to-weight ratio, emergence of engines with thrust vectoring as well as light lift engines and fans create prerequisites for the improvement of aircraft field characteristics due to the use of engine thrust directly to create lift at low speeds.

The vertical component of the thrust T_V can be obtained either by deflecting the thrust vector of the main engine or by installing special lifting devices. A combination of these methods is also possible. If T_V is available, the minimum velocity becomes equal to

$$V_{min} = \sqrt{\frac{2p_0\left(1 - \overline{T}_V\right)}{\rho_0 C_{L,max}}}, \tag{14}$$

where $\overline{T}_V = \frac{T_V}{m_0 g}$.

Thus, in this case, V_{min} is $\sqrt{1 - \overline{T}_V}$ times less than in the absence of the vertical component, and when $\overline{T}_V \approx 1$, it is close to zero, i.e., takeoff and landing with a short ground roll becomes possible.

For take-off with the thrust vector deflected by angle φ:

$$V_{min} = \sqrt{\frac{2p_0\left(1 - \overline{T}_V sin\varphi\right)}{\rho_0 C_{L,max}}}. \tag{15}$$

The effect of thrust vectoring on the unstick velocity value can be estimated using the ratio:

$$\overline{V}_{min} = \frac{V_{min,0}}{V_{min,\varphi}}.$$

The dependence of $\overline{V}_{min}$ on the angle φ for different thrust-to-weight ratios is shown in Fig. 11. It can be seen that the deflection efficiency is particularly significant for large thrust-to-weight ratios and large φ. In the presence of lift engines, the take-off ground roll distance can be determined using the following approximate equation similar to Eq. (4):

$$L_{TOGR} \approx \frac{p_0}{\rho g C_{L,unstick}\left(\overline{T}_{av}/\left(1 - \overline{T}_{av}\right) - f_1\right)}. \tag{16}$$

When deflecting the thrust vector of the main engine, the takeoff distance depends largely on the deflection program adopted. Calculations have shown that the minimum takeoff distance is obtained with the following program: the aircraft accelerates at φ = 0 to the most advantageous takeoff speed $V'_{unstick}$, which depends on q_0 and $\overline{T}_0$, and then the thrust vector is deflected to an angle

$$\varphi_{opt} = arccos\left[\left(\frac{q_{unstick}}{q_0}\right)_{opt} \frac{1}{(L/D)_{unstick} P_0}\right],$$

Fig. 11 Effect of thrust vectoring on V_{min}

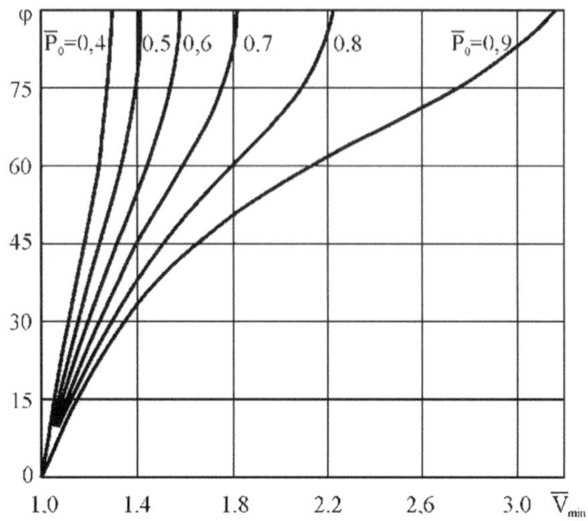

which ensures climb without trajectory acceleration. A typical dependence of $\left(\frac{q_{unstick}}{q_0}\right)_{opt}$ on the take-off thrust-to-weight ratio is shown in Fig. 12.

With this takeoff method, the ground roll distance is determined by the following equation:

$$L_{TOGR} = \frac{1}{g} \int\limits_{0}^{V'_{unstick}} \frac{VdV}{\bar{T}_0(V) - f_1 + \left(f_1 C_L \frac{\rho_0 V^2}{2p_0} - C_D \frac{\rho_0 V^2}{2p_0}\right)}. \quad (17)$$

If we accept the previously stipulated assumptions about the variation of the engine thrust with velocity, we obtain the following equations for determining the ground roll take-off distance:

Fig. 12 $\left(\frac{q_{unstick}}{q_0}\right)_{opt}$ as a function of $\overline{T}_0$

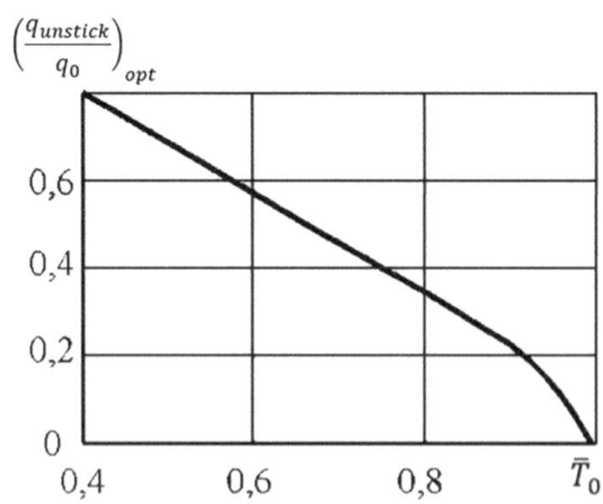

$$L_{TOGR} = \frac{p_0}{\rho_0 g(f_1 C_L - C_D)} ln\left[1 + \frac{\left(1 - \overline{T}_{aver} sin\varphi_{opt}\right)(f_1 C_L - C_D)}{\left(\overline{T}_{aver} - f_1\right) C_{L,unstick}}\right], \quad (18)$$

or roughly

$$L_{TOGR} \approx \frac{p_0\left(1 - \overline{T}_{aver} sin\varphi_{opt}\right)}{\rho_0 g\left(\overline{T}_{aver} - f_1\right) C_{L,unstick}}. \quad (19)$$

The climb is due to a constant excess of lift

$$L' = C_{L,unstick} q_{unstick} S + Tsin\varphi$$

over the aircraft's weight:

$$\Delta L = L' - m_0 g = m_0 g\left(\frac{q_{unstick}}{q_0} + \overline{T}_{aver} \sin\varphi - 1\right).$$

If we know ΔL and aircraft mass, we can calculate the time to reach a given altitude $H_{obstacle}$, and then multiplying it by the unstick speed to determine the distance of climb.

From the equation

$$\frac{d^2 H}{dt^2} = \frac{\Delta L}{m}$$

we obtain $t_{\text{climb}} = \sqrt{\frac{2m_0 H_{obstacle}}{\Delta L}}$ and the climb distance

$$L_{climb} = V_{unstick} t_{climb} = V_{unstick} \sqrt{\frac{2H_{obstacle}}{g\left(\frac{q_{unstick}}{q_0} + \overline{T}_{aver} sin\varphi - 1\right)}}.$$

Substituting φ_{opt} and rearranging the equation we obtain

$$L_{c\lim b} = 2.\sqrt{\frac{2q_{unstick} H_{obstacle}}{p_0 g\left[T_{aver}\sqrt{1 - \left(\frac{q_{unstick}}{q_0}\frac{1}{(L/D)_{unstick}\overline{T}_{aver}}\right)^2}\frac{q_{unstick}}{q_0} - 1\right]}}. \quad (20)$$

Total length of takeoff distance is $L_{TO} = L_{TOGR} + L_{climb}$. Figure 13 shows the dependency of takeoff distance on q_0 for various values of take-off thrust-to-weight ratio. The calculation was made using the approximate equations for the case of $f_1 = 0.05$, $(L/D)_{unstick} = 4$, $H_{obstacle} = 15$ m, standard atmospheric conditions. When plotting the graph, the trim penalties and thrust vector deflection penalties

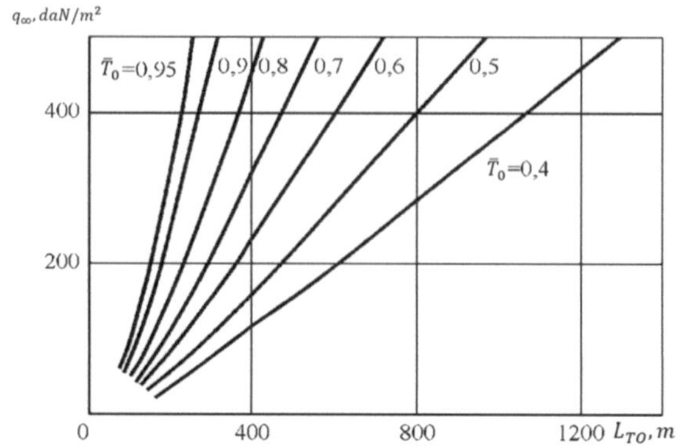

Fig. 13 Take-off distance as a function of the basic aircraft parameters

were not taken into account, therefore, the takeoff distance values obtained must be considered as the maximum achievable values.

Figure 14 shows polars obtained as a result of experimental studies of different lift augmentation systems.

1. jet flap;
2. ejector flap;
3. external blowing of wing surface and flap from above;
4. thrust vectoring;
5. external blowing of the flap from below;
6. double-slot non-blown flap;
7. efficiency = 100%, $C_t = 2$;
8. $C_D = C_{D,0} + C_L^2/\left(\pi (AR)_{eff}\right)$.

The choice of an acceptable takeoff and landing improvement system for a STOL aircraft can be made only on the basis of a comprehensive design evaluation.

3 Vertical Take-Off and Landing Aircraft

3.1 Problems of VTOL Aircraft Design and Ways to Solve Them

Vertical take-off and landing aircraft (VTOL) belong to a rather wide class of aircraft using the aerodynamic principle of flight. Their characteristic feature is the ability to take off and land from areas commensurate with the size of the aircraft. This does not mean that in order to increase payload (range) they cannot perform conventional take-off and landing, if there is such a possibility.

The generally accepted classification of vertical take-off and landing vehicles, in which the main classification attributes are based on the lifting and propulsion forces generation methods, is shown in Fig. 15.

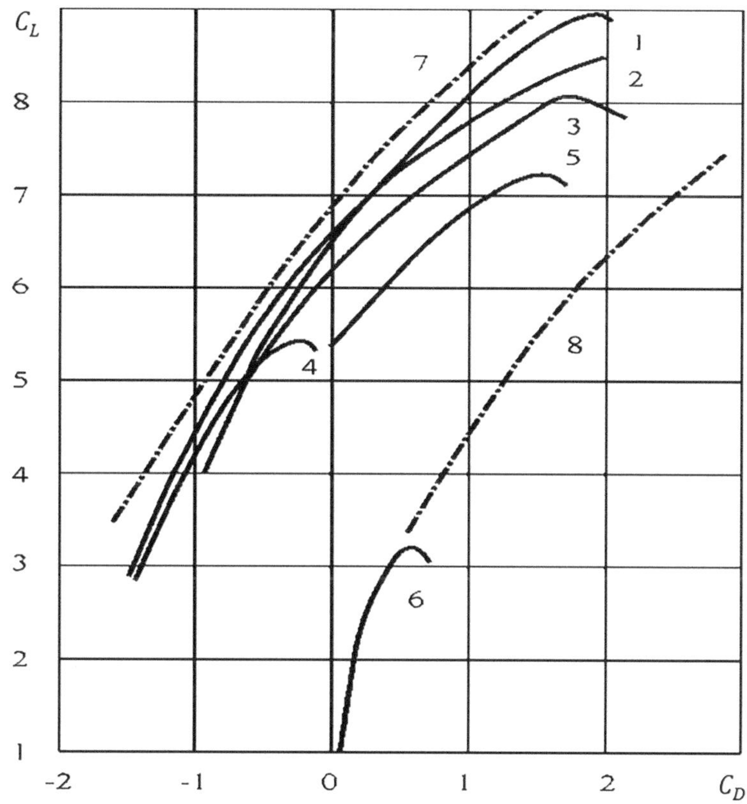

Fig. 14 Polars obtained at low speeds for different lift augmentation systems using engine power for a four-engine airplane with a BLC on the leading edge, $C_\mu = 0.08$

Depending on the method of creation of lift in horizontal flight, vertical take-off and landing vehicles are divided into four types: helicopters, rotorcraft, convertible aircraft and aircraft. They differ in the lift-to-drag ratio and, as a consequence, the achievable range and fuel efficiency. They are also characterized by different ranges of cruising speeds and vertical flight modes.

According to the method of producing thrust in horizontal flight, there are propeller, combined and jet vehicles.

The key classification attribute of the VTOL aircraft is how the lift and control forces in vertical and transitional to horizontal flight modes are produced.

The jet impulse required to generate lift can be provided by the following types of propulsion devices: large diameter rotor, propeller, fan, ejector, jet engine, etc. The method of impulse generation largely determines such important parameters of the power plant as thrust ratio, specific fuel consumption, mass ratio, jet temperature, and jet velocity. The jet area loading T_v/F_{jet} is used as a generalized parameter, since

<table>
<tr><th colspan="4">Sources of lift and thrust</th><th rowspan="3">T_v/F_{jet}, daPa</th><th rowspan="3">V_{max}, km/h</th><th rowspan="3">Vehicle layout</th><th rowspan="3">Vehicle type</th></tr>
<tr><th colspan="3">Horizontal flight</th><th>Vertical flight</th></tr>
<tr><th>Lift</th><th colspan="2">Thrust</th><th>Vertical thrust</th></tr>
<tr><td colspan="4">Main rotor with mechanical or reactive drive</td><td rowspan="3">25-50</td><td rowspan="2">280-350
450-550</td><td></td><td>Helicopter</td></tr>
<tr><td>Main rotor+wing</td><td>Propeller</td><td rowspan="8">Unitary power plant</td><td>Main rotor</td><td></td><td>Rotocraft</td></tr>
<tr><td rowspan="12">Wing</td><td>Bypass turbojet engine</td><td>Main rotor (retractable)</td><td>750-850</td><td></td><td>Convertible aircraft</td></tr>
<tr><td rowspan="3">Propeller (ducted fan)</td><td>Tilted rotor</td><td rowspan="3">100-400</td><td rowspan="3">650-750</td><td></td><td rowspan="11">VTOL aircraft</td></tr>
<tr><td>Tilted engine</td><td></td></tr>
<tr><td>Tilting wing and engine</td><td></td></tr>
<tr><td rowspan="8">Jet engine</td><td>Lift-cruise turbofan</td><td>2500-3000</td><td rowspan="5">1>M>1</td><td></td></tr>
<tr><td>Lift cruise bypass turbojet</td><td rowspan="4">5000-12000</td><td></td></tr>
<tr><td>Separate power plant</td><td>Lift turbojet</td><td></td></tr>
<tr><td rowspan="2">Combined power plant</td><td>Lift jet engine, tilted main engine</td><td></td></tr>
<tr><td>Lift (lift acceleration) jet engine, main engine thrust vectoring</td><td></td></tr>
<tr><td rowspan="3">Augmented power plant</td><td>Lifting (lift cruise) fans</td><td>1000-2500</td><td></td><td></td></tr>
<tr><td>Ejctor</td><td>1400-1600</td><td></td><td></td></tr>
<tr><td>Remote augmentor</td><td>12000-13000</td><td></td><td>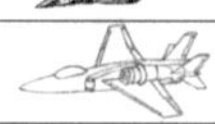</td></tr>
</table>

Fig. 15 Classification of vertical take-off and landing vehicles

this parameter also determines the above-mentioned non-dimensional parameters and dimensions of the propulsor.

As it can be seen from Fig. 16, thrust ratio is inversely correlated with $(T_v/F_{jet})^{1/2}$ and varies from 0.25–0,5 daN/kW for jet engine up to 7–15 daN/kW for helicopter main rotors.

The jet velocity increases in direct proportion to $(T_v/F_{jet})^{1/2}$, from 10–25 m/s for the rotor to 200–600 m/s for various turbojet engines (smaller values correspond to engines with a greater bypass ratio).

Specific fuel consumption C_p also increases with (T_v/F_{jet}). So, for helicopters it is 0.06–0.10 kg/daN·h, while for special lifting jet engines it reaches 1.1–1.3 kg/daN·h, i.e. it is 10 times greater.

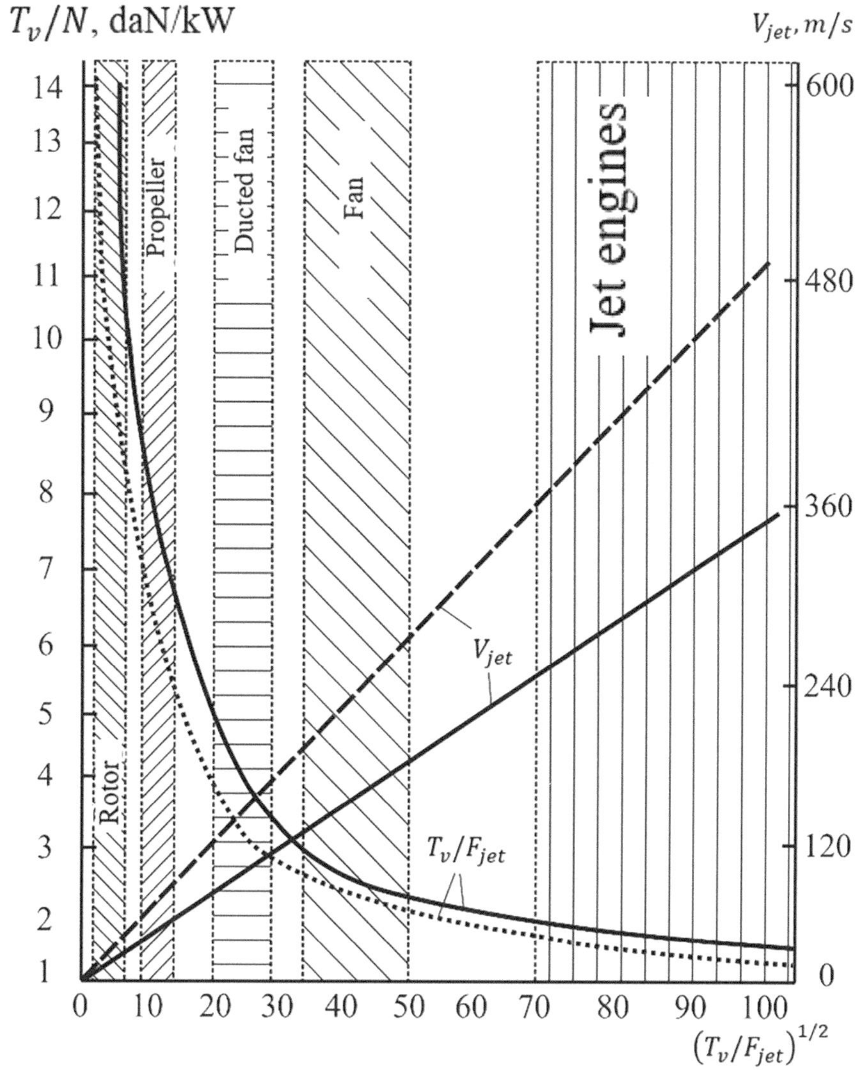

Fig. 16 Thrust ratio T_v/N and jet velocity V_{jet} as a function of the jet area loading $(T_v/F_{jet})^{1/2}$ (ideal efficiency)—propeller; — ducted fan

At the same time, the mass ratio and the size of the lift system decrease as the jet loading increases. Reducing the mass ratio allows reducing the share of the propulsion system mass, which leads to an increase in the payload mass.

In order to compare different propulsion systems in hovering mode, in addition to the power required to create a unit of thrust, it is necessary to know what mass of the propulsion system is spent to create this thrust and how much fuel is consumed per unit of hover time. The most advantageous propulsion system will be the one with the lowest value of $\left(m_{PP} + m_{fuel}\right)/T_v$. Since in the hover mode $T_v = m_0 g$, this condition is equivalent to the condition $\left(\overline{m}_{PP} + \overline{m}_{fuel}\right)_{min}$. Since at hover $\overline{T}_v = 1$, then $\overline{m}_{PP} + \overline{m}_{fuel} = r_0 + C_p t_{hover}$, r_0 is the lifting system specific mass.

The calculations show that the power plants with a high jet loading have an advantage in this parameter, losing it as the hover time increases.

The dimensions of the lifting system can be characterized by the diameter ratio D/T_v. Lifting jet engines have the smallest values of the diameter ratio (0.025–0.035 m/kN). For propeller systems this value is significantly higher (0.2–0.5 m/kN). This parameter largely determines the lift-to-drag ratio of the aircraft in the cruising mode. In addition, the use of propellers as propulsion systems in horizontal flight is limited to speeds not exceeding 600–630 km/h. This is explained by sharp drop in propeller efficiency. Velocity limit of helicopter main rotors is approximately half of the specified value.

As noted above, one of the most important functions of the lifting system of vertical take-off and landing vehicles is to ensure stability and controllability of the aircraft in the VTOL mode. This goal can be achieved by various means: from the use of thrust modulation (separate thrust change) of lift thrusters to the application of a complex system of jet controls. The ways of producing control forces depend on the type, number and location of engines and thrusters on the aircraft, i.e. on its layout. The control system must provide the required stability and controllability of the aircraft at minimum weight and minimum power consumption of the propulsion system. When air is taken away from the engines for control, their thrust decreases and fuel consumption increases. The use of thrust modulation or thrust vectoring for control is also associated with unavoidable penalties.

Let us briefly consider the specific problems arising during the design of the vertical take-off and landing vehicles. In order to implement the above mentioned additional functions, the required vertical thrust-to-weight ratio $\overline{T}_v$ is always greater than unity. The actual total vertical thrust-to-weight ratio for new vertical take-off and landing vehicles is 1.05–1.45, which is much higher than that for the most of conventional aircraft. $\overline{T}_v$ depends on the aircraft mission, the type of its lifting system, etc. Since the mass ratio of the lifting system $\overline{m}_{lift} = \overline{T}_v r_0$, where r_0 is the specific mass of the lifting system, there are two ways to reduce it.

The first way is to reduce the required vertical thrust-to-weight ratio by means of a rational choice of aircraft layout and takeoff and landing trajectories. The second way is to choose a rational propulsion system that provides the lowest value of r_0.

Since the power plant of VTOL aircraft provides stability and controllability in vertical and transition flight modes, minimization of power and mass costs for their performance is also a specific problem of VTOL aircraft design.

The power of the control system is determined by the magnitude of forces and moments required to provide the necessary linear and angular accelerations, as well as the degree of simultaneous effect of the maximum control moments. The required angular accelerations ε_x, ε_y, ε_z depend on the type of aircraft and its takeoff weight [1].

The required control moment with respect to one of the axes is determined by the product of the moment of inertia I and the angular acceleration with respect to the axis in question, i.e. $M_{control} = I\varepsilon$. Then the required thrust value of jet controls is determined as follows:

$$T_{control} = \frac{I\varepsilon}{l_{control}},$$

where $l_{control}$ is the moment arm of the jet control.

Therefore, in order to reduce the amount of air withdrawn from the engine, it is necessary, along with a rational choice of nozzle parameters, to seek to maximize the jet control moment arm and seek to reduce the moment of inertia of the aircraft by placing the heaviest loads near the center of gravity of the aircraft.

The problems of aerodynamic design of a VTOL aircraft primarily stem from the need to take into account the interaction of jets with the airframe, the airfield surface and the freestream flow in vertical and transition modes.

When hovering and flying at low speed, the jets flowing out of the lift engines, due to friction on the boundaries, entrain the surrounding air. Air flows are formed that generate additional aerodynamic forces and moments (Fig. 17). The magnitude and sign of these forces and moments depend on the parameters of the jets (primarily their speed), the number and location of the jets on the aircraft, and the distance of the nozzle from the aerodrome surface.

Air streams induced by the engine jets create a vacuum below the underside of the aircraft, causing the so-called "suction" effect, which leads to a reduction in the effective lift of the engines. On the other hand, the interaction of several jets creates an upward flow. It produces a positive lift. This phenomenon is called the "fountain effect". The resulting "fountains" are usually highly swirled and easily disintegrated. The lift generated by them quickly decreases with increasing altitude [6, 7].

Figure 18 shows the variation of the ratio of lift increment to vertical thrust component for a number of experimental VTOL aircraft (h is the ground clearance, b is the wing span).

Thus, by selecting a rational arrangement of the lift engines, it is possible to achieve a significant reduction in the loss of lifting force near the ground, and in some cases to obtain an increase in this force.

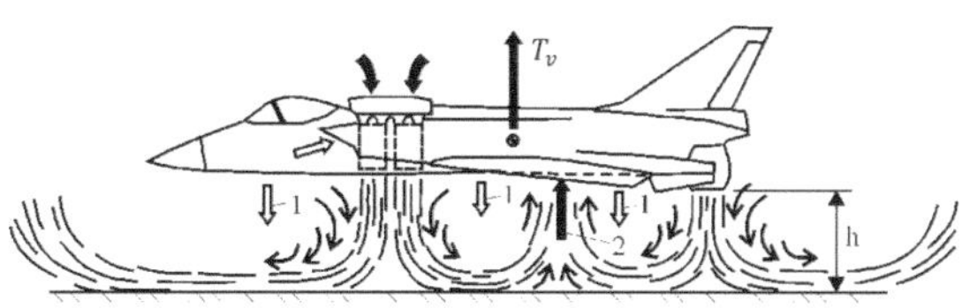

P_B	T_v

Fig. 17 Gas-air flows near the ground 1—suction, 2—spouting

Fig. 18 $\Delta Y/P$ as a function of h/b 1—Do31E; 2—Sc.1; 3—VAK-191B; 4—VJ-101CX1; 5—P.1127

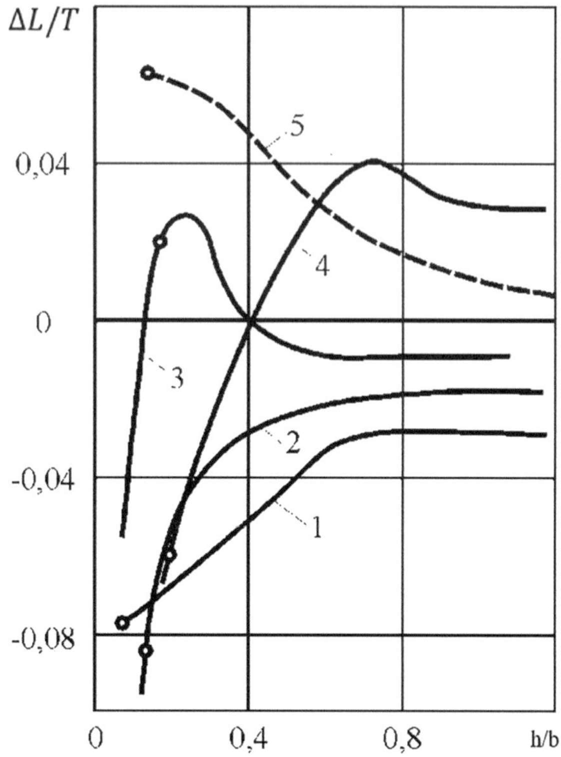

The fluxes induced by the jets from the engines also have a noticeable effect on the aircraft moment. In this case, both longitudinal and rolling moments can occur.

A jet of hot gases rising in the "fountain effect" can reach the engine inlet, causing a drop in engine thrust or even causing the engine to stall. This phenomenon of reverse flow of gases is called direct recirculation. It occurs almost as soon as the engine starts running.

Indirect recirculation—the gradual warming of the surrounding air of the aircraft and its entry to the engine inlet, is less dangerous and in the first approximation may not be taken into account.

The thrust losses due to direct recirculation can be significantly reduced or avoided altogether by a rational layout of the engines on the aircraft, when their inlets are shielded from the rising jets by the wing or the fuselage. Various shields and longitudinal ribs diverting jets from the fuselage and engine nacelles serve the same purpose.

In transition mode, jet engines also induce secondary aerodynamic forces and moments. The resulting negative lift depends primarily on the jet speed and direction, the flight speed, and the aircraft layout (number and arrangement of engines).

This negative force ($\approx$25% of thrust at the end of transient flight) reaches its greatest value at the vertical position of the jet. As the thrust vector deflection angle decreases (φ), it decreases, disappearing almost completely at $\varphi = 0°$. Aircraft which use special lift engines for takeoff cannot extinguish this thrust, so they have problems in transition flight mode (transition time and fuel consumption increase). Such problems do not arise in aircraft with thrust vectoring, where as the flight speed increases, the jet is gradually tilted backwards.

Note that, at the present level of knowledge, a rigorous quantitative assessment of the above-listed interference effects of the lift system jets in the VTOL modes based on theoretical calculations is impossible. To obtain reliable results, a significant amount of physical modeling and in-situ experiments is required. Approximate engineering methods for determining the interference forces and moments, suitable at the initial stages of design, are given in [1].

Let us also note the peculiarities of VTOL aircraft aerodynamics at cruising mode.

While having good performance in airborne modes, thrusters with low jet area loading, as noted, cannot be used as thrusters for aircraft with high flight performance due to an unfavorable balance of required and available thrust.

However, achieving high flight performance is also a challenge for VTOL aircraft using jet propulsion, as it is more difficult to achieve acceptable lift-to-drag ratio, which is known to be a function of the square of the wingspan divided by the total wetted surface of the aircraft. However, the wetted surface for most VTOL configurations tends to increase compared to conventional aircraft of the same class. This is due to the need to have additional volumes to accommodate lifting system units along with inlet and outlet devices, nozzle tilting mechanisms, etc. In addition, the VTOL aircraft designer has significantly less freedom in the placement of units due to layout constraints, as well as in the placement of engine input devices due to restrictions related to the prevention of recirculation effects. All this prevent reducing the friction drag of the aircraft.

One of the complex problems arising in the development of a VTOL aircraft is the problem of erosion: destruction of the airfield surface under the impact of jets flowing from the thrusters of the lift system during takeoff and landing. Erosion increases with increasing jet area loading, i.e. for those thrusters, the use of which is preferable for aircraft with advanced performance. The values of temperatures and jet loading for some lift thrusters of VTOL aircraft are given in Table 1.

In any case, in order to reduce erosion, the distance between the nozzle and the airfield surface should be increased as much as possible, and the jet impact time on the runway should be reduced as much as possible.

The analysis shows that, although there are serious problems encountered in the VTOL aircraft development process, designers have at their disposal a wide range of technical means and solutions to overcome these problems.

3.2 *Laying Out a VTOL Aircraft*

As follows from the above, the VTOL aircraft configuration depends on the type, number and location of thrust sources used to generate thrust in horizontal, vertical,

Table 1 Temperature and jet loading for some lifting thrusters

Type of lifting unit	Jet temperature °C	Jet pressure daPa
Lift turbojet engine	1000–1200	12,000–13,000
Vectored thrust turbofan engine	1600–1800	12,000–13,000
Lift-cruise bypass turbojet engine		12,000–13,000
– hot duct	800–900	
– cold duct	160–180	
Lift-cruise fan	100–120	2900–3000
Ejector augmentor	150–100	1500–1600

Note That the use of fans, as well as ejector augmentors makes it possible to reduce the temperature and dynamic impact of the jet by almost an order of magnitude. The type of lift system determines the requirements to the type of surface coating of takeoff and landing sites

and transition flight modes. The type of lift generation method is largely determined by the aircraft's requirements for flight speed and altitude, vertical flight time and operating conditions. The method of lifting force generation in cruising flight mode affects the lift-to-drag ratio, which determines the range and fuel efficiency of the aircraft.

In the beginning of the design process, the decision is made on the type of VTOL aircraft. The decision depends on the requirements for aircraft performance. So, if the aircraft is intended to perform operations with sufficiently long hover time, with operating on sites of limited dimensions or the deck of the ship, not imposing high requirements on speed and range, then the helicopter type will have an undeniable advantage.

However, a large unit weight of the helicopter propulsion system, poor aerodynamic efficiency, decreasing sharply with increasing cruising speed, significantly limit the field of practical application of helicopters at speed and range.

Rotorcraft, having advantages of helicopter during hover modes, can have 1.5–2 times higher speed and range, but the complexity of design of rotorcraft and large structural weight explain the fact that this type of VTOL vehicles currently has not yet left the stage of experiments. Convertible aircraft, which have advantages of helicopters at hovering modes and airplanes—at horizontal flight modes, are of certain interest, but realization of this concept requires solution of a number of complicated problems.

In contrast to the above-mentioned types of VTOL vehicles, VTOL aircraft can have flight speeds and ranges similar to those of conventional aircraft.

Depending on the method of producing thrust in horizontal flight, there are propeller-driven, combined and jet VTOL aircraft.

Approximate ranges of achievable maximum speeds and flight altitudes of VTOL aircraft with different types of control systems are shown in Fig. 19.

Let's consider the peculiarities of jet VTOL aircraft.

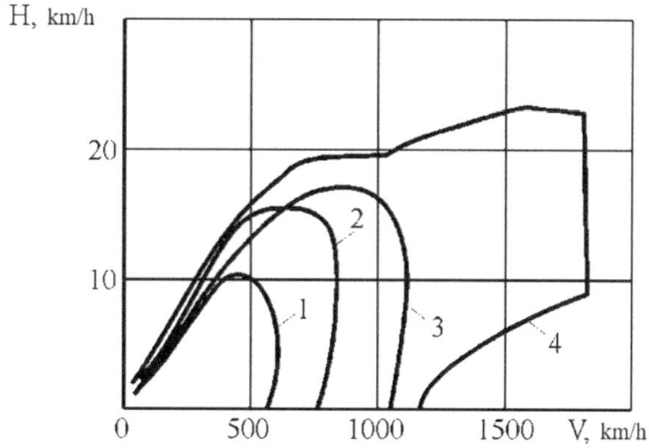

Fig. 19 Performance of VTOL aircraft with different types of power plant: 1—lift fan; 2—tilting propeller; 3—lift engine + lift-cruise jet engine; 4—bypass jet engine with vectoring nozzles

These include aircraft with unitary, separate, and combined power plant, as well as aircraft with thrust augmentors. Data on the currently operating manned jet VTOL aircraft are given in [2, 3], and on unmanned aircraft in [4].

Aircraft with a unitary power plant: the thrust required for horizontal flight, as well as at vertical and transition modes is produced by one or more engines of the same type. Production of horizontal and vertical thrust component by a single engine can be provided either by turning the aircraft, or by tilting the propulsion system relative to the aircraft, or by vectoring the engine exhaust jets. In the first case, vertical takeoff and landing are performed with the fuselage in a vertical position. By the way, the first manned UAVs were made according to this approach (Ryan X-13, Lockheed XFV-1).

A number of UAVs with tilting propeller have been built. Also a number of projects of jet VTOL aircraft with tilting engines have been proposed, but this solution has not been applied in practice.

For subsonic jet UAVs, the configuration with horizontally mounted turbojet engines equipped with devices for deflecting the thrust vector by 90° or more proved to be very promising.

Among the advantages of such a configuration is flexible control of the thrust vector direction. In transition flight, this capability leads to a reduction of lift loss due to interaction with jet and reduces the transition time and fuel consumption associated with it. In basic flight modes it allows to increase maneuvering characteristics of the aircraft. This configuration also provides an opportunity to perform a short takeoff with additional payload.

On the other hand, the need to locate the engine near the center of mass in order to trim the aircraft during VTOL, imposes significant restrictions on the permissible CG range of the aircraft. In addition, in such a layout, the cross-sectional area of the aircraft along its length is unfavorable, which leads to a sharp increase in the wave drag at transonic speeds.

The drawback of this configuration is that since the thrust-to-weight ratio required for a cruise flight, especially for subsonic non-maneuvering aircraft, is much lower than the thrust-to-weight ratio required for a vertical take-off and landing, it is difficult to match the characteristics of the power plant in vertical and horizontal modes of operation. This problem can be solved by using a complex system of air intake regulation, deep throttling, or switching off part of the lift-cruise engines. However, all of the above methods, as a rule, are associated with additional volume and an increase in the aircraft empty weight.

In *aircraft with a separate power plant,* vertical thrust is produced by special lift engines installed vertically or with a slight tilt in the fuselage or in nacelles on the wing. Horizontal thrust is produced in a conventional manner with the help of thrusters.

Short-life jet engines with relatively low thrust-to-weight ratio $(\gamma_{eng} = 0.06\ldots0.08)$ and small dimensions are used for vertical take-off and landing.

Interest in such a configuration was driven by the desire to create vertically taking-off airplanes based on modifications of existing conventional take-off and landing airplanes.

The main disadvantage of such a layout is the irrational use of the power of the propulsion system installed on the aircraft. Indeed, since on such aircraft all the thrust required in VTOL mode is produced by lift engines, more than 12% of the takeoff weight of the airplane is "dead" weight in cruise. If we add to this the increase in the volume of the aircraft associated with the installation of lift engines, then the cost of such a seemingly natural solution to the problem of vertical takeoff and landing becomes obvious.

A significant disadvantage of the configuration with a separate power plant is the large loss of thrust in the transition mode. This is explained by the fact that in such a VTOL aircraft the air velocity field at the inlet to the lift engine and the gas flow at the outlet from the nozzle are not significantly parallel (Fig. 20).

The inlet impulse vector can be estimated as follows:

Fig. 20 Thrust vector determination

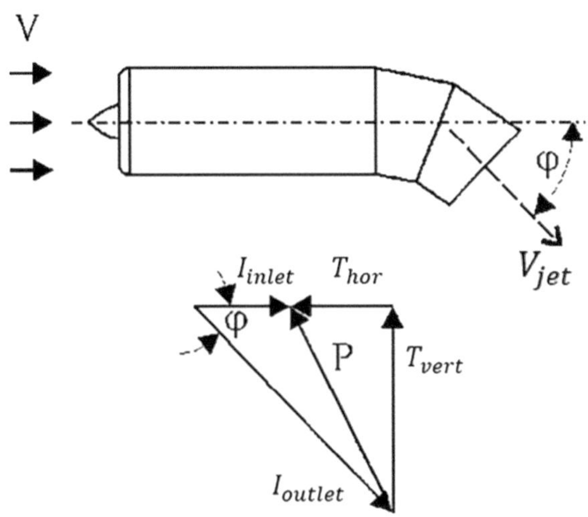

$$\overline{I}_{inlet} = G_{vert}\overline{V},$$

and the outlet impulse vector (under the condition of full expansion) is:

$$\overline{I}_{outlet} = G_{hor}\overline{V}_{jet}.$$

Then, the thrust vector is defined as follows:

$$\overline{P}_{outlet} = \overline{I}_{outlet} - \overline{I}_{inlet}.$$

Finally, the vertical and horizontal thrust can be calculated using the following equations:

$$T_{vert} = G_{hor}V_{jet}sin\varphi,$$

$$T_{hor} = G_{hor}V_{jet}cos\varphi - G_{vert}V,$$

where:

G_{hor}, G_{vert} are the gas and air flow rates respectively;
V_{jet} is the jet velocity;
V is the freestream velocity;
φ is the jet vectoring angle.

Combined-propulsion VTOL aircraft use lift-cruise engines to provide vertical thrust, along with lift engines. In turn, lift engines can be used to generate part of the horizontal thrust in order to improve transition performance. In this case, they are referred to as lift-acceleration engines. To produce horizontal thrust by lift-acceleration engines, either their installation on oscillating supports or vectoring of the engine nozzle, which provides thrust vectoring by 12–15° from the vertical, is used.

This propulsion concept is a compromise compared to a unitary and separate propulsion system. This can be explained by the possibility of selecting the thrust of the lift-cruise engine for cruising flight, which ensures its lower mass and volume as compared to the single power plant. In turn, the thrust vectoring of the cruise engines makes it possible to reduce the required thrust of the lift engines by the amount of this thrust, and hence to reduce their mass and required volumes compared to the engines of the separate power plant.

In addition to good weight characteristics, the combined-propulsion system does not have the disadvantages of unitary-propulsion systems associated with the need to locate the engines near the center of mass. In particular, this layout can be used to design an aircraft with an acceptable cross-sectional distribution over the length of the fuselage to overcome the sound barrier.

This power plant was successfully implemented on a number of aircraft, including domestic aircraft Yak-38 and Yak-141. However, it is also characterized by quite significant thrust loss during transition, caused by impulse loss at the engine inlet.

The problem of the impact of high-temperature jet of lift (lift-and-acceleration) engines on the airfield surface requires special consideration.

A summary table listing the advantages and disadvantages of VTOL aircraft with unitary, separate and combined propulsion system is presented below (Fig. 21).

The problem of matching the engine characteristics in vertical and horizontal flight modes can be solved through the use of *thrust augmentors*. Such units provide vertical takeoff and landing capability at relatively low installed power of the power plant. Fan units, ejector thrust augmentors, external afterburner can be used as augmentors. Power plants with augmentors can be based on turbojet engines. In this case, each of the engines can be used to drive one or more augmentors.

	Unitary	Separate	Combined
Configuration	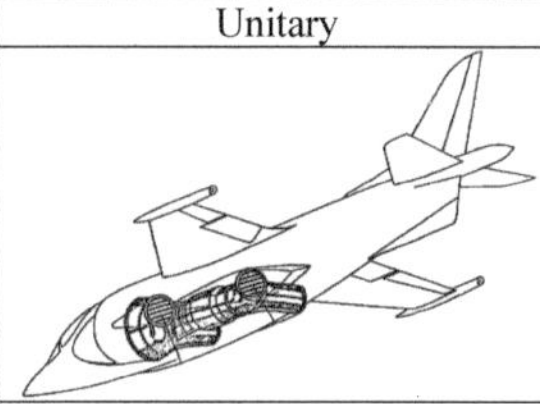	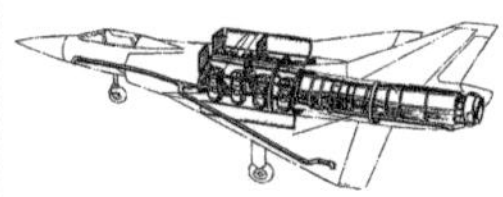	
Advantages	• Thrust vectoring • Minimal lift loss during transition • Reduced transition time • Reduced fuel consumtion • Short take-off with additional payload capability • Improved maneuverability	• The cruise engine parameters optimized for cruising flight • The use of light short-life lift engines	• Rational use of the engine power • Reduced mass • Reduced fuel consumption
Disadvantages	• Difficulty in matching engine performance for vertical and horizontal modes • Large weight and volume of the power plant • Engine must be located lear the CG • Unfavorable cross-sectional distrubution • Limited CG envelope	• Large weight of the lift engine • Large aircraft cross-sectional area • Large thrust loss during transition • Airfield erosion	• Large thrust loss during transition • Airfield erosion

Fig. 21 Comparison of VTOL aircraft with unitary, separate and combined power plant

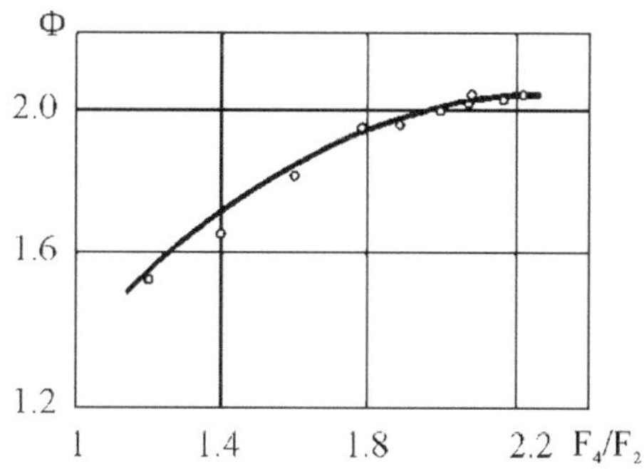

Fig. 22 Predicted and experimental values of the coefficient Φ for flat ejectors as a function of F_4/F_2(F_2 is the area of the ejected gas, F_4 is the area of the outlet section of the diffuser)

In *power plants with fan augmentors,* the lift fan can be either mechanically driven by the fan drive shaft via a gearbox from a free turbine of the main engine, or gas driven or turbo-driven. In the latter two cases, switching on and off the fan augmentor is performed by a gas deflection at the entrance to the engine exhaust nozzle, which directs the gas either to the fan augmentor turbine, located at the periphery of the lift fan, or to a special turbine, which drives the fan by means of a shaft. At that the engine operation mode does not change. Fan augmentor produces the ratio of fan thrust to engine thrust theoretically approaching three. The advantage of power plants with fan augmentor is their comparatively small weight ratio, low fuel consumption, and, that is not less important, not so high jet velocity. The disadvantage of this type of power plant is its significant size. A number of UAVs have been designed using this approach, including the American 5th generation fighter aircraft F-35B [4].

Ejector thrust augmentors are also of interest. They provide a fairly significant ratio of lift engine thrust + augmentor thrust to engine thrust. The achieved values of thrust augmentation coefficients Φ are presented in Fig. 22.

The ejectors have no rotating elements and are relatively simple in design. The low velocity and temperature of the jet flowing out of the ejector reduces the effects of erosion and the noise. On the other hand, the design of VTOL aircraft with ejector thrust augmentor raises layout problems due to the large dimensions of the unit to provide the optimum inlet and outlet area ratios and the required length of the mixing chamber.

Currently, extensive research on various aspects of aircraft ejectors has been conducted, and a number of experimental VTOL aircraft have been built, including an experimental supersonic aircraft by Rockwell XFV-12A [4, 5]. The encouraging results obtained in laboratory are very difficult to implement in a real aircraft, since the characteristics of an airplane with ejector thrust augmentor are largely determined by the features of its layout. The research conducted at the Department 101 of the Moscow Aviation Institute allowed proposing a number of innovative VTOL aircraft configurations with ejector thrust augmentors. One of them is presented in Fig. 23.

In this configuration, unlike the XFV-12A prototype, takeoff and landing are performed with the aircraft in a vertical position. The ejector thrust augmentor is made in the form of a structure consisting of a fixed wing leading-edge (1) and two

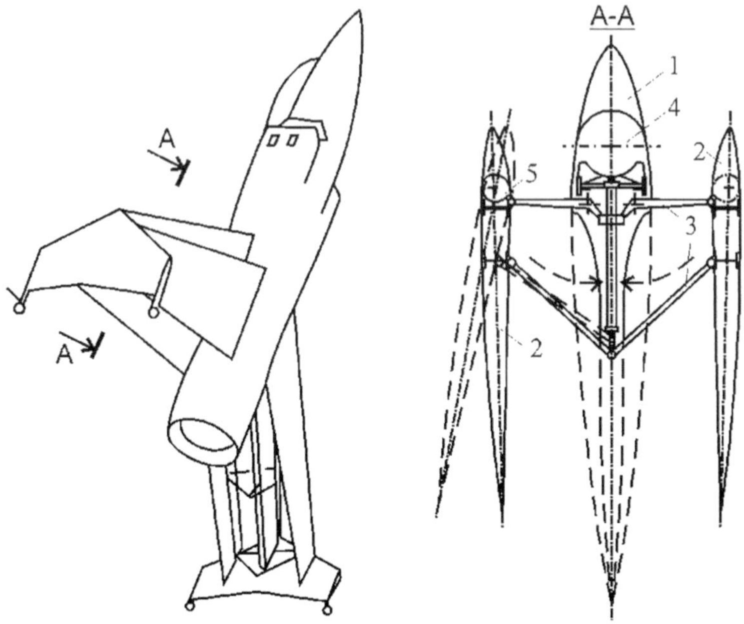

Fig. 23 Sketch of an aeroplane with ejector thrust augmentor

side panels (2) moved away in vertical and transition modes by a lever mechanism (3). The active gas flow nozzles connected by gas lines to the gas generator engine are located in the wing leading-edge (4) and movable side panels (5). The airplane is controlled in the VTOL mode by changing thrust of individual sections of the thrust augmentor and jet control located in the nose of the airplane. The configuration has significant advantages over the prototype, especially for UAVs, whose vertical position during takeoff and landing is not critical due to the absence of a pilot.

External afterburner *with adjustable outlet area can also be used as an augmentor*. It uses the air from the compressor or the fan of the main or lift-cruise engine. One of advantages of such type of thrust augmentor is a big freedom in placing the augmentor which simplifies aircraft trimming during vertical take-off and landing. Its disadvantages are associated with the impact of the jet on the airfield.

A summary table of the advantages and disadvantages of VTOL aircraft with thrust augmentors is presented in Fig. 24.

The analysis shows that there is a large variety of possible VTOL aircraft configurations that can be implemented in manned and unmanned variants. Each of the configurations has both advantages and disadvantages, so the choice of a better option for the aircraft of interest is a complex task.

<table>
<tr><th></th><th>Fan augmentor</th><th>Ejector augmentor</th><th>External afterburner</th></tr>
<tr><td>Configuration</td><td>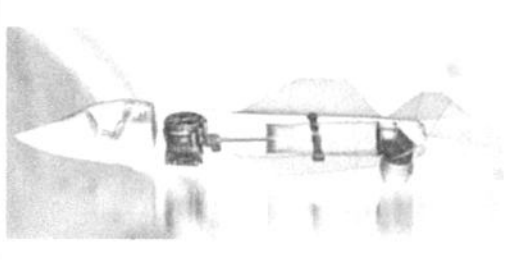</td><td></td><td>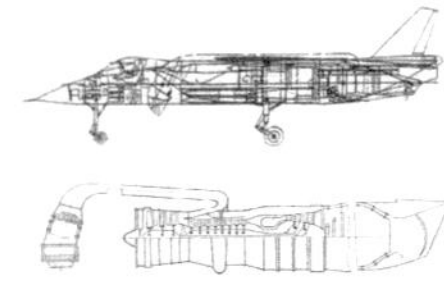</td></tr>
<tr><td rowspan="3">Advantages</td><td colspan="3">- VTOL capability at comparatively low engine power
- Reduced size and weight of the power plant
- Reduced fuel consumption
- Configurability
- High thrust increment due to augmentor</td></tr>
<tr><td>Thrust ratio
2,5...3,0</td><td>Thrust ratio
1,6...1,8</td><td rowspan="2">Thrust ratio
1,5...2,0</td></tr>
<tr><td colspan="2">Low jet temperature and velocity</td></tr>
<tr><td>Disadvantages</td><td>Large size of the augmentor</td><td>Efficiency depends on the aircraft layout and mission</td><td>High jet temperature and velocity</td></tr>
</table>

Fig. 24 Comparison of performance of VTOL aircraft with thrust augmentors

3.3 Determination of Basic Design Parameters of a VTOL Aircraft

The main difference between the design approaches used for a conventional aircraft and a VTOL aircraft is related to determining the take-off thrust-to-weight ratio and the take-off wing loading, as well as the takeoff weight of the aircraft.

3.3.1 Take-Off Thrust-to-Weight Ratio

The correct prediction of the take-off thrust-to-weight ratio is a particularly important task in designing any VTOL aircraft, since underestimating the take-off thrust-to-weight ratio may result in the aircraft not being able to take off vertically, i.e. not fulfilling its mission. Overestimating the thrust-to-weight ratio, on the other hand, results in overdesigning the aircraft. At the same time, the cost of error here is significantly higher, since the take-off thrust-to-weight ratio and the mass ratios of propulsion system and fuel, which depend on it, are usually significantly higher than that of a conventional aircraft.

The required value of $\overline{T}_v$ as a function of time to gain a given altitude can be calculated using an approximate equation:

$$\overline{T}_v \cong 1 + \frac{2H_{climb}}{g t_{climb}^2},$$

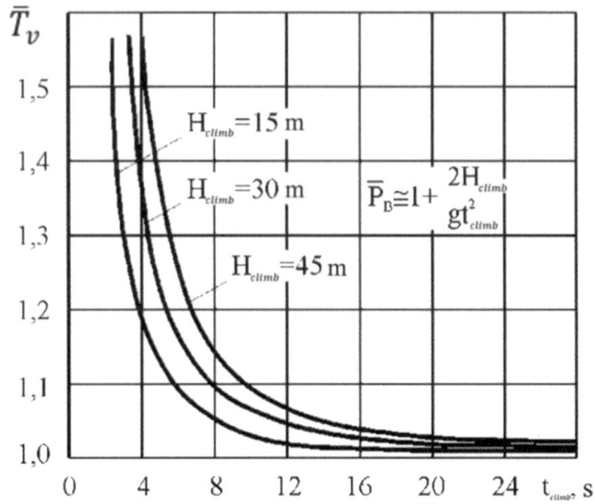

Fig. 25 Effect of climb time on minimum thrust-to-weight ratio of VTOL aircraft

where:

H_{climb} is the design climb altitude, m;

t_{climb} is the climb time, s.

As can be seen from the graph $\overline{T}_v = f(t_{climb}, H_{climb})$ (Fig. 25), plotted using this equation, with the time of climb commensurate with the time of climb of conventional aircraft, the required vertical thrust-to-weight ratio does not exceed 1.1.

However, this value does not take into account the inevitable thrust costs for the VTOL aircraft to ensure controllability and trim capability. Impulse loss, loss due to recirculation, thrust vectoring, as well as the need to overcome secondary aerodynamic forces and moments should be taken into account. Some of these factors and their influence on the take-off thrust-to-weight ratio are shown schematically in Fig. 26. The take-off thrust-to-weight ratio of a VTOL aircraft depends largely on its configuration, in particular, on the type and number of engines (thrusters) and their location on the aircraft.

It follows from this that the take-off thrust-to-weight ratio of a VTOL aircraft is a function of many variables, which depend on the initial conditions.

For the aircraft type under consideration the most typical flight modes are vertical takeoff, cruising flight, vertical landing. No special requirements are imposed on other flight modes. However, as calculations show, the vertical landing mode, as a rule, is not decisive for determining the thrust-to-weight ratio, and it can be excluded from consideration.

When determining the take-off thrust of the VTOL aircraft, one should not forget about the transition from vertical takeoff to horizontal flight and back.

Let us consider the methodology for determining the take-off thrust-to-weight ratio of a VTOL aircraft with a combined propulsion system consisting of lift-cruise and lift-acceleration engines, as well as with a propulsion system consisting of a cruise engine and a thrust augmentor. The number of engines and augmentors depends on

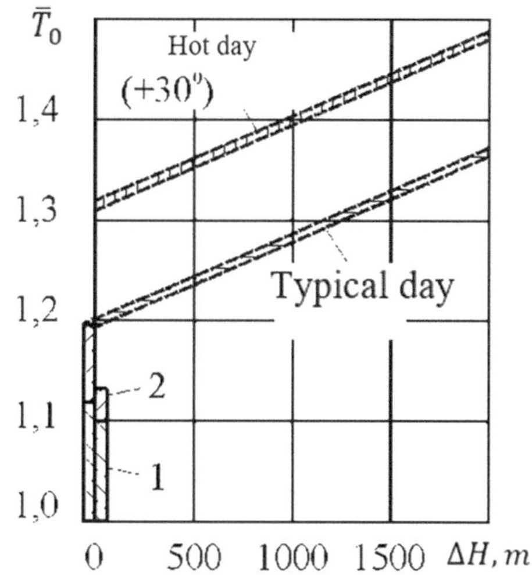

Fig. 26 Factors affecting the take-off thrust-to-weight ratio of VTOL vehicles: 1—$\Delta T = 10 - 12\%$—acceleration, stabilization, recirculation, ground effect; 2—$\Delta T = 2.5 - 7.5\%$—engine inlet thrust loss; ΔH—airfield elevation above sea level

the performance of the existing engines or those under development, as well as the peculiarities of the aircraft layout.

If the aircraft is equipped with n_{hor} lift-cruise engines and n_{vert} lift engines (lift-acceleration engines), their total design take-off thrust will be equal to $T_0 = n_{hor} T_{hor}^{des} + n_{vert} T_{vert}^{des}$. The take-off thrust ratio can be determined as follows:

$$\overline{T}_0 = \frac{n_{hor} T_{hor}^{des} + n_{vert} T_{vert}^{des}}{m_0 g} = \overline{T}_{0,hor} + \overline{T}_{0,vert}.$$

Here,

$\overline{T}_{0,hor}$ is the take-off thrust-to-weight ratio with lift-cruise engines;

$\overline{T}_{0,vert}$ is the take-off thrust-to-weight ratio with lift engines;

If $\overline{T}_{hor}$ is the design thrust of one lift-cruise engine, its effective thrust used to generate lift during VTOL is equal to:

$$T_{hor}^{eff} = T_{hor}^{des} K'_{augm} K'_{atm} K'_{cont} K'_{\varphi},$$

where:

$K'_{augm} \geq 1$ is the augmented thrust factor;

$K'_{atm} \geq 1$ is the factor taking into account the difference between the atmospheric conditions at the engine inlet and the standard ones;

$K'_{cont} \geq 1$ is the thrust control factor;

$K'_{\varphi} \leq 1$ is the factor taking into account thrust loss due to thrust vectoring.

Similarly, for lift engine (lift-acceleration) it can be written:

$$T_{vert}^{eff} = T_{vert}^{des} K''_{augm} K''_{atm} K''_{cont} K''_{\varphi}.$$

Then the effective thrust-to-weight ratio will be equal to:

$$\overline{T}_0^{eff} = \overline{T}_{0,hor} K_{total}{'} + \overline{T}_{0,vert} K_{total}^{''}.$$

Here, $K_{total} = K_{augm} K_{atm} K_{cont} K_{\varphi}$.

The following ranges of the above factors can be recommended for the preliminary design:

$K_{augm} = 1.0 \ldots 1.1$, the lower value corresponds to the lift engines without augmentor;

K_{atm} is calculated in the usual way;

$K_{cont} = 0.85 \ldots 1.0$, it depends on the type of control system;

$K_{\varphi} = 0.95 \ldots 0.98$, it depends on the design of the tilting nozzle.

$\overline{T}_0^{eff}$ must deliver the specified rate of climb in the VTOL mode. Taking into account possible thrust loss due to recirculation and secondary aerodynamic forces, it lies within $\overline{T}_0^{eff} = 1.1...1.15$. Having decided on the value of $\overline{T}_0^{eff}$, one can start determining $\overline{T}_{0,hor}$ and $\overline{T}_{0,vert}$. Since the horizontal flight in cruising mode for the majority of UAVs is the most time-consuming one, it is natural to optimize the power plant for the cruising flight condition. Based on this condition, the required take-off thrust-to-weight ratio for lift-cruise engine can be determined as follows:

$$\overline{T}_{0,hor} = \frac{\overline{m}}{(L/D)_{cr} \xi_{HV} \varphi_{th}},$$

where:

$\overline{m} = \frac{m_{av}}{m_0}$ is the coefficient taking into account the change in the takeoff weight of the aircraft in cruise flight due to fuel burn;

$(L/D)_{cr}$ is the cruise lift-to-drag ratio of the aircraft;

$\xi_{HV} = \frac{T_{HV}}{T_0}$ is the factor taking into account the change in the engine take-off thrust with elevation and speed;

φ_{th} is the throttle lever factor (engine operating mode).

$\overline{m}$ varies between 0.8–0.9. Smaller values correspond to aircraft with a longer range or flight time. This value can also be calculated using the formula $\overline{m} = 1 - 0.5\overline{m}_{fuel}$ assuming that the average weight of the aircraft is determined at half the fuel consumed.

The value of the factor ξ_{HV} can be derived from the engine altitude-velocity performance.

φ_{th} varies between 0.85–0.95.

Modern subsonic non-maneuvering jet aircraft have the take-off thrust-to-weight ratio for cruising flight conditions varying between 0.25–0.35, which is quite far from the value required for vertical takeoff.[1] If we are satisfied with this value, then the thrust-to-weight ratio for the lift-acceleration engine is determined as follows:

[1] For maneuvering aircraft, the take-off thrust-to-weight ratio depends not only on the requirements for cruise, but also a number of other conditions (see Chap. 11).

$$\overline{T}_{0,vert} = \frac{\overline{T}_0^{eff} - \overline{T}_{0,hor} K'_{total}}{K''_{total}}.$$

For UAVs with augmentors that produce all the thrust required for vertical takeoff, as is the case for XV-5A aircraft or XFV-12A aircraft, the thrust-to-weight ratio for the cruise engine is determined as follows:

$$\overline{T}_0 = \frac{\overline{T}_0^{eff}}{\Phi K'_{total}}$$

where Φ is the amplification factor.

The take-off thrust-to-weight ratio of an aircraft with an external afterburner is calculated using the same equations as for an aircraft with a combined power plant. It is only necessary to take into account that the thrust of the lift-cruise engine and the thrust of the external afterburner are physically related by the thrust enhancement factor.

3.3.2 Take-Off Wing Loading

For a non-maneuvering VTOL aircraft, the design case for the take-off wing loading estimation can be cruise and transition modes.

Let us calculate the wing loading for cruise. With the lift being equal to the weight, it can be shown that.

$$p_{0,cruise} = C_{L,cruise} \frac{\rho_H V_{cruise}^2}{2\left(1 - \overline{m}_{f,TO}\right)},$$

where:

$C_{L,cruise}$ is the cruise lift coefficient;

ρ_H is the air density at cruising altitude;

V_{cr} is the cruising speed, m/s.

$\overline{m}_{f,TO}$ is the mass ratio of fuel consumed from the beginning of takeoff to the beginning of cruise flight.

$C_{L,cruise}$ and $\overline{m}_{f,TO}$ are set at the initial stage of the design process ($C_{L,cruise} = 0.3 \ldots 0.6$; $\overline{m}_{f,TO} = 0.05 \ldots 0.1$). The other parameters must be specified in the technical specifications.

The expression for determining the take-off wing loading for transition is as follows:

$$p_{0,trans} = C_{L,allow} \frac{\rho_H V_{trans}^2}{2},$$

where:

$C_{L,allow}$ is the allowable lift coefficient during transition, $C_{L,allow} = (0.78\ldots0.85)C_{L,max}$;

ρ_H is the air density at the transition altitude (it may be assumed to be close to sea level unless a different altitude of the airfield is specified in the technical specifications);

V_{trans} is the transition speed, m/s.

For preliminary calculations it is possible to assume $V_{trans} = 50$–60 m/s. As a rule, for a non-maneuvering airplane, the main design case is cruise. For a maneuvering airplane, p_0 depends on the required normal g-force, turn radius, etc.

The lowest of all calculated values of the take-off wing loading is used as the design value.

3.4 Take-Off Weight

The take-off mass of a VTOL aircraft include systems and elements that ensure VTOL maneuver, a well as controllability and trimming capability during VTOL and transition.

The conventional equation for determining the aircraft take-of mass:

$$m_0 = \frac{m_{PL} + m_{FCEq}}{1 - \overline{m}_{struct} - \overline{m}_{PP} - \overline{m}_{fuel} - \overline{m}_{equip}}.$$

The payload mass m_{PL} depends on the aircraft mission.

As a first approximation, the mass ratio of the structure can be determined based on historical data: $\overline{m}_{struct} = 0.25\ldots0.3$. Larger values should be taken if the powerplant layout requires significant structural reinforcement, for example, due to the need to compensate for cut-outs in the most loaded parts of the wing or fuselage for placement of the augmentors.

The VTOL aircraft equipment mass ratio is, as a rule, somewhat greater than that of a conventional aircraft. As a first approximation, it can be assumed that $\overline{m}_{equip} = 0.14\ldots0.17$.

The propulsion system mass ratio for the general case of designing a VTOL aircraft with a combined propulsion system can be determined as follows:

$$\overline{m}_{PP} = \overline{T}_{0,hor} k'_{inst} \gamma'_{eng} + \overline{T}_{0,vert} k''_{inst} \gamma''_{eng},$$

Table 2 Historical data on γ_{eng} and k_{inst}

Engine type	k_{inst}	γ_{eng}
Lift turbojet engine	1.5–1.7	0.05–0.07
Lift bypass turbojet engine		0.05–0.08
Lift-cruise bypass turbojet engine	1.6–1.8	0.17–0.22
Turbofan engine	1.4–1.5	0.08–0.12

where

$\overline{T}_{0,hor}$ is the take-off thrust-to-weight ratio with lift-cruise engines;

$\overline{T}_{0,vert}$ is the take-off thrust-to-weight ratio with lift engines;

k'_{inst}, k''_{inst} are the mass ratios of additional engine installation units to the masses of lift-cruise engines and lift engines, respectively;

$\gamma'_{eng}, \gamma''_{eng}$ are the specific masses of lift-cruise engines and lift engines, respectively.

Historical data on γ_{eng} and k_{inst} are presented in Table 2.

The fuel mass ratio should be determined as the sum of the mass ratio of fuel for vertical and horizontal flight modes:

$\overline{m}_{fuel} = \overline{m}_{fuel,hor} + \overline{m}_{fuel,vert}.$

The fuel mass ratio for VTOL for an aircraft with combined power plant is determined as follows:

$$\overline{m}_{fuel,vert} = \left(C_{p,0,hor}\overline{T}_{0,hor} + C_{p,0,vert}\overline{T}_{0,vert}\right)t,$$

where:

$C_{p,0}$ is the take-off specific fuel consumption;

t is the total time of vertical and transitional modes.

For VTOL aircraft, whose mission doesn't include hovering, the time t usually does not exceed 1–3 min.

With the first approximation takeoff mass, we can calculate the other two aircraft dimensional parameters—the wing area and engine thrust:

$$S = \frac{m'_0 g}{p_0},\ T_{o,hor} = \overline{T}_{0,hor} m'_0 g,\ T_{o,vert} = \overline{T}_{0,vert} m'_0 g.$$

The second approximation of the takeoff mass of a VTOL aircraft is calculated during the laying out process. The semiempirical formulas given in [1], which were derived for VTOL aircraft of various types, can be used to estimate the mass of the aircraft components.

3.4.1 Design of the Fighter SVTOL with Combined Power Plant

In this section, the above methodology is exemplified through a brief description of the design of a short takeoff and vertical landing fighter. The aircraft uses a combined propulsion system: it incudes the domestic lift-acceleration RD-41 M engine and one lift-cruise AL-41F1 engine currently being under development.

The project was developed by A.L. Reingold, a student of Department 101 of MAI, under the requirements set out in Table 3.

Payload specification:

Artillery: Integrated cannon launcher with 30 mm GSh-301 cannon. Ammunition 120–150 rounds.

Rocket and bomb armament is located on internal and external points.

The aircraft has compartments allowing for internal suspension and the use of missile weaponry of the size corresponding to the following models:

- Medium-range air-to-air missiles: R-77(RVV-AE)
- Short-range air-to-air missiles: K-74
- Air-to-surface missiles: X-38.

The aircraft is also equipped with 4 points for external loads on the wing. External points are suitable for the loads up to 1000 kg, internal up to 1500 kg.

The general view of the developed SVTOL aircraft is shown in Fig. 27, and its enlarged layout is shown in Fig. 28.

The aircraft features a cantilever high wing and a canard layout.

The delta wing has a large leading edge sweep and a small negative trailing edge sweep.

Such a wing allows for a small thickness ratio to provide a large structural depth and to accommodate a large amount of fuel. The large sweep of the leading edge allows reducing drag at high Mach numbers and reducing signature.

Table 3 Requirements for jet VTOL aircraft

No. n/a	Name	Value, units	Note
1	Maximum speed	1,8M	
2	Cruising speed	950 km/h	
3	Payload at short take-off	1400 kg	Specific payload data is given below
4	Take-off distance at short take-off	Less than 100 m	
5	Atmospheric conditions for short take-off and vertical landing	Temperature + 27°C; Pressure 746 mm Hg	
6	Range	At least 1500 km	
7	Vertical landing conditions	Landing both with and without payload, hover time 3 min	

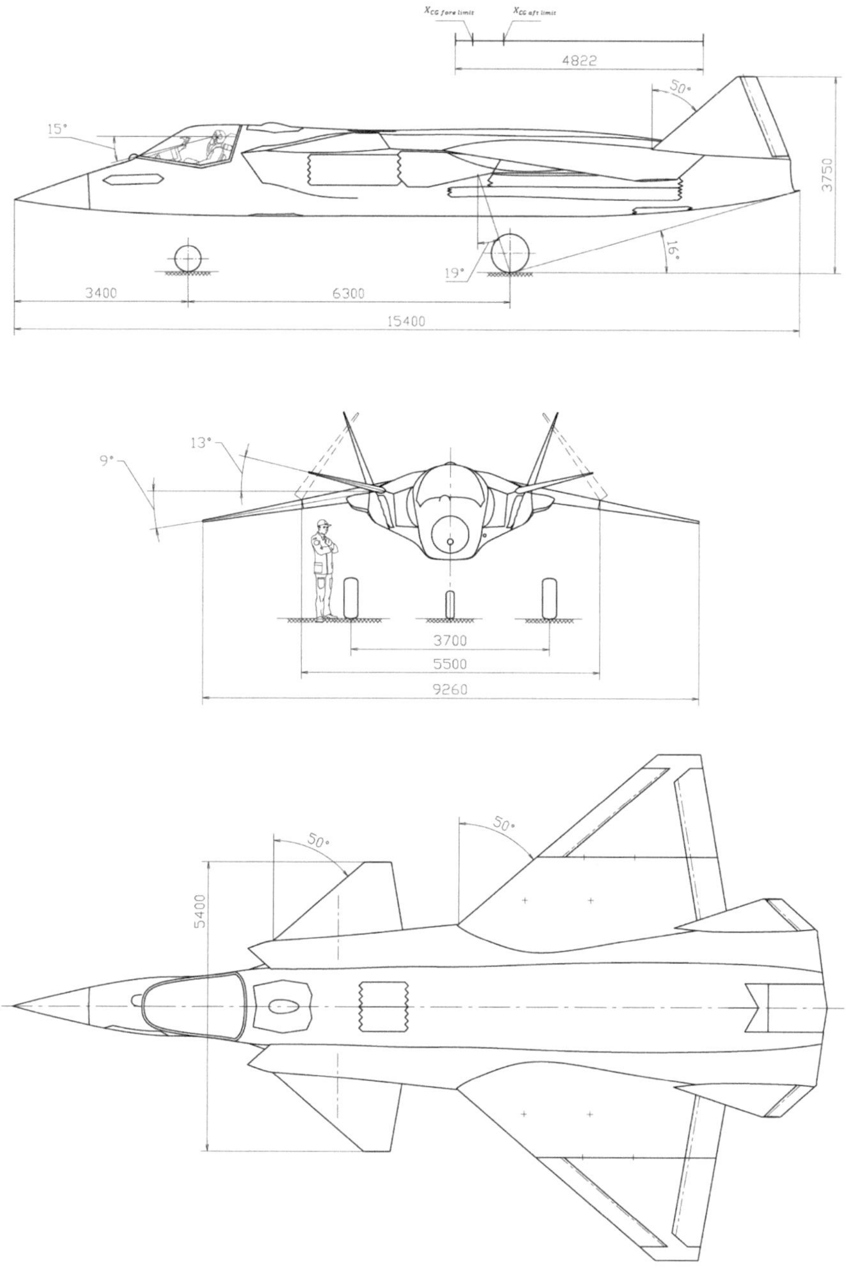

Fig. 27 General view of the aircraft

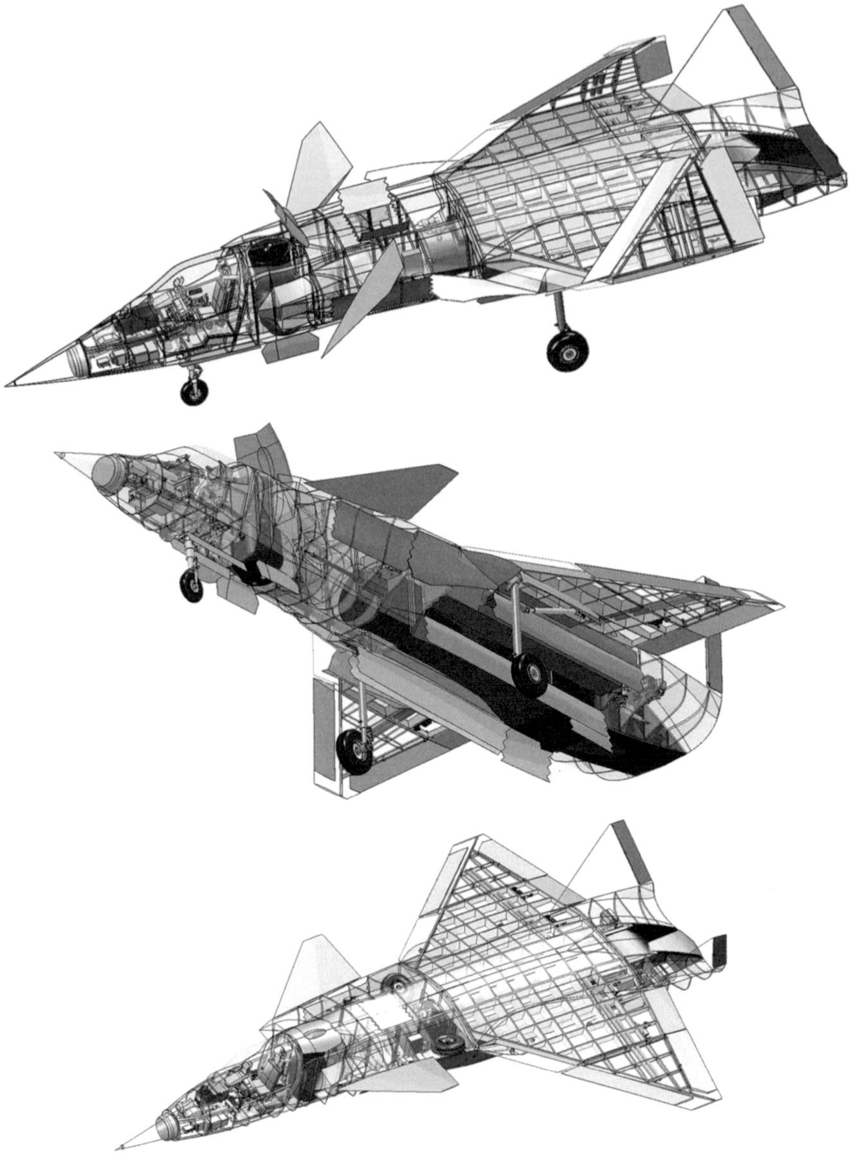

Fig. 28 STOL fighter layout

In order to reduce the influence of the secondary aerodynamic forces in the STOL modes, the distance from the wing to the runway has been increased.

The wing is equipped with slats, flaps, and ailerons.

The wing is foldable to reduce the dimensions of the aircraft when operating on the deck or using special shelters.

The fuselage has a large aspect ratio. In the forebody, equipment compartment and cockpit are located, the lift engine is mounted behind the cockpit. Side air intakes converge into a single channel behind the lift engine. Weapon compartments are located in the central part of the fuselage on each side of the engine. Flat nozzle is located in the tail of the fuselage.

Vertical Tail. For yaw control at high angles of attack and sideslip, the aircraft is equipped with a twin vertical tail. The vertical tail consists of a stabilizer and a rudder.

The **canard** is all-movable, installed on the side air intakes. The plane of the canard lies above the plane of the wing. Canard is set at a slight dihedral angle. In order to reduce the signature, the canard tips are parallel to the wingtips.

The aircraft uses a tricycle landing gear. The main supports are retracted in the upsteam direction in the fuselage-wing junction area. The nose support is retracted in the streamwise direction. The aircraft of this category is intended to operate from a concrete runway, road sections with paved surfaces or from the deck of the aircraft carrier, which allows the use of high and ultra-high pressure pneumatics (about 6.5–16 atmospheres), thereby reducing the mass of the landing gear.

The power plant of the aircraft consists of a lift-cruise engine located in the central part of the fuselage and a lift engine located behind the cockpit. Lift-cruise nozzle allows thrust vectoring to an angle of 90–100°.

During hovering, trim and control are provided by jet controls, air is taken from the lift-cruise compressor.

Advantages of the chosen configuration:

- high lift-to-drag ratio;
- high maneuverability;
- high field performance;
- reduction of nozzle signature from lateral angles and in the rear hemisphere;
- reduced engine compressor signature due to S-shaped air intake channels;
- wide CG range.

Disadvantages of the chosen configuration:

- inability to deflect lift-cruise thrust vector sideways (to increase maneuverability);
- reduced (compared to the conventional configuration) longitudinal stability;
- somewhat greater (compared to the tailless configuration) signature.

The main characteristics and parameters of the developed variant of the STOL aircraft are given in Table 4.

Table 4 Main characteristics and parameters of the developed STOL aircraft

Main parameters and characteristics	
Take-off mass, kg	13.845
Payload mass, kg	1400
Take-off thrust (STOL/ConventionalTOL)	1.26/0.94
Wing loading, daN/m^2	380
Cruising speed, km/h	950
Cruising flight altitude, m	11,000
Range, km	1517
Runway length (STOL/ConventionalTOL), m	100/1000
Wing area, m^2	36.44
Wing span, mm	9260
Length, mm	15,400
Height, mm	3750
Wheel base, mm	6300
Wheel track, mm	3700
PMD AL-41F1 take-off thrust, daN	1×14500
Takeoff thrust of RD-41 M, daN	1×4500

Variables		
Symbol	Description	Units
p_0	Wing loading	daN/m^2
L_{TOGR}	Take-off ground roll distance	m
a_x	Average acceleration during take-off ground roll	m/s
$(L/D)_{GR}$	Lift-to-drag ratio during ground roll	–
$\overline{T}_{av} = \frac{T_{av}}{m_0 g}$	Average thrust-to-weight ratio at take-off ground roll	–
$V_{unstick}$	Unstick speed	m/s
L_{LGR}	Landing ground roll distance	m
V_{land}	Landing speed	m/s
T_{rev}	Reverse thrust	N
T_{idl}	Idling thrust	N
$(L/D)_{LGR}$	Lift-to-drag ratio during landing ground roll	–
V_{st}	Stalling speed	m/s
n_{eng}	Number of engines	–
V_{app}	Approach speed	m/s
$C_{L,unstick}$	Unstick lift coefficient	–
$C_{L,max,TO}$	Maximum lift coefficient at take-off	–

(continued)

(continued)

Variables		
Symbol	Description	Units
L_{RW}	Runway length	m
$C_{L,appr}$	Approach lift coefficient	–
p_{land}	Wing loading during landing	daN/m^2
C_μ	Impulse coefficient	–
m_{air}	Blowing system mass flow rate	kg/s
V_s	Jet velocity	m/s
f_2	Rolling friction coefficient	–
δ_f	Flap deflection angle	degrees
$\nu = \frac{C_p}{C_V}$	Specific heat ratio	–
R	Gas constant	m/K
T_a	Absolute temperature of air	K
γ_{air}	Specific weight of air	daN/m^3
C_D	Drag coefficient	–
$C_{L,C_{\mu}=0}$	Wing lift coefficient without jet blowing	–
η	Static thrust recovery coefficient	–
δ_{jet}	Jet flap deflection angle	degrees
$\Delta C_{L,\Gamma}$	Lift increment due to supercirculation	–
$C_{D,0}$	Zero lift drag coefficient	–
C_t	Thrust ratio	–
T_V	Vertical component of the thrust	N
φ	Thrust vectoring angle	degrees
$\overline{T}_0$	Take-off thrust-to-weight ratio	–
$H_{obstacle}$	Elevation of obstacle above the ground	m
L_{TO}	Take-off distance	m
$\mathrm{F_{jet}}$	Jet area	m^2
C_p	Specific fuel consumption	kg/daN · h
m_{PP}	Power plant mass	kg
m_{fuel}	Fuel mass	kg
r_0	Lifting system specific mass	–
$\overline{m}_{lift} = \overline{T}_v r_0$	Lifting system mass ratio	–
$\varepsilon_x, \varepsilon_y, \varepsilon_z$	Angular accelerations	m/s^2
I	Moment of inertia	$kg \cdot m^2$
$l_{control}$	Jet control moment arm	m
b	Wing span	m

(continued)

(continued)

Variables		
Symbol	Description	Units
Φ	Thrust augmentation coefficient	–
H_{climb}	Design climb altitude	m
t_{climb}	Climb time	s
$\overline{T}_{0,hor}$	Take-off thrust-to-weight ratio with lift-cruise engines	–
$\overline{T}_{0,vert}$	Take-off thrust-to-weight ratio with lift engines	–
ξ_{HV}	Factor taking into account the change in the engine take-off thrust with elevation and speed	–
φ_{th}	Throttle lever factor (engine operating mode)	degrees
$C_{L,allow}$	Allowable lift coefficient	–
V_{trans}	Transition speed	m/s
$\overline{m}_{equip}$	Equipment mass ratio	–
$\overline{m}_{struct}$	Structures mass ratio	–
$\overline{m}_{PP}$	Power plant mass ratio	–
m_{FCEq}	Flight crew equipment mass	kg
m_{PL}	Payload mass	kg
m_0	Take-off mass	kg

References

1. Volodin VV, Liseytsev NK, Maksimovich VZ (1985) Osobennosti proektirovaniya reaktivnyh samolyotov vertikal'nogo vzlyota i posadki (Design Features of Vertical Takeoff and Landing Jet Aircraft). Mashinostroenie, Moscow
2. Ruzhickij EI (2000) Evropejskie samolyoty vertikal'nogo vzlyota (European vertical takeoff aircraft). Astrel', Moscow
3. Ruzhickij EI (2000) Amerikanskie samolyoty vertikal'nogo vzlyota (American vertical takeoff aircraft). Astrel', Moscow
4. Liseitsev NK (eds.) (2009) Bespilotnye samolyoty vertikal'nogo vzlyota i posadki (Unmanned vertical takeoff and landing aircraft.), MAI, Moscow
5. Enenkov VG, Klyachkin AL et al (1980) Aviacionnye ezhektornye usiliteli tyagi (Aircraft ejector thrust boosters). Mashinostroenie, Moscow
6. Pavlenko VF (1987) Silovye ustanovki s povorotom vektora tyagi v polyote (Power plants with in-flight thrust vectoring). Mashinostroenie, Moscow
7. Hafer X, Sachs G (1982) Senkrechtstart technik. Flugmechanik, Aerodynamik, Antriebssysteme, Springer Berlin, Heidelberg

Air-Space and Hypersonic Aircraft Design

Eduard N. Dudar and Valery P. Timoshenko

Abstract The chapter introduces the classes of aerospaceplanes (ASP) and hypersonic aircraft (HA), shows how to estimate their flight envelope, predict their flight trajectories, and how to estimate the thermal and acoustic loads imposed on these aircraft. The chapter also discusses how to design and optimize ASP and HA, based on the requirements specifications. The chapter analyses the principles of aerodynamic and thermal design of ASP and HA and discusses their thermal protection.

Keywords Aerospaceplane design · Hypersonic aircraft optimization · Flight envelope · Flight trajectory · Acoustic loads · Thermal design

1 Classification and Mission

Aerospaceplane (ASP) is a space vehicle (SV) that performs flight on Earth orbit or in the upper atmosphere and returns along a flat trajectory. It is capable of performing aerodynamic manoeuvres. Other names are orbital aircraft (OA), orbital spaceship (OS), spacecraft (SC). In most cases ASP are reusable SV that perform horizontal landing on airfield. Aerodynamic efficiency allows the ASP to reduce g-loads during descent in the atmosphere and to perform lateral manoeuvre at up to 2000 km distance from the landing path which decreases the pending time before deorbiting. Chute landing on ground or water surface can be applied for small experimental ASP (technology demo) (Bor-4, Figs. 1 and 2).

Hypersonic aircraft (HA) can also be categorised as air spacecraft because it has a space function and is capable of entering orbit using a combined cruise propulsion system (CSPS) which includes also a rocket engine for acceleration along the upper trajectory.

Depending on the cruise engines type and their arrangement, the following classes of ASP can be specified:

E. N. Dudar (✉) · V. P. Timoshenko
Scientific and Production Association "Molniya", Moscow, Lodochnaya, Russia
e-mail: dudaren1@rambler.ru

M. A. Pogosyan (ed.), *Aircraft Design*, Springer Aerospace Technology,
https://doi.org/10.1007/978-981-96-4599-2_13

Fig. 1 The Bor-4 space vehicle, a flying model of Spiral (1980 to 1987)

Fig. 2 The Spiral ASP and a hypersonic launch aircraft (1965 to 1975)

- ASP without cruise engines (Bor-4, Buran, Spiral);
- ASP with rocket cruise engines (LRE) as a component of a rocket-space (RSS) or airspace (ASS) system;
- ASP with cruise LRE, the fuel is placed in internal tanks (Horus in project ASS Sänger, single-stage launcher Venture Star);
- ASP with CSPS—hypersonic aircraft as a reusable single-stage launcher (projects NASP, Tu-2000).

Depending on the mission, the ASP can be orbital or suborbital.

Orbital ASP are designed for bringing payloads and crews into low Earth orbit and returnng them to Earth. Typical missions include: space transportation and engineering support of space station (SS), operating inspection of orbital objects, rescue of crews in disaster, remote Earth surface sensing and others.

Suborbital ASP include: experimental demo-vehicles (Fig. 1); winged rocket boosters - return rocket units (RRU) of reusable RSS (Fig. 5); suborbital space touring vehicles (SV Enterprise of suborbital system Spaceship Two). Aviation engines (TJE) can be used for reentry of the reusable RRU (SSRS first-stage boosters). TJE can be started at the altitude of 8 to 6 km. Reuse of RRU reduces injection costs and

Fig. 3 Space-rocket systems space shuttle (**a**) and Energia-Buran (**b**)

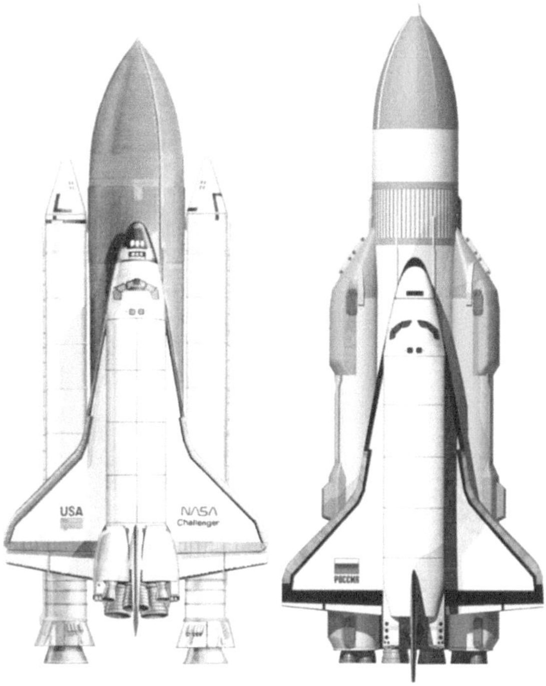

a *b*

allows eliminating restricted zones on the ground which are required for drop of disposal elements of single-use RSS. The SSRS with the 2-nd stage multiuse booster call for solving more complex problems involved with the aerodynamic heating at hypersonic speeds.

The ASS advantages (Fig. 4) are due to the airborne takeoff and operating flexibility. Simpler versions of ASS without orbital aircraft can be used for airborne takeoff of launch rockets (LR) of the light class (instances: LR Pegas with takeoff from external store of the bomber B-52G or aircraft L-1011; Airborne Takeoff project with launch aircraft An-124).

Hypersonic aircraft, when used as a single-stage ASP, could provide for a number of operating advantages. It has no drop parts, it is capable of performing a spatial manoeuvre for approaching any injection azimuth right away from take-off. It produces aerodynamic lift at ascend and descent; the air serves as an oxidiser and a working medium for the production of thrust across much of its ascent trajectory.

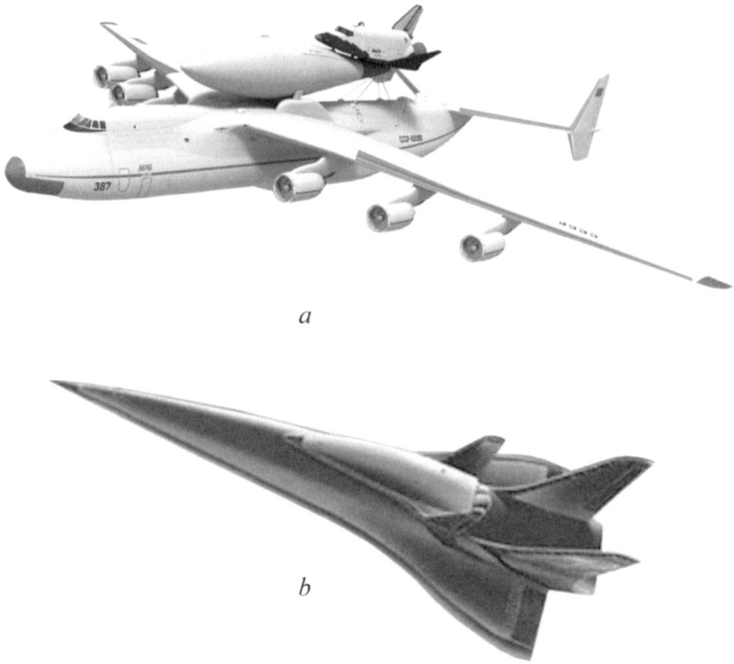

Fig. 4 Airspace system projects with reusable OA having cruise LRE: **a**—MAKS, subsonic launch aircraft (LA) An-225 and OA with an external fuel tank; **b**—Sänger, hypersonic LA, OA Horus with internal tanks for the cruise LRE

2 ASP and Hypersonic Aircraft Operation

2.1 ASP and HA Flight Envelope

The atmospheric segment of flight envelope is usually considered in terms of speed V and height H, while in the space segment the parameters of interest are the inclinations and orbit heights $(i,\ H)$ with a specified payload mass. The boundaries of such envelopes are defined by the energy capabilities of the AS and the injection system, as well as by the airframe features, characteristics of structural and thermal-protection materials, the aerodynamic features of the cruise propulsion system (CPS). Fulfilment of the specified missions in terms of the limitations and requirements to the structural strength, aerodynamic heating and flight safety are checked for at various design stages through the analysis of flight envelope. During design, analytical predictive methods are often used [2].

The most important parameters affecting the flight envelope are: (1) the ballistic coefficient $\sigma_D = C_D/(mg_0/S)$ related to the aerodynamic decceleration (where m is the mass, S is the reference area); (2) the glide parameter $\sigma_L = C_L/(mg_0/S)$ related to the aerodynamic lift; (3) the lift-to-drag ratio σ_L/σ_D; (4) the thrust-to-weight ratio

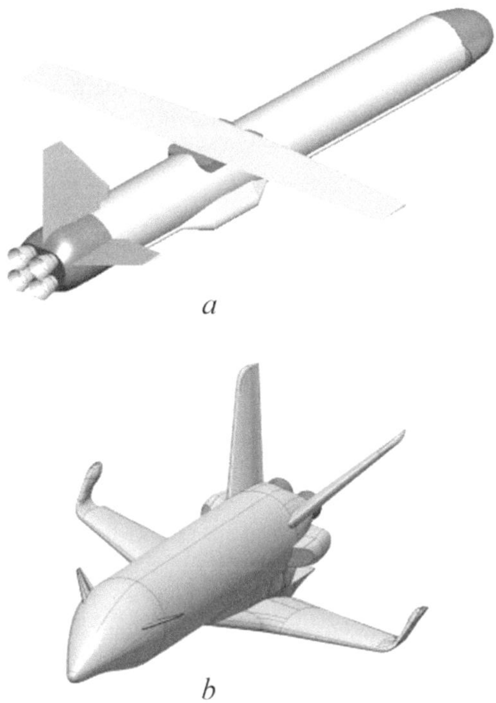

Fig. 5 Versions of RRU for the first-phase space shuttle rocket system (MRKS-1): **a**—with a straight tilting wing; **b**—with a tapered wing

of the ascent system $\overline{T} = \frac{T}{mg_0}$; (5) the specific impulse of the SPS J; (6) the fuel ratio $\overline{m}_F = \frac{m_F}{m_0}$ (where T is the thrust, m_0 is the take-off mass of the ASP or ascent system). The ASP flight trajectory in dense atmosphere and near-Earth space are interlinked through the boundary conditions for the end of active ascent and the parameter envelope for reentry.

It is convenient to perform the analysis of orbital movement using the non-dimensional variables $\overline{r} = \frac{r}{r_0}$ and $\overline{V} = \frac{V_{abs}}{V_{FC}}$, where r is the distance to the Earth centre; V_{abs} is the absolute velocity; $V_{FC} = \sqrt{g_0 r_0 = 7904.47 m/s}$ is the first cosmic velocity; $r_0 = 6371250$ m is the average Earth radius, $g_0 = 9.80665$ m/s^2 is the average gravity acceleration at $r = r_0$. The Earth surface is close in shape to an ellipsoid of revolution with the equatorial and polar radii $r_{\text{eq}} = 6378160$ m, $r_{\text{p}} = 6356775$ m. The ASP flight altitude in atmospheric segment is determined along the radius-vector $\overline{r}$ as follows:

$$H = r - r_{\text{E}}(\varphi); \quad r_{\text{E}}(\varphi) = r_{\text{eq}}\left(1 - e_{\text{E}}^2 \sin^2 \varphi\right), \tag{1}$$

where r_{E} is the distance from the Earth centre to the ellipsoid surface; $e_{\text{E}}^2 = 0.0033528$ is the square of its eccentricity; φ is the geocentric latitude (the angle of inclination of the radius-vector $\overline{r}$ to the equator plane). The flight height in orbital segment is measured from a sphere with the equatorial radius:

$$H = r - r_{\text{eq}}. \tag{2}$$

For the majority of design tasks, the Earth gravitational field may be considered as radial, with the gravity acceleration as follows:

$$g = g_0/\bar{r}^2. \tag{3}$$

Passive orbital flights of an ASP (without thrust and aerodynamic forces) is subject to the Kepler laws, the orbit is of elliptical shape (Fig. 6) and it can be represented as follows:

$$\bar{r} = \bar{p}/(1 + e\cos\kappa), \tag{4}$$

where k is the angle of rotation of the radius-vector about the perigee (true anomaly).

ASP trajectory beyond the dense atmosphere can be determined, with the accuracy sufficient for the design stage, using analytical equations. This allows obtaining explicit formulas for the flight speed, distance, endurance and the angle of inclination of trajectory at reentry in the atmosphere.

Focal parameter

$$\bar{p} = \left(\bar{r}_1\bar{V}_1\cos\theta_1\right)^2 = \left(\bar{r}\bar{V}\cos\theta\right)^2 = \text{const} \tag{5}$$

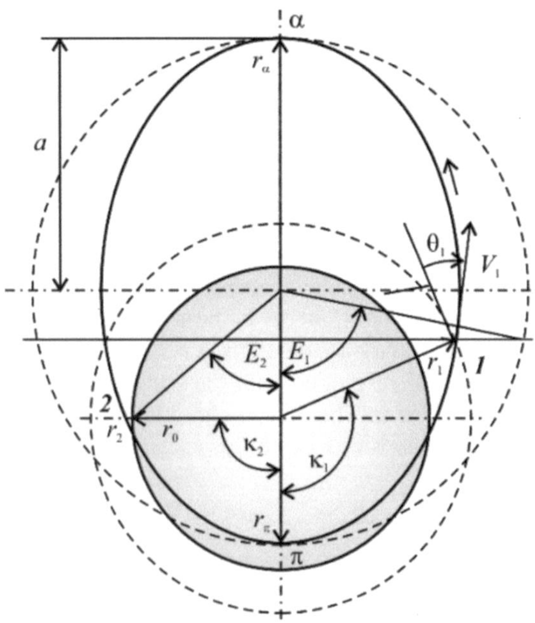

Fig. 6 The ASP movement in the Earth gravitation field: π is the perigee; α is the apogee

is proportional to the moment momentum squared and can be determined using the parameters at the end of active ascend in point *1* (Fig. 6): $\overline{r}_1$, $\overline{V}_1$ and θ_1, where θ_1 is an angle of inclination of the trajectory to the plane of the local horizon.

Orbit eccentricity

$$e = \frac{\overline{r}_\alpha - \overline{r}_\pi}{\overline{r}_\alpha + \overline{r}_\pi} = \sqrt{1 - \left(2 - \overline{r}\overline{V}_1^2\right)\overline{r}\overline{V}_1^2 \cos^2\theta_1} \tag{6}$$

characterizes the ellipsis oblongness (the difference of radius-vectors of apogee $\overline{r}_\alpha$ and perigee $\overline{r}_\pi$) which can be used to determine the focal parameter:

$$\overline{p} = 2\overline{r}_\pi\overline{r}_\alpha/(\overline{r}_\pi + \overline{r}_\alpha). \tag{7}$$

Flight distance from point *1* to point *2* in projection on the sphere of radius r_0:

$$L_{12} = r_0(2\pi - \kappa_1 - \kappa_2), \tag{8}$$

where

$$\kappa_1 = 1/2\pi - \arcsin\{(\overline{p}/\overline{r}_1 - 1)/e\}, \tag{9}$$

$$\kappa_2 = 1/2\pi - \arcsin\{(\overline{p}/\overline{r}_2 - 1)/e\}. \tag{10}$$

Flight endurance between these points:

$$t_{12} = t_{1\alpha} + t_{\alpha 2}, \tag{11}$$

where

$$t_{1\alpha} = \tau_0\overline{a}\sqrt{\overline{a}}\{\pi - (E_1 - e\sin E_1)\}, \tag{12}$$

$$t_{\alpha 2} = \tau_0\overline{a}\sqrt{\overline{a}}\{\pi - (E_2 - e\sin E_2)\}, \tag{13}$$

$$E_1 = 2\mathrm{arctg}\left\{\sqrt{\overline{r}_\pi/\overline{r}_\alpha}\mathrm{tg}(\kappa_1/2)\right\};\quad E_2 = 2\mathrm{arctg}\left\{\sqrt{\overline{r}_\pi/\overline{r}_\alpha}\mathrm{tg}(\kappa_2/2)\right\} \tag{14}$$

are the eccentric anomalies (Fig. 6); $\overline{a} = 1/2(\overline{r}_\pi + \overline{r}_\alpha)$ is the semi-major axis; $\tau_0 = \sqrt{r_0/g_0} = 806.03$ s is the time constant.

With $E_1 = E_2 = 0$ from (11), the endurance of one revolution on elliptical orbit is:

$$t_{\mathrm{rev}} = 2\pi\tau_0\overline{a}\sqrt{\overline{a}} = 2\pi\tau_0\overline{a}^{3/2}. \tag{15}$$

With $\overline{r}_\pi < 1$, the elliptical trajectory intersects the Earth surface which is typical for suborbital ASP or for fall trajectory of upper stages of LR boosters. The required condition for a multi-revolution orbital flight is $\overline{r}_{rev} \geq 1.05$ when $H_{rev} \geq 300$ km (ASP cannot remain for long on lower orbits because of deceleration in the upper atmosphere). The flight velocity on a circular orbit $\overline{V}_{cir}$ can be determined by setting the centripetal acceleration in circular motion of radius r equal to the gravitational acceleration:

$$V_{cir}^2/r = g_0\left(\overline{V}_{cir}^2/\overline{r}\right) = g_0/\overline{r}^2 \rightarrow \overline{V}_{cir} = 1/\sqrt{\overline{r}}. \quad (16)$$

Dimensional circular velocity V_{cir}: $V_{cir} = V_{cir0}\overline{V}_{cir} = \left(790447/\sqrt{\overline{r}}\right)$ m/s.

The period of revolution of ASP about Earth on a circular orbit of radius $\overline{r}$:

$$t_{rev} = 2\pi\tau_0\overline{a}^{3/2}. \quad (17)$$

For the geostationary orbit (GSO) with the angular speed of revolution around Earth equal to the angular speed of daily revolution $\omega_{Earth} = 72.921 \cdot 10^{-6} 1/s$:

$$\overline{r} = 1/(\omega_{Earth}\tau_0)^{2/3} = 6.61500;\ \overline{V}_{cir} = 0.38881;\ H = 35768\,\text{km};\ V_{cir} = 3073\,\text{m/s}$$

ASP are intended for low orbit missions, for instance, on the space station orbit. ($H_{SS} = 450$ km). Air spacecraft are brought to an intermediate elliptical orbit with the RSS or ASS where the upper stage booster is separated. Further flight is performed along the Homan ellipses with the activation of orbital manoeuvring engines (OME) in nearly horizontal segments (near points π and α, Fig. 7) which helps avoid fuel loss for counteraction to gravity. Passive flight is performed from the point of separation from the launcher to the first apogee ($H_\alpha \approx 180$ km), followed by the transit to the intermediate circular orbit ($H_1 \approx 250$ km). Two-impulse transit to the desired circular orbit ($H_2 = H_{SS}$) is performed after a hold for the appropriate time, followed by approach and docking with the SS. For an ASP to exit the orbit, deceleration is performed (the OME nozzles are directed opposite to motion) and the velocity drops by value $\Delta V_{\alpha 2}$ providing for desired perigee of entry $H_{\pi 2} = H_{\pi\ reen}$.

The corridor for entry in the atmosphere is defined by the following inequation: $H_{\pi min} \leq H_{\pi entry} \leq H_{\pi max}$, where the lower boundary $H_{\pi\ min}$ is defined by the temperature limit, the upper boundary $H_{\pi\ max}$ is defined by the condition of the ASP captured by the atmosphere. The velocities in apogee and perigee are as follows:

$$\overline{V}_\alpha = \sqrt{\frac{2\overline{r}_\pi}{\overline{r}_\alpha(\overline{r}_\pi + \overline{r}_\alpha)}}\ \overline{V}_\pi = \sqrt{\frac{2\overline{r}_\alpha}{\overline{r}_\pi(\overline{r}_\pi + \overline{r}_\alpha)}}. \quad (18)$$

The total velocity delta at transit from circular orbit of radius $\overline{r}_1 = \overline{r}_\pi$ to a higher orbit of radius $\overline{r}_2 = \overline{r}_\alpha > \overline{r}_1$ is as follows:

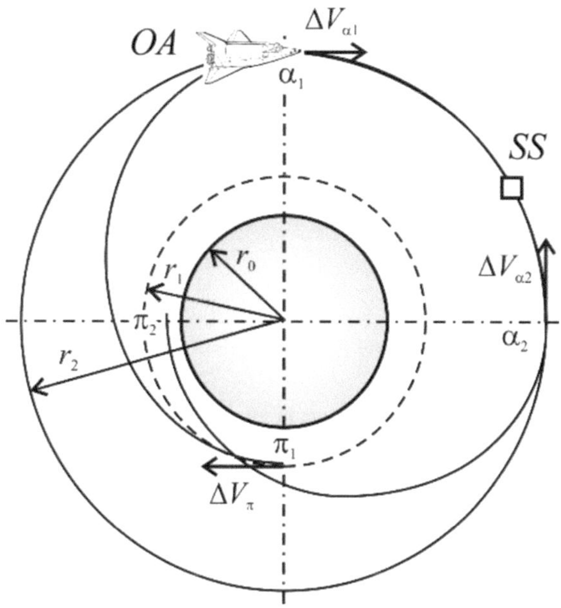

Fig. 7 Diagram of ASP transit from the holding orbit to the space station orbit and SS exiting orbit during return

$$\Delta\overline{V}_{12} = \Delta\overline{V}_{\pi} + \Delta\overline{V}_{\alpha} = \left(\Delta\overline{V}_{\pi} - 1/\sqrt{\overline{r}_{\pi}}\right) + \left(1/\sqrt{\overline{r}_{\alpha}} - \overline{V}_{\alpha}\right). \tag{19}$$

From the low of mechanical energy conservation and the moment momentum we can obtain the following equations:

$$-1/\overline{r} + \overline{V}^2/2 = \text{const}; \quad \overline{r}\overline{V}\cos\theta = \text{const} \tag{20}$$

For the desired perigee $H_{\pi\ \text{entry}}$, one determines the velocity $\overline{V}_{\text{entry}}$ and the angle of trajectory θ_{entry} when the ASP enters the atmosphere. For instance, on a conventional boundary $H_{\text{entry}} = 100$ km, the value of the deceleration impulse $\Delta V_{\alpha 2}$ is as follows:

$$\overline{V}_{\text{entry}} = \sqrt{\left(1/\overline{r}_{\overline{r}} - 1/\overline{r}^2_{\alpha 2}\right) + \overline{V}^2_{\alpha 2}/2}, \tag{21}$$

$$\theta_{\text{entry}} = -\pi/2 + \arcsin\left(\frac{\overline{r}_{\alpha 2}\overline{V}_{\alpha 2}}{\overline{r}_{\text{entry}}\overline{V}_{\text{entry}}}\right), \tag{22}$$

$$\Delta\overline{V}_{\alpha 2} = 1/\sqrt{\overline{r}_{\alpha}} - \overline{V}_{\alpha 2}, \tag{23}$$

where

$$\overline{V}_{\alpha 2} = \sqrt{\frac{2\overline{r}_{\pi\ \text{entry}}}{\overline{r}_{\alpha 2}\left(\overline{r}_{\pi\ \text{entry}} + \overline{r}_{\alpha 2}\right)}}; \ \overline{r}_{\pi\ \text{entry}} = \left(H_{\pi\ \text{entry}} + r_{\text{eq}}\right)/r_0; \ \overline{r}_{\text{entry}} = \left(H_{\text{entry}} + r_{\text{eq}}\right)/r_0. \tag{24}$$

Based on the I.V. Meschersky equation for the ideal velocity

$$V_{\mathrm{id}} = c\ln(\mathrm{m}_0/\mathrm{m}_\mathrm{f})$$

we can estimate the fuel consumed by the ASP in orbital manoeuvring as a function of the velocity delta $|\Delta\overline{V}|$:

$$\Delta\overline{\mathrm{m}}_\mathrm{F} = 1 - \exp\{-|\Delta\overline{V}|/\overline{\mathrm{c}}\}, \quad (25)$$

where $c = g_0J$ is the effective velocity of exhaust of the fuel combustion product; J is the specific OMD impulse with the unit of [s]; m_0, m_f are the initial and final ASP masses; $\overline{c} = c/V_{\mathrm{ciro}} = g_0J/\sqrt{g_0r_0} = J/\tau_0$; $\Delta\overline{m}_\mathrm{F} = m_\mathrm{F}/m_0$.

The total velocity delta in orbital flight $\Delta V_\Sigma = |\Delta V_j|$ is called the reference velocity; herewith, ΔV_Σ includes also components ΔV equivalent to the fuel consumed by the jet control system (JCS) to change the ASP angular position with respect to the centre of gravity. An example of the reference velocity estimations when the ASP flies to the space station (after the separation of the LR) and the reentry parameters are presented below.

Segment of ASP orbital flight	ΔV, m/s
Transit from the intermediate orbit ($H_\pi = 22/H_\alpha = 174$ km) to the circular waiting orbit (200/200 km)	61.4
Transit from the circular waiting orbit (200/200 km) to the specified circular SS orbit (450/450 km)	143.7
Approach and docking with the SS, control with the JCS (including reserve)	110.0
Deceleration impulse for exiting orbit ($H_{\pi\,\mathrm{entry}} = 47$ km)	117.0
Total reference velocity margin, ΔV_Σ	432.1
OME fuel capacity ratio ($J = 320$ s), $\Delta\overline{m}_\mathrm{F}$	0.129
Reentry parameters	
– flight height, km	100
– flight distance from the SS to the reentry point, km	15,455
– time of descent to the reentry point, min	35.2
– absolute reentry velocity, m/s	7926
– angle of trajectory inclination, deg	−1.18

When an ASP is used for payload transportation, the fuel consumption m_F rises with the increase in orbit height H_{rev} while the payload mass m_{pay} decreases (Fig. 8). The ASP launch mass m_0 (at separation of LR) depends on the energy of the ascend system for which the ASP is the payload. The ultimate height of the circular orbit H_{orb} is defined by the ASP reference velocity margin.

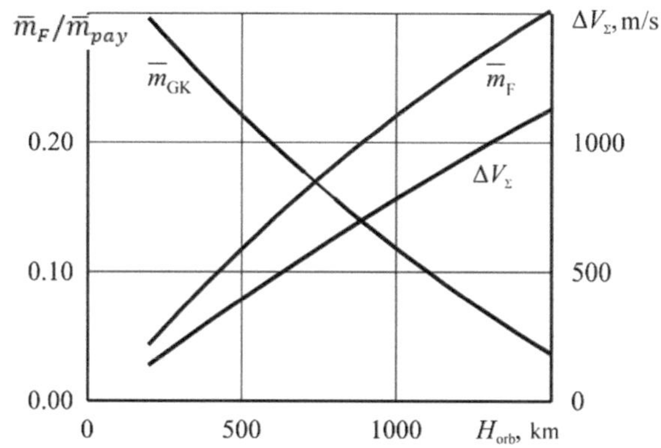

Fig. 8 Fuel mass ratio, the reference velocity spent in the ASP transits and the payload mass for various circular orbits: $\overline{m}_F = \frac{m_F}{m_0}, \overline{m}_{pay} = \frac{m_{pay}}{m_0}$, J = 320 s

Due to a large mass ratio of structures and gears for level landing (wing, landing gear) the ASP is ineffective for orbits with $H_{orb} > 1500$ km. Lighter SV of capsule type are more suitable for them. With sufficient relative fuel mass ratio $\Delta\overline{m}_F$, such SV can reach parabolic velocity V_{par} when the mechanical energy is equal to zero, and escape from the Earth gravity:

$$-1/\overline{r} + \overline{V}^2/2 = 0 \rightarrow \overline{V}_{par} = \sqrt{2/\overline{r}};$$
$$V_{par} = \overline{V}_{par}\sqrt{g_0 r_0} = V_{2c}/\sqrt{\overline{r}}, \tag{26}$$

where $V_{2c} = \sqrt{2g_0 r_0} = 11178.6$ m/s is the second cosmic velocity.

Entry in atmosphere shall meet the constraint for the maximum temperature $T_{max} \leq T_{alw}$ which corresponds to the lower boundary of the flight envelope (Fig. 9).

The thermal balance equation (see Sect. 2.3) defines the relative atmosphere density $\overline{\rho}(\mathrm{H}) = \rho(\mathrm{H})/\rho_0$ on the temperature boundary at the given velocity $\overline{V}$:

$$\overline{\rho}(H) \approx 0.1228\frac{r_n(T_{alw}/1000)^8}{10^6\overline{V}^{6.5}} \tag{27}$$

where $\rho_0 = 1.22505$ kg/m^3 is the density at sea level; T_{alw} is the maximum allowable temperature (in ° Kelvin); r_n is the radius of curvature in the ASP critical point (m); $\overline{V}=V/V_{cir0}$; V is the flight velocity relative to air. The minimum allowable flight height can be found from (27) at the given $\overline{V}$.

With a decrease in velocity $\overline{V}$, the lower temperature limit turns into the dynamic pressure limit with the non-dimentional density:

$$\overline{\rho}(H) = \overline{q}_{alw}/\overline{V}^2 \tag{28}$$

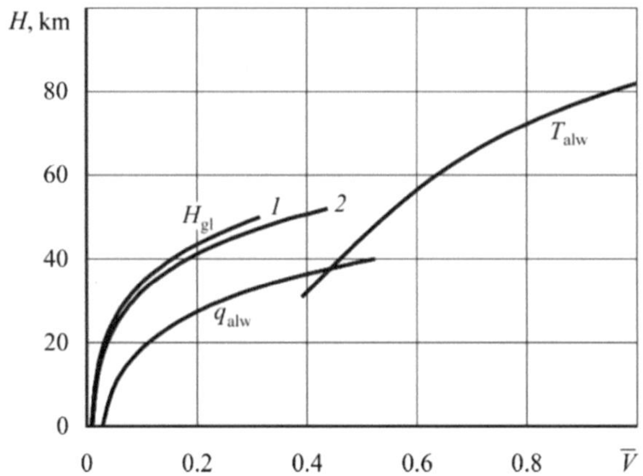

Fig. 9 ASP flight envelope: $T_{\text{alw}} = 1600\,^{\circ}\text{C}(r_{\text{n}} = 0.6\,\text{m})$; $q = 3500\,\text{daN/m}^2$; H_{cr}; H_{gl} is the upper boundary of equilibrium glide (*1*—$\sigma_L = 0.0034\ \text{m}^2/\text{daN}$; *2*—$\sigma_L = 0.0024\ \text{m}^2/\text{daN}$)

where $q_{\text{alw}} = q_{\text{alw}}/q_{\text{cir0}}$; $q_{\text{cir0}} = 1/2\rho_0 V_{\text{cir0}}^2 = 38.27084 \cdot 10^6\ \text{daN/m}^2$.

The upper boundary $H_{\text{gl}}(\overline{V})$ during descent in the atmosphere is determined as a quasistationary glide with zero roll angle:

$$\dot{\theta} = \frac{d\theta}{dt} \approx \frac{\left[\sigma_L q - \left(1 - \overline{V}^2\right) cos\theta\right] g_0}{V} = 0. \tag{29}$$

The minimum required atmosphere density on the upper boundary:

$$\overline{\rho}(H) = \left(1 - \overline{V}^2\right) cos\theta / \left(\overline{\sigma}_L \overline{V}^2\right) \approx \left(1 - \overline{V}^2\right) / \left(\overline{\sigma}_L \overline{V}^2\right), \tag{30}$$

where $\overline{\sigma}_L = \sigma_L q_{circ0}$ is the non-dimensional glide parameter.

An increase in σ_L (due to a decrease in the ASP mass or an increase in the wing area) expands the flight envelope by raising the "ceiling" H_{gl} (Fig. 9).

A generalised flight envelope for ASP and hypersonic aircraft (Fig. 10) includes the atmospheric and space segments. For an ASP of the Buran type, Space Shuttle or OA MAKS during the ascend in orbit and descent in the atmosphere, the dynamic pressure constraint is assumed at $q_{\text{alw}} = 3500\ \text{daN/m}^2$ (Fig. 9).

Hypersonic aircraft during acceleration are in more dense atmosphere because they need sufficient inflow of air mass through air intake to the ramjet engine. The structure of a HA and its PP shall allow the dynamic pressure up to $q_{\text{alw}} = 7500\ \text{daN/m}^2$ (Fig. 11). Along with the typical for ASP constraints on the external surface temperature, there exists the problem of high temperatures inside the ramjet engine path. Therefore, the limit flight speed for future HA will be defined by the applied materials and technical solutions for cooling down the most thermally stressed elements of the propulsion system (Fig. 11).

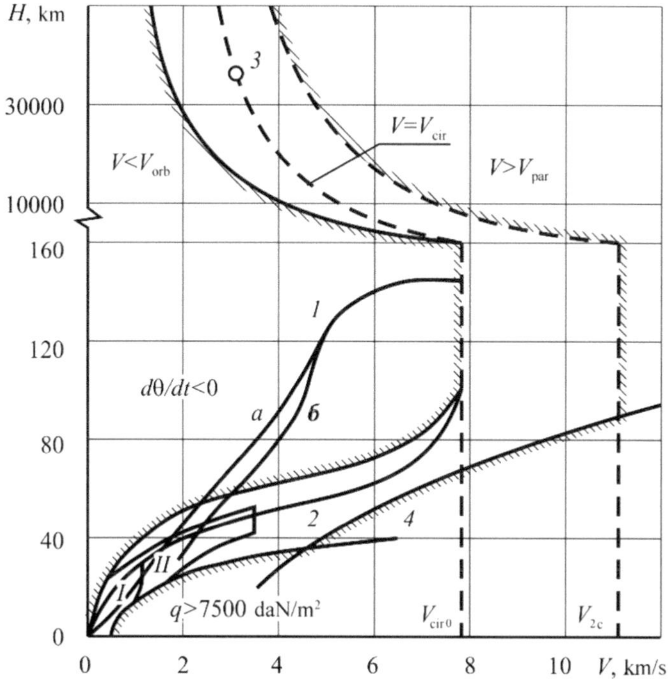

Fig. 10 The ASP flight envelope: *I*—a flight envelope for contemporary aircraft; *II*—a flight envelope for a HA; *1*—one of possible trajectories for an ASP to ascend in space *(a*—for the takeoff from the Earth, *b*—for the launch from a hypersonic LA); *2*—a possible trajectory for an ASP to return from space; *3*—geostationary orbit; *4*—the unallowable rise in the temperature of structures; V_{orb} is the minimum velocity at which an ASP can perform flight on a circular or elliptical orbit; V_{cir} is the circular velocity; V_{cir0} is the 1st cosmic velocity (circular at $H = 0$); V_{par} is the parabolic velocity; V_{2c} is the 2nd cosmic velocity (parabolic at $H = 0$)

Fig. 11 Flight envelope for hypersonic aircraft: *1*—the quasistationary glide "ceiling"; *2*—the trajectory of climb and acceleration to $M = 9$; temperature constraints: *3*—for the Rene-41 alloy; *4*—for the Inkonel-718 alloy; *5*—for titanium; 6—dynamic pressure q_{alw}= 7500 daN/m^2

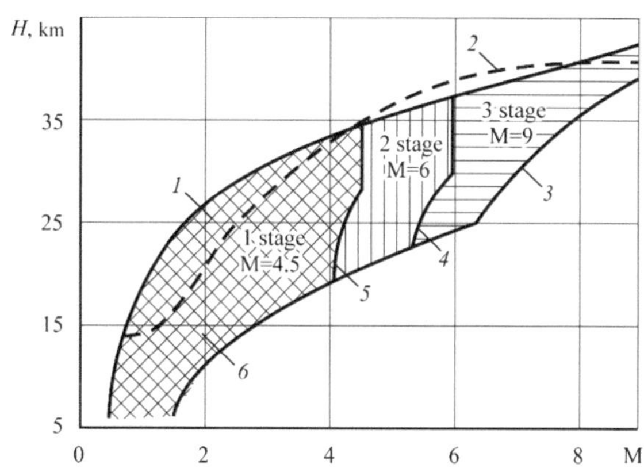

2.2 ASP and HA Flight Trajectories

The forces applied to an ASP are the gravity $mg \approx mg_0/\overline{r}^2$; the cruise thrust of engines $T = \overline{T}mg_0$, and the aerodynamic forces, namely drag $D = \sigma_D q g_0$, lift $L = \sigma_L q g_0$ (Fig. 12) and the lateral force $Y = \sigma_Y q g_0$.

The ballistic coefficient σ_D and the glide parameter σ_L depend on the angle of attack α, the parameter σ_Y depends on the sideslip angle β; these coefficients depend also on the Mach number and the current ASP mass:

$$\sigma_D = C_D(\alpha, M)/(mg_0/S);$$

$$\sigma_L = C_L(\alpha, M)/(mg_0/S);$$

$$\sigma_Y = C_Y(\beta, M)/(mg_0/S).$$

Rocket engine thrust depends on the atmospheric pressure p and decreases with the increase in altitude: $T = T_v - p(h)F_a$, where F_a is the nozzle exit area; T_v is the thrust in vacuum.

The vector dynamic (31) and kinematic (32) motion equations for the ASP centre of gravity in the atmosphere are usually considered in non-inertial coordinate systems (CS): the launch coordinate system is used for the ascend segment; the Earth spherical (Fig. 13) coordinate system is used for the descent segment; the landing coordinate system is used for manoeuvring before landing.

$$m\frac{d\vec{V}}{dt} = \vec{F}_\Sigma + m\vec{g} + 2m\left[\vec{V} \cdot \vec{\omega}_{\text{Earth}}\right] + m\left[\vec{\omega}_{\text{Earth}} \cdot (r \cdot \vec{\omega}_{\text{Earth}})\right] \tag{31}$$

$$\frac{d\vec{r}}{dt} = \vec{V} \tag{32}$$

where $\vec{F}_\Sigma$ is the sum vector of thrust and aerodynamic forces.

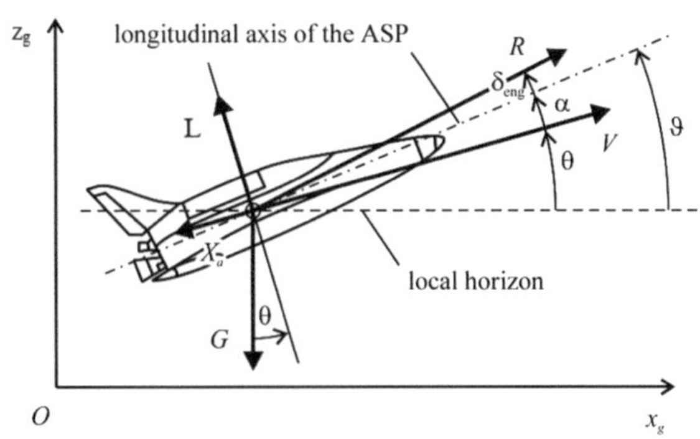

Fig. 12 The forces applied to an ASP in flight in vertical plane

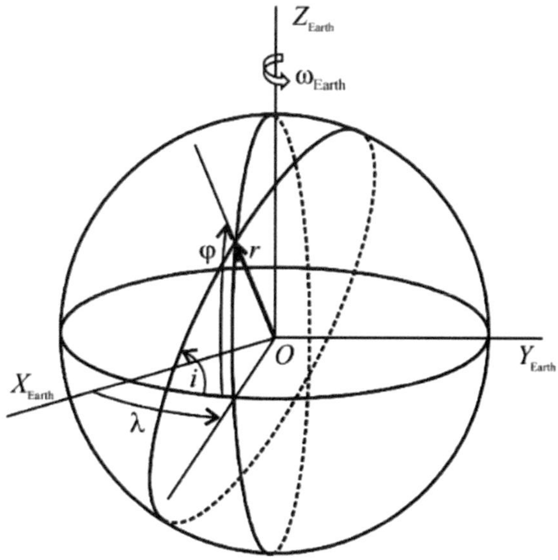

Fig. 13 The Earth spherical coordinate system: r is the distance to the Earth centre; λ, φ are the geocentric longitude and latitude; i is the orbit inclination

At the initial design stages, the ASP angular motion about its centre of gravity is usually not considered or is accounted for approximately; later, it is required to solve the vector dynamic Eq. (33) and the kinematic angular motion Eqs. (34) to (36):

$$\frac{d\vec{K}}{dt} = \vec{M}_{\Sigma} - \vec{\omega} \cdot \vec{K}, \tag{33}$$

$$\dot{\psi} = \left(\omega'_y \cos\gamma - \omega'_z \sin\gamma\right) / \cos\vartheta, \tag{34}$$

$$\dot{\vartheta} = \omega'_y \sin\gamma + \omega'_z \cos\gamma, \tag{35}$$

$$\dot{\gamma} = \omega'_x - \dot{\psi} \sin\vartheta. \tag{36}$$

Here, $\vec{\omega}\prime$ and $\vec{\omega}$ are the angular velocity vectors of the ASP relative to the moving Earth CS $O_gX_gY_gZ_g$ (connected with current meridian and current parallel) and the fixed (inertial) CS; these vectors are considered in projections on the axes of the body CS *OXYZ*:

$$\vec{\omega}' = \vec{\omega} - \vec{\omega}_{\text{Earth}} - \vec{\omega}_{\text{g}}; \tag{37}$$

$\vec{\omega}_{\text{Earth}}$ is the angular velocity vector of the daily Earth rotation;

$\vec{\omega}_{\text{g}} = \vec{\dot{\varphi}}_{\text{c}} + \vec{\dot{\lambda}}_{\text{c}}$ is the angular velocity vector of the moving CS $O_gX_gY_gZ_g$ relative to the fixed one.

The vector of the principal moment of external forces $\vec{M}_{\Sigma}$ includes the components due to the aerodynamic forces, cruise engine thrust, and the control engines of the JCS.

Yaw ψ, pitch ϑ and roll γ angles define the orientation of the ASP body relative to the Earth CS $O_g X_g Y_g Z_g$. The projections of the moment momentum vector $\vec{K}$ on the axes of the body CS $OXYZ$ are as follows:

$$\begin{aligned} K_x &= I_x \omega_x - I_{xy}\omega_y - I_{xz}\omega_z, \\ K_y &= -I_{xy}\omega_x + I_y\omega_y - I_{yz}\omega_z, \\ K_z &= -I_{xz}\omega_x - I_{yz}\omega_y + I_z\omega_z, \end{aligned} \tag{38}$$

where I_x, I_y, I_z are the axial and I_{xy}, I_{xz}, I_{yz} are the centrifugal moments of inertia of the ASP.

The equations of motion in active segment of flight with operating cruise engines (31) to (36) are added to the equation of the mass flow rate:

$$dm/dt = -\mu. \tag{39}$$

For LRE, the fuel consumption per second is

$$\mu = R_v/(g_0 J_v), \tag{40}$$

where R_v, J_v are the thrust and specific impulse in vacuum with the units of [N] and [s].

The method of balance of reference velocity is broadly used in the analysis of ascend:

$$V_{id} = (V_f - V_0) + \Delta V_{v\Sigma}, \tag{41}$$

where V_f, V_0 are the final and initial flight speeds of the SSRS or ASS in active segment of trajectory; $V_{id} = g_0 J \ln\{m_0/m_f\}$ is an ideal speed.

The following components add up to the total loss of the reference velocity $\Delta V_{v\Sigma}$: the counteraction to gravity (ΔV_{grav}), the aerodynamic deceleration (ΔV_{aer}), the trajectory curving during control (ΔV_{ctl}), as well as the compensation for the reduction in thrust due to the atmospheric pressure (ΔV_{atm}):

$$\Delta V_{grav} = g_0 \int_0^{t_f} \overline{g} \sin\theta dt;\ \Delta V_{aer} = g_0 \int_0^{t_f} \sigma_x dt;\ \Delta V_{ctl} = g_0 \int_0^{t_f} \overline{R}_v [1 - \cos(\alpha + \delta_{en})] dt, \tag{42}$$

$$\Delta V_{atm} = g_0 \int_0^{t_f} \overline{R}_v p(H) F_a \cos(\alpha + \delta_{en}) dt;\ \overline{R}_v = R_v/(m g_0). \tag{43}$$

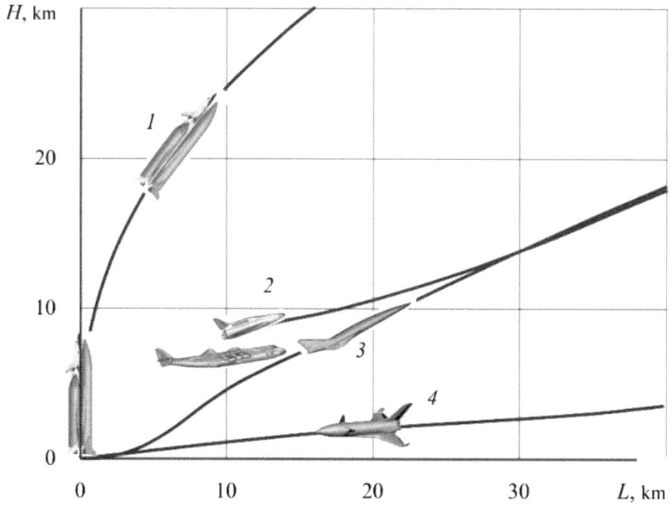

Fig. 14 Flight trajectory of ASP near takeoff point: *1*—SSRS vertical takeoff; *2*—horizontal takeoff of an ASP from a subsonic LA; *3*—the horizontal takeoff of an ASP with CSPS; *4*—return of the SSRS 1st stage winged booster using TJE

Movement of the centre of gravity and the ballistic properties of an ASP in a non-inertial coordinate system (Fig. 14) depend on the Coriolis force $2m\left[\vec{V} \cdot \vec{\omega}_{\mathrm{Earth}}\right]$ and the centrifugal force $m\left[\vec{\omega}_{\mathrm{Earth}} \cdot (\mathrm{r} \cdot \vec{\omega}_{\mathrm{Earth}})\right]$. This influence and loss of reference velocity for ascend depend on the orbit inclination i.

The ascend trajectory of a SSRS includes a vertical segment ($\theta_0 = 90°$) 15 to 35 s long; when the desired velocity is reached, the trajectory angle is reduced to $\theta_0 = 90° - \Delta\theta_0$; after that, a gravitational turn is performed with the reduction of trajectory turn at $n_z = 0$. After the separation of the 1st stage booster, the tangent of the ASP pitch angle changes linearly:

$$\mathrm{tg}\vartheta = at + b. \tag{44}$$

The constants a and b depend on the parameters of the approach to the desired point in an elliptical orbit:

$$\Delta r(a, b) = r(t_{\mathrm{f}}) - r_{1\ \mathrm{ellipsis}} = 0, \tag{45}$$

$$\Delta\theta(a, b) = \theta(t_{\mathrm{f}}) - \theta_{1\ \mathrm{ellipsis}} = 0. \tag{46}$$

The cruise engines cutoff time ($t = t_{\mathrm{f}}$) occurs when the desired velocity is reached $V_{\mathrm{abs}}(t_{\mathrm{f}}) = V_{1\ \mathrm{ellipsis}}$.

The optimised parameters of the early gravitational turn (the flight velocity V_0 and the angle of trajectory inclination θ_0) ensure the minimum fuel consumption m_{F}

Fig. 15 Example of dependency of gravitational loss on the takeoff g-load for a vertical takeoff SSRS

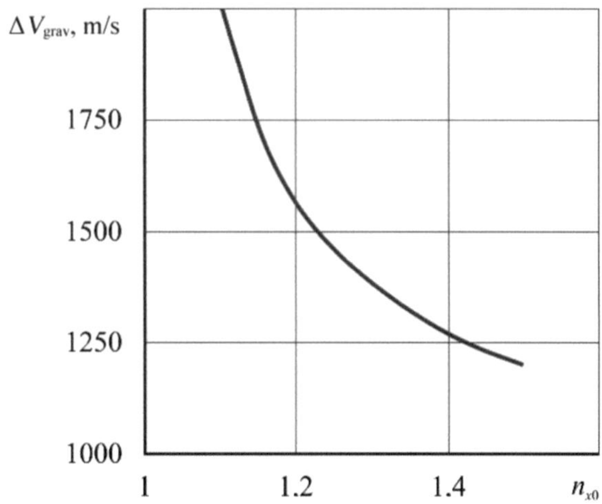

during ascend in the intermediate orbit which corresponds to the minimum V_{id} and the minimum losses $\Delta V_{v\Sigma}$. The main component is the gravitational loss. It largely depends on the takeoff g-load (Fig. 15).

The total loss of the reference velocity of a vertical takeoff system: $\Delta V_{v\Sigma} = 1650$ to $2150\,\mathrm{m/s}$; the ideal velocity in ascend to an intermediate elliptical orbit: $V_{id} = 9000$ to $9500\,\mathrm{m/s}$.

The ascend trajectory of a two-stage SSRS with reentry rocket units (RRU) is formed taking imnto account the RRU return trajectory and the fuel consumed by the air-breathing propulsion system (ABPS) (ABPS is required for the 1st stage RRU to perform cruising flight to the airfield (Fig. 16).

And the RRU reentry (the winged reusable booster of the 1st stage).

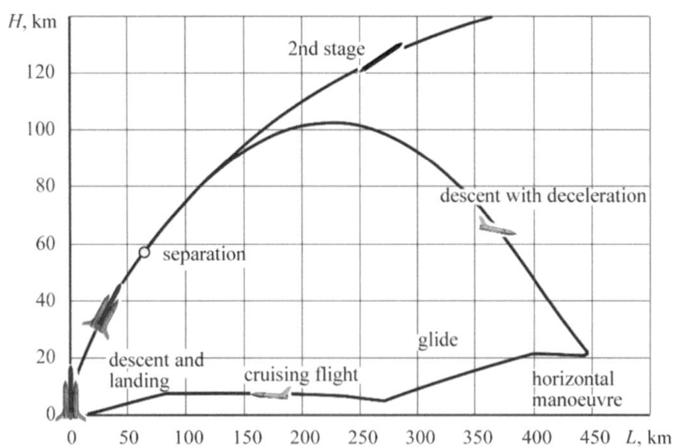

Fig. 16 Longitudinal ascend profile of a vertically starting SSRS

The use of reusable parts calls for the reduction of the angle of inclination of the trajectory in separation point.

The airborne takeoff of an ASP as a part of an air-space system (Fig. 14 (2)), significantly differs from the vertical takeoff when it is a component of a SSRS. Before the separation of the 2nd stage, the LA performs a flat glide with a growing speed and nose-up while increasing the trajectory angle up to $\theta_0 = 25$ to $38°$; the higher the subsonic lift-to-drag ratio of the ASP, the lower is the required angle θ_0.

With the airborne takeoff, as compared to the ground vertical takeoff, the gravitational losses are significantly (almost two times) lower and the total gains in the ideal velocity exceed by much the LA flight speed V_0:

Takeoff type	Vertical	Airborne from LA	
Space stage type	 ASP	 ASP + tank	 ASP
$(L/D)_{max}$	–	3	5
V_0, m/s	0	180	190
θ_0, deg	90	38	25
ΔV_{grav}, m/s	1224	700	646
ΔV_{aer}, m/s	93	312	271
ΔV_{con}, m/s	286	114	87
ΔV_{atm}, m/s	62	42	45
$\Delta V_{\text{v}\Sigma}$, m/s	1665	1168	1049
$\delta(\Delta V_{\text{v}\Sigma})$, m/s	–	–497	616
V_{id}, m/s	9070	8393	8264
$\delta(\Delta V_{\text{id}})$, m/s	–	−677	-806

Air spacecraft with CSPS (HA for space missions) feature a flat ascend trajectory at the segment of RJE operation.

The analysis of trajectories of ASP descent from orbit taking into account spatial manoeuvring can be made by integration of the following system of differential Eq. (44):

$$\frac{dV}{dt} = -\sigma_x q g_0 - g_r \sin\theta + \Delta\dot{V}_{\omega+\text{fns}};$$
$$\frac{d\theta}{dt} = \sigma_y q \cos\gamma \frac{g_0}{V} - \frac{g_r}{V}\left[1 - V^2/(g_r r)\right] + \Delta\dot{\theta}_{\omega+\text{fns}};$$
$$\frac{d\eta}{dt} = -\sigma_y q \sin\gamma \frac{g_0}{V\cos\theta} - \frac{V}{r}\cos\theta\,\text{tg}\,\varphi\cos\eta + \Delta\dot{\eta}_{\omega+\text{fns}};$$
$$\frac{dr}{dt} = V\sin\theta;$$

$$\frac{d\varphi}{dt} = \frac{V}{r}\cos\theta\sin\eta;\ \frac{d\lambda}{dt} = \frac{V}{r\cos\varphi}\cos\theta\cos\eta, \tag{47}$$

where $\sigma_y = \sigma_L$ and $\sigma_x = \sigma_D$; η is the heading angle measured counterclockwise from geographical parallel; θ is the angle of inclination of the Earth speed vector $\vec{V}$ to the local horizon plane; γ is the roll angle. The influence of the daily Earth rotation and the meridian component of the gravity acceleration is accounted for with the following corrections to the first three Eqs. (44):

$$\Delta\dot{V}_{\omega+\text{fns}} = r\omega^2_{\text{Earth}}\cos\varphi(\cos\varphi\sin\theta - \sin\varphi\sin\eta\cos\theta) + g_\varphi\cos\theta\sin\ \eta;$$

$$\Delta\dot{\theta}_{\omega+\text{fns}} = 2\omega_{\text{Earth}}\ \cos\varphi\cos\eta + \left[\text{r}\omega^2_{\text{Earth}}\ \cos\varphi(\sin\varphi\sin\theta\sin\eta + \cos\varphi\cos\theta) - \text{g}_\varphi\sin\theta\sin\eta\right]/\text{V};$$

$$\Delta\dot{\eta}_{\omega+\text{fns}} = 2\omega_{\text{Earth}}(\cos\varphi\text{tg}\theta\sin\eta - \sin\varphi) + \left[-\text{r}\omega^2_{\text{Earth}}\ \sin\varphi\cos\varphi\cos\eta + \text{g}_\varphi\cos\eta\right]/(\text{V}\cos\theta);$$

$$g_r = g\cos\varepsilon;\ \text{g}_\varphi = -\text{g}\sin\varepsilon;\ \text{g} = \text{a}_2\left(\text{g}_0/\bar{\text{r}}^2\right)\left[1 + \left(\text{a}_1/\bar{\text{r}}^2\right)\left(1 - 3\sin^2\varphi\right)\right];$$

$$a_1 = 0.001623;\ a_2 = 1.001357;\ \varepsilon = \arcsin\left\{\left(\text{a}_1/\bar{\text{r}}^2\right)\right\}\sin^2\varphi.$$

Control of longitudinal motion in the upper descent trajectory is achieved through the variation of vertical projection of normal g-load

$$n_{z,vert} = n_z cos\gamma = \sigma_L q cos\gamma. \tag{48}$$

With a preset program of the angle of attack variation α = α(M), the change in cosine of the roll angle cos γ enables the implementation of "iso-temperature entry" (Fig. 17).

The ASP temperature grows with a decrease in the glide parameter σ_L (Fig. 18), because the required density of the atmosphere grows (30) and the height of equilibrium glide drops.

Lateral control is achieved through changing the lateral projection of g-load n_z:

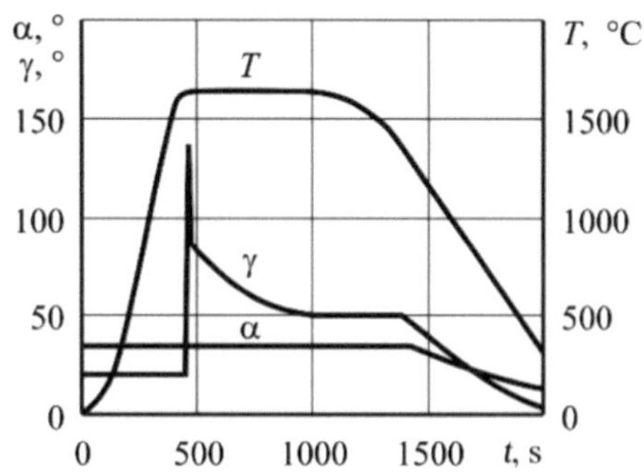

Fig. 17 Angle of attack, roll angle and temperature variation in the ASP critical point

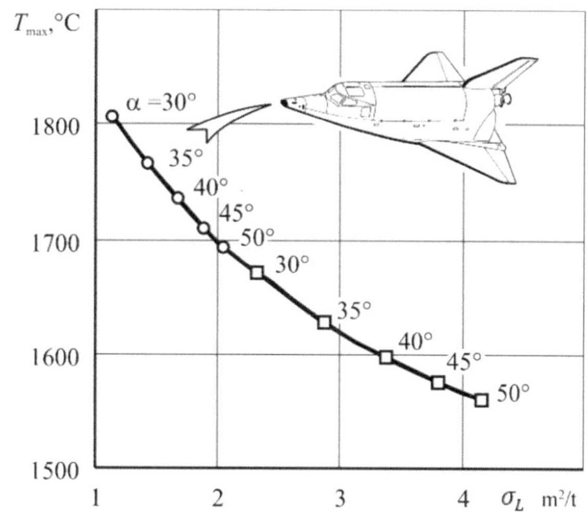

Fig. 18 Variation of temperature in the critical point as a function of ASP glide parameter: $r_n = 0.6$ m; ○—$G/S = 530$ daN/m^2; □—$G/S = 260$ daN/m^2

$$n_{z,lat} = n_z sin\gamma = \sigma_L q sin\gamma. \tag{49}$$

The size of the Earth surface available at descent manoeuvring in longitudinal and lateral directions (Fig. 19) depends on the hypersonic lift-to-drag ratio L/D. The following equations can be used for the estimation of longitudinal and maximum lateral range of descent:

$$L = 1/2r_0 L/D cos\gamma \, ln\left(\frac{1 - \overline{V}_f^2}{1 - \overline{V}_0^2}\right), \tag{50}$$

$$Y_{max} \approx r_0\left(\pi^2/48\right)(L/D)^2, \tag{51}$$

where $\overline{V}_0$, $\overline{V}_f$ are the initial and final non-dimensional flight speed at descent.

The function of control system is to steer the ASP to the landing airfield. Upon descent to an altitude of 30 km, the ASP begins manoeuvring in the horizontal plane and follows the given trajectory. It aims at one of the heading alignment cylinders (HAC) and gets in the plane of symmetry of the airfield runway. Landing is ensured for any directions of approach to the HAC (Fig. 20).

The key feature of ASP unpowered horizontal landing is the application of airbrake (AB) for control over the flight speed. The inclination of the landing glide slope for

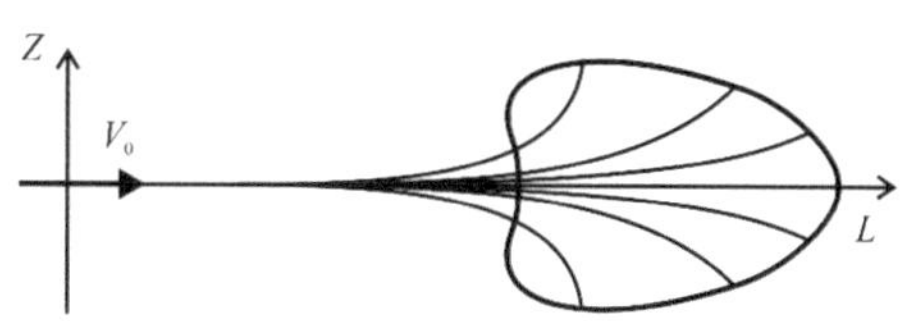

Fig. 19 The shape of the Earth surface available when an ASP descends in the atmosphere

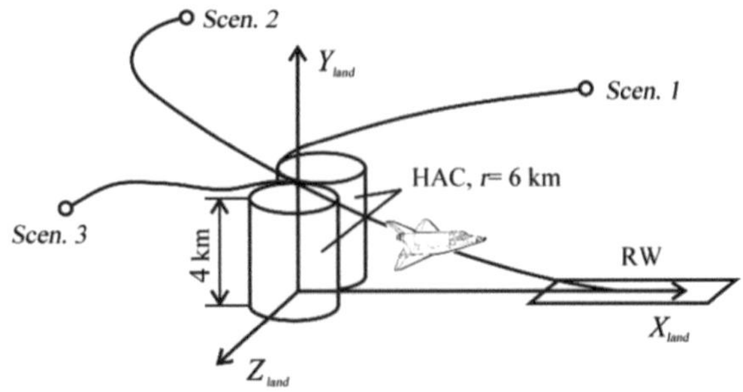

Fig. 20 Pre-landing manoeuvering and approach

ASP (θ_{glide}) is higher than for conventional aircraft. Under the nominal conditions, the airbrake is deflected to a moderate angle, herewith $\theta_{glide} = 15 \text{ to } 20^{\circ}$. At the disturbances leading to undershooting (head wind), the airbrake deflection angle is reduced which increases the lift-to-drag ratio and increases the glide distance available. At the "overshooting" disturbances, the AB is deflected to a larger angle.

The minimum landing speed for ASP under the nominal conditions can be estimated as follows (km/h):

$$V_{land,min} = 3.6\sqrt{2/(\rho_0 \sigma_L)}, \quad (52)$$

where ρ_0 is the density near the runway surface; σ_L is the trimmed glide parameter at the landing angle of attack, with the design landing mass.

The landing speed limitation for the orbit spaceship Buran: under the nominal conditions, $V_{land} \leq 320\,\text{km/h}$; in turbulence, $V_{land} \leq 360\,\text{km/h}$.

2.3 Thermal and Acoustic Loads

The flight of ASP and hypersonic aircraft is associated with a high aerodynamic heating due to the transformation of the kinetic energy of dynamic pressure to the thermal energy when it stagnates on the aircraft surface. A remarkable role is played also by the dynamic loads caused by pressure fluctuations (aerodynamic noise) in the flow separation and shockwave zones. For the typical flight conditions of ASP, the effect of thermal and dynamic loads is not simultaneous. ASP experience the highest heating at descent from orbit at an altitude of 90 to 40 km whereas the dynamic loads are high at the heights below 30 km.

When there is no heat sink from the surface heated (the wall is heat-insulated), the temperature near the wall may rise to a value close to the full flow stagnation temperature T_0:

$$T_0 = T_H\left(1 + 0.17\text{M}^2\right) \quad (53)$$

where $T_H = T_{atm}(H)$ is the air temperature at height H. For approximate estimations, we can assume $T_H = 230K$.

At moderate flight speeds ($M \leq 3$), the temperature of thermally insulated heated surface may reach the values close to T_0; but with further increase in flight speed, a major portion of convective heat flow will sink outside due to the thermal radiation of the heated surface, and the wall temperature T_w will be significantly lower than T_0.

For the estimation of the convective heat flow q_{w0} to the fore critical point of a spherically rounded body (Fig. 21), at the flight speeds from 1000 to 9000 m/s, the following equation can be used:

$$q_{w0} = \frac{2.56 \cdot 10^{-5}}{T_0}\sqrt{\frac{\rho(H)}{r_H}}V^{3,25}(T_0 - T_w),\ \ \mathrm{W/m^2}, \tag{54}$$

where $\rho(H)$ is the air density at altitude H, kg/m^3; r_H is the nose radius, m; V is the aircraft speed, m/s.

The density $\rho(H)$ can be found in the standard atmosphere tables or calculated using the following equation:

$$\rho(H) = 1.225\exp(-0.1386H),\ \ \mathrm{kg/m^3}, \tag{55}$$

where H is the flight altitude, m.

Around the critical point, with the flow assumed to be laminar, the heat flow for a specified angle θ, (Fig. 21) can be calculated from the ratio:

$$\frac{q_w}{q_{w0}} = 0.550.45\cos(2\theta),\ \left(0 \leq \theta \leq \frac{\pi}{2}\right). \tag{56}$$

The heat flow density distribution over a spherical nose under the laminar flow condition is shown in Fig. 21.

For the estimation of the heat flow on the wing leading edge (Fig. 22), rounded with radius r, in a crossflow, the following ratio can be used:

$$q_{w\ \mathrm{cyl}}/q_{w\ \mathrm{sph}} = 0.71. \tag{57}$$

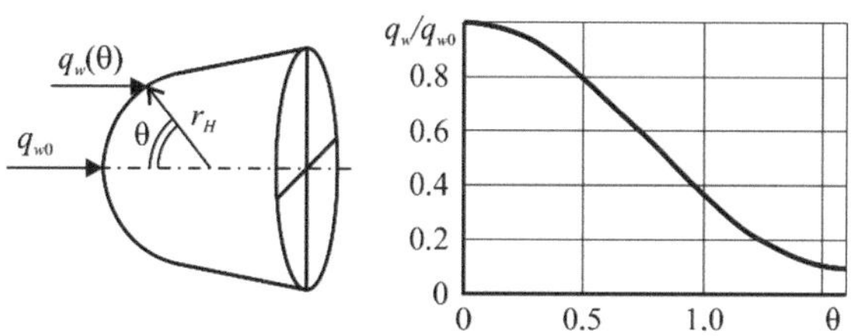

Fig. 21 Distribution of the heat flow density over the spherical nose, the flow is laminar

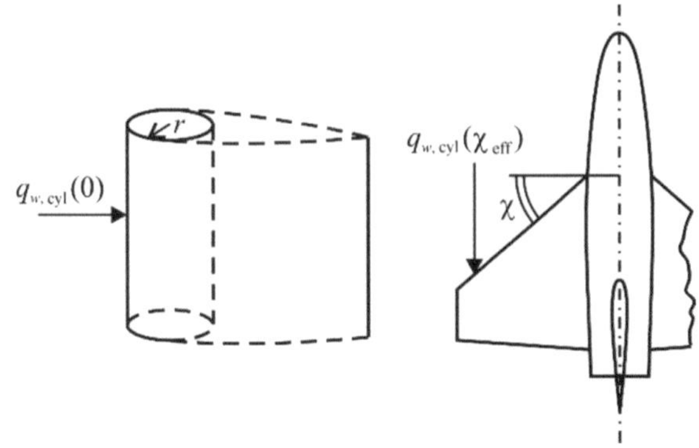

Fig. 22 Heat flow to the wing leading edge

For a swept leading-edge under angle of attack α, the heat flow is estimated as follows:

$$\begin{aligned} q_{w\,\mathrm{cyl}}(\chi_{\mathrm{eff}}) &= q_{wcyl}(0)\cos^{1.25}(\chi_{\mathrm{eff}}),\\ \sin(\chi_{\mathrm{eff}}) &= \cos\alpha\sin\chi^{+}\sin\alpha\cos\chi,\\ \cos(\chi_{\mathrm{eff}}) &= \sqrt{1-\sin^2(\chi_{\mathrm{eff}})}. \end{aligned} \tag{58}$$

Equations (50–55) are often used at early design stages for the estimation of the heat flow to the ASP most heated surfaces. Further, the convective heat flows are analysed by more complex methods, including the use of experimental data obtained during ground and flight tests of scaled ASP models.

Surface temperature T_w can be determined from the condition of heat balance on the heated ASP surface:

$$q_{\mathrm{conv}} - q_{\mathrm{rad}} = -\lambda\partial T_w/\partial x, \tag{59}$$

where q_{conv} is the convective heat flow from the boundary layer, W/m^2; q_{rad} is the heat flow radiated by the heated surface into the atmosphere; λ is the thermal conductivity of the heat shield (HS), W/(m·K); $\partial T_w/\partial x$ is the temperature gradient inside the heat shield near the heated surface; x is the linear coordinate normal to the heated surface (Fig. 23), m.

For the estimation of the surface temperature T_w under the unsteady heating conditions, one needs to solve the thermal conductivity equation:

$$c\rho\frac{\partial T}{\partial t} = \frac{\partial}{\partial x}\left(\lambda\frac{\partial T}{\partial x}\right), \tag{60}$$

for which one of the boundary conditions is (56). Here, c is the specific thermal capacity of the skin material, J/kg; ρ is the skin material density, kg/m^3; t is time, s.

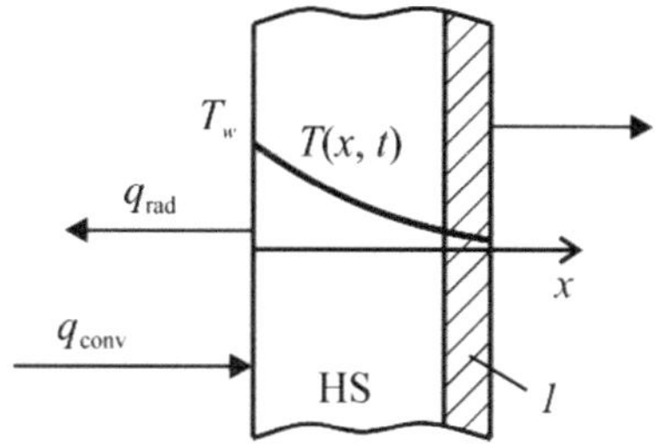

Fig. 23 Diagram of aerodynamic heating of an ASP surface

When there is no heat shield and the metallic skin of the ASP is thermally insulated on the inner side ($q_{inn} = 0$), using (56) and (57), one can obtain a simpler equation for numeric estimation of the ASP surface temperature:

$$q_{conv} - q_{rad} = c\rho\delta\partial T_w/\partial t, \tag{61}$$

where δ is the skin thickness, m.

It follows from (56) that the temperature will be the highest when $\lambda(\partial T_w/\partial t) = 0$. In this case, the temperature takes a so called radiation-equilibrium value T_{aw} which is set for a thermally insulated (adiabatic) surface in absence of heat sink into the HS. Herewith, $q_{conv} = q_{rad}$. In real applications, it is true with sufficient accuracy for most heat shielding materials with a low thermal conductivity coefficient ($\lambda < 1$ W/(m · K)).

The specific heat flow radiated by aircraft surfaces is defined by the Stefan-Boltzmann's law:

$$q_{rad} = \varepsilon\sigma T_w^4, \tag{62}$$

where ε is the surface radiation efficiency or emissivity factor, ($0 < \varepsilon < 1$); $\sigma = 5.67 \cdot 10^{-8}\mathrm{W}/\left(\mathrm{m}^2\mathrm{K}^4\right)$ is the Stefan-Boltzmann constant.

In compliance with the Newton's law, the specific convective heat flow coming to the surface from the boundary layer is defined as follows:

$$q_{conv} = \alpha_{conv}(T_0 - T_w), \tag{63}$$

where α_{conv} is the local coefficient of convective heat exchange on the air-surface boundary, W/(m^2·K). The expression for estimation of α_{conv} on the ASP rounded nose can be found by comparing (51) and (60).

Considering the equality $q_{conv} = q_{rad}$, we obtain the equation that helps determine the radiation-equilibrium temperature of adiabatically insulated surfaces of aircraft:

$$\varepsilon\sigma T_w^4 + \alpha T_0 - \alpha T_w = 0. \tag{64}$$

Figure 24 shows approximate values of the radiation-equilibrium temperature for an ASP and a hypersonic aircraft.

Fig. 24 Equilibrium temperature of a flat surface: ε = 0.8 at distance from the leading edge; for small angles of attack

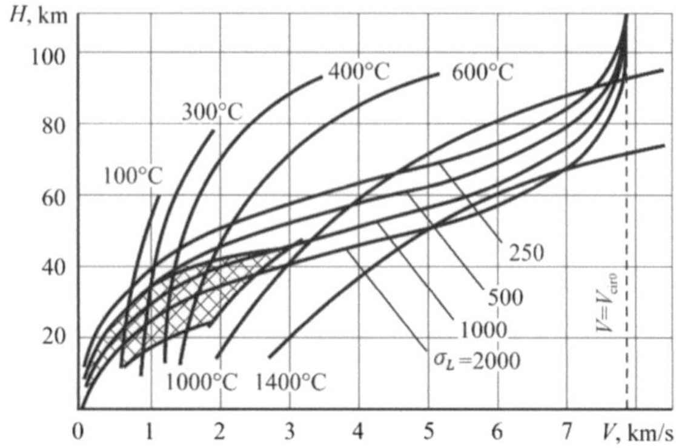

Intensity of aerodynamic heating of ASP surfaces significantly decreases with the increase in the distance from the wing leading edge (from the fuselage nose), wing (empennage) sweep angle, radius of the fuselage or wing nose. Analyses and tests showed that the surface temperature of the Russian ASP Buran during reentry in the atmosphere could reach 1650 °C on the fuselage nose and 1560 °C on the wing leading-edge. The temperature of other surfaces was significantly lower, for instance, the top surface of the fuselage experienced only 200 °C (Fig. 25). The curves show that the pressure on the major part of the ASP Buran, when exposed to the maximum heat flows, does not exceed 102 to 104 Pa (0.001 to 0.1 atm); consequently, the dynamic loads at this flight stage are insignificant.

The temperatures on the surface of hypersonic aircraft may be higher than those on the surface of the Buran type ASP.

Figure 26 shows the American and Russian experimental hypersonic aircraft (HA) under development. Such HA feature small radii of the nose fairing, wing leading edge and air intakes. Therefore, the maximum radiation-equilibrium temperatures in such surface areas may exceed 2000°, which is at least by 300 °C higher than the operating temperature limits of existing structural materials.

One of the feasible methods of thermal protection for such structures may be the active cooling, for example, using the liquid fuel components available onboard. But this concept is not yet implemented in existing HA because of great technical difficulties.

Analysis and wind-tunnel tests of the ASP Buran models (Fig. 27) revealed that the highest acoustic loads at the takeoff phase and the initial ascend will take place on the afterbody (166 dB), on the tail (165 dB), on the elevons (163 dB), on the wing top and bottom (165 dB), on the cargo compartment doors (162 dB), on the fuselage nose (163 dB). High level noise on the ASP surface reduces fatigue life of the heat shield plates which is in inverse proportion to the time of noise effect.

Acoustic loads appear again at the finishing phase of descent and they may exceed 160 dB locally.

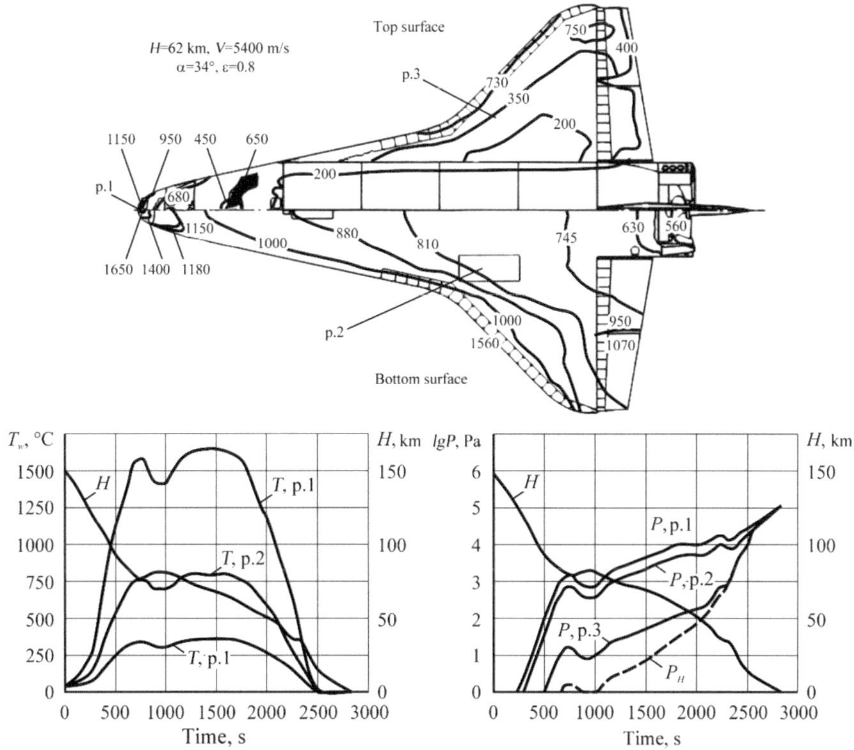

Fig. 25 The ASP Buran thermal loads during descent from orbit (temperature in degrees Celsius): 1—the distribution of the maximum values of in characteristic points of the airframe surface: p. 1—the critical point of the nose fairing; p. 2—the bottom fuselage surface; p. 3—the top wing surface; 2—the variation of flight altitude and temperature in characteristic points with time; 3 —the variation of flight altitude and pressure on the OS surface in characteristic points with time; is the static pressure of the atmosphere

Fig. 26 Experimental hypersonic aircraft

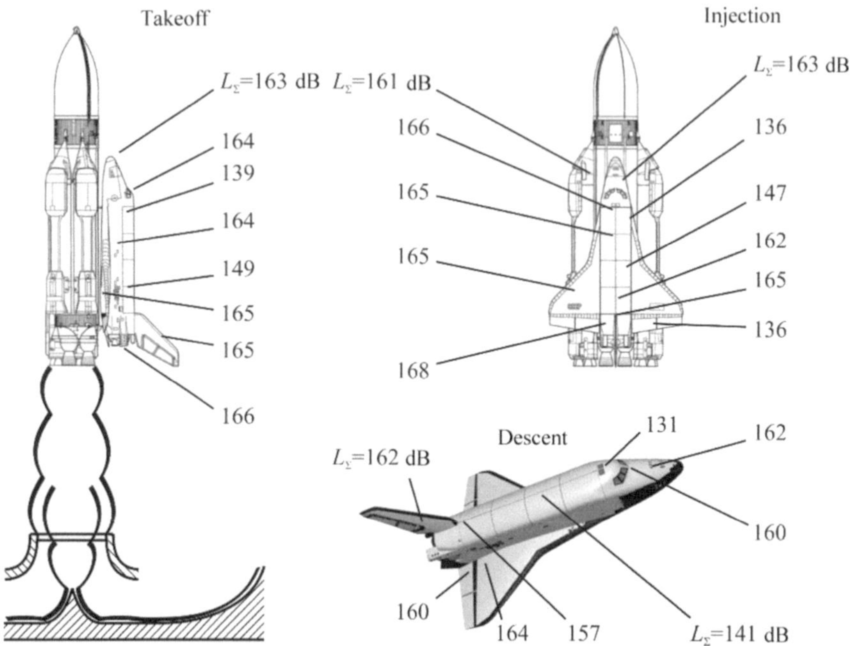

Fig. 27 Total levels of acoustic loads for the Buran ASP at various flight stages

For minimisation of aerodynamic noise, an ideal aircraft should have a shape ensuring the attached boundary layer over the major part of the vehicle surface.

3 Determining the ASP Main Parameters

3.1 ASP Requirements. SSRS and ASS Conceptual Design

The main requirements to an aerospace plane include: the mass and overall dimensions of the payload (injected in orbit and returned to the Earth); the number of crew members; the reference velocity margin ΔV_{orb}; the lateral manoeuvre relative to the landing revolution path Z_{max}; the landing speed V_{land} and the landing ground roll distance L_{roll} [1] (Fig. 28).

The dimensions of an ASP and its aerodynamic configuration, the thrust of engines and the fuel mass for orbital manoeuvring are determined according to these requirements. During the conceptual design phase, the ASP launch mass m_{0asp} is estimated, which is the payload for the ascend system—SSRS (Fig. 14 (*1*)) or ASS (Fig. 14 (*2*)). In cargo version (Fig. 16), the payload mass for the LR shuttle is assumed to be the mass of the space module $m_{0\,sm}$ injected in an intermediate elliptical orbit which

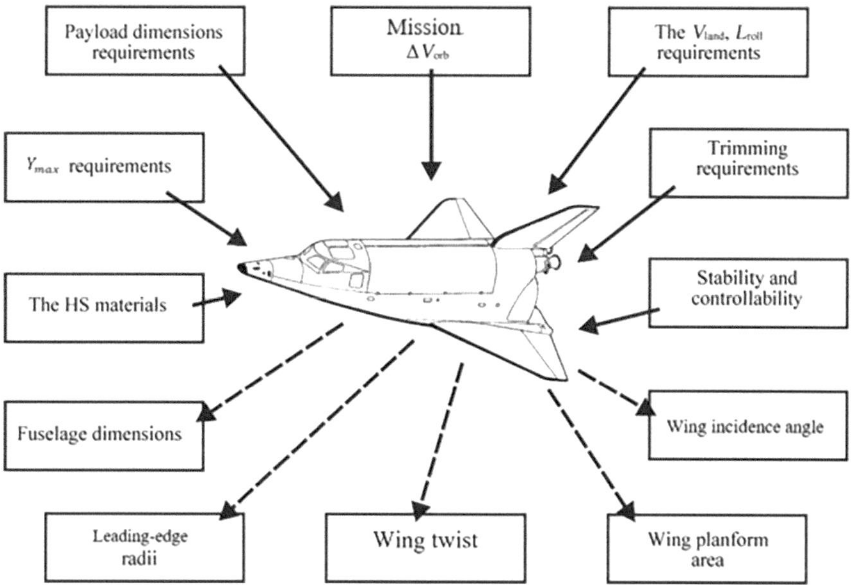

Fig. 28 ASP conceptual design

comprises a fuelled booster (space tow) and a space vehicle (satellite). The mass $m_{0\,\mathrm{sm}}$ is specified in the mission requirements.

The minimum specific cost C_{spec} of delivery of 1 kg payload to a near-Earth orbit is most often used as a criterion for future systems. With the known mass $m_{0\mathrm{asp}}$ or $m_{0\,\mathrm{sm}}$ and the specified type of engines, the minimum C_{spec}, as a rule, corresponds to the minimum of the SSRS launch mass $m_{0\,\mathrm{R}}$.

SSRS or ASS characteristics are largely defined by the engines and the fuel type; the specific impulse J_{v} influencing the efficiency of mass consumption during injection depends on them. The total engine thrust available for vertical takeoff shall provide the initial g-load $n_{x0} = 1.3$ to 1.8. Advanced systems are often subject to the requirement of the payload being delivered to a calculated or non-calculated orbit with one engine failed at takeoff; the takeoff g-load with one engine inoperative shall not be lower than $n_{x0\,\min} = 1.1$ to 1.15.

Engine function and type	Fuel components	Specific impulse in vacuum, J_{v}, s
Propulsion, Solid-Propellant Rocket Engine	Solid fuel	220 to 290
Propulsion, LRE	O_2 + HCF hydrocarbon fuel (oxygen + kerosene)	315 to 350
	$O_2 + CH_4$ (oxygen + methane)	355 to 360
	$O_2 + H_2$ (oxygen + hydrogen)	430 to 460

(continued)

(continued)

Engine function and type	Fuel components	Specific impulse in vacuum, J_v, s
Propulsion three-component two-mode, LRE	1st mode: $O_2 + HCF + H_2$	410 to 415
	2nd mode: $O_2 + H_2$	450 to 460
Propulsion, ramjet engine	Atmospheric air + HCF	1000 to 1800
	Atmospheric air + H_2	up to 3500
ASP orbital manoeuvring engine, LRE	Stable non-vaporizing components	300 to 320

Fuel components feature various density upon which the tank volume and its mass depend. Approximate data for density: HCF hydrocarbon fuel (kerosene)—830 to 860 kg/m^3; CH_4 (methane)—420 to 500 kg/m^3; H_2 (liquid hydrogen)—65 to75 kg/m^3; O_2 (liquid oxygen)—1000 to 1150 kg/m^3; grain density of solid-propellant rocket engine (SPRE)—1500 to 1700 kg/m^3.

The principle applied in selection of LRE for the Energia LR (Fig. 3b) can be considered as the most reasonable for advanced systems as well: engines of the 1st stage booster (on side units) provide most of takeoff thrust and the 2nd stage engines (on centre unit) ensure high fuel efficiency for the SPS at the end of ascend. Application of hydrogen fuel with the 2nd stage allows the reduction of fuel mass but, due to the low density of liquid hydrogen, the tank has larger volume and increased "dry" mass.

The ascend parameters can be improved with the creation of three-component two-mode LRE featuring the advantages of the 1st and 2nd stage engines. The high thrust in the 1st mode with such engines is achieved when consuming three fuel components (O_2 + HCF + H_2); in the 2nd operation mode at the end of ascend (components $O_2 + H_2$), the thrust is ~40% of the 1st mode thrust and the specific impulse increases to a level typical for the oxygen-hydrogen LRE. A decrease in the 2nd mode thrust ensures the required reduction in g-load n_x. As a result, the best combination of the LRE characteristics and the fuel tank volume is achieved. For the shuttle launch rockets, with the specified payload mass, the three-component two-mode LRE allows reducing the "dry" mass roughly by 25% against the LR equipped with oxygen-hydrogen engines.

The conceptual design phase for a two-stage injection system - with vertical takeoff (SSRS) or horizontal takeoff (SPS), that comprises an aerospace plane (ASP) or in (cargo version) a space module (SM)—includes mass specification and energy characteristics analysis. As a result, the following main design parameters are determined: the total take-off mass of a rocket-space or aerospace complex $m_{0\Sigma}$; the launch mass of the second stage $m_{0\Sigma}^{II}$ (mass of the booster and payload injected in elliptical orbit: ASP or SM); the type and number of engines of each stage. With the aerospace system (Fig. 14 2), the role of the 1st stage booster is performed by a launch aircraft (LA). In this case, the mass $m_{0\Sigma}^{II}$ is limited by the load capacity of the LA.

The take-off mass of a two-stage SSRS with reusable boosters of the stages (Fig. 14 *1*):

$$m_{0\Sigma} = m_{01} + m_{02} + m_{\mathrm{NF}} + m_{\mathrm{GK}} + m_{\mathrm{IF}}, \tag{65}$$

where m_{01}, m_{02} are the launch masses of boosters of the 1st and 2nd stages; m_{NF} is the mass of the nose fairing (the ASP version lacks the fairing); m_{GK} is the mass of the payload injected by the SSRS into elliptical orbit ($m_{\mathrm{GK}} = m_{0\mathrm{sm}}$ or $m_{\mathrm{GK}} = m_{0\mathrm{asp}}$); m_{IF} is the mass of the intermediate frame by which the SM or ASP is fixed on the 2nd stage booster;

$$m_{01} = m_{\mathrm{dry}1} + m_{\mathrm{F}1} + \Delta m_{\mathrm{F}1} + m_{\mathrm{unuse}1} + m_{\mathrm{gfs}1} + \Delta m_{\mathrm{rf}1} + m_{\mathrm{f\,JCS}1} + m_{\mathrm{f\,ABPS}}f, \tag{66}$$

$$m_{02} = m_{\mathrm{dry}2} + m_{\mathrm{F}2} + \Delta m_{\mathrm{F}2} + m_{\mathrm{unuse}2} + m_{\mathrm{gfs}2} + \Delta m_{\mathrm{rf}2} + m_{\mathrm{f\,JCS}2}, \tag{67}$$

$m_{\mathrm{dry}1}$, $\mathrm{m}_{\mathrm{dry}2}$ is the "dry" mass of the boosters (without fuel) including reserve; $m_{\mathrm{f}1}$, $\mathrm{m}_{\mathrm{f}2}$ is the mass of operating fuel of their cruise LRE; $\Delta m_{\mathrm{F}1}$, $\Delta\mathrm{m}_{\mathrm{F}2}$ is the fuel consumed at the LRE starting, due to evaporation of components in flight; $m_{\mathrm{unuse}1}$, $\mathrm{m}_{\mathrm{unuse}2}$ is the unusable fuel in tanks; $m_{\mathrm{gfs}1}$, $\mathrm{m}_{\mathrm{gfs}2}$ is the fuel reserved for the compensation of deviation of the SSRS characteristics and disturbances; $\Delta m_{\mathrm{rf}1}$, $\Delta\mathrm{m}_{\mathrm{rf}2}$ is the mass of consumable working agents of the onboard systems (oil, other fluids, explosive charge for separation) and gases; $m_{\mathrm{f\,ABPS}}$ is the mass of fuel for the air-breathing propulsion system (ABPS) that provides return of the 1st stage booster (RRU) to the launch site; $m_{\mathrm{f\,JCS}1}$, $\mathrm{m}_{\mathrm{f\,JCS}2}$ is the mass of fuel of the jet control system (JCS).

At the initial design stages, some components of the mass breakdown can be taken from statistical data for the existing systems. The "dry" mass reserve is assumed to be within 10 to 15% while the guarantee rocket fuel reserve m_{gfs}—within 0.5 to 0.8%. The operating rocket fuel reserve can be estimated on the base of the balance of reference velocity (41) for each stage of the system:

$$m_{\mathrm{F}}^{I} = m_{\mathrm{F}1} + m_{\mathrm{F}2}^{I} = m_{0\Sigma}\left[1 - \exp\left\{-\left(V_{\mathrm{f}1} + \Delta V_{\mathrm{v}\,\Sigma}^{I}\right)/\left(g_0 J_{\mathrm{v}\,\Sigma}^{I}\right)\right\}\right]; \tag{68}$$

$$m_{\mathrm{F}}^{II} = m_{\mathrm{F}2}^{II} = m_{0\Sigma}^{II}\left[1 - \exp\left\{-\left(V_{\mathrm{f}2} - V_{\mathrm{f}1} + \Delta V_{\mathrm{v}\Sigma}^{II}\right)/\left(g_0 J_{\mathrm{v}\Sigma}^{II}\right)\right\}\right], \tag{69}$$

where $m_{\mathrm{F}2}^{I} + m_{\mathrm{F}2}^{II} = m_{\mathrm{F}2}$; $m_{\mathrm{F}2}^{I}$ is the portion of the operating fuel of the 2nd stage booster used prior to separation concurrently with the 1st stage booster; $m_{\mathrm{F}2}^{II}$ is the portion used in independent flight of the 2nd stage; $J_{\mathrm{v}\,\Sigma}^{I}$, $\mathrm{J}_{\mathrm{v}\,\Sigma}^{\mathrm{II}}$ are the resulting average specific impulses of the stage SPSs; $V_{\mathrm{v}1}$, $\mathrm{V}_{\mathrm{v}2}$ is the flight velocity upon shutdown of the propulsion LRE boosters; $\Delta V_{\mathrm{v}\Sigma}^{I}$, $\Delta\mathrm{V}_{\mathrm{v}\Sigma}^{\mathrm{II}}$ is the total loss of the reference velocity of the stages.

The mass of the 1st stage booster (RRU) at separation $m_{\mathrm{f}1}$ and landing $m_{\mathrm{land}\,1}$:

$$m_{\text{f1}} = m_{01} - m_{\text{F1}} - \Delta m_{\text{F1}} - \Delta m^{I}_{\text{rf1}}; \tag{70}$$

$$m_{\text{land1}} = m_{\text{f1}} - \Delta m_{\text{f ABPS}} - \Delta m_{\text{f JCS1}} - \Delta m^{\text{ret}}_{\text{rf1}} - \Delta m_{\text{unuse1}} + m_{(\text{unuse+gfs})\text{ABPS}} + m_{(\text{unuse+gfs})\text{JCS1}}, \tag{71}$$

where $\Delta m^{I}_{\text{rf1}}$ is used during ascend and $\Delta m_{\text{f ABPS}}$, $\Delta m_{\text{f JCS1}}$, $\Delta m^{\text{ret}}_{\text{rf1}}$, Δm_{unuse1} are used during reentry; $m_{(\text{unuse+gfs})\text{ABPS}}$, $\text{m}_{(\text{unuse+gfs})\text{JCS1}}$ is the unusable fuel of the ABPS and JCS respectively.

The reserve $m_{\text{gfs ABPS}}$ is intended to compensate for the deviations of the RRU characteristics and atmospheric disturbances, including wind.

The initial and final (injected) mass in independent flight of the 2nd stage:

$$m^{II}_{0\Sigma} = m_{0\Sigma} - m^{I}_{\text{F2}} - \Delta m^{I}_{\text{F2}} - \Delta m^{I}_{\text{rf2}} - m_{01}; \tag{72}$$

$$m^{II}_{\text{f}} = m^{II}_{0\Sigma} - m^{II}_{\text{F2}} - \Delta m^{II}_{\text{F2}} - \Delta m^{II}_{\text{rf2}}, \tag{73}$$

where $\Delta m^{I}_{\text{F2}} + \Delta m^{II}_{\text{F2}} = \Delta m_{\text{F2}}$; $\Delta m^{I}_{\text{rf2}} + \Delta m^{II}_{\text{rf2}} = \Delta m_{\text{rf2}}$.

Estimation of the mass of injected load on an intermediate elliptical orbit:

$$m_{\text{GK}} = m^{II}_{\text{f}} - (m_{\text{f2}} - m_{\text{NF}} - m_{\text{IF}}), \tag{74}$$

where

$$m_{\text{f2}} = m_{\text{dry2}} + m_{\text{unuse2}} + m_{\text{gfs2}} + m_{\text{f JCS2}} \tag{75}$$

is the mass of the 2nd stage booster upon shutdown of its cruise LRE.

The mass of the injected payload m_{GK} (71) is specified in the mission requirements:

$$m_{\text{GK}} = m_{\text{pl spec}}. \tag{76}$$

With the SSRS in cargo version (Fig. 16), in the upper trajectory when the dynamic pressure decreases down to the specified value ($q < 0.5$ to 1 daN/m^2), early drop of the fairing is possible which reduces the fuel consumption m^{II}_{F2} in the segment of operation of the 2nd stage. In partially reusable SSRS with the disposable 2nd stage booster, the component $m_{\text{f JCS2}}$ of Eq. (64) is missing. When the 1st stage booster comprises several return rocket units, Eq. (63) sums up masses of all separated RRU.

3.2 Optimisation of SSRS and SPS Main Design Parameters

With the payload mass specified, the launch mass $m_{0\ \Sigma}$ should be minimum while meeting a number of constraints and requirements: the maximum g-load during ascend, the safe separation and return of the boosters, and other constraints:

$$m_{0\ \Sigma} \Longrightarrow \min\ , m_{\mathrm{GK}} = m_{\mathrm{pl\ spec}} \tag{77}$$

The mass $m_{0\ \Sigma}$ is determined iteratively with the gradual update of the rocket fuel mass and the "dry" booster mass, as well as the mass of aircraft fuel for the ABPS returning the 1st stage booster (Fig. 16). In case of a reusable 2nd stage booster (Fig. 14 (*1*)), the process of finding $m_{0\ \mathrm{R}}$ includes the analysis of its trajectories of a single-revolution return with the estimation of aerodynamic heating and duration of descent which influence the mass of heat shield and thermal insulation (Fig. 29).

In determining the SSRS parameters, selection of coefficient k_{F} is of great concern:

$$k_{\mathrm{F}} = m_{\mathrm{f2\ serv}}/m_{\mathrm{f1\ serv}}, \tag{78}$$

where $m_{\mathrm{f1\ serv}}=m_{\mathrm{F1}} + \Delta m_{\mathrm{F1}} + m_{\mathrm{unuse1}} + m_{\mathrm{gfs1}}$;

$m_{\mathrm{f2\ serv}} = m_{\mathrm{F2}} + \Delta m_{\mathrm{F2}} + m_{\mathrm{unuse2}} + m_{\mathrm{gfs2}}$ is the mass of fuel used by the 1st and 2nd stage boosters.

Frequently, cluster configurations with concurrent operation of engines of both stages until separation are considered. If the 2nd stage engines consume the fuel until separation from their own tanks only, it is advantageous to reduce their thrust down to 40 to 50% of the maximum within some time after the takeoff (~30 s).

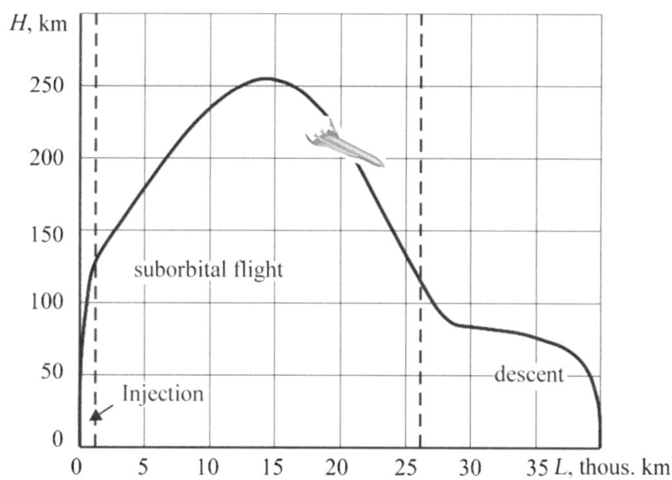

Fig. 29 The longitudinal profile of a single-revolution flight of a reusable 2nd stage booster

This will save fuel for the independent flight of the 2nd stage. For the concurrent operation of the stages with the same-type LRE the approximate fuel mass ratio factor: $k_F^{opt} \approx 1$. When the fuel supply (transfer) is provided from the 1st stage tanks to the 2nd stage engines before separation, there is no need in reduction of the thrust level with the 2nd stage booster. In this case, there is no loss of the thrust/weight ratio in the joint flight segment, the gravitational losses go down, the ballistic effect is achieved due to earlier separation of the worked out stage. The fuel mass ratio factor for the systems with "transfer": $k_F^{opt} \approx 0.5$.

Despite the difference in coefficient k_F^{opt} (Fig. 30a), the Mach number at separation for two versions of concurrent operation of the stages appears to be virtually the same: $M_{F1}^{opt} \approx 6$ (Fig. 30b).

The "transfer" with properly selected factor K_∂ increases the mass of payload injected in an intermediate elliptical orbit by approximately 18%. Moreover, the size of the 2nd stage booster decreases with the increase in the size of the 1st stage booster. When the fuel "transfer" principle is applied to a system where only "its own" fuel is consumed by the 2nd stage booster (without any change in the volumes of both stage tanks), the "transfer" effect either does not take place or is minor.

Using the same-type engines for the stages (in particular, two-mode LRE) can improve operating and economic performance. However it should be noted that a high specific impulse at the end of injection and a high takeoff thrust are attainable with different levels of the LRE nozzle area ratio. Therefore, it is reasonable to use a nozzle with smaller exit area for the 1st stage booster. Such solution will allow reducing the launch mass $m_{0\,\Sigma}$ by 4 to 5%. Another way is the application of sliding extensions in the nozzles to adjust the exit area and enable switching to a maximum nozzle area ratio at a flight altitude of 8…9 km.

For the second (space) SPS stage, the most appropriate is the application of two-mode LRE with lower thrust in the 2nd mode; and the fuel consumption ratio in such two modes should be selected such that it delivers the maximum payload mass.

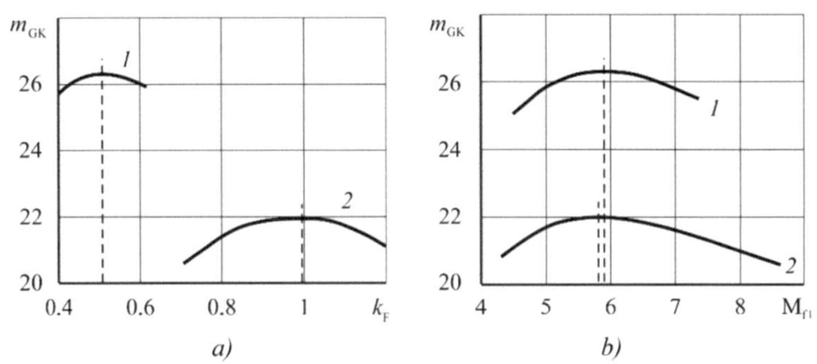

Fig. 30 Example of the effect of the fuel mass ratio of the stages and the separation Mach number on the payload mass of a SSRS with reusable boosters of the 1st and 2nd stages: *1* —the fuel supply (transfer) from the 1st stage booster tanks to the 2nd stage engines; *2*—the fuel is supplied to the 2nd stage engines from its own tanks only; $k_F = m_{f2\,serv}/m_{f1\,serv}$; $m_{f1\,serv} + m_{f2\,serv} = \text{const}$

For a vertical takeoff SSRS, reasonable trajectory kinematic parameters at separation of the 1st RRU stage are important because they have an effect on the maximum distance from the RRU launch point and the return ABPS fuel mass. Therefore, the pre-separation trajectory should be flatter as compared to the injection with a disposable RSS. Additionally, the lower angle of inclination of trajectory at separation lowers the peak dynamic pressure and aerodynamic loads in the reentry trajectory.

The advanced technical solutions reducing the total launch mass $m_{0\,\Sigma}$ with the payload mass m_{GK} are as follows:

- air-breathing engines for cruising flight of the 1st stage RRU;
- three-component two-mode LRE;
- fuel supply (transfer) from the 1st stage booster to the 2nd stage engines;
- optimum fuel mass ratio for the stages;
- regulation of the nozzle exit section area;
- high performance structural materials.

In design optimization, production-related problems, cost, reliability, flight safety and maintainability associated with the above solutions must be taken into account.

3.3 SSRS and ASS Technical Specifications

We give as example the technical characteristics of the rocket-space system Energia-Buran (Fig. 3b) and the Multirole airspace system MAKS (Fig. 4a).

Name of injection system	Energia-Buran	MAKS
Type of injection system	RSS	ASS
Type of the ASP included in the injection system	Orbital aircraft Buran	Orbital aircraft OA MAKS
Parameter or characteristic	Parameter value	
The launch mass of the RSS or take-off mass of the ASS, t	2375	630
– launch mass of the 1st stage booster (all blocks), t	1490	–
– takeoff mass of the launch aircraft, t	–	355
– launch mass of the 2nd stage, including the ASP, t	780	275
– launch mass of the ASP, t	105	26.4**
Mass of the payload injected in the datum orbit ($i = 51°$, $H = 200$ km), t	30	7.5
Mass of the payload returned from the orbit, t	20	4.6
Number of the crew members, persons	up to 10	2 (6***)

(continued)

(continued)

Name of injection system	Energia-Buran	MAKS
Type of the fuel for rocket engines:		
– 1st stage booster	O_2 + HCF	–
– 2nd stage booster	$O_2 + H_2$	O_2 + HCF + H_2
Total thrust of the LRE using boosters (in vacuum/near the Earth), ts*		–
– 1st stage booster	4 × (806/740)	2 × $(204/162)_{\mathrm{I\ m}}$
– 2nd stage booster	4 × (190/147.6)	2 × $(81/-)_{\mathrm{II\ m}}$
Specific impulse of rocket engines (in vacuum), s		–
– 1st stage booster	336.2/308.5	410/324 $_{\mathrm{I\ m}}$
– 2nd stage booster	454.7/353.2	457/–$_{\mathrm{II\ m}}$
Dimensions of the ASP (length, wing span, height), m	36.4 × 23.9 × 16.4	19.3 × 12.5 × 8.6

Notes
* the engines thrust is provided in tons of force; for conversion into kilo newtons, multiply it by coefficient 9.8
** launch mass of the orbital aircraft OA MAKS includes the mass of two propulsion LRE
*** in the base version, the crew consists of two pilots; in the version for the transport-maintenance of the space station: 2 pilots of the OA and 4 cosmonauts I m, II m are the first and second operating modes of the propulsion engine

4 ASP and HA Aerodynamic Design

The major effect on the aerodynamic configuration of ASP and hypersonic aircraft comes from such factors as the intensity and duration of aerodynamic heating, the intensity of the heat exchange in the interference zones and the transition of the boundary layer from laminar to turbulent. The requirements to the controls ensuring stability and controllability, and other requirements also play their role here. Selection of geometric parameters is complicated by the need for a compromise between aerodynamic performance of the vehicle at hypersonic and subsonic flight speeds.

Minimisation of heat flow in the stagnation point is achieved when the ASP enters the dense atmosphere at high angles of attack with the maximum lift. The heat flow on the greater part of the bottom surface is significantly lower than that on the wing leading edge. Such advantages explain the interest of ASP designers in lifing body configurations (Fig. 1). Nevertheless, a "winged" ASP (Fig. 31) features a higher lift-to-drag ratio at subsonic speeds. One of the most important changes in the configuration of the ASP Space Shuttle and Buran as compared to the earlier versions became the use of a low-mounted tapered wing with expanded extension.

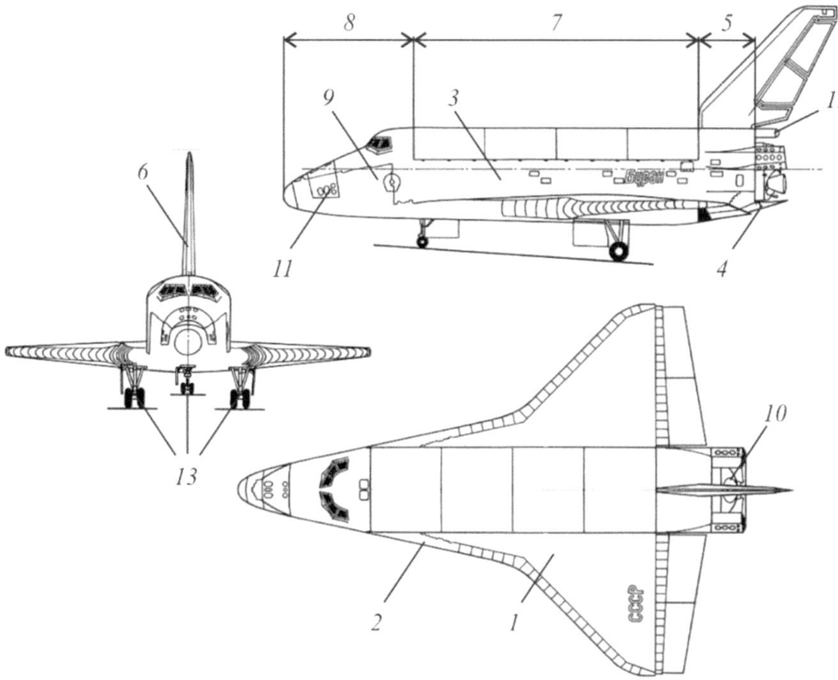

Fig. 31 Aerodynamic layout of the ASP Buran *1*—wing with elevons; *2*—wing extension; *3*—fuselage; *4*—trim panel; *5*—fuselage afterbody; *6*—vertical tail; *7*—payload compartment; *8*—fuselage forebody; *9*—cockpit; *10*—orbital manoeuvring power plant; *11*, *12*—engine units of the jet control system; *13*—landing gears

The ASP Buran (Fig. 31) is a tailless aircraft with low-aspect ratio and large sweep wings 1, wing root extensions 2, a vertical tail 6 and trim panels 4 installed on the fuselage afterbody 5.

The use of a double-sweep wing allows improving the aerodynamic characteristics of the ASP at hypersonic and supersonic speeds and ensures the required lift-to drag ratio during approach and landing of the ASP. A relatively low aerodynamic efficiency, as compared to conventional aircraft, and the absence of high-lift devices (flaps) result in an increase of the landing speed of the ASP up to $V_{\text{land}} = 320$ to 360 km/h. Such a landing speed and unpowered landing called for the construction of a special runway at the Baikonur cosmodrome with a length of about 6 km.

One of the most important geometric parameters of an ASP is the wing loading mg_0/S, as well as the glide parameter at flight angles of attack $\sigma_L = c_L(mg_0/S)$. In the hypersonic segment of descent, the parameter σ_L at high angles of attack $\alpha = 35$ to $40°$ ensures passing a trajectory above the temperature boundary; during landing at $\alpha = 12 \ldots 16°$, the value of σ_L should be sufficient to ensure the desired landing speed (49).

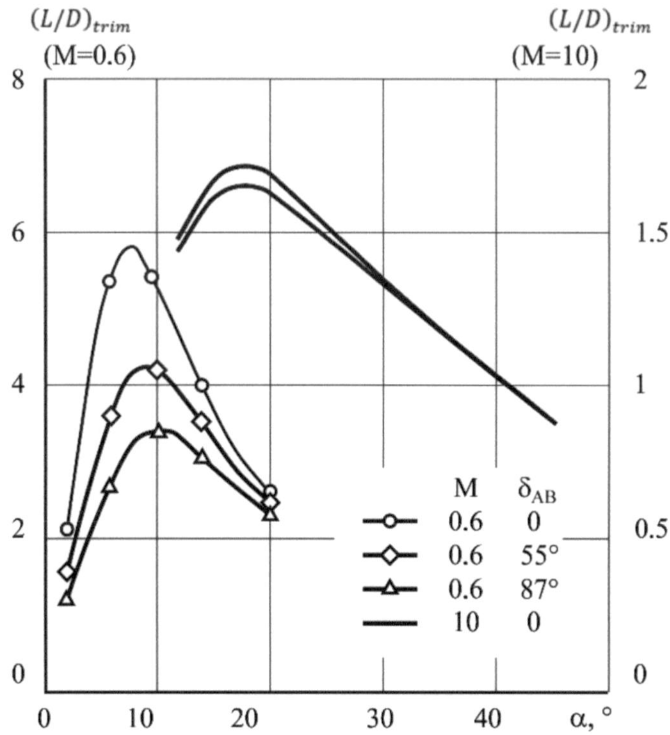

Fig. 32 The variation of trim lift-to-drag ratio with angle of attack for the ASP Buran

The trim panel deflection angle varies with the Mach number and reduces the use of elevons for longitudinal trim. The main function of the elevons is the longitudinal and lateral control and stabilisation. The rudder in the ASP Buran is made split. Additionally to the control and stabilisation in yaw, it fulfils the most important function of an airbrake (AB) that controls the rate of dissipation of mechanical energy and the flight speed at pre-landing manoeuvring and landing. Deployment of the AB at transonic speeds helps the trim panel produce the required nose-up pitching moment. On the glide slope, under the nominal conditions, the airbrake is deflected to the mid position. When open to maximum, the AB reduces the lift-to-drag ratio at landing by 35% (Fig. 32).

The orbital aircraft OA MAKS features a higher lift-to-drag ratio at hypersonic speeds and compares poorly with the orbital vehicle Buran in terms of $(L/D)_{max}$ in landing.

The minimum required lift-to-drag ratio for an ASP in unpowered landing (with a completely closed AB):$(L/D)_{max} \approx 3.5$.

Lift-to-drag ratio for two ASP configurations.

	OA MAKS	OV Buran
ASP		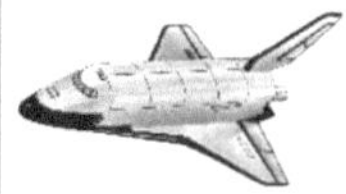
M = 10 at $\alpha = 40°$	1.2	1.0
M = 10 at $\alpha = 10°$	1.9	1.7
M = 0.6 at $\alpha = 14°$	4.0	4.0
M = 0.6 – at $\alpha_{(L/D)\max} = 9°$ – at $\alpha_{(L/D)\max} = 12°$	4.0	5.9

After suborbital flight, the winged rocket boosters of the 1st stages of a SSRS (the RRU—Fig. 5) perform a non-equilibrium descent (Fig. 16) with a peak rise in dynamic pressure and temperature. To reduce the loads, one should ensure the equilibrium at high angles of attack $\alpha = 50$ to $55°$ at the beginning of the reentry in the dense atmosphere. RRU feature high hinge moments on the aerodynamic control surfaces.

For suborbital ASP of touristic class, the problem of descent along a nonequilibrium trajectory is complicated by the forward centre of mass location due the use-up of the fuel used during acceleration. Forward horizontal fins can be used to ensure the equilibrium of such ASPs at high angles of attack.

The aerodynamic analysis of hypersonic aircraft is coupled with the gas-dynamic analysis of the ramjet engine path. Future hypersonic aircraft are for now defined in general terms only. However, it is clear now that the key feature of the aerodynamic layout of hypersonic aircraft will be implied by the need to join the HA airframe and power plant. One option is a tailless aircraft with a tapered double-sweep wing. The wing should feature a negative incidence angle of 3 to 5° which will allow increasing the fuselage angles of attack at hypersonic speeds and increasing the pre-compression of flow in the ramjet engine air intake. The shape of cross-sections of the forebody should be selected based on these considerations. The contours of the afterbody should ensure its use as an exhaust ramjet engine nozzle with supersonic combustion (Fig. 33). Future hypersonic aircraft should feature the lift-to-drag ratio in cruising flight (at $M \approx 6$) not lower than 4 to 5 which complicates the development of its aerodynamic layout. The aerodynamic heating problem will significantly influence the geometry of hypersonic aircraft.

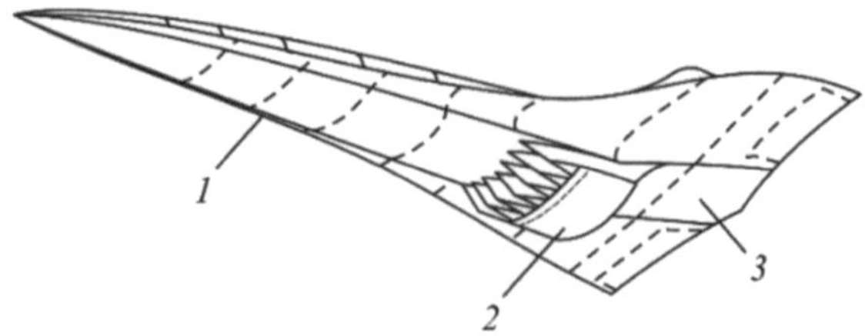

Fig. 33 Potential configuration of a future hypersonic aircraft: *1*—forebody used for air pre-compression; *2*—ramjet engine using hydrogen fuel; *3*—afterbody used as the power plant nozzle

5 ASP and Hypersonic Aircraft Thermal Design

5.1 Main Terms of Thermal Design

ASP thermal design is the structural design for minimum weight under the preset thermal constraints on the materials and equipment. Thermal design is usually split in several stages.

The thermal design of ASP layout is always performed in the first place. Its goal is deciding on the main geometric dimensions of the ASP taking into account the temperature limitations on the outside surface materials. Due to the low thermal conductivity of HS materials, the temperature of the ASP surface differs little from the radiation-equilibrium values. Therefore, the analysis of the external heat flows can be done ignoring the thermal conduct processes in the ASP structure. This stage includes the analysis of various configurations and layouts, analysis of aerodynamic, trajectory and handling characteristics, determining the external heat flows and the radiation-equilibrium temperatures for the entire ASP surface.

During the second stage, the thermal design of the heat shield and the vehicle structures is performed. This stage includes selection of the minimum required thickness for the HS taking into account the given external heat flows, the features of the structures, the properties of the materials applied, the production-related limitations and other parameters. A detailed analysis of thermal modes is performed for all main structural elements of the ASP and, when required, a modification of the structure is proposed for the reduction of its overheating or overcooling. An important part of this stage is the analysis of thermal consequences of abnormal operation of certain ASP systems (damage to the heat shield, deviation from the nominal flight mode and other emergency situations).

Thermal engineering of internal units and systems of an ASP such as the crew life support system, power plant, hardware thermal control system etc. is performed on the basis of two previous stages.

5.2 Principles of Thermal Design of ASP Heat Shield and Structures

The most part of the convective heat flow sinks outside the vehicle via radiation of the heated surface. The heat penetrating inside the HS and further in the structures comprises just a fraction of percent of the total external thermal load. The HS warmes up rather slowly and the temperature of the ASP structures begins to grow at the final flight stage only when the external heat flows are already significantly decreased and even the external HS surface cooling down by the ambient air takes place. The ASP structures acts in such conditions as an accumulation-type heat protection receiving the heat accumulated earlier in the HS layer.

The main objective of the thermal engineering of heat-protection coatings consists in minimizing the HS thickness provided that the maximum limit temperature of the structure being protected is not exceeded throughout the flight and post-flight cooling time. At that, the thermal capacity of the structures and the structure temperature before descent from the orbit are important as the input data. It is evident that the lower the initial temperature before descent and the higher the thermal capacity of the structure, the lower HS thickness is required.

As it is seen in Fig. 34, thermal conductivity of the modern low-density thermal-protection materials strongly depends on the air temperature and pressure.

This leads to the fact that, with equal levels of external heat flow, the required heat shield thickness for the top surface of the ASP Buran appears to be by 12...15% smaller than that for the bottom one (Fig. 35).

Internal thermal exchange in the ASP structures, as a rule, acts positively facilitating the temperature decrease in the hottest structural areas and smoothing the

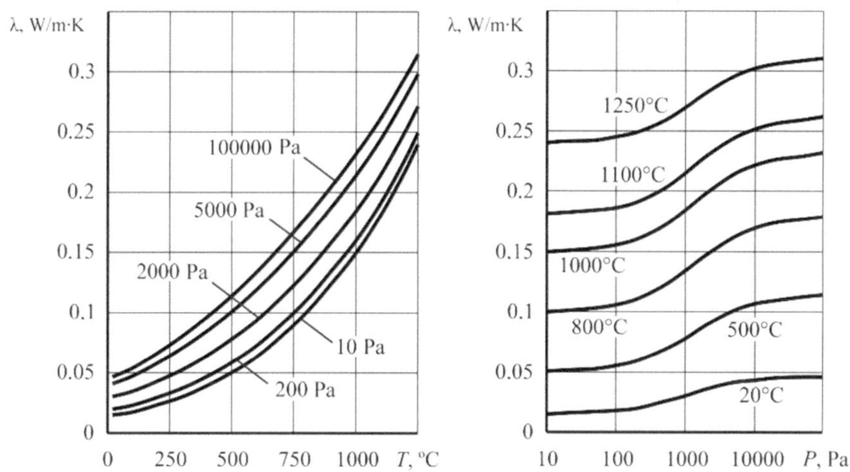

Fig. 34 Variation of thermal conductivity coefficient of the TZMK-10 material ($\rho = 150\ \text{kg/m}^3$) with temperature and pressure

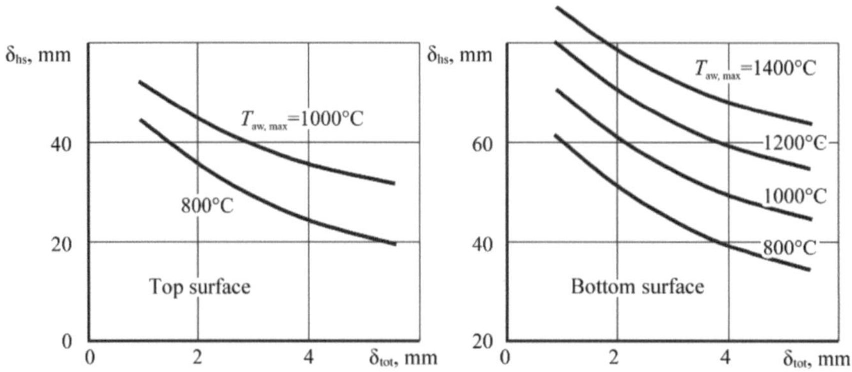

Fig. 35 Variation of the ASP Buran HS thickness with thickness of the body skin made from aluminium alloy for the top and bottom wing surface

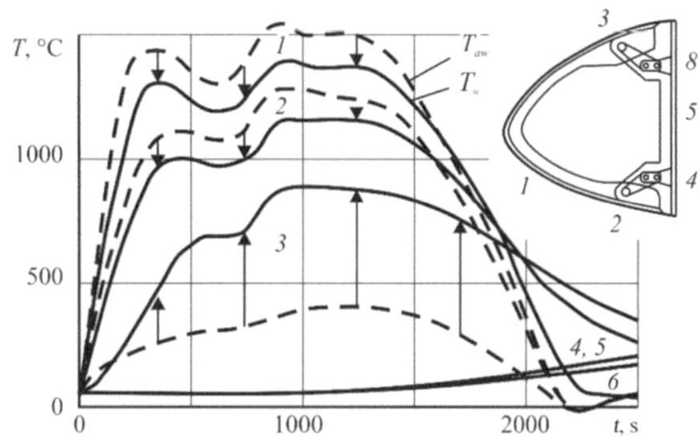

Fig. 36 Temperature variation in specific points of the ASP Buran wing leading edge

temperature gradients. For instance, in the hottest area of the carbon–carbon skin of the ASP Buran wing leading edge (point 1 on Fig. 36), the maximum temperature of the surface T_w is lower by 120…150 °C than the temperature of adiabatic wall T_{aw}, and in the lowest heating area (point 3), T_w is higher than the value T_{aw} by 500…600 °C.

Meanwhile, overheating of "cold" structural elements is possible due to thermal radiation of the "hot" areas. Hence, it comes to using rather thick and high-temperature internal heat-insulation to shield the aluminium alloy front wall of the wing (points 4 to 6) from the thermal radiation of the heated carbon–carbon edge skin.

The range of methods for modification of the ASP basic structure during its thermal design is rather limited and stays in line with physically explainable techniques for control of the heat transfer processes. For instance, in cases when the heat shield thickness cannot be increased for geometric or other reasons, it comes to increasing the skin thickness so as to lower the maximum level of its heating.

5.3 Testing the Thermal Protection

Along with the analytical engineering studies, an important part of thermal design of an ASP are ground and flight tests. The cost of tests is usually several orders higher than the cost of computations, so the adequate planning of tests with the use of analytical methods is important. In typical tests of large ASP structures with the use of radiation heating devices (Fig. 37), the external parameters of heating cannot be arranged in form of natural heat flows due to the feedback effect of the heated structure on the heat flow from radiation sources. This is why heaters are typically programmed to perform the cyclogram of temperature variation previously obtained with the use of analytical methods for actual flight conditions.

The results of flight tests of large flying models Bor-4, Bor-5 (Fig. 38) and others require a careful analysis with the use of analytical methods because the measurement accuracy in flight is significantly lower than that in ground conditions.

A series of ASP modifications follow the first flights in the finishing phase of thermal design. A detailed analysis of the collected data and its comparison against the design data shall be performed after each flight. This allows identifying both the hazardous zones in terms of heating and the areas where the actual temperature levels appear to be significantly lower than the design ones. In both cases, the required modifications of structure to improve its reliability and reduce its weight are performed.

Fig. 37 Inner view of the TPVK TsAGI and placing a fuselage compartment of the ASP Buran in the chamber for thermal tests

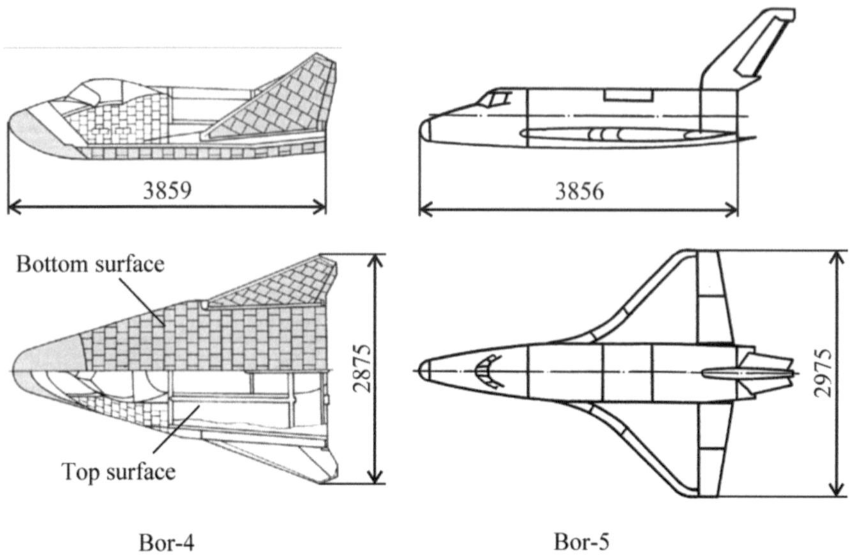

Fig. 38 Large flying models Bor-4 and Bor-5

6 ASP and HA Load-Bearing and Thermal-Protection Structures

ASP and hypersonic aircraft structures are not different from those of a conventional aircraft. However, ASP and hypersonic aircraft design should take into account the high flight speed, the intensity and duration of aerodynamic heating, as well as the specified number of flights or the service time. All these factors define the main feature of ASP and hypersonic aircraft, namely, the capability of resisting high temperatures.

The type of thermal protection system depends on certain flight conditions. For an intensive but short-term heating (experimental hypersonic aircraft), the ablation thermal-protection can be applied which is usual for single-use reentry bodies (RB) of the Soyuz type and others.

For a short-term and moderate heating, a heat-absorption design can be applied (experimental aircraft X-15). In case of prolonged intensive heating, external heat shields or heat-resistant structures are required. Materials intended for the external surfaces of various type flight vehicles is shown in Fig. 39.

The specific flight conditions of the ASP Buran (Fig. 25) impose specific requirements on the structural materials (Fig. 40). The materials must resist prolonged exposure to low temperatures (–130 °C) and high temperatures (up to 1650 °C in critical point of the nose fairing). Strict requirements in terms of mass make it necessary to use low-density materials.

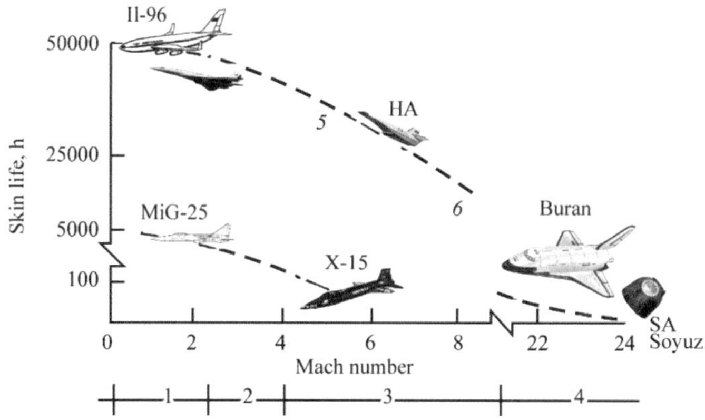

Fig. 39 Materials applicable for the external surfaces of flight vehicles for various flight speeds: *1*—aluminium; *2*—titanium; *3*—high-temperature alloys; *4*—heat-insulating, ablation and carbon-ceramic composite materials; *5*—high-temperature structure; *6*—cooled structure

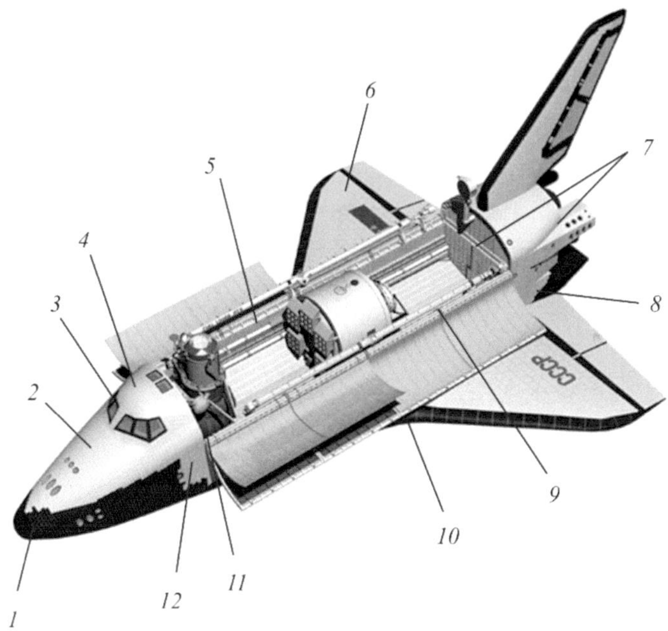

Fig. 40 Metallic materials in the Buran structure: *1*—VN3, ZhS3-VI (brackets of the cone hinge assemblies); *2*—1201T1 (cockpit); *3*—VT23 (window frames); *4*—1163T1, D16Cht1 (forebody); *5*—1163T1, D16Cht1 (frames, skin); *6*—VT23, OT4 (titanium alloys in the wing); *7*—VT23 (frames 22 and 24); *8*—VT23, VNS-5 (docking unit); *9*—VT20 (braces); *10*—VN3 (brackets of hinge assemblies); *11*—VT23 (frame 7); *12*—VT23 (docking unit)

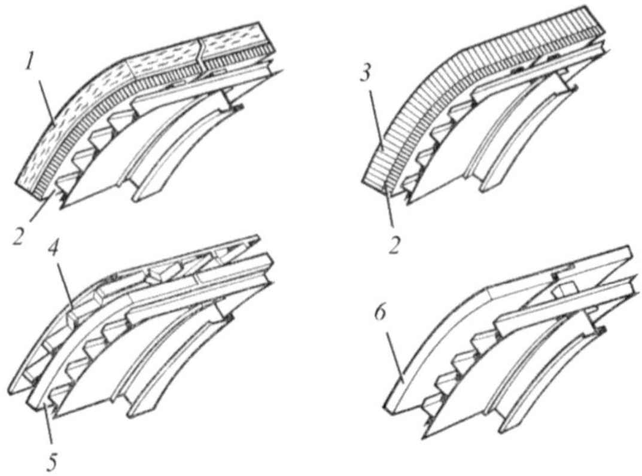

Fig. 41 Potential thermal protection systems of ASP and hypersonic aircraft: *1*—multi-use external insulation; *2*—load-bearing panel; *3*—ablation low-density coating; *4*—heat-resistant composite or metallic structure; *5*—heat-insulating layer; *6*—heat-absorption shield (molybdenum, steel or aluminium-beryllium alloy)

For the advanced winged boosters of the first stages of SSRS, high-strength aluminium alloys and polymer composite materials may be used (Fig. 5). Such materials feature relatively low maximum limit operating temperatures (below + 250 °C); therefore, the higher heat areas (fuselage nose, wing leading edges) involve structures from more thermally-resistant titanium and steel alloys.

Instances of structural design of various thermal protection systems of advanced ASP are shown in Fig. 41.

The appropriate type of material for heat shields and the areas for their installation depend on the pattern and intensity of external heat loads. With the development of carbon–carbon and carbon-ceramic heat-resistant materials, the need for external heat shields may be eliminated. Such materials are used in designing hypersonic aircraft (HA) for research and military applications (Fig. 42).

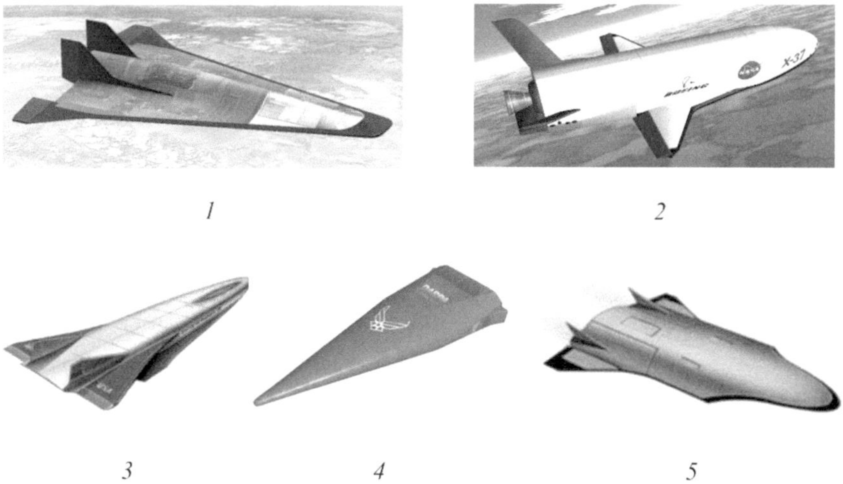

Fig. 42 Advanced HA: *1*—HyperSoar; *2*—X-37B; *3*—X-30; *4*—Falcon HTV-2; *5*—Falcon HCV

References

1. Eger SM, Mishin VF, Liseitsev NK et al (2005) Proektirovanie samolyotov (Aircraft Design). Logos, Moscow
2. Hirshel EH, Weiland C (2009) Selected aerothermodynamic design Problems of hypersonic flight vehicles. Springer, Berlin, Heidelberg

Design Features of Hydro-Aviation Aircraft

Leonid G. Fortinov and Anas M. Hairov

Abstract The chapter introduces the classes of hydro-aviation aircraft, discusses the estimation of their gross weight, their maximum lift capacity, maximum range, and the probability of their operation in the required season and site in the ocean. The chapter offers a series of design recommendations for seaplane design to overcome wind waves, to operate safely in crosswinds, intense fog, and poor visibility. The chapter introduces basic structural design concepts and some recommendations for a jet multirole bottom-amphibious aircraft of Beriev Be-200 type. Finally, the chapter presents a concept of seaworthiness and a practical algorithm for determining the geometry of the bottom of an amphibious aircraft.

Keywords Gross weight · Maximum lift capacity · Wind wave · Crosswind · Intense fog · Poor visibility

1 Introductory Remarks

Hydro aviation aircraft have the ability to operate from oceans, seas, lakes and rivers, which in total cover over 70% of the surface of planet Earth.

The hydro aviation aircraft operating only from water surface are called seaplanes, and the aircraft capable of operating also from land are called amphibious aircraft (amphibian) [1].

Seaplanes and amphibious aircraft are subdivided into contact and non-contact take-off and landing vehicles.

The vehicles with contact take-off and landing are supported by hydrodynamic forces from the moment of standing on water (passive floating) to the beginning of airborne flight. During landing, hydrodynamic force acts on the vehicle from the moment of touching water until the aircraft stops.

L. G. Fortinov (✉) · A. M. Hairov
Beriev Aircraft Company, Taganrog, Russia
e-mail: fortinovleonid@rambler.ru

M. A. Pogosyan (ed.), *Aircraft Design*, Springer Aerospace Technology,
https://doi.org/10.1007/978-981-96-4599-2_14

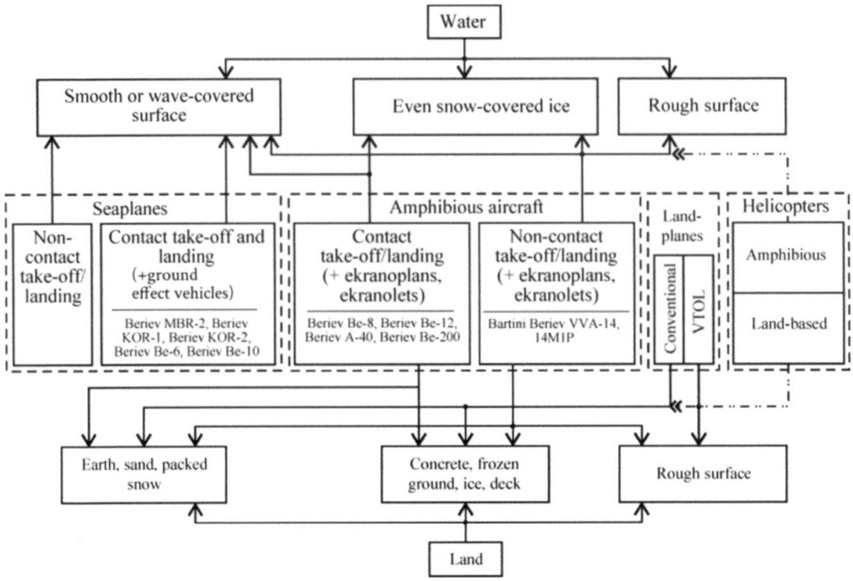

Fig. 1 Hydro aviation aircraft classification

In non-contact take-off and landing vehicles, hydrodynamic force is replaced by aerostatic or aerodynamic air cushion produced by lifting or blowing engines (VTOL aircrafts, amphibian helicopters, ground effect vehicles, etc.).

The hydro-aviation aircraft classification is shown in Fig. 1.

2 Features of Hydro-Aviation Aircraft Design

Hydro-aviation aircraft design for cruising flight does not differ from the conventional airplane design. However, the design and aerohydrodynamic features associated with takeoff from and landing on water, floating on the water surface and exposure to sea agitation and sea air require the accumulation and use of an additional set of statistical parameters that are unique to hydro-aviation.

Some of those features are the increased strength of bottom or floats of seaplanes, the top location of power units to exclude their flooding by spray and water jets, the unsinkability at penetration in the bottom by floating objects.

All this in aggregate does not make it possible to speak about sea aircraft competitiveness with land airplanes by their performance. Such competition has a scholastic character, because none of the most perfect airplanes is able to transport cargoes or people to the water surface. In other words, considering aircrafts as vehicles for delivery of payload to a given point on the ocean surface at a given distance and at a given speed, airfield (land-based) aircrafts without cargo landing have zero efficiency, while hydro-aviation aircrafts even with low seaworthiness (i.e. capability to operate on water) are able to perform such operation.

Below there will be given the recommendations developed at Beriev Aircraft Design Bureau applicable for the preliminary studies of prototypes and parameters of contact take-off and landing seaplanes. The detailed recommendations on hydrodynamics, unsinkability and stability are offered in technical literature.

Preliminary design of subsonic jet seaplanes with contact takeoff and landing begins with the technical specifications—payload mass m_{pay}, range R and average cruise speed V—and determination of the conditions of aircraft operation on the water surface (zone, area, season of operation and required probability of transport operations $\overline{T}$).

Determination of the take-off mass m_0 and identification of the prototype aircraft make it possible to answer the question about the need to design a new aircraft or possibility to modify an existing one. Preliminary research tools are designed to provide suggestions as to which of the parameters identified by the customer can be replaced in order to limit the costs of modernization. Only after a comprehensive technical and economic analysis when the need in a new aircraft is established, it is possible to proceed to further design stages.

2.1 *Take-Off Weight of a Hydro-Aviation Aircraft*

The analysis of numerous criteria for aircraft showed that none of them makes it possible to determine integrally which aircraft by its direct—transport—mission can be considered to correspond to the advanced Integral technical level, which, moreover, has not been established. Therefore, the first step was to search for a criterion that would make it possible to: firstly, set the limit for the maximum criterion values, i.e. the level achieved for a certain type of aircraft; secondly, compare the level of their excellence among themselves regardless of take-off weight and, thirdly, determine the take-off mass of the aircraft for the design payload and identify the prototype.

This step was mathematically defined based on the integral transport criterion proposed by the outstanding domestic aircraft designer and scientist Bartini, which represents the ratio of the product of three payload parameters to the take-off weight of the aircraft [2]. Initially, the criterion did not take into account the relationship between the payload weight and aircraft range. Later, this relation based on historical data (Table 1 for sea jets) was incorporated into the expression for the integral transport criterion [10] (Eq. 1).

The linearization of the dependency of payload weight on range greatly simplifies both the process of analysis and the synthesis of the appearance of aircraft.

So, the maximum value of Bartini's transport criterion, called the criterion of transport efficiency, is as follows:

Table 1 A and α

№	Aircraft class	Coefficients for $U_{\max}$		Factor for U_{nom}	
		A	α	A	α
1	Civil aircraft up to 150 t	10.4	0.794	9.16	0.775
	Civil aircraft > 150 t	70.1	0.400		
2	Passenger	87.6	0.566	33.5	0.692
3	Cargo	47.3	0.587	25.26	0.68

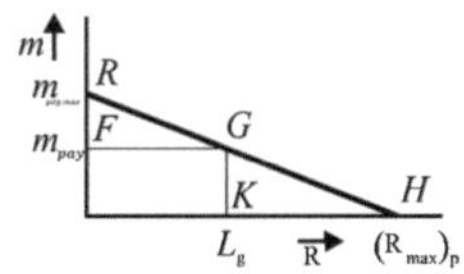

Fig. 2 Linearized dependence $m_{pay} = f(R)$

$$U = 0.25\overline{m}_0 \frac{R}{1 - \frac{m_{pay}}{m_{pay,max}}} V = 0.25\overline{m}_0 R_{max,d} V, \tag{1}$$

where: m_{pay} is the payload mass, t; $m_{pay,max}$ is the maximum payload mass, t; $\overline{m}_0$ is the aircraft mass excellence; V is the average cruising speed, km/h; R is the range, thousand km; $R_{max,d}$ is the maximum design (close to the ferry) range obtained using linearized dependence $m_{pay} = f(R)$ (Fig. 2)

Historical data for the transport efficiency criterion for aircraft of the same class (Fig. 3) are incorporated in the following equation:

$$U_{\max} = U_p = A(m_0)^{\alpha}. \tag{2}$$

The values of the coefficients A and α are given in Table 1, where, in addition to the data for U_p, the data for calculating the weighted average (nominal) values U_{nom} are also shown.

Dependencies $U(m_0)$ for some jet and turboprop seaplanes and amphibious aircraft are shown in Fig. 3. They show the advantage of heavy aircraft compared to lighter ones, so the transport efficiency criterion must take into account the aircraft takeoff weight.

To compare the technical level of aircraft using U criterion reduced by take-off weight (mass), the ratio of the estimated U_p and maximum U_{max} values of this criterion can be used.

$$\overline{U} = \frac{U}{U_{\max}} = \frac{U}{U_p} \tag{3}$$

The ratio $\overline{U}$ determines the aircraft level of transport efficiency for the time period when the historical data were processed.

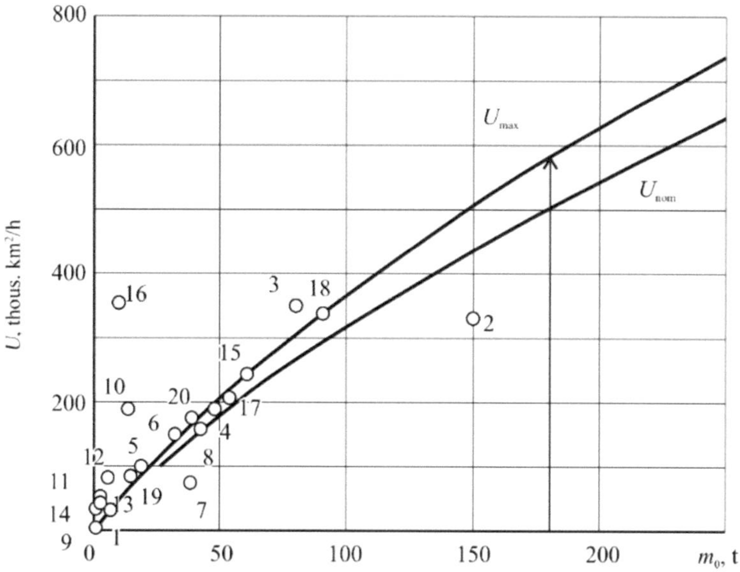

Fig. 3 Jet and turboprop see aircraft 1—SR/A1; 2—SR/45; 3—XP6M; 4—Beriev Be-10; 5—P-1; 6—Beriev Be-12; 7—PS-1; 8—US-1A; 9—TR1; 10—Do24A; 11—Avalon 680; 12—G21C; 13—Seastar; 14—Super Widgeon; 15—XP5Y-1; 16—Dolphiner; 17—SH-5; 18—Beriev A-40.

When designing a new aircraft, designers should strive to ensure that its parameters are such that the design criterion U is as close as possible to $U_{\max}$. From this consideration, a methodology was developed to determine the takeoff weight of an aircraft based on the payload requirements [3].

Its essence is to determine the ratio of payload mass to maximum payload mass of the aircraft $\overline{m}_{pay} = \frac{m_{pay}}{m_{pay,max}}$ from Eq. (4) obtained by equating (1) and (2):

$$\left[X = \frac{\overline{m}_{pay}}{\left(1 - \overline{m}_{pay}\right)^{\gamma}}\right] = \left[\frac{Cm_{pay}}{(RV)^{\gamma}} = Y\right]. \quad (4)$$

Here, C and γ are the design constants characteristic of different types of aircraft (Table 2).

The above data allows determining the range of X (Table 2) using the design values of the payload mass, range, velocity, and take-off mass. By setting $\overline{m}_{pay}$ in this range and varying it, we achieve the maximum approximation of Y to X.

The resulting value of $\overline{m}_{pay}$ allows you to find:

(a) the maximum payload mass

$$m_{pay,max} = \frac{m_{pay}}{\overline{m}_{pay}} \quad (5)$$

Table 2 Calculated values of C and γ

$\overline{m}$	0.1	0.2	0.3	0.4	0.5	0.6	0.7	0.8	0.9
X	0.114	0.264	0.470	0.760	1.190	1.898	3.180	6.050	16.26
Sea planes, over 10 t: C = 1170, $\gamma = 1.294$									
X	0.115	0.267	0.476	0.775	1.226	1.964	3.320	6.420	17.70
Sea planes, less than 10 t: C = 864, $\gamma = 1.041$									
X	0.111	0.252	0.435	0.681	1.030	1.560	2.450	4.270	9.890
Sea planes (150–300) t: C = 20.106, $\gamma = 2.5$									
X	0.130	0.349	0.732	1.430	2.830	5.930	14.20	44.72	284.6
Sea planes (600–5000) t:: C = 56.106, $\gamma = 2.12$									
X	0.130	0.321	0.639	1.181	2.173	4.190	8.990	24.200	118.6

(b) the aircraft take-off mass

$$m_0 = \frac{m_{pay,max}}{\overline{m}_{0,ex}} = \frac{m_{pay}}{\left[\overline{m}_{pay}\overline{m}_{0,ex}\right]} \tag{6}$$

where $\overline{m}_{0,ex}$ is the average mass excellence for the given take-off mass;

(c) the maximum calculated range $R_{\max}$ using the linearized dependence $m_{pay} = f(R)$ (Fig. 2).

$$R_{max} = \frac{R}{\left(1 - \overline{m}_{pay}\right)} \tag{7}$$

For calculations, $\overline{m}_{0,ex}$ is assumed to be equal to 0.375 for seaplanes.

The main historical and design parameters for some sea jets are shown in Table 3, where the masses are given in tons, ranges in thousand km, speed in km/h, and criteria U in km^2/h. It also gives the average design cruise lift-to-drag ratio. Table 3 shows that, in spite of the difference in mass excellence, the estimated take-off mass satisfactorily matches the actual mass.

Since this methodology takes into account the number of engines on existing aircraft, an adjustment of the resulting take-off mass should be made as necessary. This is done by estimating the fuel consumption per kilometer

$$C_L = \frac{m_{pay,\max}}{R_{\max}}\ (in\, t/km), \tag{8}$$

which is compared with the statistical reference value

$$(C_L)_k = \left[10^3 m_{0,av} C_{p,H}\right][VL/D]^{-1}. \tag{9}$$

Table 3 Sea jet aircraft (seaplanes and amphibious aircraft)

№	Brand	m_0	$\overline{m}_0$	$m_{pay,max}$	R_{max}	m_{pay}	R	V	U	U_p	$\overline{U}$	$m_{0,design}$	L/D
1	CL415	17.2	0.36	6.1	2.4	3.0	1.2	301	66	100	0.66	14.0	5.4
2	Do24A	18	0.25	4.5	4.0	2.3	2.0	300	76	99	0.76	15.7	7.9
3	P-1	20.3	0.30	6.0	2.0	3.1	1.0	640	96	114	0.85	18.8	5.7
4	Be-12	35	0.30	10.5	4.4	6.0	1.9	440	145	177	0.82	32.3	11.3
5	Be200	42.5	0.36	15.3	3.6	8.0	1.7	610	198	204	0.97	42.6	7.9
6	US-1A	43	0.40	17.5	3.8	8	2.1	426	163	206	0.79	40.8	7.0
7	Be-10	50	0.36	18.0	3.0	10	1.3	720	192	292	0.67	46.3	5.1
8	SH5	55	0.40	22	4.75	13	1.9	440	204	251	0.81	52.2	8.8
9	XP5Y-1	63.5	0.40	25.4	5.0	15	2.0	504	253	284	0.89	67.4	9.4
10	XP6M-1	90	0.35	28.3	5.6	13.6	2.9	744	365	340	1.07	94.8	12.2
11	A40	92.5	0.42	38.9	5.5	20	2.7	700	404	379	1.07	110.7	8.1
12	SR/45	150	0.33	49.5	6.5	36	1.8	480	257	556	0.46	138	14.4

In (9): $m_{0,av}$ is the average flight mass ($m_{0,av} = m_0(1 - 0.25\overline{m}_0)$), t; $C_{p,H}$ is the engine specific fuel consumption, t/(tf·hour); L/D is the average cruising lift-to-drag ratio; V is the average cruising speed, km/h.

However, the lift-to-drag ratio should be taken from the data on the prototype aircraft, which we have not yet defined. Therefore, it is recommended to calculate $(C_L)_k$ as follows:

$$(C_L)_K = \left(\frac{V}{V_{pr}}\right)^{0,5} E m_0^{\lambda}. \qquad (10)$$

In (10): V_{pr} is the reduced speed adopted in statistical processing, km/h; V is the cruising speed; E and λ are historical coefficients (see Table 4).

Table 4 Calculated values of E and λ are for various aircraft

Type of aircraft	Coefficients in (10)		Speed V_{pr}, to which fuel consumption $(C_L)_k$ is related
	E	λ	
Passenger with bypass turbofan engine	0.228	0.656	900
Transport with bypass turbofan engine	0.19	0.743	800
Transport with propfan engine	0.16	0.720	
Transport with turbo-propeller engine	0.352	0.720	
Sea with turbo-propeller and turbojet engines	0.337	0.658	600
Sea with bypass turbofans	0.18	0.750	
Sea with propfan engine	0.24	0.658	

If the fuel consumption calculated by (8) differs from the reference value determined by (9) by not more than 15%, the parameters $m_{pay,max}$, R_{max}, and m_0 are not corrected. Otherwise, the corrected values are calculated:

$$m_{pay,max,c} = m_{pay} + (C_L)_k R \tag{11}$$

$$R_{max,c} = R + m_{pay}/(C_L)_k \tag{12}$$

$$m_{0,c} = m_{pay,max,c}/\overline{m}_{0,av} \tag{13}$$

These are the ones used in the further synthesis process (without the lower index "c").

The tverage hourly cruising fuel consumption in t/hour:

$$\overline{m}_t = C_L 10^{-3} V. \tag{14}$$

Having determined the take-off mass of the aircraft, we find the prototypes using Fig. 3.

The given methodology responds to the payload requirements but does not prove that the aircraft provides the required seaworthiness. In this regard, unlike land aircraft, for which the methodology is also valid, one must be able to determine the most important seaworthiness parameter—the ability of the aircraft to overcome wind waves, as discussed in the next section.

2.2 Sea Plane Design to Overcome Wind Waves

The equation for the height of the wind wave which can be overcome by aircraft was introduced by the TsAGI scientist Sokolyansky [4, 5]:

$$h = A \frac{m_0^{0,67}}{V_B^2}, \tag{15}$$

where: h is the height of waves to be overcome by the aircraft, m; V_B is the take-off speed, km/h; A is a historical coefficient (for propeller aircraft it is 3960, for jet aircraft it is 4360).

It is noteworthy that the takeoff velocity depends on the height of waves.

The take-off and cruising velocities are related as follows:

$$Z = V/V_B. \tag{16}$$

It was this relation that led hydro-aviation to a deadlock in the postwar period, as the emergence of jet engines and a significant increase in cruising speeds were not compensated by the growth of the factor Z, which is associated with the efficiency of high-lift devices. For sea-planes, the value of Z ranges between 1.6 and 3.3. Blown high-lift devices and lift engines (Shin-Maiva and BBA-14 aircraft) are on the hydro-aviation agenda.

Since the value of permissible wind wave height largely determines the sea plane operation, sea-planes cannot be evaluated by the U criterion only. For example, for the amphibian A-42PE, which has an allowable maximum wind wave height of 2.2 m, the value of the criterion U is 404 km^2/h. However, if at the point of cargo delivery in the ocean a wave of less than 2.2 m makes up only 50% of time, it will be able to fly only 50% of calendar time during which it will deliver only half of cargo. This is equivalent to the criterion U being halved. Therefore, an additional criterion is introduced:

$$\overline{M}=\overline{U}h/[H]=\overline{U}\overline{h}, \tag{17}$$

where $[H]$ is the maximum wave height at delivery point.

The obtained results allow real comparison of hydro-aviation aircraft with each other (Table 5).

Having determined h, the probability of operating the designed aircraft at sea $\overline{T}$ is considered based on the turbulence of ocean (sea) surface in the given area, region (Figs. 4 and 5) and season of operation per Table 6, obtained by Leonid Fortinov by processing [6] or similar for other sections of the World Ocean.

Table 5 Jet sea aircraft

Aircraft	$\overline{U}_{LA}$	$\overline{h}_{4.5}$	$(\overline{M})_{4.5}$	$\overline{M}$ per (17)
SR/45	0.79	0.68	0.54	0.87
XP6M	1.07	0.29	0.31	0.7
Beriev Be-10	0.83	0.16	0.13	0.38
P-1	0.84	0.16	0.13	0.63
Beriev Be-12	0.83	0.24	0.20	0.72
PS-1	0.41	0.78	0.32	1.022
US-1A	0.79	0.27	0.21	0.65
Do24A	0.9	0.24	0.22	1.19
XP5Y-1	0.9	0.27	0.24	0.62
SH-5	0.97	0.31	0.30	0.84
CL-415	0.79	0.29	0.23	1.22
Beriev Be-200	0.96	0.27	0.26	0.92
Beriev A-42PE	1.04	0.45	0.48	1.01
Bartini Beriev VVA-14	0.52	2.76	1.44	4.14

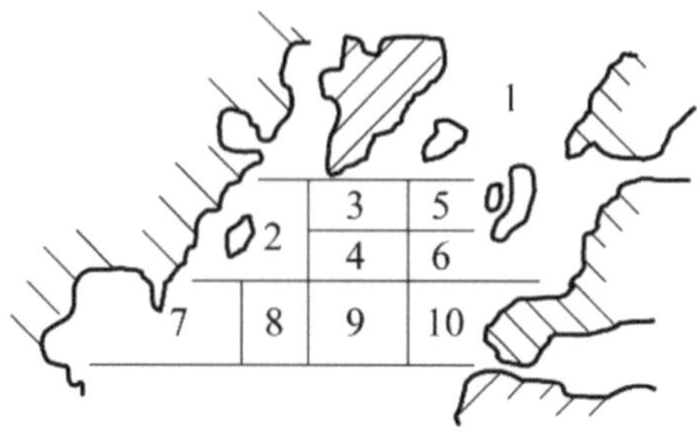

Fig. 4 North Atlantic area and regions

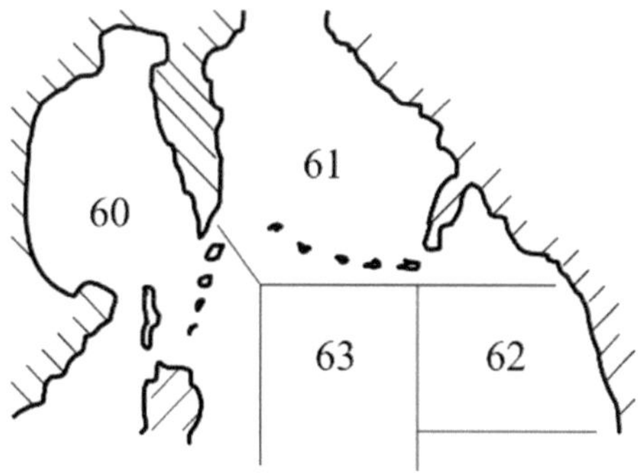

Fig. 5 North Pacific area and regions

Table 6 Probability of aircraft operation at sea

Region	Season	Analytic dependencies	
		$h(\overline{T})$, (δ%)	$\overline{T}(h)$, (δ%)
60	Winter	$1.8 + 4.4\,\overline{T}^{1,7}$ (+4.2)	0.41 (h–1.8) 0.6 (−6.0)
	Summer	$1.0 + 3.3\,\overline{T}^{2}$ (−2.0)	0.57 (h–1.0) 0.5 (+5.6)
61	Winter	$1.8 + 8.0\,\overline{T}^{2}$ (+8.5)	0.35 (h–1.8) 0.5 (−9.5)
	Summer	$0.8 + 3.3\,\overline{T}^{2}$ (+6.1)	0.57 (h–0.8) 0.5 (+2.4)
62	Winter	$0.2 + 5.3\,\overline{T}^{1,4}$ (+5.0)	0.32 (h–0.2) 0.7 (−6.5)
	Summer	$0.6 + 3.1\,\overline{T}^{2}$ (+8.0)	0.57 (h–0.6) 0.5 (+5.1)
2	Winter	$1.8 + 5.0\,\overline{T}^{2}$ (−5.7)	0.41 (h–1.8) 0.5 (−8.4)
	Summer	$0.8 + 3.2\,\overline{T}^{2}$ (+3.2)	0.57 (h–0.8) 0.5 (−5.4)
3	Winter	$1.9 + 7.0\,\overline{T}^{2}$ (+8.2)	0.38 (h–1.9) 0.5 (+6.8)
	Summer	$1.4 + 3.3\,\overline{T}^{2}$ (−6.5)	0.57 (h–1.6) 0.5 (+7.3)
6	Winter	$1.9 + 7.0\,\overline{T}^{2}$ (+8.2)	0.38 (h–1.9) 0.5 (−7.6)
	Summer	$1.6 + 3.3\,\overline{T}^{2}$ (+5.0)	0.57 (h–1.6) 0.5 (−7.6)
The Equator	Winter	$1.0 + 3.2\,\overline{T}^{2}$ (+8.5)	0.57 (h–1.0) 0.5 (−7.0)
35, south	Winter	$1.9 + 7.0\,\overline{T}^{2}$ (−7.0)	0.38 (h–1.9) 0.5 (+9.2)
	Summer	$1.8 + 5.0\,\overline{T}^{2}$ (−4.0)	0.41 (h–1.8) 0.5 (−8.4)

2.3 Determination of the Take-off Mass, Maximum Lifting Capacity, Maximum Design (Without Payload) Flight Range, Aircraft Prototype and Probability of its Operation in the Required Season of the Year, Area and Region of the Ocean

The above methodology for calculating the amphibious take-off mass as a function of the payload mass requires checking the compliance with the given operational probability using dependence (15) and 2 columns of Table 6 $h = P + Q\overline{T}^{\alpha}$. To this end, equate h under (15) and h using Table 6:

$$A\frac{m_0^{0,67}}{V_B^2} = P + Q\overline{T}^{\alpha}. \tag{18}$$

Using the ratio of cruise speed to takeoff speed Z (16), which is taken according to the prototype determined on the basis of takeoff mass analysis (13), we obtain the value of takeoff mass $(m_0)_M$, which provides necessary probability of operation on water $\overline{T}$:

$$(m_0)_M = \left[\frac{P + Q\overline{T}^{\alpha}}{A}\left(\frac{V}{Z}\right)^2\right]^{1,5} \textit{tons}. \tag{19}$$

For further calculation, a greater mass is selected out of those determined for the given payload m_{0k} (13) and for ensuring the desired seaworthiness $(m_0)_M$ (19).

In the absence of the prototype aircraft, the value of factor Z can be estimated using the empirical relationship (20):

$$Z = 1,1m_{0K}^{0,245}. \tag{20}$$

Below, you can find the calculation algorithm and some examples of calculation of parameters of amphibians for two values of probability of operation on water during summer in the North Pacific area in region 60 of the Sea of Okhotsk: $\overline{T} = 0.6$ and $\overline{T} = 0.8$. The input parameters are as follows: payload mass $m_{pay} = 20$ t, cruise speed V = 600km/h, and transportation range R = 3000 km; cruise flight altitude H = 11 km; average weight (mass) excellence $\overline{m}_0 = 0.42$.

As follows from Table 7, which is based on the theoretical and experimental dependencies obtained at the Beriev Aircraft Company by Leonid Fortinov:

(1) The increase in the required probability of the aircraft being operated on water $\overline{T}$ significantly influences the take-off mass and, consequently, the dimensions of an amphibious aircraft ($m_0 = 83.1$ t for $\overline{T} = 0.6$ and $m_0 = 114.6$ t for $\overline{T} = 0.8$).

Table 7 Algorithm for estimating the parameters of an amphibious aircraft

No	Parameter	Equation	Value	
			$\overline{T}=0.6$	$\overline{T}=0.8$
1	Speed of sound, km/h	$a*$	1062	1062
2	Mach number	$M=V/a*$	0.565	0.565
3	Coefficient of the influence of M on: – engine thrust	$\overline{p}_H=(0.4-0.17M)-\left[0.6-0.35(1-e^{-3.34M})\right](1-H/11)$ $(\overline{c}_p)_H=1+0.33M(4.03-H/11)$	0.304	0.304
	– specific fuel consumption		1.565	1.565
4	Take-off mass obtained from "$X=Y$", t	m_0	83.1	83.1
5	Prototype	Dependency $U(m_0)$ in Fig. 3	Beriev A-40	Beriev A-40
6	Efficiency of the high-lift devices of the prototype	Z from the data on the prototype aircraft (or $Z=1.1m_{0\text{k}}^{0.245}$)	3.33	3.33
7	Take-off mass from (18), t	$(m_0)_M=\left[\frac{P+Q\overline{T}^{\alpha}}{A}\left(\frac{V}{Z}\right)^2\right]^{1.5}$	67.7	114.6
8	Maximum payload mass, t	$m_m=(m_0)_M\overline{m}_0$, because $(m_0)_M$, is greater than m_0 by (4)	34.9	48.1
9	Payload ratio	$\overline{m}_{pay}=\frac{m_{pay}}{m_{pay,max}}$	0.573	0.415
10	Maximum design range, thousand km	$R_{max}=\frac{R}{(1-\overline{m}_{pay})}$	7.03	5.13
11	Specific kilometre fuel consumption, t/thousand km	$C_L=\frac{m_{pay,max}}{R_{max}}$	4.97	9.38
12	Reference specific kilometre fuel consumption, t/thousand km	$(C_L)_k=\left(\frac{V}{V_{pr}}\right)^{0.5}Em_0^{\lambda}$	4.95	6.30
13	Reference payload mass in tonnes	$m_{pay,max,c}=m_{pay}+(C_L)_kR$	Not adjusted, because C_L is equal to $(C_L)_\text{k}$	Not adjusted, because C_L is greater than $(C_L)_\text{k}$
14	Reference take-off mass in tonnes	$m_{0,c}=m_{pay,max,c}/\overline{m}_{0,av}$		
15	Reference range in thous. km	$R_{max,c}=R+m_{pay}/(C_L)_k$		
16	Number of engines i	i	2	4

(continued)

Table 7 (continued)

No	Parameter	Equation	Value	
			$\overline{T} = 0.6$	$\overline{T} = 0.8$
17	Power plant cruise thrust P_H, tf	$P_H = 2.33\left[\frac{m_{pay,\max}}{R_{\max}}\frac{V\cdot 10^{-3}}{\overline{C}_p}\right]^{1.26}\left(i\overline{p}_H\right)^{-0.258}$	5.94	11.1
18	Power plant take-off thrust P_0, tf	$P_0 = 125P_H/\overline{p}_H$	24.4	45.67
19	One engine thrust, tf	$(P_0)_1 = P_0/i$	12.2	11.4
20	Take-off specific fuel consumption, t/(tf x h)	$(c_p)_0 = 0.588(P_0)_1^{-0.205}$	0.352	0.356
21	Engine type	Bypass turbofan	PS90A-76	CFM56 - 5
22	Take-off specific fuel consumption, t/(tf x h)	$(c_p)_0$ from the data on the selected engine	0.363 (throt. 0.95)	0.320
23	Cruise specific fuel consumption, t/(tf x h)	$(c_p)_H = 0.9(c_p)_0(\overline{c}_p)_H$	0.510	0.451
24	Hourly fuel consumption, t/h	$\overline{m}_t = P_H(c_p)_H$	3.03	5.00
25	Maximum range, thous. km	$R_{max} = \frac{m_{pay,max}}{\overline{m}_{fuel}}10^{-3}V$	6.90	5.77
26	Range, thous. km	$R = R_{max}\left(1 - \frac{m_{pay}}{m_{pay,max}}\right)$	2.95	3.37
27	Criterion of transportation efficiency U, thous. km^2/h	$U = 0.25\overline{m}_0 R_{max} V$	434.7	363.5
28	Maximum criterion U_p, kkm^2/h	$U_p = (m_0)^{\alpha}(m_0)_{accept}/0.375$	389	502.7
29	Relative criterion $\overline{U}$	$\overline{U} = \frac{U}{U_p}$	1.12	0.72
30	Wind wave height, m	$h = A\frac{m_0^{0.67}}{V_{TO}^2}$	2.56	3.17
31	Maximum height of wind wave, m ($\overline{T} = 0.9$)	$h_{0.9} = P + Q0.9^2$	3.67	3.67
32	Criterion $\overline{M}$	$\overline{M} = \overline{U}h/[H]$	0.7	0.86

(continued)

Table 7 (continued)

No	Parameter	Equation	Value	
			$\overline{T} = 0.6$	$\overline{T} = 0.8$
Calculation of wing loading, wing area, cruise efficiency, wave height and probability of operation				
33	Wing loading, t/m^2	$q = \frac{m_0}{S} = 0.5\left[65.2m_0^{0.44} + 16V_{0.5}\right]$	0.407	0.442
34	Wing area, m^2	$S = \frac{m_0}{q}$	204.2	259.6
35	Average aircraft mass, t	$(m_0)_s = m_0 - 0.5\left(m_{pay,max} - m_{pay}\right)$	76.65	100.60
36	Average cruise lift-to-drag ratio	$\frac{L}{D} = \frac{(m_0)_s}{P_H}$	12.74	9.06
37	Design wave height, m	$h = Am_0^{0.667}\left(\frac{Z}{V}\right)^2$	2.18	3.12
38	Design $\overline{T}$ for the set sea region	$\overline{T} = 0.57(h-1)^{0.5}$	0.619	0.83

(2) By engineering level, both variants of amphibious aircraft are quite high. However, if they were used without landing on water, the first variant (criterion $\overline{U}$) would be more reasonable, but for the operation of amphibious aircraft as intended, i.e. for the delivery of cargo to objects on ocean surface, a heavier aircraft would have higher (criterion $\overline{M}$) performance characteristics, which is vital in selecting a transport system because it defines the time of probable operation (for example, at $\overline{T} = 0.83$ flights will have to be cancelled due to unacceptable sea turbulence only 5 days a month), and also the possibility of carrying cargo (312.6 thous.km^2/h instead of 363.5 ideally).

(3) The obtained values of average cruise lift-to-drag ratio are lower for the heavier aircraft, which is oversized due to the seaworthiness requirements.

(4) Since the $X = Y$ methodology is based on the average statistical value of weight (mass) excellence $\overline{m}_0 = 0.375$, strictly speaking, coefficients C in Table 1 need to be further verified by the dependence

$$C = \left(\frac{A}{0.25(\overline{m}_0)^{1+\alpha}}\right)^{1/\alpha}. \quad (21)$$

Computations indicate that this parameter significantly affects the take-off weight (mass) and the maximum design range, a fact which, unfortunately, is rarely regarded as significant. See, for example, Table 8 showing the parameters used as the input data in Table 7. It follows from the table that the increase in the excellence ratio decreases the aircraft take-off weight. In first approximation, it may be assumed that $\overline{m}_0 = 0.261(\overline{m}_0)^{0.096}$.

It is not always rational to ensure seaworthiness by increasing takeoff weight (and size) of seaplanes. There is yet another way of enhancing seaworthiness, namely

Table 8 Example of calculation of parameters of a seaplane

$\overline{m}_0(\%)$	m_0, t	R_{max}, thous.km	Δm_0, %	$\Delta R_{\max}$, %
0.300 (0)	130.5	6.14	0	0
0.375 (+25)	96.5	6.71	−26	+ 9.3
0.400 (+33)	88.6	6.89	−32	+ 12.2
0.42 (+36)	83.1	7.04	−36	+ 15

by installing additional lift engines (similar to this country's VVA-14 amphibious aircraft designed by Bartini [7]) or by employing powered wing high-lift devices (similar to the Japanese PS-1 seaplane).

It should be added that the calculation algorithm shown in Table 7 makes it possible to calculate parameters by varying individual data. This process becomes versatile, being programmed for computer use, in further aircraft design.

2.4 *Sea Plane Design with Respect the Height of Swell Waves*

Unfortunately, the issues of a seaplane behavior on swell waves, which are an echo of wind waves in different areas of the ocean and having a long length (distance between the crests or troughs), are not well elaborated. It is not so much the difficulty in describing the waves that affects here. The main point is that a seaplane, after passing the center of gravity vertical from the wave crest, begins to lower the bow, at a certain stage it scrambles the stern and accelerates down the slope under the influence of the thrust of the engines and the mass component. This process is still insufficiently well described, as is the process of interaction of the accelerating aircraft after passing the lowest point of the wave—its trough—with the approaching opposite wave slope. The difficulty lies in the hydrodynamic and hydrostatic interaction of the seaplane boat with water in the nose ("diving") and tail ("sinking") parts, which is sometimes definitively revealed only in full-scale tests.

The final decision on the height and length of permissible swell waves is made after water tunnel and full-scale testing.

2.5 *Special Features of Aircraft Operation on Water in Crosswind*

Take-off and landing of a seaplane or amphibious aircraft is usually carried out against the wind near the water surface. According to the data of the Registers of Seas and Oceans, the wind speed is usually not more than 8–15 m/s (up to 60 km/h), which is significantly lower than the takeoff and landing speeds. According to Fortinov data, for region 60 of the North Pacific area the wind speed in m/s equals

$V_{TO} = 0.67 + 9.17\overline{T}$ for summer, $V_{TO} = 3.17 + 14\overline{T}$ for winter. Therefore, the danger of wind impact on the aircraft arises when maneuvering it in the water, when the wind becomes lateral, and its impact on the boat, and—especially—on the keel, causes a lateral force tending to tilt the aircraft toward the leeward wing. Therefore, the aircraft stability is analyzed during advanced design stages, which allows estimating the parameters of underwing floats installed below the wings, closer to their ends. The greatest stability is demonstrated by hydroplanes with two parallel boats—catamarans.

The design crosswind is specified in the aforementioned Registers by speed, direction and distribution over the year.

The crosswind also poses a danger for the aircraft which leave water for land using a ramp because they can drift off the ramp axis, and also may become an obstacle during a turn in water if aircraft engines are not spaced wide enough relative to its longitudinal axis. This also holds special significance for manoeuvring in the water areas of sea ports, for which reason several fins used to be installed on seaplanes before WWII, which blanketed one another in crosswind thus reducing the yawing moment.

2.6 Special Features of Aircraft Operation on Water When Influenced by Negative Air Temperatures in Intense Fog and Poor Visibility

One of the serious threats to hydro-aviation aircraft operating in northern latitudes during frost is icing since the moist sea air at low temperatures can form an ice crust distorting aircraft aerodynamics, and when icing is high—shift the aircraft center of gravity. This circumstance is further aggravated by the intensive formation of sprays during take-off and landing of seaplanes. Therefore icing during operation on water is a special problem, which has been and is being tackled with little success. Nevertheless, the experience of the northern aviation of the USSR in the pre-war period showed that light non-metal aircrafts and seaplanes flew under the most unfavorable icing conditions and successfully performed their missions. This is also evidenced by the experience of operating the seaplanes MBR-2 and Be-6 in the North.

Beriev Aircraft Company together with Beriev Technical University of Southern Federal University have been developing equipment for analysis of wave and other climatic conditions at sea from an airplane flying or sitting on the water for many years and not without success.

As for intensive fogs and poor visibility, radar equipment and—mainly—equipment for careful observation from the air and afloat of water area clogging by floating objects should play a special role. The experience of Beriev Design Bureau in cooperation with the Redan shipbuilding firm in the 80 s of the last century when they were working on the Albatros amphibious support boats (A-42PE) showed that the powerful portable floodlights Problesk, installed on the launches, failed to discover a man in water in poor visibility at night even in rippled water in the Gulf of Finland. Quite unexpectedly, the possibility to search for objects on the water surface with an ordinary boat's radar enabled to see not only a dummy in a flying rescue suit, but also to observe ice splinters, the border of ice shore, floating vegetable crates, etc. This was the way the problem of rescue of people in distress floating on the water at night was solved. (Later it turned out that in the frontier marine outposts the search was carried out with the help of a cumbersome and heavy tank night vision device).

The aircraft that qualifies hands down as the amphibian with optimal protection during take-off and landing on any surface is the Bartini Beriev VVA-14 (Fig. 6, [7], which made its first flight in 1972). The aircraft is fitted with two retractable pontoons, 14 m long and 2.5 m diameter, which are pressurized at the moment of extension and retraction 0.015 to 0.025 kgf/cm^2 G (forming pressure) and 0.15 to 0.25 kgf/cm^2 (amortization pressure) in extended position. The Aircraft design take-off mass is 52 tonnes. The aircraft was manufactured and tested on the water and in the air. The aircraft cruise speed is 640 km/h, its payload mass is up to 2 tonnes, its practical flight range is 2450 km. Exactly such an aircraft is now required in connection with the global warming because the contamination of surface with floating ice floes, logs, barrels and other objects does not allow seaplanes or ground effect vehicles to land and take-off, while helicopters have insignificant range and cost effectiveness.

However, that aircraft failed to be materialised because the lift jet engines with auxiliary fans were never made.

Fig. 6 Robert Bartini's amphibious aircraft VVA-14 (USSR, 1972)

2.7 Basic Structural Design Concepts and Some Recommendations for the Jet Multirole Bottom-Amphibious Aircraft of Beriev Be-200 Type

The hydro-aviation aircraft types which have been created so far absorbed the best ideas of aerohydrodynamics. There are certainly a number of issues and problems that can be addressed in further engineering and scientific research. In this chapter, it is not possible to consider the course of further design stages, which largely coincide with the existing algorithms of aerobatic (land-based) aircraft design.

Therefore, we will limit ourselves to showing some drawings based on the general view of the Be-200 amphibious jet, the main purpose of which is to extinguish fires by scooping water into tanks during landing on a water body nearest to the wildfire, delivering it to the wildfire area and dropping it. The figures show the names of several parts of an amphibious aircraft, specific to the bottom planing aircraft. The data of the amphibious aircraft is tabulated in this chapter (Fig. 7).

The next Sect. 2.8, provides as support material the terminology and simplified recommendations on the seaworthiness and general arrangement of a bottom planing seaplane, prepared by Beriev Aircraft Company specialists Glombinsky and Kirsanov. Terms, definitions and symbols regarding the hydromechanics of aircraft of sea aviation are given in GOST 24,999–81 [8] (Fig. 8)

More complete information on the theory and practical recommendations for ensuring seaworthiness of hydroplanes and amphibian aircraft are given in special literature, which requires special training of performers in hydrostatics and hydrodynamics, as well as on regulatory materials for design, production, testing and operation of hydro aviation aircraft.

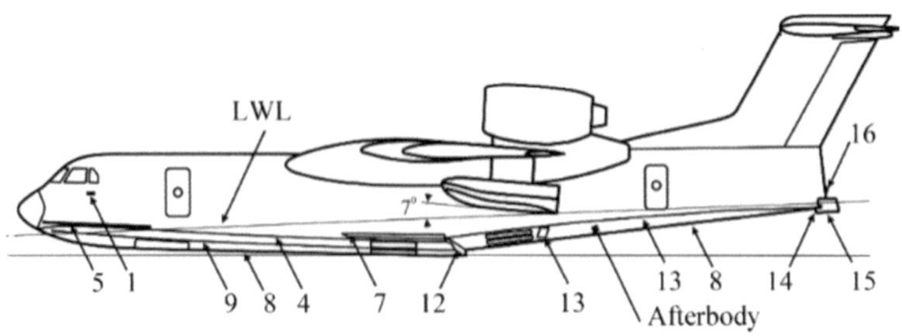

Fig. 7 Hydrodynamic surfaces and devices of the Beriev Be-200 seaplane 1—Nose ring used to tow the seaplane in water and to deploy floating anchors; 4—Chine, the line at which the boat side intersects with the bottom; 5—Spray rail; 7—Hydrodynamic flap; 8 —The line crossing the boat bottom in the centreline; 9—Longitudinal step; 12—First (main) step; 13—Deflector; 14—Rear step; 15—Water rudder; 16—Stern hook used to tow and moor the seaplane

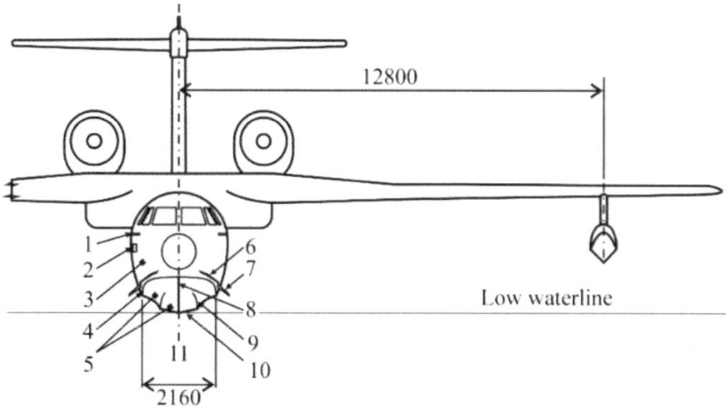

Fig. 8 Hydrodynamic surfaces and devices of a seaplane 1—Nose ring used to tow the seaplane in water and to deploy floating anchors; 2—Cleat for anchoring seaplane and mooring boats; 3—Board; 4—Chine, the line at which the boat side intersects with the bottom; 5—Bottom; 6—Spray rail; 7—Hydrodynamic flap; 8—Keel, the line crossing the boat bottom in the centreline; 9—Longitudinal step; 10—Water scoops in extended position; 11—Dynamic surface

2.8 Seaworthiness. Practical Algorithm for Determining the Geometry of the Bottom of a Planning Amphibious Aircraft

The preliminary studies set forth in the previous sections produced the following important data for the further design of an amphibious aircraft: the take-off mass, the maximum payload, the estimated range without payload and with payload, the take-off speed. However, as said above, the nomenclature of modern amphibious aircraft and seaplanes was too limited, for which reason the Beriev Aircraft Company developed methodologies for the preliminary and further design stages ensuring seaworthiness with the due account for the parameters of the prototypes of bottom seaplanes. The simplified methodology of preliminary calculations has been tabulated and is available in Table 9. In addition, the Beriev Aircraft Company and the Central Institute of Aerohydrodynamics have retained some expertise in designing sea aircraft capable of vertical and short take-off and landing.

2.8.1 About Seaworthiness

Seaworthiness is a set of measures ensuring safe operation of seaplanes under different conditions.

Seaworthiness is subdivided into hydrostatic and hydrodynamic.

The hydrostatic seaworthiness concerns buoyancy, stability, and insubmersibility.

Buoyancy is the ability to float on water without exceeding a given draft (waterline). It is ensured by the adequate displacement margin when the bottom is submerged in water.

Table 9 Some parameters of prototype aircraft

No	Parameter	Beriev Be-10	Beriev Be-12	Beriev A-42	CL-215T	US-1	SH-5	Beriev Be-200
1	Take-off mass m_0, t	50	35	92.5	20.8	43.0	55.0	42.5
2	Cruise speed V, km/h	720	440	700	290	440	450	610
3	Power plant	2 TJE	2 TPE	2 TJE	2 TPE	4 TPE	4 TPE	2 bypass turbofans
4	Wing loading p_0, kG/m^2	384	353.5	462.5	198.4	316.6	312.5	379.5
5	Boat length L, m	30.7	29.0	39.0	19.0	30.5	31.4	32.0
6	Hydrodynamic length L_h, m	28.86	21.31	37.6	19.0	20.5	29.0	30.0
7	Boat beam B, m	2.7	2.8	2.8	2.6	3.0	3.2	3.5
8	Boat height H, m	4.20	4.13	3.85	3.70	5.10	4.00	3.3
9	Hydrodynamic aspect ratio λ_h	10.7	7.6	13.4	7.3	8.85	9.0	13.5
10	Forebody length L_n, m	15.74	12.50	22.0	8.20	14.90	17.20	16.0
11	Afterbody length $L_{afterbody}$, m	15.0	10.1	17.01	10.8	13.3	14.2	14.0
12	Relative height of the first step, %	13.1	12.1	1.5/7.0	10.4	6.67	7.0	8.5
13	Forebody longitudinal deadrise γ_n, °	10 301	10 201	10.231	00	30	10	10 231
14	Afterbody longitudinal deadrise γ_h, °	80	80	6061	80	90	80	6061
15	Transverse deadrise (from keel), β_1, °	270.301	250	150/ 550	270	410	420	150/550
16	Transverse deadrise (from chine), β_2, °	220.301	200	34.40		300	220	34.40
17	Take-off/ landing speed, km/h	305	210	210	125	85	210	210
18	Z	2.36	2.10	3.33	2.32	5.15	2.14	2.90

Stability is the ability to retain a balanced position on water and return to it independently as soon as the external force ceases (for example, crosswind). There is longitudinal and lateral stability. It is ensured by wing floats, special containers by the sides (sponsons), by creating two-fuselage boats, etc.

Insubmersibility is the ability to retain buoyancy and stability with a pierced bottom. It is ensured by creating sealed compartments in the boat and floats, which do not lead (with the maximum permissible number of pierced compartments, located close to one another, and their flooding with water) to considerable consequences for the aircraft. The compartments are made either in the fuselage along its entire height, as in submarines, or under the watertight floor of the cabin.

The hydrodynamic seaworthiness includes:

- the ability to take off and land on the sea surface, including turbulent surface. It is ensured by the aerohydrodynamic configuration of the aircraft and its bottom;
- the longitudinal and lateral stability when moving on water; this is provided by bottom profiling and aerodynamic layout;
- the controllability when moving on water; this is provided by power plant layout, efficiency of aerodynamic, water or jet controls;
- the protection of vital elements of the structure against water spray; this is ensured by the aerohydrodynamic layout, bottom profiling and the presence of jet-spray deflectors;
- the ability to drift safely in the open sea; this is ensured by the adequate margin of buoyancy, stability, spray protection and configuration of the aircraft, which minimizes the formation and effect of water spray on the structure as it slams into waves when drifting;
- the ability to be towed by a sea plane and docked to an anchored container (barrel buoy); this is ensured by the appropriate bottom shape, the presence of means of communication with tug-boats (for example, radio stations, lifeline guns, controlled cleats and towline winches of tug-boats, cable cutters of emergency release, etc.) and the means of buoy docking/undocking.

Other specific requirements regarding seaworthiness parameters are given in special documents (air regulations, general specifications, airworthiness requirements, etc.).

2.8.2 The Practical Algorithm for Determining the Geometry of the Bottom of a Planning Amphibious Aircraft

The calculation algorithm presented in this chapter is exemplified by two real amphibians— the Beriev A-42 (take-off mass 92.5 t) and the Beriev Be-200 (42.5 t), which have rather similar cruise speeds of 700 km/h and 610 km/h respectively. In Table 9, some parameters of prototype aircraft are given. The outlined dimensions are given in Fig. 9.

The algorithm for calculating some of the seaworthiness parameters is shown in Table 10.

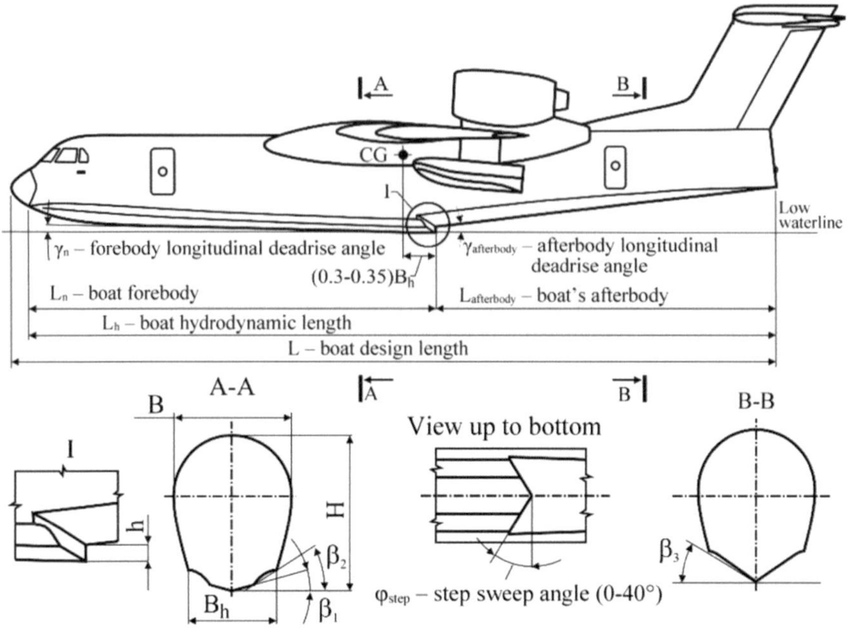

Fig. 9 Main dimensions of an amphibious aircraft

The designations of the main dimensions determined according to the algorithm in Table 10 correspond to Fig. 9.

For the values of wave height and sea condition, and also the wind speed and its condition, refer to Table 11.

Table 10 Determination of seaworthiness parameters of seaplanes

No	Parameters	Design dependencies	Results of calculation for the prototype aircraft	
			Beriev A-42 PE	Beriev Be-200
1	Take-off mass m_0, t	Calculated or adopted from the design specifications	92.5	42.5
2	Cruise speed V, km/h		700	610
3	Hydrodynamic aspect ratio λ_h	From the prototype aircraft, Table 8	13.4	13.5
4	Line efficiency coefficient,n	$n = 0.018\ to\ 0.024$	0.021	0.021
5	Cargo beam of the boat B_h, m	$B_h = (\Delta_0/\gamma\lambda_h^2 n)^{0.333}$ $\Delta_0 = m_0$; $\gamma = 1\ t/m^3$ of water	2.90	2.23
6	Hydrodynamic length of boat L_h, m	$L_h = \lambda_h B_h$	38.9	30.0
7	Design length of boat L, m	$L = (1.02\ to\ 1.1)L_h$	41.2	31.8
8	Length of boat forebody L_n, m	$Ln = (0.45\ to\ 0.55)L_h$s	20.6	15.9
9	Length of boat afterbody $L_{afterbody}$, m	$L_{afterbody} = L_h - L_n$	18.3	14.1
10	First step height h, m	$h = (0.05\ to\ 0.12)B_h$— from the prototype aircraft	0.25	0.19
11	Boat height H, m	$H = (1.3\ to\ 1.7)B_h$	4.35	3.35
12	Forebody longitudinal deadrise γ_n, °	$\gamma_n = (0\ to\ 5)°$— from the prototype aircraft	10,231	10,231
13	Afterbody longitudinal deadrise γ_h, °	$\gamma_h = (6\ to\ 9)°$— from the prototype aircraft	60 061	60 061
14	Transverse deadrise from the keel to the first step β_1, °	$\beta_1 = (15\ to\ 25)°$— from the prototype aircraft	(15/55)°	(15/55)°
15	Transverse deadrise of afterbody β_2, °	$\beta_2 = (25\ to\ 45)°$— from the prototype aircraft	34.40	34.40
16	Aircraft draft T, m	$T = V_m/L_h B_h \delta(\delta = (0.4\ to\ 0.5))$	1.9	1.4

Table 11 Statistical data on sea conditions and wave height for seaplanes

Wave height h, m	Up to 2	Up to 3.5	Up to 6	Up to 8.5
Sea condition	0 to 4	5	6	7
Wind condition	0 to 3	4 to 5	6 to 7	> 7
Speed of wind, m/s	Up to 5	Over 5 to 10	Over 10 to 15	> 15

3 Conclusion

The materials presented in this chapter make it possible to determine the appearance of a seaplane during preliminary design.

Some examples of real seaplanes, amphibious aircraft and new designs are given in Figs. 10, 11, 12, 13, 14, 15, 16, 17, 18 and 19 [9].

Variables

Symbol	Description	Units
m_{pay}	Payload mass	t
L	Range	km
V	Cruise speed	km/h
m_0	Take-off mass	t

(continued)

Fig. 10 SR/45 Princess seaplane (UK, 1952)

(continued)

Variables		
Symbol	Description	Units
U	Transport efficiency criterion	km^2/h
$m_{pay,max}$	Maximum payload mass	t
$\overline{m}_0$	Aircraft mass excellence	–
$R_{max,d}$	Maximum design range	km
U_p	Estimated value of transport efficiency criterion	km^2/h
$\overline{m}_{pay} = \frac{m_{pay}}{m_{pay,max}}$	Payload mass ratio	–
$\overline{m}_{0,ex}$	Average mass excellence for the given take-off mass	–
L/D	Lift-to-drag ratio	–
$C_L = \frac{m_{pay,max}}{R_{max}}$	Fuel consumption per kilometer	t/km
$(C_L)_k$	Statistical fuel consumption per kilometer	t/km
$C_{p,H}$	Engine specific fuel consumption	t/(tf·hour)
h	Height of waves	m
$\overline{T}$	Probability of mission success	–
M	Mach number	–
$\overline{p}_H$	Mach number factor, shows the variation of engine thrust and specific fuel consumption with Mach number	–
Z	Efficiency of high-lift devices	–
$m_{pay,c}$	Reference payload mass	t
i	Number of engines	–
P	Thrust	N
q	Wing loading	t/m^2
S	Wing area	m^2

Fig. 11 XP6M-1 SeaMaster seaplane (USA, 1955)

Fig. 12 Beriev Be-10 seaplane (USSR, 1956)

Fig. 13 Beriev Be-12 amphibious aircraft (USSR, 1960)

Fig. 14 PS1-A STOL seaplane (Japan, 1967)

Fig. 15 Harbin SH5 (China, 1976)

Fig. 16 Canadair CL-415 amphibious aircraft (Canada, 1994)

Fig. 17 Beriev A-40 Albatros amphibious aircraft (USSR, 1986)

Fig. 18 Beriev Be-200 amphibious aircraft (USSR, 1998)

Fig. 19 Beriev Be-2500–2 superheavy seaplane (Beriev Aircraft Company design 2008)

References

1. Svishchev GP (eds.) (1994) Aviaciya. Enciklopediya (Aviation. Encyclopedia), Scientific publishing house Great Russian Encyclopedia TsAGI, Moscow
2. Kaznevsky VP (1997) Robert Ludwigovich Bartini. Nauka, Moscow

3. Fortinov LG (2001) Osnovy nauchno-statisticheskogo proektirovaniya letatel'nyh apparatov gidroaviacii na predvaritel'nom etape (Fundamentals of scientific and statistical design of seaplane aircraft at the preliminary stage). Beriev Aircraft Company Publisher, Taganrog
4. Logvinovich GV (1996) Gidrodinamika gidrosamoletov (Hydrodynamics of amphibians). In: Abstracts of the conference on hydroaviation «Gelendzhik-96», TsAGI Publisher, Moscow
5. Sokolyansky VP (1997) Aero-hydrodynamics of Amphibians. In: Proceeding of the fifth Russian-Chinese symposium on aerodynamics and flight dynamics, TsAGI Publisher, Moscow
6. Myshko IG, Shorin AN (1965) Spravochnye dannye po rezhimu vetrov i volneniya v okeanah, Registr Soyuza SSR (Reference data on the wind and wave regime in the oceans, Register of the USSR). Transport, Moscow
7. Udalov KG, Panatov GS, Fortinov LG (1994) Samolet VVA-14 (Aircraft VVA-14). Aviko-Press, Moscow
8. Gidromekhanika letatel'nyh apparatov, terminy, opredeleniya i oboznacheniya GOST 24999-81 (Hydromechanics of aircraft, terms, definitions and designations GOST 24999-81) (1982), Publishing standards, Moscow
9. Fortinov LG (1998) Osnovnye gidrosamolety i samolety – amfibii mira perioda 1933…1997gg (The main seaplanes and amphibious aircraft of the world from 1933 to 1997) Beriev Aircraft Company Publisher, Taganrog
10. Eger SM, Mishin VF, Liseitsev NK et al (2005) Proektirovanie samolyotov (Aircraft Design). Logos, Moscow

Unmanned Aircraft

Nikolai A. Eremin

Abstract The chapter starts by discussing the roles of unmanned aircraft and the composition of unmanned aerial vehicle systems. The chapter explains the possible take-off and landing procedures for unmanned aircraft, as well as the layout and design parameters of unmanned aircraft of various types, including tactical low-speed unmanned aerial vehicles, unmanned long-endurance aircraft, unmanned muti-role medium-altitude long endurance aircraft, and unmanned high-altitude long endurance aircraft.

Keywords UAV role · System design · Take-off and landing · Design parameters

1 Role and Place of Unmanned Aircraft

A characteristic feature of the development of aviation technology at the turn of the 20th and 21st centuries was the rapid increase in both the volume and application areas of unmanned aircraft due to advances in robotics, including advances in electronics, automation of communication and control systems, and "artificial intelligence" technologies. Mastering technologies of automatic takeoff and landing,

---

Unmanned aircraft is a type of unmanned aerial vehicles (UAV), which are defined as the vehicles with no crew on board.

UAVs can be controlled automatically by means of onboard software devices or remotely by the operator via radio from the control station. The control station can be located either on the ground—ground control station (GCS), or on board a manned aircraft or a spacecraft. In turn, remote control can be discrete or continuous. In the discrete control mode, the operator periodically receives information about the flight progress and can make adjustments to the flight programme. In the continuous control mode, the operator actually pilots the aircraft without being on board. Unmanned aerial vehicles implementing continuous control mode are also called remotely piloted aircraft (RPA).

N. A. Eremin (✉)
Sukhoi Design Bureau, Moscow, Russia
e-mail: yeremin-nikolay@list.ru

M. A. Pogosyan (ed.), *Aircraft Design*, Springer Aerospace Technology,
https://doi.org/10.1007/978-981-96-4599-2_15

taxiing, formation flight, solving problems of autonomous route planning, identification and attack of ground targets (for combat UAVs) allow us to consider modern and advanced unmanned aircraft as robotic analogues of manned aircraft.

In terms of the mission, an unmanned aircraft can be both an alternative to a manned aircraft and a unique tool to solve problems that are inaccessible to a manned counterpart.

The unique ones include, first of all, tasks related to physiological constraints on the human body. Such limitations include:

- the long flight time, which is unacceptable due to crew performance (for example, RQ-4 "Global Hawk" unmanned high altitude reconnaissance aircraft has the endurance of about 24 h, whereas the endurance of manned high altitude reconnaissance aircraft is limited to 6–12 h);
- the long duration high speed flight in terrain-following mode associated with a prolonged exposure to alternating overloads. As practice shows, pilots find it difficult to tolerate such flight modes;
- manoeuvring with overloads and angular velocities exceeding the thresholds tolerated by the human body;
- the use of powerful radioelectronic equipment on board the aircraft, producing the fields that exclude the presence of a human in the vicinity;
- the use of the aircraft in conditions of chemical or radioactive contamination.

Also, UAVs fulfill the missions associated with a high risk of losing the aircraft and consequently endangering the life of the crew, for example:

- the combat use under conditions of unsuppressed enemy air defense system with a high probability of hitting its own aircraft;
- the application in unfavourable environmental conditions, e.g. monitoring in ecological or man-made disasters.

At present, drones are mostly used in the military sphere, first of all as reconnaissance aircrafts, starting from hand-launched micro UAVs designed to determine the tactical situation for a single platoon, and ending with strategic reconnaissance drones hovering over 24 h at altitudes up to 20 km and performing global monitoring in the interests of the general command.

Specialized combat drones are also being developed that carry a diverse range of weapons and employ low-signature technology.

Unmanned aircraft can also be used for:

- electronic warfare;
- communication relaying, including from other UAVs;
- laser "illumination" of the targets;
- as aerial decoys or as airborne targets for the training of air defence personnel.

For civilian applications, drones are mainly used for various monitoring purposes:

- monitoring of public order, traffic, state borders, nature reserves and providing information assistance in identifying and apprehending offenders;

- control over the condition of roads, road junctions, railway junctions, engineering structures, oil and gas pipelines;
- environmental monitoring, research tasks, assistance to weather forecasting services, information support during emergencies (forest fires, man-made disasters, etc.).

In addition, unmanned aircraft can find their application as:

- the relay aircraft, given the possibility of prolonged patrolling, for example, unmanned solar-powered electric aircraft can stay in the air uninterruptedly for up to several weeks;
- the agricultural aircraft that have to fly at altitudes of the order of a few meters when conducting aerochemical work, which is associated with an increased probability of collision with the ground;
- the cargo planes.

When selecting an alternative between manned and unmanned aircraft, they should be compared comprehensively, taking into account all the advantages and disadvantages of using unmanned aircraft for specific missions.

The advantage of unmanned aerial systems is the expected reduction of the life cycle cost due to:

- lower operating costs, especially for combat UAVs;
- savings on crew training (e.g., the cost of training a UAV operator is ~ 5% of the cost of training a Class 1 military pilot);
- a lower cost of the unmanned aircraft itself compared to its manned counterpart due to the absence of a number of systems and a smaller size.

No crew on board the UAV:

- reduces the influence of psycho-physiological factors on the mission;
- removes limitations, caused by physiological capabilities of human body, on maximum values of overloads, angular velocities, flight endurance;
- does not require crew rescue operations in case the aircraft is lost over enemy territory;
- ensures the absence of human losses, which, apart from the moral factor, has an important military significance, as the time cost of combat pilot training (5–8 years depending on the class) does not allow the recovery of full crews during a single military conflict.

The absence of layout limitations due to the crew and their life support systems allows the creation of an unmanned aircraft of a smaller size, with less surface inhomogeneity and, as a consequence, with less radar visibility compared to its manned counterpart.

The main disadvantage of unmanned aircraft compared to manned aircraft is less flexibility and autonomy of application, due to the currently insufficient level of development of "artificial intelligence" technologies, which does not allow to fully compensate for the absence of the crew on board. As a consequence, there are a

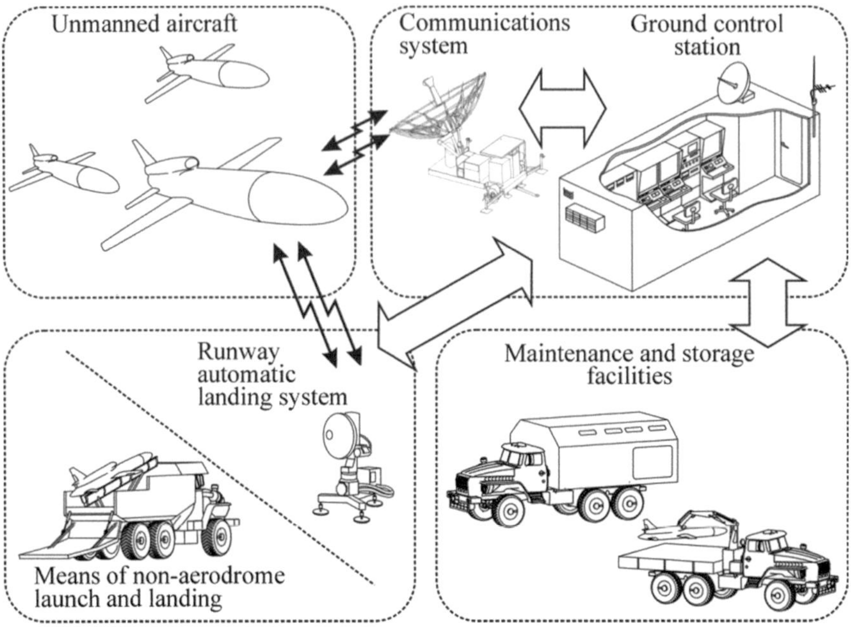

Fig. 1 Composition of an unmanned aerial system

number of tasks whose full automation may be questioned in the near future. Such tasks include:

- passenger transportation, which requires increased reliability provided by the presence of a crew that can take control in case of a malfunction of the automatic control system, as well as take non-standard actions to save the aircraft in an emergency situation;
- its use as a battlefield aircraft to provide close fire support to troops when a close visual examination of a complex and rapidly changing ground situation is required, precluding the targeting of own troops;
- combat use as a maneuvering fighter in close air combat, characterized by high dynamics, complexity, multi-variant maneuver, which can be achieved by non-standard "creative" approach of the pilot.

2 Unmanned Aerial Vehicles Within Unmanned Aerial Vehicle Systems[1]

An unmanned aerial vehicle cannot operate completely autonomously and is part of an unmanned aerial system (UAS) which includes (Fig. 1):

[1] The section is written using material from [1].

- one or more unmanned aerial vehicles (the system usually consists of 2 to 8 aircraft);
- ground control station (GCS) with UAS communication facilities, designed to control unmanned aerial vehicles, process information received from them, as well as interact with its users;
- means of automatic landing and taxiing for unmanned aircraft performing conventional takeoff and landing;
- transport, launcher and rescue equipment for the UAVs that do not require runways for take-off and landing;
- maintenance tools to ensure the preparation and operation of the UAVs;
- UAV storage facilities.

Figure 2 shows the diagram of unmanned aircraft system application with unmanned aircrafts performing takeoff and landing using airfield.

In principle, the aerodrome-based UAV operation does not differ from the application of manned aircraft systems, except for full automation of piloting an unmanned aircraft, including the takeoff, landing and taxiing phases, as well as the need to ensure stable communication channels between the UAV and the GCS at long distances, for which aircraft and relay satellites are used.

The advantages of unmanned aircraft, among others, are their smaller size compared to their manned counterparts and the capability of operating without the

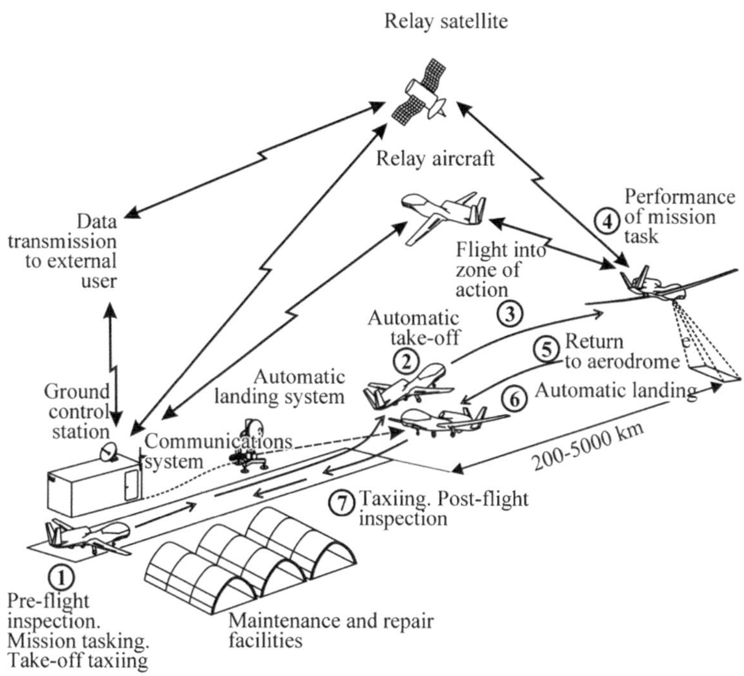

Fig. 2 Diagram of an aerodrome-based unmanned aerial system

need to use aerodrome. The diagram of application of the UAS of this type is shown in Fig. 3.

Aircraft in this case are placed and delivered to the launch site in transport/charge vehicles (TCV). The unmanned aircraft delivered to the position are installed on the launchers (L), undergo the preflight inspection, and the flight task is entered. Sometimes, TCV and L are combined in one vehicle and the UAV is launched from the transport container. Following the launch command, the aircraft takes off from the rails using boosters or ejection devices and flies into the area of operation. On completion of the mission, the UAV is automatically guided to a preset landing point and lands by means of parachute or other method. From the landing site, the aircraft is

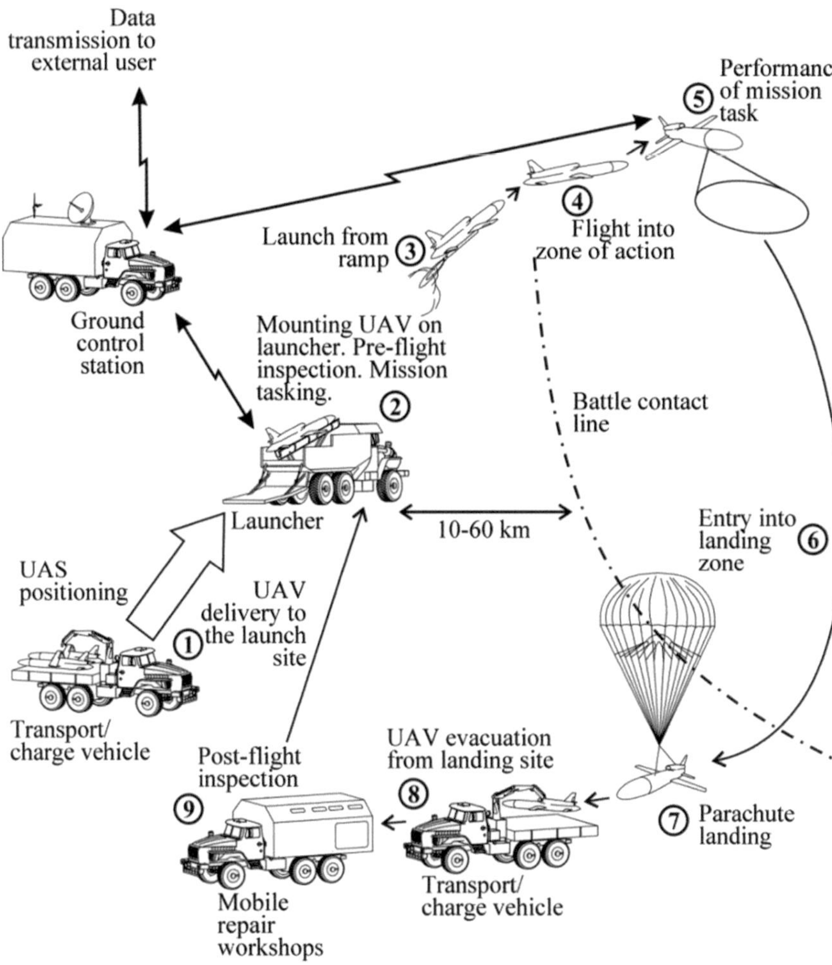

Fig. 3 Diagram of an off-aerodrome unmanned aerial system application

evacuated by the transport/charge vehicle and delivered to the so-called maintenance area for post-flight inspection.

The main components of unmanned aircraft systems are described below.

Ground control station

The main objectives of the GCS are:

- management of all elements of the system;
- development of UAV flight programs and, if necessary, their correction in the course of the flight mission;
- pre-launch check of the UAV and its onboard equipment;
- control of flight and airborne equipment of UAVs in the air, transmission of command information to and reception of telemetric information about flight parameters and system operation from the UAV, their registration and display;
- reception of target information from the UAV, its display, processing and interpretation, registration on electronic media, creation and storage of logs of the objects surveyed;
- interaction with external users of information;
- training of the UAS crew without real UAV launches.

As a rule, the GCS are made as a single chassis-mounted cabin, with the dimensions suitable for air transportation. The GCS cabin houses the workstations of UAS commander and operators (Fig. 4).

Modern unmanned aerial vehicles implement predominantly discrete control mode, which allows a single operator to control several aircraft simultaneously.

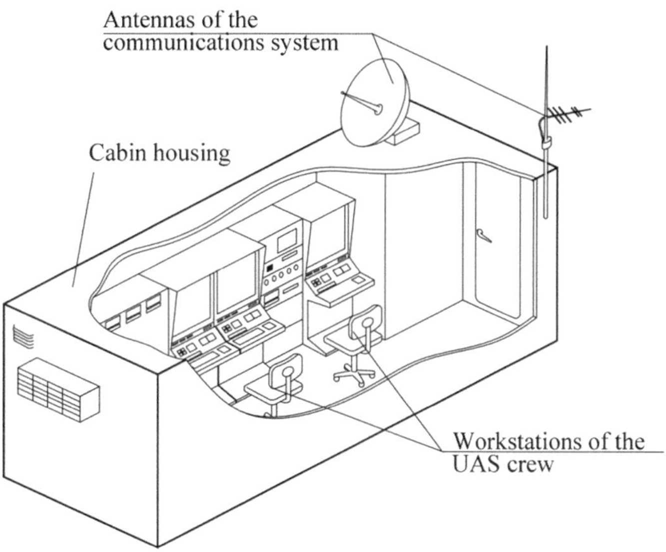

Fig. 4 The cabin of the unmanned aircraft system control station

Continuous control mode based on remote piloting of an unmanned aircraft is not widespread because, unlike the pilot, the operator on the ground is devoid of visual information in a wide range of angles given by binocular vision, subjective information on g-loads perceived by pilots, acoustic information on the operation of the power plant, etc. A remotely piloted aircraft requires a communication channel that ensures continuous transmission of large-volume video information, which at considerable distances from the GCS in a complex noisy environment (for combat aircraft) may cause known technical difficulties. Besides, continuous transmission of radio signal de-mask the aircraft emitting it in the radar range and nullifies all attempts to implement technologies to reduce radar signature on combat unmanned aircraft. However, the possibility of continuous control in some drones is retained as a backup emergency mode.

Take-off and landing of an unmanned aircraft

For implementation of off-aerodrome launch, various types of launchers are used. The most common methods of launching are those using powder boosters or catapults, the latter usually being of pneumatic or hydraulic type.

In case of ramp launch assisted by powder boosters (solid propellant rockets), they are usually mounted on the UAV such that the engine thrust passes through the centre of mass of the aircraft at a slight angle to its longitudinal axis. Since the final acceleration of the UAV by boosters can also be performed after its separation from the launcher, the advantage of this launch method is the minimum size of rails from which the aircraft starts and, therefore, the compact size of the launcher, which allows, inter alia, the launch of the UAV from the transport container (Fig. 5), thereby combining the transport vehicle and the launcher. The shortcomings of the solid propellant rocket are the dependence of the thrust of accelerators on the ambient temperature, the large signature of flame, gases and sound, as well as the cost of accelerators. Therefore, this launch method is usually used for UAVs with relatively high get-away speeds (high-speed aircraft with high wing loading), or when the dimensions of the launcher are limited, for example, when launching from a container or from the deck of a ship.

Fig. 5 Off-aerodrome launch of the Tu-143 UAV from a transport container using a powder booster

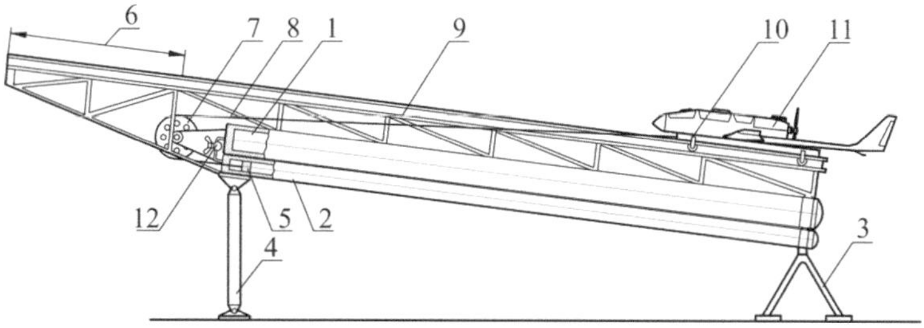

Fig. 6 UAV launch using a pneumatic catapult: 1—receiver with compressed air, 2—operating cylinder, 3—rear support, 4—front support, 5—piston, 6—braking section, 7—pulley; 8—cable; 9—ramp; 10—acceleration trolley; 11—UAV; 12—electro-pneumatic valve

In case of a pneumatic or hydraulic catapult launch (Fig. 6), the aircraft is mounted on the acceleration carriage, which is driven by compressed air or hydraulic fluid pressure along the launcher rails and accelerates the aircraft until it reaches the take-off speed. The advantage of such a start is the absence of problems associated with the use of powder boosters; the disadvantage is the large size of the launcher rails, which are composed of the length of rails needed to accelerate the aircraft and to decelerate the cart after the launch. Therefore such launching method is used for low-speed tactical UAVs with take-off mass up to 300–500 kg having low wing loading and low take-off speed.

Another method of UAV launch is using an airplane or helicopter, which is performed similarly to the launch of aviation rockets (Fig. 7).

Among the off-aerodrome landing methods, the landing net and the landing parachute energy-absorbing system are widespread.

The UAV net recovery (Fig. 8) is used to recover small-size and low-speed unmanned aircraft (take-off mass up to ~ 180 kg and landing speed up to ~ 120 km/h). This landing method requires virtually no cost to prepare the landing site, and the net itself is much smaller than any possible landing site. The net-assisted recovery of

Fig. 7 Placement of the BQM-74 UAV on board the aircraft

Fig. 8 Recovery of the Shadow 200 UAV by capture net

an unmanned aircraft is used on the UAV launched from ships because it enables the "dry" method of landing without landing on water. The latter requires that additional measures be taken to protect the structure and systems against sea water and to lift the UAV to the ship.

Landing with the use of the parachute/energy absorbing system (Fig. 9) is the most widely used method of recovery of unmanned aircraft.

The advantage of parachute landing is that it requires practically no expenses for landing site preparation.

The main disadvantage of parachute landing is a higher probability of aircraft damage. The number of flights of an unmanned aircraft may be significantly limited due to high loads applied to the aircraft structures during parachute landing. In addition, after the UAV parachute landing, it should be evacuated to the maintenance area, for which special transport vehicles equipped with a crane are used. The re-use of the aircraft is possible only after the UAV has been tested for damage during landing and its functionality is appropriately restored. All this increases the time of preparation of the unmanned aircraft for the next use.

When the UAV dimensions increase up to the values of takeoff mass over 3000 kg, off-aerodrome operation begins to lose its advantages, because for heavy and large-sized UAVs it becomes difficult to create mobile transportation and launching devices, and ensuring the evacuation of such UAV from the parachute landing site will require the use of special crane lifts. Besides, the creation of a heavy and, as a consequence, expensive unmanned aircraft with a limited (by rough parachute landing) number of flights is not economically reasonable. Therefore unmanned aircrafts with takeoff weight more than 3–5 tones perform conventional take-off and landing on airfields. The takeoff of such aircrafts is performed similarly to the piloted analogues in the automatic mode. The automatic landing system can be arranged in the following ways:

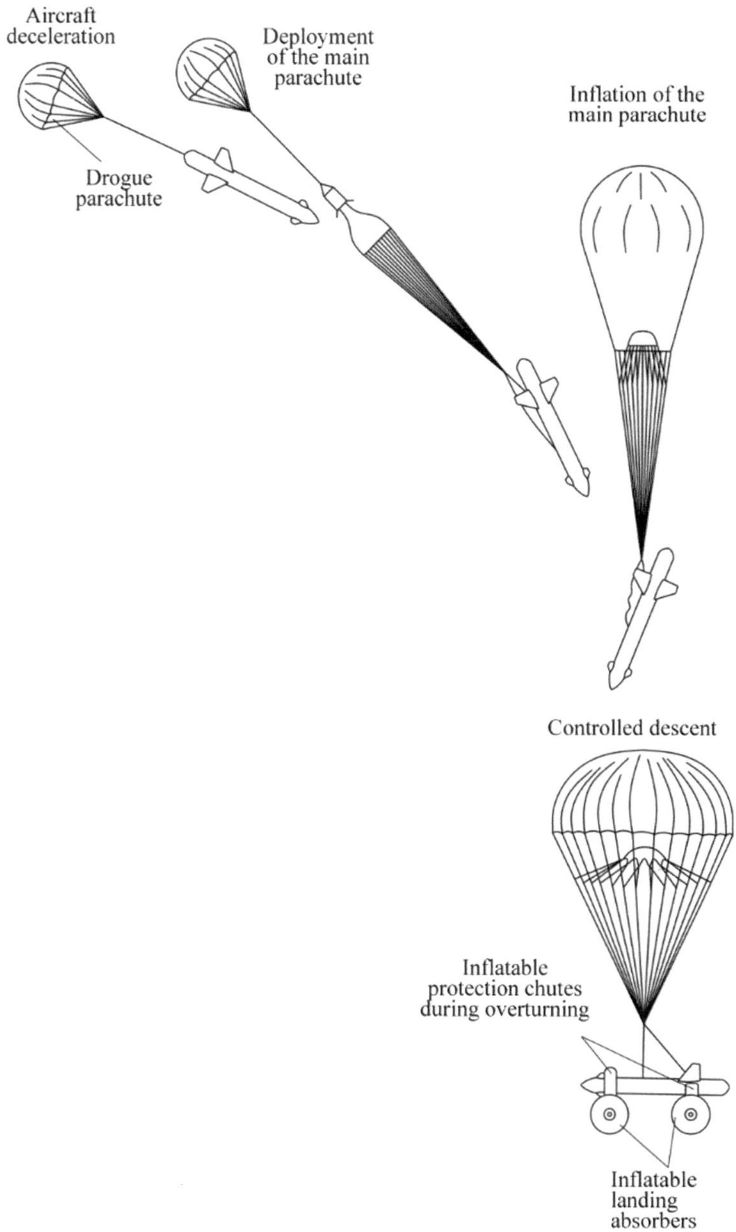

Fig. 9 Parachute landing of the CL-289 UAV

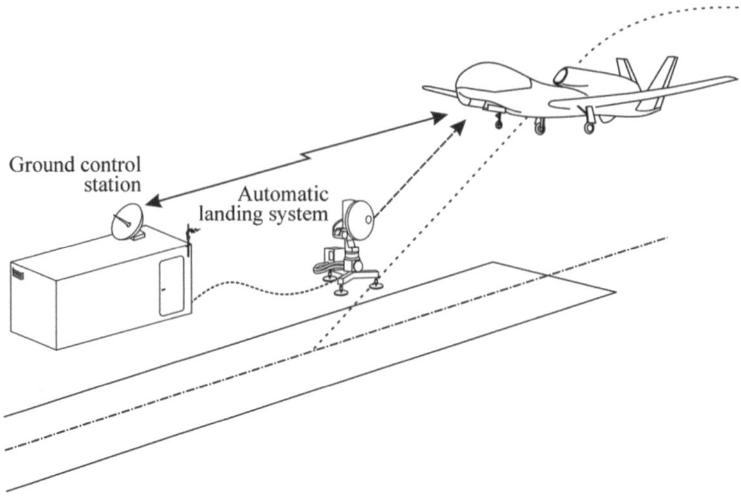

Fig. 10 Automatic landing of an UAV using ground automatic landing system

- automatic landing using special equipment installed on the ground and on the UAV. The equipment located on the ground forms a light (radio) beam, and the sensors installed on the aircraft produce signals proportional to the deviation of the vehicle's center of mass from the beam (Fig. 10);
- automatic landing using differential methods of satellite navigation, which implies placing the satellite information receiver at some reference point, the coordinates of which are well known. UAV coordinates are determined relative to this point (Fig. 11).

UAV storage facilities

Unlike manned aircraft, their unmanned counterparts do not need to perform constant flights for crew training, so for a number of unmanned aircrafts (especially combat aircraft used only for military operations or exercises) the concept of long-term storage in special containers in which UAVs are also transported to the place of application is implemented. Figure 12 shows an example of placing a demonstrator of the X-45A unmanned combat aircraft in such a container, as well as the process of loading the container with the UAV into a military transport aircraft. By regulating the temperature and humidity inside the container and providing remote control of the aircraft and its equipment, long-term storage of the UAV without opening the container for up to 10 years can be implemented.

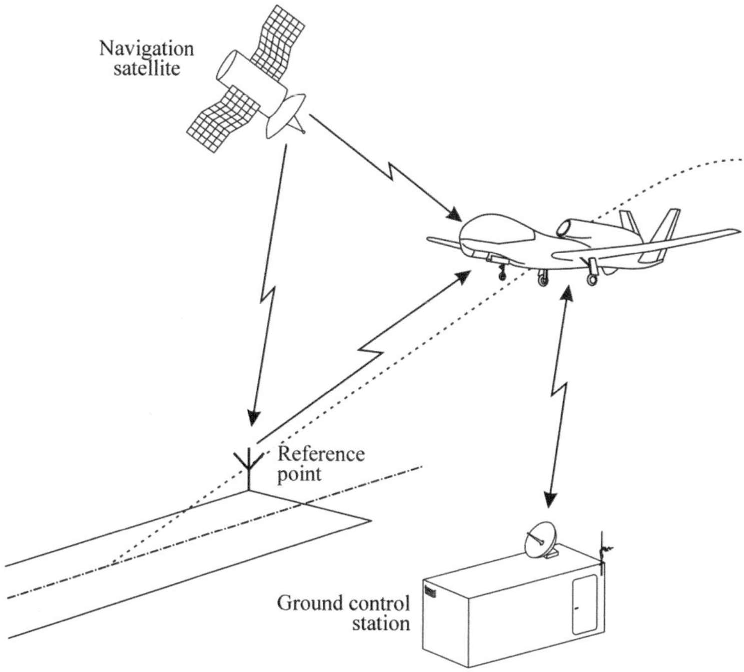

Fig. 11 Automatic landing of an UAV using satellite navigation

3 Layout and Design Parameters of Unmanned Aircraft of Various Types

The approach to determining the design parameters of an unmanned aircraft and analyzing its performance is not fundamentally different from that for manned aircraft. The differences lie in the constraints associated with the absence of a human on board.

Unmanned aircraft generally have lower strength and reliability requirements than manned aircraft. In particular, unmanned aircraft are typically designed for a safety factor of 1.3 compared to 1.5 for manned aircraft. Equipment and engines designed for the UAVs intended for a few flights may have reduced reliability characteristics and low life expectancy.

When setting the life requirements for an unmanned aircraft, it is necessary to take into account the peculiarities of its operation during its life cycle. For example, the service life of a combat unmanned aerial vehicle may be significantly less (by an order of magnitude) than that of a manned aeroplane because it is not flown in peacetime, but the service life before decommissioning may be quite long, comparable to that of its manned counterpart.

When designing an unmanned aerial launch vehicle, it is necessary to match the given limitations on the UAV (in terms of the maximum velocity head, g-loads, etc.)

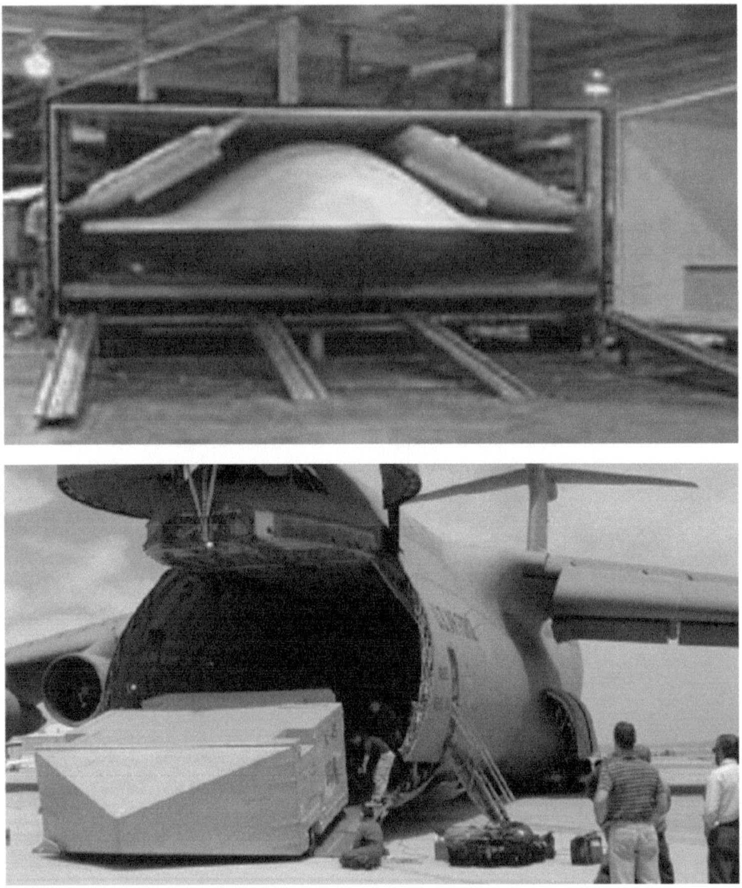

Fig. 12 Placing an unmanned aerial vehicle in a long term storage container

to the performance of the intended carrier aircraft so that the use of the unmanned aircraft does not become a rigid limitation on the carrier flight profile.

When selecting aircraft systems, it is desirable, if possible, to exclude systems with components and operating media subject to rapid aging that require replacement at short intervals. As an example, consider the concept of an aircraft with actuators built entirely on electric drives, without the use of a hydraulic system.

Particular attention should be paid to placement of antenna posts on UAVs that provide, if possible, an all-angle view of the transmitting antenna that does not depend on the mutual orientation of the UAV and the GCS, especially if large amounts of target information need to be transmitted in real time.

The main types of unmanned aircraft with the take-off weight exceeding 100 kg are briefly described below and their main characteristics and design parameters are given.

3.1 Tactical High-Speed Unmanned Aircraft

These are relatively small aircraft (Figs. 13 and 14) with the take-off mass between 300 and 3000 kg, equipped with TJE, designed to fly at a high subsonic speed (M~ 0.6–0.8), having short flight endurance (~0.5–3.5 h) and the radius of action (~100–900 km).

The flight altitude range of 100–6000 m is limited by the characteristics of the equipment and propulsion system.

High-speed UAVs find their application in environments with strong enemy air defense, where a low-speed aircraft with a long flight time can be shot down before

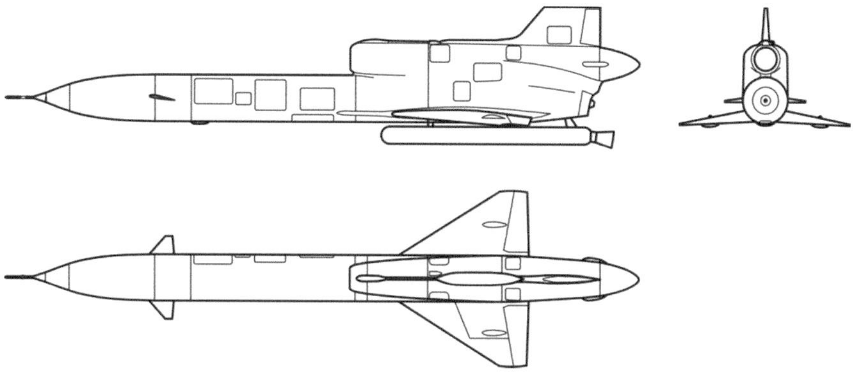

Fig. 13 Tupolev Tu-143 high-speed unmanned surveillance aircraft

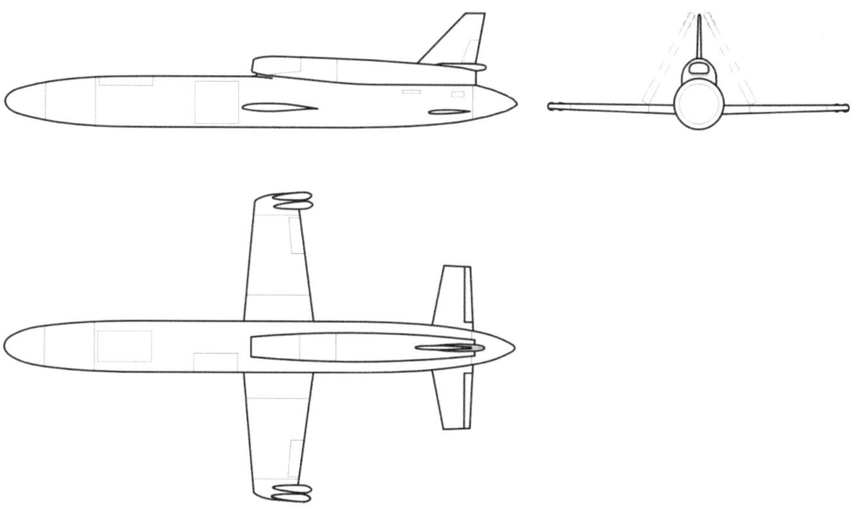

Fig. 14 *Dan* aerial target

it has time to accomplish its mission. High-speed small aircraft are more likely to overcome the air defense, and in addition, more quickly get to the target area. For reconnaissance missions, such UAVs are equipped with small radar, infrared and optical sensors, as well as real-time data transmission systems.

Another application of high-speed UAVs is their joint use with combat aircraft. In this case, some of the functions normally performed by the combat aircraft themselves are transferred to them. The use of UAVs for reconnaissance of ground targets and subsequent assessment of damage to the enemy after a strike makes it possible to improve the survivability of the piloted aircraft during its combat mission.

In addition to reconnaissance missions, high-speed drones serve as aerial targets (Fig. 14) because their flight characteristics best allow them to simulate various aerial targets.

High-speed UAVs implement the off-aerodrome concept and are launched from a ramp with the use of powder boosters or from aircraft, and land with a parachute. Shock absorbing is provided by the extendable spring-actuated supports and pneumatic shock absorbers made as soft shells consisting of canvas-covered rubber that is filled, during landing, with the gas bled during compression via special valves.

The off-aerodrome launch with subsequent acceleration by powder boosters and parachute landing remove the limitations on the wing loading associated with the low take-off and landing speeds. Therefore, the wing loading and thrust-to-weight ratio of such aircraft are determined for the cruise flight at a high subsonic speed and amount to 300–400 daN/m^2 and 0.3–0.4, respectively.

The aerodynamic layout is also optimized for cruising flight. As a rule, these aircraft feature the wing-tail layout with a straight or moderate sweep (25–35°) medium-aspect ratio wing (4–7) or the tailless layout with a delta wing. The maximum lift-to-drag ratio reaches the values of 9 to 11. For transportation purpose, the wings of such aircraft can be made foldable.

3.2 *Tactical Low-Speed Unmanned Aerial Vehicles*

Tactical low-speed UAVs are the most common class of unmanned aerial vehicles. These are small (with the take-off mass of 150–300 kg) and low-speed (the cruise speed is 100–150 km/h) aircraft equipped with piston or electric motors (Fig. 15). These vehicles have a radius of action between 50 and 150 km and are designed for battlefield reconnaissance, control of the results of combat operations, target designation for artillery. The unmanned aircraft of that class are also widely used in civil applications involving various types of monitoring.

Tactical low-speed UAVs are usually equipped with small video cameras and infrared sensors. Unlike the high-speed tactical UAVs described above, these vehicles have a longer flight time (up to 4 to 7 h) and are lighter and simpler.

The operating altitudes are 100–3000 m and are limited primarily by the range characteristics of small-size payload equipment.

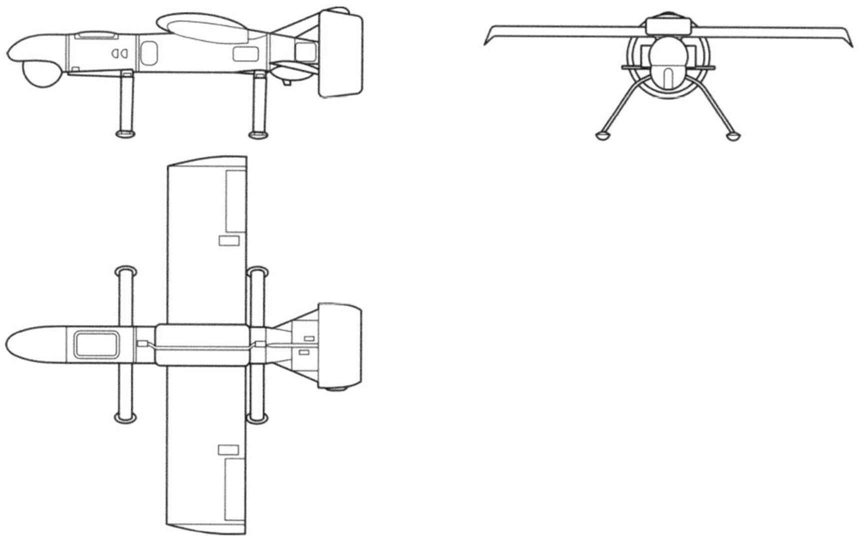

Fig. 15 *Pchela* tactical low-speed surveillance unmanned aircraft

These vehicles are usually pneumatically or hydraulically catapulted and perform either parachute or aircraft-style landing. They have spring-type landing gear, as a rule, not retractable in flight, or are made without any landing gear and land on their body.

Such aircraft predominantly have the wing-tail configuration with a straight wing of moderate AR (6…10) with the thickness/chord ratio varying between 12 and 13%.

The wing loading corresponding to the low-speed flight equals 60–70 daN/m^2, the power-to-weight ratio is 0.12–0.25 hp/daN. The payload is placed in the nose or central parts of the fuselage.

3.3 Unmanned Long-Endurance Aircraft

This type comprises unmanned aircraft capable of loitering for a long time (24 h and more). Prolonged stay of the UAV in the air makes it possible to minimize the number of aircraft involved in the task (e.g., reconnaissance in a given area), as well as to ensure a high degree of readiness of the aircraft for use from the "duty in the air" position.

With the long endurance, such UAVs can perform their missions at a great distance from the ground control station (up to several thousand kilometers). In this case, the control of the aircraft and the transmission of a large volume of reconnaissance information over long distances in real time are carried out through relays (including satellites).

The UAVs of this class take off and land as conventional aircraft. Since such UAVs have long range and the ability to communicate with the GCS at any distance, they can operate from a limited number of airfields and the design of such aircraft may not impose stringent requirements for takeoff and landing characteristics.

Several types of long-endurance UAVs can be distinguished, differing in size and loiter altitudes.

Unmanned mutirole medium-altitude long endurance aircraft

The main purpose of such UAVs is continuous (up to 30 h) aerial surveillance and real-time transmission of information to the user. In military applications, the armaments are arranged in external stores which allows quick strikes. However, these aircraft, the overwhelming majority of which have a cruising speed of 150 to 300 km/h and a high level of signature, can be successfully used only in conditions of suppressed enemy air defense.

Two main ranges of dimensions of medium-altitude UAVs can be distinguished. Let us call them conventionally "light" and "heavy" UAVs.

The light multirole medium-altitude long-endurance UAVs have a take-off mass of 700 to 1500 kg and a payload mass of 150 to 400 kg. They comprise radar and optoelectronic reconnaissance equipment weighing 100 to 250 kg and also armaments weighing up to 200 kg which are arranged externally. The aircraft of this class loiter at altitudes of 1 to 6 km at speeds between 150 and 200 km/h. Their endurance varies between 16 and 40 h without armaments, and 10 and 18 h with external store armaments.

For these flight conditions with a required power of 100 to 200 hp, piston engines will be used as a power plant. The operating ceiling of the aircraft with such engines is 5 to 9 km.

The heavy multirole medium-altitude UAVs with a take-off mass of 3500 to 9000 kg carry payloads of 600 to 1800 kg, which comprise 300 to 400 kg of radar and optoelectronic equipment, and also external store weapons weighing up to 1500 kg. The operational altitudes of such vehicles equal 6 to 12 km, the loitering speed is 250 to 300 km/h. Their endurance without weapons equals 25 to 30 h and with external store weapons equals 16 to 20 h.

For low-speed aircraft with the required power of over 500 hp, using turbo-prop engines appears preferable, therefore these aircraft are fitted with 900–1200 hp TPEs. The operating ceiling reaches 15 to 17 km.

As an example, Fig. 16 shows the MQ-9 Reaper, a multirole unmanned medium-altitude long-endurance aircraft developed by General Atomics.

Unmanned high altitude long endurance aircraft

These UAVs weighing 10,000 to 14,000 kg are designed to perform strategic surveillance at the stratospheric flight altitudes. They carry on board radar and optical reconnaissance equipment weighing 900 to 1300 kg, have the operational heights of 18 to 20 km and the flight endurance up to 30 h.

The high-altitude flight extends the direct visibility range, effectively maximising the range of action of surveillance equipment. In addition, surveillance from a great

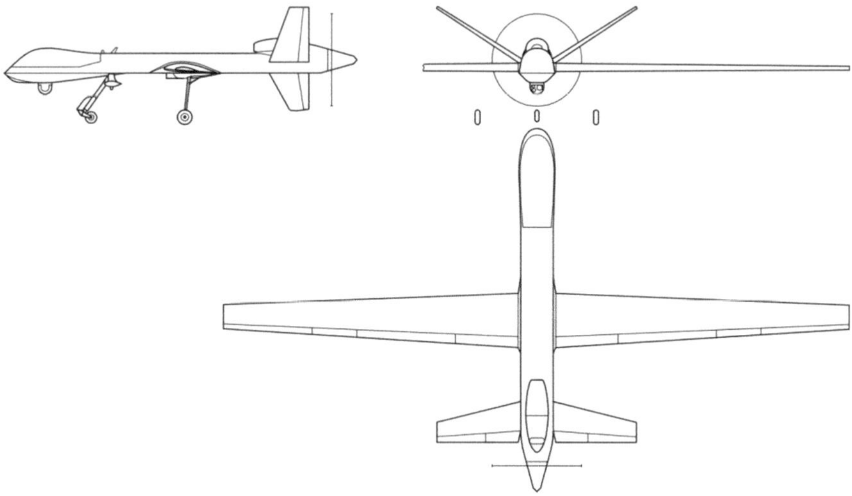

Fig. 16 MQ-9 Reaper multirole unmanned medium-altitude long-endurance aircraft

distance allows an aircraft not to fly into enemy's territory which enhances the survivability of the surveillance aircraft and eventually enhances the effectiveness of its application.

As a power plant for high-altitude long-endurance aircraft, turbojet and bypass turbojet engines are used.

The use of engine-propeller power plants in high-altitude flight is linked with the need to use large-diameter low speed propellers (for the high values of propeller efficiency), which causes certain layout difficulties. In particular, the large diameter of propellers requires a commensurate increase in the height of landing gear struts to ensure the given clearance of propeller blades relative to the ground during take-off and landing. In addition, large-diameter propellers may obstruct the view of the aircraft sensors.

The RQ-4 Global Hawk developed by Northrop Grumman (Fig. 17) is an example of a high-altitude long endurance unmanned aircraft system.

When designing long endurance unmanned aircraft the main challenge is to achieve long loiter time at operating altitudes, as well as providing the necessary visibility for the target sensors and radio communication equipment antennas.

It is known that the aircraft endurance depends on the fuel mass ratio, cruise lift-to-drag ratio and specific fuel consumption.

For medium-altitude unmanned aerial vehicles with PE and TPE which have endurance without weapons 20–25 h, the fuel ratio is 0.27–0.32, for the endurance of 30–40 h the fuel ratio is 0.35–0.4.

The wing loading of long-haul aircraft is selected to ensure the loiter at the lift-to-drag ratio close to its maximum value.

The take-off wing loading of light multipurpose long-endurance unmanned aircraft is in the range of 80–100 daN/m^2, and the take-off power/weight ratio stands

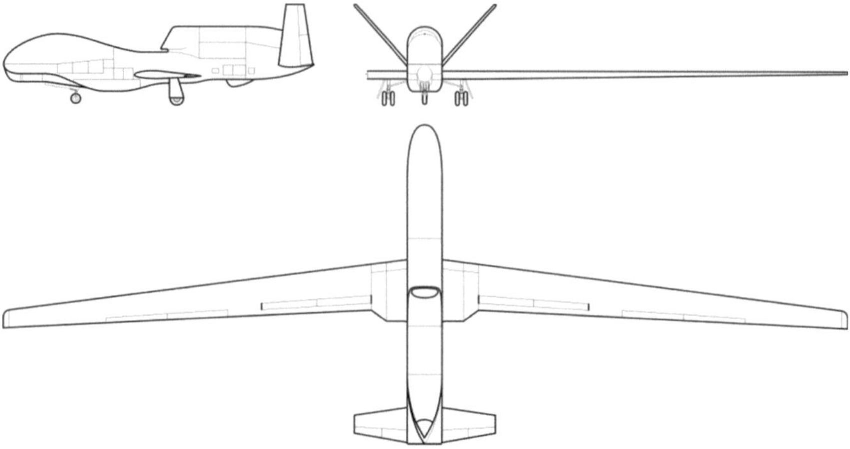

Fig. 17 RQ-4 "Global Hawk" long range unmanned aerial vehicle

at 0.08–0.15 hp/daN. For heavy multipurpose long-endurance unmanned aerial vehicles the typical wing loading is 150 daN/m^2, the take-off power/weight ratios 0.2–0.3 hp/daN.

The maximum lift-to-drag ratio of multipurpose long-endurance unmanned aircraft reaches the value of 25, which is provided by a straight wing of large aspect ratio (~20). Low cruise flight speeds allow the use of aerodynamic wing profiles with a relative thickness of 15–18%.

The requirement to loiter at altitudes of 18–20 km has a decisive influence on the design of high-altitude long-endurance aircraft. This requires high lift-to-drag ratios and thrust-to-weight ratios: the thrust-to-weight ratio of 0.3–0.4 corresponding to the lift-to-drag ratio of 25–30, the thrust-to-weight ratio 0.6–0.8 corresponding to the lift-to-drag ratio of 18–20. The take-off wing loading in this case is 200–250 daN/m^2.

To achieve the maximum lift-to-drag ratio of 30, high-altitude aircrafts use straight wings with the aspect ratio of 25…30 and the thickness ratio of 12–16%.

However, turbojet aircraft, in order to achieve a loiter time of about 24 h, have a significant fuel mass ratio of 0.5–0.6. As the fuel is burned, the aircraft mass decreases and its thrust-to-weight ratio increases. If the required thrust-to-weight ratio is determined based on reaching the design ceiling with full fuel, the engine is overdesigned with respect to thrust for the other flight segments. Consequently, when specifying the requirements for a high-altitude long endurance aircraft, it is common to specify a range of cruising altitudes (e.g., 16–22 km).

Another feature of high-altitude aircraft is the difficulty in descent and landing due to the high values of the lift-to-drag ratio.

With the high values of the thrust-to-weight ratio and aerodynamic efficiency, the thrust of a high-altitude aircraft engine, even at idle rating, may exceed the required thrust for gliding with the loss of height. Under these conditions the descent may be implemented only by an accelerated flat diving, which is impermissible given the strict limitations of the flight maximum speed.

In this circumstance, the high-altitude aircraft are fitted with effective aerodynamic means of deceleration (airbrakes and drogue chutes) and also the means of dumping the excess lifting force (spoilers).

Long endurance unmanned aerial vehicles can do without high manoeuvring characteristics, hence the decisive role of a lower manoeuvring load factor in reducing the structural mass of the aircraft with high AR wings.

In design, it is possible to regard the unmanned long endurance aerial vehicles as non-manoeuvring aircraft and choose the maximum manoeuvring load factor between 2.5 and 3.5.

At the same time, the g-load arising during flight in a turbulent atmosphere must not exceed the design maximum operational g-load.

The g-loads due to vertical gusts during flight at speed V and at altitude H may be evaluated using the equation given in [9]:

$$n_{z,gust} = 1 \pm 0.5C_L^{\alpha}\frac{\rho WV}{G/S}k,$$

where,

$k = \frac{1-e^{-\chi}}{\chi}$ is the factor taking into account the influence of the gust structure and the aerodynamic characteristics of the aircraft on the value of g-load due to gusts.

$$\chi = 0,5\frac{C_L^{\alpha}\rho gh}{G/S},$$

where,

$h = 30$ m is the length of the gust intensity rise section.

C_L^{α} has the units of [1/rad].

Gust velocity W is assumed to be equal to the effective gust velocity W_{eff} in accordance with the statistical dependence:

$$W_{eff} = 12\,to\,15\frac{V_{\max}}{V}\ [m/s].$$

Considering the high lift capacity of aircraft with straight high aspect ratio wings and low wing loading, the g-load due to gusts may prove a significant limitation for them.

To reduce this g-load, it is necessary to limit the flight maximum speed.

3.4 Unmanned Combat Aircraft

These are subsonic large aircraft (with the take-off mass between 10 and 20 tonnes), which carry a combat load of 1000 to 2000 kg to a radius up to 2000 km at a cruise speed of 0.8 M.

They perform conventional take off and landing. Some combat UAVs are also designed for operation from aircraft carriers.

The purpose of such vehicles is to stealthily penetrate into enemy territory and strike ground targets under conditions of unsuppressed air defense, where the use of manned aircraft is associated with a high risk of losing the aircraft.

Minimization of signature in the radar, infrared and visible ranges is achieved through the use of the "flying wing" layout in combination with other means of reducing the signature:

- placing the weapons in internal compartments;
- minimizing the number of antennas by combining different radio systems (e.g., combining radar and electronic countermeasures in a single device);
- minimizing the number and length of gaps and edges of controls, hatches, doors, etc. on the external surface of an aircraft (for example, provision of access to equipment via hatches located inside weapons cargo compartments and landing gear wells);
- using non-protruding pitot-static probes;
- reducing the number of directions of radio signal reflection by ensuring the maximum possible parallelism of wing edges, control surfaces, air intake, nozzle, hatches and doors, and wing pointed leading edges;
- special profiling of the channels of air intake and exhaust nozzle. For example, the S-shaped channels excluding direct visibility of the engine compressor and turbine;
- using radar absorbent coatings and materials.

Figure 18 shows the X-47B unmanned combat aerial vehicle demonstrator designed by Northrop Grumman.

The basic design parameters of combat unmanned aircraft are determined using the same approaches as those for the design of their manned counterparts.

Considering that the configuration used for unmanned combat aerial vehicles is one of the flying wing which, due to trimming difficulties, does not allow for high values of lift coefficient ($C_{L,allow} = 1.0 \ldots 1.2$), the take-off wing loading of unmanned combat aircraft is limited by the requirement to ensure the necessary values of take-off and landing approach speeds and lies within the range of 180… 230 daN/m^2.

Unmanned combat aircraft take-off thrust/weight ratio varies between 0.3 and 0.5.

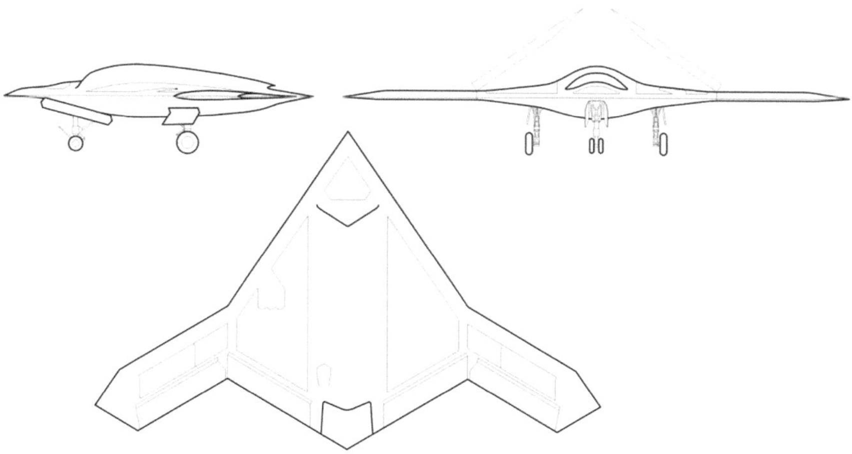

Fig. 18 The X-47B unmanned combat aerial vehicle

Variables		
Symbol	Description	Units
C_L^α	Lift curve	rad^{-1}
$n_{z,gust}$	G-load due to gust	–
ρ	Density of air	kg/m^3
W	Gust velocity	m/s
V	Aircraft velocity	m/s
G	Aircraft weight	daN
S	Wing area	m^2
g	Gravity acceleration	m/s^2
$C_{L,allow}$	Allowable aircraft lift coefficient	–

References

1. Golubev IS, Yankevich YuI (eds) (2006) Osnovy ustrojstva, proektirovaniya, konstruirovaniya i proizvodstva letatel'nyh apparatov (distancionno-pilotiruemye letatel'nye apparaty) (Fundamentals of the design, construction, construction and production of aircraft (remotely piloted aircraft)). MAI, Moscow
2. Liseitsev NK (eds.) (2009) Bespilotnye samolyoty vertikal'nogo vzlyota i posadki. Vybor skhemy, opredelenie proektnyh parametrov (Unmanned vertical takeoff and landing aircraft. Selection of the scheme, determination of design parameters), MAI, Moscow.
3. Vasilin NY (2003). Bespilotnye letatel'nye apparaty. Spravochnik (Unmanned aerial vehicles. Handbook), Popurri, Minsk.

4. Fedosov EA (ed) (2005) Razvitie bespilotnoj boevoj aviacii i rol' informacionnyh tekhnologij v eyo realizacii (Development of unmanned combat aircraft and the role of information technology in its implementation). GosNIIAS, Moscow
5. Ivanov YuA (eds.) (2004) Bespilotnye letatel'nye apparaty. Sostoyanie i tendencii razvitiya (Unmanned aerial vehicles. Status and development trends), Varyag, Moscow.
6. Grebenikov AG, Myalitsa AK, Parfenyuk VV et al (2008). Obshchie vidy i harakteristiki bespilotnyh letatel'nyh apparatov: spravochnoe posobie (General types and characteristics of unmanned aerial vehicles: a reference guide), KHAI, Kharkov.
7. Fetisov VS (eds.) (2014) Bespilotnaya aviaciya: terminologiya, klassifikaciya, sovremennoe sostoyanie (Unmanned aircraft: terminology, classification, current status), Foton, Ufa.
8. Issledovaniya vysotnyh bespilotnyh letatel'nyh apparatov s bol'shoj prodolzhitel'nost'yu poleta v SShA (Research into high-altitude, long-endurance unmanned aerial vehicles in the United States.) (1990) Tekhnicheskaya informaciya. TsAGI, Zhukovsky.
9. Zaitsev VN (eds.) (1978) Konstrukciya i prochnost' samoletov (Aircraft design and strength), Vysshaya shkola, Kyiv.

Design of Aircraft Parts

Aircraft Wing Design

Mikhail A. Pogosyan

Abstract The chapter describes the wing functions, the relationship between wing geometry and wing and aircraft characteristics, including their aerodynamic and aeroelastic performance. The chapter discusses some approaches to improve the lift capability of a swept wing, as well as how to select wing parameters at the preliminary design stage. From the chapter, the reader will learn about the aerodynamic characteristics of low-aspect ratio and variable geometry wings, as well as the design and analysis of high-lift devices and ailerons.

Keywords Aeroelastic performance · Wing sweep · Low aspect ratio · Variable geometry · High-lift devices · Aileron

1 Overview

The function of a wing is to produce the aerodynamic lift necessary to maintain flight and control of the aircraft in all modes in accordance with operational requirements.

The wing provides lateral stability and controllability. It can be used to accommodate fuel, mount and accommodate landing gear, install power plants, suspend armament (on military aircraft), etc.

The shape of wings is defined by the planform, airfoils' shape, and the front view.

The most typical wing planforms are straight, trapezoidal, straight sweepback, variable sweep, delta, and a compound planform.

The main wing geometric parameters include area S, span b, root chord c_r and tip chord c_r, mean aerodynamic chord c_{MAC}, wing aspect ratio AR, taper ratio λ, sweep angle Λ, thickness ratio $\frac{t}{c} = \frac{t_{max}}{c}$, camber ratio $\frac{f}{c} = \frac{f_{max}}{c}$, chordwise location of maximum thickness $\overline{X}_t = \frac{X_t}{c}$, leading-edge radius ratio $\overline{r}_{LE} = \frac{r_{LE}}{c}$, and trailing edge angle 2τ (Fig. 1).

M. A. Pogosyan (✉)
Moscow Aviation Institute (National Research University), Moscow, Russia
e-mail: kaf101@mai.ru

M. A. Pogosyan (ed.), *Aircraft Design*, Springer Aerospace Technology,
https://doi.org/10.1007/978-981-96-4599-2_16

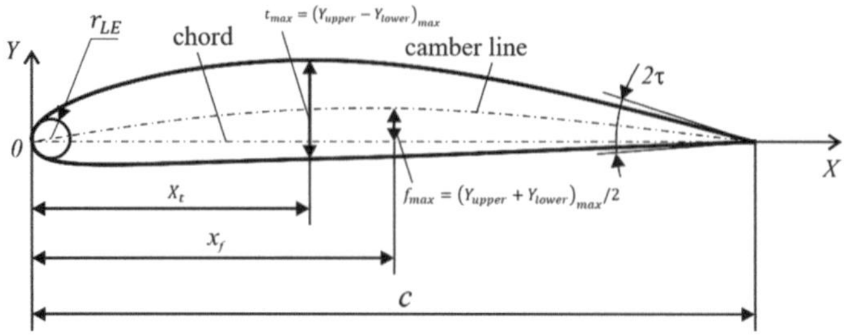

Fig. 1 Thickness and camber of an airfoil

In general, airfoils can be categorized as thin $\left(\frac{t}{c} < 8\%\right)$, moderate thickness $\left(\frac{t}{c} = 8 \text{ to } 12\%\right)$, and thick $\left(\frac{t}{c} > 12\%\right)$. The following is typical for modern airfoils: $\frac{t}{c} = 3\ldots18$, $\frac{f}{c} = 0\ldots3$, $\overline{X}_t = 30\ldots40\%$, $\overline{r}_{LE} = 0\ldots3\%$.

2 Main Geometric Parameters of a Wing

2.1 Wing Aspect Ratio

Wing aspect ratio is a dimensionless geometric parameter and is defined as follows:

$$AR = \frac{b^2}{S}, \quad (1)$$

where b is the wing span, m; S is the wing area, m^2.

When determining the wing aerodynamic characteristics, the effective aspect ratio is used rather than the geometric aspect ratio. The effective aspect ratio incorporates the effect of AR, λ, and Λ on load distribution. At the speeds less the critical Mach number when the air compressibility can be neglected,

$$\begin{aligned} AR_{eff\ incompr} &= \frac{AR}{1+\delta_{incompr}}, \\ \delta_{incompr} &= 0.02\frac{AR}{\cos\ \Lambda}\left(3.1 - \frac{14}{\lambda} + \frac{20}{\lambda^2} - \frac{8}{\lambda^3}\right), \end{aligned} \quad (2)$$

where Λ is the quarter chord sweep angle.

Using the geometric aspect ratio when predicting a wing's aerodynamic characteristics can yield serious errors. For example, for a wing with $\lambda = 4$ and $\Lambda = 35^\circ$ ($\cos\ \Lambda = 0.82$), with a geometric aspect ratio AR of 8, the effective wing aspect ratio is $AR_{eff} = 7$, and for $AR = 10$, it is $AR_{eff} = 8.5$; the error may be as high as 12.5 to 15%. An example variation of the effective aspect ratio with a trapezoidal

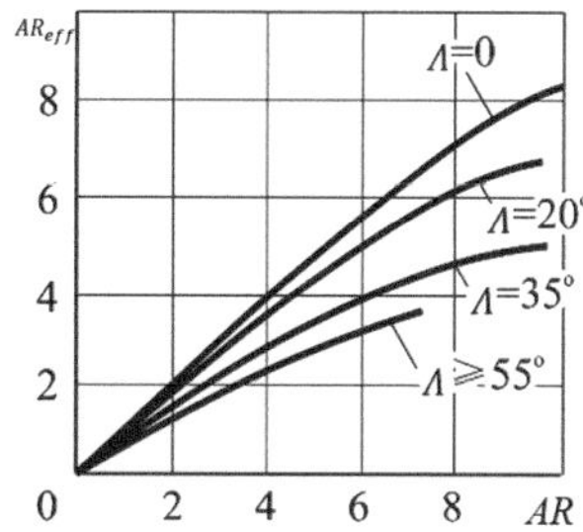

Fig. 2 Effective aspect ratio AR_{eff} as a function of AR and Λ

wing's geometric characteristics is shown in Fig. 2. It can be seen that the divergence of AR and AR_{eff} increases as the sweep angle Λ increases. The taper ratio of the wing is 2.

At a post-stall condition and transonic speed, when the air compressibility becomes significant, AR_{eff} is determined using the expression

$$AR_{eff\ compr} = \frac{AR_{eff\ icompr}}{1 + \delta_{compr}},$$
$$\delta_{compr} \approx \left\{ \begin{array}{l} 10AR\frac{t}{c}_{av}^{1/3}\left(M - M''_{crit}\right)^3 \ at\ M_{crit} < M < 1 \\ 0 \ \ at \ \ M < M_{crit} \end{array} \right\}, \quad (3)$$

where $\frac{t}{c}_{av}$ is the average thickness ratio of the wing.

At $C_L > 0$,

$$M''_{crit} = M'_{crit} - C_L^{3/2}\left(\frac{t}{c_{av}}\right)^{1/2}, \quad (4)$$

where M'_{crit} is the critical Mach number at $C_L > 0$.

$$M'_{crit} = \frac{1}{\cos \Lambda_t}\left[1 + \frac{(\chi + 1)^{4/3}\left(\frac{t}{c}_{av}\right)^{4/3}}{2(\cos \Lambda_t)^{2/3}} - \frac{(\chi + 1)^{2/3}\left(\frac{t}{c}_{av}\right)^{2/3}}{2(\cos \Lambda_t)^{1/3}}\right], \quad (5)$$

where Λ_t is the sweep angle of the maximum thickness line of the wing; $\chi = 1.4$ is the specific heat ratio for air.

2.2 Average Wing Thickness Ratio

The average wing thickness ratio is determined as follows:

$$\left(\frac{t}{c}\right)_{av} = \frac{S_{MS}}{S} = \overline{S}_{MS}, \quad (6)$$

Fig. 3 Wing longitudinal mid-section

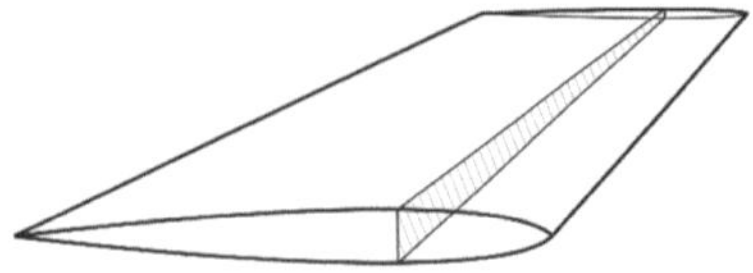

where S_{MS} is the area of wing longitudinal mid-section, m^2 (Fig. 3).

For a conical trapezoidal wing, $\left(\frac{t}{c}\right)_{av}$ is determined using the expression

$$\left(\frac{t}{c}\right)_{av} = \frac{0.5(t_{tip} + t_r)}{0.5(c_{tip} + c_r)} = \frac{t_{tip} + t_r}{c_{tip} + c_r} = \frac{c_{tip}(t/c)_{tip} + c_r(t/c)_r}{c_{tip} + c_r} = \frac{(t/c)_r\lambda + (t/c)_{tip}}{\lambda + 1}, \quad (7)$$

where t_r and t_{tip} are the absolute thicknesses of root and tip sections.

The average wing thickness cannot be determined as the arithmetic mean value between t_r and t_{tip}, since the thickness of airfoils in a conical wing varies non-linearly. For example, for a wing with a root chord of 10 m and $(t/c)_r = 15\%$, and a tip chord of 4 m and $(t/c)_{tip} = 5\%$, the average thickness ratio of wing will be 12.14%, rather than 10% (the arithmetic mean value).

2.3 *Shape of Middle Surface*

The middle surface is defined as the half-sum of ordinates of the upper and lower wing surfaces, formed by the upper and lower contours of the airfoils [3]:

$$Y_m(X, Z) = \frac{1}{2}\left[Y_{upper}(X, Z) + Y_{lower}(X, Z)\right], \quad (8)$$

The wing sections of modern aircraft are quite diverse; their middle surface is formed by a set of profiles rotated at an angle $\varphi(y)$ relative to the root section. Aerodynamic and geometric twists are widely used to obtain the desired middle surface of a wing.

Aerodynamic twist is referred to the installation of different airfoils at different spanwise stations.

Geometric twist is referred to the installation of airfoils at different angles of incidence relative to the root chord. For example, with a conical twist, the leading edges of airfoils are rotated down along the conical surface with the vertex in leading edge of the root section (Fig. 4).

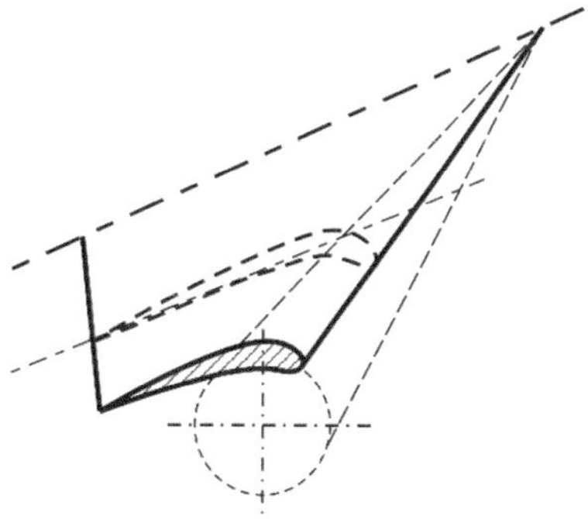

Fig. 4 Conical wing twist

The angle between the local chord and some reference plane is the local twist angle. The total twist is the twist angle of the tip section. The average twist angle and relative camber can serve as average parameters that characterise the aerodynamic and geometric wing twist:

$$\varphi_{av} = \frac{1}{S}\int_{-b/2}^{+b/2} \varphi(y)c(y)dy, \quad (9)$$

$$\left(\frac{f}{c}\right)_{av} = \frac{1}{S}\int_{-b/2}^{+b/2} \frac{f}{c}(y)c(y)dy. \quad (10)$$

A generalised twist parameter for a wing middle surface is as follows:

$$\tilde{\varphi}_{av} = \frac{1}{S}\iint \varphi(x, y)dxdy, \quad (11)$$

where $\varphi(x, y) = \frac{dy_{av}(z,y)}{dx}$ is the angle between the tangent to the middle wing surface (in ZOX-plane) and the XOY-plane.

To make a first approximation, the following expression can be used:

$$\tilde{\varphi}_{av} = \varphi_{av} - 4(f/c)_{av}. \quad (12)$$

2.4 Wing Volume

A very important geometric characteristic of a wing is its volume available for fuel. For wings with linear leading and trailing edges, the maximum theoretical volume of the entire wing (m^3) can be estimated using the formula

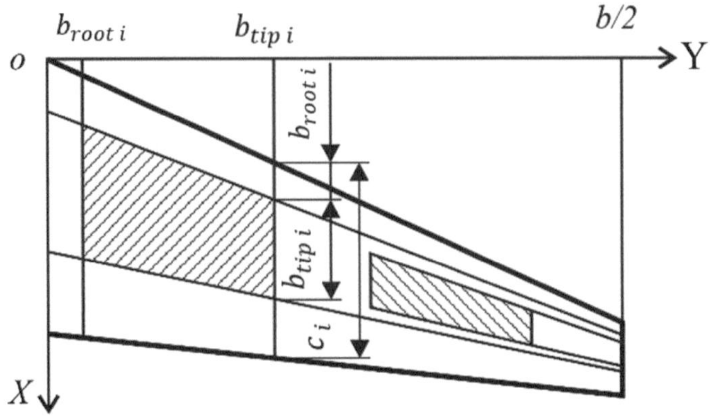

Fig. 5 Schematic explaining how to determine the volume of fuel tanks

$$W_{max} = k_{W\ max} \frac{(t/c)_{av} S^{3/2}}{AR^{1/2}}, \tag{13}$$

where

$$k_{W\ max} = \frac{4\left(2\lambda\lambda_{fr} + \lambda + \lambda_{fr} + 2\right)}{9(\lambda + 1)(\lambda + 1)};$$

$\lambda_{fr} = \frac{c_r (t/c)_r}{c_{tip} (t/c)_{tip}} = \lambda \frac{(t/c)_r}{(t/c)_{tip}}$ is the wing taper ratio as seen from the front view.

Usually, only torsion compartments out of the entire wing volume are used to accommodate fuel. Spanwise, the torsion box is also used not in full: the wing tips that have small thickness are excluded. The entire torsion tank can be divided into separate compartments or individual tanks (Fig. 5).

If $\bar{c}_{tipi}(y) = \frac{c_{tipi}}{c_i} = const$ and $\bar{c}_{rooti}(y) = \frac{c_{rooti}}{c_i} = const$, the tank volume can be estimated using the following equation:

$$W = k'_W \frac{(t/c)_{av} S^{3/2}}{AR^{1/2}}. \tag{14}$$

where:

$$k'_W = \frac{4\lambda\lambda_t}{(\lambda + 1)\left(\lambda_{fr} + 1\right)} \sum_{j=1}^{i=n} k'_{F_i} k'_{l_i},$$

$$k'_{F_i} = 2\bar{c}_{tipi}\left[\bar{c}_{tipi}\left(1 - 2\bar{c}_{rooti} - \frac{2}{3}\bar{c}_{tipi}\right) + 2\bar{c}_{rooti}(1 - \bar{c}_{rooti})\right],$$

$$k'_{l_i} = \left(\bar{b}_{tipi} - \bar{b}_{rooti}\right)\left[1 - \frac{2\lambda\lambda_{fr} - \lambda - \lambda_{fr}}{2\lambda\lambda_{fr}}\left(\bar{b}_{tipi} - \bar{b}_{rooti}\right) + \frac{(\lambda - 1)\left(\lambda_{fr} - 1\right)}{2\lambda\lambda_{fr}}\left(\bar{b}^2_{tipi} + \right.\right.$$

$$\bar{b}_{tipi}\bar{b}_{rooti} + \bar{b}^2_{rooti}\Big)\Big],$$

$\bar{b}_{tipi} = 2\frac{y_{tip}}{b}$ and $\bar{b}_{rooti} = 2\frac{y_{root}}{b}$ are the relative coordinates of the tip and root sections of the tank compartment, respectively.

The mass of fuel in the tank equals

$$m_{fuel} = \rho_{fuel} W_{fuel};$$
$$W_{fuel} = \frac{W}{k_{fuel}}; \quad m_{fuel} = \frac{W}{\rho_{fuel} k_{fuel}},$$

where $k_{fuel} \approx 1.05$ is the factor taking into account the air volume required for tank pressurization, compensation of fuel expansion when heated and the volume of the equipment located in the fuel tank; W is the wing volume, m^3; ρ_{fuel} is the fuel mass density, kg/m^3.

3 Aerodynamic Characteristics of Moderate and High-Aspect Wings (AR > 3)

Wing lifting properties, which are determined by the magnitude of wing lift curve slope C_L^α at subsonic and transonic speeds, significantly depend on the wing sweep angle. The second parameter determining the C_L^α value at subsonic and transonic speeds is the wing aspect ratio. The theoretical maximum incompressible value of C_L^α for a straight infinite aspect-ratio wing equals 2π.

When the wing aspect ratio decreases, the lift slope of the wing decreases as well. However, at aspect ratios of 15 to 20, C_L^α is close to the theoretical value. Other wing parameters (wing taper, airfoil thickness and shape) have a secondary influence on C_L^α. The lift slope also significantly depends on the Mach number (Fig. 6).

The influence of air compressibility begins to show itself at Mach numbers of 0.4 to 0.5. When M increases within the transonic speed range, C_L^α increases as the suction on the upper surface of the wing grows faster than on the lower one. C_L^α

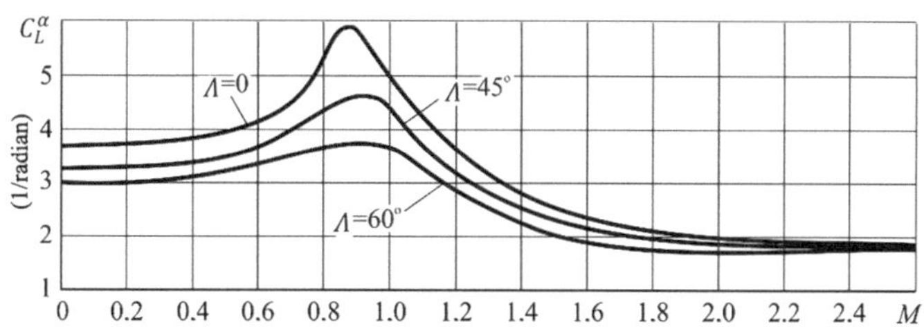

Fig. 6 Lift slope of a swept wing as a function of Mach number and sweep angle

increases as $C_L^\alpha = \frac{C^\alpha_{Lincompr}}{\sqrt{1-M^2}}$. C_L^α reaches its maximum at the critical Mach number. This is explained by the fact that the suction zone, as a result of the shift of the shock wave to the trailing edge, occupies an increasingly large surface of the wing, which leads to an increase in C_L^α. When $M = 1.1 \ldots 1.3$, C_L^α begins to reduce smoothly. At supersonic speeds, when the wing leading edge becomes supersonic, the influence of the wing planform on C_L^α becomes insignificant.

The bearing capacity of moderate and high-aspect wings at low subsonic flight speeds is defined as follows:

$$C_L = C_L^\alpha(\alpha - \alpha_0), \tag{15}$$

where α_0 is the angle of attack at $C_L = 0$;

$$C_L^\alpha = \frac{\partial C_L}{\partial \alpha} = 2\pi \frac{AR}{\overline{p} + 2}, \tag{16}$$

where $\overline{p}$ is the ratio of the wing half-perimeter to its span; C_L^α is the slope of the linear portion of the lift curve.

For wings with straight leading and trailing edges and with tip chords parallel to the air flow,

$$\overline{p} = \frac{1}{2}\left(\frac{1}{\cos\,\Lambda_{LE}} + \frac{1}{\cos\,\Lambda_{TE}}\right) + \frac{AR}{\overline{p} + 2}, \tag{17}$$

where $\Lambda_{LE},\ \Lambda_{TE}$ are the sweep angles of the leading and trailing edges of the wing, respectively. They are interrelated as follows:

$$\tan\,\Lambda_{TE} = \tan\,\Lambda_{LE} - \frac{4}{AR}\frac{\lambda - 1}{\lambda - 2}. \tag{18}$$

Formulas (15) and (16) can be applied to any wing planform.

The effect of air compressibility at subsonic speeds on the wing lift can be considered using the subsonic similarity law. In Eq. (16), the effect of compressibility is taken into account by the dimensionless parameter $\overline{p}$. For wings with straight edges, instead of $\overline{p}$ we take

$$\overline{p}_{compr} = \frac{1}{2}\left(\sqrt{1 - M^2 + (\tan\,\Lambda_{LE})^2} + \sqrt{1 - M^2 + (\tan\,\Lambda_{TE})^2}\right) + \frac{2}{AR(\lambda + 1)}. \tag{19}$$

Equations (16, 17, 18, 19) are not suitable for calculations within the transonic speed range ($M = 0.85 \ldots 1.15$). At such flight speeds, the flow over a wing is mixed, i.e. there are zones of local subsonic and supersonic speeds on the wing surface. There is no generic method for calculating the wing bearing properties within this speed range (especially at $M \approx 1$). Based on experimental data, using the similarity rules, with sufficient accuracy for practice, it is possible to determine the magnitude of wing lift slope at $M \approx 1$ as follows:

$$C_L^\alpha = \frac{2\pi AR}{\overline{p}AR(t/c)^{1/3} + 2}. \tag{20}$$

For wings with straight leading and trailing edges and with a tip chord parallel to the air flow (for swept and tapered wings),

$$\overline{p}AR(t/c)^{1/3} \cong AR(t/c)^{1/3}\sqrt{1+\tan\Lambda_{LE}} + \frac{2}{\lambda+1}\left[1-(\pi-1)\left(\frac{\tan\Lambda_{LE}}{\sqrt{1+M^2+(\tan\Lambda_{LE})^2}} - \frac{2(\lambda-1)}{AR(t/c)^{1/3}(\lambda+1)\sqrt{1+M^2+(\tan\Lambda_{LE})^2}}\right)\right]. \tag{21}$$

The Mach number, at which C_L^α will have the maximum value, is determined by the following equation:

$$M_{C_{L\,max}^\alpha} = \sqrt{1-(t/c)^{1/3}\left(2-\frac{1}{AR(t/c)^{1/3}}\right)}. \tag{22}$$

$C_{Lmax}^\alpha / C_L^\alpha$ at $M \approx 1$ can be predicted using the following equation:

$$\left(C_{Lmax}^\alpha / C_{LM=1}^\alpha\right)M_{C_{Lmax}^\alpha} = 1 + (t/c)^{1/3}\left(2-\frac{1}{AR(t/c)^{1/3}}\right)^2. \tag{23}$$

Depending on whether the wing edges (leading and trailing) are subsonic $\left(\tan\Lambda_i > \sqrt{M^2-1}\right)$ or supersonic $\left(\tan\Lambda_i < \sqrt{M^2-1}\right)$ [1], particular equations for the lift slope $C_L^\alpha(M)$ can be obtained. For a straight wing with $AR\sqrt{M^2-1} > 1$, we have

$$C_L^\alpha = \frac{4}{\sqrt{M^2-1}}\left(1-\frac{1}{2AR}\sqrt{M^2-1}\right). \tag{24}$$

For $R\sqrt{M^2-1} < 1$, there is a more complex and lengthy dependence. For a straight wing with oblique tips, if they are subsonic $|\tan\Lambda_{tip}| > \sqrt{M^2-1}$, at $\sqrt{M^2-1} > 1$ we obtain

$$C_L^\alpha = \frac{4}{\sqrt{M^2-1}}\left(1-\frac{\sqrt{M^2-1}-1-\tan\Lambda_{tip}}{2AR}\right). \tag{25}$$

If they are supersonic,

$$C_L^\alpha = \frac{4}{\sqrt{M^2 - 1}}, \tag{26}$$

i.e. C_L^α is the same as that of an infinite aspect-ratio wing or a two-dimensional flat-plate.

Characteristics of swept wings at supersonic speeds are of great interest for aircraft engineers. However, the theoretical definition of their bearing capacities is very difficult: the variation of C_L^α with M is quite lengthy. Approximately, these dependences can be represented as follows:

– for subsonic leading edges:

$$C_L^\alpha = \frac{4}{E(k)\tan \Lambda_{LE}}\left[\frac{1-\frac{1}{\lambda}}{1+\xi}\left(\xi + \frac{2\ \mathrm{acos}(-\xi)}{(2+\xi)(1+\xi)^{1/2}}\right)\right] + \frac{1+\pi\sqrt{2}}{2\lambda}; \tag{27}$$

– for supersonic leading edges,

$$C_L^\alpha = \frac{8}{\pi\ \tan \Lambda_{LE}}\left[\frac{1-\frac{1}{\lambda}}{1+\xi}\left(\xi + \frac{2\mathrm{acos}\left(\frac{-\xi}{m}\right)}{\sqrt{m^2-\xi^2}}\right) + \frac{\xi\,\mathrm{acos}\left(\frac{1}{m}\right)}{\sqrt{m^2-1}}\right] + \frac{\mathrm{acos}\left(\frac{1}{m}\right)}{\lambda\sqrt{m^2-1}}. \tag{28}$$

Here, $E(k)$ is the complete elliptic integral of the second kind with module $k = \sqrt{1-m^2}$, where $m = \frac{\sqrt{M^2-1}}{\tan \Lambda_{LE}}$, $\xi = \frac{\tan \Lambda_{TE}}{\tan \Lambda_{LE}}$.

The above equations cover a wide class of wing planforms. They are the most commonly used in practice. For these wings, $C_L^\alpha(M)$ can be plotted for a speed range from subsonic to supersonic, including the transonic zone. Curve $C_L^\alpha(M)$ at transonic Mach numbers must pass through the points $C_L^\alpha(M < 1)$, $C_L^\alpha(M > 1 \text{and } M < 1)$ and $C_L^\alpha(M = 1)$, smoothly mating with the curves $C_L^\alpha(M)$ for subsonic and supersonic speeds.

The data on $C_L^\alpha(M)$ is necessary to determine the limit and ultimate g-loads in case of vertical gust.

3.1 The Maximum Lift of a Wing

The maximum lift coefficient of a wing $C_{L,max}$ depends on the three-dimensional geometry of a wing.

Let us examine the flow around trapezoidal swept wings (Fig. 7).

In zone *I*, the stream tubes expand; in zones *II* and *III*, they narrow; and in zone *III*, the narrowing occurs more intensively due to the influence of less deformed

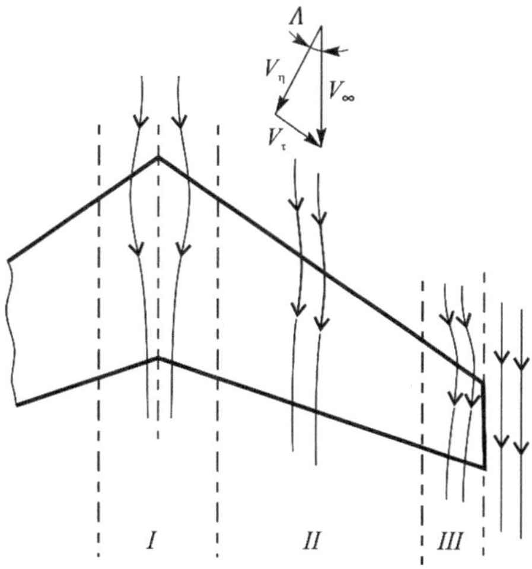

Fig. 7 Flow around a trapezoidal swept wing

stream tubes downstream the wing tip. The trajectories of streamlines bend because the tangential component of speed along the leading edge remains almost unchanged, and the normal component is reduced due to slowing down of the air flow. Then, with the narrowing of the stream tubes, the air flow accelerates and the normal component of the speed increases. Examples of chordwise distributions of pressure coefficients and spanwise distributions of lift coefficients are shown in Figs. 8 and 9.

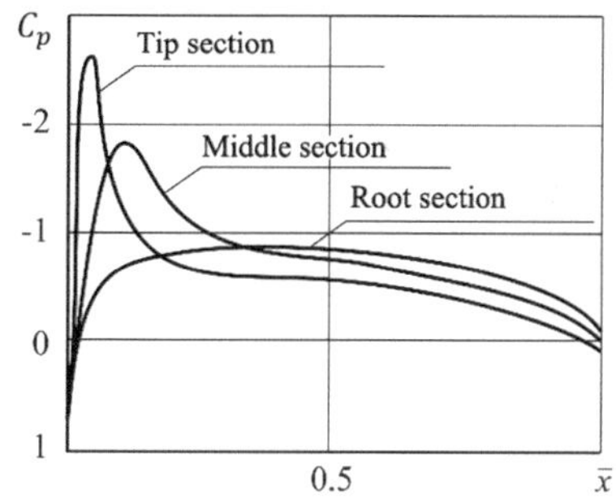

Fig. 8 Chordwise distributions of pressure coefficients in different sections of a swept wing

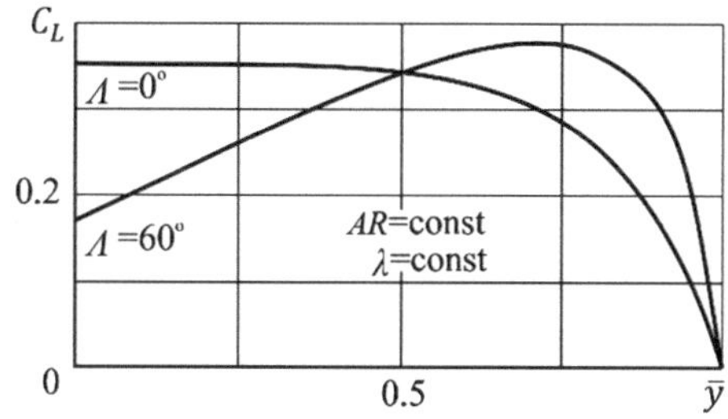

Fig. 9 Spanwise distributions of lift coefficients for a straight and a swept wing

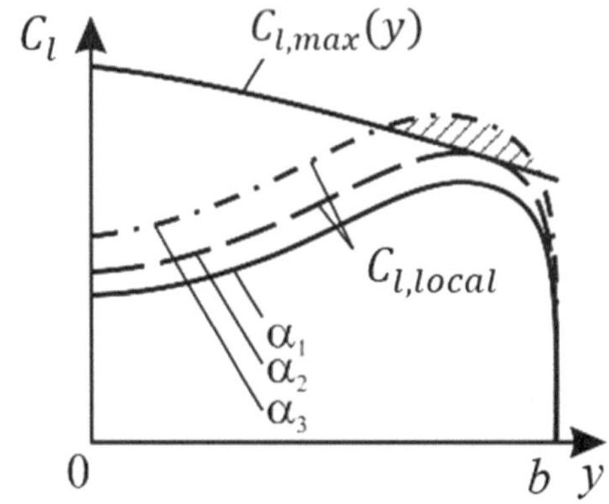

Fig. 10 The variation of $C_{l,max}$ and $C_{l,local}$ for a swept wing

When the angle of attack increases, $C_{l,max}$ is first reached locally in some sections of the wing (at α_2). If the angle of attack increases further, the flow begins to separate in these sections. From Fig. 10, it is clear that the stall onset occurs at the tips.

Figure 11 demonstrates that the spanwise distribution of lift coefficients of a swept wing depends on the taper ratio λ and sweep angle Λ of the wing.

Figure 12 shows that the pitching moment coefficient of a swept wing varies with the angle of attack. As the wing angle of attack increases to α_1, $C_m(\alpha)$ curve takes a negative slope (which characterises a longitudinally statically stable aircraft).

With the onset of stall at the wing tips ($\alpha = \alpha_1$), the lift of the wing tips ceases to increase with the angle of attack, while the lift of the wing root increases (there is no flow separation); therefore, the wing pitching-down moment begins to decrease.

The $C_m(\alpha)$ curve exhibits a positive slope (pitch instability region) and the aircraft becomes statically unstable. With a further increase in angle of attack, when stall covers most of the wing, the negative curve slope can be recovered.

When stall occurs on a straight wing, it leads to a sharp drop in C_L beyond $C_{L,max}$. The linear segment of $C_L(\alpha)$ is retained up to C_L close to $C_{L,max}$. On a swept wing, the $C_L(\alpha)$ curve begins to deviate from the linear one after the stall onset on tips,

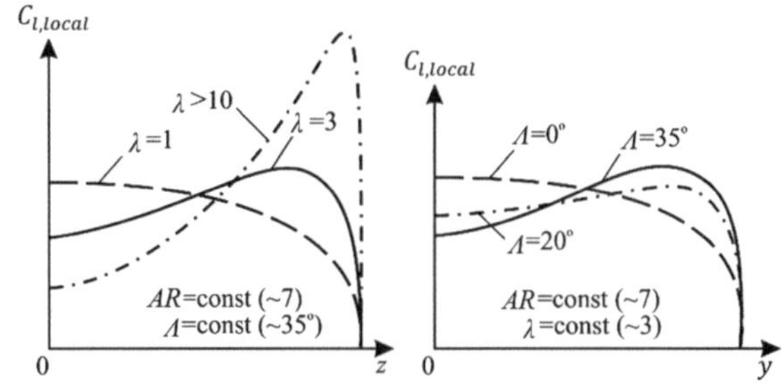

Fig. 11 Spanwise distribution of lift coefficients $C_{l,local}$ for a swept wing as a function of the taper ratio λ and sweep angle Λ

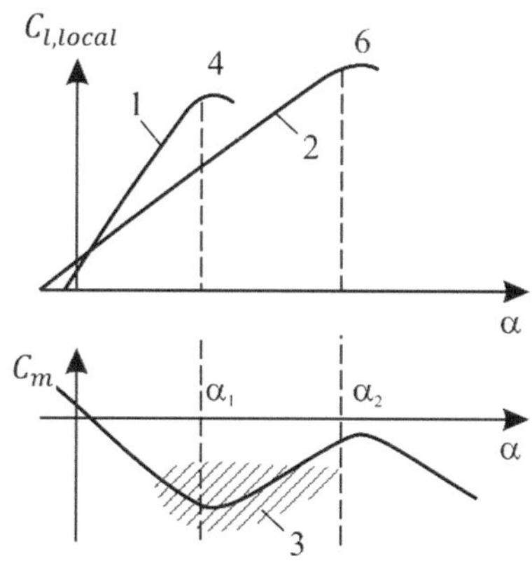

Fig. 12 Pitch instability region in moment curve $C_m(\alpha)$: 1—tip section (~ 8%); 2—root section (~14%); 3—pitch instability region; 4, 6—stall

but C_L continues to grow until it reaches the value of $C_{L,max}$. The greater the sweep angle Λ, the earlier the $C_L(\alpha)$ curve begins to deviate from the linear one, and the smoother the change is in C_L around $C_{L,max}$ zone.

$C_{L,max}$ is greatly influenced by the air compressibility. For modern moderately thick airfoils ($t/c = 10 \ldots 15\%$), the influence of compressibility begins to manifest itself at the Mach numbers of 0.2 to 0.3.

In general, for swept wings, the airflow features are as follows:

- the wing tips carry a larger load than the root sections, which leads to that the tips stall first with an increasing angle of attack;
- the air flowing from root sections to the tips also increases the tendency of a swept wing to stall at the tips first;
- the flow separation at tips causes a pitching-up longitudinal moment (a decrease in the pitching-down moment) and a non-linearity in the moment curve that makes it difficult to control the aircraft.

3.2 Approaches to Improve the Lift Capability of a Swept Wing

To increase the $C_{L,max}$ of a swept wing, the following measures can be taken.

1. Twist the wing by installing the airfoils with a high $C_{L,max}$ at the tips, and the airflois with a flattened upper surface and even with a negative camber at the root section. With such an arrangement, the wing C_L increases, the stall begins to develop in the middle wing at significantly larger values of $C_{L,max}$, the instability region in the moment curve decreases and (or) shifts towards larger angles of attack.

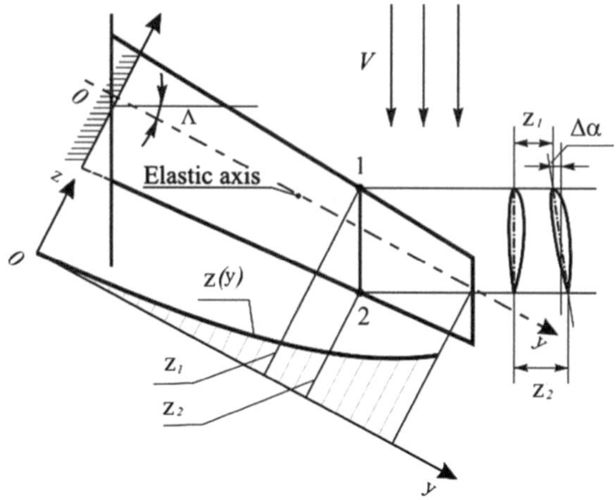

Fig. 13 Change in local angles of attack of a swept wing sections due to bending

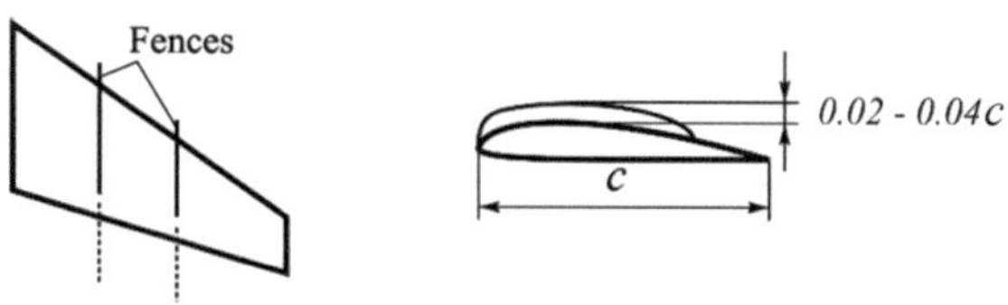

Fig. 14 Aerodynamic fences on a swept wing

2. Twist the wing by rotating the tip sections to some negative angle (the leading edges of the airfoils rotate downward) relative to the root section. Due to this, the stall on the tip sections occurs later which increases the wing $C_{L,max}$. It should be noted that in flight, the wing tips are deflected due to the air load as shown in Fig. 13., and that leads to a decrease in the local angles of attack.
3. To reduce the negative effect of the boundary layer flowing along the swept leading edge to the wing tips, wing fences are installed on the upper surface. This device prevents premature separation of the boundary layer (Fig. 14).
4. Instead of wing fences, a step change in chord length and special airfoils may be used. In this portion of the wing, a vortex is produced, which also prevents the premature separation of the boundary layer; at the same time, the wing drag (compared to the configuration with fences) decreases, since the friction dag due to fences is eliminated.
5. The detrimental effect of the boundary layer separation can be reduced by installing one or two rows of vortex generators, which are small devices mounted perpendicular to the wing surface. Vortex generators mix the boundary layer with fresh air, increasing its kinetic energy, which prevents premature stall.

3.3 Wing Drag

The total wing drag can be approximated by a parabolic equation:

$$C_D = C_{D,0} + C_{D,i} = C_{D,0} + AC_L^2, \tag{29}$$

where $C_{D,0}$ is the drag coefficient at zero lift; $C_{D,i}$ is the induced drag coefficient; A is the lift-induced drag constant.

The drag coefficient at zero lift can be presented in the following form:

$$C_{D,0} = C_{D,p} + C_{D,w}, \tag{30}$$

where $C_{D,p}$ is the pressure drag coefficient; $C_{D,w}$ is the wave drag coefficient emerging at flight speeds exceeding M_{cr};

$$C_{D,p} = 2C_{f,pl}(1+3t)\left[1 + \frac{t}{c(5M-3)\overline{S}_w} + 0.001\overline{l}_{slot}\right], \tag{31}$$

where $C_{f,pl}$ is the flat plate friction coefficient,

$$C_{f,pl} = \frac{0.455}{\left(1+0.1M^2\right)^{2/3}(\lg Re)^{2.58}}\left(1 - \overline{x}_{tr} + \frac{40}{Re^{2/3}}\overline{x}_{tr}^{5/8}\right)^{4/5}, \tag{32}$$

$\overline{S}_w$ is the area ratio of wing wetted surface, $\overline{S}_w = 1 - k_{int}\overline{S}_{belly}$;

$\overline{S}_{belly}$ is the ratio of the wing area inside the fuselage;

k_{int} is the interference factor:

Wing vertical position	High-wing monoplane	Mid-wing monoplane	Low-wing monoplane
k_{int}	~9	~0.7	~0.5

$_{slot}$ is the total length (span) of wing slots (between wing and aileron, wing and flaps, etc.) divided by the wing span;

Re is the Reynolds number;

$\overline{x}_{tr}$ is the chordwise coordinate divided by the chord length where the laminar boundary layer transitions to the turbulent one,

$$\overline{x}_{tr} = min\begin{cases} \frac{10^n}{Re} \\ \sqrt{\overline{x}_t^2 + \overline{x}_f^2 at\sqrt{\overline{x}_t\overline{x}_f}}, \\ \overline{c}_{slat} \end{cases}$$

where $\overline{x}_t$ and $\overline{x}_f$ are the relative coordinates of the maximum airfoil thickness and camber; $\overline{c}_{slat}$ is the slat chord ratio;

$$n = 5 + \left[1.3 + 0.6M\left(1 + 0.25M^2\right)\right]\sqrt{1 - \left[\frac{\lg\left(\frac{h_r}{c_{av}}Re\right) - 1}{2.2 - \frac{0.08M^2}{1+0.312M}}\right]^2},$$

where h_r is the average height of the wing surface roughness, $h_r \approx (5\ldots15)10^{-6}$m.

The wave drag coefficient at transonic speeds can be determined using the equation

$$C_{D,w} = \frac{2\pi AR(t/c)\cos\,\Lambda_t}{2 + AR(t/c)^{1/3}(\cos\,\Lambda_t)^{5/3}}\left(\frac{M - M'_{crit}}{M_{C_{D,max}} - M'_{crit}}\right)^3\left(4 - 3\frac{M - M'_{crit}}{M_{C_{D,max}} - M'_{crit}}\right), \tag{33}$$

where Λ_t is the sweep angle of the line of maximum thicknesses; M'_{crit} is the critical Mach number for the wing at $C_L = 0$ (refer to 14.5); $M_{C_{D,max}}$ is the Mach number corresponding to $C_{D,wmax}$.

$$M_{C_{D,max}} = \frac{1}{\cos\,\Lambda_t}\left\{1 + 0.4\frac{(t/c)^{3/2}}{(\cos\,\Lambda_t)^{3/2}}\left[2 - AR(t/c)^{1/3}(\cos\,\Lambda_t)^{3/2}\right]\right\}. \tag{34}$$

Wave drag coefficient at supersonic speeds ($M > 1.2$):

$$C_{D,w} = \frac{4\left[(t/c)^2 + 2(f/c)^2\right]}{\sqrt{M^2 - 1}}\left[1 + \left(k_{pr} - 1\right)\varphi\right] \tag{35}$$

where $k_{pr} = 1$ for diamond-shaped airfoils; $k_{pr} = 4/3$ for the airfoils formed by circular arcs; $k_{pr} = 3$ *to* 4 for subsonic airfoils;

$$\varphi = \begin{cases} \frac{0.16\left(AR\sqrt{M^2-1} - AR\ tan\ \Lambda_t\right)^2}{1+0.16\left(AR\sqrt{M^2-1} - AR\ tan\ \Lambda_t\right)^2} & for\ \sqrt{M^2 - 1} > \ tan\ \Lambda_t \\ 0 & for\ \sqrt{M^2 - 1} \le tan\ \Lambda_t \end{cases}.$$

Thus, the drag coefficient can be determined using Eqs. (32) and (33) or (37) for a whole range of Mach numbers. Moreover, these equations include wing geometrical parameters, which can be used for optimizing an aircraft's geometry.

The wing induced drag coefficient can also be regarded as consisting of two components: vortex and wave drag. The first component is produced at all flight speeds, the second one—at transonic and supersonic speeds.

At low speeds, the induced drag coefficient can be predicted as follows:

$$C_{D,i} = \frac{C_L^2}{\pi AR_{eff,incompr}}, \tag{36}$$

where $AR_{eff,incompr}$ is effective aspect ratio of wing in an incompressible flow.

Under the post-stall conditions (with respect to the Mach number) within the transonic speed range, when air compressibility must be taken into account:

$$C_{D,i} = \frac{C_L^2}{\pi AR_{eff,compr}}, \tag{37}$$

where $AR_{eff,compr}$ is determined using Eq. (3).

For wings with supersonic leading edges, the induced drag coefficient equals

$$C_{D,i} = \frac{C_L^2}{C_L^\alpha}. \tag{38}$$

In the case of subsonic leading edges for sweptback and other wing planforms, the induced drag coefficient can be determined using the expression

$$C_{D,i} = \frac{C_L^2}{C_L^\alpha} - \xi_T C_T, \tag{39}$$

where C_T is the suction force coefficient; ξ_T is the suction force factor (shows what portion of the theoretical suction force is produced). For a sharp leading edge, $\xi_T = 0$; for a rounded leading-edge, $\xi_T = 0.8 \ldots 1.15$.

Suction force coefficient equals

$$C_T = \frac{\pi AR}{8C_L^\alpha} \frac{\sqrt{1-m^2}}{E^2(k) C_L^\alpha},$$

where $E(k)$ and m are the parameters from expression (30).

The elliptic integral $E(k)$ can be approximated by the expression $E^{-1}(k) \approx \frac{2}{\pi}\left(\frac{2}{\pi}\text{a} \cos k + \text{a} \sin k\right)$.

Then, the induced drag coefficient of wings with subsonic leading edges when flying at supersonic speeds will be determined by the expression

$$C_{D,i} = \frac{C_L^2}{C_L^\alpha}\left[1 - \xi_T \frac{\pi AR}{8C_L^\alpha} \frac{\sqrt{1-m^2}}{E^2(k)}\right]. \tag{40}$$

It should be noted that some suction force can be produced by an airfoil with a sharp leading-edge as well, provided that it is deflected down to an angle slightly exceeding the wing angle of attack in this flight mode.

To increase the maximum lift-to-drag ratio of swept back and delta wings, the so-called "conical" wing twist is applied (refer to Fig. 4). Such a twist is effective at transonic speeds; at high supersonic speeds, it can give some reduction in the maximum lift-to-drag ratio. In order to reduce the induced drag, wings must be designed to have a non-planar middle surface. The optimal wing twist is determined

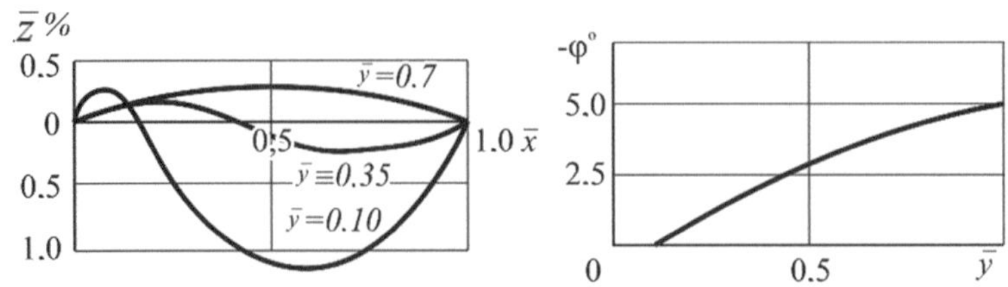

Fig. 15 Aerodynamic (right plot) and geometric (left plot) wing twist

through solving the variational or extremum problem on the minimum of induced drag under the given isoperimetric and relations conditions.

Figure 15 shows an example of aerodynamic twist ($f/t(y)$, left plot) and geometric twist ($\varphi(y)$, right plot) for a non-planar middle surface.

From the equation for the maximum lift-to-drag ratio $L/D_{max} = \frac{1}{2}\sqrt{\frac{\pi AR_{eff,compr}}{C_{D,0}}}$ follows that in order to increase it, it is necessary to optimize the wing aspect ratio, planform and geometric twist, as well as the aerodynamic twist in terms of the shape of airfoils and their spanwise location, since this can significantly reduce $C_{D,0}$.

3.4 Wing Aerodynamic Moment

Wing pitching moment coefficient C_m depends on the wing's aerodynamic configuration and flight mode (C_L and M).

To first approximation, $C_m = C_{m,0} + C_m^{C_L} C_L$. Here, $C_{m,0}$ is the pitching moment coefficient about the aerodynamic center; it depends on the aerodynamic and geometric wing twist, as well as on the Mach number. The longitudinal static stability $C_m^{C_L} = \overline{X}_{CG} - \overline{X}_{AC,w}$ depends on the center of gravity position $\overline{X}_{CG}$, the position of the wing aerodynamic centre $\overline{X}_{AC,w}$, which in turn depends on the wing planform and Mach number.

For subsonic speeds, $\overline{X}_{AC,w}$ can be approximately determined as follows:

$$\overline{X}_{AC,w} = \overline{X}_{AC,airfoil} + 0.033\left(AR\ tan\ \Lambda^{1/2}\frac{\lambda - 1}{\lambda + 1}\right)\frac{\lambda - 1.7}{\lambda}, \tag{41}$$

$\overline{X}_{AC,airfoil} = \frac{1}{4}\left[1 - 2(t/c_{av})^2\right]$ is the aerodynamic centre of the airfoil.

The shift of the aerodynamic centre caused by air compressibility for $M_{crit} < M < 1.2$ can also be determined approximately:

$$\Delta\overline{X}_{AC} = \Delta\overline{X}_{AC,max} k_F(M), \tag{42}$$

where

$$\Delta\overline{X}_{AC,max} = \frac{3}{4}\left(0.12 + 0.03\frac{ARtan\ \Lambda^{1/2}}{\lambda}\right)\frac{(\lambda+1)^2}{\lambda^2+\lambda+1}, \tag{43}$$

$$k_F(M) = \begin{cases} 0 & atM < M'_{crit} \\ \left(3 - 2\frac{M-M'_{crit}}{1.2-M'_{crit}}\right)\left(\frac{M-M'_{crit}}{1.2-M'_{crit}}\right)^2 & at\,M'_{crit} < M < 1.2 \\ 1 & atM > 1.2 \end{cases} \tag{44}$$

Equations (43) and (44) can be used for a wide range of wing shapes.

The wing contributes the most to the yawing stability of aircraft, C_n^β. The wing rolling moment coefficient with respect to sideslip angle is characterised by the derivative $\frac{\partial C_{l,w}}{\partial \beta} = C_{l,w}^\beta$. The value of this derivative can be estimated as follows:

$$C_{l,w}^\beta = \frac{1}{6}\left[\left(1 + \frac{C_L^{\alpha,M}M}{C_L^\alpha}\right)C_{L,\ \Lambda_{1/2}} + C_L^\alpha\Gamma\right]\frac{\lambda+2}{\lambda+1}, \tag{45}$$

where $C_L^{\alpha,M} = \frac{\partial^2 C_L}{\partial\alpha\partial M}$; $\Lambda_{1/2}$ is the wing mid-chord sweep angle; Γ is the dihedral angle (all angles are in radians).

It can be seen from (45) that at supersonic speeds (where $C_L^{\alpha,M} < 0$) or at post-stall angles of attack ($C_L^\alpha < 0$), a roll reversal can occur.

In terms of aircraft as a whole, C_n and C_l are affected by the interference of wing with fuselage and other parts of the aircraft, as well as their elastic deformations under load.

3.5 *Selection of Wing Parameters at the Preliminary Design Stage*

The wing geometric parameters are selected on the basis of compromises between the aerodynamic, weight and volume requirements.

When establishing the wing shape for a subsonic aircraft, the major consideration is to ensure an efficient flight at a cruising speed of $M \approx M_{cr}$. Expressions (4) and (5) show that M_{cr} depends on the average thickness ratio of the wing $(t/c)_{av}$ and on the wing sweep angle Λ. Since $(t/c)_{av}$ can vary within a very narrow range and it is determined by the conditions of static and dynamic strength, as well as the capability of the wing to accommodate the required volume of fuel, the selection of the sweep angle is critical when establishing the wing planform.

We may state approximately that $M_{cr,\ \Lambda>0} = \frac{M_{cr,\ \Lambda=0}}{\cos\ \Lambda}$ or $\cos\ \Lambda = M_{cr,\ \Lambda=0} = const$.

This makes it possible, assuming certain M_{cr} values for a straight wing ($\Lambda = 0$), to select Λ based on the given cruising speed and the corresponding Mach number. To first approximation, it can be assumed that, depending on the wing airfoils, $M_{cr,\ \Lambda=0}$ has the following values:

Fig. 16 Fairings and wing root extensions: **a**—fairings; **b**—extensions; 1—forward fairing; 2—aft fairing

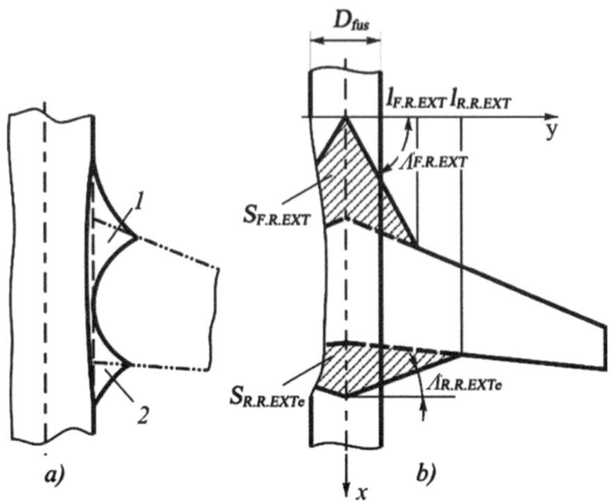

- conventional high-lift airfoils: 0.65
- high-speed airfoils: 0.70
- supercritical airfoils: 0.75 or greater.

However, note that $C_{L,max,\ \Lambda>0} \approx C_{L,max,\ \Lambda=0}\cos\Lambda$ and as Λ increases, $C_{L,max}$ decreases; the lift-to-drag ratio decreases as well. This fact directly affects the aircraft field performance and the range.

When configuring an aircraft, fairings and root extensions are often used in the wing root section (Fig. 16). On high-aspect-ratio wings, rear root extensions as a rule are also used to increase the cross-section area in the wing inboard part and to accommodate the high-lift devices.

On a subsonic swept back aircraft:

(1) fairings decrease the detrimental interference between wing and fuselage, especially for a low-wing aircraft;
(2) forward extensions reduce the wing drag due to a decrease in the "middle effect" (a swept back wing in its central part has the property of a straight wing with a corresponding C_D increase);
(3) forward and rear extensions increase the wing's structural depth and volume to accommodate fuel, units, etc., while maintaining the required wing thickness ratio.

The relative increase in the wing area due to the extensions can be estimated as follows:

(1) for the forward extension:

$$\overline{S}_{FRext} = \frac{S_{fFRext}}{S} = \overline{y}^2_{FRext}(tan\ \Lambda_{FR\ ext} - tan\ \Lambda_{LE}); \tag{46}$$

(2) for the rear extension:

$$\overline{S}_{RRext} = \frac{S_{RRext}}{S} = \overline{y}^{2}_{RRext}(tan\ \Lambda_{RRext} - tan\ \Lambda_{TE}), \tag{47}$$

where

$\overline{y}_{FRext}$ and $\overline{y}_{RRext}$ are the span ratios of extensions;

Λ_{FRext} and Λ_{RRext} are the sweepback angles of the leading and trailing edges of extensions;

Λ_{LE} and Λ_{TE} are the sweepback angles of the leading and trailing edges of the baseline wing.

For a subsonic aircraft, if $\overline{S}_{FRext} + \overline{S}_{RRext} < 0.15$, usually extensions are not considered when defining the wing area and the mean aerodynamic chord.

If $\overline{S}_{FRext} + \overline{S}_{RRext} \geq 0.15$, then the following equation is used:

$$S = S_{baseline}\left(\overline{S}_{FRext} + \overline{S}_{RRext}\right), \tag{48}$$

where $S_{baseline}$ is the area of the baseline wing.

In this case, the MAC (c_{MAC}) is estimated using the equation for the compound wing.

For a supersonic aircraft, forward root extensions are of great importance. At subsonic flight speeds, the position of the wing aerodynamic centre ($\overline{X}_{AC}$) is calculated using the baseline wing MAC, because the extensions have little effect on this position. During the transition to supersonic flight speeds, the entire wing area operates equally, and the aerodynamic center shifts backward to a lesser degree if compared to its shift without extension. It is possible to select the area and sweep of the front extension so that the position of the baseline wing aerodynamic centre without extensions at subsonic flight speeds and the position of wing aerodynamic centre with the extensions at supersonic speeds will almost coincide and the static stability will almost remain constant.

It should be borne in mind that the use of front extensions increases the instability region in the moment curve and shifts it to lower angles of attack.

The effect of Λ, AR and $(t/c)_{av}$ on the wing mass characteristics should be considered as well. In general, for a straight wing, the dependence of the wing mass on its geometric parameters can be presented in the following form:

$$m_w = A\frac{AR^{3/2}}{(t/c)_r}k_\lambda + BAR^{1/2} + C\frac{1}{AR^{1/2}} + D, \tag{49}$$

where $k_\lambda = 45.5\left(1 + \frac{1}{\lambda}\right) - (\lambda + 1)$ is the coefficient that considers the wing taper influence; A, B, C, D are the coefficients that do not depend on $(t/c)_r$, AR, and λ.

Equation (49) makes it possible to assess the influence of $(t/c)_r$, AR, and λ on the wing mass. If we substitute AR with the expression $\frac{AR}{(\cos\ \Lambda)^2}$ in equation (49), we arrive at the following relationship to determine the mass of a swept wing:

$$m_w = A\frac{AR^{3/2}}{(\cos\,\Lambda)^2}\frac{k_\lambda}{(t/c)_r} + B\frac{AR^{1/2}}{\cos\,\Lambda} + C\frac{\cos\,\Lambda}{AR^{1/2}} + D. \tag{50}$$

We can make the following conclusions from Eqs. (49) and (50):

(1) an increase in AR leads to a significant increase in m_w;
(2) an increase in $(t/c)_r$ leads to a decrease in m_w;
(3) an increase in λ leads to a decrease in m_w;
(4) an increase in Λ entails an increase in m_w.

4 Aerodynamic Characteristics of Low-Aspect Ratio Wings

4.1 Flow Field Around a Low-Aspect Ratio Wing

The flow field around low-aspect wings ($AR \leq 3$) of various planforms is essentially three-dimensional, even at small angles of attack. It is characterised by an intense leakage of air from the bottom surface to the upper one around the tip or leading (with a large sweep) edge. The spatial vortices formed on these edges roll up into powerful vortex cores. Further downstream, they connect with tip vortices formed by the main vortex sheet of the wing (Fig. 17). Such a three-dimensional vortex system induces additional speed on the upper surface of the wing in the chordwise direction. Therefore, the suction increases on the wing upper surface, which produces additional lift, compensating with excess for the local flow separation at the wing tips.

For a low-aspect ratio wing, this aerodynamic effect increases when the angle of attack increases. It delays the full wing stall to higher angles of attack.

The wing lift coefficient with respect to the additional vortex lift produced by low-aspect ratio wings equals

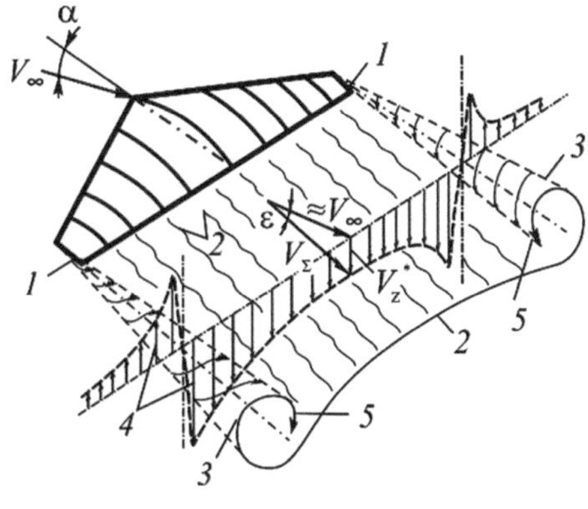

Fig. 17 The flow pattern downstream of a finite wing: 1—Wing tips; 2—Vortex sheet; 3—Vortex cores; 4—Vertical components of speed vector; 5—Direction of rotation of vortex cores

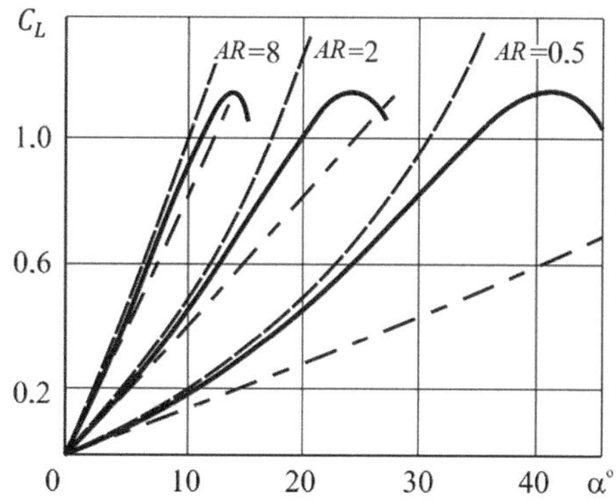

Fig. 18 Lift curves for wings with various aspect ratios: —·—·— linear theory; — — — non-linear theory; ——— viscous flow

$$C_L = 2\pi AR\frac{\alpha - \alpha_0}{\overline{p}AR + 2}\left(1 + 2\frac{\alpha - \alpha_0}{\overline{p}AR + 2}\right), \tag{51}$$

where $\overline{p}$ is the ratio of the wing half-perimeter to its span and it is defined by expressions (19) and (21).

This equation can be applied to wings of moderately low aspect ratio ($1 < AR < 3$). For very low-aspect ratio wings ($AR < 1$), the lift curve at small angles of attack has the form $C_L = \left(\frac{\pi AR}{2}\right)(\alpha - \alpha_0)$ and $C_L^\alpha = \frac{\pi AR}{2}$.

The non-linearity of the lift curve for such wings can be modelled by means of expression (51); with the view that at $AR \rightarrow 0, \overline{p} \rightarrow \frac{2}{AR}$, then

$$C_L = \frac{\pi AR}{2}(\alpha - \alpha_0)[1 + 2(\alpha - \alpha_0)]. \tag{52}$$

It should be noted that, provided the airfoils are the same, the maximum lift coefficient will also be approximately equal for wings of any aspect ratio (Fig. 18).

For delta wings with supersonic leading edges and at supersonic speeds, C_L^α is calculated using the following equation:

$$c_{y\max\,\chi>0} = c_{y\max\,\chi=0}\cos\chi.$$

For subsonic leading edges $\left(tan\ \Lambda_{LE} < \sqrt{M^2 - 1}\right)$, the following expression can be used:

$$C_L^\alpha = \frac{2\pi}{E(k)tan\ \Lambda_{LE}}, \tag{53}$$

where $E(k)$ is the complete elliptic integral of the second kind with module

$$k = \sqrt{1 - m^2}, m = \frac{\sqrt{M^2 - 1}}{tan\ \Lambda_{LE}}.$$

We can use a simple semi-empirical equation as a universal model for the C_L^α of delta wings:

$$C_L^\alpha = \frac{4AR}{AR\sqrt{M^2 - 1}} + 0.3. \tag{54}$$

It should be noted that the non-linearity of the lift curve of low aspect ratio wings almost disappears at supersonic speeds.

4.2 Aerodynamic Drag and Moment of Low Aspect Ratio Wings

At subsonic speeds, the drag coefficient at zero lift $C_{D,0}$ of low aspect ratio wings can be estimated using the same equations as for wings of moderate and high aspect ratio. Vortex-induced drag for low-aspect wings ($AR < 3$) is determined as follows:

$$C_{D,i} = \frac{3}{\pi AR} C_L^2. \tag{55}$$

At supersonic speeds, the total induced drag (vortex and wave) of these wings at optimal (elliptic) load distribution along span and chords is determined by the following expression:

$$C_{D,i} = \frac{C_L^2}{\pi AR}\left(1 - 2\frac{M^2 - 1}{tan\,\Lambda_{LE}}\right). \tag{56}$$

To reduce the drag of low aspect ratio wings, as for the high aspect ratio wings, aerodynamic and geometric twist is applied.

The moment curve at subsonic speeds for low aspect ratio wings depends on their planform:

(1) ogival wings feature an instability region;
(2) gothic wings, on the contrary, display an increased longitudinal static stability;
(3) delta wings have almost linear dependence of the pitching moment coefficient of the lift coefficient.

At supersonic speeds, the instability region in the moment curve is almost eliminated for wings of any shape.

5 Wing Aeroelastic Characteristics

5.1 Static Aeroelasticity

In flight, aircraft structure is exposed to aerodynamic forces, elastic forces and mass forces interacting with each other. This must be taken into account when designing the wing, that is, sizing the wing, selecting an appropriate structural arrangement for it, and sizing the structural elements of the wing. Static aeroelasticity deals with the interaction of only aerodynamic and elastic forces and its consequences. These include wing divergence, aileron reversal, longitudinal static instability due to wing flex.

When inertia forces are added to the analysis, problems of dynamic aeroelasticity arise: flutter (dynamic instability of structure), buffeting (dynamic behaviour of structure during flow separation), dynamic reactions in structure to external disturbances (wind gust, landing, etc.).

At speeds typical for modern aircraft, the wing stiffness at bending and torsion often determines the rational values of wing parameters. The general expression for the maximum deflection ratio of a cantilever wing is as follows:

$$\overline{f} = \frac{f}{b} = B_1 \frac{AR}{(t/c)_r} \frac{\lambda + 1}{\lambda}. \qquad (57)$$

where B_1 is the coefficient depending on the spanwise load distribution and moments of inertia of the spar cross-sections.

The general expression for the twist angle of the wing (the moment is applied at the tip) has the following form:

$$\varphi_{tip} = \frac{Tb}{4GF^2} \oint \frac{dS}{\delta},$$

where φ is the twist angle at the wing tip; T is the torsional moment applied to the wing tip; G is the modulus of elasticity in shear; F is the area of cross-section; dS is the linear segment of the cross-section; δ is the shell thickness.

The twist angle ratio at the wing tip with $\lambda = 1$ will be

$$\frac{\varphi_{tip}}{b} = B_2 T \frac{AR}{(t/c)_r^2} \qquad (58)$$

Thus, the twist angle increases with increasing wing aspect ratio AR and decreases with increasing root thickness ratio $(t/c)_r$.

Inadequate wing stiffness in bending and torsion affects the flight in the most unfavourable way. Large flexing and twisting of wing change its aerodynamic characteristics (due to the change in the angles of attack of the wing sections and the dihedral).

Very large deformation due to torsion at low torsional stiffness leads to a partial and even complete loss of aileron efficiency, which is called the aileron reversal.

Let us examine this phenomenon applied to a wing element of length dy. When the aileron is deflected on an inelastic wing to angle δ_a, the lift coefficient of a wing element of length dy can be expressed as follows:

$$C_{l'} = \frac{dC_l}{d\alpha}(\alpha_0 + \alpha) + \Delta C_l(\delta_a).$$

When the aileron is deflected on an elastic wing to the same angle δ_a, the lift coefficient of a wing element of length dy will be (Fig. 19):

$$C_l' = \frac{dC_l}{d\alpha}(\alpha_0 + \alpha) + \Delta C_l(\delta_a) - \frac{dC_l}{d\alpha}\varphi = \frac{dC_l}{d\alpha}[(\alpha_0 + \alpha) - \varphi] + \Delta C_l(\delta_a\,), \quad (59)$$

where φ is the angle to which the wing element is rotated as a consequence of the moment due to the aileron deflection.

The angle φ can be estimated by equating the torsional moment due to the aileron deflection with the moment due to the wing's internal elastic forces; it depends on the dynamic pressure q.

At some value of the dynamic pressure, called the aileron reversal dynamic pressure $q = q_r$, φ is be so large that

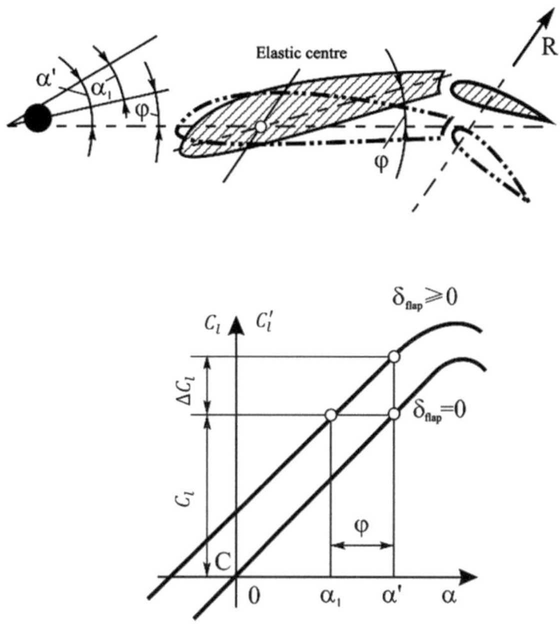

Fig. 19 The effect of twisting an elastic wing by angle φ. R is the force due to aileron deflection; $\Delta C_l = \frac{dC_l}{d\alpha}\varphi$

Fig. 20 The variation of the moment due to aileron deflection L_{ail} as a function of dynamic pressure for an inelastic (1) and elastic (2) wings

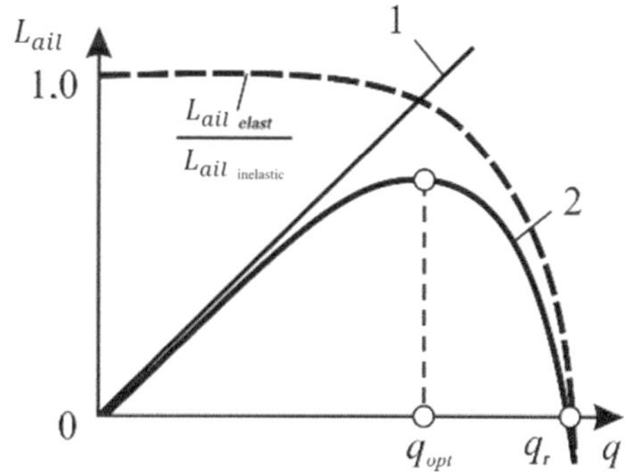

$$\frac{dC_l}{d\alpha}\varphi = \Delta C_l(\delta_a)$$

and the lift increment of the wing segment dy due to aileron deflection will be cancelled by the reduction of lift due to the wing segment's twisting by the angle φ. The variation of the moment due to aileron deflection L_{ail} as a function of dynamic pressure is shown in Fig. 20.

At $q = q_r$, as can be seen in the plot, $L_{ail} = 0$. Consequently, at $q = q_r$, the aileron reversal begins, the ailerons completely lose their authority, and with a further increase in q the aircraft will start banking in the opposite direction.

The sweep angle strongly affects the bending and torsional stiffness of a wing. The greater the sweepback angle, other things being equal, the greater the wing flex. Fig. 21 shows the variation of the maximum displacement of the tip of a torsion box wing f_{sw} and the corresponding twist angles φ_{sw} as a function of sweep angle Λ. Here, f_{sw} and φ_{sw} are related to the displacement and twist angle of straight wings with all other parameters equal to those of swept back wings.

Elastic deformations of a wing's structure must be taken into account during the design process.

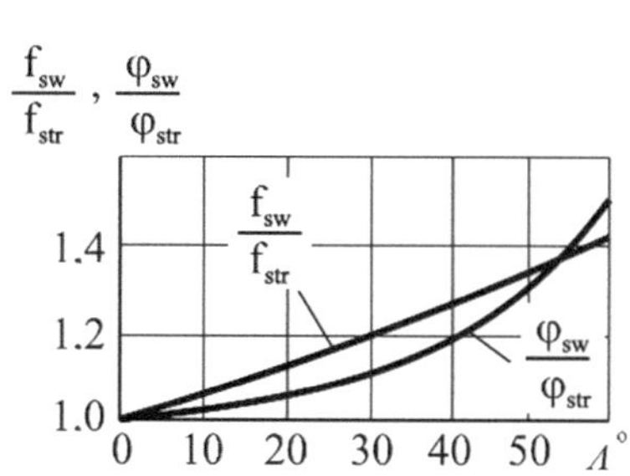

Fig. 21 The variation of the maximum displacement of the tip of a torsion box wing f_{sw} and the corresponding twist angles φ_{sw} as a function of sweep angle Λ (f_{sw} and φ_{sw} are the displacement and twist angle of a sweepback wing's tip; f_{str} and φ_{str} are the displacement and twist angle of a straight wing's tip)

For a long-range aircraft, in order to maintain a high lift-to-drag ratio (as a consequence of the elliptical spanwise load distribution in cruise), it is necessary to add the predicted twist angles due elastic deformations $\varphi_{elastic}^{pred}(\overline{y})$ (with a negative sign) to the optimal aerodynamic (geometric) wing twist $\varphi_{\alpha}(\overline{y})$. The resultant twist of the unloaded wing is $\Delta\varphi_{\Sigma}(\overline{y}) = \varphi_{\alpha}(\overline{y}) - \varphi_{elastic}^{pred}(\overline{y})$. This should be reflected in the drawings of the wing and in the drawings of the manufacturing equipment. Then the actual wing twist in flight can be equal (or almost equal to) to the given optimal twist, i.e.

$$\Delta\varphi_{\Sigma}^{actual}(\overline{y}) = \varphi_{\alpha}(\overline{y}) - \varphi_{elastic}^{pred}(\overline{y}) + \varphi_{elastic}^{actual}(\overline{y}) \approx \varphi_{\alpha}(\overline{y}),$$

under the assumption that $\varphi_{elastic}^{pred}(\overline{y}) \approx \varphi_{elastic}^{actual}(\overline{y})$.

For a manoeuvring aircraft, the wing flex in flight affects its aerodynamic efficiency, longitudinal stability and controllability.

The below equation is applicable to the wing of an arbitrary planform:

$$\Delta C_{L,elastic} \approx \frac{2}{S}\int_0^{b/2} \big[\varphi_{twist}(\overline{y})\cos\,\Lambda_{elcentre} - \Theta_{man}(\overline{y})\sin\,\Lambda_{elcentre}\big] C_L^{\alpha}(\overline{y}) c(\overline{y}) dy, \tag{60}$$

or

$$\Delta C_{L,elastic} \approx \frac{2}{S}\int_0^{b/2} \big[\varphi_{twist}(\overline{y})\cos\,\Lambda_{elcentre} - \Theta_{man}(\overline{y})\sin\,\Lambda_{elcentre}\big] C_L^{\alpha}(\overline{y}) c(\overline{y}) \times \big[X_{CP}(\overline{y}) - X_{MAC,LE}\big] dy, \tag{61}$$

where $\varphi_{twist}(\overline{y})$ and $\Theta_{man}(\overline{y})$ are the twist angle and the slope angle of the elastic axis of the wing at station $\overline{y}$; $\Lambda_{elcentre}$ is the sweep angle of the elastic axis at station $\overline{y}$; $C_L^{\alpha}(\overline{y})$ is the derivative of the lift coefficient of the section with respect to the angle of attack; $c(\overline{y})$ is the spanwise chord distribution; $X_{CP}(\overline{y})$ is the chordwise centre-of pressure location as a function of spanwise station; $X_{MAC,LE}$ is the coordinate of the MAC leading-edge.

5.2 *Dynamic Elasticity*

The general relationship between the main wing design parameters and the flutter limit $V_{flutter}$ is provided by the following equation (wing taper ratio $\lambda = 1$):

$$V_{flutter} = \frac{\pi}{S}\sqrt{\frac{2GJ_w}{\overline{r}C_L^{\alpha}\rho_H\cos\,\Lambda}}, \tag{62}$$

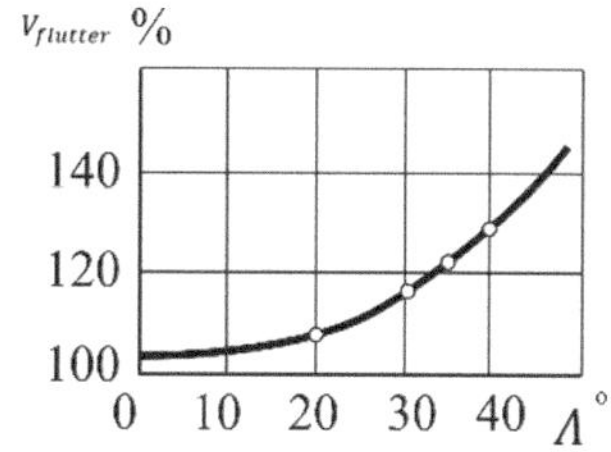

Fig. 22 Variation of $V_{flutter}$ with the sweep angle Λ; $\overline{X}_{CG} = 0.45$, $\lambda = 4$

where S is the wing area; $\overline{r} = \frac{r}{c \cos \Lambda}$ is the distance between the section's aerodynamic centre and the centre of mass, with respect to $c \cos \Lambda$ (c is the wing chord measured parallel to the free stream velocity vector).

The flutter limit $V_{flutter}$ depends on the wing torsional stiffness GJ_w which in its turn depends on $\overline{r}$ and the air density at the flight altitude ρ_H.

$V_{flutter}$ increases with a decrease in C_L^α. Recall that C_L^α increases with increasing aspect ratio AR; it decreases with increasing sweep angle Λ. As a result, $V_{flutter}$ increases with decreasing AR and increasing Λ.

Figure 22 shows an experiment-based variation of $V_{flutter}$ with the sweep angle Λ.

The experiment was carried out using a model in which the sweepback angle of the wing could vary from 0 to 50°.

The following equation relates the taper ratio λ of a wing to the flutter limit $V_{flutter}$:

$$V_{flutter} = \frac{\lambda^2 + 0.8\lambda + 0.4}{c_{MGC}\lambda^2\left(\overline{X}_{CG} - 0.1\right)}\sqrt{\frac{GJ'_w}{0.6bf(M)p_{11}}}, \tag{63}$$

where $\overline{X}_{CG}$ is the distance ratio from the leading edge to the centre of mass axis; GJ'_w is the average torsional stiffness of the wing portion between the root section and 75% of semi-span; b is the wing span; $f(M) = \sqrt{1-M}$ if $0 < M < 0.75$; $f(M) = 0.67$ if $M > 0.75$.

It should be borne in mind that an increase in taper also entails an increase in torsional stiffness GJ'_w.

$V_{flutter}$ strongly depends on the mutual location of the aerodynamic centre axis and the centre of mass axis in the wing, as well as on the torsional stiffness of the wing structure (it depends very little on the bending stiffness). A shift of the centre of mass axis along the chord with respect to the aerodynamic centre axis (Fig. 23) can dramatically change $V_{flutter}$.

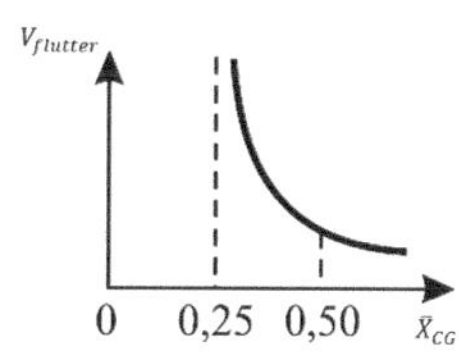

Fig. 23 Variation of $V_{flutter}$ as a function of $\overline{X}_{CG}$ for the given position of the wing aerodynamic centre, $\overline{X}_{AC} = 0.25$

Table 1 Relationship between main wing parameters and characteristics

Variables	Aerodynamic characteristics				Weight	Stiffness	Vibration
	$C_{D,pr}$	$C_{D,i}$	$C_{D,w}$	$C_{L,max}$	Wing mass	Stiffness	Flutter limit
Increasing AR	*	↓	**	—	↑↑	↓↓	↓
Decreasing *t/c*	↑	—	↑	↑	↓	↑	—
Increasing λ	—	***	—	****	↓	↑	↑
Increasing Λ	—	—	↓	↓	↑	↓	↑

* $C_{D,pr}$ increases at high aspect ratios, very small chords and speeds (at low Re)
** At very small AR (<2.5), $C_{D,pr}$ decreases at transonic speeds
*** $C_{D,i}$ at $\lambda = 2.0 \ldots 2.5$ is minimized
*** For λ up to $2.0 \ldots 2.5$, $C_{L,max}$ increases, and then decreases

If the aerodynamic centre axis and the centre of mass axis are coincident, the flutter limit goes to infinity, $V_{flutter} = \infty$. However, this is very difficult to implement in practice. In engineering practice, balancing weights are often used to shift the centres of mass of the cross-sections towards the aerodynamic centres (i.e. forward). These weights are installed in the wing leading-edges (closer to the wing tips) or arranged in special streamlined brackets fixed at the tips. However, a significant increase in flutter limit can only be obtained at the cost of a quite significant increment in the total wing mass that reaches 10% or more of the mass of its structure.

The main points of the section are summarized in Table. 1.

6 Variable-Geometry Wings

The altitudes and speeds of flight can be increased through the use of variable-geometry wings and, in particular, variable-sweep wings. There is an advantage to these wings: as the sweep angle increases from $\Lambda \approx 0^{\circ}$ to $\Lambda \approx 90^{\circ}$, all other geometric parameters change in a wide range, resulting in a desirable change in the wing's and aircraft's main aerodynamic characteristics.

Some important geometric parameters of a variable-sweep wing are the total theoretical area S and the wing span b of one of its configurations. In general, it is difficult to derive analytical equations describing the variation of geometrical parameters of such a wing as a function of the outer wing's sweep angle. Given the importance of these equations for aircraft design, they can be obtained approximately. The chords of the outer movable wing will change as follows:

$$c_{outer}(\Lambda) = \frac{c_{outer}}{\cos\,\Lambda} = c'_{outer}\frac{\cos\,\Lambda'}{\cos\,\Lambda}, \tag{64}$$

where $c\prime_{outer}$ is the chord of the outer movable wing at $\Lambda = \Lambda' = \Lambda_{min}$.

Since structural depth of the wing does not change with sweep angle, the thickness ratio of airfoils changes as follows:

$$(t/c)\prime_{outer}(\Lambda) = \frac{t\prime_{r,outer}}{c_{outer}(\Lambda)} = (t/c)_{r,outer}c\prime_{outer}\cos\ \Lambda = (t/c)\prime_{outer}\frac{\cos\ \Lambda\prime}{\cos\ \Lambda}, \tag{65}$$

where $(t/c)\prime_{outer}$ is the thickness ratio of the airfoils of the outer movable wing at $\Lambda = \Lambda' = \Lambda_{min}$. The camber ratio of airfoils will change in the same way.

The main geometric properties of a variable-sweep wing are as follows:

$$S = S_{Inboard} + S_{outer}, \tag{66}$$

$$b = 2y_{joint+} + b_{outer}\cos\ \Lambda, \tag{67}$$

$$c_{MAC} = \frac{1}{S}\left(S_{Inboard}c_{MAC,inboard} + S_{outer}c_{MAC,outer}\right), \tag{68}$$

where $S_{Inboard}$ is the inboard wing; S_{outer} is the outer wing; $2y_{joint}$ is the distance between left and right hinged joints; b_{outer} is the span of two outer wings; $c_{MAC,inboard}$ and $c_{MAC,outer}$ are the mean aerodynamic chords of the inboard and outer wings, respectively; S_{outer} and $c_{MAC,outer}$ are the functions of sweep angle of outer wing.

Another important wing property is the shift of its centre of mass as a function of sweep angle. This shift in terms of MAC of the baseline wing configuration can be estimated as follows:

$$\Delta\overline{X}_{CG,outer} = \frac{\overline{m}_{outer}}{c_{MAC}}y_{CG,outer}\left(\sin\ \Lambda - \sin\ \Lambda'\right), \tag{69}$$

where $\overline{m}_{outer} = \frac{m_{outer}}{m_w} \approx \overline{S}_{outer} \approx 1 - \overline{S}_{inboard}$ is the mass of the outer wings referred to the mass of the entire wing; $\overline{X}_{CG,outer} \approx \frac{1}{6}b_{outer}$ is the distance from the centre of mass of the outer wings to the hinge axis.

It is evident from Eq. (69) that at small Λ the wing span and aspect ratio change little and there is no great need for decreasing Λ' to more than 15 to 20°. In terms of structural design, this provides a significant simplification. The maximum sweep angle of the outboard wing Λ'' is in the range of 65 to 72°.

C_L^{α}, $C_{D,0}$, and C_D for variable-sweep wings can be calculated as shown below.

The spanwise position of the hinge axis y_{joint} and front root extensions affect strongly the longitudinal stability of a variable-sweep aircraft.

Figure 24 demonstrates the influence of extensions on $C_m^{C_L}$ when the sweep is changed.

It should be noted that the elastic wing flex in flight (wingtip vertical displacement and twist) tends to shift the wing's aerodynamic centre forward, as shown in Fig. 25.

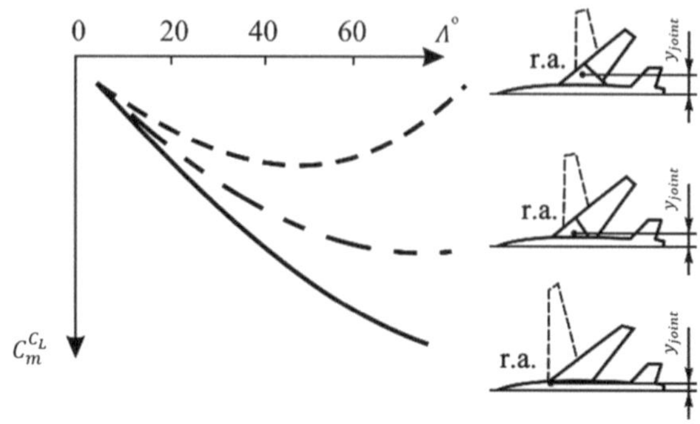

Fig. 24 Effect of wing root extensions on $C_m^{C_L}$ when the sweep angle of the outer wing is changed; r.a = rotation axis

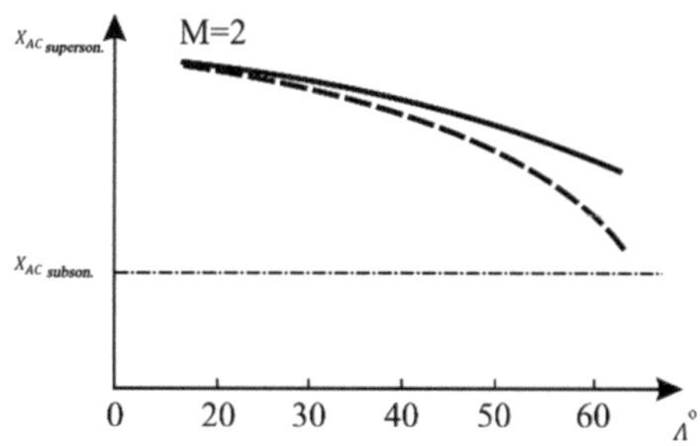

Fig. 25 The effect of the elastic wing flex on the position of the wing aerodynamic centre as a function of the wing sweep angle at $M = 2$: — inelastic wing; - - - elastic wing.

As a result, an appropriate wing geometric layout with root extensions and wing stiffness can ensure that the aircraft has an almost constant $C_m^{C_L}$ over a wide speed range (subsonic and supersonic). In accordance with Eq. (67), the thickness ratio and camber ratio of the outer wing decrease when the outer wing rotates. This leads to a decrease in $C_{D,0}$. As a result, the aircraft's drag polars change as shown in Fig. 26.

The maximum sweepback angle provides a variable-sweep aircraft with the best speed performance at supersonic flight, while the minimum sweep angle ensures the best field performance.

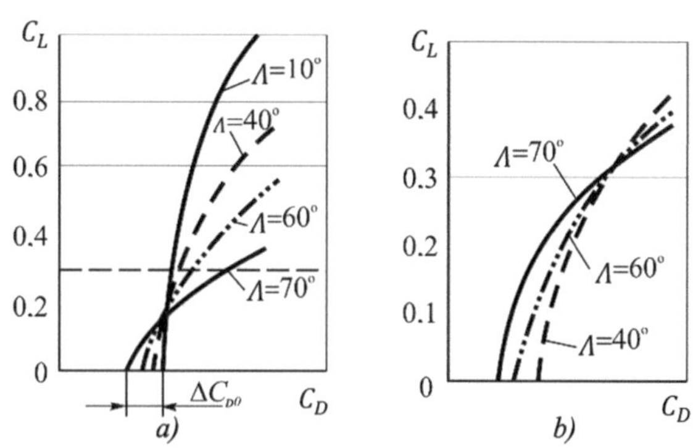

Fig. 26 Drag polars for different outer wing sweep angles: **a**—at subsonic speeds; **b**—at supersonic speeds

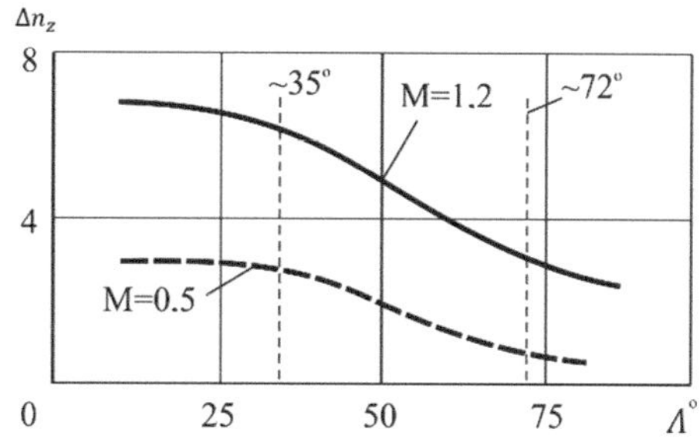

Fig. 27 Effect of the outer wing sweep angle on an aircraft's g-loads when exposed to a vertical gust

During flight at extremely low altitudes the number of vertical gusts encountered by aircraft is 103 times greater than when flying at altitudes of 10,000 to 14,000 m. When an aircraft encounters a vertical gust, it is exposed to a g-load that equals

$$\Delta n = k_0 \frac{\Delta L}{mg} = k_0 \frac{\Delta C_L \rho_0 V_i^2 S}{2mg} = k_0 \frac{\rho_0 C_L^\alpha (W_i/V_i) V_i^2}{2 - mg/S} = k_0 \frac{C_L^\alpha W_i V_i}{16.33p}, \tag{70}$$

where k_0 is the vertical gust factor (usually $k_0 = 0.88 \ldots 0.95$); W_i is the indicated speed of vertical gust, m/s; V_i is the indicated horizontal speed, m/s; p is the wing loading, daN/m^2.

It was shown above that the aspect ratio of variable-sweep wings decreases with an increasing sweep angle, while the lift slope C_L^α decreases. This allows significantly reducing the g-loads when flying at low altitudes and the maximum sweep angle (Fig. 27), while also providing an increase in crew performance and a significant increase in the aircraft's fatigue life.

7 High-Lift Devices and Ailerons

7.1 Purpose and Types of High-Lift Devices

The purpose of high-lift devices is to change the aerodynamic characteristics of an aircraft in flight and during take-off and landing.

High-lift devices solve the following tasks:

(1) control the lift and drag of an aircraft during take-off and landing;
(2) increase the manoeuvring capabilities of an aircraft;
(3) improve the lateral stability and control of an aircraft at high angles of attack;
(4) prevent the lateral stability divergence, as well as improve the aileron efficiency at high angles of attack;
(5) increase $C_{L,max}$ at insignificant change in critical angle of attack α_{crit}.

To solve these tasks, various types of wing high-lift devices are used. Some of them are given in Table 2.

Table 2 Wing high-lift devices and their properties

Type of high-lift devices	Sketch	Increase in lift coefficient, $\frac{C_{L,HLon}}{C_{L,HLoff}}$
Plain flap with a chord of *0.3c* deflected to 45°	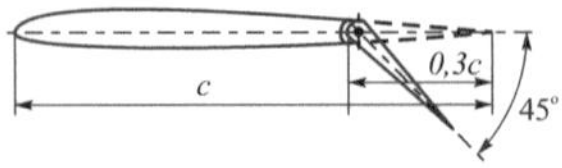	1.45–1.55
Split flap with a chord of *0.3c* deflected to 45°	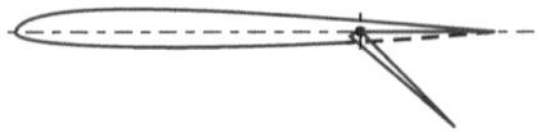	1.65–1.75
Full-span slat	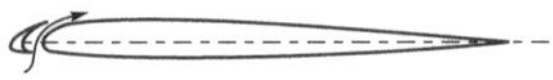	1.35–1.40
Boundary layer blowing	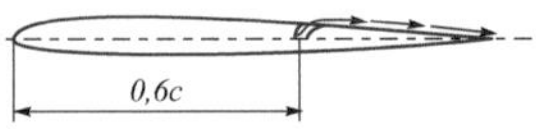	3.00–4.00
Boundary layer suction	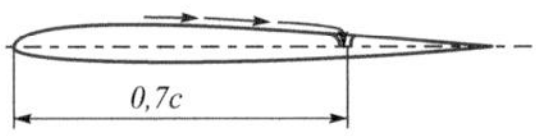	3.00–4.00
Flap coupled with boundary layer blowing	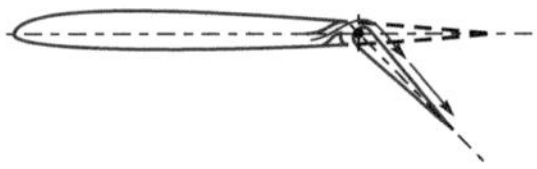	2.50–3.00
Slotted flap with a chord of *0.3c* deflected to 45°	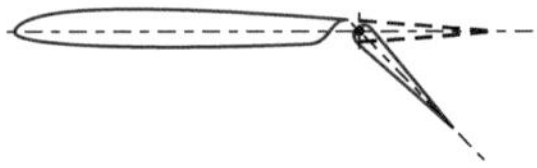	1.50–1.60
Split flap with a sliding hinge axis and a chord of *0.3c* deflected to 45°	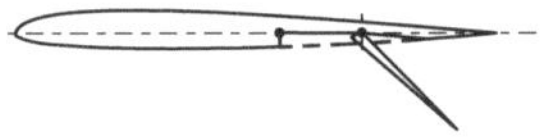	1.75–2.85
Fowler flap with a chord of *0.3c* deflected to 40°	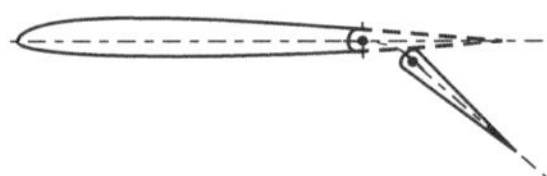	2.15–2.25
Double-slotted fowler flap deflected to 40°	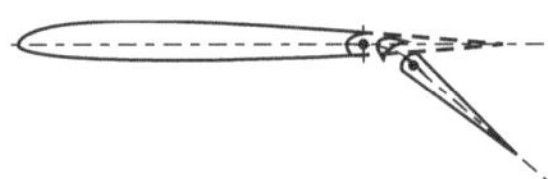	2.40–2.50
Triple-slotted Fowler flap	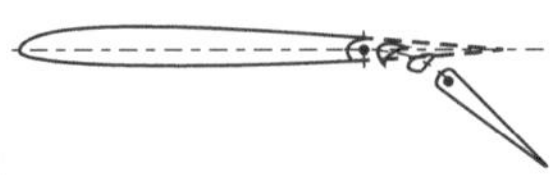	3.00–4.00
Slat and flap with a chord of *0.3c* deflected to 45°	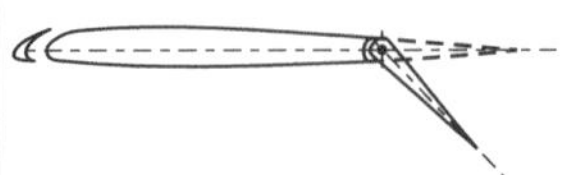	1.55–1.65

(continued)

Table 2 (continued)

Type of high-lift devices	Sketch	Increase in lift coefficient, $\frac{C_{L,HLon}}{C_{L,HLoff}}$
Slat and slotted flap with a chord of $0.3c$ deflected to 45°	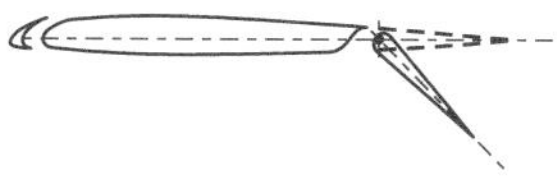	1.75–1.85
Slat and Fowler flap with a chord of $0.3c$ deflected to 45°	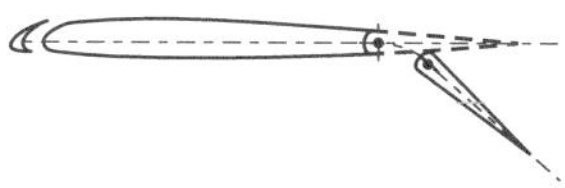	2.55–2.65
Slat and double-slot Fowler flap deflected to 40°	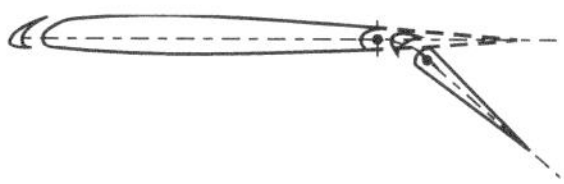	2.80–3.20
Flap with a chord of $0.3c$ deflected to 45° and hinged leading-edge deflected to 10… 15°	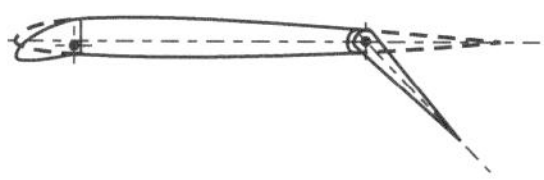	1.50–1.60
Jet flap	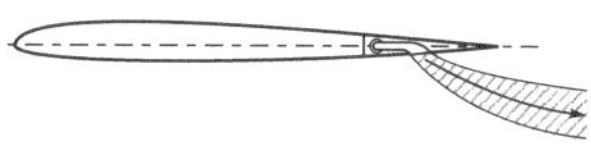	8.00–10.00

As an example, Fig. 28 shows the high-lift devices used in Boeing 727.

7.2 *Trailing Edge High-Lift Devices*

In modern aircraft, trailing edge high-lift devices are quite diverse. The most common types are shown in Fig. 29.

Lift curves for various types of high-lift devices is given in Fig. 30.

The increase in lift due to high-lift devices $\Delta C_{L,max}$ depends on the following factors:

(1) the type of the device;
(2) the span of the device;
(3) the wing taper λ;
(4) the wing aspect ratio AR;
(5) the wing sweep angle Λ;
(6) the chord length of the device;
(7) the shape of the wing's airfoil, including its thickness ratio t/c;
(8) the deflection angle of the device;
(9) the backward extension of the Fowler flap;
(10) the shape of slot between flap and wing.

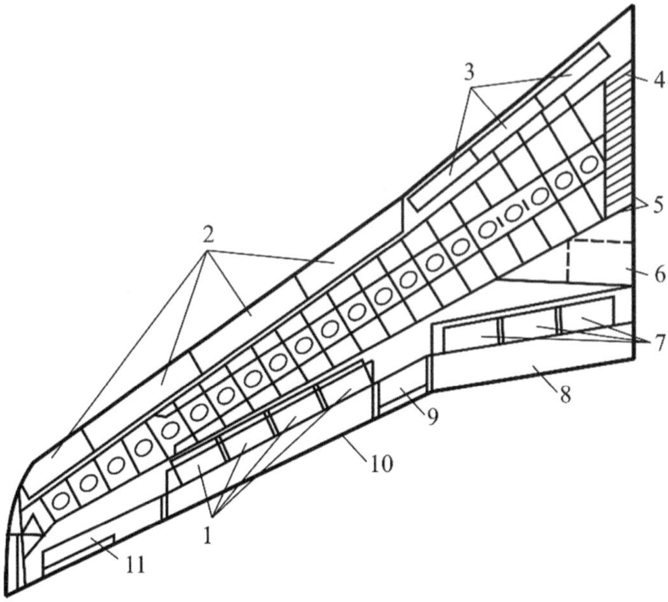

Fig. 28 The Boeing 727 wing and its high-lift devices: 1—spoilers; 2—slotted slats; 3—Krueger flaps; 4—front spar; 5—rear spar; 6—main landing gear access door; 7—spoilers—airbrakes; 8—inboard double-slotted flap; 9—inboard aileron; 10—outboard double-slotted flap; 11—aileron

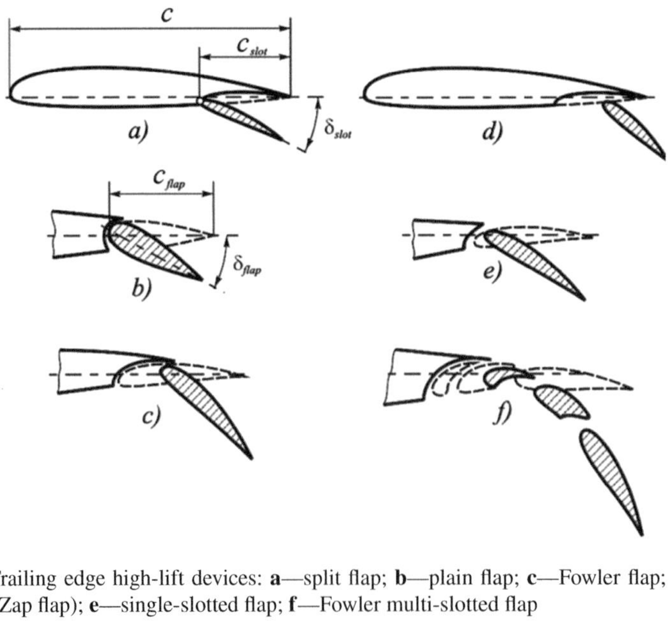

Fig. 29 Trailing edge high-lift devices: **a**—split flap; **b**—plain flap; **c**—Fowler flap; **d**—sliding split flap (Zap flap); **e**—single-slotted flap; **f**—Fowler multi-slotted flap

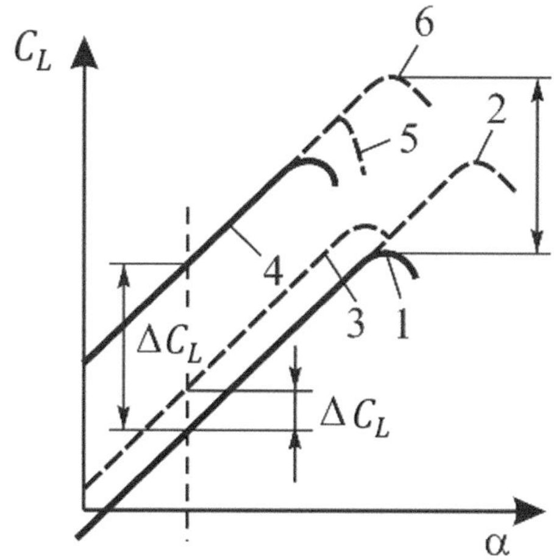

Fig. 30 Lift curves for various types of high-lift devices: 1—clean wing; 2—wing with a slat; 3—wing with a split flap; 4—wing with a multi-slotted Fowler flap; 5—the same with a Krüeger flap; 6—the same with a slotted slat.

High-lift devices reduce an aircraft's lift-to-drag ratio, which must be considered to ensure the aircraft's sufficient thrust-to-weight ratio during the initial segment of climb. The relationship between the required thrust and the lift-to-drag ratio can be expressed as follows:

$$T_{req} = mg\frac{n_{eng}}{n_{eng} - 1}\left(\frac{1}{L/D_{TO}} + tan\ \Lambda\right), \tag{71}$$

The dependency of the lift-to-drag ratio on the type of flap is given in Fig. 31.

High λ and AR enhance the effect of high-lift devices. It should be borne in mind that when the flaps are deployed, the wing pitching moment increases sharply. For trim, a negative lift must be produced by the horizontal tail, which significantly reduces the increase in wing lift due to high-lift devices. To reduce the absolute

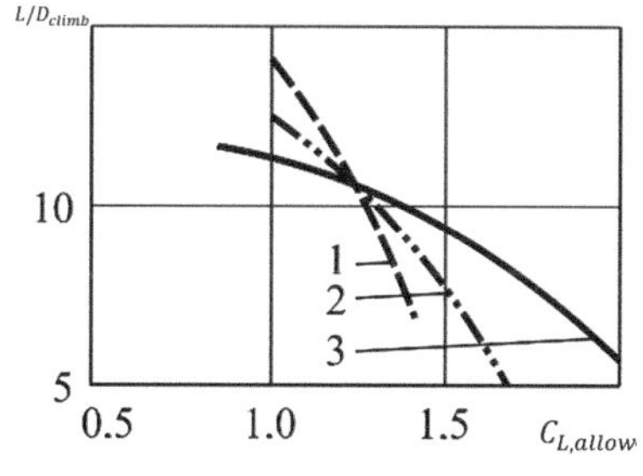

Fig. 31 Variation of lift-to-drag ratio as a function of $C_{L,allow} = \frac{C_{L,max,TO}}{1.44}$ at climb: 1—single-slotted flaps; 2—triple-slotted flaps; 3—triple-slotted flaps and slats

value of pitching moment, the flaps cover only up to 65–70% of the wing trailing edge.

$C_{L,max}$ of a wing with high-lift devices can be estimated as follows:

$$\overline{C\prime}_{L,max} = \frac{S_{HL}}{S} C_{L,max},$$

where $\overline{C\prime}_{L,max}$ is the maximum lift coefficient of the wing with high-lift devices; S_{HL} is the wing area serviced by the wing high-lift devices (refer to Fig. 32); $C_{L,max}$ is the maximum lift coefficient of a wing with the full-span high-lift devices.

Below, some typical chord ratio values for flaps are given:

(1) split flaps: $\frac{c_{SFl}}{c_w} \approx 0.25$;
(2) plain flaps: $\frac{c_{PFl}}{c_w} \approx 0.3$;
(3) Fowler flaps: $\frac{c_{FFl}}{c_w} \approx 0.3 \ldots 0.4$.

The maximum deflection angles:

(1) split flaps: $\delta_{SFl} = 55 \ldots 60°$ (measured parallel to free stream flow);
(2) plain flaps: $\delta_{PFl} = 40 \ldots 50°$ (measured parallel to free stream flow);
(3) multi-slotted Fowler flaps $\delta_{FFl} = 50 \ldots 60°$ (measured parallel to the last section).

Sweep significantly reduces the effect of wing high-lift devices. The greater the sweep angle Λ, the smaller the ratio $\frac{C_{L,max,\ \Lambda}}{C_{L,max,\ \Lambda=0}}$, where $C_{L,max,\ \Lambda}$ is the maximum lift coefficient of a swept wing with deployed high-lift devices; $C_{L,max,\ \Lambda=0}$ is the maximum lift coefficient of a straight wing with deployed high-lift devices. For split and plain flaps, the reduction in $C_{L,max}$ due to the sweep is much more pronounced than for extendable ones.

The effect of high-lift devices largely depends on the sweep angle of the wing trailing edge Λ_{TE} (Fig. 33). The greater Λ_{TE}, the smaller $C_{L,max}$.

The following data is provided to compare plain flaps and Fowler flaps:

– for $\Lambda = 35°$, multi-slotted Fowler flaps with $\frac{c_{FFl}}{c_w} \approx 0.3 \ldots 0.35$ and $\frac{b_{FFl}}{b} \approx 0.6$ provide $\Delta C_{L,max} \approx 0.9 \ldots 1$;

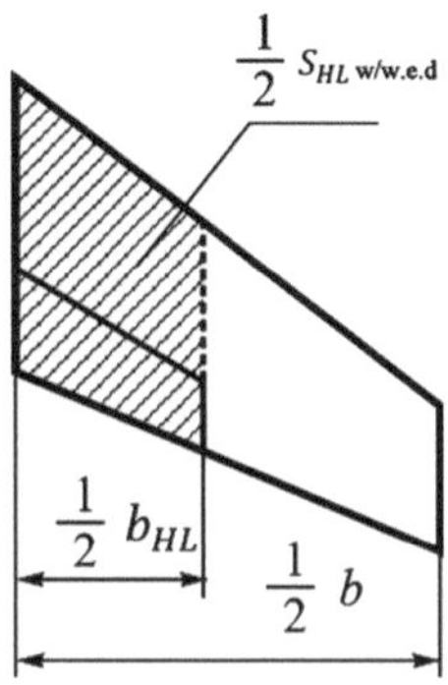

Fig. 32 Wing area serviced by the wing high-lift devices explained

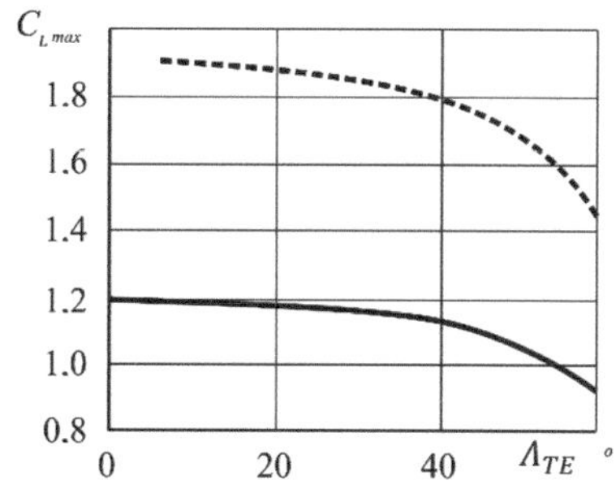

Fig. 33 $C_{L,max}$ of wings with $\Lambda_{LE} = 60°$ and $\Lambda_{TE} = 0 \ldots 60°$: ---- wing with Fowler flaps; — clean wing

- for the same wing with the same size plain flaps, $\Delta C_{L,max} \approx 0.35 \ldots 4$.

In preliminary calculations for a straight wing, the following historical data on flap properties can be used:

- slotted flaps: $\Delta C_{L,max} = 0.8 \ldots 1.0$; $\alpha = 13 \ldots 14°$;
- Fowler single-slotted flaps: $\Delta C_{L,max} = 1.4 \ldots 1.5$; $\alpha = 12°$;
- multi-slotted Fowler flaps: $\Delta C_{L,max} = 1.6 \ldots 1.8$; $\alpha = 12 \ldots 13°$.

7.3 Leading-Edge High-Lift Devices

Leading-edge high-lift devices are aimed at eliminating or delaying the flow separation from the wing and thereby at increasing $C_{L,max}$. The additional effects are the enhanced lateral stability, control, and aileron authority at high angles of attack.

Some types of the leading-edge high-lift devices are: slotted slats, Krüeger flaps, hinged leading-edges (Fig. 34).

Hinged leading-edges are used in aircraft with thin leading edges and small thickness ratios, because it is difficult to install slats in such wings. Hinged leading-edges provide smaller increase in $C_{L,max}$ than slats or Krüeger flaps and they cause a very sharp stall.

Slotted slats have the most complex structure and the greatest mass. They delay flow separation on the wing to higher angles of attack and thus provide the greatest $\Delta C_{L,max}$ (up to 1.0–1.2) at $\alpha = 25 \ldots 30°$. The main disadvantage of slotted slats is that the main wing and slats flex differently in flight. As a consequence, a step-like slot is formed between the slat trailing edge and the upper wing surface when the slat is retracted. That slot significantly increases $C_{D,0}$ of aircraft and reduces the lift-to-drag ratio.

Krüeger flaps delay flow separation to smaller angles of attack than slotted slats and trigger an abrupt stall beginning at the leading edge of the wing. This property is used to prevent swept-wing aircraft from reaching dangerously high supercritical

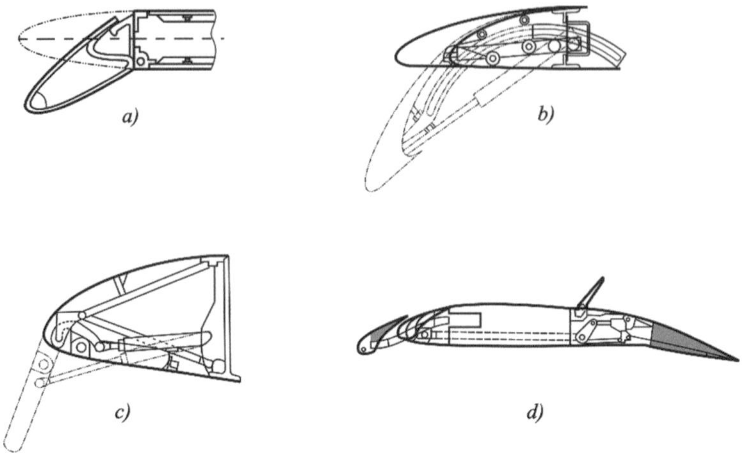

Fig. 34 Some types of the leading-edge high-lift devices: **a**—hinged leading-edge; **b**—slotted slat; **c**—Krüeger flap; **d**—a wing of manoeuvring aircraft, equipped with a slat, a plain flap and an spoiler

angles of attack. When Krüeger flaps are installed in the wing root section and slotted slats are in the middle and tip sections, the stall begins at the wing root section as the angle of attack increases. Since the flow on the middle wing and tip sections is still attached, a pitching moment is generated that returns the aircraft to lower angles of attack.

The leading-edge high-lift devices used in conjunction with the trailing-edge high-lift devices improve the flow over a wing, provide better field performance and enhance aircraft manoeuvrability.

7.4 Ensuring Aircraft Lateral Control

The most common means of aircraft lateral control are ailerons located at the wing tips (Fig. 36). As a rule, the area of ailerons is 0.05…0.07 the wing area.

During preliminary design, the aileron area can be established using an approximate equation for the aileron volume coefficient A_{ail} (Fig. 35):

$$A_{ail} = \frac{S_{ail,serv}a}{Sb} = 0.05\ldots0.07,$$

where $S_{ail,serv}$ is the wing area serviced by ailerons; a is the distance between the centres of mass of $S_{ail,serv}/2$; S is the wing area; b is the wing span.

The derivative of the rolling moment coefficient with respect to the aileron deflection angle or aileron efficiency (Fig. 36) as a function of $\frac{b_{ail}}{b}$ and $\frac{c_{ail}}{c}$ can be expressed approximately as follows [2]:

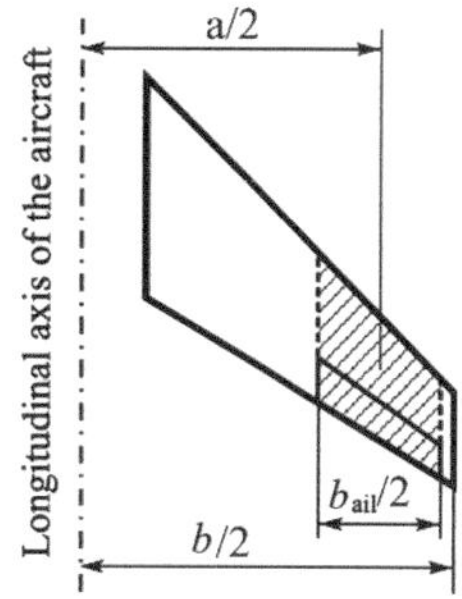

Fig. 35 Aileron arrangement in a wing

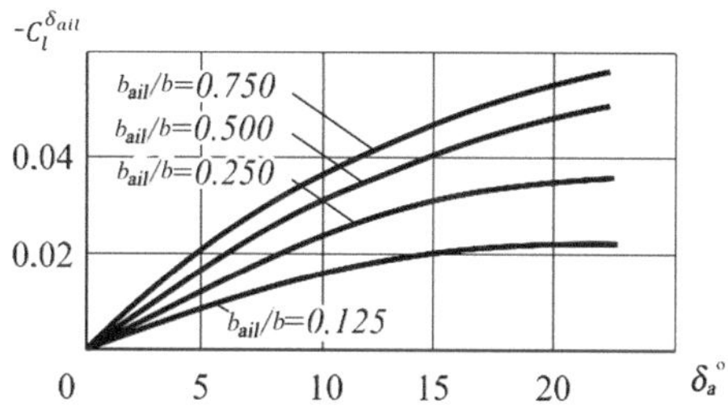

Fig. 36 Variation of the rolling moment coefficient derivative $C_l^{\delta_{ail}}$ with the aileron deflection angle δ_{ail} and aileron span ratio $\frac{b_{ail}}{b}$

$$C_l^{\delta_{ail}} = \frac{b_{ail}}{b}\sqrt{\frac{c_{ail}}{c}},$$

where b_{ail} is the aileron span; c_{ail} is the aileron chord length.

With $\frac{c_{ail}}{c} > 0.25$ $C_l^{\delta_{ail}}$ grows very slowly, therefore, $\frac{c_{ail}}{c} = 0.2\ldots0.25$ is considered to be reasonable values.

When the aileron span ratio $\frac{b_{ail}}{b}$ approaches 0.5, the sensitivity of $C_l^{\delta_{ail}}$ to $\frac{b_{ail}}{b}$ reduces, and since the span of ailerons affects their $C_{L,max}$, b_{ail} is usually 0.3 …0.4 wing span.

When the aileron deflection angles are close to 25°, $C_l^{\delta_{ail}}$ grows very slowly, therefore, it is advisable that the maximum aileron deflection angles do not exceed $25-30°$.

It should be borne in mind that downward aileron deflection increases the wing angle of attack. If the aircraft already flies at a fairly high angle of attack, a large downward aileron deflection can bring this wing to supercritical angles of attack and cause flow separation on this wing. Instead of rolling out, the aircraft will begin to increase the bank angle. In order to prevent such a phenomenon, aileron differential deflection is applied: upwards ($\delta_{ail} = -25°$) and downwards ($\delta_{ail} = +15°$).

When the aileron size and flight speeds increase, the hinge moment grows rapidly, and, as a result, the stick force increases as well. On high-speed aircraft, this force may exceed the manpower capabilities. In order to reduce this force, various types of aerodynamic balancing and servo tabs are used. "Axial" and "internal" means of aerodynamic balancing are the most common options (Fig. 37), producing a moment opposite to the hinge moment of the aileron.

Fig. 37 "Axial" and "internal" means of aerodynamic balancing an aileron: **a**—axial balancing; **b**—internal balancing; **c**—"internal" means of aerodynamic balancing ; 1—upper and lower chambers; 2—narrow slots; 3—flexible fence; 4—chamber wall

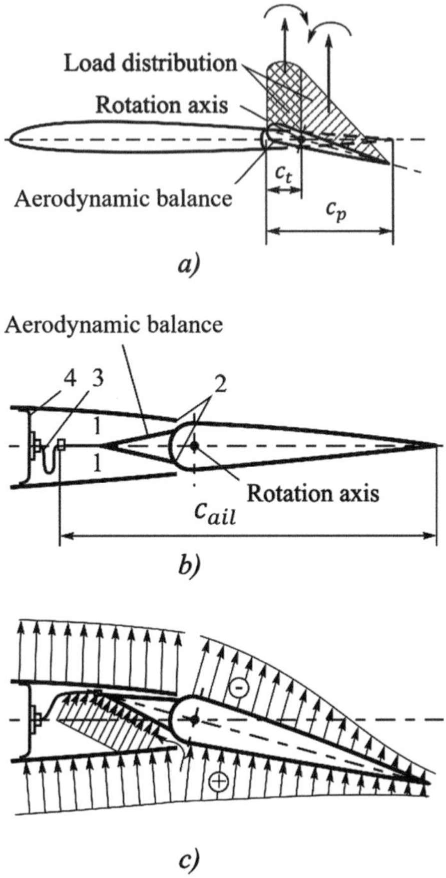

For "axial" aerodynamic balancing, $\frac{S_{AxBal}}{S_{ail}} = 0.25$ is recommended, where S_{AxBal} is the area of the portion of the aileron in front of the hinge axis; S_{ail} is the aileron area.

For "internal" aerodynamic balancing, this ratio is $\frac{S_{IntBal}}{S_{ail}} = 0.3 \ldots 0.31$.

Internal aerodynamic balancing is especially desirable at high speeds; however, its use in thin airfoils is difficult, since it greatly limits the aileron deflection angles ($\delta_{ail,max} = +15 \ldots + 18°$).

Servo tab is a part of the aileron that deflects simultaneously with the aileron but in the opposite direction (Fig. 38). When deflected, it slightly increases the aircraft's drag; however, it reduces the aileron hinge moment and sometimes can cause its vibration. Therefore, it is used more often as an additional device.

In a steady flight, trim tabs are used to eliminate stick force. Trim tabs are distinct from servo tabs in that they are controlled by the pilot (Fig. 39).

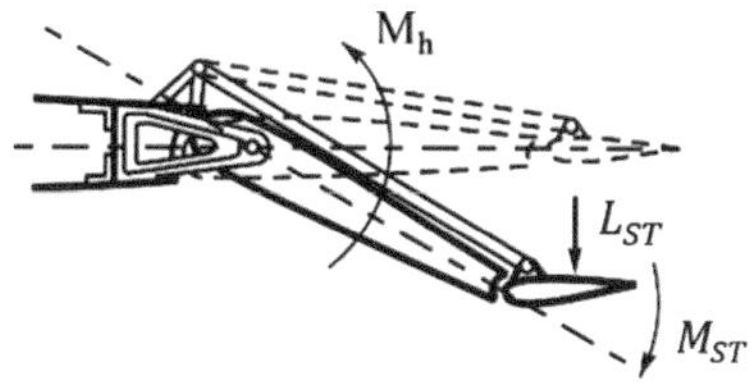

Fig. 38 A servo tab

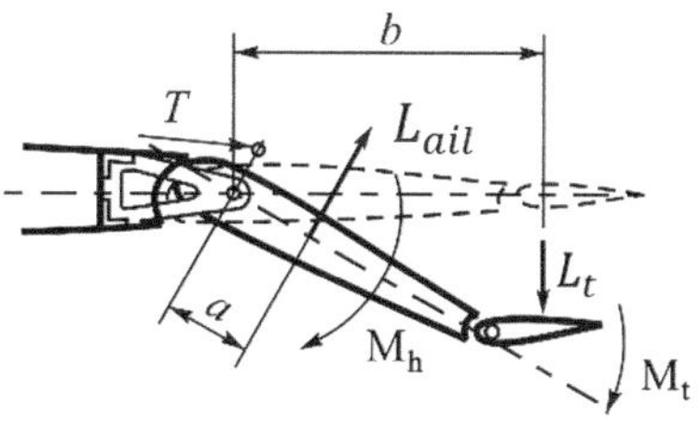

Fig. 39 A trim tab

Increasing the bending and torsional stiffness of swept wings in order to prevent aileron reversal leads to a very large increase in wing mass. The following is used on heavy aircraft to eliminate or reduce the effect of aileron reversal:

- the inboard aileron, which operates during the entire flight, and the outboard aileron, which is applied only during take-off and landing or when flaps are deployed;
- the spoilers (Fig. 40), operating either in conjunction with the ailerons or completely replacing them.

Spoilers, when deflected, cause intense flow separation and reduce wing lift dramatically. Typically, they are used in conjunction with ailerons, which increases their effectiveness (Fig. 41).

Spoilers can be installed in the root and central (middle) wing sections as well. However, they have a different purpose in this case. They are applied when the aircraft

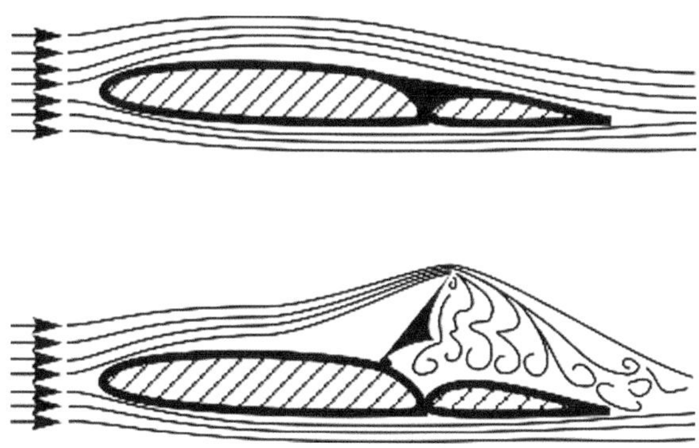

Fig. 40 A spoiler

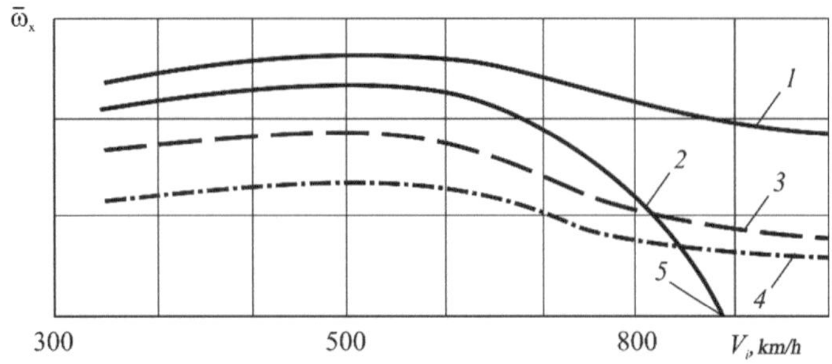

Fig. 41 Authority of various roll control devices as a function of the indicated flight speed V_i: $\overline{\omega}_x = \frac{\omega_x}{2V} b$ is the dimensionless angular rolling speed; 1—spoilers and inboard ailerons; 2—outboard ailerons; 3—middle flaps—ailerons; 4—inboard ailerons; 5—the onset of outboard aileron reversal

touches the ground during landing, which sharply decreases the lift and reduces the braking distance.

8 Some Technical Solutions Implemented in Recent Years

Ogival-shaped wing extensions with a sharp leading edge

Vortices (Fig. 42) generated by such extension produce additional lift at high angles of attack, which makes it possible to maintain the wing bearing properties far beyond the linear part of the lift curve.

Vortex effectiveness increases with increasing angles of attack. In addition, (Fig. 43) the extensions reduce wave drag at supersonic speeds. Due to large root chords and, as a consequence, large Reynolds numbers, the friction drag decreases. The large sweepback of the leading edge extension allows it to remain subsonic, i.e. be inside the Mach cone. This offers a pronounced gain in the maximum lift-to-drag ratio at supersonic speeds.

Fig. 42 Vortex pattern around a wing with extensions

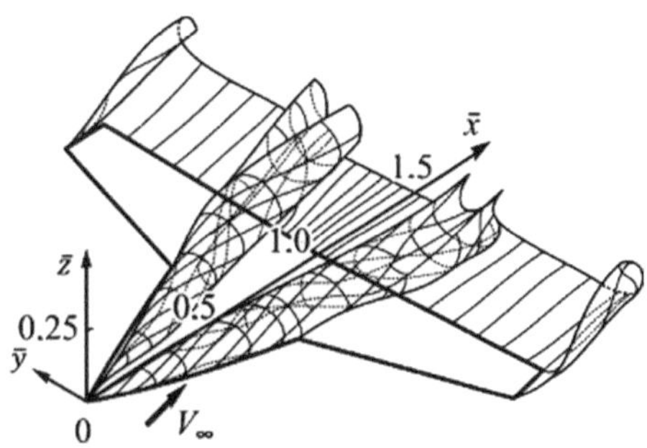

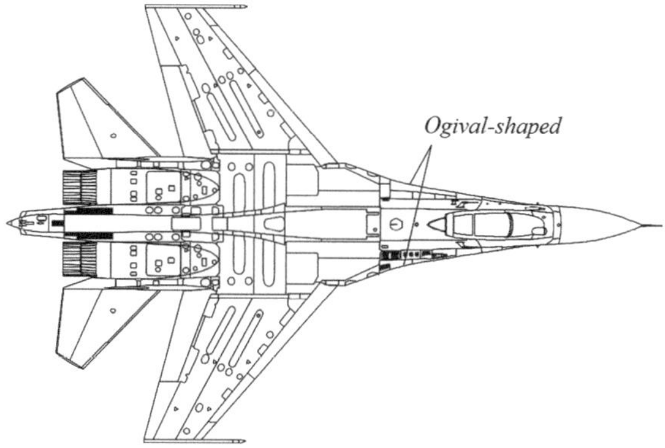

Fig. 43 Ogival-shaped extensions

Wing/fuselage integration

Wing/fuselage integration (Fig. 44) suggests a smooth transition from wing to fuselage, aiming at:

(1) reducing the induced drag;
(2) lowering the zero-lift drag by reducing the wetted area;
(3) increasing the internal volume for fuel and equipment;
(4) decreasing the wing mass by increasing the structural depth of the root section;
(5) applying the transonic area rule, which reduces aircraft wave drag.

The integrated configuration also contributes to reducing the diffuser effect, which decreases an aircraft's induced drag.

On the other hand, the increase in fuselage midsection leads to higher wave drag at supersonic speeds.

Flaperons

Flaperons are the devices used at take-off and landing as flaps, and throughout the flight—as ailerons (Fig. 45).

The use of flaperons suggests that a fly-by-wire control system must be available on the aircraft. It simplifies the wing structure and reduces its weight.

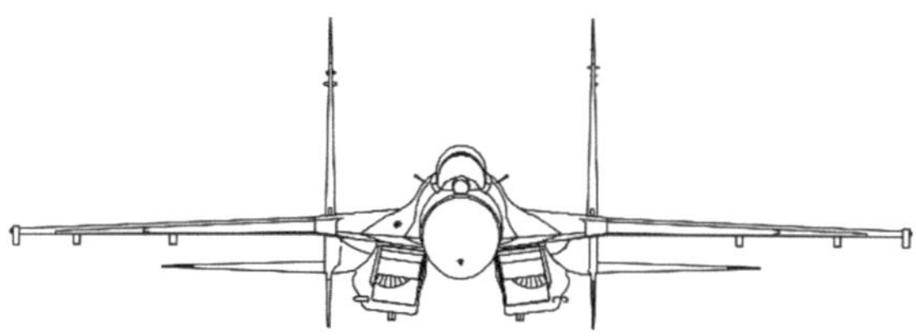

Fig. 44 Wing/fuselage integration

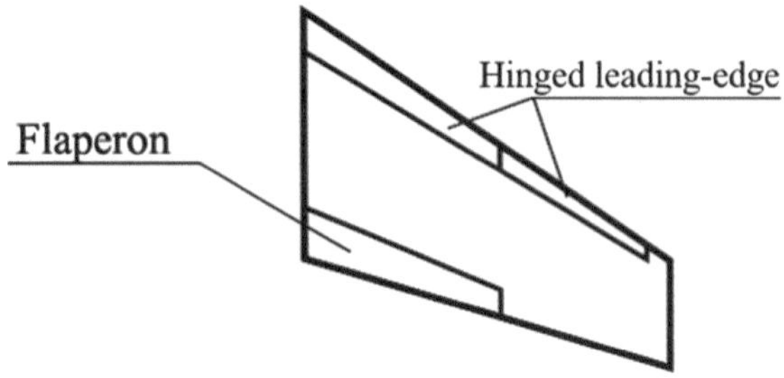

Fig. 45 A flaperon and hinged leading-edge arrangement

Wing tip devices

Wing tip devices come in different shapes and sizes (Figs. 46 and 47).

The wing lift is the consequence of the pressure differential under the wing and above the wing. A wing tip vortex is produced when the flow underneath the wing curls around the tip. The generation of vortices requires energy, which causes induced drag. Wing tip vortices also alter the wing spanwise load distribution, reducing its effective aspect ratio and decreasing the lift-to-drag ratio.

Wing tip devices prevent the leakage of air around the wing tip, which helps achieve a more optimal lift distribution. In addition, they increase the wing effective aspect ratio (Fig. 48).

Wing tip devices increase the lift-to-drag ratio and improve the fuel efficiency of aircraft by 3.5–5.5%.

In addition to the obvious advantages, wing tip devices have several disadvantages:

- wing tip devices shift the wing centre of mass backwards and reduce the flutter limit. As a result, the wing loading must be reduced;
- lateral stability somewhat worsens, which leads to additional restrictions on the crosswind during landing;
- wing mass increases.

Fig. 46 Winglets on Boing-737-800

Fig. 47 Vertical stabilisers on An-225 and Rutan long-EZ

Wing dihedral

Dihedral increases an aircraft's lateral stability, while anhedral reduces it. Typically, this angle varies between −7 and +7 degrees. The mechanism behind the effect of dihedral on an aircraft's rolling stability is quite simple (Fig. 49). If an aircraft with dihedral is rolled to a small bank angle in horizontal flight, then the upper wing will experience a smaller angle of attack than the lower wing. Therefore, the lift produced by the two wings will be different as well. This difference generates a stabilizing moment, returning the aircraft to its initial flight path. When wing lateral stability is too high, the aircraft experiences undamped yaw and roll oscillations (oscillatory instability). If wing lateral stability is low, the aircraft enters a sharp spiral in sideslip (spiral instability).

As a rule, straight wings have dihedral. Swept-back wings use some anhedral to reduce the excessive lateral stability at high angles of attack caused by the sweep angle. Gull-wings and inverted-gull wings reduce drag through the favorable interference of the wing with the fuselage, but are more difficult to manufacture.

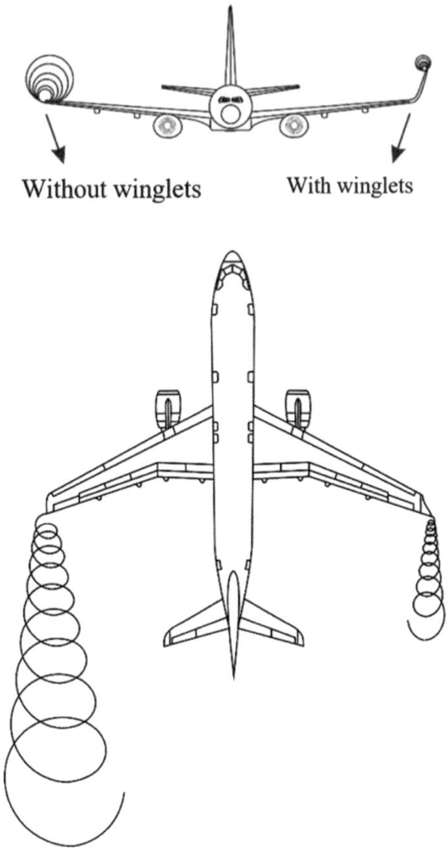

Fig. 48 Wing tip vortices on swept back wings

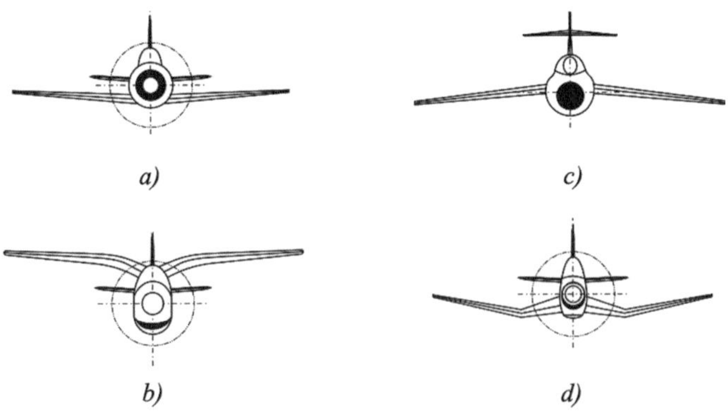

Fig. 49 Wing dihedral and anhedral: **a**—simple dihedral; **b**—compound dihedral (inverted gull-wing); **c**—simple anhedral; **d**—dihedral/anhedral (gull-wing)

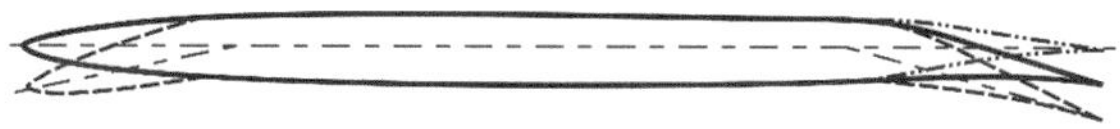

Fig. 50 Adaptive wing

Adaptive wings

Adaptive wings ensure continuity of aerodynamic derivatives within the entire range of parameters. An important distinguishing feature of such a wing is the smoothness of its shape in flight when the wing configuration is adjusted to the flight conditions to ensure a high lift-to-drag ratio. In cruise, a wing of this type becomes nearly flat. In maneuvers, the wing shape is changed in order to retain an elliptical lift distribution for drag reduction (Fig. 50).

For example, wing adaptation was implemented on the Su-27 aircraft; the technology involved the application of flaperons and hinged leading edges. This technology makes it possible to increase the lift-to-drag ratio and critical angles of attack, which improves field performance, manoeuvrability, and endurance at the same fuel mass.

Supercritical airfoils

The distinctive features of supercritical airfoils are a relatively flat upper surface and a curved rear portion of the lower surface (Figs. 51 and 52).

Supercritical airfoils provide a more uniform pressure distribution on the wing and increase the critical Mach number by 10-15%. This is equivalent to an increase in an aircraft's speed by almost 100-150 km/h without the onset of shock waves (Fig. 53).

In other respects, the benefits of using supercritical airfoils are insignificant, while they are more difficult to produce. However, for a transonic airliner, the use of supercritical airfoils can have a significant impact on their fuel efficiency, which depends on the factors indicated in Fig. 54.

Forward-swept wings

In aircraft design, forward-swept wings receive great interest (Fig. 55). These wings have a number of distinctive features:

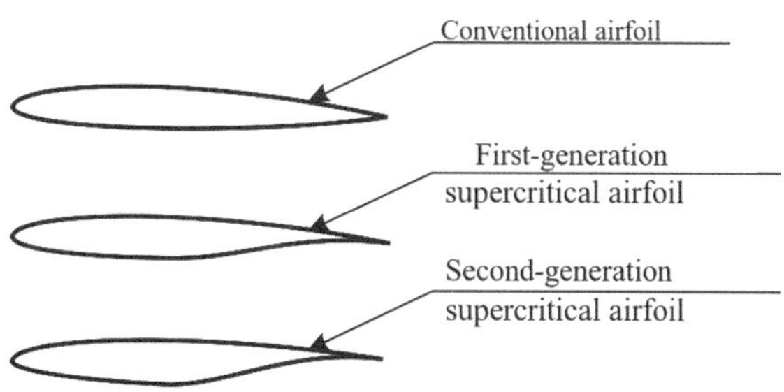

Fig. 51 Supercritical airfoils of the first and second generation

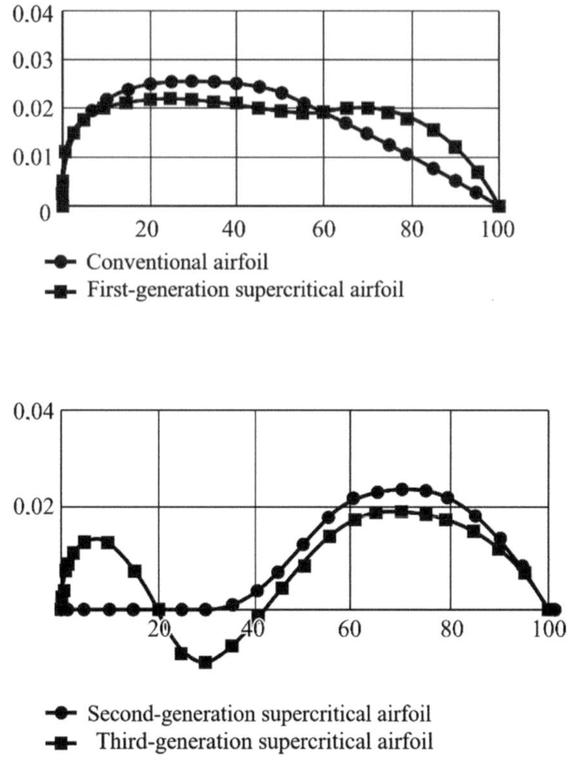

Fig. 52 Camber lines of supercritical airfoils of the first, second and third generations

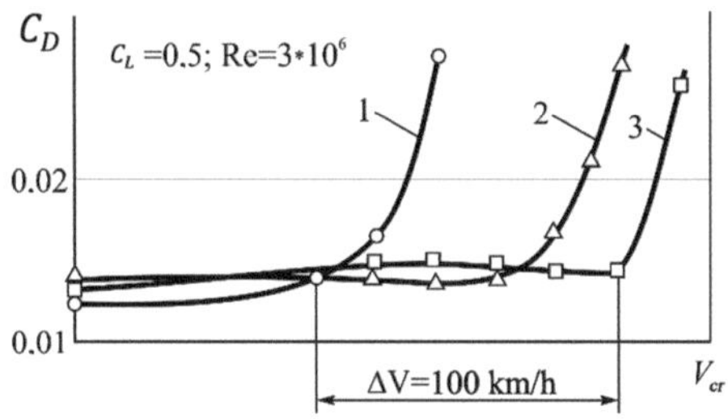

Fig. 53 Effect of supercritical airfoils on cruise speed: 1—swept wing with conventional low-speed airfoils; 2, 3—swept wings with supercritical airfoils

- vortex cores in the root and tip sections rotate in different directions, and the suction zone between them above the upper wing surface increases the lift;
- when compared to swept back wings, this effect is retained at higher angles of attack;
- improved controllability at low speeds.

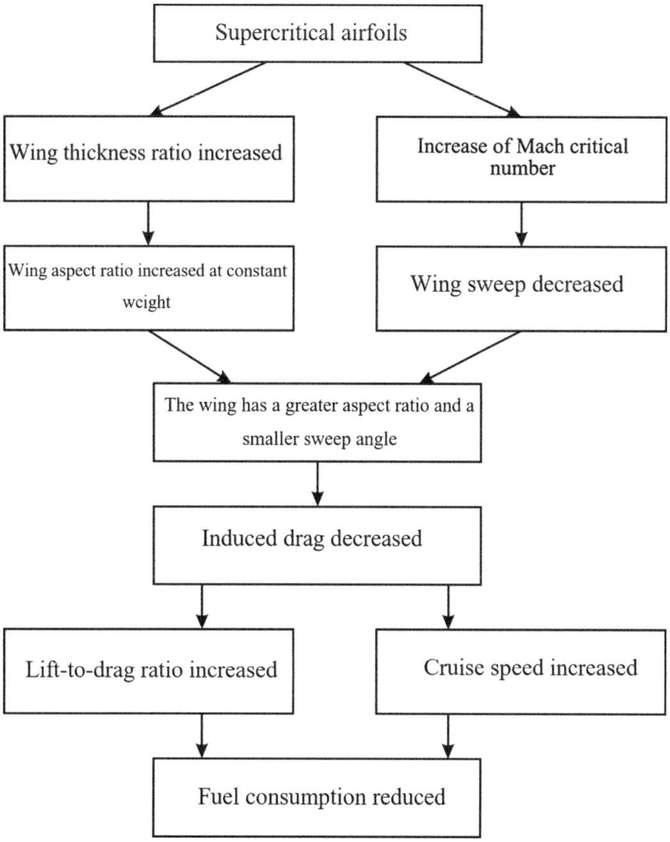

Fig. 54 The effect of supercritical airfoils on wing characteristics

Fig. 55 The Berkut forward-swept aircraft

Forward-swept wings make it possible to increase the lift-to-drag ratio (by 10 … 15%), critical angle of attack (by 15 to 20%), improve field performance, and reduce minimum turn radii. In addition, forward-swept wings simplify the technical realization of the transonic area rule.

However, their wide use in practice is restrained by the tendency to divergence which results in a substantial increase in structural mass. An acceptable solution to this problem is to use composite materials with a carefully designed orientation of fibres.

Variables		
Symbol	Description	Units
S	Wing area	m^2
b	Wing span	m
c_r	Root chord	m
c_t	Tip chord	m
c_{MAC}	Mean aerodynamic chord	m
AR	Aspect ratio	–
λ	Taper ratio	–
Λ	Sweep angle	Degrees
$\frac{t}{c}$	Thickness ratio	%
$\frac{f}{c}$	Camber ratio	%
$\overline{X}_t$	Chordwise location of maximum thickness	%
$\overline{r}_{LE}$	Leading-edge radius ratio	%
2τ	Trailing edge angle	Degrees
$AR_{effincompr}$	Incompressible effective aspect ratio	–
$AR_{effcompr}$	Compressible aspect ratio	–
$\frac{t}{c}_{av}$	Average wing thickness ratio	–
C_L	Aircraft lift coefficient	–
M'_{crit}	Critical Mach number at zero lift	–
S_{MS}	Area of wing longitudinal mid-section	m^2
t_r	Root section thickness	m
t_{tip}	Tip section thickness	m
φ	Twist angle	Degrees
W	Wing volume	m^3
λ_{fr}	Wing taper ratio as seen from the front view	–
m_{fuel}	Fuel mass	kg
ρ_{fuel}	Fuel density	kg/m^3
W_{fuel}	Fuel volume	m^3
C_L^α	Wing lift curve slope	1/rad
Λ_{LE}	Leading edge sweep angle	Degrees

(continued)

(continued)

Variables		
Symbol	Description	Units
Λ_{TE}	Trailing edge sweep angle	Degrees
α_0	Angle of attack at zero lift	Degrees
M	Mach number	–
$\overline{p}$	Ratio of wing half-perimeter to its span	–
C_m	Pitching moment coefficient	–
$C_{D,0}$	Drag coefficient at zero lift	–
$C_{D,i}$	Induced drag coefficient	–
A	Induced drag constant	–
$C_{D,p}$	Pressure drag coefficient	–
$C_{D,w}$	Wave drag coefficient	–
$C_{f,pl}$	Flat plate friction coefficient	–
$\overline{S}_w$	Wing wetted surface area ratio	–
$\overline{x}_{tr}$	Laminar-turbulent transition chordwise coordinate divided by chord length	–
$\overline{x}_t$	Relative coordinate of the maximum airfoil thickness	–
$\overline{x}_f$	Relative coordinate of the maximum airfoil camber	–
h_r	Average height of wing surface roughness	mm
Λ_t	Sweep angle of the line of maximum thicknesses	Degrees
M'_{crit}	Wing critical Mach number at $C_L = 0$	–
C_T	Suction force coefficient	–
ξ_T	Suction force factor	–
φ	Twist angle	Degrees
$C_{m,0}$	Pitching moment coefficient about aerodynamic center	–
$\overline{X}_{CG}$	Center of gravity position measured along MAC	–
$\overline{X}_{AC}$	Wing aerodynamic centre position measured along MAC	–
$\Lambda_{1/2}$	Wing mid-chord sweep angle	Degrees
$C^{\beta}_{l,w}$	Rolling moment coefficient with respect to sideslip angle (rolling stability)	1/rad
C^{β}_{n}	Yawing moment coefficient with respect to sideslip angle (yawing stability)	1/rad
Γ	Dihedral	Degrees
C_n	Yawing moment coefficient	–
C_l	Rolling moment coefficient	–
m_w	Wing mass	kg
$\overline{f}$	Maximum deflection ratio of a cantilever wing	–
G	Modulus of elasticity in shear	Pa

(continued)

(continued)

Variables		
Symbol	Description	Units
δ_a	Aileron deflection angle	Degrees
q_r	Aileron reversal dynamic pressure	N/m^2
L_{ail}	Moment due to aileron deflection	Nm
$\varphi_{elastic}^{pred}$	Predicted twist angle due elastic deformations	Degrees
φ_α	Optimal aerodynamic (geometric) twist angle	Degrees
φ_Σ^{actual}	Actual twist angle in flight	Degrees
Θ_{man}	Slope angle of wing elastic axis	Degrees
$\Lambda_{elcentre}$	Elastic axis sweep angle	Degrees
$X_{MAC,LE}$	X-coordinate of MAC leading-edge	m
$V_{flutter}$	Flutter limit	m/s
ρ_H	Air density at flight altitude	kg/m^3
$C_m^{C_L}$	Pitching moment coefficient derivative with respect to lift coefficient (longitudinal static stability)	–
W_i	Indicated speed of vertical gust	m/s
p	Wing loading	daN/m^2
T_{req}	Required thrust	N
S_{HL}	Wing area serviced by high-lift devices	m^2
$\overline{C'}_{L,max}$	Maximum lift coefficient of wing with high-lift devices	–
c_{SFl}	Split flap chord length	m
c_{PFl}	Plain flap chord length	m
c_{FFl}	Fowler flap chord length	m
$C_{L,max,\ \Lambda}$	Maximum lift coefficient of a swept wing with deployed high-lift devices	–
$C_{L,max,\ \Lambda=0}$	Maximum lift coefficient of a straight wing with deployed high-lift devices	–
Λ_{TE}	Trailing edge sweep angle	Degrees
A_{ail}	Aileron volume coefficient	–
$S_{ail,serv}$	Wing area serviced by ailerons	m^2
$C_l^{\delta_{ail}}$	Aileron efficiency	1/rad
b_{ail}	Aileron span	m
c_{ail}	Aileron chord length	m
δ_{ail}	Aileron deflection angle	Degrees
S_{AxBal}	Aileron area in front of the hinge axis	m^2
S_{ail}	Aileron area	m^2
V_i	Indicated flight speed	m/s
$\overline{\omega}_x$	Angular rolling speed ratio	–

(continued)

(continued)

Variables		
Symbol	Description	Units
ω_x	Angular rolling speed	rad/s

References

1. Byushgens GS (ed) (1998) Aerodinamika, ustojchivost' i upravlyaemost' sverhzvukovyh samolyotov (Aerodynamics, stability and controllability of supersonic aircraft). Fizmatlit, Moscow
2. Byushgens GS, Azarov YuA, Amiryants GA et al (2002) Aerodinamika, dinamika polyota i prochnost' (Aerodynamics, flight dynamics and strength). In: Byushgens GS (ed.) Mechanical Engineering. Encyclopedia, Masinostroenie, Moscow
3. Matveenko AM, Akimov AI, Akopov MG et al (2004) Proektirovanie, konstrukcii i sistemy samoletov i vertoletov (Design, construction and systems of aircraft and helicopters). In: Matveenko AM (ed) Mechanical Engineering. Encyclopedia, Masinostroenie, Moscow

Fuselage Design

Valentin A. Kiselev and Evgeniy A. Pen'kov

Abstract The chapter discusses the fuselage function, the fuselage parameters and characteristics, the loading of a fuselage in flight and its structural design (including structural optimization), as well as the shapes of a fuselage cross-section. The chapter provides historical data on fuselages of various aircraft and shows how to design the forward and aft parts of a fuselage. Finally, the chapter provides detailed recommendations for the design of passenger cabins.

Keywords Fuselage parameters · Fuselage characteristics · Structural design · Cross-section · Passenger cabin

1 Fuselage Function and Design Features

In terms of purpose and function, the fuselage is one of the most complex parts of the aircraft. Typically, the fuselage accommodates payload, crew, and equipment; sometimes it carries the power plant and fuel; on the other hand, the fuselage integrates the most important parts of the aircraft: the wing, empennage, landing gear, and engines. This functional complexity causes some difficulties in the design process when establishing the fuselage geometry and defining the external loads. The fuselage is exposed not only to the weight of its own structure and the items it accommodates, but also to the loads transferred from other parts of the aircraft.

The main fuselage parameters and characteristics should be defined in parallel with the parametric analysis of other parts of the aircraft. It is convenient to represent this analysis as a part of the iterative aircraft sizing process. The schematic of this process is given in Fig. 1. Herewith, the input data are the design range R_{design} (or radius of action), payload mass m_{PL}, design speed V (maximum, cruise) and flight

V. A. Kiselev (✉)
Moscow Aviation Institute (National Research University), Moscow, Russia
e-mail: valentinkiselev36@mail.ru

E. A. Pen'kov
Sukhoi Design Bureau, Moscow, Russia

M. A. Pogosyan (ed.), *Aircraft Design*, Springer Aerospace Technology,
https://doi.org/10.1007/978-981-96-4599-2_17

altitude H, airfield class, and runway length L_{RW}; the dimensions and masses of the fuselage contents, which are loads, equipment, engines, etc., are more or less known. As a rule, the structural materials to be used are also known.

Using the initial data, different versions of the fuselage and the entire aircraft are developed. Their basic dimensions and parameters are established based on accommodating all the required items in the fuselage and integrating other parts of the aircraft.

For each configuration, in view of all the requirements, the forward and aft parts, the cross-sectional shape, the geometric parameters, and the structural layout must be adjusted. Each configuration must meet the requirements for the pilot's view, safety in emergency situations, as well as the operational, production, and other requirements.

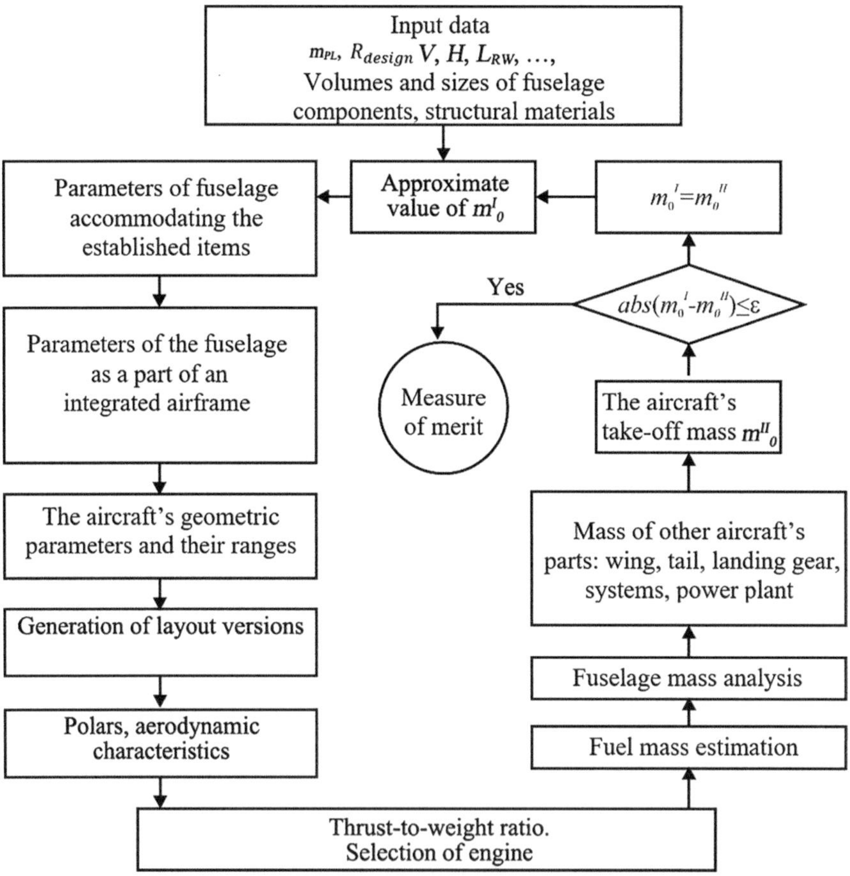

Fig. 1 Schematic of fuselage parametric analysis

2 Fuselage Parameters and Characteristics

Fuselage geometric parameters are (Fig. 2) the length l_{fus}, diameter d_{fus}, the area of middle (maximum) section $S_{f,m-s}$; fuselage dimensionless parameters are the fuselage fineness ratio $\lambda_{fus} = l_{fus}/d_{fus}$; the forebody fineness ratio $\lambda_{FB} = l_{FB}/d_{FB}$; the afterbody fineness ratio $\lambda_{AB} = l_{AB}/d_{AB}$. If the cross-section is a non-circular shape, its characteristic dimensions are the maximum width B and height H. Often, the diameter equivalent to the area of the fuselage midsection $d_{eq} = 2\sqrt{S_{f,m-s}/\pi}$ is also used.

It should be noted that the fuselage parameters λ_{fus}, l_{fus}, d_{fus} are not independent. As a rule, one parameter is subject to optimization, while the others are established as input.

If the area of midsection or the fuselage diameter is fixed, λ_{fus} changes as its length changes. In this case, the fuselage diameter and mid-section can be unambiguously determined based on the engine arrangement in the fuselage.

Another approach is that used in airliners, where a given number of passengers can be accommodated in different ways: a fuselage can be wide and short, or narrow and long.

At subsonic Mach numbers, the fuselage profile drag is composed of friction drag and pressure drag. Fuselage drag due to friction, which is a function of the fuselage wetted area, is much greater than the fuselage drag due to pressure. That is why if we increase the fuselage length keeping the diameter fixed, the fuselage wetted area F_{FWA} and, therefore, its drag D_{fus} and drag coefficient $C_{D,fus} = D_{fus}/(qS_{f,m-s})$ will also increase (Fig. 3).

However, when $\lambda_{fus} < 3.5$, the fuselage drag increases again because the pressure drag increases.

Therefore, with the given $S_{f,m-s}$ and d_{fus}, and at subsonic speeds, a fuselage with a very small fineness ratio has better aerodynamic characteristics. The aircraft IAI Arava with $\lambda_{fus} \approx 3.5$ serves as an example.

If we increase λ_{fus} by reducing d_{fus} and increasing l_{fus}, keeping the number of seats or the area of cargo floor constant, $F_{FWA} \approx const$. Therefore, friction drag remains constant, while pressure drag and total drag decrease. $C_{D,fus}$ increases because it is based on the reduced midsection area. Thus, in this case, for a better aerodynamic

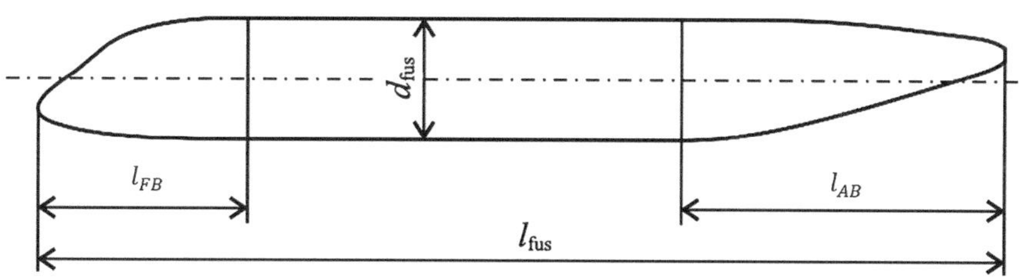

Fig. 2 Fuselage dimensions

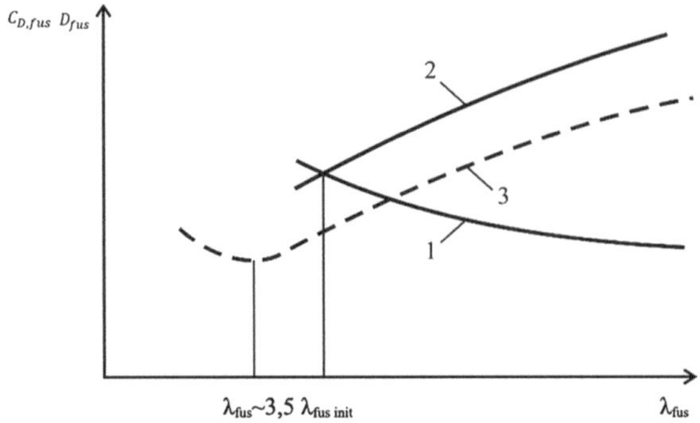

Fig. 3 Fuselage drag as a function of fineness ratio λ_{fus}; $M, H = const, M < M_{cr}$; 1—D_{fus} at $F_{FWA} = const$; 2—D_{fus} at $S_{f,m-s}, d_{fus} = const$; 3—$C_{D,fus}$

performance, it is advantageous to increase λ_{fus} (the upper limit of λ_{fus} is established by the fuselage mass limit).

Figure 3 shows the change in fuselage drag as a function of fuselage fineness ratio and illustrates the two aforementioned scenarios (the dashed $C_{D,fus}$ curve roughly corresponds to the two cases).

The fuselage wave drag can be reduced by increasing the forebody fineness ratio λ_{FB}. For this reason, supersonic aircraft have a slender fuselage and pointed forebody. For example, the supersonic airliner Boeing 2707-300 was designed with $\lambda_{fus} = 23$! Fuselage critical Mach number increases when the fineness ratio of the fuselage or its parts increases. To first approximation, $M_{cr,fus} = \lambda_{fus}(0.17 - 0.0076\lambda_{fus})$.

For the fuselage, the critical Mach number problem is less acute than for the wing. This is because the fuselage, which is a body of rotation or close to it, accelerates the airflow to a lesser degree than a wing. However, the development and application of supercritical airfoils have made the critical Mach number issue relevant for the fuselage as well.

The fuselage mass depends on the parameters of the fuselage and other parts of the aircraft. In strength analysis, the fuselage is considered a beam lying on two supports, which are the front and rear spars of the wing (Figs. 4 and 5). The fuselage sections close to the wing spars take the largest load. The forebody and afterbody are like cantilever beams, constrained at the front and rear wing spars. If the fuselage length increases at $S_{f,m-s}, d_{fus} = const$, the length of these cantilever beams increases as well, which increases the overall dimensions and bending moments experienced by the fuselage, and, consequently, the size and mass of the fuselage structural elements.

If we increase l_{fus} while reducing d_{fus} or $S_{f,m-s}$ for an airplane with the fixed number of passengers, there is an optimum value of λ_{fus} delivering the minimum structural mass of the fuselage (refer to Sect. 5). If a fuselage is very narrow and long, its weight increases due to an increase in arms and bending moments. If a fuselage is short and wide, the load applied to the fuselage structure decreases, while the durability and mass increase. Moreover, with an increase in diameter, the structural mass ratio increases rapidly due to the pressure differential in the pressurized cabin.

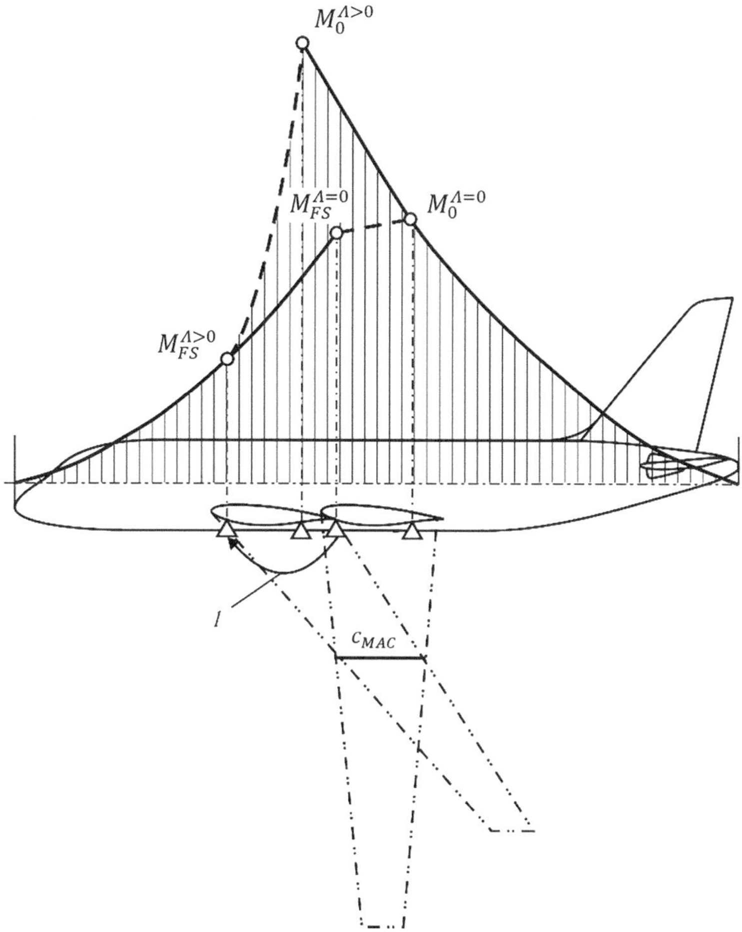

Fig. 4 Fuselage bending moments in a configuration with a straight and a swept back wing: 1—the wing centre section is shifted forward when the wing sweep angle is increased

Swept wings significantly increase the fuselage mass. In order to retain the aircraft center of gravity with an increasing wing sweep, it is necessary to move the wing center section and, consequently, the reinforced sections of the fuselage forward. This will call for an increase in the length of the afterbody which will cause an increase in fuselage mass (refer to Fig. 4). If the wing is swept, the fuselage sections close to the rear wing spar are loaded with a much greater bending moment M_0 than the sections at the front wing spar M_{FS} These moments are close if the wing is straight.

The afterbody mass significantly increases with aft-podded engines. This is because in this configuration, the afterbody reacts to the engine weight, which is not compensated by the reduced moment arm (the wing moves backwards to retain the

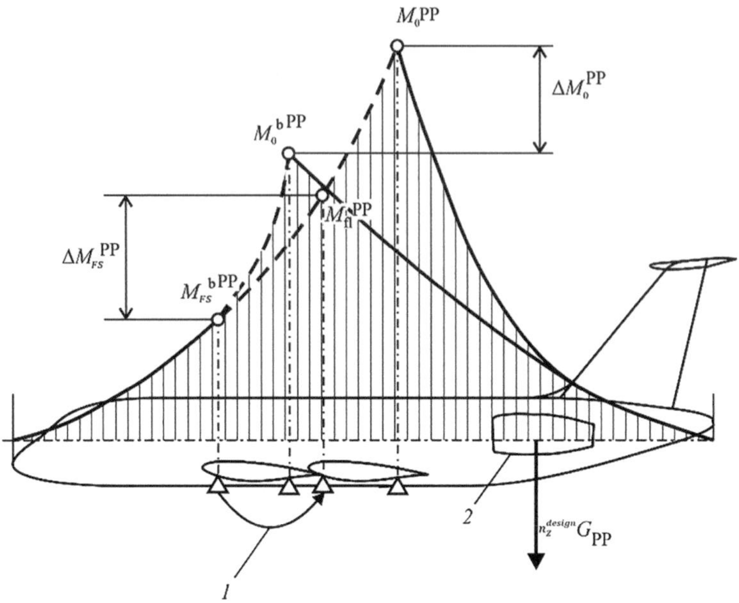

Fig. 5 Change in bending moment distribution along the fuselage when engines are moved from the wing to the afterbody: 1—the wing centre section is shifted backwards when the engines are moved; 2—power plant

CG envelope—refer to Fig. 5). Simultaneously, the length and mass of the forebody increase.

Statistical data on fuselage weight as a function of geometric parameters is given in [1–3]. Table 1 shows statistical data on fuselage fineness ratio.

Below, some statistical equations for fuselage parameters are given:

$$l_{fus} = (0.25\ldots0.3)\frac{\lambda_{fus} b}{\sqrt{AR}},$$

where b and AR are the wing span and aspect ratio;

$$l_{fus} \approx \sqrt[3]{1.3 V_{fus} \lambda_{fus}^2 / k_{fus}},$$

Table 1 Fuselage fineness ratio

Fineness ratio	Subsonic aircraft ($M \leq 0.7$)	Transonic aircraft ($M = 0.8$ to 0.9)	Supersonic aircraft
λ_{fus}	6 to 9	8 to 13	10 to 20 (up to 23)
λ_{FB}	1.2 to 2.0	1.7 to 2.5	4 to 6
λ_{AB}	2 to 3	3 to 4	5 to 7

where

k_{fus} is the shape coefficient, $k_{fus} = 0.75\ldots0.8$ for subsonic aircraft and $k_{fus} = 0.70\ldots0.75$ for supersonic aircraft;

V_{fus} is the fuselage volume, $V_{fus} \approx 0.2\pi d_{fus}^2 \lambda_{fus}$;

fuselage surface area: $F_{FWA} \approx 2.85 l_{fus}\sqrt{S_{f,m-s}}$.

3 Shape of Fuselage Cross-Section

A circular cross-section seems to be the most advantageous for a fuselage since it provides the minimum perimeter for a constant cross-section area or the minimum surface area of the fuselage with constant volume. This results in the lowest friction drag. The circular shape is also preferred for pressurized compartments since it eliminates significant stresses due to bending and, therefore, provides the smallest structural mass. If the layout requirements do not allow a circular section to be used or make it disadvantageous, the section shape should be close to the circular one, or it should be formed by intersecting circles. In this case, the points of the circles' intersection on opposite sides of the contour are interconnected by reinforced structural members (for example, floor beams), which reduce the fuselage bending. The structural elements react tension or compression loads depending on the specific fuselage section shape (Fig. 6; also refer to Fig. 16).

Bulb section shape is used to increase the volume of freight compartments under the floor of the passenger cabin.

However, for unpressurised fuselages, other shapes may be advantageous (Fig. 7).

Table 2 shows statistical data on the dimensions of fuselage cross-sections in passenger aircraft.

The different levels of comfort are what cause the variances in dimensions for the same number of seats.

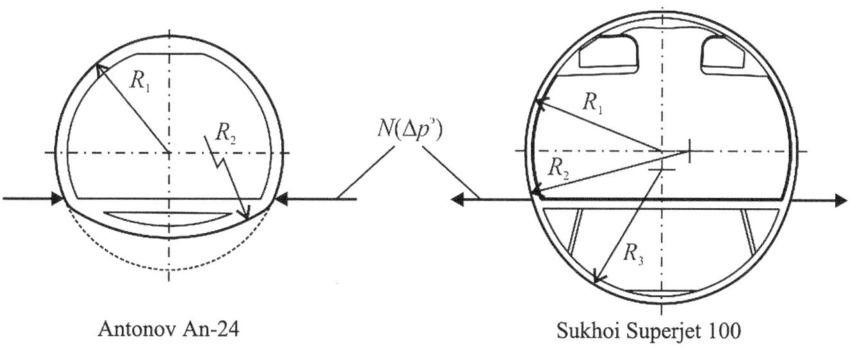

Fig. 6 Fuselage cross-sections formed by intersecting circles, with the floor beams reacting tension (SuperJet 100) and compression (An-24) due to pressure differential Δp^e

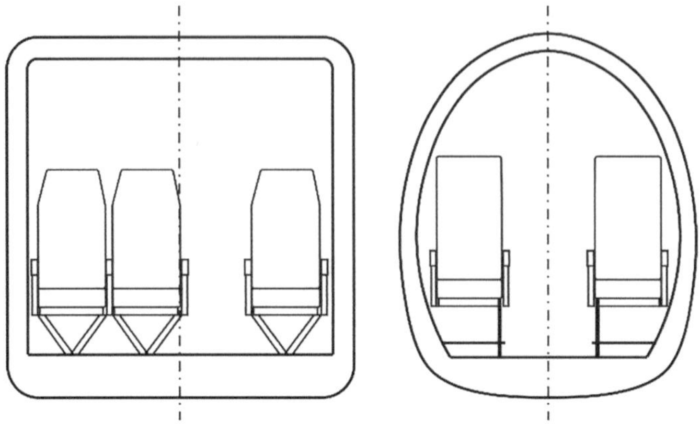

Fig. 7 Fuselage cross-sections of the Short Skyvan aircraft (United Kingdom) and the light passenger Be-30 (USSR)

As a rule, with four seats in a row or less, it is not possible to accommodate luggage under the floor of a circular cross-section passenger cabin, so it is arranged in the luggage compartments on the passenger deck. One aisle is allowed when there are no more than six seats in a row because a flight attendant can serve no more than three passengers on each side of the aisle, while a passenger who sits at the window seat should not disturb more than two neighbors when leaving. Obviously, with two aisles, the maximum number of seats in a row increases to 12.

The fuselage cross-section has a very unique shape when several engines or air intakes are arranged in the fuselage (Fig. 8).

4 Shapes of the Forward and Aft Parts of the Fuselage

In terms of aerodynamics, the fuselage should be an axisymmetric body of rotation with smoothly tapering forebody and afterbody. In a subsonic aircraft, the afterbody is longer and straighter than the forebody.

The forebody outlines are defined by the pilot visibility requirements. This feature is inherent in both civil and military aircraft of various sizes. It is due to the fact that below the cockpit, there is a zone where the extended forward portion of the fuselage can be placed with no obstruction to the pilot's field of view (Fig. 9). This is advantageous in terms of aerodynamics (the nose is pointed) and layout (radar antennas or other equipment can be arranged in the stretched nose).

To reduce the aerodynamic drag, it is necessary to tilt the cockpit windscreen as far as possible from the vertical line (refer to Fig. 9). However, if the windscreen angle is greater than 70°, the effect of total internal reflection emerges (the light falling at such an angle does not pass through the windshield, i.e., the pilot will see

Table 2 Dimensions of fuselage cross-sections of passenger aircraft

Aircraft	D_{fus} or $B \times H$*, m	Number of seats/number of aisles
Be-30 (USSR)	1.7 × 2.0	2/1
HFB.320 (Federal Republic of Germany)	2.06	
L. 410 (Czechoslovak Socialist Republic)	2.38	3/1
Yak-40 (USSR)	2.4	
Fokker F.27 (Netherlands)	2.8	4/1
An-24 (USSR)	2.9	
Tu-134 (USSR)	2.9	
Caravelle (France)	3.2	5/1
BAC One-Eleven (United Kingdom)	3.4	
DC.9 (USA)	3.378 × 3.634	
Il-18 (USSR)	3.5	
DH.121 Trident (United Kingdom)	3.7	6/1
Il-62 (USSR)	3.75 × 4.1	
Boeing 707 (USA)	3.75 × 4.33	
VC.10 (United Kingdom)	3.76 × 4.3	
Tu-154 (USSR)	3.8	
Boeing 767 (USA)	5.03	7/2
A.300B, A.330/340 (Airbus Industry)	5.64	8/2
Boeing 787 (USA)	5.74	
Lockheed L-1011 (USA)	5.97	8–9/2
DC.10 (USA)	6.02	9–10/2
Il-86/96 (USSR)	6.08	9/2
Boeing 777 (USA)	6.2	9–10/2
Boeing 747 (USA)	6.5	10/2
A.380 (Airbus Industry)	7.14 × 8.41	10/2—lower deck 8/2—upper deck

* B is the section width, H is the section height

only a reflection of the cockpit interior). That is why the recommended windscreen angle for subsonic aircraft is 50…55°; for supersonic aircraft, it may be as high as 60…65°.

For a supersonic passenger aircraft, the reduction of drag in cruise and the required field performance can be ensured by tilting down the forward part of the fuselage during take-off and landing (Fig. 10).

To reduce drag, cockpits are made with an aspect ratio of 4…6 (this is the ratio of the cockpit length to its width or height). In terms of structural weight, forming the canopy's cross-section by an arc of a circle is advantageous.

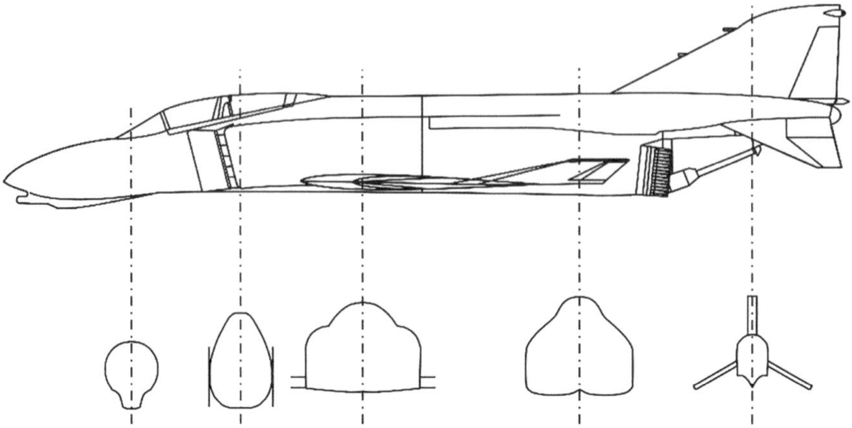

Fig. 8 Fuselage cross-sections of the F-4 Phantom aircraft (USA)

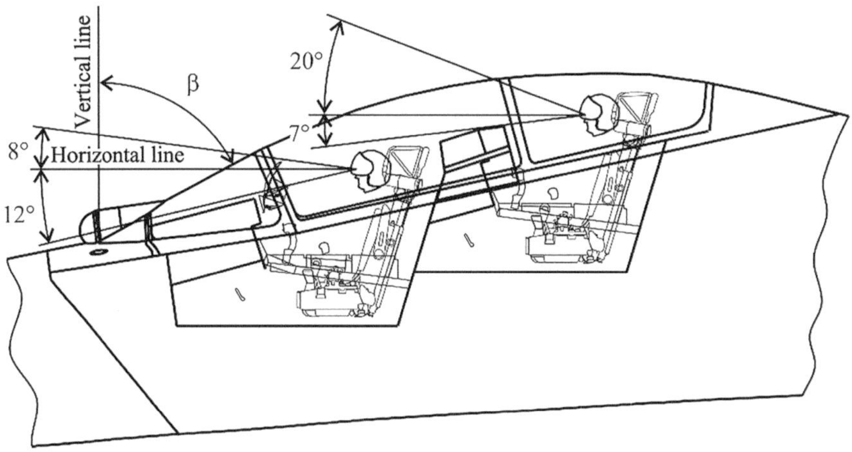

Fig. 9 Pilot's field of view

The tail upsweep angle must guarantee that the required angle of attack during takeoff and landing can be ensured at the shortest length of the main landing gear (Fig. 11).

Many freight aircraft use a large hatch in the aft fuselage and a loading ramp lowered onto the ground for loading and unloading of various cargoes without the use of ground support equipment. The flat bottom of the fuselage tail may be deflected up in order to ensure the required dimensions of the cargo opening at the shortest cutout length and to use a significant part of the hatch cover as a ramp.

The downside of this solution is an increase (by 10 to 15%) in fuselage drag. A similar shape of the fuselage tail is typical, for example, for the military transport aircraft Lockheed S-130 (USA), Transall C-160 (France, German Federal Republic,

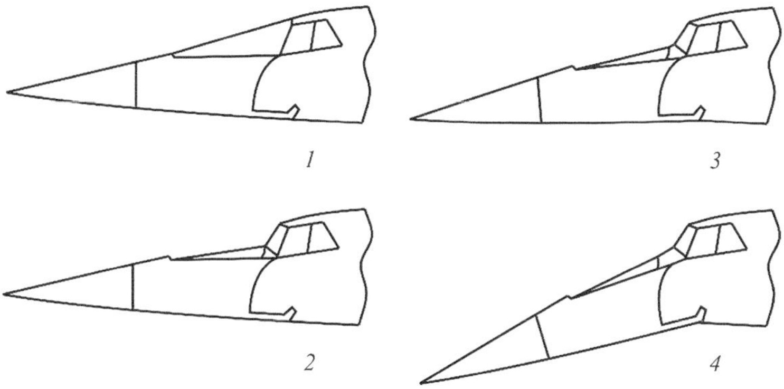

Fig. 10 The forward part of the Concorde supersonic airliner: (1)—the fuselage nose and windscreen are up; (2)—the fuselage nose is up and the windscreen is retracted; (3)—the fuselage nose is deflected down to 5°, the windscreen is retracted; (4)—the fuselage nose is deflected down to 12.5°, the windscreen is retracted

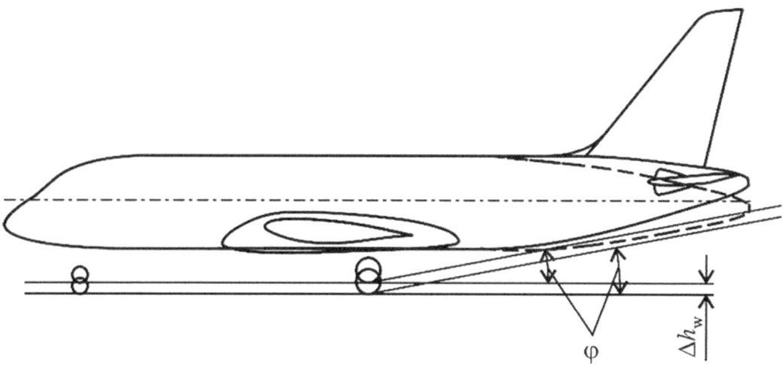

Fig. 11 How an increase in the tail upsweep decreases the landing gear length Δh_w

the Netherlands). On more modern aircraft (IL-76, An-72, Lockheed S-141A, C5A, Boeing YC.14, McDonnell Douglas C-17), the afterbody features increased length and curvature to reduce aerodynamic drag. As a result, its bottom surface is upswept at a minimum angle, drag decreases, and the tail moment arm increases; however, the cutout length increases as well (Fig. 12).

Some of the largest transport and cargo aircraft (C.5A, Boeing 747F) are provided with a cargo hatch in the forward fuselage as well, and for that reason, its bottom part (below the cockpit) is designed to hinge upward (in some projects, it hinges sideways). Such a solution affects the configuration of the entire forebody. Thus, the cockpit is raised on the Boeing 747 aircraft to ensure the required dimensions of the cargo opening in the forward fuselage (Fig. 13).

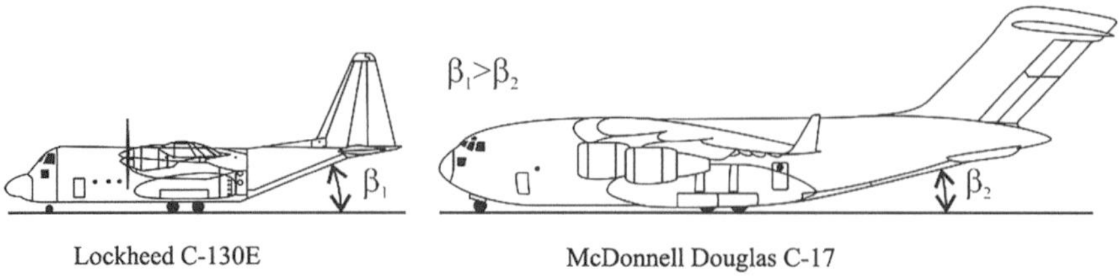

Fig. 12 Comparison of the fuselage tail configuration of the military transport aircraft Lockheed C-130 and the McDonnell Douglas C-17

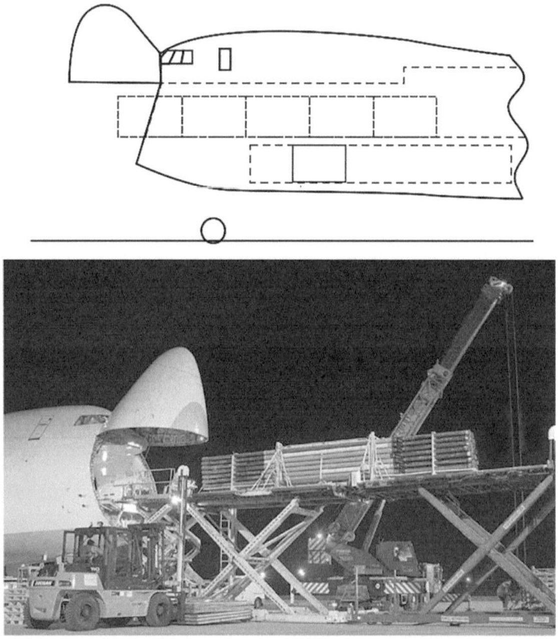

Fig. 13 Cargo door in the forward part of the Boeing 747F fuselage

5 Sizing the Fuselage of a Passenger Aircraft

With an increase in the dimensions and capacity of passenger aircraft, the variety of passenger cabin configurations (size and pitch of seats, width of aisles, etc.) has increased. Since the configuration of the passenger cabin is closely related to the fuselage configuration, the selection of the most advantageous dimensions and cross-section shape of the fuselage, its length, and the number of passenger decks have become very crucial. As it was mentioned in Sect. 2, in this case, changes are due to the simultaneous variation in the fuselage length and diameter with a constant number of passenger seats. Consequently, various configurations will call for different fuselage lengths, shapes, and areas of their cross-sections, as well as the structural mass and the fuselage drag.

Let us consider a subsonic passenger aircraft with the following properties: its capacity is 500 passengers; the pitch of seats is 810 mm; the design range is 2500 km; the cruise Mach number is 0.85 at $H_{cruise} = 10$ km; the runway length corresponds to an airfield of class B; the take-off thrust-to-weight ratio $(T/W)_0$ is 0.3; the take-off wing loading $p_0 = \frac{m_0 g}{S}$ is 550 daN/m^2; it must feature a low wing, four bypass jet engines, $AR = 7.5$, $\lambda = 3.5$, $\Lambda = 30°$, $(t/c)_r = 0.12$ (the maximum structural depth of root section is 1.3 m); triple-slotted flaps; slats; and spoilers.

To identify advantageous combinations of length and diameter for a single-deck fuselage with a circular section, several options for passenger cabin layouts have been analyzed (the obtained relations between fuselage diameter and length are given in Fig. 14). For all configurations, the fineness ratios of the fuselage nose and tail are assumed to be constant (2 and 3, respectively). Since the arms of the horizontal L_{HT} and vertical L_{VT} tail change with the fuselage length, their areas S_{HT} and S_{VT} are determined by their volume coefficients $S_{HT}L_{HT}/(Sc_{MAC}) = 1.0$ and $S_{VT}L_{VT}/(Sb) = 0.1$, where S, b, c_{MAC} are the area, span, and the wing mean aerodynamic chord, respectively. The change in fuselage length affects the length of the landing gear if the fuselage tail upsweep angle is approximately constant, $\varphi = 10$ to 12°. With an increase in the fuselage diameter, the wing inside the fuselage extends, and the length of the net wings decreases at a fixed span. For each configuration, the following is determined: the fuselage, wing, empennage, landing gear, fuel masses, the aerodynamic drag $C_{D,min}$, the lift-to-drag ratio L/D , and the cost per tonne-kilometer. To simplify the calculations (and avoid iterations), we assume that the takeoff mass of the aircraft $m_0 = 186.5$ t is constant. This takeoff mass corresponds to a passenger capacity of 500 passengers, with 10 passengers in a row with two aisles at a fuselage diameter of 6.4 m and a length of 70 m; the changes in the components of the takeoff mass are compensated for by small changes in the payload mass m_{PL} that are possible with the variations in the masses of baggage, mail, and cargo accommodated under the cockpit floor.

Cost per tonne-kilometer a, rub/(t · km) is a measure of merit for the considered layouts. We can write down the analysis algorithm as follows:

$$C_L(M_{cr}, H_{cr}) \rightarrow C_p(M_{cr}, H_{cr}) \rightarrow C_{D,min} \rightarrow L/D \rightarrow m_{fuel} \rightarrow m_i \rightarrow a,$$

where c_p is the specific fuel consumption; m_{fuel} is the fuel mass; m_i are the components of the aircraft's take-off mass, including the fuselage mass. Note that narrow long fuselages ($\lambda_{\mathrm{fus}} > 14$), featuring too low stiffness, will tend to flex excessively under load. We should take into account the influence of fuselage deflection on stability and control of aircraft using the following factor:

$$c_{\mathrm{fus}} = \frac{P_{\mathrm{HT}}}{X} > 0.1,$$

where P_{HT} is the force produced by the horizontal tail that bends the fuselage; X is the force applied to the aerodynamic centre of the horizontal tail and causing the angle of attack of the tail to change by 1° due to fuselage bending.

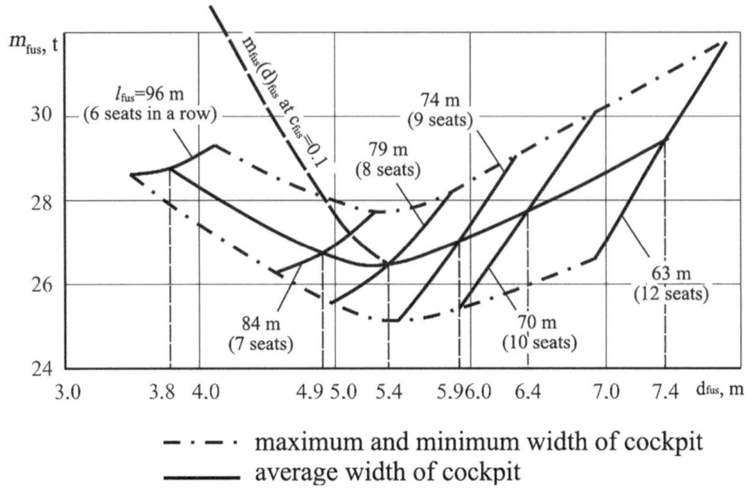

Fig. 14 Change in fuselage structural mass with $n_{pass} = 500$, $d_{fus}, l_{fus} = var$; – · – maximum and minimum width of cabin; —— average width of cockpit

If $C_{fus} > 0.1$, the fuselage stiffness must be increased. With the given geometric parameters of the fuselage and tail, this can be achieved only by increasing the cross section and mass of the fuselage structural members. We can use statistical equations to calculate the fuselage mass, which provides a reduction of C_{fus} to 0.1 (Fig. 14 shows this increased mass with a dotted curve).

The equations that can be used within the framework of the above algorithm are given in [4].

As can be seen in Fig. 14, the smallest mass of the fuselage structure $m_{\text{fus}} = 26.5$ t is obtained with $d_{\text{fus}} = 5.4$ m, $l_{\text{fus}} = 79$ m (8 seats in a row with two aisles).

The same number of seats in a row can be accommodated in different fuselage diameters, ensuring different levels of comfort. That is why Fig. 14 depicts the diameter ranges. The lines that intersect the main curve (corresponding to an average width of the passenger cabin) and that are plotted for a constant number of seats in a row and l_{fus} = const show the variation of the fuselage diameter as a function of the width of seats and aisles.

The collection of curves in Fig. 14 gives a fuller picture that allows the designer to select the most favorable parameters of the fuselage. So, although the minimum m_{fus} for the main curve occurs at $d_{\text{fus}} = 5.4$ m (8 seats in a row), we can consider a diameter of 5.75 m with a slightly tighter arrangement of 9 passengers in a row. This reduces the fuselage mass and length (the latter may be advantageous for future fuselage modifications to increase the passenger capacity).

Figure 15 shows the mass ratios of the take-off mass components as functions of the fuselage diameter. The minimum mass of the aircraft structure is achieved by the same configuration as the minimum fuselage mass. The minimum cost per tonne-kilometer and the maximum payload are almost the same for the layouts with $d_{fus} = 5.4$ m($l_{fus} = 79$ m) and with $d_{fus} = 4.9$ m ($l_{fus} = 84$ m).

The above results correspond to the specific input data, including the passenger capacity, range, section shape, and the number of decks. We can identify the most advantageous structural and design solutions in a wide range of passenger capacities and ranges of an airliner, as well as for various fuselage cross-section shapes with different numbers of passenger decks. Let us examine the configurations that accommodate between 300 and 700 passengers with a design range between 2000 and 8000 km. Various types of fuselage cross-sections are given in Fig. 16. For all combinations of passenger capacity, range, and cross-section shapes, it is necessary to determine the mass of the fuselage, wings, empennage, landing gear, and fuel. It is also necessary to predict the aerodynamic drag and aircraft lift-to-drag ratio, and the aircraft take-off mass; the cost per tonne-kilometer can be used as a measure of merit.

Fig. 15 Optimization of fuselage parameters with respect to mass and cost performance

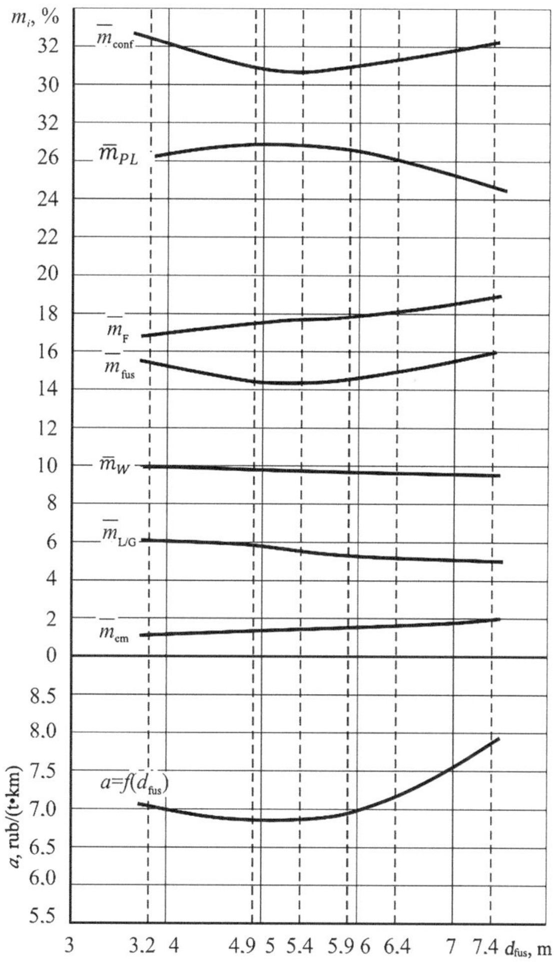

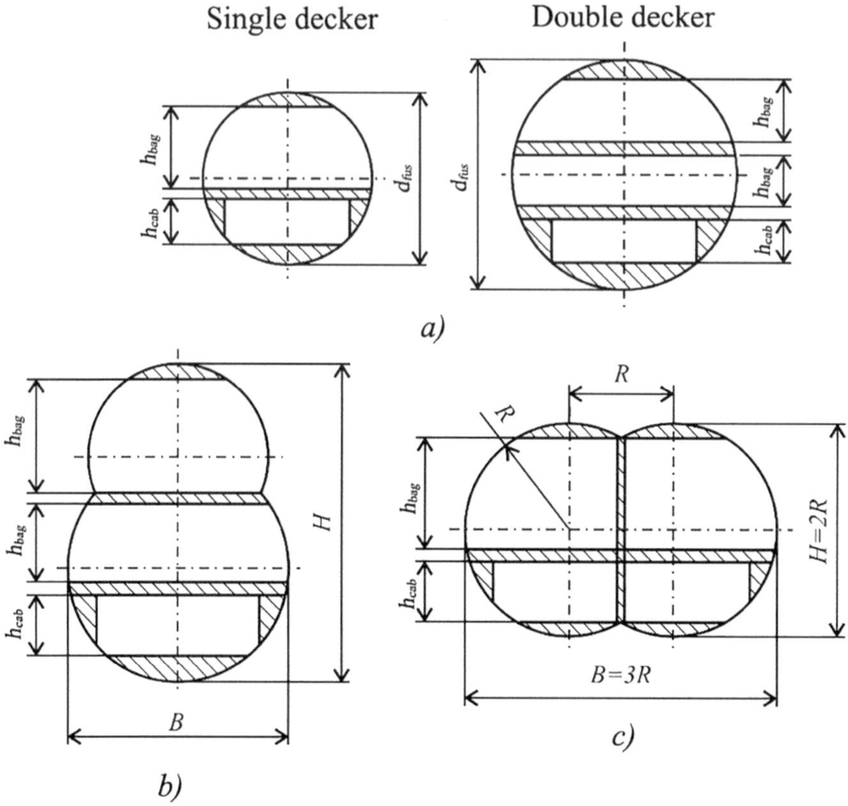

Fig. 16 An example of fuselage cross-sections: **a** and **b** are the circular single and double-deck sections; **b** is a double-deck section in the bulb shape; **c** is a single-deck section in the shape of a horizontal eight; h_{cab} is the cabin height; h_{bag} is the height of the cargo compartment

For an increased range, the effect of the aerodynamic efficiency on the measure of merit (cost per ton-kilometer) becomes more important than the effect of the fuselage mass; vice versa, for a shorter range, the aerodynamic efficiency is less important than the fuselage mass. This explains the decrease in the optimal value of the fuselage midsection with the increasing design range. Thus, for $n_{pass} = 300$ at the estimated range $R_{design} = 2000$ to 4000 km, single-deck fuselages of a circular cross-section with 7 to 8 seats in a row with two aisles ($d_{fus} = 5.02$ to 5.51 m) turn out to be advantageous; with $R_{design} = 8000$ km, the arrangement of 6 seats in a row with one aisle becomes advantageous. This rule can be illustrated by the variation of the efficiency of single and double-deck fuselages with the passenger capacity and range. Single-deck fuselages have lower aerodynamic drag, but they are longer and heavier than double-deck fuselages of equal capacity. The mass advantage of the latter increases with an increasing passenger capacity. This is confirmed by the curves $a_{min} = f(n_{pass})$ given in Fig. 17, where only the optimal fuselage configurations for each type of cross-section are shown. The advantages of the single-deck fuselage of a circular cross-section are maximum at the least capacity $n_{\text{pass}} = 300$ and the longest ranges of 6000 to 8000 km (refer to Fig. 17b), while the advantages of the double-deck fuselage are maximum at the largest capacity $n_{\text{pass}} = 700$ and the least design ranges of 2000 to 3000 km (refer to Fig. 17a). However, for small, medium and large values of both passenger capacity and range, it is not clear which of the fuselage layouts will be advantageous, since their efficiencies are becoming close. The graphs show that with an increase in the design range, single-deck fuselages of a circular cross-section become optimal for ever-increasing passenger capacity. Thus, with $R_{design} = 2000$ to 3000 km, they are advantageous for the capacity $n_{\text{pass}} = 350$ (refer to Fig. 17a), with $R_{design} = 6000$ km, the passenger capacity may be increased up to $n_{\text{pass}} = 450$, and with $R_{design} = 8000$ km, $n_{\text{pass}} = 575$ (Fig. 17b). Double-deck fuselages are the most advantageous for larger passenger capacities. As a rule, the cross-sections in the shape of a horizontal eight feature insufficient stiffness ($c_{\text{fus}} > 0.1$) and become disadvantageous in terms of mass with an increase in stiffness. The efficiencies of double-deck bulb-shaped fuselages of a circular and a horizontal eight cross-sections become closer with an increasing range.

The efficiency border between single-deck and double-deck fuselages depends on the number of passenger seats and the design range. Single-deck fuselages have less aerodynamic drag, but they are longer and heavier than double-deck fuselages of equal capacity. The weight advantage of the latter increases with increasing passenger capacity. Therefore, single-deck fuselages are preferable for aircraft with smaller passenger capacities and longer ranges, while an increased passenger capacity and decreased range make the double-deck configuration a better option.

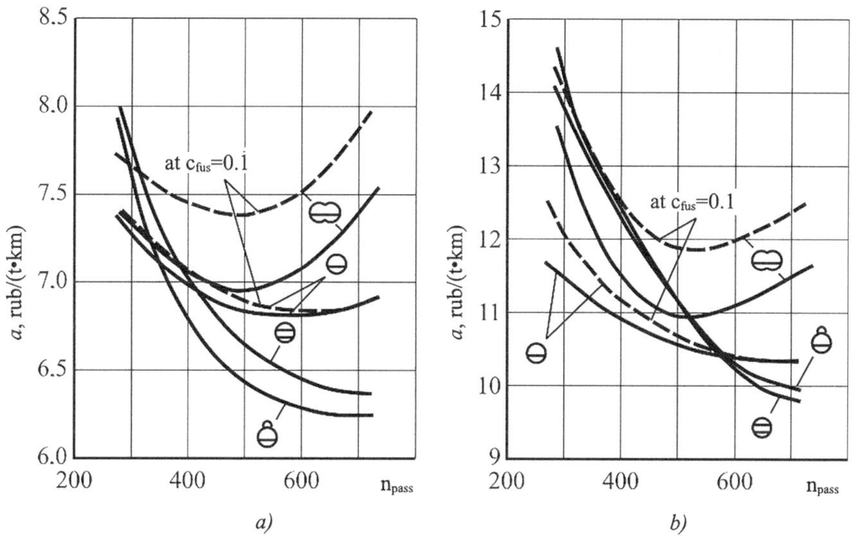

Fig. 17 Comparison of fuselages with the optimized geometry: **a** $R_{design} = 3000$ km; **b** $R_{design} = 8000$ km.

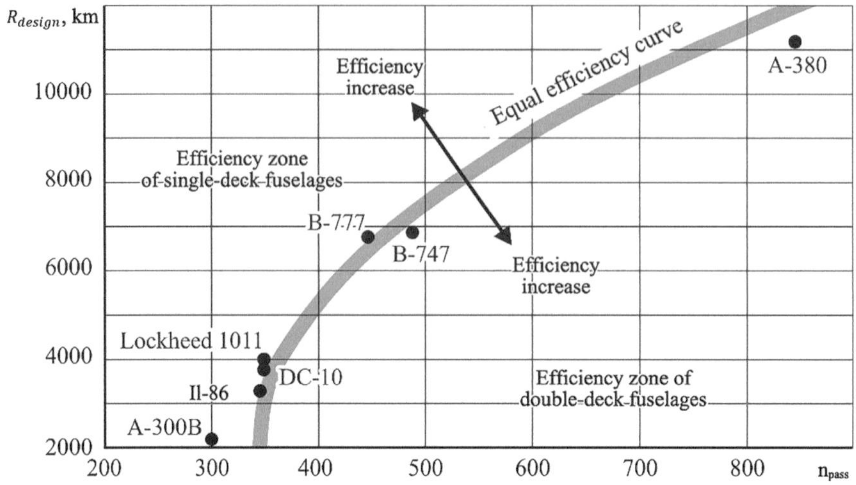

Fig. 18 Efficiencies of single- and double-deck fuselages

In addition to the border separating the efficiency zones of single- and double-deck fuselages, Fig. 18 shows data on some modern airliners (Il-86, A 300B, DC.10, Lockheed L-1011 TriStar, Boeing 777, Boeing 747, A.380). The points lying on the border correspond to the cases where single- and double-deck fuselages are equally efficient. An offset from the borders increases efficiency. In 2006, the full-length double-deck aircraft A380 began to operate (the second deck of the Boeing 747 occupies only the forward part of the fuselage). In spite of the large passenger capacity of the A380 ($n_{\text{pass max}} = 850$), its efficiency is deteriorated due to the extremely long range.

6 Fuselage Structural Design

6.1 Loads Applied to the Fuselage

In flight, the fuselage is exposed to the following types of loads:

- the weight of the fuselage structure, as well as the weight of the payload, equipment, and systems, accommodated in the fuselage;
- the forces transmitted from other parts of the aircraft; these can be the weights of these parts and the forces they produce, such as the empennage weight, its aerodynamic forces, the engine thrust and weight, the landing gear weight and reaction during take-off and landing;
- the aerodynamic forces produced by the fuselage;
- the pressure differential in the pressurized compartments.

Also, loads are classified as distributed (structural weight, aerodynamic forces) or concentrated (forces transmitted to the fuselage from other parts via joints and the weight of the items accommodated inside the fuselage).

The forces acting on a fuselage are considered in two planes: in the center line of the aircraft (the plane of symmetry) and in the horizontal plane perpendicular to the center line. The greatest loads applied to the fuselage are parallel to the center line, for example, the loads experienced by the fuselage during pull-out or landing. An example of loads acting on the fuselage in the horizontal plane is the force produced by the vertical tail (when rudders are deflected).

Exposed to the forces acting parallel to the aircraft plane of symmetry, the fuselage bends in the vertical plane. Due to the forces applied in the horizontal plane, the fuselage bends in the horizontal plane. Moreover, the fuselage reacts the torsional moment due to the forces applied at a significant arm (for example, the load due to the vertical tail or the lateral component of the force transmitted from the nose landing gear when landing with lateral impact).

Therefore, the fuselage experiences a transverse (shear) force Q, bending moments M in the vertical and horizontal planes, and a torsional moment about the fuselage axis M_t. All design loads, as well as the corresponding safety coefficients are specified in the strength standards for typical design cases. In general, the fuselage loaded in the vertical plane can be considered a beam supported by the wing spars.

In Fig. 19, the diagrams of distributed loadq_z, shear forceQ_z, bending M_y and torsional M_t moments are shown. The reactions in the fuselage/wing joints are R_1 and R_2, the reactions in the fuselage/empennage joints are R_3 and R_4; the weight of the items placed in the fuselage is also shown ($P_i = m_i g n_z$,m_i is the item mass, n_z is the design load factor). The torsional moment due to the load on the vertical tail P_{VT} and the lateral component of the reaction in the nose landing gear (landing with the lateral impact) for the fuselage nose section are also shown. In flight, the fuselage torsional moment is balanced by a couple of forces at the wing attachment points and then by the wings when the ailerons deflect. During landing with a lateral impact experienced by the nose gear, the fuselage's torsional moment is balanced by a couple of forces in the right and left main landing gear legs.

6.2 *Fuselage Structural Layouts*

The fuselages of the early aircraft, starting with the Wright's Flyer, were of the truss structural layout. Truss is a system of rods, with the rods reacting only axial loads (tension and compression).

Later, in order to streamline the fuselage, its structure was covered with skins that increased the structural mass of the aircraft. In addition, truss elements cluttered up the fuselage interior and made it difficult to arrange payload there. For these reasons, truss fuselages were replaced with the stress-skin beam-type structures. It should be noted that truss fuselages are still used for aerobatic low-speed aircraft.

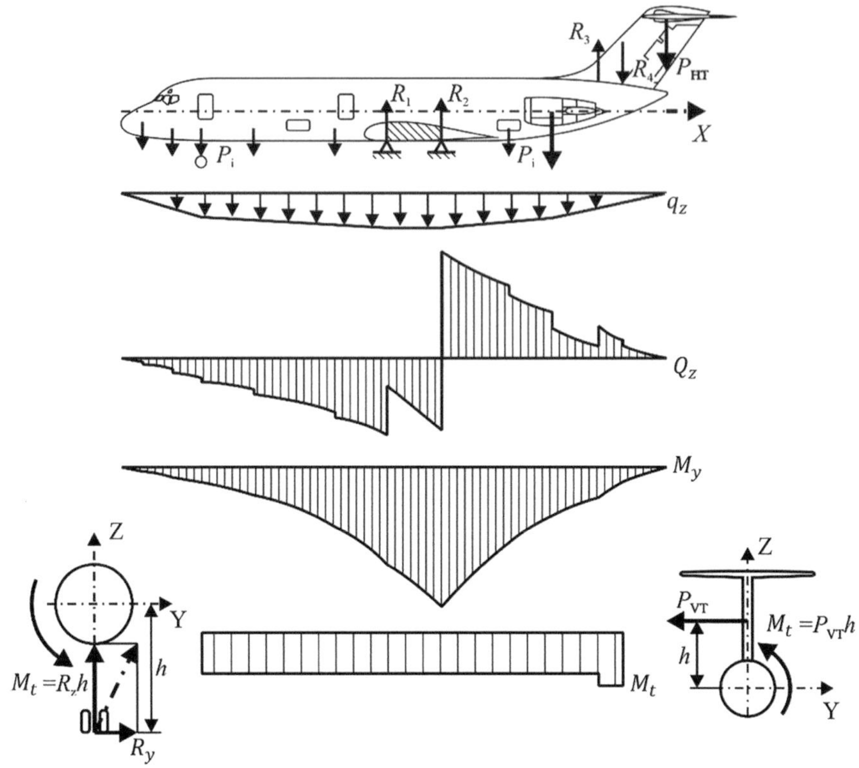

Fig. 19 Shear force and moments acting on a fuselage

Predominantly, the fuselages of modern aircraft feature the beam layout. The components of the beam fuselage structure are shown in Fig. 20. The fuselage structure is divided into:

(1) the primary structural members, which transmit the main loads along the fuselage, and
(2) the secondary structural members.

The primary structure includes longitudinal and transverse members, which is a distinctive feature of the beam fuselages.

Longitudinal members. The longitudinal fuselage members react the bending moment M.

Beam fuselages are divided into three types as to how they react the bending moment: spar-based, stringer-based, and skin-based layouts (Fig. 21.).

The distinctive feature of the spar-based structural layout is that the bending moment is reacted by the longitudinal members (metal sections of different shapes), i.e., spars, as a force couple. Stringers and skin either do not participate at all, or participate very weakly in reacting bending. Due to fuselage bending, the spars

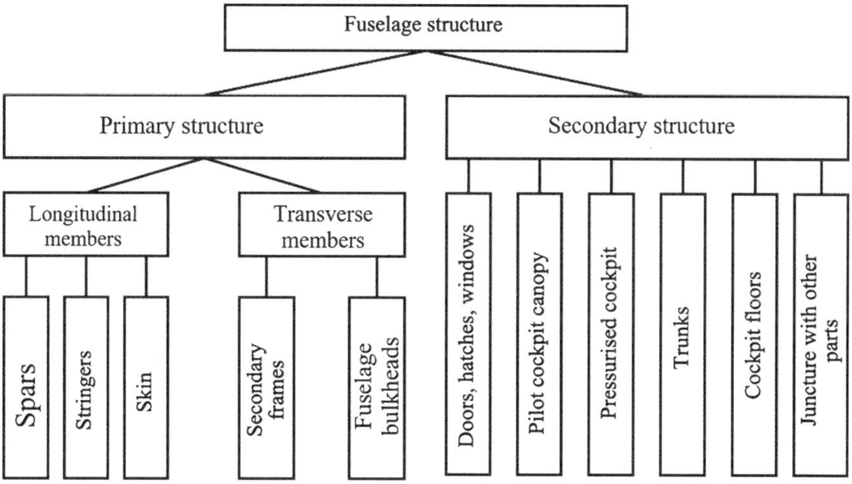

Fig. 20 Components of the beam fuselage structure

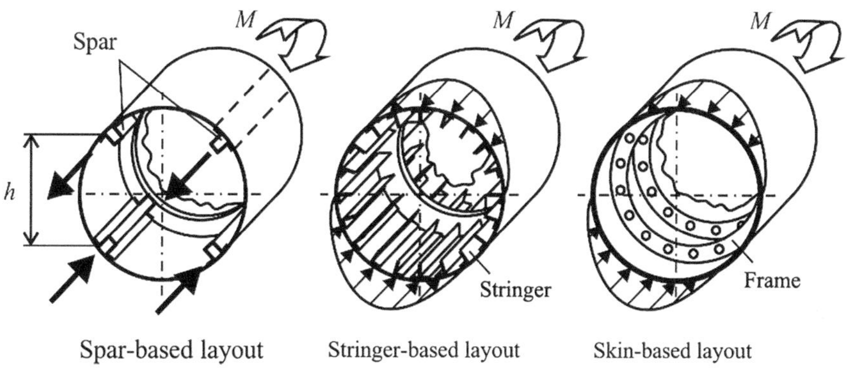

Fig. 21 Types of the beam fuselage

carry axial loads (compression and tension). The couple arm h refers to the distance between the spars (perpendicular to the axis about which the bending moment acts).

In the stringer-based structural layout, the bending moment is reacted by stringers and skin through tension and compression. The greatest compression or tension occurs in stringers with attached skin that are the farthest from the neutral axis of the fuselage section.

There is also a mixed spar-stringer-based layout incorporating spars, stringers, and skin participating in reacting tension and compression arising from the fuselage bending moment.

The skin-based structural layout consists of skin only. Due to fuselage bending, one part of the cross-section resists tension, and the other one resists compression. Since a thin skin features low stability when exposed to compression or shear, numerous frames must be used to reinforce it. For this reason, it is more common to use sandwich panels consisting of two sheets of skin with a filler between them in the form of honeycomb, corrugation, or foam plastic. Such panels are highly resistant to compression and shear.

The selection of a particular structural layout depends on the loading of the fuselage structure, the size, and type of aircraft, as well as the configuration of the particular compartment. Therefore, often the fuselage features different structural layouts along its length. Fuselage forward and aft parts that are slightly loaded with bending frequently feature a skin-based layout. The spar-based fuselages allow making cutouts and hatches between spars at a low structural weight, which is typical of many fighter aircraft. Large-diameter passenger aircraft usually feature the stringer-based structural layout.

Since the type of structural layout depends on how the structure reacts the bending moment, the remaining types of load (shear force Q and torsional moment M_t) are reacted by each layout equally. The shear force is reacted by the shear of the lateral (relative to the direction of Q) fuselage panels. If Q acts vertically, then these are the right and left fuselage panels; if Q acts horizontally, then these are the upper and lower panels. Panels can be of the stringer-based or skin-based type.

The most rational type of structure for reacting the torsional moment is a closed-contour structure. That is why the fuselage of any structural layout reacts the torsional moment by skin forming a closed contour. The skin is supported by longitudinal and transverse structural members.

Transverse members. Fuselage transverse members can be divided into frames and bulkheads. Bulkheads resist the concentrated loads transmitted from other parts of the aircraft and from items inside the fuselage. Frames maintain the prescribed shape for the fuselage sections and support the longitudinal members (skin, stringers, spars) when they resist compression and shear.

By supporting the compressed longitudinal members, frames prevent the buckling of the fuselage shell due to bending.

There are two forms of buckling. The general fuselage buckling is characterised by a significant distortion of its cross-section shape, i.e. the frame shape (Fig. 22a). When the local fuselage buckling occurs, the compressed longitudinal members arranged between the frames, such as panels (stringers and skin), lose their stability, but the frame shape is maintained (Fig. 22b).

To prevent the general buckling of the fuselage, the frames must be stiff. However, it is not advisable to overdesign the frames because of the weight penalty. Similarly, the longitudinal members must be designed to prevent the local buckling of the entire fuselage.

The ratio between the stiffness of longitudinal and transverse members of the fuselage exposed to bending can be optimized for the weight reduction. In an optimized structure, both forms of buckling occur simultaneously. On the grounds of experimental studies, N.D. Hoff proposed a criterion that serves as an indicator of the form of the fuselage buckling:

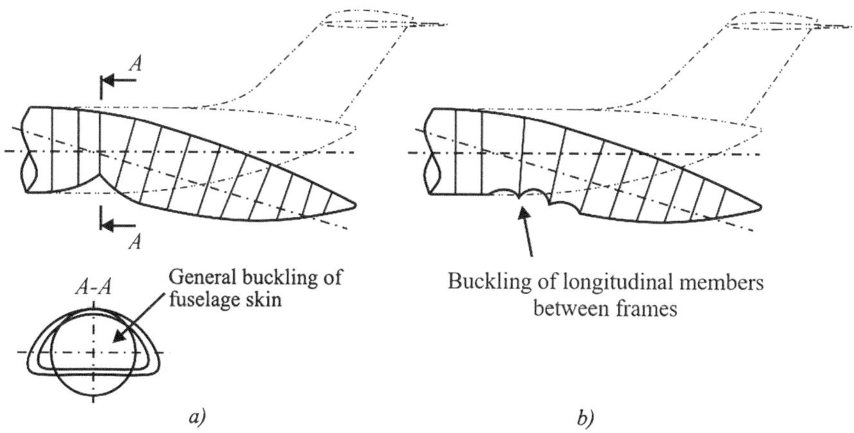

Fig. 22 Fuselage general and local buckling due to bending

$$\Lambda = \left(\frac{D/2}{L}\right)^3 \left(\frac{D/2}{b_{\text{str}}}\right) \frac{(EJ)_{\text{str}}}{(EJ)_{\text{fr}}},$$

where $(EJ)_{\text{str}}$, $(EJ)_{\text{fr}}$ are the bending stiffness of the stringer and frame, respectively, in the radial direction; L is the distance between frames; b_{str} is the distance between stringers; D is the fuselage diameter.

At $\Lambda \approx 50$, the loads at which the two forms of buckling occur are equal; at $\Lambda < 50$, the local buckling occurs; and at $\Lambda > 50$, the general fuselage buckling occurs. For the minimum fuselage structural mass, $\Lambda \approx 50$. Then the required stiffness of the frames can be determined from the stiffness of the stringer panel (which in turn is based on the structural analysis of the fuselage longitudinal members):

$$(EJ)_{\text{fr}} = \frac{(EJ)_{\text{str}}}{50} \left(\frac{D/2}{L}\right)^3 \left(\frac{D/2}{b_{\text{str}}}\right). \tag{1}$$

If the same material is used for longitudinal and transverse members, Eq. (1) determines the moment of inertia of the frame section J_{fr}.

Typically, frames feature a cap and a shear web to resist bending and shear. A frame can be composed of parts joined together by cover plates (Fig. 23), which facilitates manufacture and assembly. The height of the frame cross-section usually exceeds the height of stringers, that is why stringers, as a rule, go through the frames via openings in them (Fig. 23a). In order not to weaken the frame, stringers are sometimes passed under it and joined with the skin by means of compensators (Fig. 23b).

Unlike frames, bulkheads often go through stringers, which calls for joining stringers to transfer the loads.

Bulkheads perform all functions of frames, as well as react the concentrated loads. In order to be reacted, the load must act in the plane of the frame. If a concentrated load P is directed at an angle to the frame plane, a reinforced structure must be

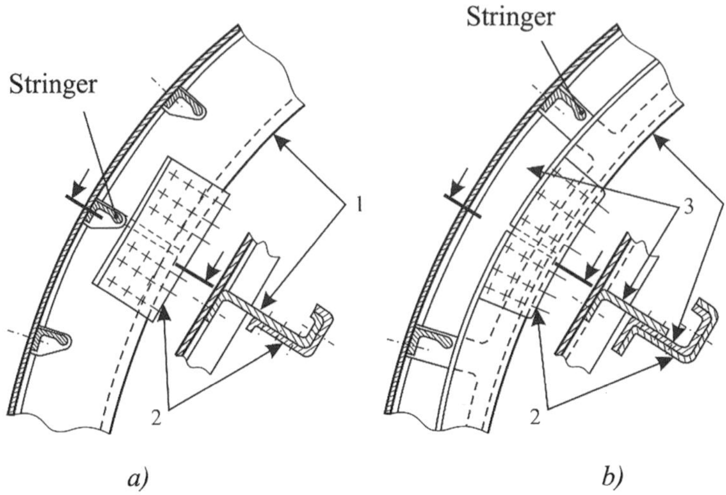

Fig. 23 Fuselage frames. **a** A frame with openings for stringers; **b** A frame with compensators; 1—frame; 2—cover plate; 3—compensator

used, which resolves the load into two components: P_{fr} acting in the frame plane and reacted by it, and P_{str} acting perpendicularly or at an angle to it, reacted by the reinforced longitudinal members, such as a spar or a reinforced stringer (Fig. 24a).

In some cases, the frame plane is tilted to fully react this load, which renders it unnecessary to reinforce the longitudinal members. Such "slanting" frames (similar to the "slanting" ribs in a swept-back wing) can be found, for example, in the aft part of the fuselage where the spar of a swept-back stabilizer is joined with the fuselage (Fig. 24b).

The variety of loads and design features result in a variety of bulkhead designs. Figure 25 shows the attachment point of a bulkhead that transfers the concentrated load P (or its component in the frame plane) to the bulkhead members, for example, the column, which converts the concentrated load into the shear stress in the shear webq_{SW}. In turn, the shear web transfers the load to the bulkhead caps q_{skin}, and q_{SW} eventually to the side fuselage panels (relative to the force P).

7 Structural Design of the Fuselages of Integrated Aircraft

Since the early 1980s, there has been a clear tendency to significantly complicate the aerodynamic configurations of aircraft. New requirements for aircraft called for novel aerodynamic layouts. Wing/fuselage integration became widely used in engineering practice. To a greater extent, this tendency has affected combat aviation. Such aircraft as the Su-27, MiG-29 (maneuvering aircraft), and Tu-160 (heavy strategic aircraft) were created. A requirement to significantly reduce the signature of combat aircraft

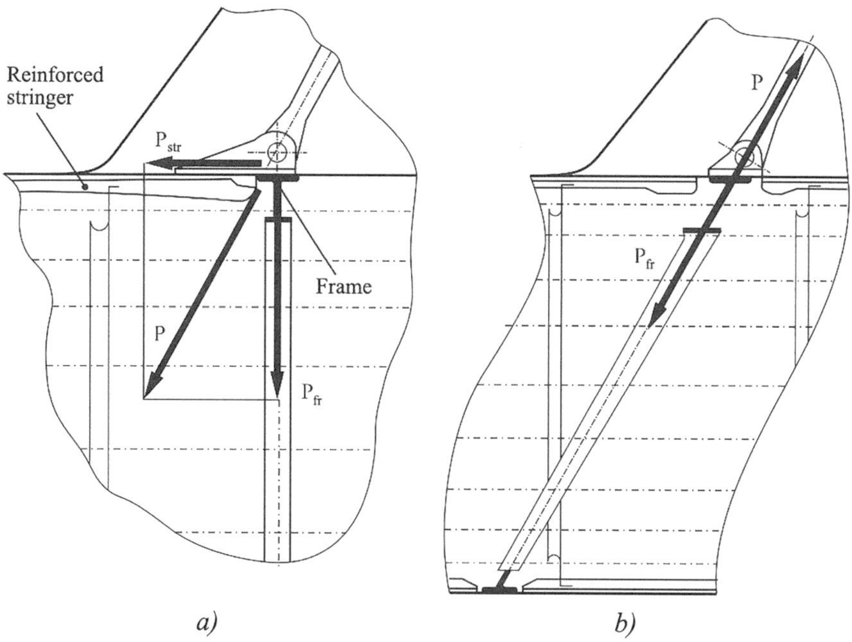

Fig. 24 Bulkheads

within all wavelength ranges, leading to the need to arrange the entire payload inside the fuselage, increased the integration of wings and fuselage in new projects (Fig. 26).

A flying wing can also be considered a logical development of an integrated aerodynamic configuration (Fig. 27). The B-2 aircraft features this layout. Currently, there are numerous civil aircraft projects employing this configuration.

Any integration complicates the traditional division of aircraft airframe into parts: wing, fuselage, and empennage. Here, the aircraft central part can only conventionally be called fuselage. Externally, it bears little resemblance to the traditional fuselage, while combining the functions of both the fuselage (accommodation of payload, crew, equipment, power plant, and other) and the wing (generation of lift, sometimes 50% or more of the total lift produced by the aircraft).

The lift generated by such a fuselage can balance the inertia loads arising from the cargo, crew, equipment, power plant, and other items, directly in those places where it is exerted, or close to them. Consequently, the length of load paths measured is significantly reduced. This reduces the structural weight and, in this respect, offers an advantage over conventional aerodynamic layouts.

However, to realize the potential of this configuration during the design process, it is important to consider the specific features and the structural solutions inherent in such layouts. Special attention should be paid to the structural design.

The structural design of the integrated fuselages compared to conventional configurations has become much more complicated, both in geometrical and topological

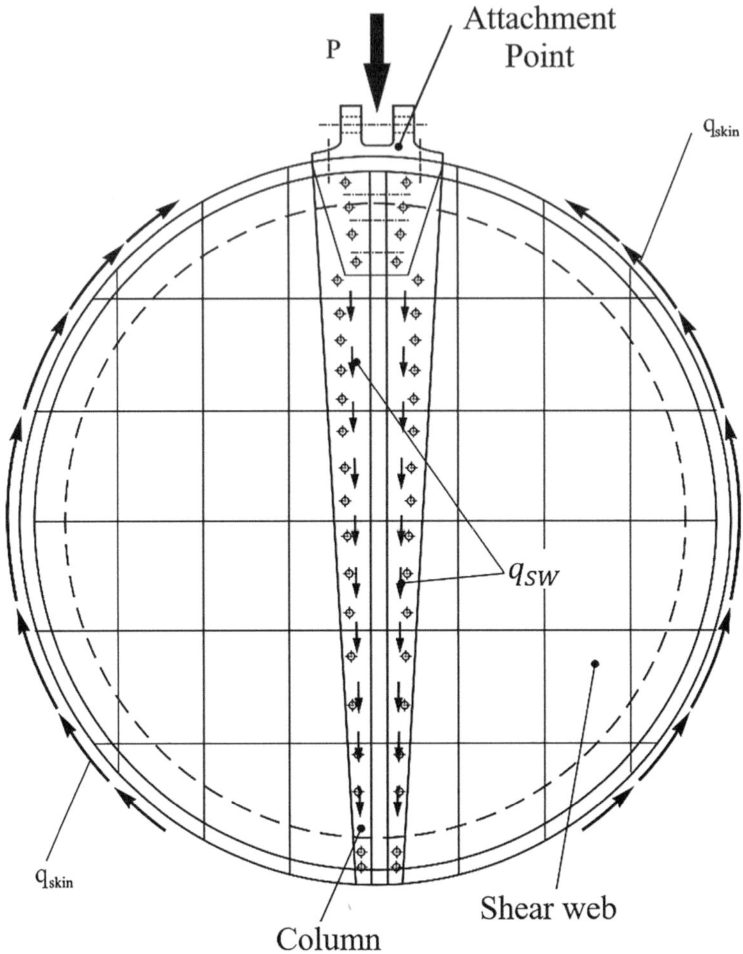

Fig. 25 Load paths in a bulkhead

terms. The most loaded parts of the fuselage became particularly irregular. Significant cut-outs may be placed on the paths of loads arising on the wings, empennage, and control surfaces. For this reason, the structural members of an integrated fuselage, even in the most loaded zones, can have large steps in structural depth and numerous stress concentrations. Because neither the method of reacting and transmitting loads nor the rational layout for such structures are obvious, their selection always necessitates extensive preliminary research and optimization. Figure 28 demonstrates a fragment of such a structure.

Since it is impossible to separate the longitudinal and transverse load paths in the integrated airframe structure (which is typical for conventional configurations), the most stressed zones of such a fuselage are loaded simultaneously in both the

Fig. 26 Aircraft with integral aerodynamic layout

Fig. 27 Flying wing aircraft

longitudinal and transverse directions and react both bending and torsion. The internal stress state of these elements becomes complex (multi-component). All this results in the fact that the analysis of the total stress–strain state (SSS) of such structures using analytical methods becomes difficult and sometimes even impossible. Neither a plate-and-beam analogy nor a classical theory of shells is able to strictly and correctly model the operation of such structures. Often, only the finite element method (FEM) is suitable for modeling complex structures.

Fig. 28 A fragment of an integrated aircraft structure

To exploit the weight advantages of an integrated configuration, the loads must be analysed by means of methods that provide results in a distributed form. It is unadvisable to simplify the loads to integral form (circulation along the elastic axis of the aircraft's part), since in this case the loads must be re-distributed over the area or volume of interest by means of approximate laws (trapezoid, triangle, etc.). When replacing the actual load distribution with an approximate one, the solution accuracy may be compromised.

The design and analysis of conventional structures are often made in practice based on engineering experience. However, for integrated configurations, this approach is inappropriate. Firstly, the engineering practice has not accumulated sufficient experience here. Secondly, the collected experience may turn out to be too conservative, since the tasks and conditions in which they had been solved before have changed significantly with the transition to other layouts. Therefore, extrapolating the previously obtained solutions to new aircraft layouts as well as to new engineering conditions and new production technologies may not always be entirely correct.

The next engineering feature of integrated aircraft is that their parameters and shapes have a much greater influence on the mechanical characteristics of structures (their strength, stiffness, etc.) than in traditional layouts. Some of these parameters are determined during an aircraft's aerodynamic design, while others are determined during its weight design. That is why the aerodynamic, weight and structural design of an aircraft should be carried out concurrently to obtain an efficient structure. Herewith, the structural design should be based on a general measure of merit related to the aircraft as a whole, and on reaching a compromise between numerous contradictory requirements.

Since the engineering of a complex structure, as a rule, involves examining many possible configurations, it is obvious that aircraft structural design is impossible without the use of information technologies. Along with the experience and qualifications of the designer, the use of numerical tools is a prerequisite for the creation of new integral fuselage structures.

8 Modern Methods of Aircraft Structural Design

In general, structural design is an iterative process. In this process, a successive approximation of structure characteristics to optimal values is performed. Its essence is shown in Fig. 29.

Analysis methods establish the relationships between model parameters, external conditions, and model characteristics. The variables are chosen from among the total number of parameters in the computational model. Their numerical values should provide, in the iterative optimization process, the extremal values for the corresponding characteristics of the model selected as criteria. When the characteristics reach such values, the iteration process ends.

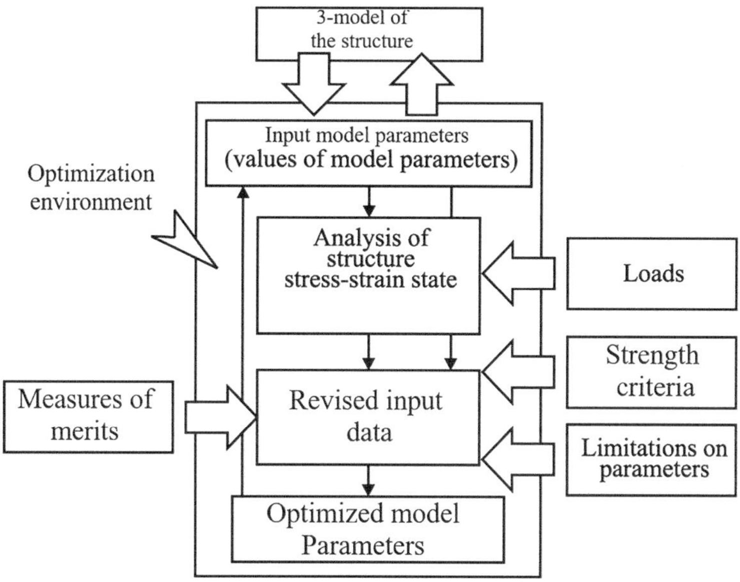

Fig. 29 Flowchart of a structure's optimisation process

9 A Method for Designing Theoretically Optimal Structures

In the initial stage of structural design, a method for designing theoretically optimal structures can be used. As a rule, the minimum of the total weight of structural members is used as a measure of merit. The method makes it possible to determine the magnitude and direction of the main load paths and to evaluate the mass of the meat necessary to react and transfer these loads. A large role in the development of design methods for theoretically optimal structures is attributed to Professor V. A. Komarov.

With regard to the fuselage structure, the essence of this method is as follows. We should first analyse the external loading of the fuselage (both the aerodynamic load acting directly on the fuselage and the loads from the wings, empennage, control surfaces, etc.) using panel methods. Then, the aircraft part (fuselage as a whole or its part) is replaced by a computational mesh. Initially, this area may be shaped as a rectangular parallelepiped without any limitations. Later, the fuselage outline will serve as the external boundary for the mesh. Internal boundaries can also be established. The external load is applied to the nodes of the mesh, while the aerodynamic load can be reduced to the concentrated forces and moments at the attachment points (Fig. 30).

Next, an iterative procedure is applied to search for a rational distribution of the material over the fuselage using the ratio of the limit and actual loads:

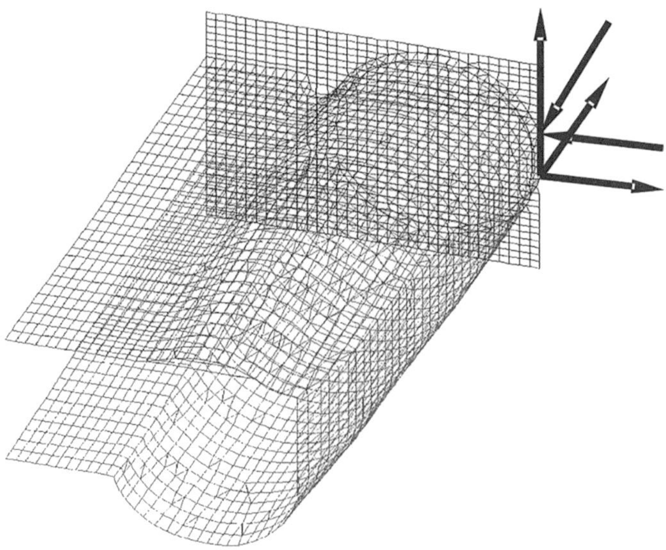

Fig. 30 Formation of a theoretical optimal structure

$$\delta_i^K = \delta_i^{K-1} \frac{\sigma_i}{[\sigma_{limit}]},$$

where: δ_i^K is the thickness of a finite element i at a step k;
σ_i is the actual stress in a finite element i;
σ_{limit} is the limit stress.

If the stress in some finite elements exceeds σ_{limit}, their thickness increases in proportion to the stress, whereas some weakly loaded elements degenerate. Therefore, the actual stress field of the structure is levelled. The result of this procedure is a fully stressed structure that is close to the statically determined structure. Herewith, its total weight reaches its minimum and can be expressed as

$$M = \sum_{i=1}^{n} \rho_i S_i \delta_i,$$

where ρ_i and S_i are the material density and area of a finite element i. The orientation of the main stresses in the finite elements shows the direction of the main load paths in the future structure and thereby indicates the rational orientation of its structural members (Fig. 31).

The main advantage of this method is that it enables the elaborated solution of a theoretically optimal structure. However, the real physical structure almost always differs from the theoretically optimal one. The first reason for that are various limitations related to fabrication and production. Secondly, in order to meet the survivability and fail-safe requirements, a certain minimum value of redundancy is typically built into the structure, even if a statically determined structure under the same loading conditions would be strong enough at a smaller mass. Therefore, the practical significance of theoretically optimal structures is as follows: as a source of objective information, they allow a designer to explore the physics of the optimal force transfer process in an ideal structure, to draw the necessary conclusions, and to make the right decisions when designing the real physical structures.

Fig. 31 Main stresses in an integrated fuselage panel

Modern algorithms of the method are capable of incorporating additional requirements in the analysis and thereby bringing the result closer to the physical one. During the iterative optimization procedure, it is possible to set limitations on the minimum thickness of individual structural elements, such as skins and other longitudinal and transverse structural members.

10 The Method of Improving Variations

One of the heuristic approaches used in structural design is the method of improving variations. If the preliminary structural layout is known, this method allows for the identification of an optimal strategy for its further improvement. This is achieved through the assessment of various possible changes and the selection of those that ensure an optimized design. The variables used in optimization may be geometric, topological (additional structural members, etc.), or combined. They can be related to the general structural layout, sizing the structural members, or both.

The aggregate impact of the changes to the generalized parameters of aircraft structural layout $\{\delta_i\}$ on the quality criterion of the structure $Q_j = Q_j(\delta_1 ... \delta_N)$ can be written down as follows:

$$\Delta Q_j = \sum_{i=1}^{N} \frac{\partial Q_i}{\partial \delta_i} \Delta \delta_i, \ (j = 1, ..., M),$$

where N is the number of considered changes to the aircraft structural layout, M is the number of criteria applied. The expression for partial increments of criterion Q_j for each of the expected changes in project parameters δ_i

$$q_{ji} = \frac{\partial Q_i}{\partial \delta_i} \Delta \delta_i$$

represents the contribution of this change to the overall increment in quality criteria for structure Q_j. The physical meaning of q_{ji} is the sensitivity of Q_j to a change i. Thus, when developing a strategy for making changes to a structure using q_{ji}, it is possible to assess which change influences Q_j to a greater extent.

Since not all variations of generalized parameters $\{\delta_i\}$ out of the investigated parameters ($i = 1,...,N$) will lead to the expected results (some of them will be rejected due to a lack of efficiency, and some will be rejected due to a negative result), the overall change in the structural layout can be represented by N' sampling the most effective variations $\{\delta_i\} i = 1,...,N'$ out of the whole $N (N' \subset N)$ set, providing a sufficient level of improvement in the quality indicators of structure, i.e.

$$\Delta U = \sum_{i=1}^{N'} \Delta \delta_i.$$

ΔU depends on the problem at hand. For a correct comparison of q_{ji} (increment of j-criterion for all investigated changes $i = 1,...,N$), their calculation for each fixed j should be performed under the same initial conditions using the same analysis algorithm. The initial conditions imply the use of the same material, the same sizes of structural members (thickness, cross-section areas, etc.), as well as the same loading conditions.

The analysis algorithm includes linear stress–strain state analysis as well as the models to determine the required stiffness characteristics of the structural members at hand.

11 Direct-Comparison Method

Since variations $\Delta\delta_i$ in the expression for ΔQ_j are generalised, and some of them (for example, the changes to the topology of the structure) can only be qualitative, it is not always possible to calculate the total increment of j-criterion by using the formula for ΔQ_j. Neither ΔQ_j-formula can be applied in the evaluation of a new aircraft layout version; this case is quite important. In these situations, a direct comparison of various versions of aircraft structural layouts is relevant. Moreover, a direct comparison is always carried out for the final assessment of alternative aircraft structural layouts and selecting the best one.

For the direct comparison, the same system of quality criteria $\{Q_j\}$ can be used as when using the method of improving variations. And here it is also important to ensure that the comparison of each Q_j criterion for the investigated aircraft structural layout is correct. Therefore, their calculation for the layouts at hand should also be carried out under the same initial conditions using the same analysis algorithm. The requirements for the initial conditions and the algorithms are identical to those imposed on the previous method.

Variables		
Symbol	Description	Units
R_{design}	Design range	km
m_{PL}	Payload mass	kg
L_{RW}	Runway length	km
l_{fus}	Fuselage length	m
d_{fus}	Fuselage diameter	m
$S_{f,m-s}$	Area of fuselage middle (maximum) section	m^2
λ_{fus}	Fuselage fineness ratio	–
λ_{FB}	Forebody fineness ratio	–
λ_{AB}	Afterbody fineness ratio	–
B	Fuselage maximum width	m
H	Fuselage maximum height	m

(continued)

(continued)

Variables		
Symbol	Description	Units
d_{eq}	Fuselage equivalent diameter	m
F_{FWA}	Fuselage wetted area	m^2
$M_{cr.fus}$	Fuselage critical Mach number	–
c_{MAC}	Mean aerodynamic chord	m
M_o	Fuselage bending moment at the aft spar of the wing	N•m
M_{FS}	Fuselage bending moment at the front spar of the wing	N•m
n_Z^{design}	Z-component of the design load factor	–
b	Wing span	m
AR	Wing aspect ratio	–
V_{fus}	Fuselage volume	m^3
H_{cruise}	Cruise altitude	km
$(T/W)_0$	Take-off thrust-to-weight ratio	–
p_0	Take-off wing loading	daN/m^2;
λ	Wing taper ratio	–
$(t/c)_r$	Wing root thickness ratio	–
n_{pass}	Number of passengers (seats)	–
L_{HT}	Moment arm of horizontal tail	m
L_{VT}	Moment arm of vertical tail	m
S_{HT}	Horizontal tail area	m^2
S_{VT}	Vertical tail area	m^2
φ	Fuselage tail upsweep angle	Degrees
L/D	Lift-to-drag ratio	–
m_0	Aircraft take-off mass	t
m_{PL}	Payload mass	t
a	Cost per tonne-kilometer	Rub/(t·km)
m_{fuel}	Fuel mass	kg
m_i	Component of the aircraft's take-off mass	kg
C_{fus}	Fuselage stiffness factor	–
$\overline{m}_W$	Wing mass ratio	–
$\overline{m}_{PL}$	Payload mass ratio	–
M_t	Torsional moment	N·m
M	Bending moment	N·m
Q	Shear force	N
R	Reaction force	N
σ	Stress	Pa

References

1. Eger SM, Mishin VF, Liseitsev NK et al (2005) Proektirovanie samolyotov (Aircraft design). Logos, Moscow
2. Sheinin VM, Kozlovsky VI (1984) Vesovoe proektirovanie i effektivnost' passazhirskih samoletov (Weight design and efficiency of passenger aircraft). Mashinostroenie, Moscow
3. Kiselev VA (1970) Proektirovochnyj raschyot vesa i prochnosti fyuzelyazhej passazhirskih samolyotov (Design calculation of weight and strength of fuselages of passenger aircraft). TsAGI, Moscow
4. Kiselev VA (1977) Voprosy komponovki passazhirskih samolyotov (Passenger aircraft layout issues). MAI, Moscow

Aircraft Power Plant Design

Nikolai K. Liseitsev

Abstract The chapter explains the composition of aircraft power plants of various types and shows the properties of aircraft engines (with the focus on jet engines). The chapter also covers the features of air intakes, both subsonic and supersonic, the arrangement of the intake on an aircraft, as well as the design of exhaust systems and thrust reversers. The chapter explains the installation of engines on aircraft, the noise of jet engines, and its reduction. Finally, the chapter presents the composition and arrangement of fuel systems on aircraft.

Keywords Aircraft engine · Jet engine · Air intake · Exhaust system · Thrust reverser · Fuel system

1 Power Plant as an Aircraft Subsystem

The power plant is a major structural and functional component of an aircraft.

The power plant of an aircraft comprises:

- engines with their parts and systems;
- intake and exhaust devices;
- fuel system, etc. (see appendix).

The theory and design of individual components of a power plant are considered in detail in special literature. This chapter focuses on power plant design issues that are directly related to the general design of a modern aircraft, and mainly on turbojet engines. The reader may become familiar with the design aspects of a turboprop aircraft in [1, 2].

The main requirements that the power plant has to satisfy are the lowest specific fuel consumption (especially for long-range aircraft), the lowest specific weight (especially for aircraft with a high thrust-to-weight ratio), and also sufficient reliability and cost effectiveness.

N. K. Liseitsev (✉)
Moscow Aviation Institute (National Research University), Moscow, Russia
e-mail: nikolai.liseitsev@yandex.ru

M. A. Pogosyan (ed.), *Aircraft Design*, Springer Aerospace Technology,
https://doi.org/10.1007/978-981-96-4599-2_18

The successful outcome of the design of a high-performance modern aircraft largely depends on how well the airframe and the power plant are integrated. It is not enough today to design an airframe that would be highly efficient in terms of aerodynamics and weight and to fit it with an engine with a small specific fuel consumption and a small specific weight. It is essential for airframe and power plant characteristics to match each other, i.e. to select optimal conditions of aircraft flight and power plant operation according to a certain criterion underlying the efficiency of an aircraft of a given type. In other words, alongside the optimization of airframe characteristics, the optimal engine layout and gas-dynamic parameters defining its basic characteristics must be selected.

The matching characteristics of the airframe and power plant deliver the maximum (or minimum) of the objective function and, as a result, a more complete compliance with the military operational requirements that the aircraft in development must satisfy [3, 4].

In the design of a modern aircraft, two situations may occur:

(1) by the time the aircraft design begins, the engine has been completed (the aircraft is designed to fit a completed engine);
(2) a new engine needs to be developed to fit the aircraft in development.

The issue of selecting the engine type and its parameters is not raised if the aircraft is designed for a completed engine. In that case, the required thrust-to-weight ratio of the aircraft, the required number of engines, etc. are determined.

If a new engine is developed for the aircraft in development, then in the process of the aircraft's detailed design, all necessary characteristics of the future engine are elaborated (efforts are made to match the airframe and power plant characteristics). Based on the results of the work done, the aircraft designers team up with engine designers to draft mission requirements for the design of the engine (and nozzle).

2 Properties of Aircraft Engines

The main properties of aircraft engines are: altitude-speed performance $P = f(M, H)$ and $c_{fc} = f(M, H)$, air mass flow rate m_{air}, specific thrust P_{sp}, engine specific weight γ_{eng} and its overall dimensions (length and maximum diameter)*.

Engine specific thrust (thrust per 1 kg of air mass flow) is not only a measure of power plant efficiency, but also a key parameter defining engine noise and a good criterion for comparing the power plant's aerodynamic drag (and mass). The higher the specific thrust, the lower the air flow for the required engine thrust, and the smaller the dimensions of the engine and air intake.

The less the specific fuel consumption c_{fc}, the less the fuel mass required for the flight, and the greater the cost efficiency of the aircraft.

It is also obvious that reducing the engine specific weight γ_{eng} directly translates into reducing the mass of the aircraft power plant and its share in the aircraft weight balance equation.

The above properties are mainly determined by the following important gas-dynamic characteristics of the engine:

- engine bypass ratio m;
- gas inlet temperature T^*_{gas}, K;
- total compressor pressure ratio (compression ratio) π^*_{compr}.

The engine bypass ratio m is the ratio between the mass flow rate of the bypass stream and the mass flow rate entering the core. It affects specific fuel consumption, engine dimensions, temperature, and jet noise. Except for the power plant dimensions, an increase in m has a positive effect on all of the above characteristics. The bypass engines arose from a desire to reduce the cost of passenger aircraft operations, which is heavily reliant on fuel consumption (fuel costs account for 50 to 65% of aircraft flying hour costs).

T^*_{gas} and π^*_{compr} also affect the specific fuel consumption and the specific engine weight and dimensions (see below).

The GTE gas-dynamic parameters have radically changed between 1950 and 2000. If initially they were characterised by the following values: $T^*_{\text{gas}} =$ 1000 to 1200K; $\pi^*_{\text{compr}} =$ 7 to 12; $m =$ 0.7 to 1.5, today $T^*_{\text{gas}} =$ 1500 to 1800K; $\pi^*_{\text{compr}} =$ 20 to 40; $m =$ 0.3 to 0.7 for the ABTFE; $m =$ 3 to 6 and more for modern TFE.

During the preliminary design of an aircraft, the above characteristics of the engine for take-off conditions ($H = 0, M = 0$) may be determined in the following way.

For TFE and also ABTFE, afterburning off:

Specific thrust, daN*s/kg,

$$P_{sp_0} = \xi_T T^*_{gas}\left(1 + 0.05\text{m} - \sqrt{0.5\text{m}}\right);$$

where $\xi_T =$ (0.052 to 0.054) is a statistical factor that takes account of the effect of the relative fuel consumption, the relative mass flow of the turbine-cooling air, the nozzle exit velocity, the nozzle loss factor, the atmospheric air pressure at the flight altitude, the compressor efficiency, the turbine efficiency and other gas-dynamic parameters;

specific fuel consumption, $kg/(daN \bullet h)$,

$$c_{f.c0} = \xi_\tau \frac{\sqrt{T^*_{gas}}}{\pi^{*0.25}_{compr_0}}\left(1 + 0.05m - \sqrt{0.14m}\right),$$

where $\xi_\tau =$ (0.052 to 0.054) is the statistical factor; T^*_{gas} in K; $\pi^*_{\text{compr}_0} = \pi^*_{\text{compr}}$ at ($H = 0, M = 0$).

For ABTFE, full afterburning:

Specific thrust,

$$P_{\text{sp}_0} = \xi_\text{T}\left(T^*_{\text{AB}} + T^*_{\text{gas}}\right)(1 - 0.12m),$$

specific fuel consumption,

$$c_{f.c_{0AB}} = \xi_{\tau} \frac{T^*_{AB} - 02T^*_{gas}}{T^*_{AB} + T^*_{gas}} \left(1 + 0.18m + 0.01m^3\right),$$

where ξ_T = (0.029 to 0.031); ξ_{AB} = (3.8 to 3.9); T^*_{gas}, T^*_{AB} is the gas temperature at the exhaust of the common afterburner and at the inlet of the turbine of the core duct accordingly; for modern ABTFE $T^*_{gas} < T^*_{AB} < 2400\text{K}$.

It is noteworthy that actuating the ABTFE afterburner significantly raises the specific thrust and consequently the engine full thrust by 30 to 60%; however, the specific fuel consumption rises by over 200 to 250%. For example, at $H = 0$, $M = 0$ for the ABTFE with parameters $T^*_{gas} = 1060\text{K}$; π^*_{compr} = 7 to 12; $T^*_{AB} = 2100\text{K}$ and $m = 0.5$ we have specific thrusts $P_{sp_0} = 65$ daN s/kg; $P_{sp_{0AB}} = 105$ daN s/kg; ($\Delta \approx 60\%$) and specific consumption $c_{f.c_0} = 0.7$ kg/daN h;$c_{f.c_{0AB}} = 105$kg/daN h; ($\Delta \approx 270\%$).

For this reason, the engine full afterburning is activated for a short period of time: take-off, climb, and air combat.

The compression ratio of the compressor $\pi^*_{compr_0}$ has no effect on $P_{sp_{0AB}}$ and $c_{f.c_{0AB}}$. The effect of gas-dynamic parameters T^*_{gas} and m on P_{sp_0} and $c_{f.c_0}$ is shown in Fig. 1.

When an aircraft transitions from climb to cruising flight, the engine is switched to cruising mode: rotor speed n_{cru} and specific fuel consumption $c_{f.c}$ decrease (Fig. 2).

The effect of the bypass ratio on the specific fuel consumption of certain known TFEs is shown in Fig. 3.

An increase in π^*_{compr} leads to a decrease in specific fuel consumption, but the same parameter has a weak effect on the specific thrust (Fig. 4).

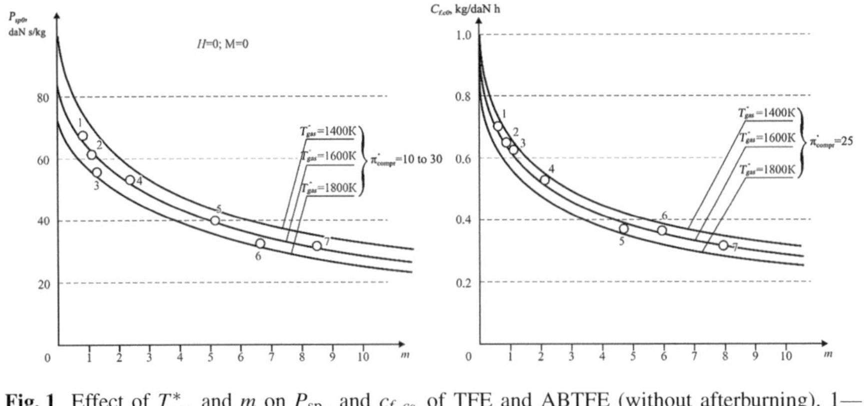

Fig. 1 Effect of T^*_{gas} and m on P_{sp_0} and $c_{f.c_0}$ of TFE and ABTFE (without afterburning). 1—F-100-PW ($T^*_{gas} = 1640$ K,, $\pi^*_{compr} = 25$, $m = 0.7$), F-15, F-16; 2—RB-199 ($T^*_{gas} = 1600$ K,, $\pi^*_{compr} = 23$, $m = 1$), Tornado; 3—TF-30P ($T^*_{gas} = 1410$ K,, $\pi^*_{compr} = 17$, $m = 1.1$), F-111; 4—F-101-GE ($T^*_{gas} = 1647$ K,, $\pi^*_{compr} = 27$, $m = 2.1$), B-1; 5—CA6-80C ($T^*_{gas} = 1595$ K, , $\pi^*_{compr} = 26$, $m = 4.7$), B747-200B; 6—TF-34-GE ($T^*_{gas} = 1150$ K,, $\pi^*_{compr} = 21$, $m = 6.2$), A-10; 7—TF-39 ($T^*_{gas} = 1580$ K,, $\pi^*_{compr} = 26$, $m = 8$), C-5A

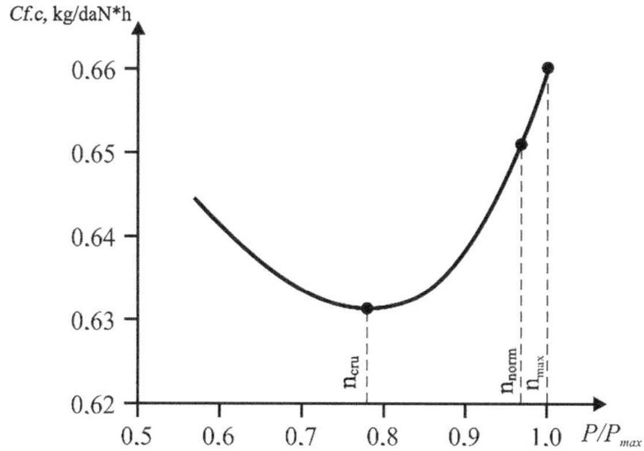

Fig. 2 TFE throttle characteristics ($m = 5$, $H - 11$ km, $M = 0.85$); n_{cru} is the rotor speed in cruise; n_{norm} is the rotor speed in the normal mode; n_{max} is the rotor maximum speed

Fig. 3 Effect of the TFE bypass ratio on the specific fuel consumption at take-off and under cruising conditions at altitude $H - 11$ km

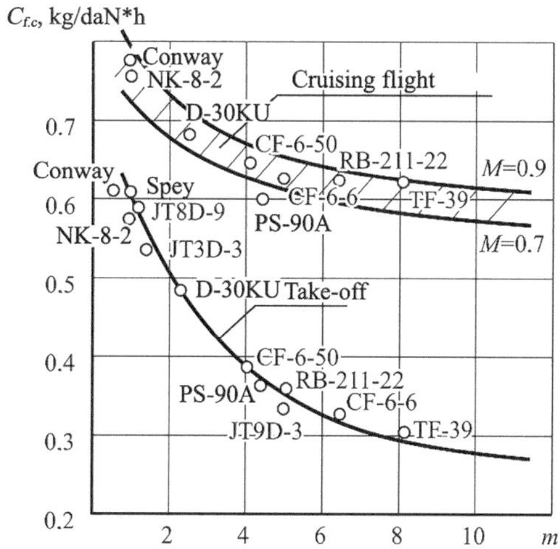

Engine altitude-velocity characteristics are used to calculate the take-off/landing characteristics of an aircraft, the climb, manoeuvrability and other characteristics of the aircraft in development.

At the early stages of aircraft design, when a particular engine has not been chosen yet, the altitude-velocity characteristics may be determined by the following dependencies (for hypothetical ABEs):

For TFE and ABTFE (afterburning off):

Specific thrust:

Fig. 4 TFE cruising characteristics at $M = 0.85$, $H = 11$ km

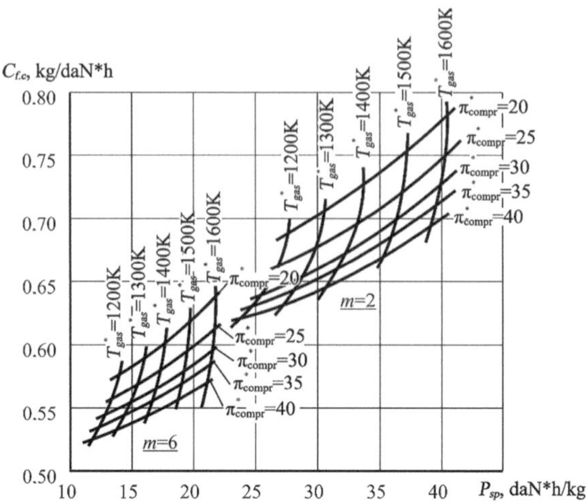

$$\frac{P}{P_0} = \frac{\sigma_{\text{inlet}}^{1.25}\left(1 - 0.18M^{1.2}m^{0.65}\right)}{\left(\frac{T_H}{T_{H=0}}\right)^{0.35M}}\left(1 - \frac{\pi^{*0.2}_{\text{compr}0}M}{\kappa_1\, T^{*0.4}_{\text{gas}}}\right)\left(1 + \frac{k-1}{2}M^2 - M^{5.3}\,10^{-3}\right)^3 \Delta\kappa_4^*;$$

specific fuel consumption:

$$\frac{c_{f.c}}{c_{f.c_0}} = \frac{\left(\kappa_2 - T_H^*\,10^{-3}\right)\left(1 - M^3\,10^{-3}\right)}{\kappa_3\,\sigma_{\text{to}}^{0.25}\left(1 - \frac{\pi^{*0.2}_{\text{compr}0}}{\kappa_1\,T^{*0.4}_{\text{gas}}}M\right)}\left(1 + 0.162m\,\Delta^{0.4}M^{1.7}\right)\left(\frac{T_H}{T_{H=0}}\right)^{0.5M},$$

where σ_{inlet} is the total pressure recovery coefficient at engine inlet; M is the flight Mach number; T_H is the air temperature at flight altitude; $T_H^* = T_{H1} + 0.2M^2$ is the stagnation temperature (at compressor inlet) at the flight altitude H, K; $k_1 = 0.36$ to 0.362 is a statistical factor; $k = 14$ is the adiabatic index of air; $\Delta = p/p_0$ is the air density ratio at flight altitude; $\kappa_2 = 1.23$ to 1.32 is a statistical factor; $\kappa_3 = 0.92$ to 0.94 is a statistical factor; $\kappa_4^* = \frac{1-0.1mM\,(1-0.6)}{1-0.74M^{1.5}(1-\Delta)}$.

When throttling the TFE from the maximum $n_{\max}$ to the idling speed n_{idle}, the specific fuel consumption may be determined by the following dependence:

$$\overline{c}_{f.c_{\text{thr}}} = 0.5\overline{P}^2_{\text{thr}} + \frac{0.006}{\overline{P}^2_{\text{thr}}} - 0.656\overline{P}_{\text{thr}} + 1.15,$$

where $\overline{c}_{f.c_{\text{thr}}} = \frac{c_{f.c_{\text{thr}}}}{c_{f.c(n_{\max})}}$; $\overline{P}_{\text{thr}} = \frac{P_{\text{thr}}}{P_{P(n_{\max})}}$; at $\overline{P}_{\text{thr}} = 1$; $n = n_{\max}$; at $\overline{P}_{\text{thr}} = 0.8$ to 0.9; $n = n_{\text{norm}}$;

at $\overline{P}_{\text{thr}} = 0.6$ to 0.8; $n = n_{\text{cru}}$; at $\overline{P}_{\text{thr}} \approx 0.05$; $n = n_{\text{idle}}$.

For ABTFE, full afterburning:

Specific thrust:

$$\frac{P_{\text{AB}}}{P_{0\text{AB}}} = \sigma_{\text{inlet}}^{1.25}\left(1 + 0.2M^2 - M^{5.3}10^{-3}\right)^3 \Delta \frac{\sqrt{T_{\text{AB}}^*}\left(8.17 + M^{1.5} - 0.05M^{3.5}\right)\kappa_4 - Ma}{g\xi_{\text{T}}\left(T_{\text{AB}}^* + T_{\text{gas}}^*\right)(1 - 0.04\,mM)};$$

specific fuel consumption:

$$\frac{c_{f.c0}}{c_{f.c0\text{AB}}} = \frac{g\xi_T\left(T_{\text{AB}}^* + T_{\text{gas}}^*\right)\left(T_{\text{AB}}^* + T_H^*\right)}{\sigma_{\text{inlet}}^{0.25}\left(T_{\text{AB}}^* - 288\text{K}\right)\left(1 + 0.05m^{1.2}M\right)} \frac{1}{\left[\sqrt{T_{\text{AB}}^*}\left(8.17 + M^{1.5} - 0.05M^{3.5}\right)\kappa_4 - M\,a\right]},$$

where T_{AB}^* is the afterburner exit temperature, K; $\kappa_4 = 2.95$ to 3.05 is a statistical factor; a is the speed of sound at flight altitude, m/s; $g = 9.81\,\text{m/s}^2$; $\xi_{\text{T}} = 0.029$ to 0.031 is a statistical factor.

When the engine is throttled from full to low afterburning, the specific fuel consumption ratio may be determined by the following dependence:

$$\overline{c}_{f.c_{\text{AB thr}}} = 0.75\overline{P}_{\text{AB thr}} + 0.25,$$

where $\overline{c}_{f.c_{\text{AB thr}}} = \frac{c_{f.c_{\text{AB thr}}}}{c_{f.c_{\text{AB full}}}}$; $\overline{P}_{\text{AB thr}} = \frac{P_{\text{AB thr}}}{P_{\text{AB full}}}$; $\overline{P}_{\text{AB full}} = 1$ at full afterburning; $\overline{P}_{\text{AB thr}} \approx 0.9$ at 80% afterburning; $\overline{P}_{\text{AB thr}} \approx 0.75$ at low afterburning.

Engine specific weight is defined as follows:

$$\gamma_{\text{eng}} = \frac{G_{\text{eng}}}{P_{0(\text{AB})}},$$

where G_{eng} is the engine dry weight, $P_{0(\text{AB})}$ is engine maximum take-off thrust; for ABTFE—at full afterburning.

The effect of the engine gas-dynamic parameters on its specific weight is determined by the equations given below.

For engines of TJE(TFE) type:

$$\gamma_{\text{eng}} = \frac{k_{\text{nozzle}} k_\gamma \sqrt{\pi_{k_0}^*}}{P_{\text{sp}_0}\sqrt{T_{\text{gas}}^{*3}}(10 + m)},$$

where k_{nozzle} is a statistical factor that takes into account the type of nozzle, for the nozzle without thrust reverser $k_{\text{nozzle}} = 0.95$ to 1.05; for the nozzle with thrust reverser $k_{\text{nozzle}} = 1.15$ to 1.25; $k_\gamma = (1.0$ to $1.05)10^6$ is a statistical factor.

It is noteworthy that parameters P_{sp_0}, m and T_{gas}^* are interconnected.

For engines of AFTJ(ABTFE) type:

$$\gamma_{\text{eng}} = \frac{k_{\text{nozzle}} k_\gamma \left(25 + \sqrt{\pi_{k_0}^*}\right)}{P_{\text{sp}_0}\sqrt{T_{\text{gas}}^{*3}}(7.5 + m)},$$

where $k_\gamma = (2.3 \text{ to } 2.5)10^5 \frac{\text{daN}K^{3/2}}{\text{kg/s}}$; for convergent variable geometry nozzle $k_{\text{nozzle}} = 0.85$ to 0.95; for convergent-divergent variable geometry nozzle $k_{\text{nozzle}} = 0.95$ to 1.05; for variable geometry nozzle with thrust reverser $k_{\text{nozzle}} = 1.0$ to 1.1; for variable geometry two-dimensional nozzle with vectored thrust $k_{\text{nozzle}} = 1.1$ to 1.2.

For modern TFEs, γ_{eng}=0.16 to 0.18, and for ABTFE γ_{eng}=0.1 to 0.2.

Engine overall dimensions. The maximum length and the maximum diameter of the engine determine dimensions of the power plant nacelle or the fuselage aft section (if the engine is mounted in the fuselage).

The full length L_{eng} of the modern aircraft engine of the ABTFE type may be represented as the sum of the lengths of its main parts (Fig. 5).

It is obvious that the largest number of assembly units will be in the engine of ABTFE type:

$$L_{\text{eng}}=l_{\text{fan}}+l_{\text{compressor}}+l_{\text{combustor}}+l_{\text{turbine}}+l_{\text{afterburner}}+l_{\text{nozzle}},$$

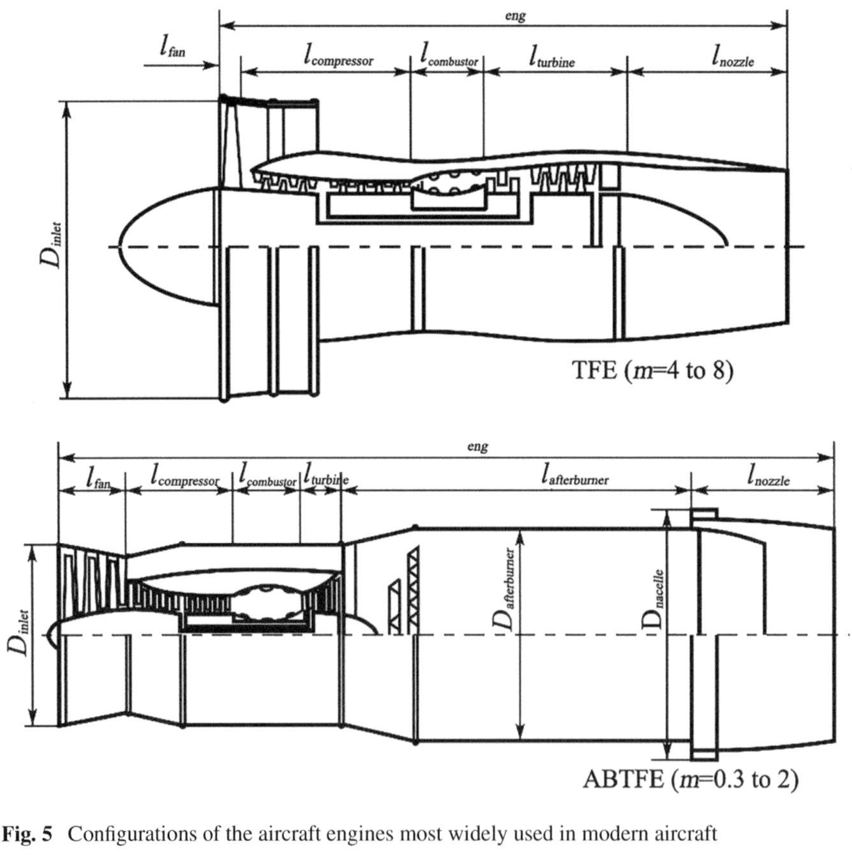

Fig. 5 Configurations of the aircraft engines most widely used in modern aircraft

where l_{fan} is the length of fan (LP compressor); $l_{compressor}$ is the length of compressor (intermediate and high pressure); $l_{combustor}$ is the length of combustor; $l_{turbine}$ is the length of turbine (total); $l_{afterburner}$ is the length of afterburner; l_{nozzle} is the length of nozzle.

For non-afterburning engines of the TFE type, the engine length depends on the maximum take-off thrust, turbine inlet temperature and the engine bypass ratio and may be determined by the following formula:

$$L_{eng} = k_l k_{nozzle} \frac{\sqrt{P_0}}{T^*_{gas}\left(10 + \sqrt{m}\right)},$$

where $k_l = (5.58 \text{ to } 5.62)10^2$ is a statistical factor linking the gas-dynamic parameters with the engine length; for a nozzle without thrust reverser, $k_{nozzle} = 0.95$ to 1.05; for a nozzle with thrust reverser, $k_{nozzle} = 1.15$ to 1.25.

For engines of the ABTFE type, the formula is written as:

$$L_{eng} = k_l k_{aug} k_{nozzle} \frac{P_0^{2/3}}{T^*_{gas}(10 + m)},$$

where $k_l = (1.4 \text{ to } 1.5)10^2$ is a statistical factor linking the gas-dynamic parameters with the engine length; $k_{aug} = P_{0AB}/P_0$ is the engine augmentation ratio; at $T^*_{gas} = 1200$ to $1400\,K$, $k_{AB} = 1.4$ to 1.6; at $T^*_{gas} = 1600$ to $1800\,K$, $k_{aug} = 1.6$ to 1.8; k_{nozzle} is a statistical factor taking account of the nozzle type:

for a variable geometry nozzle, $k_{nozzle} = 0.95$ to 1.05; for a variable geometry nozzle with a thrust reverser, $k_{nozzle} = 1.05$ to 1.15; for a vectored thrust nozzle $k_{nozzle} = 1.15$ to 1.25.

Estimating an aircraft's center of gravity position requires the coordinates of its engine's centre of gravity. The position of CG (measured from the inlet) may be taken to be:

$$x_{CG} = k_{CG} L_{eng},$$

where k_{CG} is a statistical factor; for a THE, $k_{CG} = 0.4$ to 0.5; for AFTJ, TFE, ABTFE, $k_{CG} = 0.3$ to 0.4.

The maximum diameter of TJE and TFE engines D_{inlet} is located at the compressor (fan) inlet, the maximum diameter of AFTJ and ABTFE is the afterburner diameter $D_{afterburner}$. D_{inlet} and $D_{afterburner}$ define not only the diameter of the engine nacelle, but also the diameter of the air duct inside the fuselage, if the engine is located in the fuselage (Fig. 6).

For engines of TJE, AFTJ, TFE, and ABTFE type, the expression determining the compressor (fan) inlet diameter is written as:

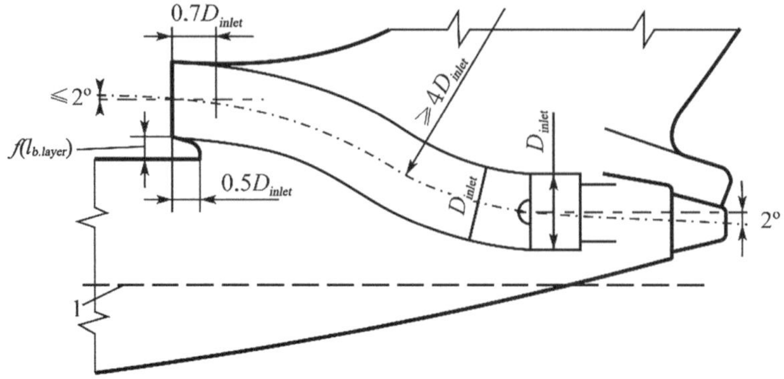

Fig. 6 Example of an engine (TFE) mounted inside the fuselage afterbody (D_{inlet} is the inlet diameter; $f\left(l_{\text{b.layer}}\right)$ is the length of the surface on which the boundary layer is formed): 1—passenger cabin floor

$$D_{\text{inlet}} = k_{\text{inlet}} \sqrt{\frac{P_0}{T^*_{\text{gas}}\left(1 + 0.05m - \sqrt{0.15m}\right)}},$$

where $k_{\text{inlet}} = 0.375$ to 0.395 is a statistical factor taking into account the effect on D_{inlet} of parameters P^*_H, T^*_H, σ_{inlet} etc.; P_0 is the take-off thrust of engine (without afterburning); T^*_{gas} is the turbine inlet temperature.

At the preliminary phase of aircraft development, D_{inlet} may be determined by the chart drawn in accordance with the given dependence (Fig. 7).

For AFTJ and ABTFE engines, the diameter of the afterburner is

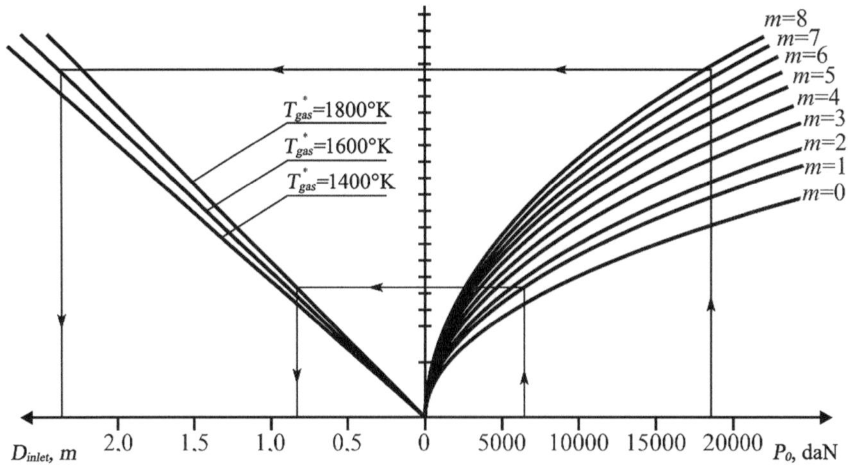

Fig. 7 A chart to determine an ABTFE D_{inlet} by take-off thrust (without afterburning)

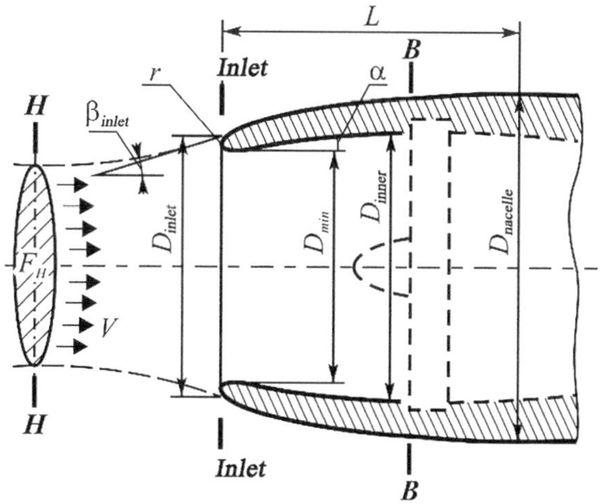

Fig. 8 Configuration of a subsonic air intake: D_{inlet} is the diameter of air intake inlet; $D_{\text{inner}} = D_{\text{eng}}$ is the diameter of compressor; D_{nacelle} is the maximum diameter of nacelle; L is the distance from the nacelle nose to the cylindrical part

$$D_{\text{afterburner}} = k_{\text{afterburner}}\sqrt{P_{0\text{AB}}\left(\frac{1}{P_{\text{sp}_{0\text{AB}}}} + \frac{c_{P_{0\text{AB}}}}{3600}\right)},$$

where $k_{\text{afterburner}} = 0.103$ to 0.107 is a statistical factor; $P_{0\text{AB}}$ is the take-off thrust at full afterburning; $P_{\text{sp}_{0\text{AB}}}$ is the take-off specific thrust; $c_{P_{0\text{AB}}}$ is the take-off specific fuel consumption.

3 Air Intakes of Modern GTEs

The functions of the air intake within a modern aircraft's power plant system include ensuring the following:

- steady operation of the engine under all flight conditions that are mandatory for the aircraft in development;
- the compression of air entering the air intake is accomplished by converting the kinetic energy of ram air into pressure.

If the increase in air pressure in the engine duct occurs primarily in the GTE compressor (approximately five times that of the air intake) at subsonic flight speeds, then as flight speed increases, the compressor functions gradually pass to the air intake, at $M \approx 2$ the air intake and compressor compress the flow equally. At high supersonic speeds ($M > 3$), the compressor role becomes negligent, while the compression degree in the inlet device reaches almost 40:1, i.e. an air-breathing engine without a compressor and a turbine becomes a better proposition, i.e. a ramjet engine.

When the air flow gets compressed in the air intake, there will always be pressure losses due to friction, vortex (flow separation with the non-uniform speed envelope), heat exchange; and during the deceleration of the supersonic flow there are wave losses due to the emergence of shock waves.

Pressure losses during compression in the air intake are evaluated by the total pressure recovery coefficient:

$$\sigma_{inlet} = \frac{p^*_{inlet}}{p^*_{ram\ air}},$$

where p^*_{inlet} is the total pressure at compressor inlet; $p^*_{ram\ air}$ is the total pressure of the ram air.

For the air intake of a modern aircraft to perform its functions effectively, it should:

- ensure the possibly high values of the total pressure recovery coefficient σ_{inlet};
- create a sufficiently uniform speed field at the compressor inlet;
- maintain steady operation (without pronounced stalls and pressure pulsations) in all operation modes;
- have the lowest possible external drag.

3.1 Subsonic Air Intakes

The experience of designing and operating subsonic air intakes accrued so far makes it possible to obtain in them rather high values of the total pressure recovery coefficient $\sigma_{inlet} = 0.97$ to 0.98.

During the design of subsonic air intakes, their parameters are selected for the main flight conditions.

The dimensions of the inlet of the diffuser are determined by the air mass flow through the inlet area. According to the conservation-of-mass theorem, the air flow rate in sections $H - H$ and Inlet – Inlet (Fig. 8) will be equal:

$$m_{air} = F_H V \rho_H = F_{inlet} V_{inlet} \rho_{inlet},$$

where V is the flight speed at altitude H.

The inlet area may be written as:

$$F_{inlet} = \frac{m_{air}}{V_{inlet} \rho_{inlet}},$$

where m_{air} is the air mass flow through the engine, for which the air intake is developed; V_{inlet} is the air speed at the air intake inlet; ρ_{inlet} is the air density at the inlet.

V_{inlet} may be written to first approximation as $V_{inlet} = V\overline{V}_{inlet}$, where $\overline{V}_{inlet} = 0.3$ to 0.7 is the relative air speed at the air intake inlet. Lower values of $\overline{V}_{inlet}$ are taken for long and bent channels (in order to keep hydraulic losses low), large values $\overline{V}_{inlet}$ are used for short channels and GTE with large speeds at compressor inlet.

An increase in the air density ρ_{inlet} during deceleration from V to V_{inlet} is determined by special gas-dynamic tables.

For preliminary calculations, the dimensions of inlet may be determined through the relative diameter of inlet

$$\overline{D}_{inlet} = 1.1\sqrt{\frac{1}{\left(1-\overline{V}_{inlet}^{2}\right)/\left(1-M^{2}\right)}+1}.$$

Here, $\overline{D}_{inlet} = D_{inlet}/D_{eng\ nacelle}$; D_{inlet} is the diameter of air intake inlet; $D_{eng\ nacelle}$ is the maximum outer diameter of engine nacelle, with $D_{eng\ nacelle} \approx (1.1 \text{ to } 1.2)D_{inner}$,

where D_{inner} is the maximum inner diameter of nacelle; M is the Mach number. It is possible to assume $D_{inner} \approx D_{eng}$ (D_{inlet} is engine diameter by compressor).

Inlet lip profiling is undertaken in order to obtain a smooth streamlining of the nacelle and to avoid flow separation at inlet.

The angle of inflow of the stream outer boundary to the inlet of the air intake in relation to its axis β_{inlet} (degree) to first approximation may be determined through the relative inlet velocity:

$$\beta_{inlet} = 22\sqrt{1/\overline{V}_{inlet} - 1}.$$

The radius of curvature of the inlet lip (cowl lip) of the air intake ensuring smooth streamlining may be taken to be

$$r_{min} = (0.04 \text{ to } 0.05)\sqrt{F_{inlet}}.$$

The air duct profiling is provided for to obtain the largest value of σ_{inlet} and a uniform speed field at the inlet of compressor (section $B-B$), attainable if the flow is attached in the duct.

The attached flow may be ensured by choosing a profile such that the pressure ratio along the length of the duct has a constant value. In that case, the half-angle of duct width is $\alpha \leq 4$ to 5°.

If the duct has turns and bends, then the variation of the area of duct cross sections should be smooth and in the last portion (before the engine) the duct axis should coincide with the compressor axis. This cylindrical part of the duct is made to have length at least $(0.5 \text{ to } 1.0)D_{inlet}$.

The profiling of air intake outer lines should ensure the minimum drag for it. Therefore, the outer lines of the air intake are profiled independently of internal lines.

The relative outer length of the inlet, feasible under the condition of minimizing the drag of the engine nacelle, may be expressed as a function of the Mach number:

$$\overline{L} = L/D_{inlet} \approx 1.5M^{2},$$

where L is the distance measured from the nacelle nose to its cylindrical part. It should be noted that the dependencies given above are approximate. Theoretically, it is highly difficult to account for all circumstances of the real flow; therefore, the ultimate recommendations in regard to air intake profiling are established experimentally.

3.2 Supersonic Air Intakes

When the supersonic flow passes around three-dimensional bodies, *shock waves are known to arise.* When the air flow passes through the shock wave, the flow parameters (speed, pressure, density, and temperature) sharply change with a varying intensity depending on the shock inclination angle. The supersonic air intakes utilize that phenomenon.

The increases in pressure and temperature in the shock waves that form at the inlet of the air intake are part of the compression cycle in the power plant of a supersonic aircraft. The losses resulting from the air compression are the sum of wave losses (in the shock wave system), vortex losses, and losses due to friction. However, by far the largest are the wave losses:

$$\sigma_{\text{inlet}} = (0.9\ \text{to}\ 0.95)\sigma_{\text{shock}},$$

where $\sigma_{\text{shock}} = \sigma_1\sigma_2\cdots\sigma_n = \prod_{i=1}^{n}\sigma_i$ is the total pressure recovery coefficient in the shock system; σ_i is the total pressure recovery coefficient in one shock.

Depending on the shape of the inlet cross section, the supersonic air intakes may be broken down into two types: two-dimensional (flat) and three-dimensional (round, semiround, etc.)

Depending on the location of oblique shocks, the air intakes are broken down into external, internal, and mixed compression; all three types may be two-dimensional or three-dimensional (Fig. 9).

External compression air intakes have slightly higher aerodynamic drag and lower efficiency than internal compression air intakes; for example, at a flight speed corresponding to $M = 2, 2$, it is possible to obtain:

Compression	External	Internal	Mixed
σ_{inlet}	0.91	0.95	0.93

However, the external compression air intakes are quite simple to regulate, do not require a complex startup system, have weight advantages; therefore, they have found wide application in supersonic aviation.

Internal compression air intakes require a highly complex regulation system, are hard to start up (design mode is hard to achieve), etc. Therefore, despite higher values of coefficient σ_{inlet}, they have yet to find practical application.

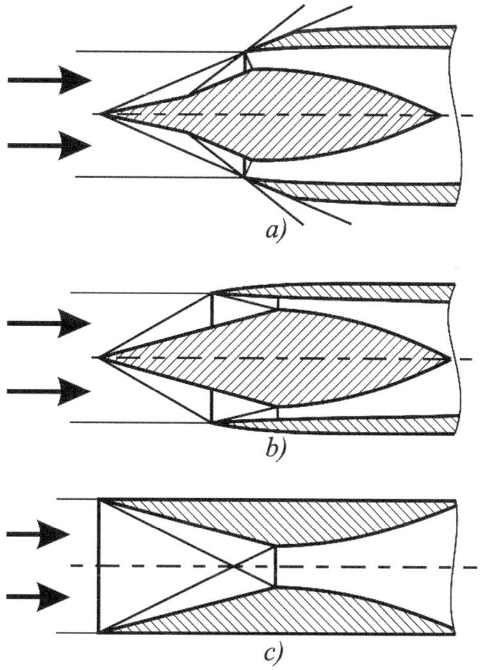

Fig. 9 Methods of generation of shock waves. **a**—external compression air intake (all oblique shocks are located externally); **b**—mixed compression air intake (oblique shocks are located externally and internally to the air intake); internal compression air intake (all oblique shocks are located internally)

Mixed-compression air intakes are also very sensitive to the variation of operation mode and various kinds of perturbations, which may be generated, for example, by a significant variation of the angle of attack during the aircraft's manoeuvre or by the activation (deactivation) of the engine afterburner, etc. This type of air intake is best suited for single-mode supersonic aircraft.

The parameters and dimensions of a supersonic air intake are selected with the goal of ensuring the uninterrupted operation of the engine throughout the flight speed envelope, for which a particular aircraft is designed.

Inlet area. The capacity of the air intake is evaluated by the mass flow coefficient φ, which is a ratio of the actual air mass flow to the maximum possible mass flow, i.e.

$$\varphi = m_{\text{air}}/m_{\text{air max}}.$$

The mass flow coefficient φ is numerically equal to the ratio of the air jet area in the free stream (Fig. 10, section $H - H$) to the air intake inlet area:

$$\varphi = F_H/F_{\text{inlet}},$$

with the inlet area of the supersonic air intake considered to be the complete area of section Inlet - Inlet. The area of the inlet slot *per se*:

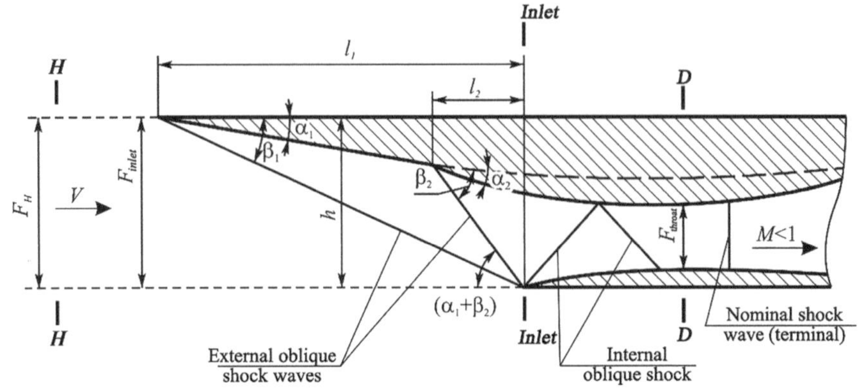

Fig. 10 Layout of a supersonic mixed compression air intake (design air flow regime)

$$F_{\text{slot}} = F_{\text{inlet}} - F_{\text{body}},$$

where F_{body} is the area of the section of the body generating the shock wave system.

To create the shock wave system, the air intakes may use both two-dimensional bodies (wedge) for two-dimensional air intakes and round bodies (cone, half-cone, quarter-cone) for three-dimensional air intakes of appropriate shape. However, the geometric parameters of all air intakes are similar, so the longitudinal sections, for example, of half-round and two-dimensional air intakes will be practically similar (see Fig. 10), and the longitudinal sections of the round air intake will differ only by the symmetry of the lower and upper parts of the section.

At the design operation mode of the air intake, the external oblique shock waves are focused on the cowl front lip, i.e.

$$\varphi = m_{\text{air}}/m_{\text{air max}} = \frac{F_H V \rho_H}{F_{\text{inlet}} V \rho_H} = 1,$$

hence

$$F_{\text{inlet}} = m_{\text{air}}/V\rho_H,$$

where m_{air} is the air flow rate; V is the aircraft flight speed; ρ_H is the air density at the flight altitude. The total air flow through the air intake:

$$m_{\text{air }\Sigma} = m_{\text{air}} + m_{\text{b.layer air}} + m_{\text{air bypass}}$$

where m_{air} is the air flow through engine; $m_{\text{b.layer air}}$ is the mass of the boundary layer air bled from compression surfaces (wedge, cone, etc.); $m_{\text{air bypass}}$ is the mass of the air bypassed from the diffuser to the atmosphere through the bypass (anti-surge) doors. However, with an accuracy sufficient for the detailed design, it is possible to take $m_{\text{air}\,\Sigma} \approx m_{\text{air}}$. The air mass flow through the engine can be estimated as follows:

$$m_{\text{air}} = m_{\text{red. air flow}} \frac{p^*_{\text{inlet}}}{p_0} \sqrt{\frac{T_0}{T^*_H}},$$

where $m_{\text{red. air flow}}$ is the reduced air flow; $p^*_{\text{inlet}} = p^*_H \sigma_{\text{inlet}}$; p_0, T_0 are the pressure and temperature of the ambient air at $H = 0$, $V = 0$.

Let us express the stagnation pressure and temperature through the flight Mach number:

$$p^*_H = p_H \left(1 + 0.2M^2\right)^{3.5}; \; T^*_H = T_H \left(1 + 0.2M^2\right),$$

where p^*_H, T^*_H are the stagnation pressure and temperature of the ram air at the flight altitude; p_H, T_H are the pressure (static) and temperature of the ambient air at the flight altitude. We will obtain

$$m_{\text{air}} = m_{\text{red. air flow}} \frac{\sigma_{\text{inlet}} p_H \left(1 + 0.2M^2\right)^3}{1.033} \sqrt{\frac{288}{T_H}}.$$

The expression for the determination of the inlet area of a supersonic air intake will be written as

$$F_{\text{inlet}} = m_{\text{red. air flow}} \frac{\sigma_{\text{inlet}} p_H \left(1 + 0.2M^2\right)^3}{10.14 V \rho_H} \sqrt{\frac{288}{T_H}},$$

where F_{inlet} is in m^2; $m_{\text{red. air flow}}$ is in kg/s (the reduced air mass flow is given in the appendix); p_H is in daN/cm^2; V is in m/s; T_H is in K.

The number of compression surfaces on the wedge (cone), i.e., the number of oblique shock waves in the air intake, should ensure that the shock wave system achieves the maximum total pressure recovery coefficient (value is greater with a greater number of oblique shock waves; see Fig. 11).

Angles of the step wedge (cone) $\alpha_1, \alpha_2, \alpha_3$ etc. (see Fig. 10) are selected such that in the design operation mode of the air intake, the external oblique shock waves are focused on the cowl's front lips. Only shock waves of equal intensity produce the highest value of σ_{shock}, which is calculated as a ratio of stream speed before the shock wave to stream speed after the shock wave. Therefore, angles $\alpha_1, \alpha_2, \alpha_3$, etc. should ensure equality:

$$V/V_1 = V_1/V_2 = V_2/V_3 = \ldots,$$

where V is the free stream speed (flight speed); V_1 is the free stream speed behind the first oblique shock wave; V_2 is the free stream speed behind the second oblique shock wave, etc.

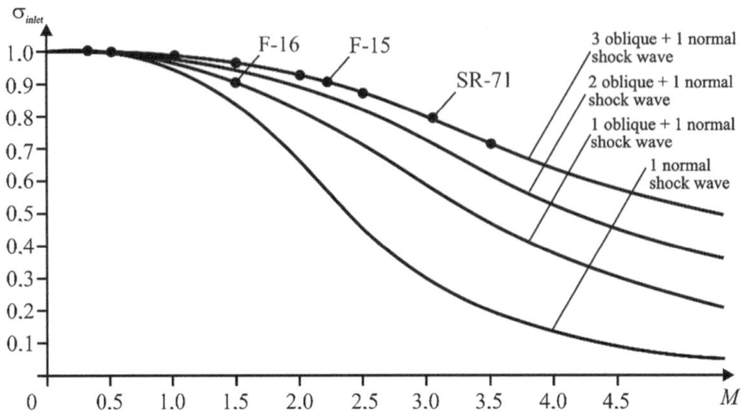

Fig. 11 The effect of flight speed on σ_{inlet}

For the total pressure recovery coefficient in the shock wave system to have a greater value, the stream speed before the normal (terminal) shock wave should not exceed $M = 1.2$ to 1.25.

The stream speed behind i oblique shock wave V_i is linked to the speed of the stream before the shock wave by the following dependence:

$$V_i = V_{i-1} \frac{\cos \beta_i}{\cos(\beta_i - \alpha_i)},$$

where V_{i-1} is the stream speed before i oblique shock wave.

For the first oblique shock wave, we will have accordingly

$$V_i = V \frac{\cos \beta_1}{\cos(\beta_1 - \alpha_1)},$$

where V is the flight speed.

The Mach number of the stream behind i oblique shock wave is determined as follows:

$$M_i^2 = \frac{5 + M_{i-1}^2}{7M_{i-1}^2 \sin^2 \beta_{i-1}} + \frac{5M_{i-1}^2 \cos^2 \beta_i}{5 + M_{i-1}^2 \sin^2 \beta_i},$$

where M_{i-1} is the Mach number of the stream before i oblique shock wave (for the first oblique shock wave—the flight Mach number).

Knowing the number of oblique shock waves, it is possible to determine from the equations the required values of step wedge (cone) angles, ensuring equal intensity in the shock waves.

Special attention should be paid to the inclination angle of the first compression surface α_1 because it actually determines the cone (wedge) projection, which is the distance from the wedge apex to the inlet plane.

For modern supersonic air intakes (depending on the design Mach number), angle α_1 assumes the following values:

M	<2.5	2.5… 3.5
Wedge	~9°	~7°
Cone	~15°	~11°

The length of the cone (wedge) steps is easily determined if angles α and β and dimension *h* are known.

The distance along the axis from the inlet plane to the cone (wedge) apex:

$$l_1 = h/\mathrm{tg}\beta_1.$$

We will find the distance from the inlet plane to the beginning of the second step (in Fig. 10—dimension l_2) by drawing from the focus point of the second shock wave a line at an angle $(\alpha_1 + \beta_2)$ to the oncoming air flow. Dimensions l_3, l_4, etc. are determined similarly.

The air intake throat area F_{throat} (in section $D - D$) will decrease as the flight speed increases. Physically, it is quite obvious: with an increase in the Mach number the air pressure ratio in the shock wave system rises and consequently the air pressure and density in the throat increase prompting the need to reduce its area (otherwise the air in the throat will expand and π_{inlet} will decrease).

Usually, the required throat area ratio is calculated as follows:

$$\overline{F}_{\mathrm{throat}} = F_{\mathrm{throat}}/F_{\mathrm{inlet}}.$$

The value of $\overline{F}_{\mathrm{throat}}$ as a function of the Mach number to first approximation may be taken to be:

M	1.5	2.0	2.5	3.0	3.5
$\overline{F}_{\mathrm{throat}}$	0.5	0.42	0.35	0.32	0.3

The supersonic air intake parameters arrived at as a result of detailed design are obligatorily checked and adjusted in the course of experimental tests.

3.3 Air Intake and Engine Compatibility

An important task in matching the airframe and power plant characteristics is to ensure air intake and engine compatibility. Compatibility in this case means steady operation of the engine with preset characteristics in sustained and transitional modes in the presence of various destabilizing factors.

Variation of engine speed, angles of attack and sideslip, afterburner actuation (deactivation), etc. cause a variation of the capacity of the air intake and engine and also disrupt the uniformity of the air flow speed field at the engine inlet, which, naturally, disrupts its stable operation. An appreciable distortion of the air flow speed field, for example, may cause vibration and damage to compressor blades. When the air intake and engine capacity are mismatched, the power plant operates in an unsteady pulsating manner (surge, "buzz" of the air intake). Also, the total pressure recovery coefficient σ_{inlet} drops appreciably, i.e., the engine thrust lessens, and the specific fuel consumption rises.

A characteristic feature of the operation of subsonic air intakes is the automatic coordination of the air flow passing through the engine and the inlet. For this reason, the subsonic air intakes do not need the air flow regulation system. Occasionally, for better uniformity of the speed field at maximum thrust ratings, additional inlet doors are installed on the peripheral circumference of the air intake cowl (Boeing 747, Tupolev Tu-154, etc.). An operation diagram of inlet doors is shown below (see Fig. 26).

Air intakes of aircraft designed for high supersonic flight speeds should have a special regulation system (despite the higher complexity, mass, and cost of the air intake) in order to avoid the harmful influence of the above destabilizing factors on the operation of the power plant and to preserve the high value of σ_{inlet} for a large speed range.

The task of the regulation system to ensure the coordinated operation of air intake and engine consists of keeping the shock wave system (especially the terminal normal shock wave behind the throat) in a preset position at supersonic speeds. This may be done by varying the throat area and bypassing excess air into the atmosphere. The air is bypassed into the atmosphere by opening special doors installed on the surface of the air intake duct (behind the throat). These doors are referred to as "anti-surge" or "bypass" doors. Under supersonic cruise flight conditions, the air bypass doors are continuously open, and part of the air is vented into the atmosphere, precluding the occurrence of a "surge" of the air intake.

At take-off and small subsonic speeds the required throat area turns out to be higher than the $F_{\text{throat max}}$, which is determined by structural capabilities (this is explained by the relatively small air density in throat). Therefore, despite the fully open throat, the engine is short of air for normal operation. To avoid disrupting engine operation, auxiliary (take-off) doors are opened at take-off and subsonic speeds to allow more air into the engine, bypassing the throat. To vary the throat area, the structure may have a provision for moving the cone fore and aft (round air intakes) or moving moveable ramps (two-dimensional air intakes). Additional holes open in the duct behind the throat for additional air suction or bypassing (additional and bypass doors in supersonic air intakes).

An automatic system regulates the operation of the air intake.

In supersonic, highly manoeuvrable aircraft (which are characterized by a considerable variation of angles of attack), the forward part of the air intake may turn by

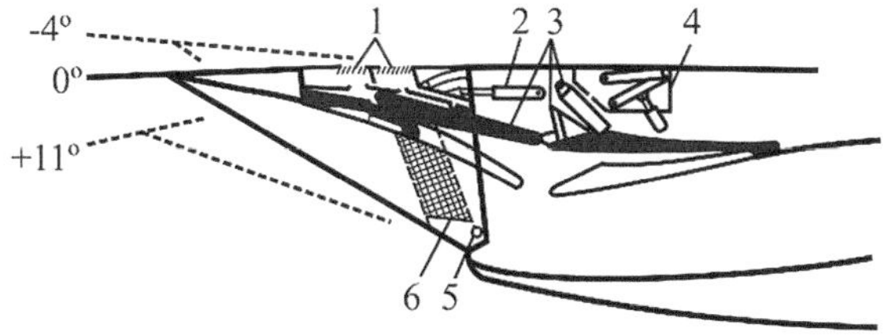

Fig. 12 Air intake regulation diagram of the McDonnell Douglas F-15 fighter. 1—boundary layer bleed holes; 2—the power cylinder of the air intake forward part deflection; 3—pivotally-connected ramps regulated with the aid of one hydraulic drive; 4—air bypass doors; 5—the hinge axis of the air intake forward part; 6—the air suction holes from the side surfaces of the air intake duct

13° to 15° as one unit (Fig. 12). Such air intake enables an effective variation of the area of the section of captured air flow..

3.4 Arrangement of Air Intakes on an Aircraft

Selecting the location of air intake installation on the aircraft plays an important role in dealing with the issues of compatibility between the air intake and the engine (eventually matching airframe and power plant characteristics), because the characteristics of an isolated inlet may vary appreciably once it is installed in the aircraft.

The air intake configuration is largely determined by the aircraft's general aerodynamic configuration and flight characteristics (manoeuvrable or non-manoeuvrable aircraft, operating envelope of flight altitudes and speeds, limit angles of attack and sideslip, ground or concrete runway, and so on). If the air intake arrangement is not right, the interaction between the wing and the fuselage may become one of the main causes of distortion of the flow field, especially when the angles of attack and sideslip vary.

Another source of speed field nonuniformity in the air intake is friction (due to air viscosity). The presence of friction is known to cause the boundary layer to appear on the streamlined surface, whose speed sharply falls from the free-stream speed to zero.

The boundary layer interacts with shock waves in the supersonic flow pattern, distorting their regularity: there are local detachments of the flow from the walls; the boundary layer passes through shocks, increasing its thickness further; in places where the boundary layer thickens, new weak oblique shock waves (shock waves) form, and so on.

The deviation from the design flow pattern, caused by air viscosity, eventually leads to the non-uniform velocity field and the reduction of σ_{inlet}. Therefore, the air intakes of all modern aircraft should have a system of boundary layer diversion (bleed). Both the boundary layer forming on the fuselage (or wing) surface and the boundary layer forming on compression surfaces—the cone (wedge) and internal surface of the cowl (Fig. 13)—are removed.

The thickness of the boundary layer depends on the flow speed, the air viscosity coefficient and the length of the zone of the flow contact with the wetted surface. When arranging the air intake to remove the boundary layer, the height of bleed slots (h_1,h_2...) to first approximation may be taken by ratio $h \approx 0,01l$, where l is the length of the surface on which the boundary layer forms. If the air intake snugs tightly the fuselage surface (i.e. $h_1 = 0$), then at $M = 2$, for example, the total pressure recovery coefficient will decrease by 25 to 30%, which will eventually lead to the reduction of the engine thrust by approximately 45% and to an increase in specific fuel consumption by 15%.

When selecting a location for the installation of the air intake on the aircraft, it is also necessary to take into account the positive influence of aircraft surfaces on the performance of the inlet section. Under certain flight conditions, the wing, and the fuselage may be used to significantly attenuate the flow and to mitigate the nonconformity of flow speed field before the air intake. The air intake at a sufficient distance from the front lip finds itself in the flow, whose downwash angle is much less than the aircraft's angle of attack due to the effect of wing (fuselage), which leads to the smoothing of the flow before the air intake. This effect is of great importance for highly manoeuvrable aircraft, capable of angles of attack in excess of 40°. Figure 14 shows for this air intake arrangement the effect of the wing on the downwash angle at subsonic speeds (a similar effect also happens at supersonic speeds).

At supersonic speeds, the wing (fuselage) decelerates the flow acting as the first stage of supersonic flow compression. As a result of such pre-compression, the total pressure recovery coefficient σ_{inlet} increases in the air intake. This effect is especially significant at $M > 1.2$ which allows the successful use of a fixed geometry air intake with subsonic geometry at $M = 1.5$ to 1.7 (Fig. 15).

Depending on their location on the aircraft, the air intakes may be broken down into the following basic types: nose (mainly round), side and underwing (or under fuselage).

The nose air intakes are arranged either in the nose portion of the fuselage on light aircraft, or in the nose portion of an engine nacelle mounted on the wing of heavy aircraft.

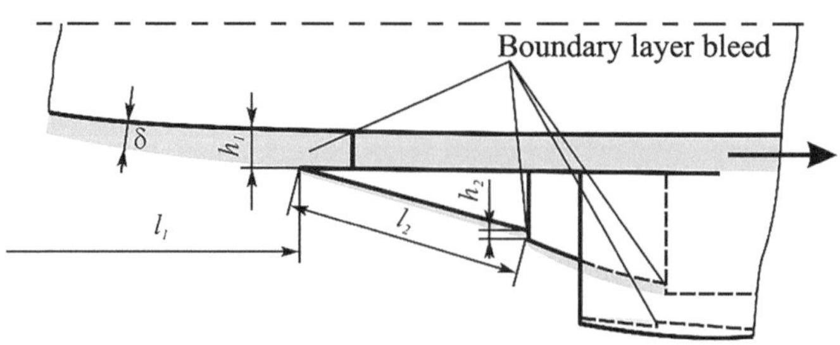

Fig. 13 Boundary layer diversion diagram

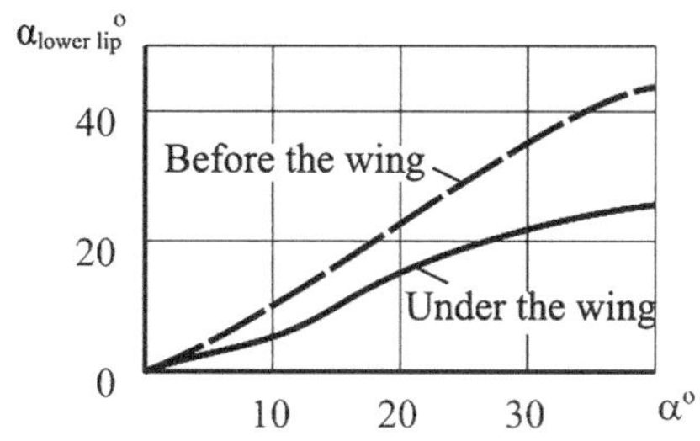

Fig. 14 The angle of attack of the air intake lower lip $\alpha_{lower\ lip}$ as a function of the aircraft's angle of attack $\alpha(M = 0.8)$

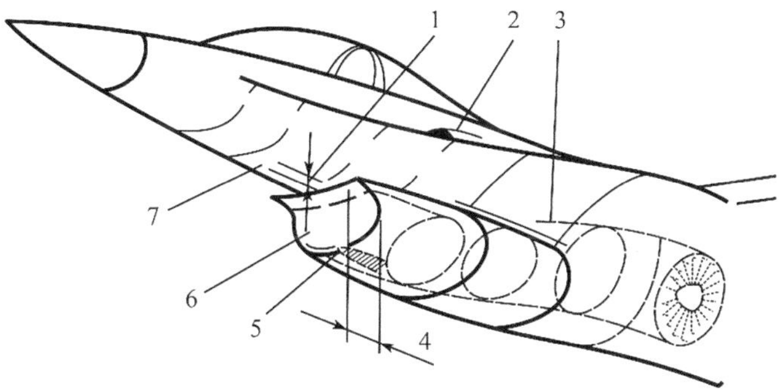

Fig. 15 Basic features of the arrangement of F-16 fixed geometry air intake (the air intake is made as a separate module and may be replaced with another air intake). 1—gap between the fuselage and air intake to bleed the boundary layer; 2—the gun muzzle is positioned away from the air intake face; 3—gradual increase in the areas of air ducts (ensures a nearly linear variation of M); 4—protruding plate to prevent the boundary layer from entering the air intake; 5—cowl blunt lip (a separate component that can be replaced if the air intake performance needs to be improved); 6—normal shock wave position; 7—bottom surface of the fuselage acting as the pre-compression surface

The main advantage of the nose air intakes is their high uniformity of speed zero when they encounter the free stream, and in a supersonic cruise flight under design conditions, they also strictly maintain the preset position of the shock wave system.

However, the nose air intakes are not without shortcomings. Whereas in a heavy, non-manoeuvring aircraft, the angle of attack practically does not change during the cruise flight, and thus the shock system at the inlet of the air intake maintains the preset position, in a light aircraft, when the angle of attack rises appreciably during a high-loading manoeuvre, the focusing of shock waves is disrupted. It leads to the non-uniformity of the speed field and the reduction of the total pressure recovery coefficient. For this reason, the nose air intakes perform less effectively at large angles of attack (Fig. 16).

The second shortcoming: with the air intake located in the forward portion of the fuselage, large inner volumes in the fuselage have to be set aside for the air duct (the air and propulsion trains cut through almost the entire fuselage from nose to tail),

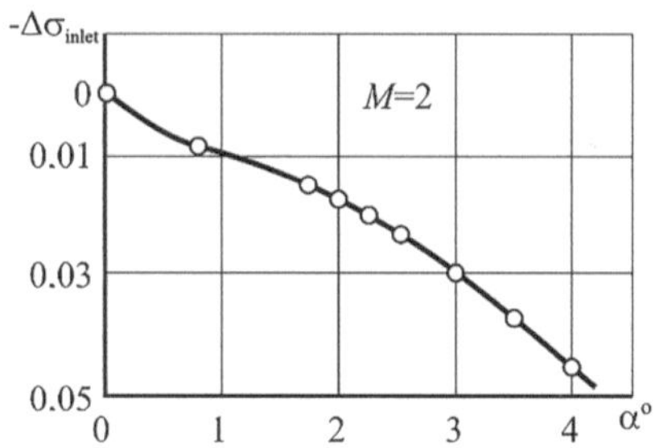

Fig. 16 The effect of angle of attack on total pressure recovery coefficient in axisymmetric nose air intake

which naturally complicates the aircraft layout. In addition, the nose air intake makes it impossible to place a large-diameter radar antenna in the nose of the fuselage (the antenna is constrained by the dimensions of the air intake cone).

The side air intakes have a wide variety of inlet shapes: round, semi round, rectangular, etc. Subsonic aircraft are mainly fitted with round air intakes.

The presence of a boundary layer on the fuselage surface, as noted above, calls for creating bleed slots during the arrangement of air intakes on the aircraft, and if the engines are located in the fuselage afterbody, the entire engine nacelle is moved aft and mounted on a pylon to the fuselage (Fig. 17a).

It should be noted the pylon in this case makes it possible only to organize the boundary layer bleed without, however, ensuring the conditions for the operation of the nose air intake.

a) *b)*

Fig. 17 An example of a subsonic (**a**) and supersonic (**b**) air intake arrangement that prevents the boundary layer from entering it

Supersonic aircraft mainly use side air intakes of a rectangular or similar shape (Fig. 17b) (admittedly, other inlet shapes are also encountered).

Placing the air intakes on either side of the fuselage significantly reduced the length of the air duct, as well as completely cleared the fuselage nose to accommodate the radar station. Given a well-organized boundary layer bleed, the side air intakes perform quite efficiently.

The underwing (under fuselage) air intakes on existing aircraft have a two-dimensional or oval inlet (see Fig. 15). In the heavy supersonic aircraft, the air intakes, as a rule, constitute the forward portion of the nacelle of underwing engines. Figure 18 illustrates the longitudinal section of the Concorde nacelle.

v

The shortcoming of the underwing air intakes is their poor performance at large negative angles of attack.

4 Exhaust System

4.1 Mitigation of Base Drag

The aft part of the fuselage (or of the engine nacelle) should be profiled such that the drag at subsonic speed is minimal while meeting the supersonic flight requirements.

If the integration of the nozzle with the fuselage afterbody is not made properly (especially when several closely located engines are installed), the large base area will cause an appreciable base drag to appear. Under certain flight conditions (with the maximum convergence of the nozzle), this drag may reach 30% of the aircraft's total drag, which leads to a dramatic degradation of flight performance.

Base drag is primarily determined by the shape of the aircraft afterbody's outer lines. The smoother the tail surface lines, the more uniform the pressure field on it, and the external flow will remain attached to the streamlined surface, i.e., the base drag will be small.

Figure 19 illustrates the effect of the fineness ratio of the fuselage (or engine nacelle) afterbody on the distribution of the pressure coefficient, written as follows:

$$c_{f.c}=\frac{p-p_H}{\rho_H V^2/2},$$

where p is the pressure on the tail surface; p_H is the static pressure at the flight altitude; $\rho_H V^2/2$ is the dynamic pressure at the flight altitude.

Coordinates of the generating line of a well streamlined afterbody surface may be determined using the following dependence:

$$r=0.5\left[D_{\text{mid-section}}-(D_{\text{mid-section}}-D_{\text{tail section}})x^2/l_{\text{tail}}^2\right],$$

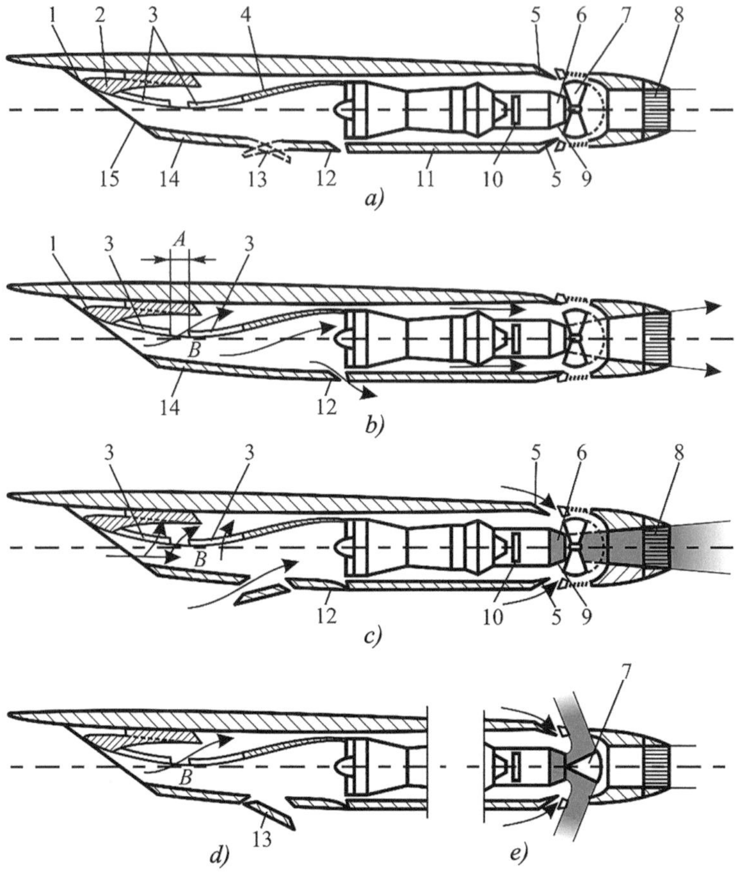

Fig. 18 Layout of the engine nacelle of a heavy supersonic aircraft: **a**—general arrangement; **b**—cruising flight; **c**—take-off; **d**—engine shutdown; **e**—thrust reverser; 1—a slot to bleed the boundary layer from the wing surface; 2—a non-movable wedge plane; 3—movable ramps; 4—a subsonic air intake duct; and 5—additional air scoops. 6—primary nozzle; 7—thrust reverser; 8—variable geometry secondary nozzle; 9—suppressor; 10—afterburner nozzles; 11—web under the engine; 12—bypass (antisurge) door; 13—auxiliary (take-off) door; 14—cowl; 15—vertical partition separating the scoops of two engines; **A**—slot to bleed the boundary layer from the wedge plane; **B**—boundary layer air

where r is the radius of the afterbody current section; $D_{\text{mid-section}}$ is the diameter of the afterbody mid-section; $D_{\text{tail section}}$ is the diameter of the tail part; x is the distance from the current section to the mid-section (see Fig. 19); l_{tail} is the length of the afterbody. At $M < 0.7$, the gas jet emanating from the jet engine nozzle has almost no effect on the distribution of pressure on the external surface of the afterbody of the fuselage (or engine nacelle). At $M \approx 0.8$ and a small length of the profiled afterbody, a flow separation occurs, in whose zone a significant expansion is observed. The base drag in this case increases appreciably.

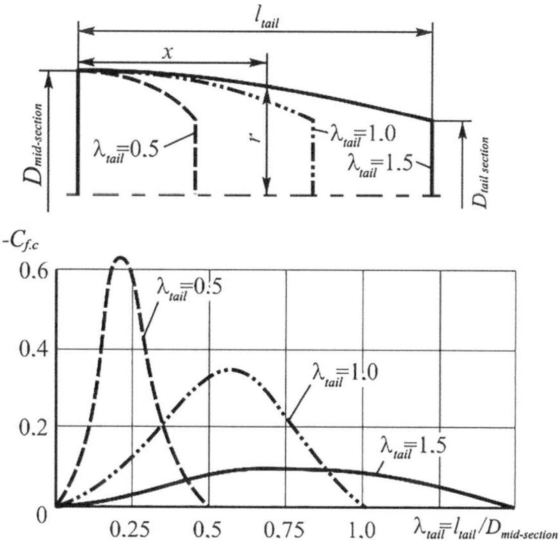

Fig. 19 The pressure coefficient distribution along solids of revolution at $M = 0.9$

The fixed-geometry nozzles of the engines of subsonic aircraft do not cause great problems when the base drag is mitigated. However, modern supersonic aircraft need more effective exhaust systems. These aircraft have exhaust nozzles with variable geometry that are consistent with the cycles of operation of engines with an after-burner and a high compression ratio, which were developed for fast acceleration and flight at high Mach numbers.

For the jet aircraft nozzle to have a small base drag at its dry power rating, it should have secondary doors that are as long as possible (Fig. 20). The external stream flowing around the fuselage (nacelle) afterbody, at a small angle of slope of the secondary doors (10 to 15°), remains attached to them, which also reduces the base drag. In a subsonic flight, however, almost all types of supersonic variable geometry nozzles have a higher drag than a variable geometry convergent nozzle designed specifically for a subsonic flight. Figure 21 illustrates the additional drag of possible types of nozzles on modern supersonic aircraft, written as follows:

$$\frac{\Delta D}{P_{ideal}} = \frac{D}{P_{ideal}} - \left(\frac{D}{P_{ideal}}\right)_{fgcn},$$

where D is the drag of a supersonic variable-geometry nozzle; P_{ideal} is the ideal thrust of an isentropic nozzle; D_{fgcn} is the drag of a fixed-geometry convergent nozzle.

The base drag is influenced to a great extent also by the mutual position of several engines on an aircraft: the drag of all types of nozzles decreases as the relative distance between them increases.

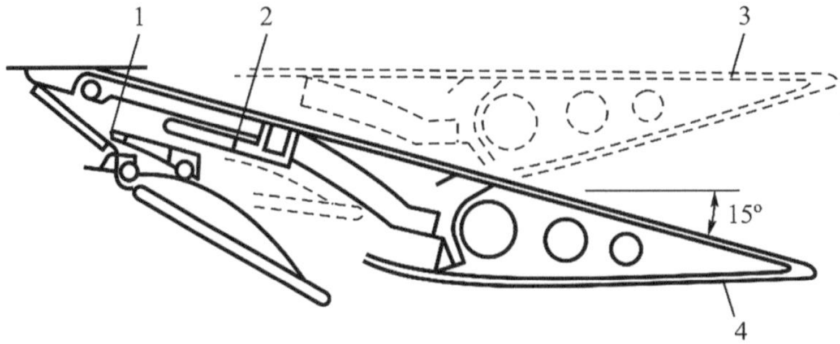

Fig. 20 Section of a nozzle with a small base drag: 1—drive ring of the primary nozzle; 2—drive ring of the secondary nozzle; 3—position during afterburning; 4—position without afterburning

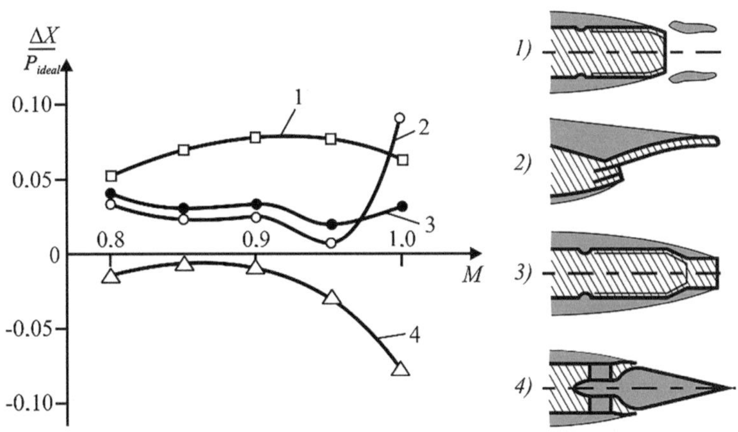

Fig. 21 Additional drag of various supersonic nozzles: 1—aerodynamic ejector; 2—isentropic ramp; 3—convergent-divergent nozzle with long doors; 4—centre body

4.2 ABE Thrust Reverser

The exhaust systems of many modern aircraft are equipped with a device to reverse engine thrust. The requirement for a negative thrust appeared in connection with the increase in the length of the landing run of modern jet aircraft (aircraft with TPE obtain negative thrust with the help of reverse-pitch propellers; on military jet aircraft, a drogue parachute is usually used for this purpose).

Thrust reversal is especially effective when landing on a wet or icy airfield when the friction coefficient is small and the landing run increases significantly.

Depending on the bypass ratio of the TFE and the reverser design, the jet from the jet engine is used totally or partially to create thrust reversal (with a high TFE bypass ratio, only the fan duct may be used to reverse thrust).

The efficiency of thrust reverser is determined not only by the amount of engine negative thrust but also by factors such as t_{init}, the time of initiation of the maximum negative thrust; P, the aircraft's thrust-to-weight ratio; V_{land}, the aircraft's landing speed, f_{fric}, the tire/runway friction coefficient during run.

In general, the negative thrust on actuation of the reverser may be written as follows:

$$P_{rev} = n_{eng}\left[m_{rev} W_{rev} \cos\varphi - \left(m_{gas} - m_{rev}\right) W_c\right],$$

where n_{eng} is the number of aircraft engines equipped with a thrust reversal device; m_{rev} is the gas mass flow at the reversal device exit; W_{rev} is the thrust reversal device exhaust velocity; φ is the angle between the gas velocity vector and the engine axis; m_{gas} is the total gas mass flow; W_c is the nozzle exhaust velocity.

Efficiency of the reversal device design is evaluated by the reverse thrust ratio $\overline{P}_{rev} = P_{rev}/P_0$, where P_0 is the maximum take-off thrust of engine (without afterburning).

For modern TFE, $\overline{P}_{rev} = 0.3$ to 0.6. Obtaining quantity $\overline{P}_{rev} > 0.6$ is practically impossible.

The time of initiation of the maximum negative thrust t_{init} is determined by factors such as the time of making decision to turn on thrust reverse (pilot's response), the time of actuation of the reversal device mechanism, and the time of engine's transition from idle to maximum thrust (engine acceleration). In an ideal case $t_{init} = 0$, i.e. the engine's maximum negative thrust is formed at the moment of LG wheels touchdown; however, in real conditions, $t_{init} = 5$ to 10 s (for example, for the NK-8–2 engine installed on the Tupolev Tu-154, the time of obtaining the maximum negative thrust is equal to 6 s).

As Fig. 22 shows, the thrust reversal has the highest effect on the run distance at small t_{init} (even at $\overline{P}_{rev} = 0.3$). At a greater t_{init}, the effect of $\overline{P}_{rev}$ on the run distance is reduced. In other words, the efficiency of the reversal device may be raised mainly at the expense of a reduced t_{init} (at the expense of a reduced engine acceleration, which exerts a defining influence on t_H).

Design-wise, all reversal devices may be subdivided into two groups: reversal devices that turn the gas jet before the main nozzle exit and reversal devices that divert the jet behind the main nozzle exit. In the reversal devices of the first type, in the thrust reversal regime, special doors block the gas path to the main nozzle, directing it to the rotating cascade (see Fig. 18). Instead of the rotating cascade, other design solutions are possible in order to turn the gas jet (Fig. 23).

In the reversal devices of the second type, the reverser doors in the forward thrust regime are positioned over the nozzle, forming the outer lines of the engine nacelle (Fig. 24).

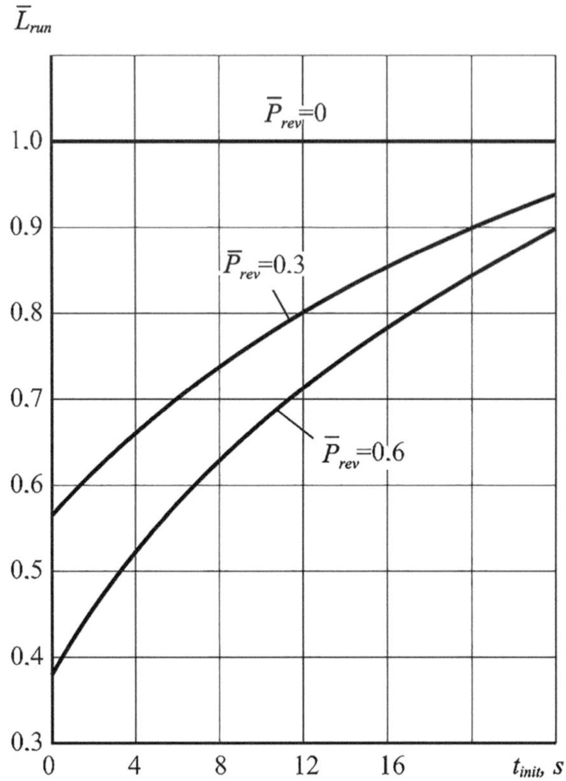

Fig. 22 Influence of the time of obtaining the maximum negative thrust on actuation of the reversal device on the run distance $\left(V_{\text{land}} = 250 \text{ km/h}, \overline{P}_0 = 0.25 \text{ to } 0.30, f_{\text{fric}} = 0.2\right)$

5 Installation of Engines on Aircraft

The installation of engines on modern aircraft should meet the following basic requirements:

- ensure the minimum increment of the aircraft's mass and aerodynamic drag;
- ensure easy engine mounting and dismantling, as well as easy access to all assembly units during technical inspection and maintenance;
- ensure the possibility of localizing and promptly extinguishing the fire in the engine.

The engine mount design will ensure the compensation of temperature deformations and the absorption of engine (propeller) vibrations. All attachment fixtures' vibration loading will not exceed rated values. The reduction of vibrations is especially vital when engines are mounted on the fuselage (or inside the fuselage) of the aircraft when the resonant vibration of insufficiently balanced engines may be transmitted via engine supports to cockpit skin where, due to the "membrane effect",

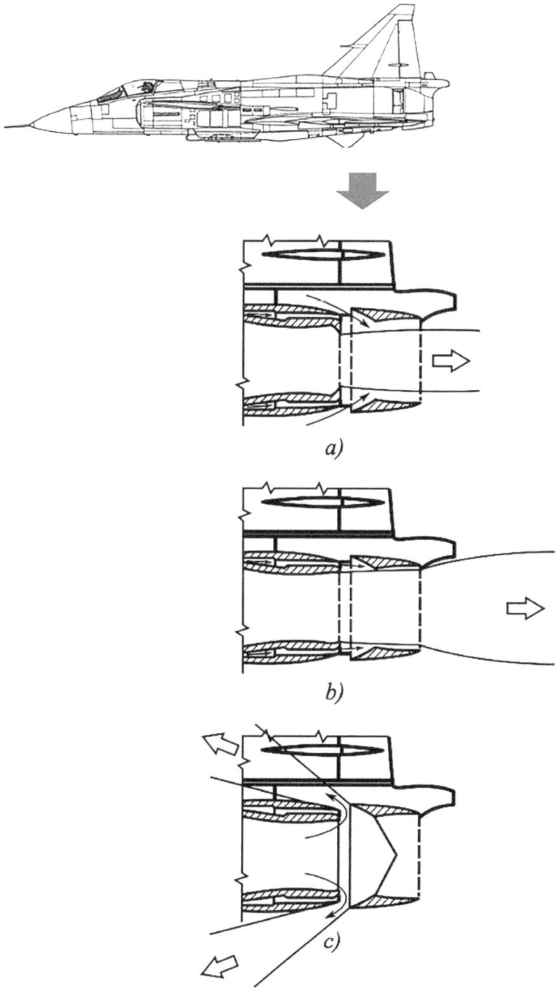

Fig. 23 Aerodynamic ejector integrated with the thrust reverser of the RM-8A engine installed on the Viggen multirole fighter (Sweden): **a**—inlet duct open; **b**—inlet duct closed; **c**—reverser actuated

it may turn into intense sound oscillations (this happened on some Douglas DC-9 production planes).

The designs of attachments for aircraft engines are quite diverse and depend on the type of engines and their location on the aircraft.

The turboprop engines are usually installed on the wing and attached to the wing structure with the help of a grid framework (truss) through dampers (Fig. 25).

On modern aircraft, the jet engines are often arranged in special nacelles, which are thin-walled structures (with longitudinal and lateral formers) similar to fuselage structures.

Engine nacelles are installed on the wing or fuselage with the help of pylons. All communications in the pylon plane are detachable. The pylon has a firewall, which isolates the wing (or fuselage) structure from the zones of fuel and oil systems.

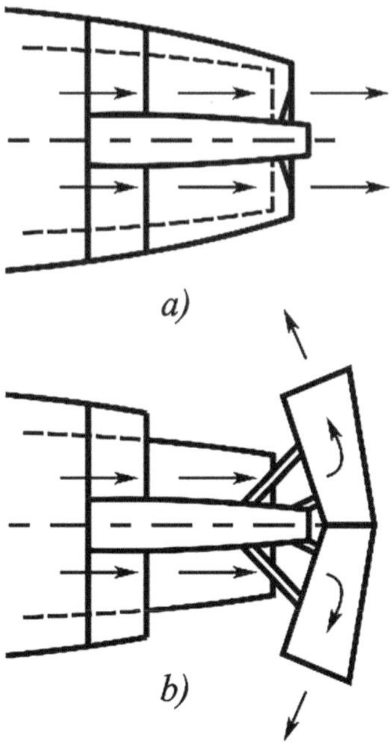

Fig. 24 Golfstream II thrust reverser. **a** Doors retracted; **b** doors in operating position

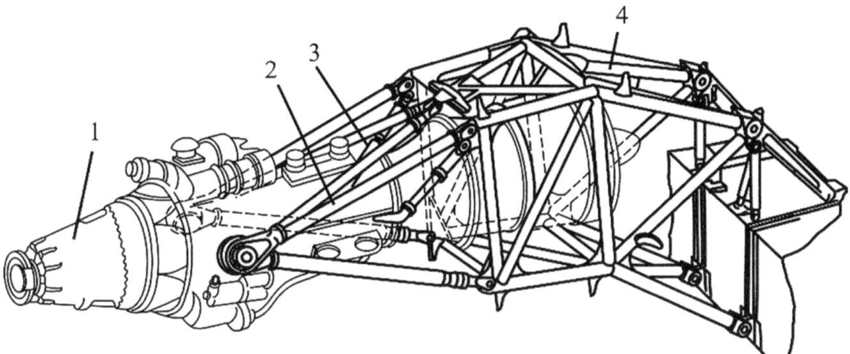

Fig. 25 TPE attachment to aircraft wings design: 1—engine, 2—struts, 3—damper, 4—engine attachment truss

The basic dimensions and mass of the nacelle are determined by the dimensions of the engine, air intake, and exhaust nozzle; the structure of the acoustic lining also has some effect on the nacelle's dimensions and mass.

Figure 26 illustrates the installation of the high-bypass TFE nacelle on the wing of a modern aircraft.

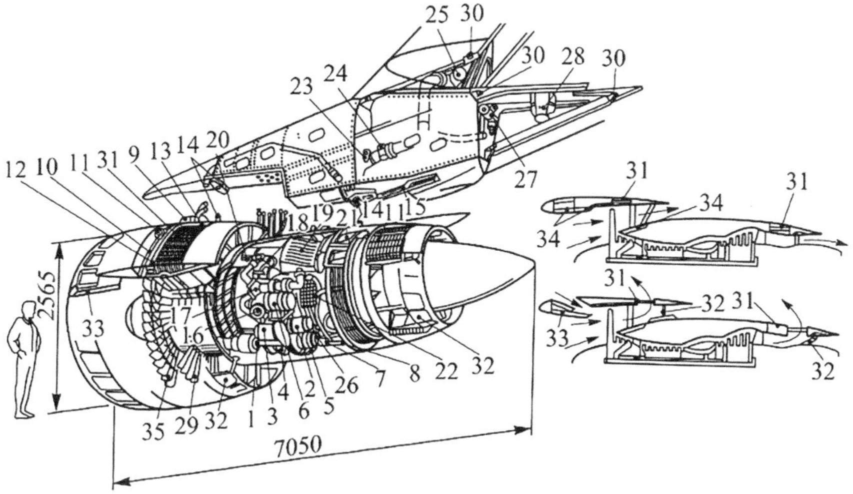

Fig. 26 Pratt Whitney J T9D TFE installation and operation diagram of thrust reversers and inlet doors (Boeing 747): 1—drive shaft of accessories; 2—constant speed coaxial drive; 3—60 kV·A generator; 4—pump and control unit; 5—centrifuge; 6—oil tank; 7—gas generator thrust reverser drive; 8—cable to the reverser; 9 - pneumatic motor powered by a reversed fan thrust; 10—flexible drive shaft; 11 – reverser cascade; 12—a rotating wheel; 13—roller rails; 14—engine mount; 15—thrust link; 16—air bleed; 17—8th stage check bleed valve; 18—15th stage shut-off bleed valve; 19—primary heat exchanger; 20—air from fan to heat exchanger; 21—fan air exhaust valve; 22—pressure discharge valve; 23—bleed air temperature control unit; 24—shut-off and control valve; 25—air system line; 26—pneumatic starter; 27—pneumatically powered hydraulic pump; 28—hydraulic tank; 29—stator; 30—bolts; 31—reverser cascade; 32—reverser doors; 33—inlet doors; 34—acoustic (sound-absorbing) panels; 35—single-stage fan

A possible design of engine mount on underwing pylon suspension is shown in Fig. 27.

If the aircraft's required thrust-to-weight ratio is ensured by three engines, then the external engines are mounted either on pylons under the wing, or on horizontal pylons in the afterbody, while the center engine is inside the fuselage afterbody. The installation of the center engine with an S-shaped duct calls for carefully integrating it with the fuselage.

An engine installed inside the fuselage is usually mounted and dismantled simply by lifting (lowering) the engine via a large secondary structural hatch, which is also used for access to engine accessories during in-service maintenance of an aircraft.

The possible design of an engine mount during installation in the fuselage afterbody is shown in Fig. 28.

Engine attachment in the nacelle and fuselage is accomplished at three points and represents a combination of several struts in the nacelle (or in the fuselage). This attachment structure compensates for variations in engine length caused by temperature expansions. Engine position may be varied by adjusting the length of struts, etc.

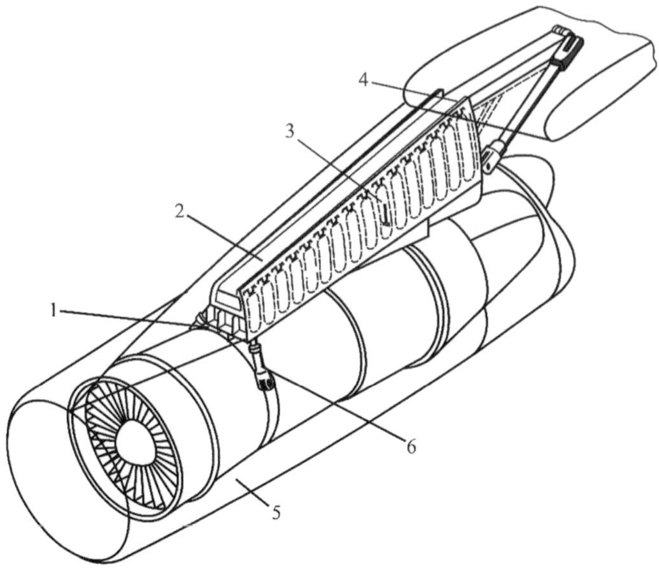

Fig. 27 Underwing engine pylon suspension: 1—engine forward mounts; 2—torsion box; 3—engine aft mount; 4—pylon-to-wing attachment fittings; 5—engine nacelle; 6—strut

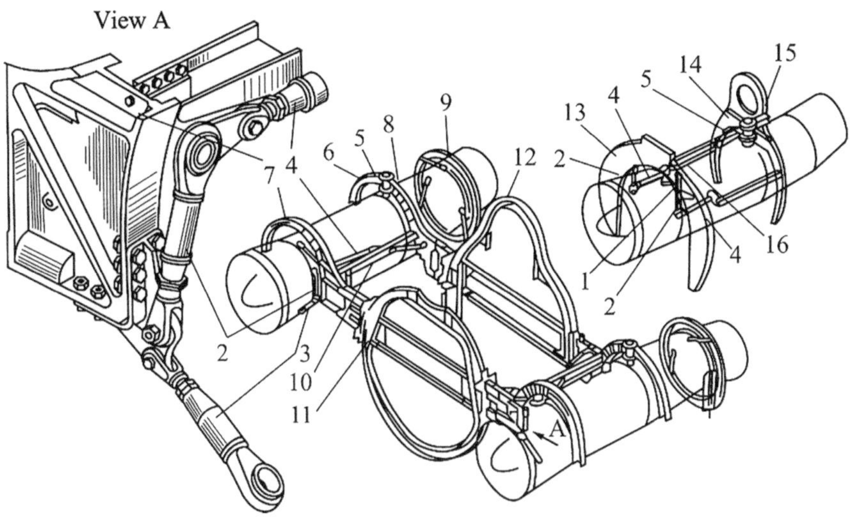

Fig. 28 Mounting a TFE in the Tupolev Tu-154 tail with the aid of struts: 1, 2, 3, 4, 5, 6—struts; 7, 8, 9—reinforced frames of the engine nacelle; 10—side beam; 11, 12, 13, 14—reinforced frames of the fuselage; 15, 16—beam

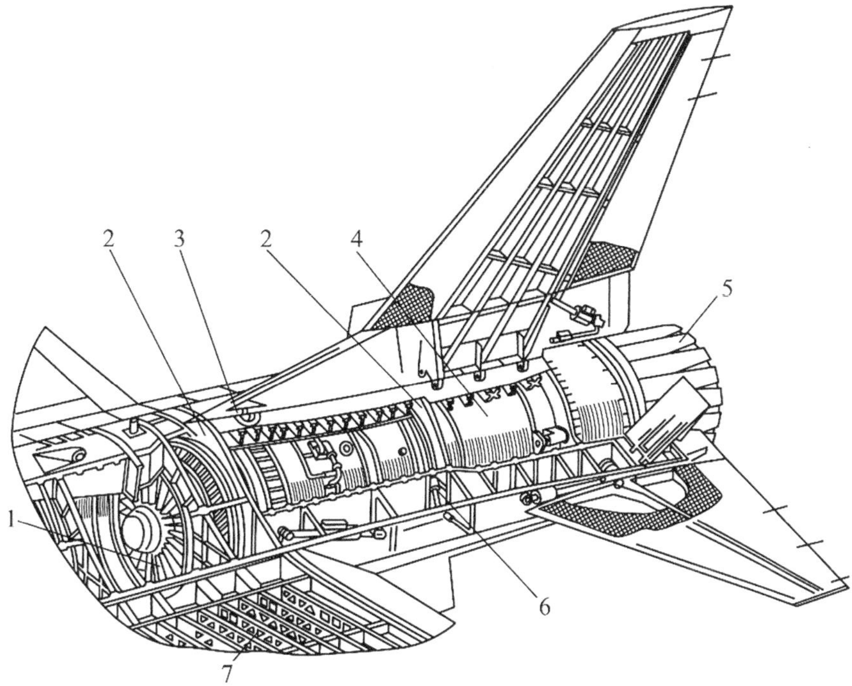

Fig. 29 Pratt-Whitney F-100 ABTFE installation on the F-16 fighter: 1—engine fan; 2—engine reinforced frame; 3—forward engine mount; 4—afterburner; 5—variable geometry exhaust nozzle; 6—main engine mount; 7—multispar wing structure

On modern fighters, one or two turbojet engines are installed inside the fuselage tail and attached to reinforced frames (Fig. 29).

To improve fighter maintenance, engine accessories may be arranged in the airframe rather than on engines, which simplifies installation and ensures interchangeability of right and left engines (for example, in the McDonnell Douglas F-15 fighter).

6 Noise of Jet Aircraft

6.1 Noise Level

The increasing intensity of air traffic and the increasing dimensions of passenger aircraft have created a problem with reducing the noise of passenger aircraft power plants. The large capacity of major airports (over 1000 aircraft per day) has led to noise from turbojet engines inflicting a harmful physiological effect on the population of areas adjacent to airports and the personnel of aerodromes. Therefore, when passenger aircraft are designed, the task of power plant noise abatement is today considered as important as achieving good flight performance characteristics and high cost-effectiveness of an aircraft in development.

The strength (intensity) of the monophonic sound, i.e., the sound of one frequency, is determined by the amount of energy carried by the sound wave over 1 s through a 1 cm^2 area, perpendicular to the direction of wave movement. The relationship between sound strength and sound pressure (excessive in relation to atmospheric mean quadratic pressure, generated by sound waves) is as follows:

$$I = p_{\text{sound}}^2/(\rho a),$$

where I is the sound strength in a given point in space; p_{sound} is the sound pressure; ρ is the air density; a is the air speed of sound.

The level of the mono-sound strength L is measured in decibels (dB) and is written as

$$L = 10\ \lg(I/I_0),$$

where I_0 is the strength of sound at the audibility threshold (at $L = 0$).

The noise produced by a jet engine represents chaotic air oscillations that consist of several simple sound oscillations of varying intensity and frequency.

Since the human ear has different sensitivity to sounds of varying frequencies, the level of noise perceived by a person is admittedly a subjective notion and not always characterized precisely by objective acoustic indicators. For example, the noise with level $L = 101$ dB and frequency $f = 100$ Hz and the noise with $L = 82$ dB and $f = 4000$ Hz are perceived as equally unpleasant, despite the 80-fold difference in intensity.

Therefore, for a more precise assessment of the level of perceived noise, a PNdB unit (English—Perceived Noise) was introduced. The general level of noise from a turbojet engine is converted into PNdB units for individual sections of the noise frequency band using special tables, which are based on the equal noise lines.

However, the PNdB unit has proved insufficiently refined for the correct evaluation of the noise level of jet aircraft. As further investigations showed, at the same level of perceived noise, additional irritation effect on man is produced by the duration of noise, which depends on aircraft speed and flight path. In addition to the duration of noise, the PNdB unit fails to take account of the effect of discrete tones in the noise spectrum (separate spikes of noise level in the narrow frequency band), which also exert an additional irritation effect in comparison to the broadband noise (noise with the gradual variation of level by frequency). To take account of these factors, a new unit of assessment of the noise of jet aircraft, the EPNdB (effective PNdB), has been proposed, based on the PNdB unit adjusted for the duration of noise effect and presence of discrete tones in the noise spectrum.

6.2 Acceptable Noise Level Norms

The range of noise levels perceived by the human ear is quite broad: L = 0 to 140 EPNdB, which corresponds to a 10^{14}-fold change in noise intensity. The region of perceived frequencies ranges from 20 to 20,000 Hz (approximately ten octaves). Practical analyses are usually limited to a narrower range of frequencies f = 45 to 11000 Hz (eight octaves). It is noteworthy that not all perceived noises are fatiguing for humans. Noises in excess of 80 EPNdB cause distress and fatigue. A particularly strong effect on man is exerted by the noise at $L >$ 110 EPNdB with frequencies f = 2000 to 5000 Hz (higher and lower frequencies have a less distressing effect).

Levels of perceived noise from certain conventional sources in EPNdB are as follows:

Audibility threshold	0
Woods at zero wind	10
Conversation in a low voice	40
Typing pool	80
Subway car	90
High-speed train (vestibule)	110
Pain threshold	140

The Boeing 707 passenger aircraft (take-off mass 151 tonnes, entry in service 1957) during climb created noise of level L = 114 EPNdB, and during landing approach L = 120 EPNdB. If account is taken of the range of frequencies of the noise of modern TFE (f = 50 to 10000 Hz) and the fact that the take-off mass of passenger aircraft is over 350 tonnes, then it becomes clear why special standards of acceptable noise level of jet passenger aircraft have appeared.

Acceptable noise levels are given in GOST 17,228-78. Similar norms have been established in the United States, Britain, and several other countries. All of them are close to ICAO norms, under which the noise created by passenger aircraft is measured at three reference points:

- during take-off—at a distance of 0.65 (0.45) km from RW (0.65 km—for existing aircraft and 0.45 km—for newly developed aircraft);
- during climb—6.5 km from the start of the run;
- during landing approach—at a distance of 2 km from the end of RW (Fig. 30).

The acceptable noise level is expressed through the aircraft's take-off mass. During take-off and landing approach:

$$L = 6.6 \lg m_0 + 91.6 \text{ EPNdB};$$

during climb:

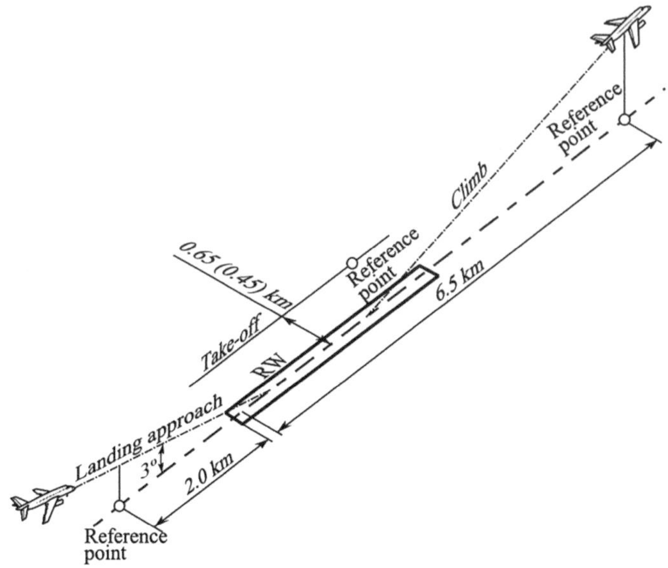

Fig. 30 Location of reference points for measuring noise level

$$L = 16.6 \lg m_0 + 65.8 \text{ EPNdB},$$

where m_0 is the take-off mass of the aircraft (t).

With the reduction of the aircraft's take-off mass, noise requirements become stricter: the acceptable noise level is lowered, which is linked to a higher frequency of flights of dimensionally smaller aircraft. However, for aircraft with a take-off mass less than 37.5 t or over 350 t, the acceptable noise level is taken to be constant (Fig. 31).

It is worth noting that reducing the noise of a jet aircraft by 10 EPNdB corresponds to a perceived reduction of noise of approximately 50% for the observer on the ground.

6.3 *Possible Methods of Noise Abatement*

On modern jet aircraft, the main sources of noise are the jet engine and the unsteady stream flowing around the aircraft airframe (aerodynamic noise). In the first case, the noise is generated by the jet exhaust, fan, turbine, and compressor (Fig. 32).

One of the more intensive sources of noise from the jet engine is the jet itself. At present, the noise of the jet exhaust has been thoroughly studied. The presence on the border of the jet of a large speed gradient leads to air ejection from surrounding space and the formation of vortices in the mixing zone, which generates noise. The speed of the jet exhaust has the greatest effect on the noise level (noise level is proportional

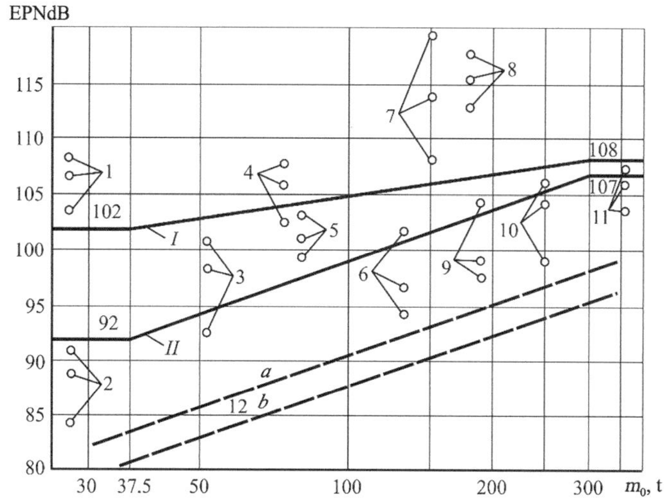

Fig. 31 Acceptable noise levels and actual noise levels of some modern passenger aircraft: I – takeoff and landing approach; II—the climb; 1—Jet Commander; 2—Learjet; 3—DC-9-30; 4—Boeing 727-100; 5—Boeing 727-200; 6—A300B; 7—Boeing 707; 8—Concorde; 9—L-1011; 10—DC-10; 11—Boeing 747; 12—aircraft's aerodynamic noise (**a**—wing extendable devices and landing gear extended; **b**—flight configuration)

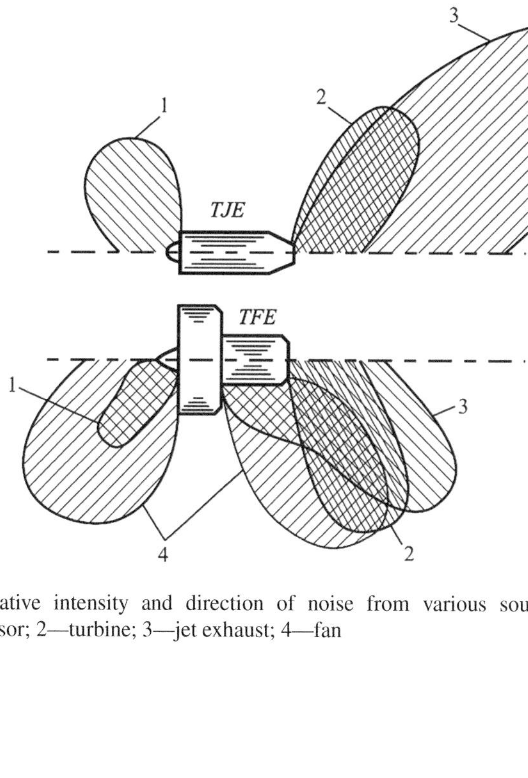

Fig. 32 Relative intensity and direction of noise from various sources on TJE and TFE: 1—compressor; 2—turbine; 3—jet exhaust; 4—fan

to $\sim W_c^8$). Consequently, the more effective method of jet noise abatement is the reduction of its speed.

Successes in abating the noise of jet exhaust are primarily related to the creation of bypass turbojet engines, whose jet speed is much less than that of single jet engines. When the first TFEs with a small bypass ratio ($m \approx 1$) were installed on passenger aircraft instead of TJEs, the noise level was reduced by approximately 15 EPNdB. The development of the high-bypass TFE ($m = 4$ to 6) further reduced jet exhaust noise by 13–17 EPNdB. With a further increase in the bypass ratio, the jet noise practically does not change, but the level of fan noise rises progressively, significantly surpassing the noise level of jet exhaust. In the high-bypass TFE, the fan has become the principal noise source, which is distributed both forward and aft, like the noise of the TJE compressor and turbine. The intermediate pressure compressor, turbine, and two jets (from the fan and the gas generator) are additional noise sources of the TFE.

Further reduction of the noise of the fan, compressor, and turbine of the TFE is effected by utilizing external noise suppression. To this end, the internal surfaces of engine nacelles are covered with sound-absorbing (acoustic) lining. The sound-absorbing lining represents a layered structure, whose outer layer consists of porous material, and a layer of cell structure underlies the outer layer. In this case, cells act as acoustic cavities.

Modern TFE noise levels were reduced by 10 to 13 EPNdB by using an acoustic nacelle liner. Elongating the air intake and exhaust nozzles (Fig. 33) can provide additional noise suppression of 3 to 5 EPNdB (though this is associated with an increase in nacelle mass).

The efficiency of using a noise reduction system depends on the bypass ratio of the jet engine. For example, the acoustic lining of the TJE and low-bypass TFE nacelles

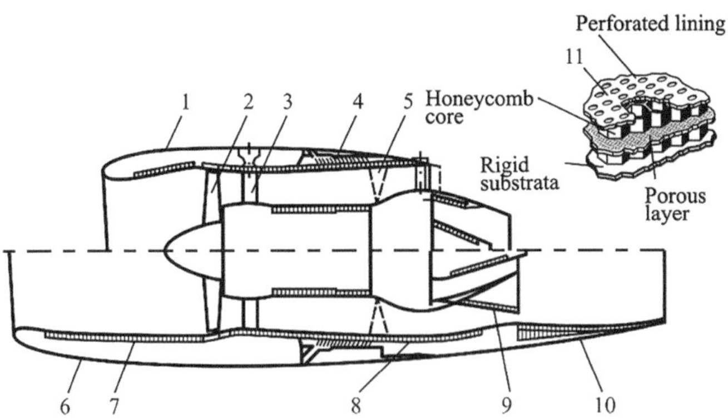

Fig. 33 Layout and location of the acoustic lining of TFE nacelle: 1—high-bypass TFE nacelle; 2—single-stage fan without inlet guide vanes 3—a rectifier; 4—an engine thrust reverser cascade; 5—thrust reverser doors; 6—possible modified nacelle; 7—acoustically lined, elongated air intake; 8—elongated bypass duct lining; 9—mixer lining; 10—elongated exhaust nozzle lining 11—acoustic lining structure

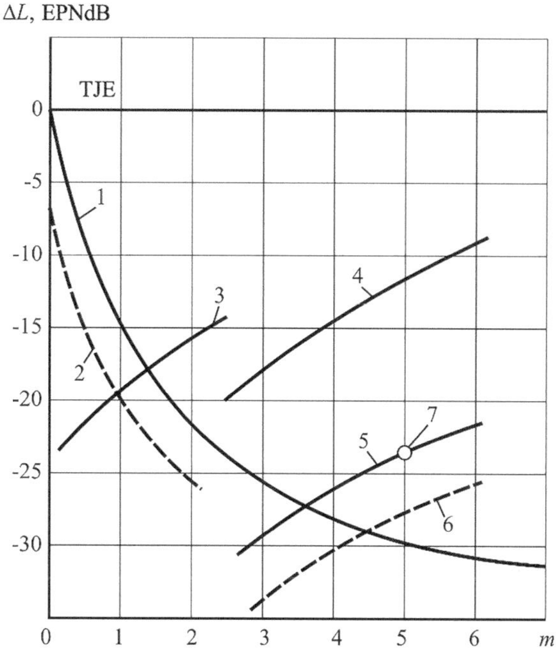

Fig. 34 Effect of the TFE bypass ratio on the level of perceived noise: 1—noise of jet exhaust; 2—nozzle with suppressor; 3—noise of multistage fan, compressor, and turbine; 4—single-stage fan; 5—engine nacelle with sound absorbing lining; 6—elongated air intake and exhaust nozzle; 7—the Rolls-Royce RB-211 TFE

is meaningless because the noise of the exhaust jet is dominant for those engines. In this case, a special noise suppression nozzle (notched, lobe, multitube, etc.) is used for noise reduction, which has found wide application in the first passenger jet aircraft. Using the suppressor nozzle becomes inefficient with a bypass ratio of $m > 1$ (Fig. 34).

One more method of noise reduction needs to be mentioned. It will help reduce aircraft noise during landing approaches. The use of microwave landing systems will enable a landing approach along the gliding path with a slide angle of 6° instead of 3°, along which modern aircraft are gliding currently. The increase in flight altitude over territory adjacent to the aerodrome will naturally reduce the noise level generated by an aircraft.

Further noise reduction at the power plant will reduce the noise of the entire aircraft only up to a certain limit, defined by the airframe airflow noise, or, as it is often called, the aerodynamic noise. The airframe noise of the modern aircraft is approximately 6 to 10 EPNdB below the noise of the power plant (see Fig. 31). This noise is defined by factors such as the turbulent boundary layer, the airflow of extended LG, flaps, and slats, and also the turbulent wake behind the wing and fuselage.

The intensity of airframe noise is proportional to the flight speed to the power of five or six. In the next few years, the aerodynamic noise may turn out to be the main component of the aircraft's general noise during landing approach.

7 Fuel System

7.1 Fuel Storage on Aircraft

The fuel system of a modern aircraft comprises the following principal elements: fuel tanks, plumbing, pumps, valves, cocks, filters, a system of diverse automatic devices, sensors, measuring instruments, etc.

The purpose of the fuel system is to ensure the supply of fuel to engines under all possible flight conditions for a particular aircraft (by altitude, speed, and load) in the required amount and at the requisite pressure. In supersonic aircraft, especially those designed for long-duration supersonic flight, the fuel system also performs several other important functions, including cooling of certain parts and systems of the aircraft (hydraulic, conditioning, etc.). In addition, fuel transfer (forward-aft) may be used to shift the aircraft balance, countering the release of weapon payload, the shift of the aerodynamic centre when transitioning from a subsonic to a supersonic flight, etc. Naturally, the aircraft's stability margin and, as a result, trim drag are reduced.

Figure 35 illustrates the layout of the fuel system of a modern aircraft.

Accommodation of fuel largely determines the overall weight arrangement of the aircraft because the fuel reserve of modern subsonic and supersonic aircraft may exceed 50% of the aircraft's take-off mass (on the Concorde value $\overline{m}_{\mathrm{f\ max}} = 0.52$).

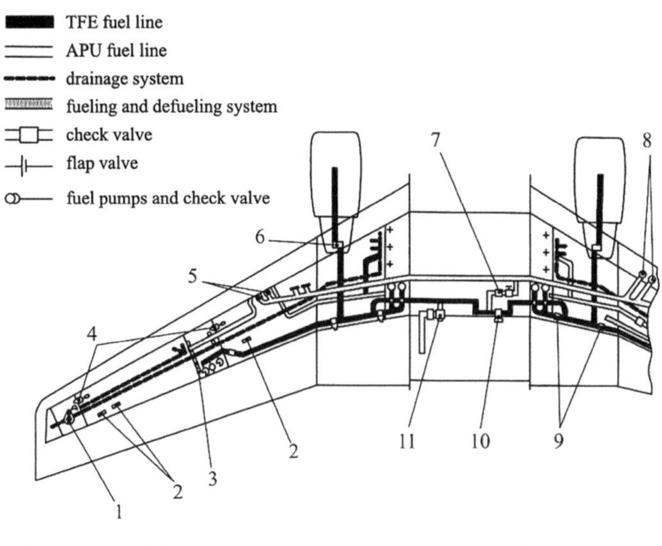

Fig. 35 Block schematic of the fuel system of a modern aircraft (fuel tanks are not shown): 1—drain inlet; 2—fueling sensor; 3—ejector; 4—drain float-type valves; 5—fuel-defuel valves; 6—low pressure cock; 7—relief valve (used only on the ground); 8—defueling system connections; 9—tank isolation cocks; 10—cross-feed cock; 11—APU fuel valve and pump

Fuel is stored in special fuel tanks that are subdivided by purpose into main, feeder, and balancing (depending on the type of aircraft, the fuel system may have no provision for balancing tanks and occasionally feeder tanks).

Fuel may be supplied to engines from any fuel tanks, but more often special feeder (collector) tanks, into which fuel is transferred in a certain sequence from the main tanks, are used for this.

The fuel system with the feeder tank has certain advantages (especially for military aircraft): it is more reliable because, should the main tanks become inoperative, there still is the feeder tank (up to 20% of all fuel). Its protection may be provided by armoring; one (or several) feeder tanks are easier to fit with special devices for the uninterrupted supply of the engine during manoeuvres with acceleration and deceleration, etc.

Fuel tanks may be arranged in the fuselage and in the wing of an aircraft (symmetrically in relation to the centre of mass). Both arrangements have certain advantages and shortcomings. The wing fuel tanks have a greater surface area that is susceptible to damage, which makes the fuel system less robust. This is the key shortcoming of such an arrangement of fuel tanks. However, when fuel is stored in the wing, its mass unloads the wing in flight from the bending moment, which yields a certain gain in the mass of wing the structure. In addition, when fuel is stored in the wing, the fuselage can almost completely be taken up by payload, which has great significance primarily for passenger and transport aircraft. Therefore, on all passenger, transport, and heavy military aircraft, the fuel tanks are mainly arranged in the wing (Fig. 36).

Storing fuel in the fuselage may prove more expedient for military aircraft (especially in combat operations at low altitudes). Placing fuel tanks in the fuselage over the wing centre section, engine, landing gear, etc., which serve as a kind of protection screen against anti-aircraft fire, can enhance the aircraft's survivability.

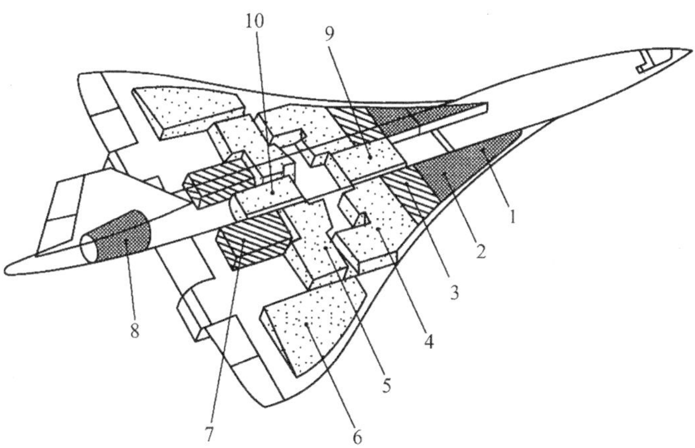

Fig. 36 Arrangement of fuel tanks on the Concorde supersonic passenger airliner: 1, 2, 8—balancing tanks; 3, 7—feeder tanks; 4, 5, 6, 9, 10—main tanks

Design-wise, the fuel tanks are of three types: rigid, flexible, and structural tanks, the latter being part of the aircraft structure. The main advantage of flexible rubber tanks is that they are more volume efficient than rigid tanks and that they are easier and more convenient to manufacture and operate (they are foldable and mountable through small hatches). Flexible tanks are also impervious to vibration, have no torn edges when shot through (small holes even seal with a special rubber layer), and have good heat insulation qualities.

Rigid tanks are lighter than flexible ones (with containers), and they can be comparatively easily dismantled.

Leakproof structural fuel tanks make for a more rational use of the aircraft's internal volume because there is no fuel tank as such, but the fuel is filled into the wing or fuselage section lined on the inside with a kerosene- and temperature-resistant sealant. The use of structural fuel tanks helps increase the fuel reserve on board an aircraft. However, the structural fuel tanks are more prone to damage, which degrades the reliability of the fuel system; they are also difficult to repair. Fuel in the structural tanks is directly subjected to the effects of low temperatures during a flight at subsonic speeds and high temperatures during a flight at large supersonic speeds.

Subject to aircraft purpose, the advantages of said types of fuel tanks manifest themselves to a greater or lesser degree. On modern aircraft, several types of fuel tanks are often installed simultaneously.

7.2 Fuel Flow Control

Retaining the aircraft's centre of mass within acceptable limits as fuel is depleted is ensured not only by the symmetrical layout of fuel tanks. For the stability margin of the aircraft to remain unchanged within considerable limits, a special automatic system is installed within the fuel system of modern aircraft to measure fuel reserve and control flow in accordance with a preset program without crew interference. Automatic control is attained with the help of pressure-reducing valves, by turning on and off booster pumps, etc.

As a rule, the fuel is first depleted from the internal and then from the external (along the wingspan) tanks. This operation diagram of the flow control system delivers maximum reduction of the wing bending moment.

If the aircraft is designed for long-duration supersonic flight, then the fuel system, to reduce the stability margin at supersonic speed, will usually have special balancing tanks installed (tanks located at the possible largest distance forward and aft of the aircraft's centre of mass; see Fig. 36). Fuel transfer from the forward balancing tanks to the aft tanks may be used to bring the centre of mass closer to the aircraft's aerodynamic centre.

Figure 37 illustrates the typical behavior of the position of the aircraft's centre of mass and aerodynamic centre depending on speed (or flight time).

A question may arise: is it possible to ensure a fuel depletion sequence for the reduction of $C_m^{C_L}$ such that the centre of mass at a supersonic speed passes the preset aft

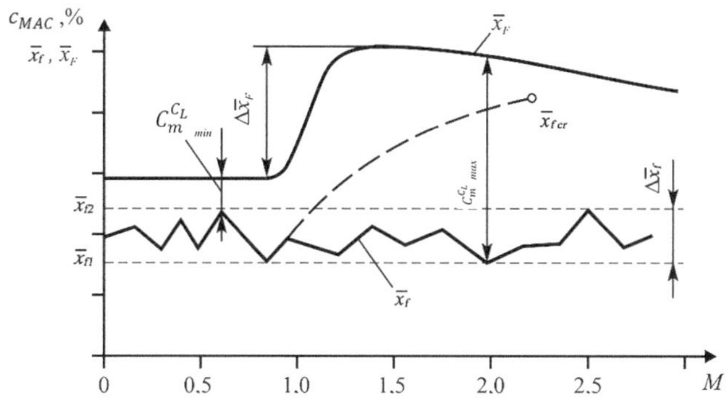

Fig. 37 The possible change of the position of the aircraft's centre of mass and aerodynamic centre during flight: $\bar{x}_{f1}$—forward limit of the centre of mass; $\bar{x}_{f2}$—aft limit of the centre of mass; $\Delta\bar{x}_f$—aircraft's acceptable CG range; $\Delta\bar{x}_F$—shift of aerodynamic centre during transition from subsonic to supersonic flight; $\bar{x}_{f\,cr}$—possible position of the centre of mass in a supersonic cruise flight (after fuel transfer from forward to aft balancing tanks)

limit $\bar{x}_{f2}$ and approaches the aircraft's aerodynamic centre, for example, by depleting fuel only from forward tanks? This cannot be done for two reasons: first, if the fuel in the forward tanks is depleted (rather than transferred to the rear tanks), the centre of mass will shift backward very slowly, and the effect of a degraded stability margin will be significantly lower. Secondly, when the centre of mass is behind position $\bar{x}_{f2}$ and for any reason it is required to reduce the speed quickly and go back to subsonic flight, it will be necessary to also bring the aircraft's centre of mass back into subsonic position quickly (otherwise, the aircraft will become unstable), but the quick return of the centre of mass to subsonic position can be attained only by fuel transfer to forward tanks.

7.3 Characteristics of Aviation Fuel

Hydrocarbon fuel is used as the main fuel for the engines of modern aircraft: kerosene for air-breathing engines and gasoline for reciprocating engines.

The mass (weight) calorific value of kerosene is of the same order as that of gasoline, but the density of kerosene is higher, which explains why its volumetric calorific value is 10 to 15% higher than gasoline's. The flashpoints of kerosene and gasoline are respectively around +30 °C and 35 to 40 °C (as known, the lower the flashpoint, the easier the engine starts at subzero temperatures).

For hypersonic aircraft and in the future for supersonic and subsonic aircraft, cryogenic fuel will be the order of the day (for example, liquid hydrogen), which will enable longer flight ranges (proportional under other equal conditions to the fuel calorific value).

Because hydrogen's flashpoint does not exceed −253 °C, its use is fraught with complications. At the same time, hydrogen has the highest calorific value and specific heat, i.e. the highest cooling capacity. These qualities qualify hydrogen as one of the more promising fuels for the engines of aircraft flying at high speeds, which require intensive cooling of the aircraft structure.

Liquid methane is also considered a promising fuel for ABE. It has a slightly higher mass (weight) calorific value and three times the cooling capacity of kerosene.

At high flight speeds, when the direction of heat flow in the aircraft structure changes to the opposite, the fuel (especially placed in wing structural tanks) is heated. For example, at a cruise speed of $M = 3$, the heat flow emanating from the skin becomes so significant (average skin temperature is around 260 °C) that it could heat kerosene to flashpoint. The use of heat insulation (a laminated structure with a honeycomb core that significantly reduces the heat flow) makes it possible to avoid excessive fuel heating; however, at high speeds ($M > 3$), its temperature rises so high that the hydrocarbon fuel cannot act as a cooling agent to cool the aircraft's structure and systems. In addition, the disadvantage of hydrocarbon fuel is its thermal instability at high temperatures (for example, for kerosene, this limit is equal to approximately 200 °C). If said limit is reached even for a short period of time, then a certain amount of solid depositions may form in heat exchangers, filters, nozzles of engines, etc., with the possibility of their blockage. Thus, one can speak about the advantage of cryogenic fuel for hypersonic aircraft.

At hypersonic speeds hydrogen before being fed to the engine may be used to cool aircraft structures (specific heat of hydrogen is 15 times the kerosene's). Then the considerable difficulties connected with its use will be compensated to some degree. Liquid hydrogen, while being undoubtedly superior to kerosene in terms of mass calorific capacity and specific heat, has a density which constitutes merely 0.086 of kerosene density. Hydrogen energy efficiency per volume unit is 4 times less than that of kerosene. Therefore, the aircraft using that fuel must have significantly larger volumes, with fuel stored in cryogenic tanks which will further increase the aircraft's dimensions and, consequently, its mass. This significantly complicates the design of an aircraft using hydrogen fuel.

The first creditable practical results of using cryogenic fuel in aerospace technology were obtained in this country.

In the mid-1970s, amidst a grave economic crisis, the USSR Academy of Sciences and several research institutes and design bureaus came up with a R&D program envisaging a mass-scale introduction of a hydrogen-based power industry into the national economy. Under that program, Moscow's engineering factory Opyt was to create a flying test bed using liquid hydrogen as fuel on the basis of the Tupolev Tu-154. It was found at the design stage that the research had to be expanded to include the use of liquefied natural gas (LNG) as an alternative to kerosene and gasoline, which are conventional fuels for the means of conveyance. LNG, like liquid hydrogen, has numerous valuable advantages compared with conventional types of fuel: its calorific value is 15% higher, its cost is roughly half that of kerosene, its use dramatically reduces the environmental footprint of a flying aircraft.

At the same time, the LNG density is 1.83 times less than the density of kerosene. Therefore, its storage requires an equally large volume. The LNG is stored on board only in a liquid form, which requires special measures to be taken when it is stored on aircraft. In addition, a new ground infrastructure is required to ensure the normal operation of LNG-fueled transport aircraft, which will accumulate, store, and service aircraft with liquefied gas [5].

Based on the research and developments in the USSR in the 1980s, the first flying test bed, the Tupolev Tu-155, was built for the first time in the world. It had an experimental engine that used alternative types of fuel. Its maiden flight took place on April 15, 1988. The large volume of research conducted on it has enriched this country's science and engineering with valuable experience and potential in the field of the development of advanced cryogenic technologies in the aircraft industry and secured Russia's leading positions in that field.

The Tupolev company is currently working on LNG-fueled versions of all aircraft, both in production and in development. Estimates show that the operation of the aircraft with cryogenic fuel, the Tupolev Tu-204 K, compared with the basic version, the Tupolev Tu-204 will significantly reduce toxic effluents, achieve a 2.5-fold decrease in fuel costs, and cut direct operational costs by 4 to 5% per hour of flight. One of the options for locating the cryogenic tank over the fuselage on the Tupolev Tu-204 K (Tu-206) is shown in Fig. 38.

Variables

Symbol	Description	Units
P	Engine thrust	N

(continued)

Fig. 38 Modification of the Tupolev Tu-204 (Tu-206) with the cryogenic fuel tank over the fuselage

(continued)

Variables		
Symbol	Description	Units
c_{fc}	Specific fuel consumption	kg/(daN•h)
m_{air}	Air mass flow rate	kg/s
P_{sp}	Specific thrust	N/kg
γ_{eng}	Engine specific weight	–
m	Engine bypass ratio	–
T^*_{gas}	Gas inlet temperature	K
π^*_{compr}	Total compressor pressure ratio (compression ratio)	–
$P_{sp,0}$	Specific thrust at take-off	–
n	Rotor rotational speed	1/min
σ_{inlet}	Total pressure recovery coefficient at engine inlet	–
T^*_H	Stagnation temperature (at compressor inlet) at flight altitude H	K
Δ	Air density ratio at flight altitude	–
T^*_{AB}	Afterburner exit temperature	K
a	Speed of sound	m/s
G_{eng}	Engine dry weight	N
L_{eng}	Engine length	m
F_{inlet}	Inlet area	m^2
ρ	Air density	kg/m^3
σ_{shock}	Total pressure recovery coefficient in the shock system	–
σ_i	Total pressure recovery coefficient in one shock	–
φ	Mass flow coefficient	–
D	Drag	N
V_{land}	Aircraft landing speed	m/s
f_{fric}	Tire/runway friction coefficient	–
$\overline{P}$	Reverse thrust ratio	–
I	Strength of sound	dB
L	Level of mono-sound strength	dB
m_0	Aircraft take-off mass	t

References

1. Badyagin AA, Mukhammedov FA (1978) Proektirovanie lyogkih samolyotov (Design of light aircraft). Mashinostroenie, Moscow
2. Arep'ev AN (2005) Vybor proektnyh parametrov i ocenka lyotnyh harakteristik passazhirskih samolyotov s turbovintovymi dvigatelyami (Selection of design parameters and evaluation of flight characteristics of passenger aircraft with turboprop engines). MAI, Moscow
3. Yugov OK, Selivanov OD (1980) Soglasovanie harakteristik samolyota i dvigatelya (Coordination of aircraft and engine characteristics). Masinostroenie, Moscow
4. Yugov OK, Selivanov OD (1989) Osnovy integracii samolyota i dvigatelya (Fundamentals of Aircraft and Engine Integration). Masinostroenie, Moscow
5. Andreev V, Borisov V, Klimov V et al (2001) Vnimanie: gazy. Kriogennoe toplivo dlya aviacii (Warning: gases. Cryogenic fuel for aviation). Moskovskij rabochij, Moscow.

Empennage Design

Nikolai K. Liseitsev

Abstract The chapter starts by explaining the functions, composition, and the performance indicators of empennage of different types. The chapter shows how to size the empennage of an aircraft for the desired stability and control. The reader will learn about the approaches to designing the structures of the empennage, as well as some design features of the attachment points to the airframe.

Keywords Tail sizing · Structural design · Attachment point · Performance indicator

1 General Concepts. Objectives of Empennage Design

The purpose of an empennage is to ensure aircraft's stability and controllability. So far, the typical configuration of empennage is the so-called "normal" or "classic" (wing-tail) configuration, where the horizontal and vertical empennage surfaces are mounted in the fuselage's aft part, forming the aircraft tail assembly. With the "canard" configuration, the horizontal empennage is moved to the fuselage's forward part, while the tail retains only the vertical unit. The evolution of supersonic aviation and the search for methods to improve supersonic aircraft manoeuvrability resulted in a high level of interest among aviation specialists in the direct control of aerodynamic forces without changing the aircraft's angles of attack or sideslip.

Direct control of aerodynamic forces (lift and lateral) can be achieved by deflection of special aerodynamic surfaces. These include wing flaps and additional horizontal and vertical surfaces in the nose and tail parts of the aircraft, which provide variation in aerodynamic forces without aircraft rotation (Fig. 1).

Success in solving the technical issues related to direct control of the aerodynamic forces will allow for significantly improved performance for the aircraft featuring such a system because it will provide for:

N. K. Liseitsev (✉)
Moscow Aviation Institute (National Research University), Moscow, Russia
e-mail: nikolai.liseitsev@yandex.ru

M. A. Pogosyan (ed.), *Aircraft Design*, Springer Aerospace Technology,
https://doi.org/10.1007/978-981-96-4599-2_19

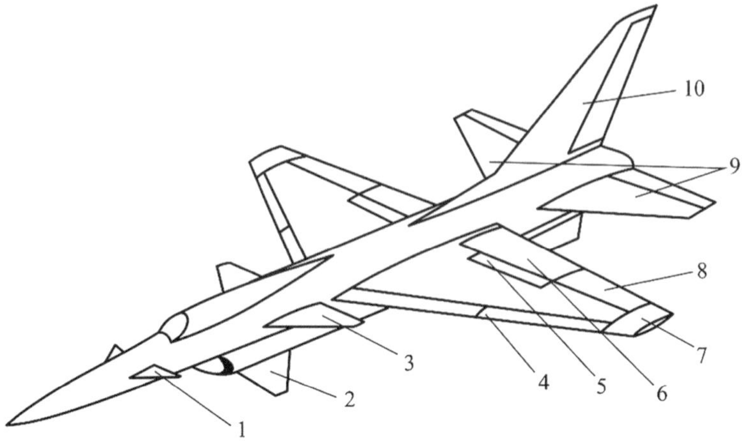

Fig. 1 The empennage and control surfaces conventionally combined in a theoretical aircraft are: 1—the horizontal surfaces for cockpit vibration stabilization; 2—the nose vertical stabilizer for automatic directional stability and direct control of lateral force; 3—the canard for longitudinal trim through direct control of lift force and automatic stabilization; 4—the hinged wing leading edge for direct control and increased lift force; 5—a spoiler (for roll control and braking); 6—the flaperon for direct control of lift force; 7—an outboard aileron; 8—an aileron; 9—a stabilizer capable of differential deflection of the right and left parts for roll control; 10—the vertical empennage with rudder or all-moving empennage (for directional static stability, sideslip angle control, and direct lateral force control)

(1) significantly more aggressive maneuverability due to a faster change in aerodynamic forces than the "classic" method, which involves turning the entire aircraft to change the angles of attack, roll, or sideslip;
(2) the capability of changing the aircraft's flight path without changing its spatial orientation (angular attitude);
(3) the capability of reducing the aerodynamic loads caused by the wind gusts ("buffeting");
(4) the capability of distributing the manoeuvring aerodynamic loads across the wing for lower bending moments at the wing root.

The latter two factors will allow for an improved aircraft performance due to improvements in both the load distribution and the airframe life.

The empennage design should provide the desired characteristics of the aircraft's stability and controllability in all flight modes. It also should provide sufficient efficiency of controls for stabilizing the aircraft in the normal flight mode after unintended exceeding of the critical angle of attack, stall, and the aircraft pitching down after stall or spinning dive. Therefore, the main relative parameters of the empennage and wing, as well as the required efficiency of the control surfaces should be selected in the course of general aircraft engineering after the layout, configuration, and centre of mass of the aircraft are established (see chapters "Analysis and Selection of Aicraft Configuration and Type of Power Plant" and "Aircraft Layout").

The efficiency of a tail can be assessed per variation of non-dimensional moment coefficients C_l, C_m, C_n with the unit change of α and β, i.e., the following derivatives:

$$\begin{aligned} C_{m,HT}^{\alpha_{HT}} &= C_{L,HT}^{\alpha_{HT}} K_{HT} \overline{S}_{HT} \overline{L}_{HT} = C_{L,HT}^{\alpha_{HT}} K_{HT} A_{HT}; \\ C_{n,VT}^{\beta} &= C_{Y,VT}^{\beta} K_{VT} \overline{S}_{VT} \overline{L}_{VT} = C_{Y,VT}^{\beta} K_{VT} A_{VT}; \\ C_{l,VT}^{\beta} &= C_{Y,VT}^{\beta} K_{VT} \overline{S}_{VT} \overline{z}_{VT}, \end{aligned} \tag{1}$$

where $\overline{S} = S_{\mathrm{VT}} / S$ and $\overline{z}_{VT} = z_{VT}/b$.

The effectiveness of the elevators and rudders mounted on a tail is measured using the derivatives of the respective moment coefficients with the unit deflection of such control surfaces:

$$\begin{aligned} C_m^{\delta_{el}} &= C_{L,HT}^{\delta_{el}} K_{HT} \overline{S}_{HT} \overline{L}_{HT} = C_{L,HT}^{\delta_{el}} K_{HT} A_{HT}; \\ C_n^{\delta_{rud}} &= C_{Y,VT}^{\delta_{rud}} K_{VT} \overline{S}_{VT} \overline{L}_{VT} = C_{Y,VT}^{\delta_{rud}} K_{VT} A_{VT}; \\ C_l^{\delta_{rud}} &= C_{Y,VT}^{\delta_{rud}} K_{VT} \overline{S}_{VT} \overline{z}_{VT}. \end{aligned} \tag{2}$$

If the "classic" control surfaces arranged at the tail surfaces are unable to provide the aircraft with the required manoeuvrability, which is typical for supersonic aircraft at high altitudes, then all-movable empennage surfaces are used, whose efficiency is much higher (Fig. 2) and is measured by the variation of their respective moment coefficients with the unit angle of their deflection ($C_m^{\varphi_{HSt}}$, $C_n^{\varphi_{VSt}}$, $C_n^{\varphi_{VSt}}$).

As it follows from expressions (1) and (2), the empennage efficiency is defined not only by its area ratio $\overline{S}_{\mathrm{emp}}$ and the length $\overline{L}_{\mathrm{emp}}$ but also by the aerodynamic characteristics $C_{L,HT}^{\alpha_{HT}}$, $C_{L,HT}^{\delta_{el}}$, $C_{Y,VT}^{\beta}$, $C_{Y,VT}^{\delta_{el}}$ that much depend on the empennage shape, its aspect ratio *AR*, sweep *Λ*, taper ratio λ, airfoil shape and thickness ratio *t/c*, area and chord ratio of control surfaces arranged on the empennage S_{EL}, S_{R}, $\overline{c}_{el}$, $\overline{c}_{rud}$. The values of maximum control moments depend also on the maximum allowable angles of deflection of the control surfaces.

Hence, upon deciding on the aircraft layout, the empennage's arrangement, and its required relative parameters, the objectives of the empennage design as such include: tailoring its optimum shape, determining the dimensions, determining the

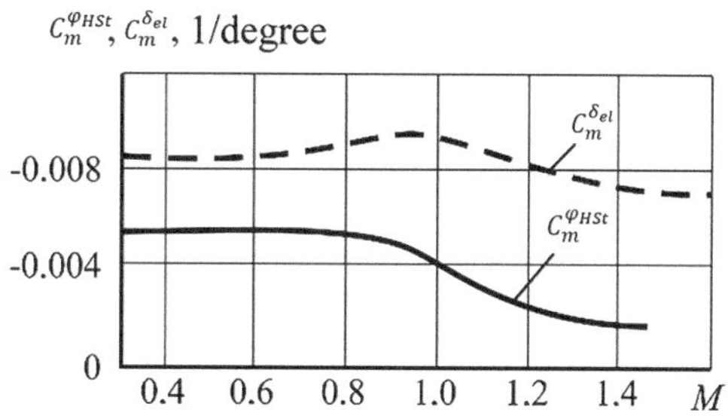

Fig. 2 Variation in elevator efficiency $C_m^{\delta_{el}}$ versus all-movable stabilizer efficiency $C_m^{\varphi_{HSt}}$ with the Mach number for similar size empennages

effective loads, designing the structural layout and its linkage to the fuselage structural elements, and performing a preliminary evaluation of the empennage mass and its effect on the aircraft center of gravity.

2 Selection of the Empennage Shape

As it follows from expressions (1) and (2), the efficiency of empennage and control surfaces arranged on it depend not only on the area of empennage S_{emp} and distances from the empennage centres of pressure to the aircraft centre of mass (L_{emp}) but also on its aerodynamic characteristics.

The most important characteristics thereof are the derivative of the empennage lift coefficient with respect to its angle of attack $a_{HT} = C_{L,HT}^{\alpha_{HT}}$ and the derivative of the vertical empennage lateral force coefficient with respect to the sideslip angle $a_{VT} = C_{Y,VT}^{\beta}$. These derivatives greatly depend on the empennage shape and size. Empennage planform and airfoil depend to a certain extent on the shape and airfoil thickness of the wing. Currently, the typical empennage shape is a trapezoidal straight or swept shape; sometimes it is a delta shape.

For transonic and supersonic wing-tail aircraft, a swept empennage is usually adopted. Such a shape allows the empennage to achieve sufficiently high values of the critical Mach number with a comparatively low gain in drag, high values of a_{HT} (due to a relatively high aspect ratio), and some increase in the empennage arm (Fig. 3a).

In establishing the empennage sweepback angle Λ, aspect ratio AR, and empennage airfoil thickness ratio t/c, one should try to ensure sufficient empennage efficiency at any angles of attack of the wing, as well as $M_{\text{crit emp}} > M_{\text{crit wg}}$. Typically, empennage uses symmetric airfoils with a thickness ratio somewhat lower than that of the wing, a low aspect ratio, and a high sweepback angle.

The empennage thickness ratio for low-speed aircraft with straight wings is typically 10–12%. For high-speed aircraft with a swept-back empennage, the thickness ratio is 1–2% smaller than the wing airfoil thickness ratio and usually amounts to 5–6% (the chords are measured parallel to the free stream velocity).When a straight back empennage is used on a supersonic aircraft, then sufficiently high values of M_{crit} are ensured by a low aspect ratio (1–2) and very small airfoil thickness ratio (3–4%). Approximately the same airfoil thickness ratio (4–5%) is used for the delta-shape empennages.

Rather significant effect on increase in values of M_{crit} for the airfoils and, consequently, the entire empennage, is made by the maximum airfoil thickness chordwise location. The highest values of M_{crit} are for those airfoils where the maximum thickness falls approximately in the middle of the chord. It is precisely these airfoils that are adopted for empennages.

Airfoil modifications that reduce the trailing edge angle have a significant effect on the efficiency of empennage and controls located near its trailing edge, especially at transonic and supersonic flight speeds (Fig. 4). For this purpose, it is not uncommon

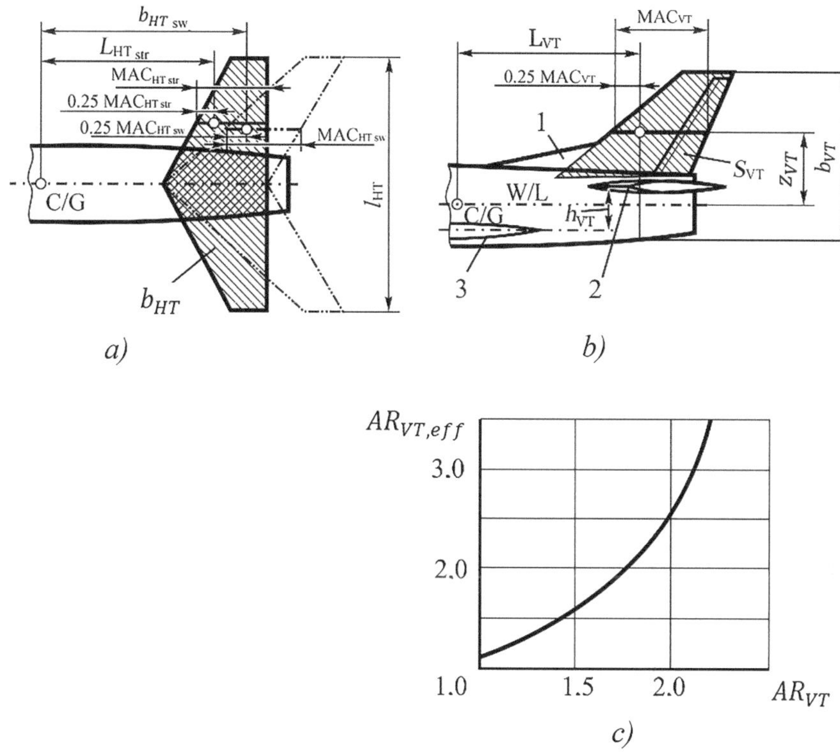

Fig. 3 Diagram for determining the tail size: **a** the variation of the horizontal tail parameters with an increase in its sweep (crosshatching shows the fuselage portion of the tail $S_{\mathrm{HT\,fus}}$; the wetted area of the horizontal tail $S_{\mathrm{HT\,req}} = S - S_{\mathrm{HT\,fus}}$; **b** parameters of the vertical tail; **c** the variation of the effective aspect ratio of the vertical tail $AR_{VT,eff}$ with the geometric aspect ratio AR_{VT}; 1—dorsal fin; 2—leading edge $\mathrm{MAC_{HT}}$; 3—inboard wing rib; $AR_{HT} = b_{HT}^2/S_{HT}$; $AR_{VT} = b_{VT}^2/S_{VT}$

for the empennage airfoils to be modified to replace the arc-trailing elements of the airfoil with straight ones (shown on top of Fig. 4).

To improve efficiency, the empennage on high supersonic and especially hypersonic aircraft is sometimes designed with blunt trailing edges (as the elevons on the Swedish aircraft SAAB-35) or tapered edges (as the vertical tail airfoils on the experimental hypersonic aircraft North-American-15).

The following empirical expression can be used in determining M_{crit} for the empennage:

$$M_{\mathrm{crit\,emp}} \approx M_{\mathrm{crit\,airf}} \frac{2}{1 + \cos \chi_{\mathrm{emp}}},$$

where χ_{emp} is the sweepback angle of the empennage along the line of aerodynamic centres (approximately, it can be assumed to coincide with 1/4 airfoil chord line).

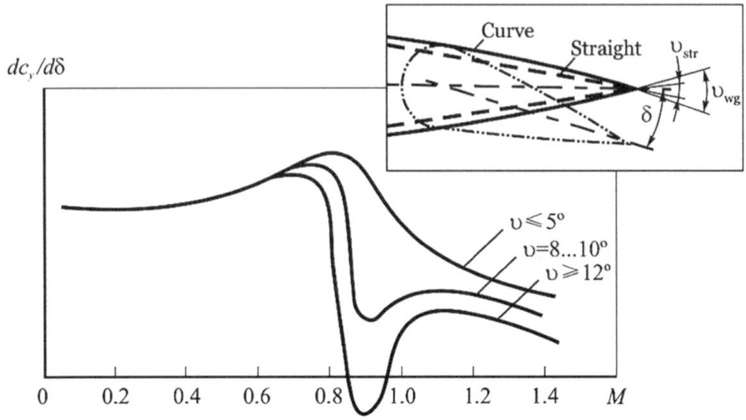

Fig. 4 Variation of the control surface efficiency with Mach numbers at various airfoil trailing edge angles ϑ; the dashed line at the top right shows the airfoil modification for a lower trailing edge angle

$M_{crit,airf}$ is determined by the following equation:

$$M_{crit,airf} \approx 1 - 0.7\sqrt{t/c_{emp}} - 3.2tc_{emp}/t_{emp}^{1.5},$$

where $c_{y\,\mathrm{emp}}$ is taken corresponding to the flight modes for which M_{crit} is determined for the wing.

For supersonic aircraft, one usually tries to achieve a "subsonic leading edge" for empennage, i.e. the edge located inside the shock wave cone. This condition is met at $\chi_{\mathrm{L/E}} > \arcsin\frac{1}{M_{\max}}$. The sweepback angle of empennage usually exceeds the wing sweepback angle by 3°–5°.

The aspect ratio *AR* and taper ratio λ for horizontal and vertical tail are chosen with the effect of such parameters on the structural mass and empennage efficiency in mind.

When the aspect ratio increases, the empennage efficiency increases as a consequence of an increase in a_{HT} (Fig. 5a) and a smaller portion of the empennage area is blanketed by the wing wake (Fig. 5b). Hence, an increase in the empennage aspect ratio makes it possible to reduce the area required for the stability and controllability [see (1) and (2)]. However, with a higher aspect ratio, the mass of the empennage structure and its attachment assemblies grows due to a higher bending moment and torque (with the same load on the empennage). Moreover, with a higher aspect ratio of the empennage, the adverse effect of the empennage elastic deformations on the stability and controllability characteristics significantly grows, and the flutter characteristics worsen (the flutter limit decreases).

Evidently, there is an optimum value $AR_{emp,opt}$ with which the structural mass is the lowest (Fig. 5c).

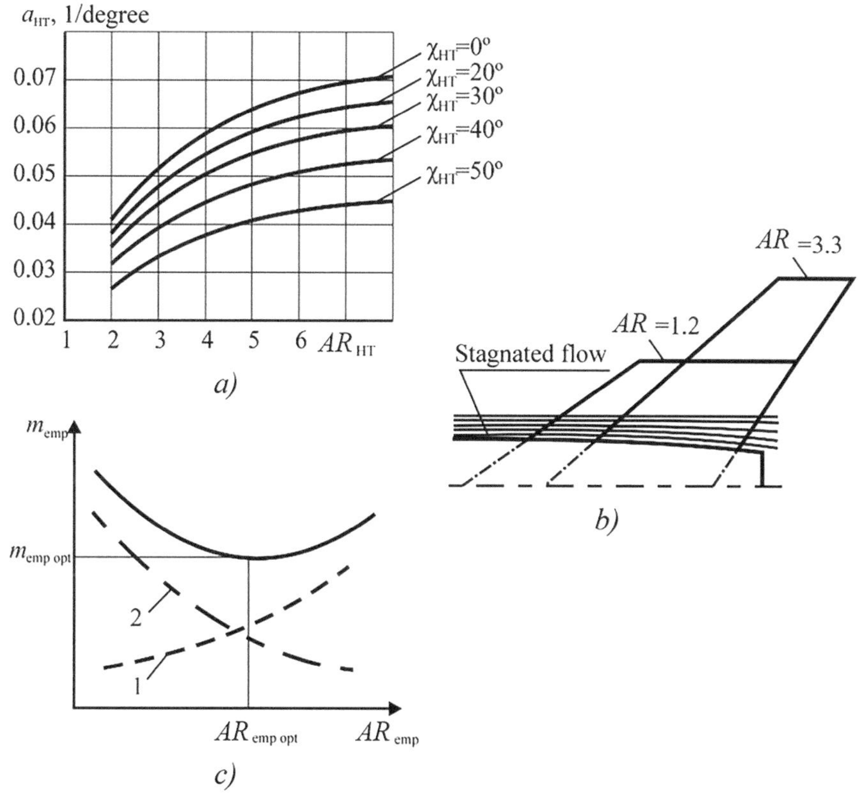

Fig. 5 Effect of the empennage's parameters on its efficiency: **a** dependency of a_{HT} on the empennage aspect ratio at subcritical Mach numbers; **b** decrease in tail area exposed to stagnated flow as the aspect ratio increases (the empennage areas are equal); **c** origin of the optimum empennage aspect ratio at which the empennage mass is the lowest (with the given efficiency). 1—increase in structural mass due to an increase in the empennage aspect ratio; 2—decrease in the empennage mass due to a smaller area required

With an increased empennage taper ratio, its structural mass decreases due to lower bending moments, but the efficiency of the part of the empennage exposed to the wing wake decreases as well. The latter prompts us to increase the empennage area, which, of course, results in an increase in its mass. Hence, it is possible to properly select the combination of empennage parameters with which its structural mass appears to be the lowest (similarly to the selection of the wing parameters) only through optimization.

Processing statistical data from modern aircraft yields the following values for the aspect ratio and the taper ratio for the horizontal and vertical tails:

$AR_{HT} = 3.5 \ldots 4.5$—for low-speed aircraft with a high wing aspect ratio ($AR > 4.5$);

$AR_{HT} = 2 \ldots 3$—for high-speed aircraft with a low wing aspect ratio;

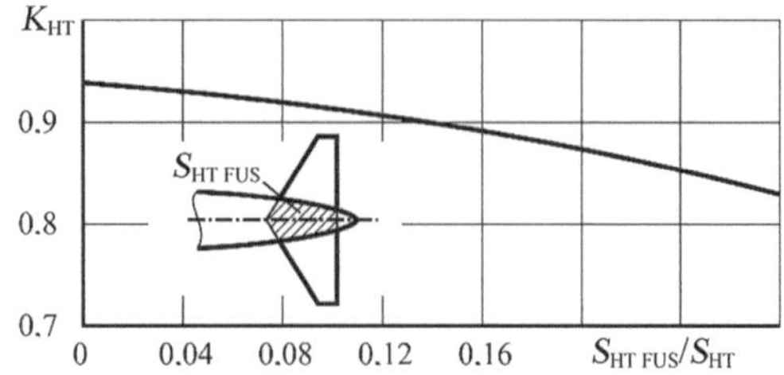

Fig. 6 Horizontal tail efficiency factor as a function of $S_{HT\ FUS}/S_{HT}$ (at subcritical Mach numbers)

$\lambda_{HT} = 2.0 \ldots 3.5$ (for aircraft with delta-shaped empennage, $\lambda_{HT} = \infty$);
$AR_{VT} = 0.8 \ldots 1.2$;
$\lambda_{VT} = 2.0 \ldots 3.5$—for aircraft with the horizontal tail mounted on the fuselage or at the root of the vertical tail.

If the horizontal tail is mounted in the tip of the vertical stabilizer, $\lambda_{VT} \approx 1$, i.e., the vertical tail is designed to have no taper ratio. This is done to avail of the sufficiently large structural height of the fin airfoil at its tip to accommodate attachment units for the horizontal tail and actuators.

The tail efficiency factor for flights at subcritical values of M can be approximately determined as per the curve (Fig. 6). For aircraft with a T-tail, one can adopt $K_{\text{HT}} \approx 0.95$.

The coefficient a_{HT} for straight and swept back empennages with the airfoil thickness ratio $t/c = 8 \ldots 10\%$ at subcritical Mach numbers can be approximately determined as per the curves provided in Fig. 5a.

3 Design of Control Surfaces

In tailoring parameters of the control surfaces ($\overline{S}_{\text{ctl}} = S_{\text{ctl}}/S_{\text{emp}}$, $\overline{c}_{ctr} = c_{ctr}/c_{emp}$, $\overline{S}_{\text{a.c.}} = S_{\text{a.c.}}/S_{\text{ctl}}$), one should note that, with an increase in the control surface size, the hinge moment grows faster than its efficiency.

The ratios $C_{m,HT}^{\delta_{el}}/C_{m,HT}^{\alpha_{HT}} = C_{L,HT}^{\delta_{el}}/C_{L,HT}^{\alpha_{HT}}$ and $C_{n,VT}^{\delta_{rud}}/C_{n,VT}^{\beta} = C_{Y,VT}^{\delta_{rud}}/C_{Y,VT}^{\beta}$ usually denote the elevator and rudder efficiency and are designated as n_{elev} and n_{rud}.[1]

At subsonic speeds, $n_{\text{elev}} \approx \sqrt{S_{\text{elev}}/S_{\text{HT}}}$ and $n_{\text{rud}} \approx \sqrt{S_{\text{rud}}/S_{\text{VT}}}$. At supersonic speeds, effectiveness of the control surfaces significantly drops, and $n_{\text{elev}} \approx S_{\text{elev}}/S_{\text{HT}}$, $n_{\text{rud}} \approx S_{\text{rud}}/S_{\text{VT}}$.

The hinge moment of a control surface is defined by the following expression:

$$M_{hin} = \left(m_{hin,0} + m_{hin}^{\alpha}\alpha + m_{hin}^{\delta}\delta_{ctl} m_{hin}^{\tau}\tau\right) S_{ctl} c_{ctl,av} q. \tag{3}$$

[1] For all-movable control surfaces (stabilizer or fin) the concept of efficiency coefficient is still relevant since $C_{L,HT}^{\alpha_{HT}} \neq C_{L,HT}^{\varphi_{el}}$ and $C_{Y,VT}^{\beta} \neq C_{Y,VT}^{\varphi_{rud}}$.

For certain flight conditions, one can assume $\left(m_{\text{hin}0}+m_{\text{hin}}^{\alpha}\alpha+m_{\text{hin}}^{\delta}\delta_{\text{ctl}}+m_{\text{hin}}^{\tau}\tau\right)q = A = \text{const}$. Then, $M_{hin} = AS_{ctl}c_{ctl,av}$. Taking into consideration that with S_{ctl} = const, the control surface area can only change if the chord is changed, and the control surface span remains constant $\left(S_{ctl} = c_{ctl,av}b_{ctl}\right)$, then $M_{hin} = Ab_{ctl}c_{ctl}^{2}$.

Hence, the control surface efficiency at subsonic speeds increases in proportion to $\sqrt{c_{ctl,av}}$ and the hinge moment increases in proportion to c_{ctl}^{2}.

If the control surface area ratio $\left(\overline{S}_{ctl} = \frac{S_{ctl}}{S_{emp}} = \frac{c_{ctl,av}b_{ctl}}{c_{emp,av}b_{emp}} = \overline{b}_{ctl}\overline{c}_{ctl,av}\right)$ increases, the force transferred from the control surface to the fixed empennage rapidly increases as a consequence of the air load and hinge moment on it. The structure of the fixed stabilizer and fin should be strengthened in order to react such a growing force, which results in a fast increase in its mass. When tailoring the empennage parameters and the area ratio of control surfaces, compare the weight effectiveness of increasing the empennage area vs. increasing the control surface area. This means that an optimum value of $\overline{S}_{ctl}$ should be decided on.

It is desirable that the chord ratios of the control surface be constant along the span of the empennage, $\overline{c}_{ctl} = \frac{c_{ctl}}{c_{emp}} = const$. Otherwise, first, the hinge moment coefficient grows significantly (primarily due to one of its components, m_{h}^{α}—see expression (3)), and second, the caps of the rear spar of the fixed empennage (stabiliser or fin) to which the control surface is attached feature variable twisted bevels, which complicate manufacturing.

For low-speed aircraft, the optimum area ratios of control surfaces are usually $\overline{S}_{\text{elev}} = 0.3$–$0.4$; $\overline{S}_{\text{rud}} = 0.35$–$0.45$.

For modern high-speed (transonic) aircraft, the optimum area ratios of control surfaces are somewhat lower: $\overline{S}_{\text{ctl}} = 0.2$–$0.3$.[2]

An important issue in designing control surfaces is deciding on the degree of its axial compensation $\overline{S}_{\text{a.c.}} = S_{\text{a.c.}}/S_{\text{ctl}}$.

For small low-speed aircraft with manual control system, $\overline{S}_{\text{a.c.}} = 0.2-0.25$ is usually selected based on ensuring the required characteristics of manoeuvrability and controllability, as well as considering that with $\overline{S}_{\text{a.c.}} \geq 0.28$ overcompensation takes place. For instance, the required axial aerodynamic compensation of the elevator for such aircraft is limited by the trim force in level flight $P_{\text{elev tr}}$ and the force per unit load factor $P_{elev}^{n_z}$.

Power-controlled systems (see chapter "Analysis and Selection of Aicraft Configuration and Type of Power Plant") provide the opportunity to decrease the axial aerodynamic compensation of control surfaces to the so called "designed" degree (Fig. 7). This significantly reduces the drag due to the slot between the fixed empennage and control surface (especially at supersonic speeds) and eliminates the drag of the leading portion of the control surface protruding into the airflow at high angles of deflection with higher axial compensation.

The augmentation of the lift coefficient of the horizontal tail due to deflection of the elevator and the side-force coefficient of the vertical tail due to deflection of the

[2] In aircraft with long fuselage tail parts (and consequently, with larger empennage length), the relative areas of control surfaces can decrease to values $\overline{S}_{\text{ctl}} = 0.1-1.15$ as with the Boeing B-52.

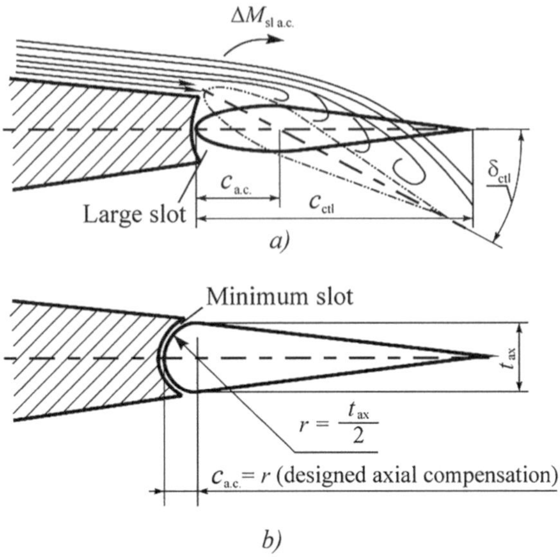

Fig. 7 Origin of the overcompensation hinge moment $\Delta M_{sl\ a.c.}$ on a control surface with a high axial compensation due to protrusion of its nose into the air flow at high angles of deflection (**a**) and formation of the "designed" axial compensation of control surfaces on a high-speed aircraft (**b**)

rudder gradually slows down and then drops because of the flow separation at large deflection angles. Simultaneously, the hinge moment and drag steeply rise. Therefore, the maximum deflection angle of control surfaces is limited. For rudders, $\delta_{rud\ max} = \pm 20°$ to $\pm 25°$. The maximum required rudder deflection angles are established for:

(1) takeoff with one engine inoperative and side wind on the same side, and
(2) landing with non-symmetric thrust with side wind, i.e., the same modes as for tailoring the area of the vertical tail or A_{VT}.

The negative angles of deflection (for pitching up) are much greater than the positive ones (for pitching down) because for trimming at high angles of attack (at a low speed or when manoeuvring at a positive load factor), the elevator should create much greater nose-up moments than the dive moments. Therefore, the maximum elevator up settings usually lie within $\delta_{rud\ max} = \pm 25°$ to $\pm 30°$, and down settings lie within $\delta_{rud\ max} = +15°$ to $+20°$.

The required elevator deflection angles can be determined using the expression $\delta_{el} = \frac{C_m(\alpha, M)}{C_m^{\delta_{el}}(M)}$.

Dependences $C_m = f(\alpha, M)$ for the present-day aircraft with swept back and delta wings feature pitch instability shown in Fig. 8a. This longitudinal static instability occurs at angles between 12° and 15° and is caused by the flow separation from the tips of the wing and an unfavourable change in the side wash near the horizontal tail.

Such flow separation and the related longitudinal static instability usually occur at angles of attack below the critical one α_{crit}. An exceedance of the angles of attack where the longitudinal static instability occurs results in the aircraft starting to inadvertently move to an even higher angle of attack (so called "pickup" phenomenon) due to an augmentation of the nose-up moment, and if the pilot (or the automatic

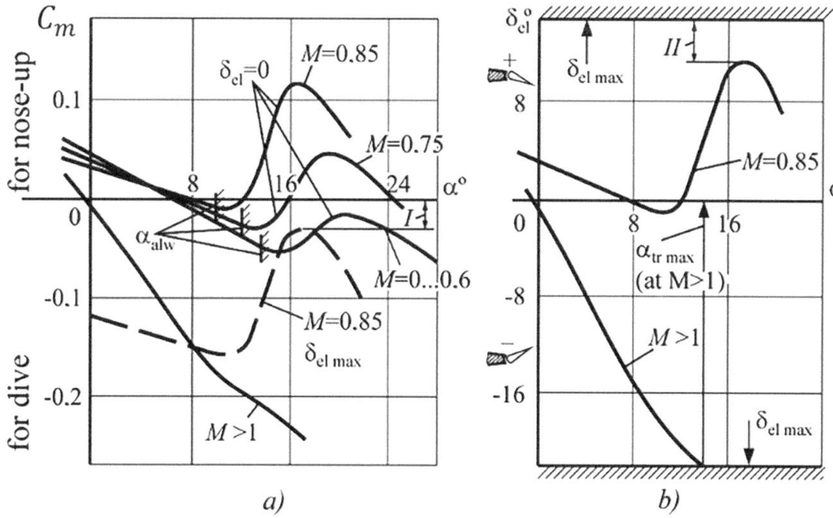

Fig. 8 Longitudinal moment coefficient versus angle of attack for a swept-back aircraft at various Mach numbers (**a**) and corresponding curves of the elevator deflection for $M = 0.85$ and $M > 1$ (**b**): $\delta_{el} = \frac{C_m}{C_m^{\delta_{el}}}$, C_m for the maximum deflection of the elevator down $\delta_{\text{el max}} = 15°$; I—rated moment margin for the dive; II—minimum margin for elevator down setting with $M = 0.85$

stall barrier system) fails to act in time the critical angles of attack can be exceeded, followed by stall.

As a result, today's high-speed aircraft are limited in their allowable angles of attack α_{alw}, ensuring normal operation with the preset longitudinal static stability.[3] As a result, the current aircraft's wing bearing capacity cannot be fully utilized.

When α_{alw} is exceeded (due to air turbulence or as a consequence of the pilot's error), the elevator must provide the dive moment for the aircraft to recover to safe angles of attack $\alpha < \alpha_{\text{alw}}$. This is the design case for the selection of the maximum elevator deflection down (Fig. 8b).

When selecting the maximum elevator-up deflection (or for the stabilizer—nose-down), the design cases are take-off and landing with the forward centre-of-gravity position for non-manoeuvering aircraft, and high-altitude and supersonic-speed flight for manoeuvering aircraft.

The required vertical manoeuvrability is ensured by a high wing bearing capacity and sufficient longitudinal control. It is defined by the ratio

$$n_{z,avail} = C_{L,avail}/C_{L,HFl}, \tag{4}$$

where $C_{L,HFl} = \frac{2G}{\rho V^2 S}$ is the aircraft lift coefficient in level flight, and $C_{L,avail}$ is the available lift coefficient in the given flight mode.

[3] According to the airworthiness directives, α_{alw} in this case shall be lower than the angle of attack at initial loss of longitudinal static stability by 3°.

The lift coefficient at which the aircraft can be balanced at full deflection of the longitudinal control surface $\left(C_{L,\delta_{el,max}}\right)$ is determined through the longitudinal balance equation:

$$C_{L,\delta_{el,max}} = -\frac{C_{m,0} + C_m^{\omega}\omega_y + C_m^{\dot{\alpha}}\dot{\alpha} + C_m^{\delta_y}\delta_{el,max}}{C_m^{C_L}}. \quad (5)$$

With the transition to supersonic speeds, the longitudinal static stability of aircraft $C_m^{C_L}$ usually grows and the pitch authority (especially of the elevators, $C_m^{\delta_y}$) significantly lowers (see Fig. 2). As a result, the wing bearing capacity cannot be fully utilized for maneuvering in such flight modes. As a result, if the aircraft mission requires some load factor at the given Mach numbers and flight altitudes, then when using (4), one should first determine the required $C_{L,avail}$ in this mode (5) and then determine $\delta_{el,max}$ or $C_m^{\delta_y}$. If the required pitch authority of the elevator $C_m^{\delta_y}$ is not ensured there, a stabilizer should be applied, providing for more complete use of the wing bearing capacity and substantial growth of the aircraft's manoeuvrability (Fig. 9).

In determining the maximum required trim elevator-up deflection angle $\delta_{el_{max}}$ by the longitudinal balance equation for an aircraft during takeoff and landing, one should remember to ensure the manoeuvrability margin. The margin needed to ensure the minimum allowed aircraft maneuverability during a path correction and air disturbance attenuation can be calculated using the following expression:

$$\Delta\delta_{el,max} = \frac{I_y (d\omega_y/dt)_{req}}{C_m^{\delta_y}} Sc_{MAC}q,$$

where $\left(d\omega_y/dt\right)_{req}$ is the minimum allowed angular acceleration to be provided by the longitudinal control surface. It is specified in the aircraft mission requirements. Hence, $\delta_{\text{el max}} = \delta_{\text{el tr max}} + \Delta\delta_{\text{el mar}}$.

Similar manoeuvrability margins for the deflection angles should be ensured for the rudder and ailerons.

For an easier trim of heavy subsonic aircraft at high angles of attack (at takeoff and landing), their stabilizer is usually designed to have a variable incidence angle. Changing the incidence angles enhances the CG range, shifting ahead the forward centre-of-gravity limit with the same value of A_{HT} (see Fig. 7, dash-and-dot line). The stabilizer incidence angle varies between $+3°$ and $-12°$.

The area of the control surface tabs (aerodynamic servo tabs) with the manual control system is tailored to ensure a complete elimination of the forces from controls (stick-free) in all long flight modes, including the flight modes with asymmetric power and side wind. This condition is met at

$$m_{hin} = m_{hin,0} + m_{hin}^{\alpha}\alpha + m_{hin}^{\delta}\delta + m_{hin}^{\tau}\tau = 0. \quad (6)$$

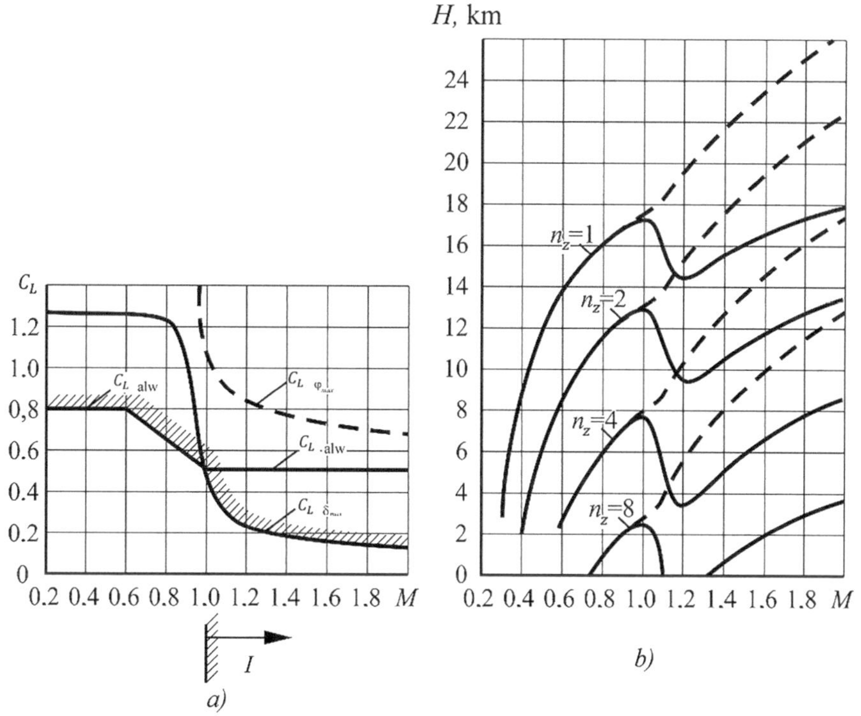

Fig. 9 Pitch authority of elevator versus all-movable stabilizer: **a** available C_L when elevator and stabilizer are used for trim ($C_{L,alw}$ is the limit defined by the wing bearing capacity, the start of the aircraft buffet or entry in the longitudinal static instability zone; I—range of Mach numbers where the wing bearing capacity cannot be fully utilized; $\delta_{el\varphi_{max}}$—stabilizer-operated; $\delta_{el\delta_{max}}$—elevator-operated); **b** shows the limit altitudes where the aircraft can be trimmed by the elevator at a given speed (solid lines) and increase in such altitudes at supersonic speeds if a stabilizer is used (dashed lines)

Therefore, $\left(m_{hin}^{\tau}\tau\right)_{req} = -\left(m_{hin,0} + m_{hin}^{\alpha}\alpha + m_{hin}^{\delta}\delta\right)$, where $m_{hin}^{\tau} = \sigma m_h/\sigma\tau$ depends on the tab area ratio $\overline{S}_{tr} = S_{tr}/S_{ctl}$ and its distance from the hinge axis of the control surface.

The control surface tabs area is usually $S_{tr} = (0.06-0.12)S_{ctl}$, and the higher values are used for heavy transport and military aircraft with widely spaced engines and a large CG range.

Control surface tabs deflection angles are usually $\tau_{max} = \pm(8-12)°$.

4 Structural Design of Empennage and Attachment Points

Structural design of empennage and attachment points depends on the fuselage aft part configuration (or the fuselage forward part configuration, for a canard).

Placing the engines in the fuselage aft part brings up a new arrangement requirement for the main fuselage and empennage structural elements involved with the flight safety in the event of destruction of the engine turbine discs. All the main fuselage and empennage structural elements shall be designed to either prevent them from damage from high speed fragments or minimize the probability of such damage in the event of destruction of the engine turbine. Such requirement is especially difficult to comply with in the presence of a third (central) engine in the fuselage tail which is installed in a number of passenger aircraft (Tu-154, Yak-40, Boeing 727, Lockheed Corporation L-1011 Tristar, DH-121 Trident, Fig. 10).

First, let us consider the structural layout and pattern of joints of the empennage comprising fixed parts (stabiliser and fin) and hinged control surfaces.

In such configuration, the structural layouts of the stabiliser and fin scarcely differ from the corresponding wing layouts considered above. In general, they feature two-spar configurations in light aircraft and torsion-box or multi-spar configurations in heavy aircraft.

Control surfaces, as a rule, feature a single-spar configuration with ribs perpendicular to the spar. The trailing edges of control surfaces often use laminated or welded structures with honeycomb or foam core instead of ribs.

When control surfaces are divided into sections with independent drives, as is done in heavy aircraft for increased flight safety, each section is hinged on two mounts and is not connected to adjacent sections. The number of sections in this case is decided

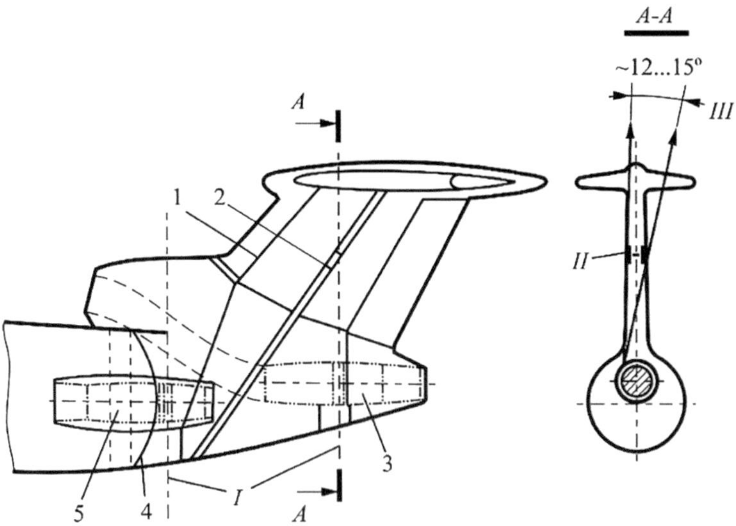

Fig. 10 Arrangement of engines and main structural elements in the aft fuselage part of the passenger aircraft DH-121 Trident ensuring the minimum probability of destruction of the main structural elements with flying fragments of the engine turbine discs: 1—forward fin spar; 2—main fin spar; 3—central engine; 4—pressurised cabin bottom; 5—side engines; I—planes of engine turbine discs; II—spar cross-section; III—area of probable spar destruction with turbine fragments

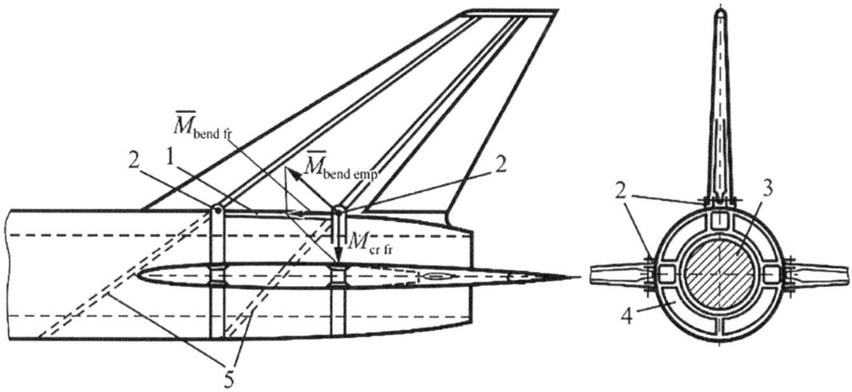

Fig. 11 Structural layout of an empennage attachment to the fuselage's reinforced frames, with the engine installed inside: 1—reinforced longitudinal elements; 2—attachment units; 3—engine; 4—circular bulkhead; 5—optional installation of sloping frames

on the basis of the degree of redundancy and the allowable number of failures of boosters adopted in the project.

The lightest structure is attained when the designer manages to run the main structural elements of the empennage reacting bending (spar or torsion box) through the fuselage. Though such a structure is not easily applied to light aircraft with engines placed in the fuselage tail.

If the horizontal tail of such aircraft cannot be mounted on the top or bottom of the fuselage cross-section, then the stabiliser spars as well as the fin spars are fixed with joint assemblies to the reinforced fuselage frames. To avoid overweight of structure, we should try to arrange the fin and stabiliser joint assemblies on the same reinforced fuselage frames (Fig. 11).

In such a configuration, the bending moment reacted by a longitudinal member of empennage (shown as vector $\overline{M}_{bend\ fr}$) is resolved near the fuselage sidewall into the components acting in the plane of the highest frame stiffness $M_{cr\ fr}$ and perpendicular to it $M_{bend\ fr}$.

When a stabiliser uses not-split longitudinal members, the mid-part of the stabiliser (torsion box) is placed on the fuselage top between two reinforced frames (Fig. 12). The rear spar of the stabilizer is attached to the rear frame, the forward spar is attached to the forward frame. If the stabiliser is of resettable design, then, as a rule, the rear attachments around which the stabiliser is rotated are made fixed and the forward attachment assemblies are replaced with the drive actuator attachment bracket as shown in Fig. 12.

For a more smooth load distribution on the forward and rear reinforced fuselage frames when the right and left halves of the horizontal tail are loaded asymmetrically, the drive actuator attachment assembly can be made with a guide railing or with torque links (the scissor type applied in the landing gear struts) to take a part of the bending moments. Instances of such a design are shown in Fig. 13b, c. For comparison,

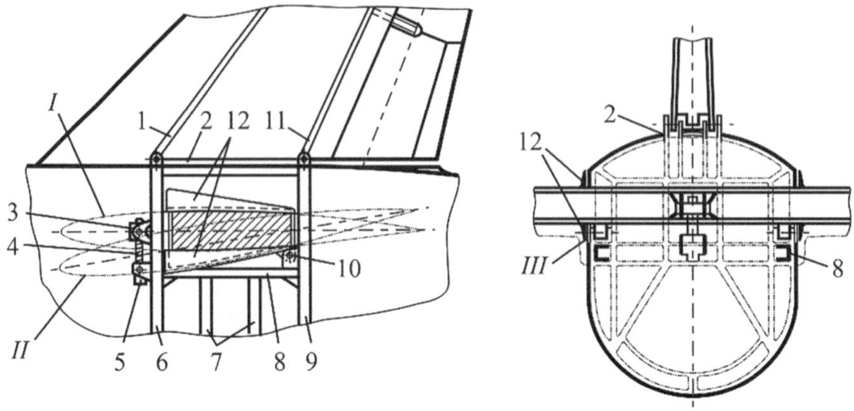

Fig. 12 Installation of variable setting stabilizers on the fuselage top: 1—forward fin spar; 2—longitudinal elements (reinforced stringers) reacting the component $M_{cr\ fr}$ that occurs when the bending moment is transferred from the swept back fin to the straight fuselage frames (see Fig. 11); 3—nut with a hinge fitting; 4—stabiliser setting actuator leadscrew; 5—setting actuator (hydraulic or electric motor with gearbox); 6—forward reinforced fuselage frame (symbolically shown as a contour in cross-section); 7—semi-frames (top portion removed); 8—sidewall longitudinal elements (reinforced stringers edging the cutout for the stabiliser torsion box); 9—rear reinforced fuselage frame; 10—stabiliser-to-fuselage attachment points; 11—rear fin spar; 12—sectorial fairing-plates covering the fuselage sidewall cutouts (the contour of the cutout in the fuselage sidewall skin ensuring the change of the stabilizer's incidence angle is shown by the dashed line, and the stabilizer's torsion box is cross-hatched); I—cruise incidence angle; II—takeoff-landing incidence angle; III—cutout in fuselage sidewall

Fig. 13a shows a schematic where the bending moment from asymmetric loads on the right and left horizontal tail panels is reacted by only the rear attachment units of the stabiliser torsion box; the trim and manoeuvring loads (Y_{HT}) are reacted by the rear attachment units and the actuator lead screw.

The schematic on Fig 13b shows a rear attachment unit of the stabiliser's torsion box partially relieved from the bending moment due to the application of a two-link scissor in the forward unit; force L_{HT} similarly to Fig. 13a is reacted by the rear attachment units and the actuator lead screw.

Changing the stabiliser incidence angle is usually made by a helical mechanism (lifter) rotated by hydraulic or electric motors, or sometimes manually using a mechanical emergency system linking the lifter with a special wheel in the cockpit. The helical lifter can be attached immediately to the stabiliser bracket. However, in this case, the entire load from the horizontal tail (excluding the part that falls on the rear attachment points of stabiliser) shall be reacted by the lifter (Fig. 13a, b). To relieve the lifter at high speeds when the stabiliser is nose-up (positive setting angles) and aerodynamic loads are at their maximum, a special scissor with a kinematic lock can be applied to react these loads and relieve the lifter. The links of this scissor partially relieve the rear attachment points of the stabiliser from the bending moments as well (Fig. 13c).

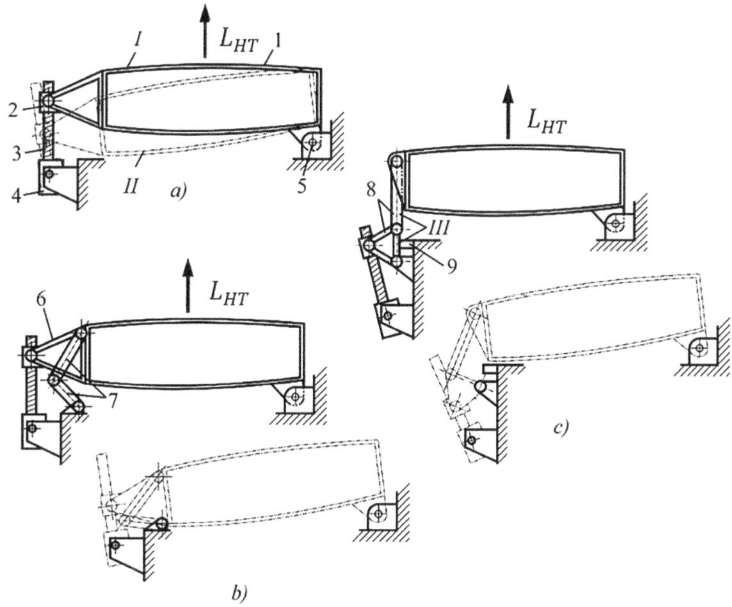

Fig. 13 Structural design of a forward stabiliser actuator assembly with a variable setting angle: **a** without additional structural elements; **b** with a three-link scissor; **c** with a three-link scissor with a kinematic lock; 1—stabiliser torsion box; 2—nut with a hinge fitting; 3—lead screw; 4—setting actuator drive; 5—rear stabiliser attachment units (to fuselage or fin); 6—fitting bracket for the nut 2; 7—scissor links; 8—kinematic lock links made as a scissor; 9—safety stop of the kinematic lock; I—cruise setting; II—takeoff-landing setting; III—kinematic lock links are straight, load is not transferred to the lead screw

With the T-tail configuration, a number of unique design challenges arise that differ from the challenges of arranging the horizontal tail on the fuselage.

To decrease the horizontal and vertical tail interference, it is recommended to position the horizontal tail in such a manner that its maximum airfoil thickness line be shifted backward relative to the same line of the vertical tail by $e \approx (0.01 \ldots 0.15)c_{VT}$.

Installation of the horizontal tail on the vertical tail will result in a substantially higher load on the latter than with the above configurations. Such additional loads on the vertical tail occur both at sideslip (due to an uneven flow over the right and left horizontal tail, the vertical tail is affected by the bending moments perpendicular to its plane) and in flight with no sideslip angle (the trim and manoeuvring loads on the horizontal tail produce for the fixed vertical tail additional bending moments in its plane). To react these additional loads, the vertical tail structure shall be made stronger, which increases its mass (by 50–70%) relative to the mass of a free vertical tail with the same area. Though, an increase in efficiency of the vertical tail due to the installation of the horizontal tail on it acting as an end plate allows for the reduction of the required area S_{VT}. As a consequence, the total increase in mass of the T-tail drops down to 15–20%.

When deciding where to place the fin spars with the horizontal tail arranged on them, the goal is to find the optimum distance between the fin spars with the least amount of structure mass. This is usually accomplished when the forward spar is located within $(0.15\ldots0.20)c_{VT}$ and the rear spar is within $(0.6\ldots0.65)c_{VT}$.

The hinge unit for stabilizer rotation is typically installed on the rear fin spar. The stabilizer actuator is attached to the forward spar and reinforced ribs.

A fairing surrounds the intersection of the horizontal and vertical tails. Its design falls into three types. If the fairing is rather large (Fig. 14a), the design of the joint is similar to that of the attachment point of a movable stabiliser to the fuselage. The design shown in Fig. 14b features a moving trailing part of the fairing because it is connected with the stabiliser, the forward part is fixed to the fin.

On heavy aircraft where the loads are rather high due to the large area of the horizontal tail, the design of a joint with a movable fairing is applied, the forward part of which encloses the nose of the fin tip (Fig. 14c).

Let us consider now the structural layouts, configurations of joints and drives of all-moving control surfaces. Such structures use two patterns of bearings on the hinge axis, which define to a great extent the structural layouts of such surfaces. The bearings are mounted on the fuselage load-bearing frames (or the fin spars) in the first pattern (Fig. 15a), and the axle is attached to the stabilizer structure. The second pattern (Fig. 15b) features the axle fixed to the fuselage structure, the bearings mounted on the stabiliser itself.

The design features of the first pattern relate to the fact that all types of loads (bending moment and torque, shear, and normal forces) applied to the stabilizer are transferred to the fuselage via only one axle, which is therefore made tubular outside the stabilizer. Near the root rib inside the stabilizer, the tubular cross-section of the axle changes to the H-section of the stabilizer spar, which reacts bending much better while the stabilizer contour reacts the torque. Obviously, the axle with the least mass and highest stiffness should have the largest diameter possible. Therefore, it should be installed where the stabilizer airfoil thickness is the largest. As it follows from Fig. 16a, with such an attitude of the straight axle (A), the distance from it to the centre of pressure of the swept back stabilizer, especially at supersonic speeds, is very large, and, consequently, the hinge moment of the stabilizer will be large as well. If the straight axle of a swept-back stabilizer is installed so that it is approximately between the centers of pressure at subsonic and supersonic speeds (B), the stabilizer hinge moments will be significantly reduced, but ensuring axle strength (and especially stiffness) is practically impossible due to the small structural depth in the trailing part of the airfoil.

Therefore, when designing all-movable control surfaces with a large sweepback angle, their axle is often mounted at some angle Λ_{as} as shown in Fig. 16a (C) and on the diagrams shown in Fig. 15. Such an axle position provides for both its arrangement along the line of maximum airfoil thickness and the least hinge moments.

Though selecting a swept hinge axis for all-movable control surfaces is not as simple as the preceding elementary considerations suggest, the difficulty in making

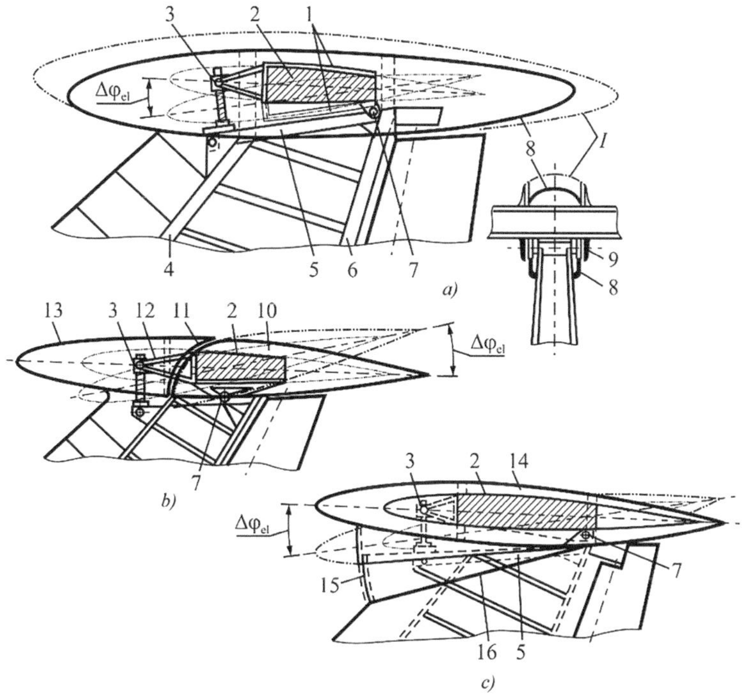

Fig. 14 Various types of the outer contours of a resettable stabiliser installed on a fin: **a** fixed fairing for small variations of the stabiliser incidence angle; **b** fairing on the juncture of horizontal and vertical tails, with fixed forward and moving rear parts; **c** movable fairing connected to the stabiliser (this configuration allows the chord of the non-split stabiliser-torsion box to be increased to its maximum); 1—contour of the skin cutouts on the fairing sides; 2—stabiliser torsion box; 3—stabiliser resetting actuator; 4—forward fin spar; 5—reinforced rib at the tip of the fin; 6—rear fin spar; 7—hinge unit for the stabiliser to fin attachment; 8—fairing; 9—movable sliding plates to overlap the cutouts in the side fairing panels; 10—moving (connected with stabiliser) rear part of the fairing; 11—joint line of the movable and fixed parts of the fairing; 12—stabiliser actuator bracket; 13—forward fixed fairing part secured to the fin; 14—movable fairing linked to stabilizer; 15—fin tip enclosed by fairing in takeoff-landing stabiliser setting; 16—contour line of the fin merging with the contour of the tip enclosed by fairing in takeoff-landing stabiliser setting; I—the required fairing expansion for covering the top slots during takeoff and landing

such a choice is that other important factors must be considered (the required actuating power, the efficiency of the control surface, drag produced by it, structural complexity of the embodied attachment units and actuator, and so on).

The required actuating power is defined by the following expression:

$$N_{\text{req}} = k_M M_{\text{h max}} \dot{\delta}_{\max}, \tag{7}$$

where $M_{\text{h max}}$ is the maximum hinge moment of the control surface about its axis of rotation and $\dot{\delta}_{\max} = (d\partial/dt)_{\max}$ is the maximum required rate of rotation of this axis defined by the aircraft controllability requirements.

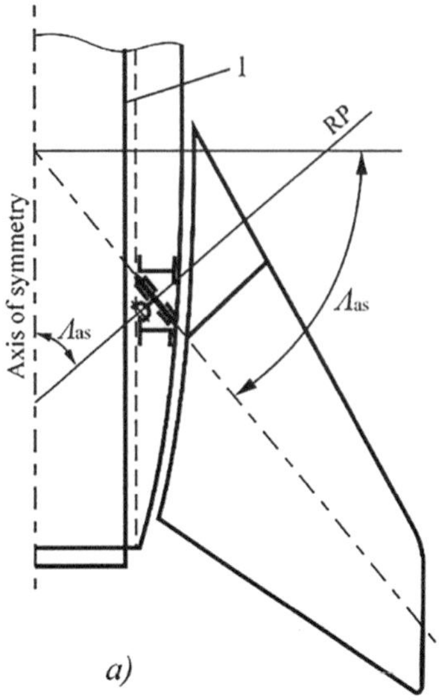

Fig. 15 Patterns of bearings for the hinge axis of a swept back stabiliser: **a** the bearings mounted in the fuselage reinforced frames, the axle fixed to the stabiliser structure; **b** the bearings mounted in the stabiliser, the axle fixed to the fuselage structure; 1—engine contour; 2—axle bearing; 3—drive lever; RP—rocking plane of the drive lever

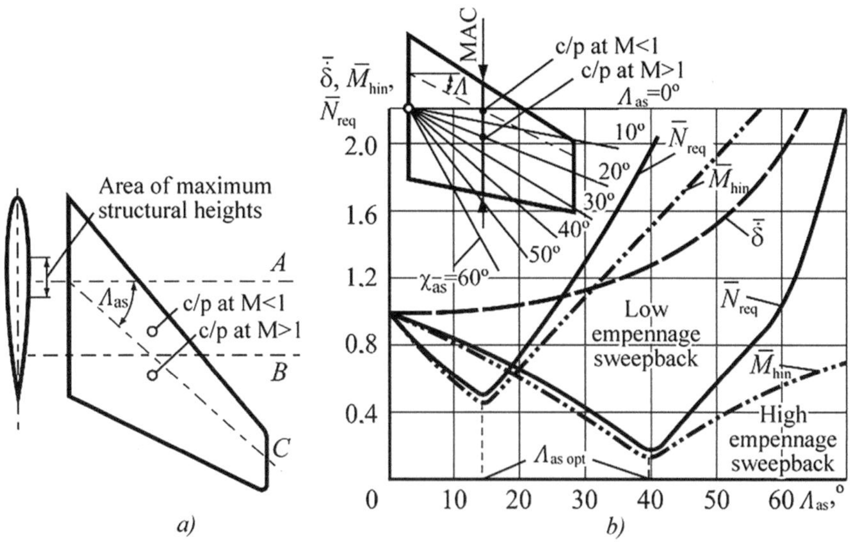

Fig. 16 Optional positions of the sweepback stabiliser hinge axis (**a**) and selection of the optimum sweepback of the hinge axis for the minimum required actuating power (**b**)

If the sweep angle of the axis increases, as shown in Fig. 16b on the left top, the distance from this axis to the centres of pressure of the control surface decreases at first and then begins to increase again. The hinge moments change as well (dash-and-dot lines in Fig. 16b).

The change in aerodynamic forces applied to a control surface and, consequently, its authority, are determined by the change of the incidence angle of this surface relative to the free stream ($\Delta\varphi$) rather than by the angle of rotation of the axis ($\Delta\delta$). These angles are interrelated by the following ratio:

$$tg\,\Delta\varphi \approx tg\,\Delta\delta \cos\Lambda_{as}. \tag{8}$$

As a result, for the controllability to remain constant as the sweep angle of the axis Λ_{as}, increases, the angles of rotation of the axis $\Delta\delta$ and the rate at which they change δ must also increase. The consequence is that the required actuating power increases, as shown by the dashed line in Fig. 16b.

As a result, an optimum sweep angle of the hinge axis ($\Lambda_{as\,opt}$ in Fig. 16b) is determined for each shape of a swept back control surface based on the condition of the minimum required actuating power.

The lift-to-drag ratio of a control surface deteriorates significantly as the axis sweep increases, because as the axis sweep increases, the control surface becomes more "cross-flow" than "in-flow."

In addition to reducing aerodynamic efficiency, the use of control surfaces with swept hinge axes introduces a number of structural issues. These include the necessity to separate the axes for the right and left stabilizer halves and the increase of the required angles of rotation for such axes with the growth of their sweep.

If the axis is common, then bending moments from the right and left stabilizer halves are balanced on it and are not transferred to the fuselage. Therefore, when the axis is straight, the structure of joint assemblies and frames results in the lightest weight. Moreover, configuring the stabilizer drive in this case is easily implemented because the lever mounted on the axle is rocking in a plane parallel to the aircraft axis. Therefore, the stabilizer actuators (hydraulic servos) are easily accommodated even in the rather narrow space between the fuselage sidewall and the internal engine, which is typical for light supersonic aircraft.

Making the swept axes non-split is impossible; this being the case, the resulting structure of joint units and frames is heavier than that with the straight axes due to the transfer of bending moments from each of the two stabilizer halves to the frames. The configuration of actuators of the stabilizer halves is further complicated by the fact that the levers rock in interfering planes at angles to the longitudinal aircraft axis (see Fig. 15). Therefore, the hydraulic servos can be installed horizontally either above or below the engines installed in the fuselage tail section (Fig. 17a). If the stabilizer is installed in mid position, accommodation of the hydraulic servos in rather narrow space between the fuselage sidewall and engines is complicated and the only feasible attitude for them is vertical (Fig. 17b).

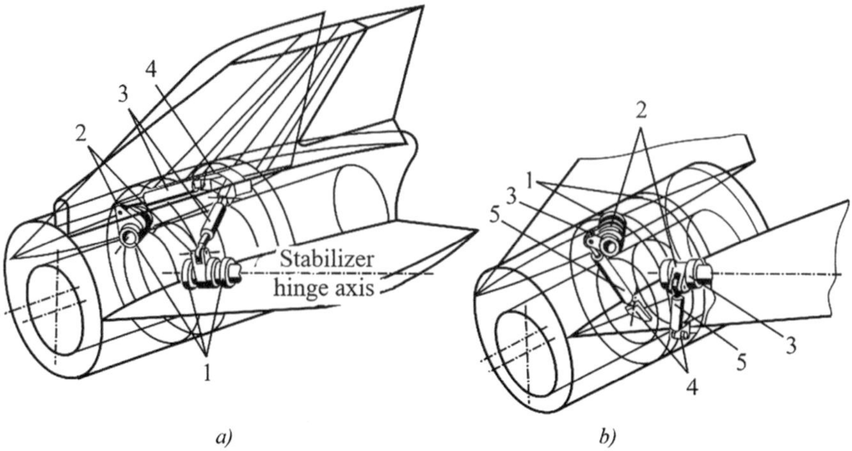

Fig. 17 Layout of a controlled stabilizer actuator's hydraulic servos with swept axes: **a** accommodation of hydraulic servos above the engines and their attachment to a common bracket connected with reinforced longitudinal members between the reinforced frames of the fuselage; 1—bearing on axle-arms for the right and left stabilizer halves; 2—levers of actuators of stabilizer halves; 3—hydraulic servos; 4—attachment bracket of hydraulic servos; **b** optional installation of hydraulic servos in a vertical position with their attachment on the load-bearing frame; 1—load-bearing frames for embedding bearings on axle-arms for the right and left stabilizer halves; 2—actuating levers for the right and left halves; 3—axle-arm bearings; 4—hydraulic servo attachment bracket on a reinforced frame; 5—hydraulic servos

The above disadvantages of all-movable control surfaces with swept hinge axes often outweigh the benefit of lower hinge moments. As a result, new supersonic aircraft are increasingly adopting the all-moving control surface configuration with a straight hinge axis (for example, the F-15 and F-16).To reduce hinge moments for the straight hinge axis of all-movable control surfaces, a special planform (delta or low-aspect-ratio and tapered, as shown in Fig. 18c–f) is used.

Since the axis of rotation of all-movable control surfaces is usually placed between the centers of pressure at subsonic and supersonic speeds for lower hinge moments, in subsonic flight modes such surfaces become overcompensated (statically unstable because the hinge moment tries to deflect them still farther from their initial attitude). This eliminates the availability of change for emergency manual control and forces the use of higher redundancy-level actuators to improve flight safety.

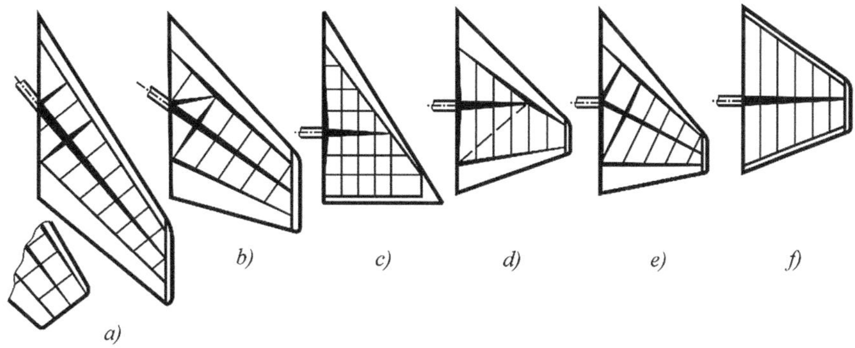

Fig. 18 Structural layouts of all-movable control surfaces with an embedded axle (bearings in fuselage): **a** a large sweep angle control surface (shown below is an optional tip design for insufficient stiffness and vibrations); **b** a relatively low sweep layout; **c** a delta-shape empennage; **d** and **e** optional structural layouts of low aspect ratio empennage with low sweep (in diagram **d**, an oblique rib sometimes used in such layouts is shown by the dashed line)

Structural layouts most often adopted in all-movable control surfaces with embedded axles (bearings in the fuselage) are shown in Fig. 18.

Variables

Symbol	Description	Units
C_l	Rolling moment coefficient	–
C_m	Pitching moment coefficient	–
C_n	Yawing moment coefficient	–
b	Wing span	m
α	Angle of attack	degrees
β	Angle of sideslip	degrees
$C_m^{C_L}$	Longitudinal static stability	–
$C_{m,HT}^{\alpha_{HT}}$	Derivative of HT pitching moment coefficient with respect to HT angle of attack	1/deg
$C_{n,VT}^{\beta_{VT}}$	Derivative of VT yawing moment coefficient with respect to angle of sideslip	1/deg
$C_{l,VT}^{\beta}$	Derivative of VT rolling moment coefficient with respect to angle of sideslip	1/deg
$a_{HT} = C_{L,HT}^{\alpha_{HT}}$	Derivative of HT lift coefficient with respect to HT angle of attack	1/deg
$a_{VT} = C_{Y,VT}^{\beta}$	Derivative of VT side force coefficient with respect to sideslip angle	1/deg
S_{VT}	Vertical tail area	m^2
S_{HT}	Horizontal tail area	m^2
A_{HT}	Horizontal tail volume coefficient	–

(continued)

(continued)

Variables		
Symbol	Description	Units
A_{VT}	Vertical tail volume coefficient	–
K_{HT}	Horizontal tail efficiency factor	–
K_{VT}	Vertical tail efficiency factor	–
$C_m^{\delta_{el}}$	Derivative of pitching moment coefficient with respect to elevator deflection angle	1/deg
$C_n^{\delta_{rud}}$	Derivative of yawing moment coefficient with respect to rudder deflection angle	1/deg
$C_{l,VT}^{\delta_{rud}}$	Derivative of rolling moment coefficient with respect to rudder deflection angle	1/deg
$C_m^{\varphi_{HSt}}$	Derivative of pitching moment coefficient with respect to horizontal stabilizer deflection angle	1/deg
$C_n^{\varphi_{VSt}}$	Derivative of yawing moment coefficient with respect to vertical stabilizer deflection angle	1/deg
$C_n^{\varphi_{VSt}}$	Derivative of rolling moment coefficient with respect to vertical stabilizer deflection angle	1/deg
AR	Aspect ratio	–
Λ, χ	Sweep angle	degrees
λ	Taper ratio	–
t/c	Thickness ratio	–
$\overline{c}_{el}$	Elevator chord ratio	–
$\overline{c}_{rud}$	Rudder chord ratio	–
n_{elev}	Elevator efficiency	–
n_{rud}	Rudder efficiency	–
M_{hin}	Control surface hinge moment	N m
τ	Control surface trim tab deflection angle	degrees
m_{hin}	Hinge moment coefficient	–
S_{ctl}	Control surface area	m^2
$c_{ctl,av}$	Control surface average chord	m
q	Dynamic pressure	N/m^2
$\overline{S}_{ctl}$	Control surface area ratio	–
$S_{a.c.}$	Axial compensation area	m^2
$C_{L,HFl}$	Aircraft lift coefficient in level flight	–
$C_{L,avail}$	Available lift coefficient in the given flight mode	–
$C_{L,\delta_{el,max}}$	Lift coefficient at which the aircraft can be balanced at full deflection of the longitudinal control surface	–
Λ_{as}	Hinge axis sweep angle	degrees
S_{tr}	Control surface trim tab area	m^2

Aircraft Control System Design

Oleg S. Dolgov

Abstract The chapter starts by highlighting the role of the human factor in the design of airplane control systems. The chapter explains the composition and design objectives of manual, automatic, powered, as well as remote control systems. The chapter introduces the concept of a control loop and discusses its main elements and their characteristics. In the chapter, the reader will find information about stability and control of modern high-speed aircraft and methods for their improvement, as well as stability and control augmentation systems. Finally, the chapter provides data on direct lift control systems and aerodynamic load reduction systems.

Keywords Human factor · Control loop · Control augmentation system · Direct lift control system · Aerodynamic load reduction system

In contrast to other aircraft systems, when designing the control system for an aircraft, the human factor must be taken into account [1]. Moreover, to ensure the control system provides the required aircraft controllability, it is important to take into account not only the hinge moments, mass-inertia, and stiffness of the aircraft control surfaces, but also the aerodynamic, inertial, and dynamic characteristics of the aircraft itself.

In order for the aircraft to fulfill the technical requirements for controllability, the available control moment must exceed the required one, which is determined by the aircraft's moment of inertia and the required angular acceleration relative to the axes under consideration. In general, this condition can be written as follows [2]:

$$\left.\begin{aligned} M_0^{\text{req}} &\le M_0^{\text{avail}} \\ M_0^{\text{avail}} &= \sum_i P_i(\delta^i) r_0^i \\ M_0^{\text{req}} &= I_0 \dot{\omega}_0 \end{aligned}\right\} \Rightarrow \iiint\limits_{xyz} r_0^2 dm \le \sum_i P_i(\delta^i) r_0^i / \dot{\omega}_0,$$

O. S. Dolgov (✉)
Moscow Aviation Institute (National Research University), Moscow, Russia

M. A. Pogosyan (ed.), *Aircraft Design*, Springer Aerospace Technology,
https://doi.org/10.1007/978-981-96-4599-2_20

where M_0^{avail} is the available stabilizing or control moment; M_0^{req} is the required stabilizing or control moment; I_0 is the aircraft moment of inertia relative to the axis under consideration; r_0^i is the distance between the resultant aerodynamic force applied to the control surface and the aircraft centre of mass; P_i is the force applied to the control surface; $\dot{\omega}_0$ is the required angular acceleration.

This system of equations reflects the interrelationship between the parameters of the aircraft and the characteristics of the control system.

The values of the required angular accelerations are given in the aircraft specifications. For roll control, for example, they can be determined to first approximation from an empirical dependence developed by the Central Institute of Aerohydrodynamics [3]. It is based on statistical data and a subjective assessment of the controllability reported by the pilots. This dependence is presented in terms of the roll angular acceleration and the roll mode time (T_{roll}):

$$\dot{\omega}_{0x} = \frac{M_{0x}^{\text{avail}}}{I_{0x}},$$

$$T_{roll} = -\frac{I_{0x}}{m_{0x}^{\overline{\omega}_{0x}} qSb} \frac{2V}{b}.$$

It can be seen from the expression above that the roll mode time constant determines the damping rate in rolling motion and depends on the roll damping moment and flight speed.

A graphic interpretation of the control system conceptual design is given in Fig. 1. Nominally, the first level of the hierarchical structure corresponds to the control system. From the system perspective, it is enough to consider two adjacent levels for correct consideration of the problem when solving tasks at a certain hierarchical level.

The conceptual design phase of the control system (Fig. 2) is a complex, multi-iteration process. As part of this process, several stages are distinguished by specific goals, methods, research models, and their results.

The control system conceptual configuration is formed by analyzing the results of multiple calculations of the effectiveness of various options and the conditions for their use. The first step in designing a control system is to develop a hierarchical system of models (Fig. 2).

It should be noted that the strategies for the control system design provide fundamentally different levels of controllability for an aircraft at the design flight conditions.

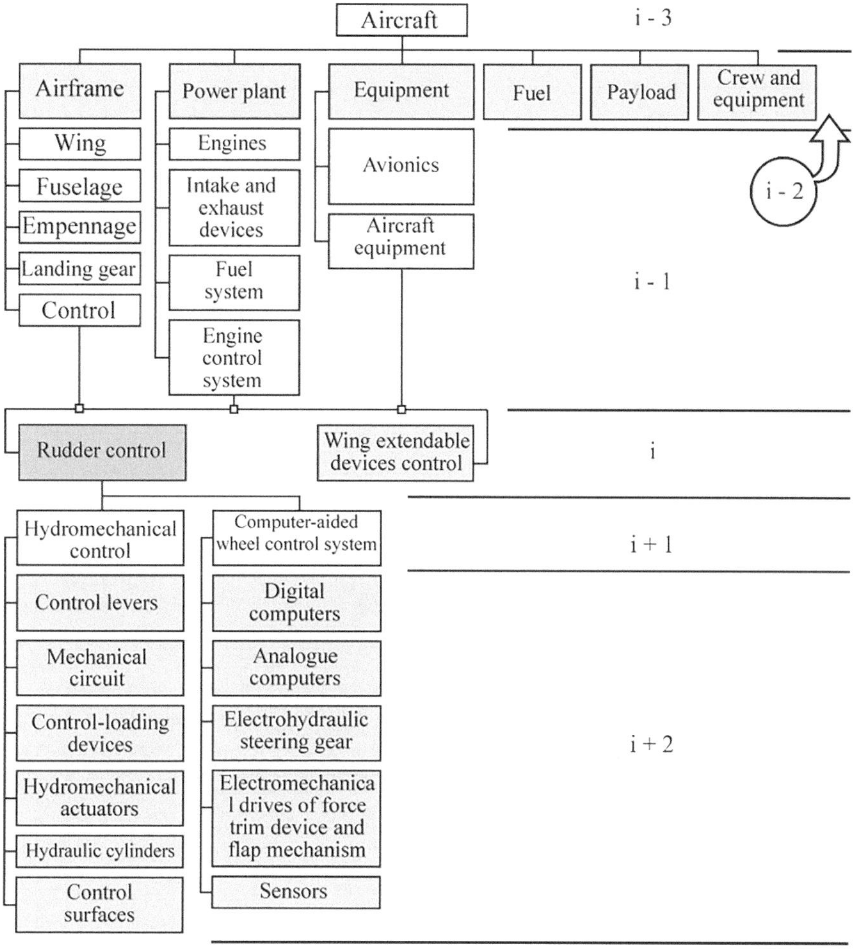

Fig. 1 The hierarchical structure of an aircraft in the conceptual design phase of the control system

1 Composition of the Aerodynamic Control System and Its Design Objectives

The control system of an aircraft is a complex of mechanical, hydromechanical, electrohydraulic, electromechanical, and electrical equipment that can be nominally decomposed according to the nomenclature accepted in the industry [4–6] (Fig. 3):

– roll control. It controls the aircraft's lateral movement and trim by deflecting the ailerons and spoilers;
– pitch control. It controls and trims the aircraft longitudinally by deflecting the elevator and stabilizer;

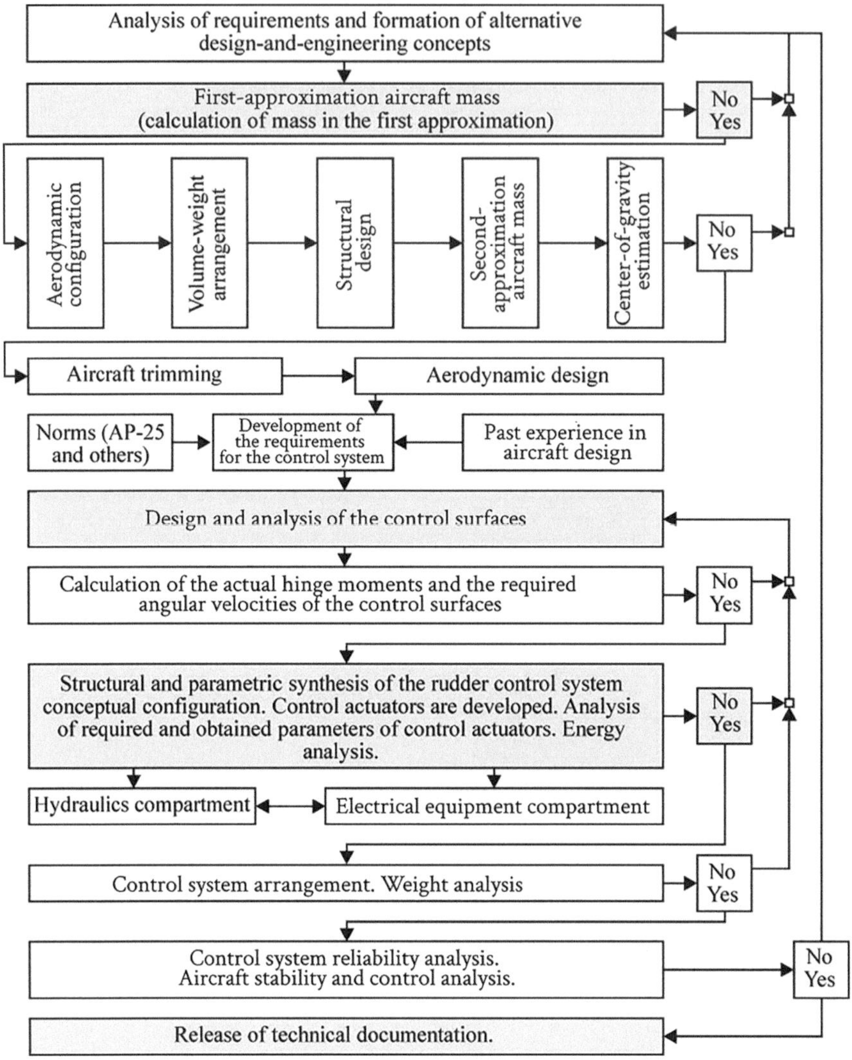

Fig. 2 Design process of an aerodynamic control system

- yaw control. It realizes the directional control and trim of the aircraft through the rudder deflection;
- air brake control. It provides braking and reduces aircraft lift in flight by deflecting spoilers, and when running on the ground by deflecting spoilers and air brakes.

Figure 4 shows the structure of the control system of an airliner.

The control system on aircraft designed to fly at low subsonic speeds is only concerned with ensuring the transmission of control signals from the pilot (or autopilot) to the control surfaces (Fig. 5).

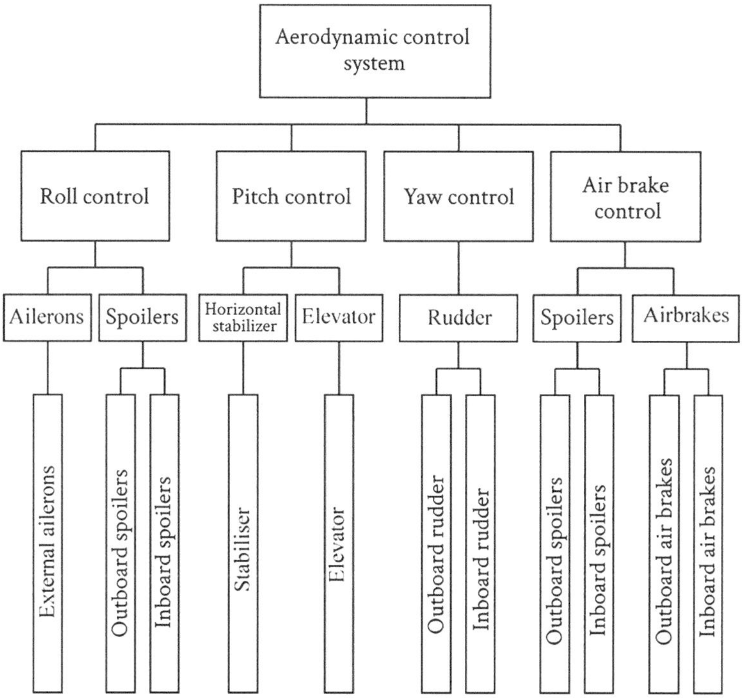

Fig. 3 Structural decomposition of the aerodynamic control system

On a high-speed aircraft, the tasks of the control system become much more complicated, and its composition is significantly extended. The control systems on these aircraft solve the following tasks:

- transmit control signals from the pilot to the controls (flying mode by the pilot);
- transmit control signals from the actuators of the automatic control system to the control surfaces;
- provide the required deflection power;
- provide static and dynamic stability;
- generate the required control characteristics;
- generate control signals for automatic trajectory control;
- generate signals for directors in semi-automatic flight;
- increase pilot safety by using appropriate signaling and limiting control deflections when approaching the maximum permissible values of flight parameters (V, M, n_z, α, β, γ, and others).

On high-speed aircraft of the first generation, these tasks were solved by introducing into the control system a whole range of new mechanical, hydraulic, electromechanical, and automatic systems, each of which solved a specific task [7]. Therefore, along with the mechanical circuit, the control systems of these aircraft included (Fig. 6):

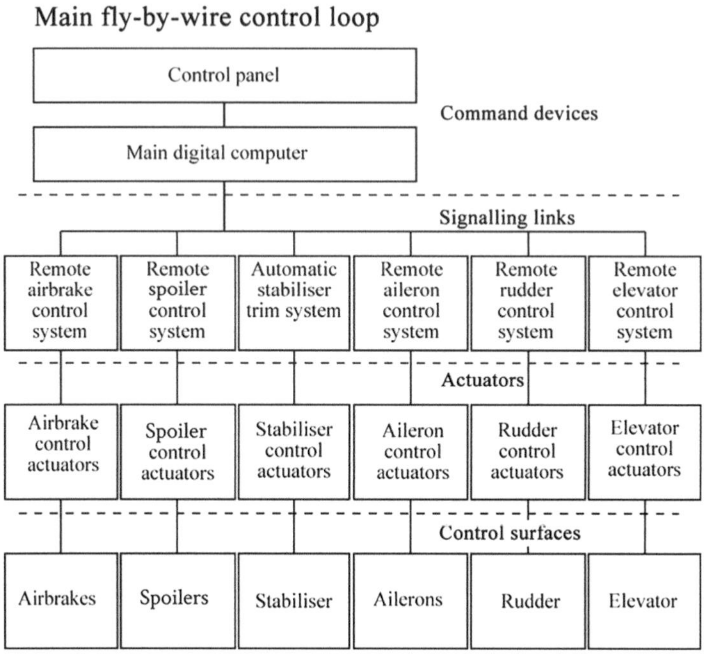

Fig. 4 Structural decomposition of the aerodynamic control system of an airliner

- hydraulic power amplifiers—power control actuators (boosters) and hydraulic systems for their power supply;
- control lever feel systems (with full-powered control);
- systems to ensure the necessary characteristics of static and dynamic stability and control (gain changers in the wiring from control levers to the rudders, vibration dampers, stability controllers, slide controllers, trim controllers, autothrottles, etc.);
- director system with command signal computers;
- auto-steering system (autopilot or other automatic systems that perform tasks such as automatic search and sighting, automatic landing, etc.);
- alarm systems and permissible mode limitation systems.

Not all of the systems listed above could be installed on a specific aircraft. The choice of a particular system was made in the process of designing the aircraft and its control system, based on the characteristics of the aircraft and its purpose. An example of how the elements of a mechanical circuit and the actuators of automatic systems were arranged on a light, maneuverable aircraft is shown in Fig. 7. When installing various automatic devices (in control laws of which the current flight parameters are used) into the control system to ensure the solution of the above tasks, the control loop (refer to Fig. 3a) of high-speed aircraft becomes significantly complicated due to the additional correction circuits that occur there (refer to Fig. 6b). The operation of

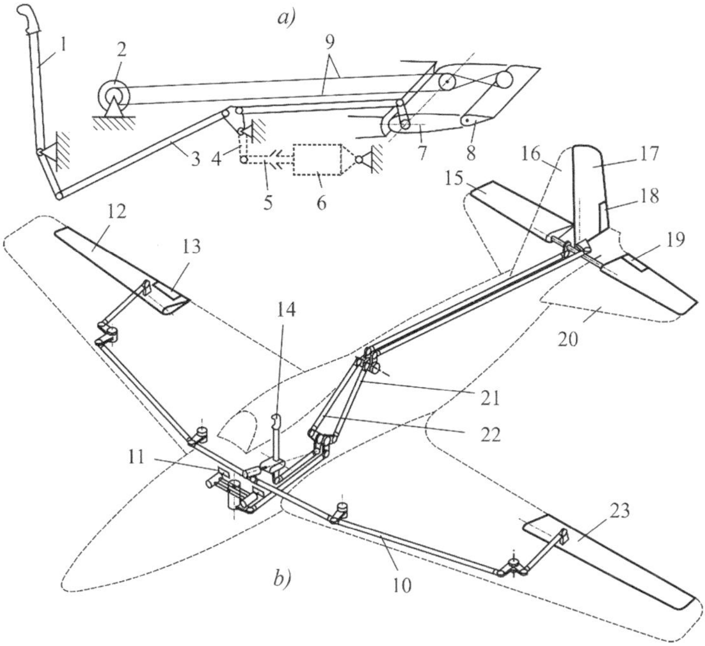

Fig. 5 Manual control in a low-speed aircraft: **a** diagram of the mechanical drive of a control surface; **b** control circuit arrangement and controls of a light low-speed aircraft; 1—actuating bellcrank; 2—tab control wheel; 3—rudder control circuit; 4—bellcrank of the autopilot actuator connection; 5—release clutch of actuator from circuit; 6—autopilot actuator; 7—control surface; 8—aerodynamic tab; 9—tab control circuit; 10—rods of the aileron steering linkage; 11—rudder pedals; 12—aileron (right); 13—aileron tab; 14—aileron and elevator control column; 15—elevator; 16—fin; 17—rudder; 18—rudder tab; 19—elevator tab; 20—stabiliser; 21—rods of rudder control circuit; 22—rods of elevator control; 23—aileron (left)

these compensating circuits ensures optimal controllability characteristics, stability of the entire control circuit in all flight modes, and increased flight safety [8].

The evolution of electronic equipment and the development of high-reliability multi-channel drives in recent years have created a technical basis for replacing numerous automatic and semi-automatic systems (with their sensors, computers, and actuators) with a single automatic flight control system (powerful, multiple redundant) (Fig. 8) that performs all of their functions.

The actuator of the automatic flight control system is a multiple redundant (modular) electrohydraulic device that connects differentially to the control wiring, replacing the actuators of the old systems (Fig. 9). When using such drives that should be able to deflect flight control surfaces at sufficiently high angles to perform their tasks, flight safety is ensured by applying a new redundancy principle.

This principle of force-summing redundant subchannels (modules) and "majority voting" (i.e., "overriding" the failed subchannel with the remaining serviceable ones

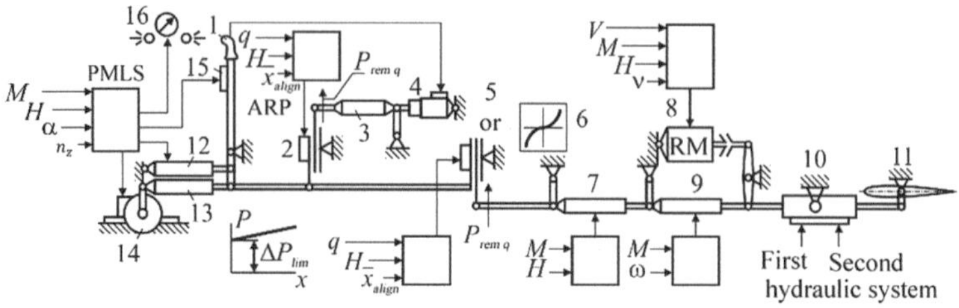

PMLS – Permissible mode limitation system

ARP – Aircraft repair plant

a)

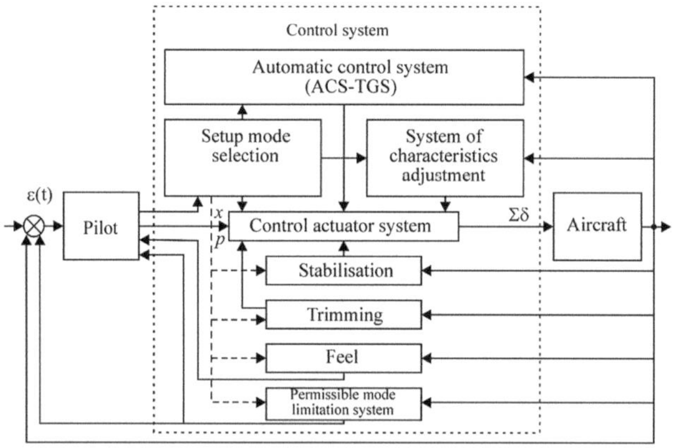

b)

Fig. 6 Composition of a control system in the first generation supersonic aircraft (**a**) and generation of additional links and correction circuits in the control loop (**b**): 1—control lever; 2—feel control actuator; 3—feel mechanism; 4—force trim device with electrical control; 5—gearing adjuster from control lever to rudder; 6—nonlinear transmission mechanism; 7—trim or stability controller; 8—automatic trajectory guidance system (TGS) or autopilot (actuator with clutch connecting to the circuit upon activation); 9—damper; 10—twin-chamber booster with supply hydraulics; 11—rudder; 12—control lever velocity limiter (it can be combined with the pusher of the permissible mode limitation system); 13—limiting loader of the permissible mode limitation system and its characteristics; 14—lever locking mechanism for the permissible mode limitation system; 15—lever shaking mechanism for the permissible mode limitation system; 16—indicator of current values of flight parameters and available ranges for manoeuvres from the light warning system

in majority) replaced the old principle of position-summing redundant subchannels (Fig. 10).

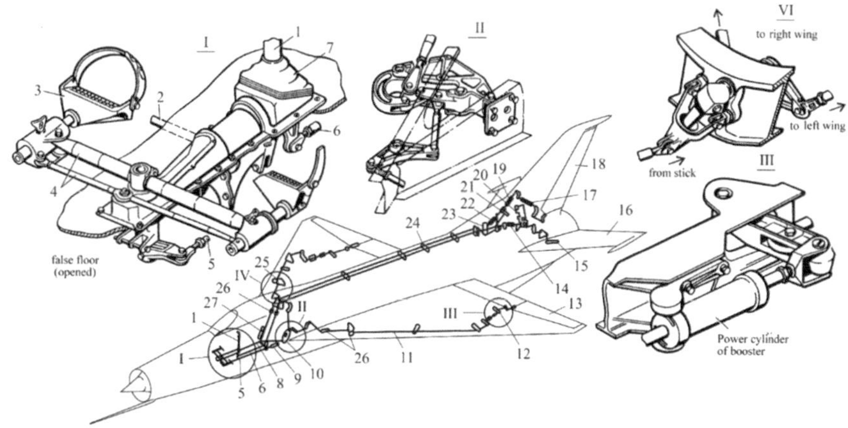

Fig. 7 Layout of the main elements of the mechanical part of the power-controlled system in a light, high-speed aircraft: 1—longitudinal and lateral control stick; 2—aileron control rod; 3—rudder pedal; 4—parallel-link motion; 5—rudder actuation rod; 6—stabiliser actuation rod; 7—protective cover from the ingress of foreign objects into wiring; 8—aileron control rod; 9—feel pedal mechanism; 10—central unit for distributing aileron control to the right and left wings; 11—aileron control circuit in the wing; 12—aileron drive booster; 13—left aileron; 14—adjustable rod of pitch damper; 15—stabiliser hinge axis; 16—all-movable stabiliser (left half); 17—rudder drive booster; 18—rudder; 19—adjustable rod of yaw damper; 20—stabiliser drive booster; 21—force trim device of longitudinal control; 22—feel mechanism of the longitudinal control surface; 23—gain changers in longitudinal control; 24—roller bearings of stabiliser and rudder control rods; 25—nonlinear transmission mechanism in aileron control; 26—rod pressure seals from the pressurised cabin; 27—feel mechanism of lateral control

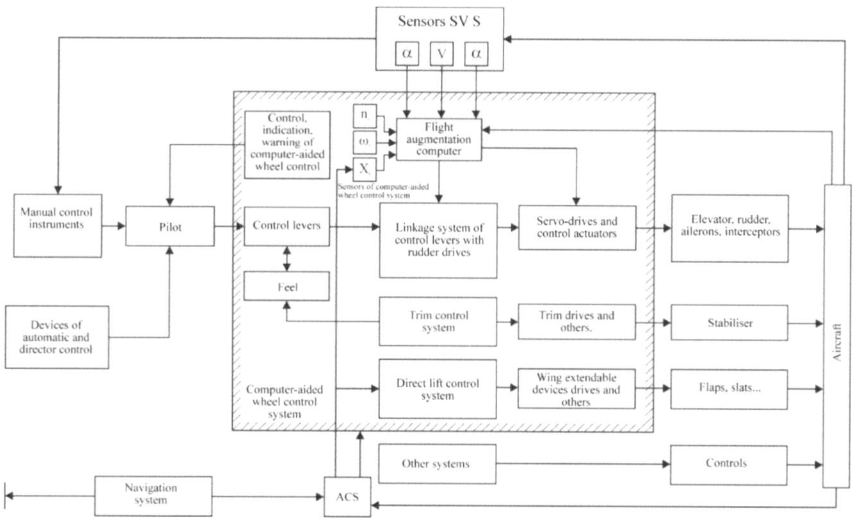

Fig. 8 Structure of an airliner's automated control system [3]

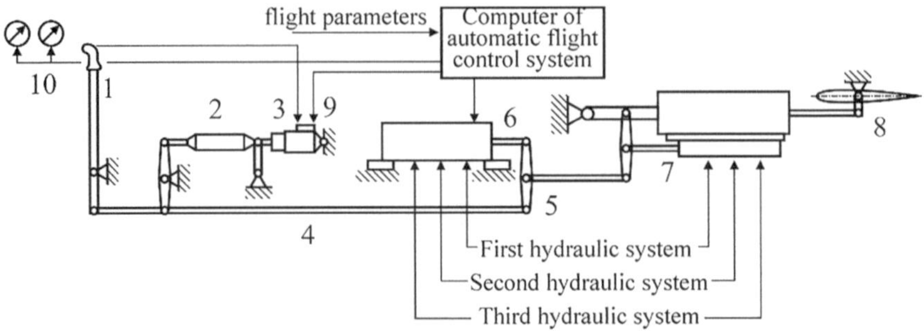

Fig. 9 Composition of control systems in a modern high-speed aircraft. 1—control lever; 2—feel mechanism; 3—force trim device; 4—mechanical circuit; 5—differential summing bellcrank; 6—multichannel drive of automatic flight control system; 7—many-chambered control actuator; 8—flight control surface; 9—lever and rudder coupling signal ("transfer" via force trim device); 10—directors and warning

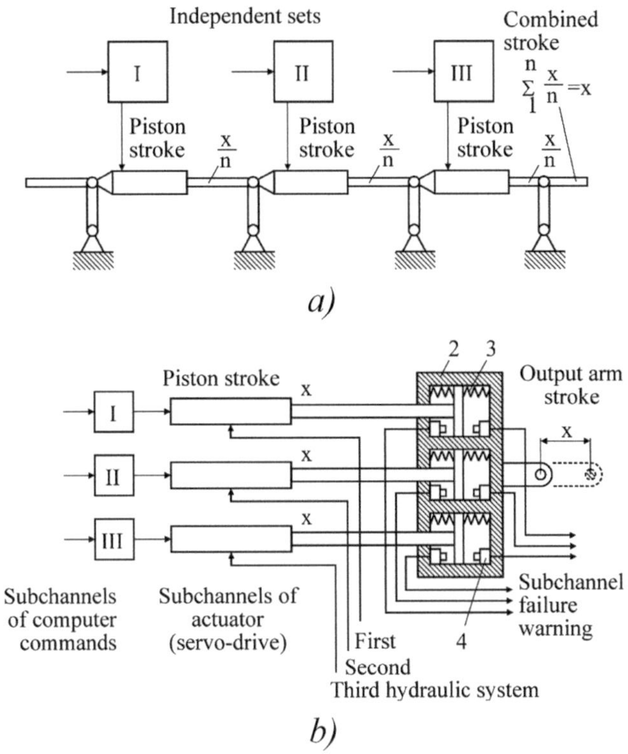

Fig. 10 Two principles of automated control system redundancy: **a** combined strokes of electromechanical actuator pistons to obtain the required signal (X); **b** combined forces on pistons of electrohydraulic subchannels (modules) at common output arm 2; 1—power hydraulic cylinders of subchannels; 2—common output arm of servo-drive in cross-section; 3—pre-loaded springs; 4—limit switches of subchannel failure warning

1.1 Remote-Control Systems

The incorporation of various automatic and computing devices in the control system of modern aircraft (Fig. 11), as well as difficulties in ensuring the necessary accuracy in the mechanical transmission of control signals from control levers to control surfaces, has increased aviation experts' interest in fly-by-wire systems.

According to the principle of transmitting control signals, remote systems can be divided into several groups, for example, fly-by-wire, fly-by-light, radio systems (relevant for unmanned aerial vehicles), etc. At the present stage of scientific and technological development, the most widespread are fly-by-wire systems, which, in turn, can be divided into two main groups depending on the use of analogue or digital equipment [3].

Analog systems were used (as main loops) in the early development of remote control systems. At a later stage, the dominant positions were taken by digital systems, enabling them to implement quite complex adaptive control modes and promptly change them depending on the situation. However, analog systems are still used today as fly-by-wire backup control loops based on alternative physical principles of operation.

The primary advantage of fly-by-wire systems is that they allow for the circumvention of the constraints associated with elastic deformations. Such systems shape the required law of control to further modify it in accordance with the current situation without the need for a substantial change in the system. It is quite easy to correct the pilot's commands in accordance with the estimated maneuver and current parameters of the aircraft. Fly-by-wire systems are far easier to automate using algorithms for augmented stability and control, as well as feeling control. Modern fly-by-wire systems enable the change of groups of control surfaces during flight in the event of the failure of several engines or the seizure of individual control surfaces.

Fly-by-wire systems come with better dynamic properties; they are characterized by lower inertia of the circuit, the absence of adverse friction in signaling links, as well as the influence of elasticity and backlash characteristic of hydromechanical systems. They enable the use of cockpit controls designed in full compliance with the pilots' requirements.

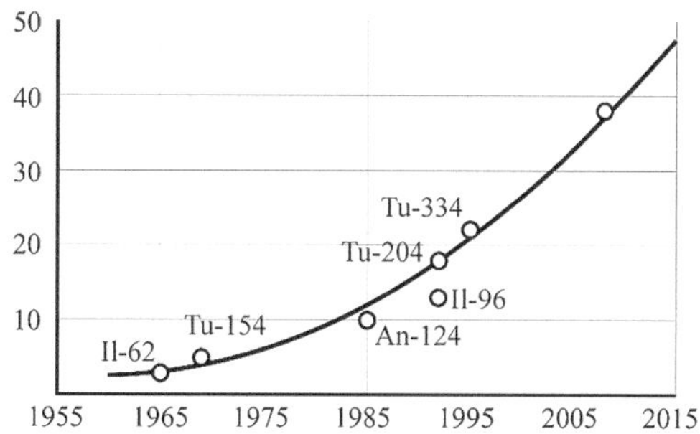

Fig. 11 Increase in automated functions of a computer-aided wheel control system [3]

On the other hand, fly-by-wire systems are liable to failures of a general type, the so-called "avalanche failures." It is hardly possible to detect such failures using the monitoring system. This is due, on the one hand, to the presence of "common latent points" (for example, energy systems, control systems, etc.), and, on the other hand, to low jamming resistance from both external electromagnetic effects and internal interference. Fly-by-light systems are considered advanced technologies that can dramatically increase the jamming resistance of the signaling links.

When using digital fly-by-wire systems, digital-system-specific failures of the software suite are present in addition to equipment failures. That is why the alternative software suites quite often operate in parallel. Their operation is based on equipment of different types. It should be borne in mind that the complexity of creating software that meets modern requirements for reliability and functionality is commensurate with the complexity of developing the control system hardware.

Despite these drawbacks, fly-by-wire systems are the most advanced ones and are used as the main control systems in most modern aircraft at the present stage of scientific and technological development. Fly-by-wire systems force out alternative types of control systems by taking into account continuous improvement of the element base and best practices in operating mixed-type control systems.

However, the previously listed drawbacks hinder the use of purely fly-by-wire systems on aircraft of certain classes. So, fly-by-wire control systems are used along with hydromechanical control systems on modern passenger aircraft. At the same time, normal operation is attached to the fly-by-wire circuit, while the emergency landing function is assigned to the hydromechanical circuit, which is traditionally considered to be more reliable.

The failure of the main fly-by-wire system on these aircraft, as a rule, leads to the need to limit the number of flight modes in which the mechanical system cannot provide satisfactory characteristics of control loop stability and controllability.

Saturation of the control systems in high-speed aircraft with automatic stabilization and stability systems using the signals of angular velocities and g-loads created a new problem for their design: it is necessary to consider the elastic vibration effect of the fuselage design on the stability of the stabilization circuits and to increase the aircraft's stability. This problem is particularly relevant for large aircraft, in which the natural frequencies of elastic vibrations of the fuselage can be quite close to the vibration frequencies of the entire aircraft (as a solid) in the air. With bending vibrations of the fuselage, depending on their tones, antinodes and nodes are generated at different sites along its length (Fig. 12).

Maximum local angular velocities occur in nodes, whereas g-loads occur in antinodes. The g-load and angular velocity sensors will respond to these g-loads and angular velocities if the sensors are not installed correctly. It is extremely difficult to filter out these false signals when their frequencies are close to the vibration frequencies of the entire aircraft. As a result, the automatic system will begin to induce oscillations in the aircraft instead of stabilizing it. Therefore, searching for the mounting location of the angular velocity and g-load sensors on the aircraft is an important task in designing an automated control system. Approximate mounting locations of these sensors are shown in Fig. 12.

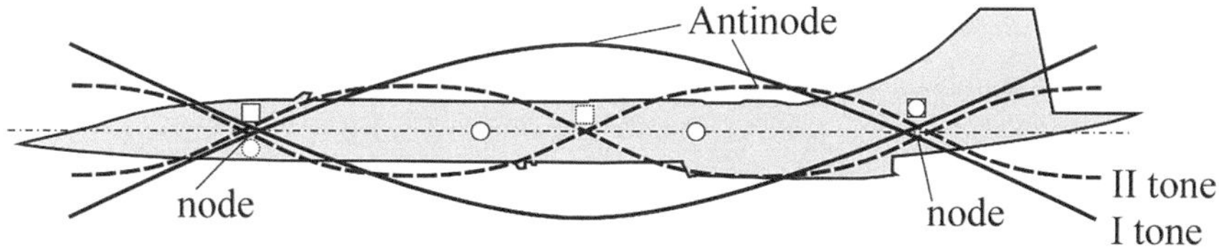

Fig. 12 The arrangement of angular velocity sensors and accelerometers to ensure the absence of structure vibrations and deflections: ○—angular velocity sensor; □—g-load sensor; grey colour indicates incorrect arrangement of sensors

1.2 Features of Fly-by-Wire Systems in Modern Fighters of the Fourth Generation

The majority of fourth-generation fighters have multiple redundant fly-by-wire (FBW) systems on board. It is designed for:

- manual control of the control surfaces and movable nozzles of the aircraft by means of electrohydraulic actuators in response to command signals from the controls and signals from the flight parameter sensors;
- automatic or combined control of the aircraft along with the avionics system;
- measurement and calculation of the altitude-velocity parameters;
- measurement and calculation of the true angles of attack and sideslip, normal g-load, limit values of angle of attack, indicated airspeed, and Mach number in the entire operational range of aircraft weights and the design center-of-gravity envelope, taking into account its current configuration and distribution of loads;
- pedal control of aircraft motion on the ground by means of fly-by-wire control of hydraulic actuators for turning the wheels of the nose landing gear;
- braking the main landing gear wheels with anti-skid control.

A fly-by-wire system controls the aircraft relative to the three axes. It provides the required stability and control characteristics over the entire range of altitudes, flight speeds, and the required center-of-gravity envelope.

Control is accomplished through the use of a control stick and pedals located in the aircraft cockpit. The control system provides automatic trimming of the longitudinal and lateral forces on the control stick and the pilot's manual trimming of the effort on the pedals.

Longitudinal control is carried out by the in-phase deflection of the stabilizer, flaperons, canard, and wing leading edges. Lateral control is carried out by the differential deflection of flaperons, ailerons, and horizontal stabilizer, or elevons. Directional control is carried out through rudders and, on some aircraft, through control of engine thrust.

The control structure in the longitudinal direction is ensured by the constancy of forces on the control stick to create a 1-g load or angle of attack; when the control stick is in a neutral position, a 1-g load flight is provided.

A special piloting mode is introduced into the FBW system to improve longitudinal and lateral control while refueling in flight from a tanker aircraft.

To ensure the required reliability of computers, the FBW system is provided with multiple redundancies. All control signals in modern fighters are generated by digital algorithmic computers. To control the flight control actuators, a digital drive computer is additionally used in most aircraft.

Control means (that monitor the operation of the drives and computational part of the FBW system and provide control) detect, localize, and switch off the failed part of the control channel.

The fly-by-wire system solves the tasks of automatic control of the aircraft in the entire operational range of altitudes and flight speeds, weights, and the design center-of-gravity envelope by generating control signals.

The automatic control mode of the FBW system in modern fighters of the fourth generation provides the following:

- mode of angular position stabilization of the aircraft in pitch, roll, and yaw;
- mode of altitude stabilization;
- mode of straight-line horizontal flight from any aircraft spatial position in roll and pitch;
- mode of stabilization of the indicated airspeed by generating control signals in the engine thrust control channel;
- automatic landing approach to altitudes of 40–70 m.

Moreover, the FBW system can provide the following:

- calculation of the current aircraft mass;
- calculation of the vertical and horizontal accelerations;
- generation of the current altitude H_{flight};
- generation of the available lift coefficients $C_{L,avail}$ and g-load $n_{z,avail}$;
- generation of the limit parameter values α, $n_{z,avail}$, V_{ias}, M;
- generation of the minimum speed for landing approach $V_{\text{min 1.appr}}$.

These values are used in the FBW system algorithms and are given to the pilot as an indication of the maximum permissible values of pilot-centered parameters.

2 Control Loop, Its Main Elements and Their Characteristics

As discussed, the aircraft can be controlled either by a pilot or by automatic systems.

A large role in the pilot's information about changing the flight mode and actions of the controls is played by physical sensations of g-loads and changes in forces on the controls. By comparing the current values of flight parameters with the required ones, the pilot is able to generate control signals in the form of lever motion based on certain piloting skills. These motions are converted by the control system into the corresponding deflections of the aircraft control surfaces, which leads to a change in flight parameters. The pilot controls the results of his forces on control surfaces (i.e., changes in flight parameters) using sense organs, seeking to eliminate misalignment between their current and desired values, and ceases to force on controls if these values match.

Thus, a closed control loop is generated in flight that consists of three main elements connected to each other (via links): the pilot, the control system, and the aircraft (Fig. 13a).

The pilot, as a component of the control loop, can be thought of as an automatic adjusting system composed of three major elements (Fig. 13b): sense organs (sensory receptor organs are "sensors"); the central nervous system, which performs information processing and decision-making functions; and executive organs—muscles of the arms, legs, and back. The movements of the pilot's executive organs and forces developed by them are the "output signals" ("output") of a human as a link in the control loop.

The possibility to measure out the forces and motions of the executive organs allows the pilot, who is familiar with the control characteristics of a given aircraft, to make the necessary changes in the flight mode by appropriately moving the control levers and applying certain forces to them.

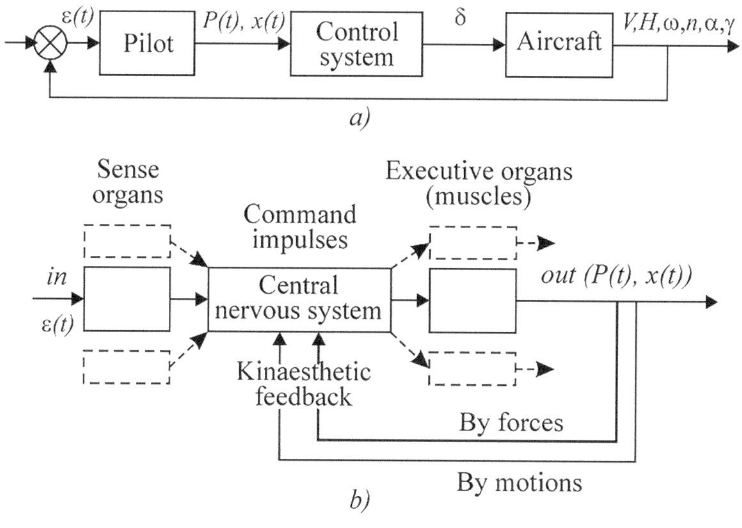

Fig. 13 Scheme of a closed control loop "pilot—control system—aircraft" (**a**) and block diagram of the pilot operator as a command link of the control loop, eliminating misalignment between the required and current values of a parameter by acting on the control lever (**b**): $\varepsilon(t)$—misalignment between the current and required values of flight parameters; $P(t)$, $x(t)$—forces applied to control levers and their displacement; δ—deflection of controls; V, H, ω, n, α, γ—flight parameters; input—input perceptions (visual, audio, acceleration sensation); output—output effects on control levers

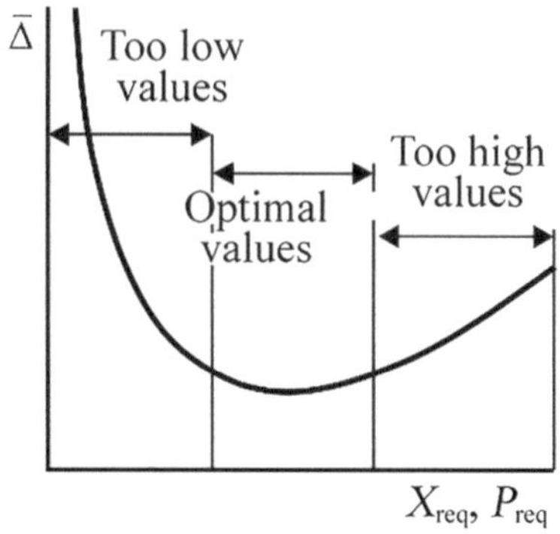

Fig. 14 The relative error in displacements of a control lever and the forces applied to it as a function of their required values

The pilots feel the changes in forces better than the motion, and the accuracy of measuring the forces is significantly higher than the accuracy of measuring the motion.

Studies show that the loss of control loop stability occurs not only in the complete absence of change in force applied to the control lever when the adjustable flight parameter is changed but also when such a change is too low or too high. This is due to the fact that the pilots most accurately measure out the displacements of control levers and forces applied to them in a certain range of their values, as can be seen from Fig. 14. In this figure, the $\overline{\Delta}$ symbol indicates the relative error ($\overline{\Delta} = (\Delta X)/X_{req}$ or $\overline{\Delta} = (\Delta P)/P_{req}$) in the displacement of a control lever and the force applied to it at various required values of (X_{req}, P_{req}).

Long-term operating practice is used to determine the optimal values of control lever displacements and forces required to perform certain maneuvers ($X_{el}^{n_z}$ and $P_{el}^{n_z}$). They are specified in the relevant requirements for the given aircraft or determined in the process of its design with the help of modeling control processes on special test stands using computers. These values of controllability parameters should be ensured when designing control systems.

The pilot, as a commanding (or controlling) element of the control loop, has a number of properties that affect the control process. The following can be distinguished from these properties:

- delay in response to external signals. Its magnitude greatly depends on the pilot's training level and his psychological and physical condition. For standard conditions, it is possible to consider a delay in reaction of a pilot with an average qualification of $\tau = 0.2$–0.3 s[1];
- inertia, which is manifested in the fact that the pilot's actions cannot instantly (i.e., "stepwise") reach the necessary level of response (Fig. 15a);

[1] The delay of the pilot's reaction to the misalignment signal that is clear to him between the current and required values of the adjustable flight parameter should not be confused with the delay in response to an unexpected change in the aircraft's behavior, the reason for which is not immediately clear to the pilot. In such a way, for example, as evidenced in practice, the pilots begin to take countermeasures only 3–5 s after they detect the failure of a multi-engine aircraft.

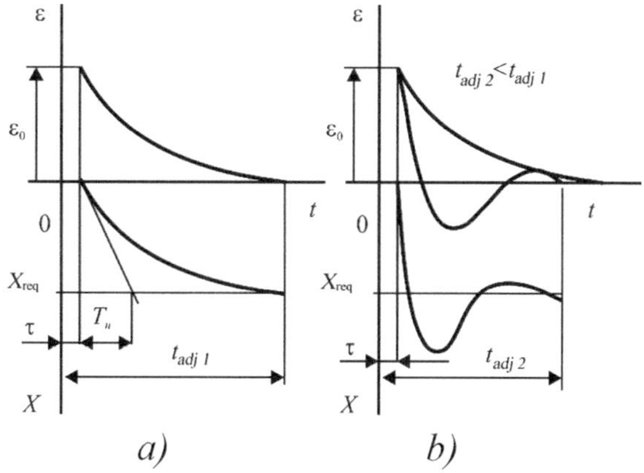

Fig. 15 Pilot's reaction to a step change in an adjustable parameter (misalignment ε_0) when operating with a simple amplifier with inertia and delay (**a**) and when using an afterburner to reduce the adjustable time (**b**)

- some dead zones;
- ability to filter external signals;
- the ability of the pilot to change his own transfer function over a wide range (refer to the example in Fig. 15b), including differentiation and integration, i.e., the ability to respond not only to deviations of any flight parameter from the required value, but also to the first and second derivatives (speed and acceleration) of this deviation, as well as to the integral of this deviation;
- ability to generate output command signals with a certain accuracy depending on the magnitude of these signals and their frequency;
- ability to track signals arriving with a frequency of not more than 2.5–3.0 Hz (presence of bandwidth).

Despite the ability to change his own transfer function over a wide range, the pilot most accurately works in the control loop as a single-channel amplifier, consistently eliminating the $\varepsilon(t)$ misalignment between the current and required values of any flight parameter. However, the delay in the response onset and inertia in the completion of a corrective action on the external signal increase the time to eliminate the arising misalignment and, with a large frequency of the error signal, create a significant delay in the phase of the pilot's output signal as compared with the required one. In such a way, for example, at an alternating frequency of the input misalignment signal of 0.5 Hz, a delay of the pilot's onset response of only 0.2 s causes a delay in his response in the phase of 36°, and at a frequency of 1 Hz, it amounts to 72°. If we take into account that the natural oscillation frequencies of modern aircraft, especially in some flight modes, increase significantly and can even reach 1 Hz or more in heavy aircraft, then the reason for pilot-induced oscillation in these modes becomes obvious. In actual piloting conditions, the pilot, knowing that he has his own delay and inertia, begins to speed up, as shown, for example, in Fig. 15b.

The aircraft as an object of control has six degrees of freedom in space, and the motion of the aircraft is described by six Euler differential equations. The solution of these equations in the general case would allow one to determine the nature of the spatial movement of the aircraft at any time, in particular after the pilot's effect on the controls. Also, it would allow us to judge the stability of this movement. However, the direct solution of these equations poses certain difficulties, even with the use of modern computers.

In general, this allows you to create a system of equations for an aircraft that reflects limitations on all six degrees of freedom. Accordingly, the translational degrees of freedom are reflected in the mass balance equation, while the rotational degrees are reflected in the moment-inertia balance equations:

$$\begin{cases} 1 = \sum_i \overline{m}_i \\ 1 = \sum_j \overline{I}_{jOX} \\ 1 = \sum_k \overline{I}_{kOZ} \\ 1 = \sum_n \overline{I}_{nOY} \end{cases}$$

Contrary to the equation of mass balance, the equations of moment-inertia balance do have not only obvious limitations on maximum values but also the limitations on minimum moments of inertia, which are dictated by the need to ensure stability and controllability for some types of aircraft.

If we take the steady level flight with no sideslip as the reference flight mode and consider the deviations of the motion parameters from the initial values to be rather low (apply the method of small perturbations), then due to the symmetry of the aircraft, the system of six motion equations can be divided into two independent systems of equations with a certain degree of accuracy describing the aircraft motion in the plane of symmetry (the so-called longitudinal motion) and in two other planes (lateral motion).

When using kinematic equations, each of the motions (longitudinal and lateral) is described by a system of four differential equations. The system of four equations of longitudinal motion describes two oscillatory motions that develop once the external disturbance does not affect the aircraft anymore (atmospheric actions, deflections of control surfaces, changes in engine thrust, etc.). One of these motions (called short-period motion) develops rapidly and has a short period (about 1–5 s), while the other develops relatively slowly and has a significantly longer period (about several tens of seconds). This motion is called long-period.

The solution of the lateral motion equation system shows that in modern aircraft, lateral motion is, as a rule, the sum of two aperiodic and one oscillatory motion.

The long-period longitudinal motion is easily counteracted by the pilot and does not arouse much interest. Aperiodic lateral movements, one of which is a rapidly convergent roll-subsidence mode and the other a rather slow accelerating spiral mode, are also of little interest to the study because they have no effect on the pilot's assessment of aircraft controllability.

However, it is another matter with short-period longitudinal and oscillatory lateral motion. The nature of the development of these motions after the action of external disturbances or control surface deflection is a criterion for the pilots to assess the stability and controllability of the aircraft. These motions are thoroughly studied during the aircraft design process in order to develop appropriate measures in advance if the parameters of these motions are unsatisfactory. At this stage of the aircraft design process, these studies are carried out using computers that solve the motion equations at varying design parameters.

Examples of changes in flight parameters during the longitudinal short-period motion due to an external disturbance (wind gust) and with the longitudinal control deflection are shown in Fig. 16. Transient processes are similar to lateral oscillatory motion. Since the flight parameters in these motions are associated with differential equations of the second order, the nature of the transient processes is entirely determined by the coefficients with the corresponding derivatives of the considered parameters. Thus, the coefficient of the first derivative determines oscillation attenuation, and the coefficient of the zero derivative determines frequency.

If we consider the change in the angle of attack in longitudinal short-period motion as an example, the damping of these oscillations is determined by coefficient α_1:

$$\alpha_1 \approx C_L^{\alpha} - \frac{C_m^{\omega_y} + C_m^{\alpha}}{\overline{r}_y^2}, \tag{1}$$

and the frequency is determined by the coefficient α_2:

$$\alpha_2 \approx C_L^{\alpha} - \frac{\mu\left(C_m^{C_L} + C_m^{\frac{\omega_y}{\mu}}\right)}{\overline{r}_y^2}. \tag{2}$$

In these expressions, $\mu = \frac{2m}{\rho S c_{MAC}}$ is the aircraft density ratio; m is the aircraft mass; $\overline{r}_y^2 = \frac{I_y}{m c_{MAC}}$ is the square of the dimensionless radius of aircraft inertia relative to the lateral axis (OY_1); I_y is the aircraft moment of inertia relative to the same axis.

Aircraft stability during a short-period motion, i.e. the oscillation attenuation, is determined by the conditions $a_1 > 0$ and $a_2 > 0$.

The condition $a_1 > 0$ is always met since it is impossible to create an aircraft without positive damping. This fact is explained by the following: rotation of the lifting surfaces (wing, empennage, and even the fuselage) in air creates additional aerodynamic forces, moments of which will always impede this rotation. However, in some flight modes, particularly at high altitudes, this damping is insufficient to provide good feedback to pilots on the aircraft's dynamic stability because the oscillations of the aircraft do not attenuate for a long time after disturbance.

Condition $a_2 > 0$ is met, if

$$\sigma_n = \frac{C_m^{C_L} + C_m^{\omega_y}}{\mu} < 0. \tag{3}$$

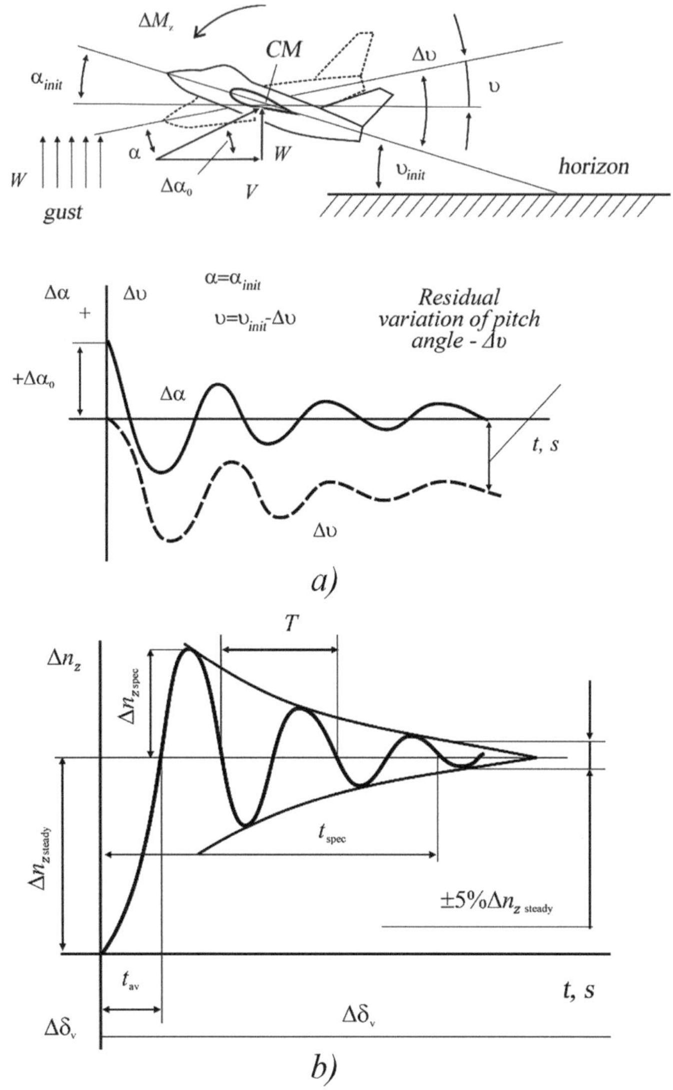

Fig. 16 Examples of changes in flight parameters during the longitudinal short-period motion: **a** response of a stable aircraft to an upward gust; **b** transition to a new normal g-load after a stepwise deflection of the elevator and the longitudinal dynamic stability and controllability parameters; T is the period of natural short-period oscillations; t_{av} is the response time; $\Delta n_{z\,overs h}$ is the initial overshoot of the adjustable parameter (in this case, — g-load); $\Delta n_{z\,steady}$ is the steady-state value of the adjustable parameter; $v = \frac{\Delta n_{z\,oversh}}{\Delta n_{z\,steady}}$ is the initial overshoot ratio (readjustment); t_{atten} is the transition process's attenuation time; n_{atten} is the number of oscillation periods prior to attenuation

The coefficient of longitudinal static stability with respect to g-load σ_n affects the aircraft's longitudinal static controllability in terms of the pitch control lever displacement per unit of normal g-load:

$$X_{el}^{n_z} = -\frac{1}{k_{\lg\, el}} \frac{C_{Lif}}{C_m^{\delta_{el}}} \sigma_n, \tag{4}$$

and the pitch control lever force per unit of normal g-load:

$$P_{el}^{n_z} = P_{el}^{X} x_{el}^{n_z}. \tag{5}$$

Here, $k_{\lg.\,\mathrm{el}} = d\delta_{\mathrm{el}}/dx_{\mathrm{el}}$ is the coefficient of kinematic transmission from the lever to the pitch control surfaces; $P_{\mathrm{el}}^{X} = dP_{\mathrm{el}}/dx_{\mathrm{el}}$ is the force gradient applied to the pitch control lever with respect to its displacement.

It follows from the expression (4) that the force applied to the pitch control lever will be zero at $\sigma_n = 0$, i.e. if $\bar{x}_{CG} = \bar{x}_{AC} - C_m^{\omega_y}/\mu$. It is called the manoeuvre (neutral) point and designated by $\bar{x}_{CG,MP}$. Since piloting at $\sigma_n = 0$ is impossible, it is necessary to limit the movement of the center of gravity backwards to a certain point called the aft center of gravity limit, which ensures the minimum permissible values of $|\sigma_n|_{\min}$ and $\left|X_{el}^{n_z}\right|_{\min}$ for this class of aircraft (Fig. 17).

The meeting of condition (3), as well as obtaining the necessary control characteristics [for example, refer to expressions (4) and (5)] on aircraft with a manually

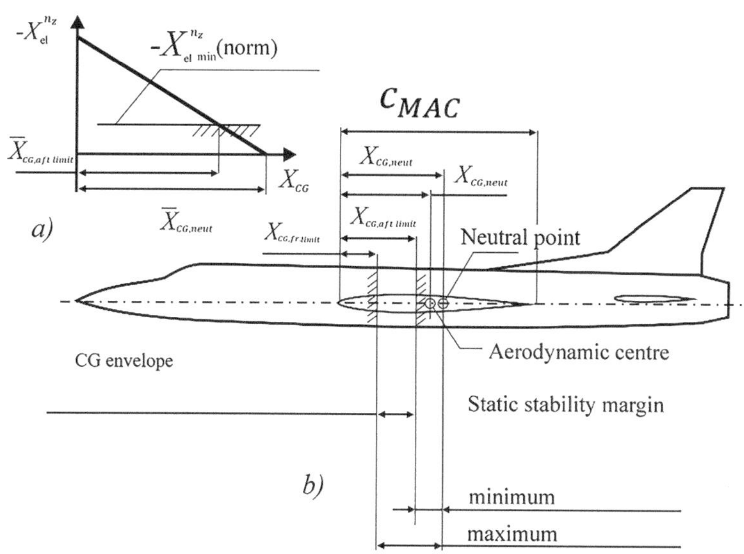

Fig. 17 Pitch control lever force per unit of g-load as a function of the aircraft center of gravity (**a**) and the positions of the centre of gravity, aerodynamic centre, and the neutral point of a longitudinally statically stable aircraft (**b**): $x_{CG,aftlimit}$ is the aft centre-of-gravity limit; $x_{CG,fr.limit}$ is the forward centre-of-gravity limit; $C_m^{C_L} = \bar{x}_{CG} - \bar{x}_{AC}$; $\sigma_n = \bar{x}_{CG} - \bar{x}_{CG,neut}$

operated control system, is achieved exclusively by providing the required longitudinal static stability $C_m^{C_L} = \overline{x}_{CG} - \overline{x}_{AC}$ during the design process since the values of the term $C_m^{\omega_y}/\mu$ are quite small in modern aircraft (about 0.02–0.03 at low altitudes, while with ascent to greater altitudes they decrease to negligibly small values).

The longitudinal static stability margin $C_m^{C_L}$ largely determines the trim displacement of the pitch control lever with respect to the speed and flight altitude obtained from the known equation of longitudinal balance ($C_m = 0$):

$$X_{el\,trim} = \frac{C_{m,0} + C_L}{k_{lg,el} C_m^{\delta_{el}}}.$$

Summarising the above, it can be noted that the characteristics of longitudinal stability and control of an aircraft are mainly determined by the following parameters: $mg/S, \overline{r}_y, C_L^{\alpha}, C_m^{C_L}, C_m^{\omega_y}, C_{m.}^{\alpha}$

The derivative $C_m^{\omega_y} = \frac{\partial C_m}{\partial \overline{\omega}_y}$ (where $\overline{\omega}_y = \omega_y c_{MAC}/V$), which characterises the longitudinal damping, depends on the wing, fuselage, engine nacelles, and the horizontal tail geometry. The main contribution to the longitudinal damping moment is due to the horizontal tail; taking into account the influence of the fuselage, this part can be determined by the approximate formula

$$C_m^{\omega_y} \approx -1.2 C_{L,HT}^{\alpha} \frac{S_{HT} L_{HT}^2}{S c_{MAC}^2}. \tag{6}$$

The contribution of the wing to the derivative of the longitudinal damping moment of an aircraft increases rapidly with an increasing wing sweep:

$$C_{m,WG}^{\omega_y} \approx -C_L^{\alpha}\left(A + B\,ARtg\alpha + C\,AR^2 tg^2 \Lambda\right) - D^*. \tag{7}$$

For a tailless configuration, the entire longitudinal damping moment is created by the wing and fuselage. Therefore, only backswept (or delta) wings can be used to obtain sufficient damping in aircraft of this kind.

The longitudinal damping at subsonic speeds increases remarkably due to the delay of downwash at the horizontal tail. This increase is estimated by the derivative

$$C_m^{\dot{\alpha}} = \frac{\partial C_m}{\partial \dot{\alpha}} \approx -C_{L,HT}^{\alpha} \frac{\partial \varepsilon}{\partial C_L^{\alpha}} \frac{S_{HT} L_{HT}^2}{S c_{MAC}^2}. \tag{8}$$

where $\dot{\alpha} = \frac{d\alpha}{dt}\frac{c_{MAC}}{V}$ is the dimensionless speed of change in the angle of attack during the aircraft rotation.

For subsonic aircraft with moderately-swept wings, this derivative has values of the order of (0.4–0.6) $C_m^{\omega_y}$.

Studies of lateral oscillatory motion, which is characterised by a close relationship between yaw and roll motions, show that, as previously stated, the nature of this

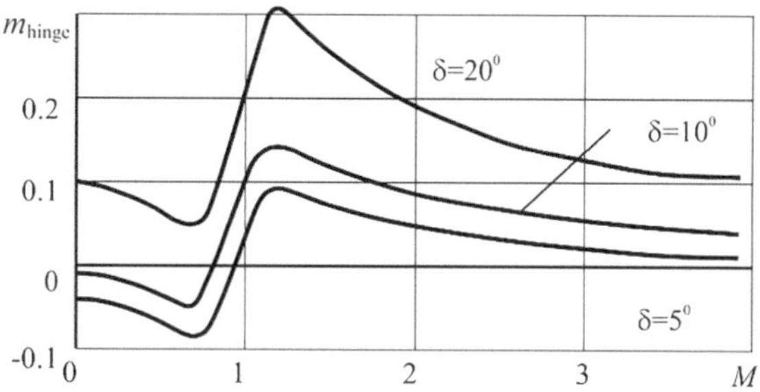

Fig. 18 Variation of the hinge moment coefficient of a control surface with the Mach number

motion is mainly determined by the derivatives[2] C_m^{β}, C_n^{β}, C_l^{β}, as well as by the mass-inertia characteristics $\overline{r}_x^2 = \frac{4I_x}{mb^2}$, $\overline{r}_y^2 = \frac{4I_y}{mb^2}$ and $\frac{I_x}{I_y}$.

3 Stability and Control of Modern High-Speed Aircraft and Methods for Their Improvement

A common feature of all high-speed aircraft is a sharp increase in the hinge moments of control surfaces at supersonic speeds, associated both with an increase in dynamic pressure and redistribution of pressure over the control surface when M_{crit} is exceeded, and with an increase in the size of these control surfaces. An example of the variation of the hinge moment coefficient of a control surface with the Mach number is shown in Fig. 18.

The sharp increase in the hinge moments of high-speed aircraft control surfaces led to a significant increase in the power required to drive these controls. The pilots' physical capabilities were insufficient for this, and that is why hydraulic power amplifiers of control signals—boosters—were installed in the control wiring. Initially, boosters were installed according to the so-called reversible scheme, in which they helped the pilot by taking part of the hinge moment. However, the further growth of flight speeds and the development of boosters led to the transition to the irreversible scheme. It has a number of undeniable advantages and can solve a variety of additional tasks in addition to increasing power. The use of an irreversible scheme for installing boosters (i.e., transition from purely mechanical to hydromechanical drive of control surfaces) required the introduction of special feel devices into the aircraft control system in order to create lever forces and restore the "control feel."

The most important feature of supersonic aircraft is the substantial dependence of longitudinal static controllability with respect to g-load on flight mode (Fig. 19). At subsonic speeds (especially at low altitudes), σ_n decreases significantly with increasing speed due to elastic deformations of the wing. This decrease becomes more noticeable as the wing sweep angle and aspect ratio increase (the absolute

[2] Definitions of these derivatives are given in chapter "Relationships Between Aircraft Parameters and Performance. Ways to Improve Performance".

value of σ_n may decrease by 0.06–0.08 with a wing sweep angle of 50°–60°). This decrease is especially significant for aircraft with variable-geometry wings, due to their lower stiffness (the decrease in the absolute value of σ_n may exceed 0.10). When the critical Mach number is exceeded, the σ_n coefficient begins to grow rapidly due to the redistribution of pressure along the wing chords, which shifts the aerodynamic centre backwards. This shift occurs before reaching the Mach numbers of 1.15–1.25. The aerodynamic centre shift and the corresponding increase in the absolute value of σ_n can be as high as (0.15–0.25). For aircraft with variable-geometry wings, the aerodynamic centre shift as a function of the sweep angle can be significantly greater.

It has already been noted that a sharp increase in longitudinal static controllability with respect to g-load during a transition to supersonic speeds leads to a deterioration in manoeuvrability and increase in trim penalties. Therefore, when designing supersonic aircraft, in addition to the transition to an all-movable stabilizer, as already mentioned, some means are usually used to mitigate the increase in stability (Fig. 20). Since the longitudinal stability is determined by $\overline{x}_{CG} - \overline{x}_{AC}$, these means must affect the position of the aircraft aerodynamic centre (shift it forward at $M > 1$) or the position of the centre of mass (shift it backwards at $M > 1$).

The first group of measures (Fig. 20a) includes the use of the leading-edge extensions in the wing root (for example, on Tu-144, Concorde, F-16, F-18, YF-12A, etc.) and extendable destabilisers in the forward part of the fuselage (for example, on F-14A Tomcat aircraft).

It is possible to shift the aircraft's centre of mass backwards when switching to supersonic flight speed by pumping fuel from central tanks to special trim tanks located in the aft part of the aircraft. This technique is used on almost all supersonic aircraft with a long flight range. When fuel is pumped to tail trim tanks, they must be equipped with an emergency drain system since the aircraft may become unstable in the event of a reverse pumping system failure when braking to subsonic speed (refer to Fig. 20b).

The transonic speed instability, which manifests itself as nose-heaviness during supersonic acceleration and an arbitrary increase in g-load (g overshoot) during supersonic to subsonic braking, is caused by an intensive backward shift of the aircraft aerodynamic center in the transonic speed range. The transonic speed instability is

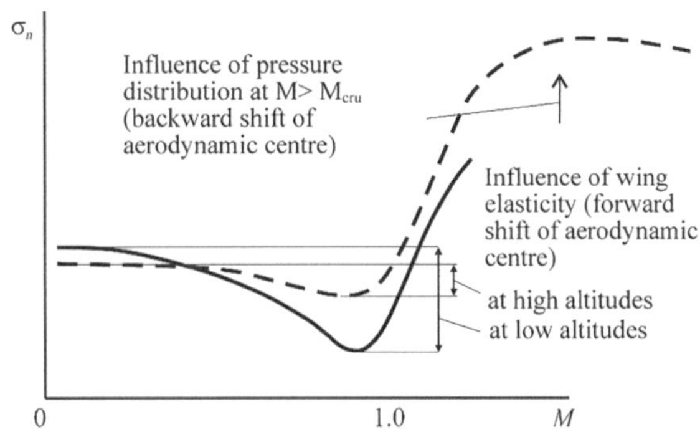

Fig. 19 The variation of the longitudinal static controllability coefficient with respect to g-load as a function of the Mach number and altitude: $\sigma_n = C_m^{C_L} + C_m^{\omega_y}/\mu$; —— low altitudes; - - - high altitudes (the figure shows the negative values of σ_n)

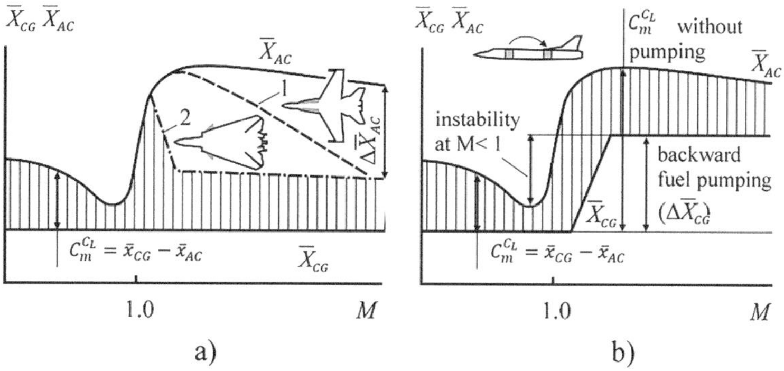

Fig. 20 Measures to reduce an aircraft's longitudinal static controllability with respect to g-load at supersonic speeds: **a** aerodynamic means of shifting the aerodynamic centre forward; **b** displacement of the centre of mass when pumping fuel backwards; 1—the use of the wing leading-edge extensions; 2—nose destabiliser exposed to the flow (or the retraction of horizontal tail)

Fig. 21 "Pitching-up region" in trim curve in the transonic range of speeds and its correction with the help of trim controller or automatic flight control system: $\Delta\delta_{el} = f(M, H)$—misalignment introduced by the trim controller or automatic flight control system

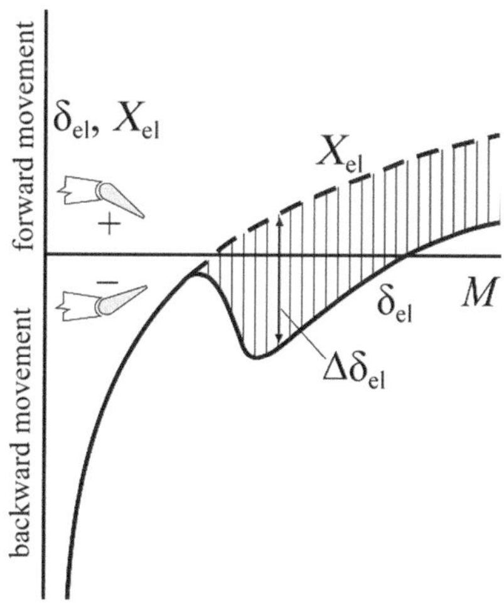

illustrated by the change of the positive gradient of the trim curve slope $x_{el} = f(M)$ to the negative in the near-sonic speed range, i.e., the emergence of the pitch-up tendency in the trim curve (Fig. 21). In the zone of transonic speed instability, movements of the longitudinal control lever to trim the aircraft are reversed: "pull-up"—during acceleration (instead of the usual one—"pushover"), and "pushover"—when braking (instead of the usual ones—"pull-up").[3]

In some first-generation supersonic aircraft, this controllability defect was rectified by installing a so-called trim controller in the control system. It introduced

[3] Do not confuse this change in movement during speed control with the remaining unchanged directions of lever movement during g-load control: "pull-up" is to increase the g-load, "push-over" is to reduce it.

misalignment between the positions of the lever and longitudinal controls in a certain range of Mach numbers in which transonic speed instability manifested itself (usually in the $M \approx 0.9$–1.2 range).

During speed control, the constant nature of lever movement was ensured, while the longitudinal controls deflected according to the "curve with a pitch-up region" law. On modern supersonic aircraft, near-sonic speed instability manifests itself less noticeably due to the "smoothing" of intensity in stability growth by other means discussed above. If necessary, speed controllability is improved by introducing the appropriate control laws [$\Delta\delta_{el} = f(M, H)$ type] in the automatic flight control system.

The fourth feature (the loss of longitudinal static controllability with respect to g-load when reaching angles of attack of the order of 12°–15°) was discussed in detail in chapter "Empennage Design" (refer to Fig. 8). Here it is worth adding only that several means are used to prevent reaching such angles of attack in normal operation in order to increase flight safety. These means include specific instructions, alarm systems (sound, light, shaking of the pitch control lever), as well as special automatic devices (permissible mode limitation systems) that limit the backward movement of the lever or even forcibly deflect it forward to escape the dangerous mode ("pushers"). Permissible mode limitation systems, as a rule, provide an appropriate indication on the pilot's instrument panel displaying what margins for g-load, sideslip angles and angles of attack are acceptable for this flight mode. One of the tasks solved by the automatic flight control system is to prevent the aircraft from entering dangerous modes.

The systems for artificial longitudinal stability provide for a significant expansion of the available range of angles of attack. As a result, the aircraft's maneuverability improves in flight modes where it was previously limited by wing lift properties.

So, for example, the maximum operational angle of attack for the F-16 aircraft is 30°. When this angle of attack is reached, the stabilizer automatically deflects the leading edge up to create a pitching-down moment, and in this way, the aircraft avoids dangerous mode. On the F-18, normal controllability characteristics and air intake operation are provided up to 50°.

A common problem of all high-speed aircraft is the deterioration of the natural oscillations damping relative to all three axes of the aircraft. This is due to two factors. The first one is the reduction in C_L^{α} and C_Y^{β} (especially at $M > 1$—(Fig. 22a)) typical of all high-speed aircraft, which largely determine the values of rotational derivatives $C_l^{\omega_x}$, $C_m^{\omega_y}$ and $C_n^{\omega_z}$ characterising the aircraft damping properties [for example, refer to expressions (6) and (7)]. The second one concerns flights at high altitudes characteristic for high-speed aircraft, where the density of air is low.

The deterioration in damping of natural oscillations leads to large initial overshoots of adjustable flight parameters with deflected control surfaces and long-lasting vibrations of the aircraft that hinder precise piloting control (Fig. 22c). In the longitudinal control channel, the deterioration in damping is accompanied at supersonic speeds by a significant increase in the frequency of these oscillations, which is associated with an increase in static stability. Mathematically, these phenomena

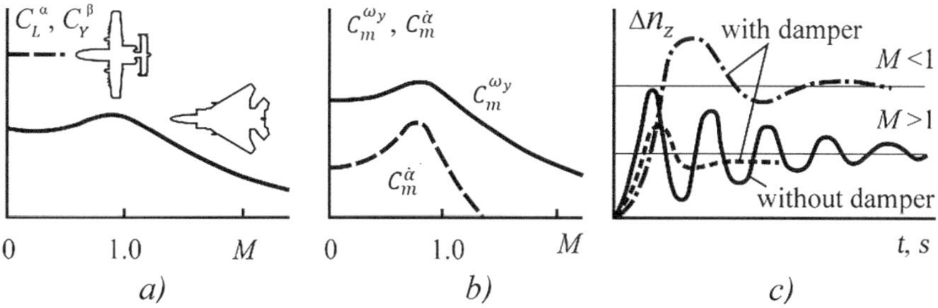

Fig. 22 The variation of C_L^α and C_Y^β as a function of the Mach number for supersonic aircraft in comparison with low-speed aircraft (**a**), rotational derivatives $C_m^{\omega_y}$ and $C_m^{\dot{\alpha}}$ (**b**), as well as the transient processes of establishing a new g-load value with the same deflection of longitudinal controls without damper and with a damper of longitudinal oscillations at subsonic and supersonic speeds (**c**)

are illustrated by a decrease in coefficient a_1 that determines attenuation of oscillations [refer to formula (1)] and an increase in coefficient a_2 that determines the oscillation frequencies [refer to formula (2)] at the first and zero derivatives in the equation of short-period motion. Since the frequency of natural oscillations in even heavy modern supersonic aircraft in the longitudinal channel may approach 1 Hz and even exceed this value in the transverse channel, the pilot's delay and inertia create prerequisites for induced oscillations in aircraft. It is confirmed by practice that in real-world conditions of piloting in these flight modes, the pilots do not cope with the task of suppressing the aircraft oscillations and begin to induce them.

Surely, during the aircraft design stage, it would be possible to obtain the required values of rotational derivatives by increasing the areas and relative arms of the damping surfaces, primarily of the empennage [for example refer to expressions (6) for $C_m^{\omega_y}$]. However, this would require an excessive increase of A_{HT} and A_{VT} that, on the one hand, would entail a significant weighting of the structure, while, on the other, it would cause difficulties in ensuring acceptable controllability characteristics in other flight modes (at low altitudes), where these surfaces would be too effective [9]. The latter would create prerequisites for the loss of control loop stability and pilot-induced oscillations in these flight modes.

Therefore, since the early days of the first serial supersonic aircraft, special automatic devices installed in the control system, i.e., dampers operating in parallel with pilots and independently of them, have provided the necessary degree of damping of their oscillations (Fig. 22c). The principle of vibration damper operation is based on measuring the angular velocity of aircraft rotation and the deflection of the corresponding control surfaces in the counter-direction of this rotation. The damper actuators on first generation high-speed aircraft were manufactured as electrically adjustable rods. On new aircraft, these mechanisms are manufactured as multiple redundant electrohydraulic devices, being a part of the automatic flight control system.

Depending on the axis relative to which the aircraft oscillations are to be dampened, dampers of pitch, roll, and yaw are distinguished. Their configuration is the same, and dampers differ only in the installation of their sensitive elements, i.e. gyroscopes.

Their simplest operation laws are the deflections of the control surfaces in proportion to the corresponding angular velocities of the aircraft's rotation:

$$\begin{aligned} (\Delta\delta_{el})_{p.d.} &= k_{\omega_y}\omega_y \quad \text{for the pitch damper;} \\ (\Delta\delta_{ail})_{r.d.} &= k_{\omega_x}\omega_x \quad \text{for the roll damper;} \\ (\Delta\delta_{rud})_{y.d.} &= k_{\omega_z}\omega_z \quad \text{for the yaw damper.} \end{aligned} \tag{9}$$

In these expressions, $k_{\omega_x}, k_{\omega_y}, k_{\omega_z}$ are the damper gear ratios that determine how many degrees the control surface deflects per 1 degree/s of the angular velocity of the aircraft rotation (the unit of the gear ratio is [s]). When dampers operate according to such laws, the increment of rotational derivatives defining the damping is determined by the following expressions:

$$\left(\Delta C_m^{\omega_y}\right)_{p.d.} = C_m^{\delta_{el}} k_{\omega_y}\frac{V}{c_{MAC}},$$

$$\left(\Delta C_l^{\omega_x}\right)_{r.d.} = C_l^{\delta_{ail}} k_{\omega_x}\frac{2V}{b},$$

$$\left(\Delta C_n^{\omega_z}\right)_{y.d.} = C_m^{\delta_{rud}} k_{\omega_z}\frac{2V}{b}.$$

When the pitch damper operates according to (3), the increment in $C_m^{\omega_y}$ leads to the corresponding increment in the σ_n coefficient and the controllability properties defined by it.

Thus, the displacement of the neutral point (Fig. 23a) is determined by the expression $\left(\Delta\overline{x}_{CG,neut}\right)_{p.d.} = \frac{C_m^{\omega_y} k_{\omega_y}\rho SV}{2m}$, while the change in the pitch control lever force per unit of normal g-load (refer to Fig. 23b) is determined by the expression $\left(\Delta\overline{x}_{CG,neut}\right)_{p.d.} = -\frac{k_\omega}{k_{lg.el}}\frac{g}{V}57.3$.

When $C_m^{\omega_y}$ increases, a_1 increases as well [refer to expression (1)].

Lateral stability and controllability also increase with the dampers operating according to the above laws. The property of dampers to increase the control lever force during manoeuvres can be used in those flight modes where these forces become unacceptably low (high subsonic speeds at low altitudes).

The reduction of the control force and the longitudinal control lever with respect to g-load with an increase in flight speed is typical for all aircraft [refer to expression (4)]. However, on aircraft with manual control system, this decrease in the deflections of the elevators and pitch control levers required for manoeuvring was accompanied by an increase in the hinge moments of the rudders in direct proportion to the velocity squared. The hinge moment of the elevator increasing the pitch control lever

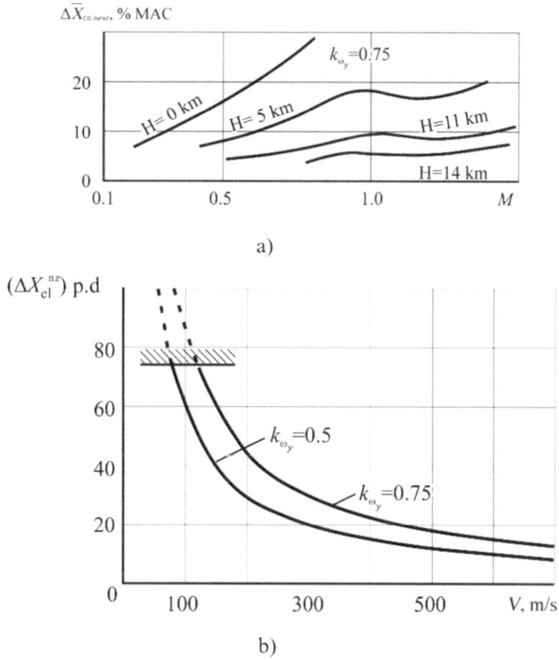

Fig. 23 Variation of the longitudinal static controllability properties when the pitch damper operates as $(\Delta\delta_{el})_{p.d.} = k_{\omega_y}\omega_y$: **a** displacement of the neutral point with respect to g-load in different flight modes; **b** $X_{el}^{n_y}$; the hatching shows the $X_{el}^{n_y}$ increment limitation by the maximum possible stroke of the adjustable rod

force when it was deflected to perform manoeuvres with the same g-load remained approximately constant in all flight modes, as well as, according to expression (4), the coefficient k_1 in $\delta_{el}^{n_z} = k_1/q$ is roughly constant at subsonic speeds.

Consequently, regardless of the flight mode, the more important parameter of longitudinal controllability—the control force per g-load unit—remained approximately constant on aircraft with manual control systems $P_{rud}^{n_z}$ (Fig. 24).[4]

In other words, with a manual control system in expression (5), the gradient of force to deflect the longitudinal control lever increased in proportion to the dynamic pressure $P_{\text{el}}^{x} = k_2 q$, where k_2 is also approximately constant coefficient.

When switched to power-controlled systems with the simplest spring loaders of control levers with a constant characteristic, the P_{el}^{x} gradient became constant (regardless of the flight mode), while the nature of $x_{el}^{n_z}$ remained the same. This determined a rapid decrease in the stick force per unit of g-load $P_{\text{el}}^{n_y}$ (that was unusual for pilots) with an increase in the dynamic pressure (indicated airspeed). At subsonic speed, the magnitudes of $x_{el}^{n_z}$ and $P_{el}^{n_z}$ significantly decreased and fell into the zone of the greatest control errors (refer to Fig. 14).

This situation was further aggravated by the fact that in the longitudinal control, the stabilizer began to be used to increase manoeuvrability at supersonic speeds at

[4] Some increase in this controllability parameter with an increasing speed for aircraft with a manual control system is due to the influence of elastic deformations of the structure, as well as the onset of the compressibility effects.

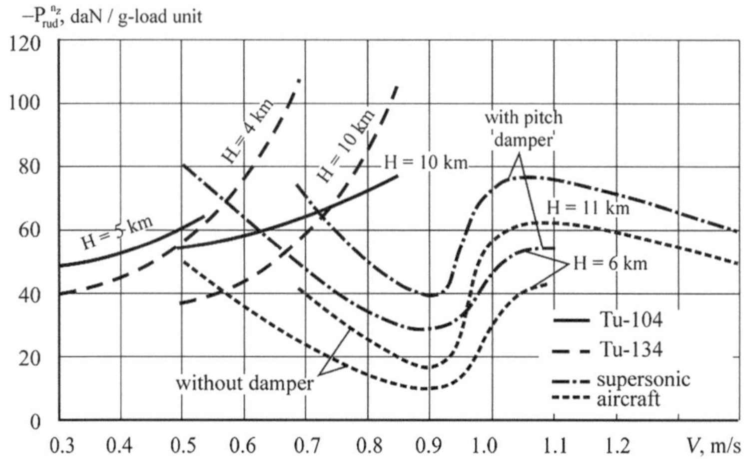

Fig. 24 Variations in $P^{n_z}_{rud}$ with altitude and velocity for aircraft with a manual control system (Tu-104 and Tu-134) and power-controlled systems with a simple spring loader (without damper and with pitch damper)

high altitudes. Its effectiveness was found to be excessively high at high subsonic speeds, particularly at low altitudes (limit modes indicated airspeed).

Under these conditions, the values of $x^{n_z}_{el}$ decreased to 2…3 mm (they became commensurate with the backlash in the control wiring). In accordance with the expression (5), $P^{n_z}_{el}$ also decreased. This created real conditions for the loss of control loop stability[5] that was common in early supersonic aircraft testing and development.

Consideration of expressions (3), (4), and (5) suggests technical ways to increase the values of $x^{n_z}_{el}$ and $P^{n_z}_{el}$, i.e. reduce the gain coefficient of the pilot-aircraft system to ensure the stability of the control loop. Even without increasing the longitudinal static stability $C^{C_L}_m$, an increase in $x^{n_z}_{el}$ can be achieved in three ways:

- through a decrease in the transmission coefficient from the longitudinal control lever to elevator (or stabiliser) $k_{\text{lg. el}}$;
- through a reduction in the longitudinal controls effectiveness $C^{\delta_{el}}_m$;
- through an increase in the longitudinal damping $C^{\omega_x}_m$.

$P^{n_z}_{el}$ can be reduced to the required level, in addition to the above three methods, by increasing the force gradient P^x_{el}.

All these methods are technically implemented on a broad scale in modern aircraft construction:

- $k_{\text{lg. el}}$ is adjusted to flight modes (it decreases with increasing air speed);
- a combined-type pitch control is employed, in which for pitch control, elevator ($M < 1$), and all-movable stabilizer ($M > 1$) are used;

[5] The loss of stability of the control loop can also occur in the autosteering modes and it is expressed in self-oscillations of the "automatic control system—aircraft" system.

– $m_x^{\omega_x}$ is increased by using pitch dampers;
– P_{el}^x is adjusted to flight modes.

In each case, one or another method, or a combination of them is applied depending on the type of aircraft, its characteristics and requirements.

The peculiarity of lateral stability and controllability characteristics in modern aircraft is associated with C_l^β, C_n^β, $C_n^{\omega_z}$ and their mass-inertia characteristics.

Rolling static stability C_l^β significantly depends on the angle of attack and Mach number. An excessive increase in the absolute value of C_l^β at the angles of attack as low as 12°–15° is characteristic for aircraft with swept back and delta wings (Fig. 25a).

Because this derivative characterizes the value of the roll moment that occurs in sideslip, its extreme increase causes excessive dynamic rolls when even minor sideslip occurs, such as from a gust of crosswind. In flying practice, there were cases when aircraft with such characteristics flipped on their backs due to crosswind gusts or one-sided engine failure before the pilot was able to take countermeasures. The roll hover is also a result of excessive transverse stability: a slower rollout after deflecting the ailerons to eliminate the roll indicates a lack of ailerons effectiveness at high lateral stability.

As noted above, in the first high-speed aircraft with swept back and delta wings, the wingtips were lowered to reduce C_l^β at high angles of attack, i.e. an anhedral was used. This measure, however, resulted in some structural complications and an increase in structure mass, as well as the roll reversal to rudder deflection at low angles of attack and high indicated airspeed. The disadvantages of lateral motion at high angles of attack (intense rolls from wing to wing—"roll overshoot") at elevated C_l^β values and the onset of roll reversal in sideslip at decreased values of this derivative made it necessary to limit both maximum and minimum flight speeds of the aircraft by reducing the operating speed range.

At present, if a special wing airfoils do not give a noticeable improvement in lateral stability, acceptable characteristics of lateral controllability are obtained by introducing special automatic devices (roll and yaw dampers, lateral stability controllers, etc.) into the control system. The operation laws of these automatic devices can be very diverse. For example, ailerons can be deflected in proportion to the sideslip angle ($\delta_{\text{ail}} = k_\beta \beta$), or rudder is deflected in proportion to the angular velocity in roll ($\delta_{\text{contact}} = k_{\omega_x} \omega_x$). For example, the function $(\Delta\delta_{rud})_{y.d.} = k_{\omega_z} \frac{T_{urn}}{T_{urn}+1} \omega_z + k_{\omega_x} \omega_x$ is used for the yaw damper with flaps retracted on the Tu-134 aircraft. In this law, the second term compensates for an increased stability by increasing the efficiency of lateral control, while the filter $\frac{T_{\text{turn}}}{T_{\text{turn}}+1}$ eliminates the influence of damper on pedal movements at steady turns.

The operation laws of automatic means of improving lateral stability and controllability are chosen based on the aircraft's properties and requirements. One of these requirements concerns the limitation of the rate of roll onset in sideslip or one engine inoperative, which is set by the limit angle of roll during the delay in the pilot's response (usually 5 s).

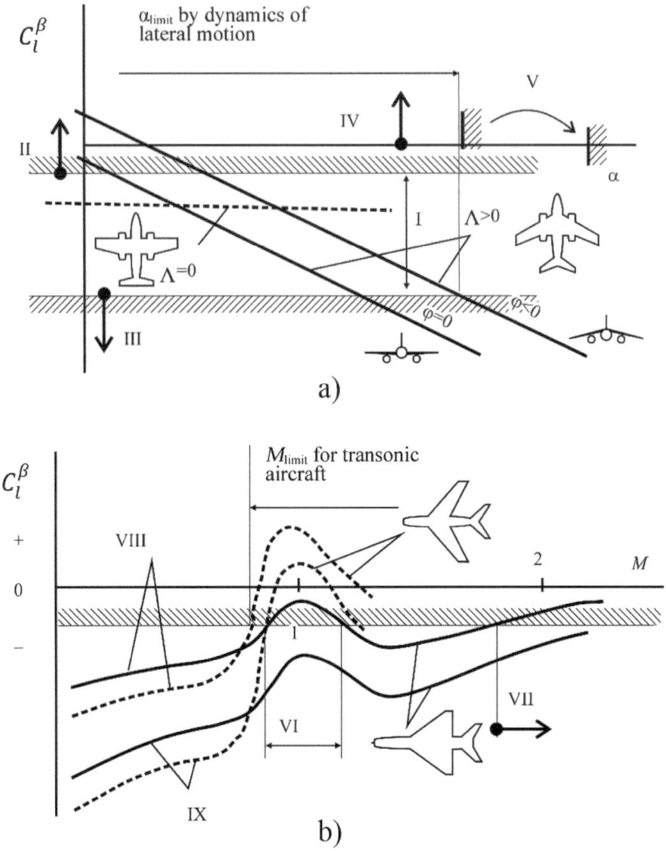

Fig. 25 Rolling static stability of aircraft of various types: **a** as a function of the angle of attack; **b** as a function of the Mach number: I—allowable limits of C_l^β variation by dynamics of lateral motion; II—zone of roll reversal in sideslip; III—excessive roll stability; IV—roll instability; V—potential increase in α_{limit} when using high-lift devices; VI—zone of transonic roll reversal in sideslip of supersonic aircraft; VII—supersonic zone; VIII—low angles of attack; IX—high angles of attack; in figure (**b**), the hatching shows m_x^β above which a roll reversal in sideslip occurs

The simplest and most common of these automatic devices is the yaw damper. By counteracting the resulting angular velocity in yaw, the damper prevents the development of high sideslip angles. Automation of the control system in this case enables the operation at large angles of attack, i.e., the reduction of the minimum speed. This leads not only to expanding the speed range, but also to reducing the required length of the runway due to a decrease in take-off and landing approach speeds.

Transonic and supersonic decreases in C_l^β (refer to Fig. 25b) led to the aircraft's roll reversal in sideslip. It often forced to impose limitations on the maximum allowable flight speed. It is now compensated automatically as well.

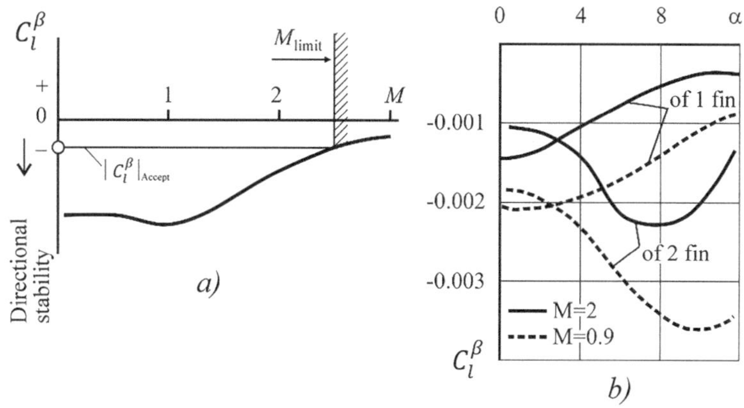

Fig. 26 Variation of the directional stability of supersonic aircraft as a function of the Mach number (**a**) and as a function of the angle of attack at different Mach numbers for configurations with one and two vertical stabilizers (**b**); the hatching shows the limit value of the Mach number (by directional stability); $\left|C_l^\beta\right|_{allow}$ is the lower limit of $\left|C_l^\beta\right|$ decrease

A characteristic feature of all supersonic aircraft is the considerable difficulty in providing directional stability $\left|C_l^\beta\right|_{allow} < 0$ (Fig. 26). The reasons for that were already discussed in previous chapters. Serious difficulties in providing acceptable lateral stability and controllability arise at supersonic speeds and high altitudes. To fly in these modes, due to the strong rarefaction of the atmosphere, even at supersonic speeds, it is necessary to use relatively high angles of attack. Such flight parameters lead to the most unfavourable combination of high lateral stability with insufficient directional stability. This is aggravated by $\frac{I_z}{I_x} > 10$ which is characteristic for supersonic aircraft and is associated with their large length. As a result, the intensity of rolling motion development in relation to yaw motion is large. The aircraft intensively rolls from wing to wing, and the pilot's attempts to correct these rolling, as a rule, lead to a progressive oscillation.

The directional stability of supersonic aircraft is now ensured both by classical methods of selecting the appropriate parameters of vertical tail and by using special automatic devices.

The need to trim the aircraft in flight with one engine inoperative, as well as during takeoff and landing in cross-wind conditions (with the possibility of engine failure in mind), necessitates a significant increase in the total area of the vertical tail on supersonic aircraft. For example, the YF-12A aircraft designed for flying at altitudes of up to 30 km at $M = 3$, features five vertical stabilizing surfaces. The choice of the number, size, shape, and arrangement of these surfaces is quite a challenge for each particular aircraft.

An artificial increase in directional stability can be provided by automatic means that deflect the rudder, for example, by $(\Delta\delta_R)_{adir.stab} = k_\beta \beta$ law. In this case, the change in the directional static stability is determined by the expression

$\left(\Delta C_l^{\beta}\right)_{a.dir.stab} = k_{\beta} C_n^{\delta_{rud}}$. However, the conditions for ensuring the stability of the "controller-aircraft" loop, as in the case of using the dampers, do not allow for increasing infinitely the gear ratios of high-performance automatic stabilization systems and increasing stability. In such systems, as a rule, the characteristics of the hydraulic drive (booster) are decisive. Therefore, a significant increase in directional stability or damping of yaw oscillations at high supersonic speeds can be achieved by increasing the effectiveness of directional control ($\Delta C_l^{\delta_{rud}}$) by using all-movable vertical surfaces (as it is used, for example, on YF-12A and SR-71 aircraft).

One of the characteristics of high-speed aircraft lateral controllability is a noticeable decrease in lateral control effectiveness at high flight speeds. This phenomenon, quite often inaccurately called the aileron reversal, is caused by elastic bending and torsion deformations at the aileron deflected, and it is manifested, as a rule, more and more noticeably with an increasing aspect ratio and sweep and with a decreasing wing thickness ratio. On the variable-geometry wings, ailerons are practically not used, since the limit speed of their reversal would be very low due to the reduced stiffness of such wings. Lateral control in such aircraft is, as a rule, carried out by a combination of spoilers and a differentially-deflected stabiliser.

However, using the differential deflection of the right and left halves of the stabilizer gave rise to a new problem of lateral controllability: the nose of the aircraft skidded in the opposite direction to roll when turning. This is caused by the yaw moment, which arises due to an asymmetrical change in pressure from different sides of the vertical tail when the right and left halves of the stabilizer are deflected in different directions. A similar problem was observed in a number of tailless aircraft with elevons deflected in opposite directions for roll control.

To improve the efficiency of lateral control on modern aircraft, the following are used: spoilers (Fig. 27), flap-ailerons (flaperons), root ailerons, droop noses, and others, in addition to the differential deflection of the stabilizer halves. Drooped ailerons (flap-ailerons or flaperons) are used to ensure that overdeveloped lateral controls do not reduce the effectiveness of high lift devices due to the reduction in the length of the wing trailing edge.

The use of power-controlled systems has also affected a number of lateral controllability characteristics. Thus, the change in the control force per unit of angular velocity in roll ($P_{\mathrm{a}}^{\omega_x}$) as a function of the speed of an aircraft with a power-controlled system with a spring loader having a linear characteristic as compared to an aircraft with a manual control, is shown in Fig. 28. A decrease in this property with increasing speed for an aircraft with a power-controlled system at high speeds is somewhat compensated by the influence of the wing elastic deformations. Nevertheless, at high wing stiffness (for example, at a delta wing), such flight modes may appear where this parameter will be too low while the rate of roll onset, even with a slight deflection of the control lever, is too high for the pilot to cope with its correct adjustment.

In other words, similar to the previously discussed longitudinal control, the "pilot-aircraft" system gains in some flight modes, and in the lateral control becomes too high, resulting in stability loss by the system. The aircraft begins to oscillate in roll motion. Methods to suppress this phenomenon are similar to those discussed above

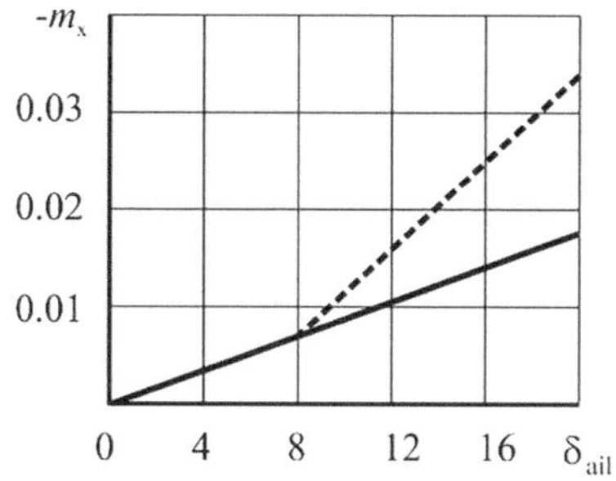

Fig. 27 Variation of the rolling moment coefficient as a function of the angle of aileron deflection and its increase when the spoilers are deflected: —— aileron is deflected, $m_x^{\delta_{ail}} = 0.0009\ \frac{1}{\text{degree}}$; - - - - - aileron and spoilers are deflected, $m_x^{\delta_{ail}} = 0.0023\ \frac{1}{\text{degree}}$

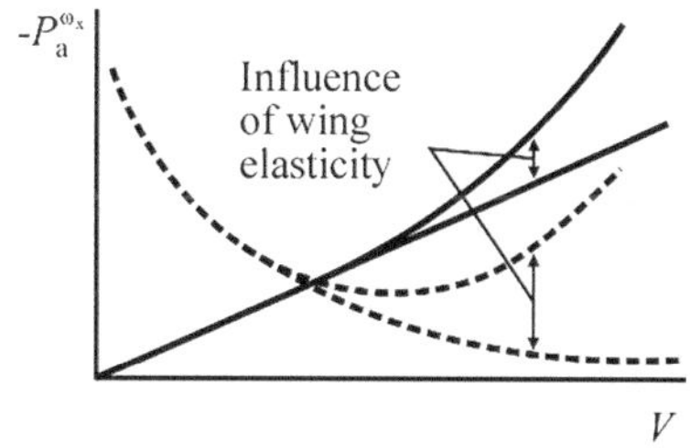

Fig. 28 Variation of the control force per unit of angular velocity in roll as a function of speed: —— without power boosters; - - - - - with an irreversible hydraulic booster

for longitudinal control. We can only add that in some cases (this depends on the specific characteristics of this type of aircraft), a nonlinear power transmission in lateral control wiring may be sufficient to reduce the "pilot-aircraft" system gain by reducing the transverse control efficiency (Fig. 29).

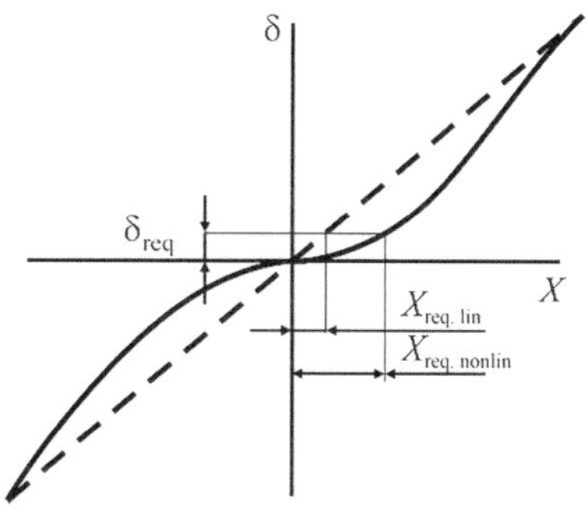

Fig. 29 Increase in the control lever deflection required for a manoeuvre when applying a nonlinear transmission in the control wiring to the rudder: X—control lever travel; δ—rudder deflection; - - - - - linear transmission; —— nonlinear transmission; δ_{req}—required rudder deflection; $X_{req.\ lin}$, $X_{req.\ nonlin}$—required rudder deflections for linear and nonlinear transmissions

In this case, special kinematic mechanisms are installed in the wiring (for example, refer to pos. 25 in Fig. 7) that provide a significant reduction in the output movement (to the ailerons) with low displacements of the control lever. Such mechanisms, due to their simplicity and high reliability, are often used in other control channels (in pitch and yaw).

3.1 Stability and Control Augmentation Systems

Modern stability and control augmentation systems solve a number of problems to ensure the aircraft's stability and controllability, with specific tasks based on the aircraft's weight and centre of gravity:

- to improve aircraft damping performance;
- to increase the degree of longitudinal stability;
- to improve overall controllability performance;
- to prevent flight parameters from exceeding operational limits.

In total, this enables the aircraft to operate with large margins of longitudinal static stability, which leads to a reduction in trim penalties and makes it possible to improve the technical and cost performance of the aircraft. At the same time, the pilot should perceive the flight characteristics of automated aircraft provided with a stability and control augmentation system in the same way as if these characteristics were provided by traditional means. In other words, the pilot should not feel operation of the stability and control augmentation system.

To this end, certain conditions for its functioning should be provided. With mechanical wiring, the stability and control augmentation system is activated in a so-called sequential circuit, i.e. the system signal is summed with the pilot's mechanical signal. In the operation of the stability and control augmentation system, the control levers should remain stationary, i.e. during system operation there should be no "blowback" to the control levers. As is known, blowback arises from the increased friction in the wiring and drive slide that the servo-drive must overcome. The following measures are usually taken to eliminate blowback: to create preload in the control lever feel system (Fig. 30) and to install an auxiliary drive in the mechanical system.

Increasing the preload leads to a heavier control and lower piloting accuracy. The auxiliary drive eliminates the blowback to the control levers, and it is necessary to clench a spring link; the control lever forces are reduced due to the friction in drive slides; the accuracy of the stability and control augmentation system improves due to the cut-off of some wiring by the auxiliary drive; at manual control, the reversal rate is limited.

The main problem with high-performance stability and control augmentation systems is ensuring high reliability. For such a system, first of all, an uncontrolled skid to the extreme position should be excluded, since the pilot practically has no effective means to compensate for this skid. In this case, the use of a redundancy system to

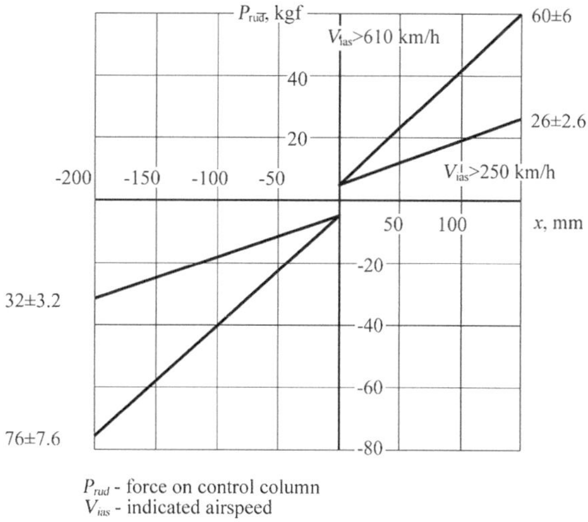

Fig. 30 Characteristics of the control lever feel system

improve stability and controllability with full channel separation, including channel distribution on the two aircraft sides, should be considered the main safety measure in such a way that this failure does not involve more than one side at the worst, while the other side could counteract this failure. Such an approach was applied, in particular, to the An-124 and Il-96-300 aircraft.

3.2 *Direct Lift Control Systems*

On heavy aircraft of conventional configuration, a system consisting of slats and flaps is used to control the lift. This system is applied mainly during takeoff and landing. Separate rotational electro-hydraulic drives control the slats and flaps. These drives transmit the moment through rotational transmission to flap and slat actuators. Each of the systems can operate jointly or independently of each other, according to a given program.

The operation of slats and flaps is synchronized through a mechanical transmission. The deflection of the wing high-lift devices on such aircraft is limited by flight speed to protect these surfaces from destruction.

The considered high-lift devices usually have low deflection rates; for example, the time necessary for complete a reversal from one extreme position to another reaches 30–40 s. During this time, the pilot can compensate for the change in aircraft trim caused by the change in its configuration. The deflection range of slats and flaps is usually 0°–20° and 0°–35°, respectively.

The flaps are multi-slotted to obtain the greatest increment in lift. Such flaps are extended in several stages during the landing approach. This procedure makes it easier for the pilot to compensate for the flap extension’s out-of-trim effect.

Lift controls on light aircraft (slats and flaps) are used at takeoff and landing speeds to reduce the landing speed and decrease the takeoff distance, as well as at high flight speeds, thus improving the aircraft's maneuverability. In this regard, the deflection rate of the high-lift devices approaches the rate of the main control surfaces deflection.

Accordingly, the control system of these elements is organized in much the same way as the control system of the main control surfaces (sectioning, redundancy without sacrificing system performance after failure). For example, the Su-27 aircraft is equipped with control systems for drooped leading-edges and flaperons that use translational drives as a group of cylinders mounted along the span of the leading-edges and flaperons.

On some aircraft (F-5, F-16, F-18, etc.), the drooped leading edges are deflected by a light-sized mechanical transmission containing planetary gearboxes with a large gear ratio. A hydro-motor drive mounted in the fuselage is used to drive the transmission.

When performing maneuvers, the leading edges are usually deflected automatically in response to signals from sensors of the dynamic pressure, the Mach number, and the angle of attack. In this case, the high-lift devices are deflected both in tracking mode and in discrete mode.

Often, spoilers are used to control the lift. Spoilers are extended to an intermediate angle (10°), relative to which they are symmetrically deflected in response to signals from the direct lift control system within angles of $\pm 10°$ to correct the flight trajectory or glide path. This type of lift control is mainly used during landing approaches (on an airfield or on an aircraft carrier). Spoilers are usually deflected by electro-hydraulic actuators that can receive air deceleration signals in addition to the signals of the direct lift control system. In this case, the rates of spoiler deflection reach 100 deg/s, while the angles of spoiler deflection are 50°–60°.

Four to twelve sections per wing may be used in the spoiler control system, depending on the aircraft class. A smaller number of sections is mainly used on manoeuvring aircraft (MiG-23, Su-24, F-14, F-111, Tornado, Jaguar), while the number of sections per wing is up to 12 on heavy aircraft (Tu-160, V-1, Tu-22M). Each section of spoilers is usually deflected by a single-chamber drive with electric or mechanical control.

Usually, a dead zone is introduced in the spoiler control system by aileron deflection. This zone does not allow the spoilers to deflect in response to small signals, for example, when the roll damping system is running. However, this zone does not affect the operation of the direct lift control system, since this system works with respect to the extended position of spoilers (approximately 10°).

The variable wing sweep can also be referred to as the lift control system. It includes: a redundant rotary hydraulic drive; a rotational mechanical transmission that links the drive with the ball screw power mechanisms; ball screws that rotate the outer wings to the given angle; a synchronizing shaft that provides simultaneous rotation of both outer wings; a feedback mechanism; a unit coordinating the change in sweep and deflection of the wing high-lift devices; and other components. The variable wing sweep system is a fairly powerful system. For example, the wing sweep drive of the F-111 aircraft is provided with two hydraulic motors with a capacity of 100 hp each. A screw jack can withstand 230 tons of force. For the B-1 aircraft, this force amounts to 450 tons, while the time for complete sweep reversal from one position to another is 20 s.

3.3 Automatic Control Systems (ACS)

First automatic control systems, or autopilots, performed an auxiliary function of the pilot's relief during continuous aircraft piloting, mainly in cruise. In particular, the autopilots stabilized the aircraft's angular positions, altitude, and flight speed. Minor deflection angles of the control surfaces and, accordingly, low efforts of the actuators were required to perform these functions on aerodynamically stable aircraft. In regards to their significance for flight safety, such autopilots belonged to the category of "comfort" or "auxiliary" systems, i.e., in the event of an autopilot fault and disabling, the aircraft flight could continue without a change in the flight plan. Such an autopilot was usually made as an autonomous system that contained the necessary functional elements (sensors, computers, actuators, consoles, indicators, etc.).

The principles of the automatic control system's inclusion in the manual control system were changing during the evolution of the control system. In general, they correspond to those discussed in the section on stability and control augmentation systems.

3.4 The Inclusion of an Automatic Control System in the Fly-by-Wire System

On aircraft equipped with a fly-by-wire system, there are several options for including automatic control systems (Fig. 31). The choice of one or another option is associated with the peculiarities of the FBW system implementation, the level of redundancy of automatic control systems, the importance of automatic modes for aircraft piloting, etc. In general, the inclusion of signals produced by automatic control systems in the FBW system should not lead to a decrease in the reliability of the main control of the FBW system since usually these systems have different categories of significance from the point of view of flight safety.

In one option (T-4, Concorde), the signals produced by the automatic control system are entered into the FBW system using a special autopilot servo drive that directly affects the control levers. When the control levers are deflected, the signals from the lever movement sensors arrive at the FBW system input. To ensure override, the follow-up drive of the automatic control system is connected to the control levers via a clutch or spring rod with a preload. When the automatic control system is

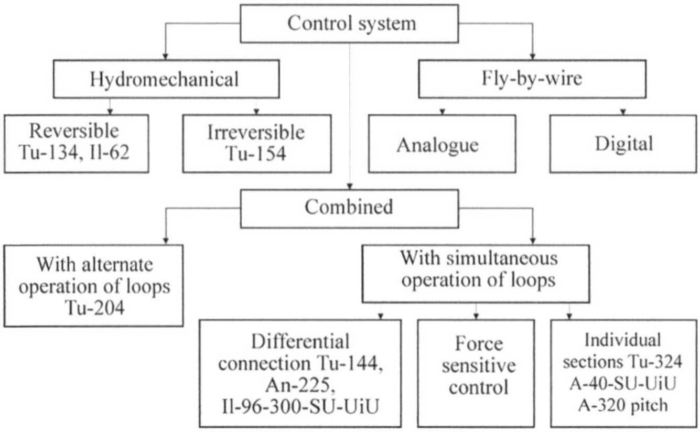

Fig. 31 FBW system implementation options

running, its follow-up drive overcomes the force of the feel device of the FBW system control lever. A self-balancing of forces may be required to ensure a bumpless switching in this inclusion scheme.

Another option to include autopilot signals relates to the FBW system, in which only control actuators with electrical inputs are used. The FBW system has the highest level of redundancy out of all aircraft systems. When the autopilot signals are included directly in the FBW system, security measures must be taken in the event of possible failures in the automatic control system. Moreover, the computers of the automatic control system must have reliable controls that exclude false signals in the FBW system. In an automatic control system, the control levers do not work according to the signals of the automatic control system, the need for which in the FBW system of the type considered is controversial. At some companies (for example, Boeing), according to tradition, the control levers are actuated by special drives; other companies (for example, Airbus) do not find it expedient.

3.5 Aerodynamic Load Reduction System

Aerodynamic load reduction systems are designed to reduce the external loads applied to the airframe structure and to reduce the internal stresses resulting from the redistribution of the aerodynamic forces. What is taken into account here is the structural layout, the features of the structure loading, and the natural frequencies of the airframe oscillation that, in turn, are implemented through the active use of control surfaces and wing high-lift devices [3].

Depending on the purpose, which is set during the implementation of the conceptual configuration and algorithm of the aerodynamic load control system, there are several priority areas:

- the reduction of loading of the structure during maneuvers with the maximum vertical g-load, that entails a decrease in the wing bending moment due to the redistribution of circulation along the wing span at higher normal g-loads;
- the reduction of the gust and turbulence-induced loading;
- the increase in flutter limit;
- the improvement in passenger comfort that entails a decrease in g-loads acting on crew and passengers.

The greatest effect from applying the aerodynamic load reduction system is achieved by those types of aircraft for which aerodynamic loads are one of the determining factors in the choice of the design cases (for example, a long-range aircraft with a high passenger capacity [10, 11]). In this case, it is possible to obtain a reduction in the structure's mass and an increase in the aircraft's flight range of the order of 5%.

Most often, it seems advisable to apply load reduction systems in an integrated manner. At the same time, in each specific case, it is necessary to prioritize between design cases, taking into account the possible cost effect at the expense of increasing passenger comfort. On the other hand, increasing the life of the structure and reducing its mass can be limited by some factors. It may be difficult to directly influence those factors using aerodynamic load reduction systems. For example, the factor determining the service life for a short-range aircraft may be the loads experienced on the ground including those associated with the takeoff and landing. In this case, the possible cost effect of increasing passenger comfort comes to the fore in deciding on the composition of the load reduction system.

The effective use of aerodynamic load reduction systems requires a detailed study of the frequency characteristics and elastic deformations of the structure. It also necessitates a significant increase in the performance of the aerodynamic control system, as well as a fundamental shift in the conceptual configuration of a number of adjacent systems (for example, the control system of the wing high-lift devices).

The use of aerodynamic load reduction systems for aircraft not originally equipped with these systems may be appropriate if one of the design cases "significantly dominates" and limits the aircraft's performance.

Variables

Symbol	Description	Units
M_0^{avail}	Available stabilizing or control moment	N m
M_0^{req}	Required stabilizing or control moment	N m
C_l	Rolling moment coefficient	–
C_m	Pitching moment coefficient	–
C_n	Yawing moment coefficient	–
I	Aircraft moment of inertia	kg m^2
r_0^i	Distance between the resultant aerodynamic force applied to the control surface and the aircraft centre of mass	m

(continued)

(continued)

Variables		
Symbol	Description	Units
P_i	Force applied to control surface	N
$\dot{\omega}$	Angular acceleration	$1/s^2$
T_{roll}	Roll mode time	s
q	Dynamic pressure	N/m^2
S	Wing area	m^2
b	Wing span	m
α	Angle of attack	degrees
β	Angle of sideslip	degrees
M	Mach number	–
n_z	Load factor in the z-direction	–
$C_{L,avail}$	Available lift coefficient	–
C_L^α	Lift slope	1/deg
C_m^α	Pitching moment coefficient with respect to angle of attack	1/deg
$\overline{r}$	Dimensionless radius of aircraft inertia	–
σ_n	Coefficient of longitudinal static stability with respect to g-load	–
$X_{el}^{n_z}$	Pitch control lever displacement per unit of normal g-load	mm
$P_{el}^{n_z}$	Pitch control lever force per unit of normal g-load	N
$\overline{x}_{CG}$	Center of gravity X-position in terms of the MAC length	–
$\overline{x}_{AC}$	Aerodynamic center X-position in terms of the MAC length	–
P	Force applied to control lever	N
x	Control lever displacement	mm
c_{MAC}	Mean aerodynamic chord	m
$C_m^{C_L}$	Longitudinal static stability	–
$x_{CG,fr.limit}$	Forward center of gravity limit	m
$x_{CG,aftlimit}$	Aft center of gravity limit	m
$x_{CG,neut}$	Neutral point position	m
μ	Aircraft density ratio	–
δ	Control surface deflection angle	Degrees

References

1. Kestelman VN, Fedorov AV (1987) Mekhanizmy upravleniya samoletom (Aircraft control mechanisms). Mashinostroenie, Moscow
2. Ostoslavsky IV, Strazheva IV (1969) Dinamika poleta. Ustojchivost' i upravlyaemost' letatel'nyh apparatov (Flight dynamics. Stability and controllability of aircraft), Mashinostroenie, Moscow

3. Byushgens GS (ed) (1995) Aerodinamika i dinamika poleta magistral'nyh samoletov (Aerodynamics and flight dynamics of mainline aircraft). TsAGI, Moscow
4. Lebedev AA, Adzhimamudov GG, Baranov VN et al (1996) Osnovy sinteza sistem letatel'nyh apparatov (Fundamentals of synthesis of aircraft systems). MAI, Moscow
5. Rudis VI (1978) Poluavtomaticheskoe upravlenie samoletom (Semi-automatic control of the aircraft). Mashinostroenie, Moscow
6. Shumilov IS (2006) Sistemy upravleniya rulyami samolyotov (Aircraft rudder control systems), MGTU im. Baumana, Moscow
7. Boris YuA, Bocharov VYa (1978) Nadezhnost' sistem upravleniya samoletov i ih struktura (Reliability of aircraft control systems and their structure). Mashinostrenie, Moscow
8. Goniodsky VI, Sklyansky FI, Shumilov IS (1974) Privod rulevyh poverhnostej samoletov (Aircraft control surfaces drive). Mashinostroenie, Moscow
9. Sheinin VM, Kozlovsky VI (1984) Vesovoe proektirovanie i effektivnost' passazhirskih samoletov (Weight design and efficiency of passenger aircraft). Mashinostroenie, Moscow
10. Eger SM, Mishin VF, Liseitsev NK et al (2005) Proektirovanie samolyotov (Aircraft design). Logos, Moscow
11. Novozhilov GV (ed) (1976) Teoriya i praktika proektirovaniya passazhirskih samoletov (Theory and practice of designing passenger aircraft). Nauka, Moscow

Landing Gear Design

Nikolai K. Liseitsev

Abstract The chapter starts with some recommendations for the selection of the landing gear arrangement for an aircraft (including the number of wheels and gears), discusses in detail the geometric parameters and performance of aircraft with a tail-dragger, tricycle, tandem landing gear with outriggers, and a multiple-wheel landing gear. The chapter shows how to estimate aircraft ground roll maneuverability, as well as how to design structural and kinematic diagrams of the retractable landing gear. At the end of the chapter, the reader will learn about possible design solutions to reduce the volume of the landing gear in the retracted position.

Keywords Landing gear arrangement · Wheel · Gear · Ground roll maneuverability · Retractable landing gear

1 Selecting the Landing Gear Arrangement

Landing gear is the device that enables an aircraft's ground run, take-off, landing, and manoeuvring on the ground. It reacts the loads acting on the aircraft and dissipates most of its landing impact kinetic energy.

The landing gear arrangement (type) refers to the number of landing gear legs and their location with respect to the aircraft's centre of mass. There are currently four arrangements of the landing gear used on aircraft: taildragger (Fig. 21.1a); tricycle (Fig. 21.1b); tandem with outriggers (Fig. 21.1c); multiple-wheel.

1.1 Taildragger

The taildragger landing gear arrangement has been widely used before the mid-1940s. In this arrangement, the main wheels are located forward of the aircraft's

N. K. Liseitsev (✉)
Moscow Aviation Institute (National Research University), Moscow, Russia
e-mail: nikolai.liseitsev@yandex.ru

M. A. Pogosyan (ed.), *Aircraft Design*, Springer Aerospace Technology,
https://doi.org/10.1007/978-981-96-4599-2_21

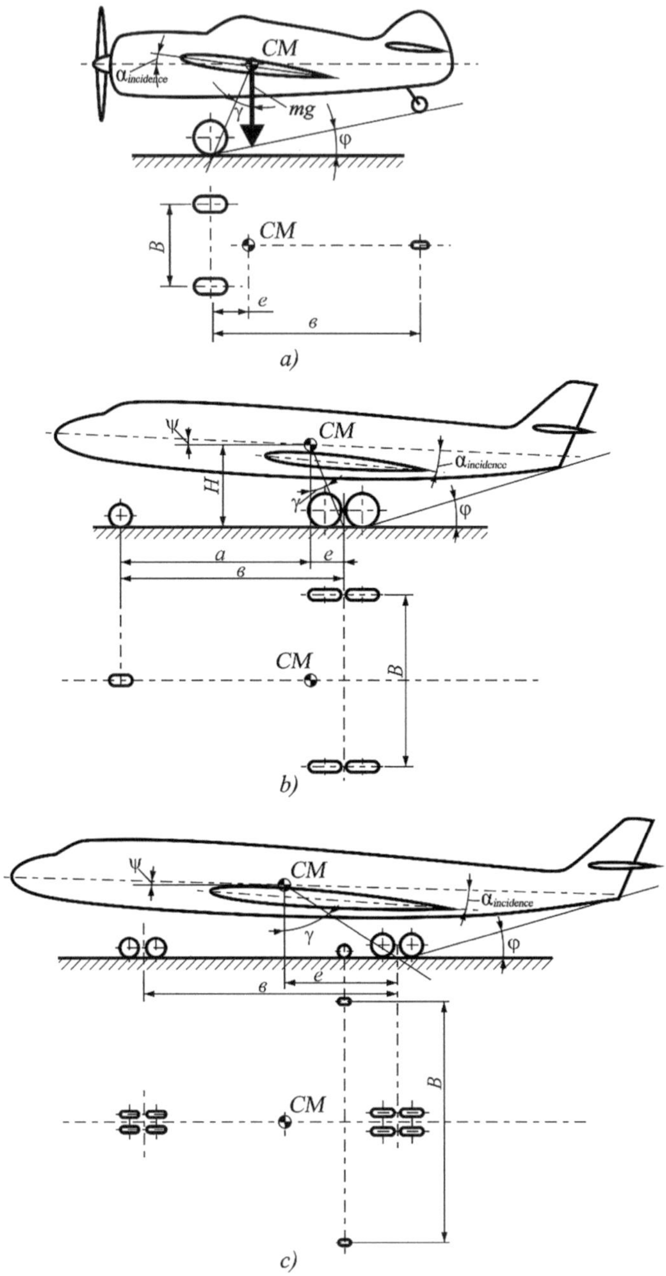

Fig. 21.1 Landing gear arrangements: **a**—taildragger; **b**—tricycle; **c**—tandem

centre of mass, while the tailwheel is located aft. This landing gear arrangement has the following shortcomings.

1. The aircraft's landing speed has to be limited: for passenger aircraft, it is 80–120 km/h, for fighters and bombers −130–150 km/h, for agricultural aircraft −60–90 km/h. Increasing these speeds is impossible. As V_{land} increases, the pilot finds it harder to keep the landing trajectory, the landing pancake altitude increases, while during flare or holding the aircraft may prematurely, at a speed exceeding the landing speed, touch the RW with its main wheels. The drag due to landing gear at that time creates a negative pitching moment relative to the aircraft's centre of mass. If this moment is higher than the nose-up pitching moment created by the empennage and the aircraft's weight, the aircraft will turn over (nose over), resting on the main landing gear. If the moment due to drag is insufficient to cause the nose over, the moment due to weight will cause the airplane to lower the tail and increase the angle of attack; because the aircraft speed is higher than the landing speed, the wing lift will become greater than the aircraft weight, and the aircraft will soar up to a certain height, from which it will then pancake due to the speed loss. This aircraft behaviour is called "bounce". If bounce occurs at a speed significantly higher than that of landing, the altitude of pancaking may be significant and the aircraft's drop from it will result in damage.

This means that for a taildragger aircraft, touching down on the RW with the main wheels at a speed exceeding the landing speed is impermissible. The normal landing is ensured only if the three wheels touch down on the RW simultaneously at a speed equal to the landing speed.

2. Tailwheel aircraft features dynamic ground instability. During take-off run, the difference between the weight of the aircraft and the lift of the wings and horizontal tail is reacted by the main landing gear, with respect to which the aircraft is trimmed. In the course of this motion, various disturbances (crosswind, uneven runway, etc.) cause the aircraft to turn by angle β, while the friction forces applied to the landing gear create a moment with respect to the aircraft centre of mass, which tends to turn it even to a greater angle (Fig. 21.2).

The destabilising moment is determined by the following equations:

$$M_{dest} = F_{MGL}\, a \sin\beta, \tag{21.1}$$

$$F_{MGL} = (mg - L_{WG} - L_{HT})\mu, \tag{21.2}$$

$$M_{dest} = (mg - L_{WG} - L_{HT})\mu\, a \sin\beta, \tag{21.3}$$

where μ is the ground roll friction coefficient.

The pilot counteracts the aircraft's turns caused by disturbances by applying the rudder, which is immersed in the propeller slipstream. In addition, the friction forces

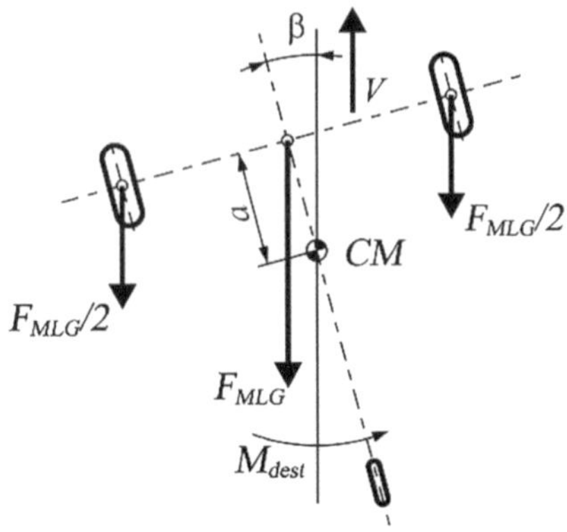

Fig. 21.2 Formation of destabilising moment during aircraft operation on the RW (the tailwheel is not in contact with the RW)

also create a nose-down pitching moment, which disrupts the aircraft's balance and complicates steering.

To counteract disturbances on the ground, the pilot creates control moments by breaking the wheel or wheels. The wheel braking forces are an effective means of counteracting disturbances; however, this increases the landing ground run distance.

Some directional stability of such aircraft while moving on the RW is attained by forming a planform wheel camber (Fig. 21.3). The planform wheel camber usually amounts to 3–4° and is intended to make the aircraft dynamically stable on the ground since the friction force on the landing gear acts against the turn caused by the disturbance. However, the planform wheel camber results in an increased tire wear and increases the landing run distance, as the friction forces directed aft are constantly applied to the wheels.

To reduce the effect of the destabilising moment on the ground, the tailwheel is locked in the aircraft's plane of symmetry. When the aircraft begins to turn, the locked tail wheel produces a stabilizing moment. At the beginning of the landing ground run, the force created by the horizontal tail presses the tailwheel down to the RW, thus contributing to an increased ground stability. At the final section of the run, the horizontal tail force keeping the tailwheel down disappears. Therefore, if at the end of the run during the initiation of the turn the pilot fails to brake and produce the required controlling moments on the main landing gear, it becomes impossible to put the initiated turn under control.

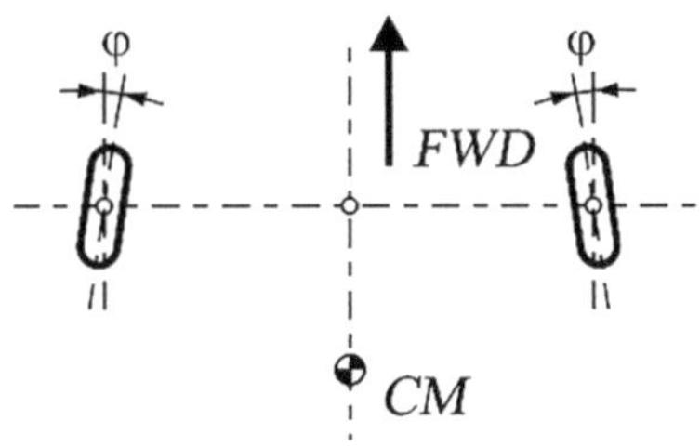

Fig. 21.3 Planform wheel camber diagram: φ is the planform wheel camber angle

The instability of taildragger aircraft is another factor that contributes to a lower landing speed.

During manoeuvring of a taildragger aircraft, the tailwheel needs to be unlocked and made to caster, otherwise, during the aircraft's turn, forces will arise on the tailwheel that will hamper the turn and destroy the tires, tearing them off the rim.

3. On a taildragger aircraft, jet engines are not used (take-off run is hampered). Under the speed at which the horizontal tail becomes effective, the pilot can neither control the aircraft's pitch nor accomplish its trim with respect to the main landing gear, hence the possibility of the aircraft's nose-over.

Therefore, the taildragger arrangement is used only on propeller or turboprop aircraft. The horizontal tail of such an aircraft is immersed in the propeller slipstream, creating the necessary moments for aircraft trim and control.

1.2 Tricycle Landing Gear

Landing on the main wheels has no negative consequences for aircraft equipped with tricycle landing gear. Due to the drag of the main wheels, the aircraft's nose dips downward and comes to rest on all wheels. At the same time, the angle of attack and the lift of the wings decrease. If the shock absorber of the landing gear has absorbed the aircraft energy properly, the aircraft will no longer be able to take off from the runway.

This circumstance allowed the aircraft with tricycle landing gear to sharply increase the landing speed, which (on military as well as passenger planes operated off concrete runways) reaches 240–280 km/h and more.

The possible steering mistakes on such aircraft during landing caused by increased landing speed are not dangerous.

High landing speeds lead to an increase in the landing gear's loads and consequently its mass; however, this appeared justified because of the possibility to significantly increase the wing loading, reduce its area, and consequently increase the flight speed.

The aircraft with tricycle landing gear arrangement feature smooth ground run. Their nose wheels rotate clear, and the wheels and skis of the gear are capable of rotating with respect to the vertical axis located in the plane of symmetry of the aircraft. As a result, when the aircraft turns to an angle β due to disturbances, the ground friction force does not act on the nose wheel because the nose wheel casters in the direction of movement. Meanwhile, the friction force F_{MGL} that acts on the main landing gear creates a stabilising moment with respect to the aircraft's centre of mass (Fig. 21.4). This moment can be estimated by the following equations:

$$M_{\text{dest}} = F_{\text{MLG}} e \sin \beta, \ or \quad (21.4)$$

Fig. 21.4 Formation of stabilising moment during aircraft motion on the RW

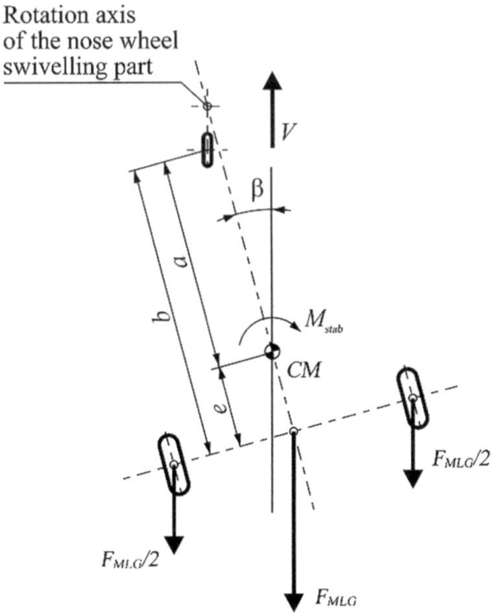

$$M_{\mathrm{dest}} = mg(a - \mu_{\mathrm{aircraft}} h)(e/b)\mu_{\mathrm{MLG}} \sin \beta, \tag{21.5}$$

where μ_{aircraft} is the friction coefficient of the aircraft's landing gear; μ_{MLG} is the friction coefficient of the main landing gear. Quantities a, b, e and h are given in Fig. 21.1b.

The nose wheel, positioned forward of the aircraft's centre of mass, allows the use of jet engines on it. Additionally, aircraft with tricycle landing gear have the following advantages:

(a) a simpler steering technique during take-off and landing;
(b) a better pilot's view during take-off, landing, and taxi because the aircraft fuselage has a zero deck angle or is close to it;
(c) a more intensive braking reducing the ground roll distance and the possibility to land with the wheels not fully braked. As for a taildragger aircraft, the braking forces at the beginning of the take-off run are limited because of the nose-over possibility, which increases the ground roll distance;
(d) the possibility to land without flare. This landing pattern is used on the carrier-based aircraft and STOLs allowing a shorter runway because the deceleration by wheel braking is more effective in comparison with the deceleration by air drag during flare.

The tricycle's shortcoming is that self-induced oscillations of the castering nose wheel may emerge. They are referred to as "shimmy." The oscillations are damped with the aid of a specially mounted hydraulic damper. If the aircraft has a nose wheel steering system, the shimmy damper also serves as the system's hydraulic power drive. The hydraulic system providing energy for turning the nose wheel compensates for possible liquid leaks from the damper, and this considerably enhances the reliability of its operation.

1.3 Tandem Landing Gear with Outriggers

In the tandem arrangement, the nose wheel is located forward of the centre of mass, and, therefore, the aircraft can land like a tricycle-geared aircraft at high landing speeds and feature directional stability during take-off run and landing roll.

The use of the tandem arrangement with auxiliary wing outrigger wheels is mainly caused by aircraft configuration. The need for a large-volume bay in the region of the aircraft's centre of mass results in the main landing gear being moved away from it by a considerable distance. Furthermore, the load on it accounts for up to 55% of the aircraft weight. However, the use of the tandem landing gear is also the cause of the following serious shortcomings.

1. It calls for a more experienced pilot to perform a successful take-off.

   Once the aircraft reaches the take-off speed, it can only increase the take-off angle of attack with the aid of a special system ensuring that the nose is lifted (its height is increased) or the tail is tipped back (its height is decreased), because the horizontal tail at the take-off speed neither has the necessary effectiveness nor ensures the longitudinal trim with respect to the main landing gear, which is well off the centre of mass.
2. The aircraft's roll distance increases due to the limitation of the braking force created by the NLG wheels to 60–70% of the braking forces of the main landing gear, because the friction forces on the nose wheel during disturbances along the track create a destabilising moment with respect to the centre of mass.
3. The fuselage mass rises by 15–20% due to the need to reinforce the frames and beams in the region of the central compartment and the landing gear attachment points.
4. The mass of the tandem landing gear, despite its low height, is larger than that of the tricycle landing gear because of the wing outrigger wheels and a more powerful turn control system.
5. In the retracted position, the outriggers usually do not fit into the aircraft outlines, necessitating the installation of fairings on them, which increase the aircraft's drag.

Due to the above shortcomings, the tandem landing gear arrangement has not been used widely.

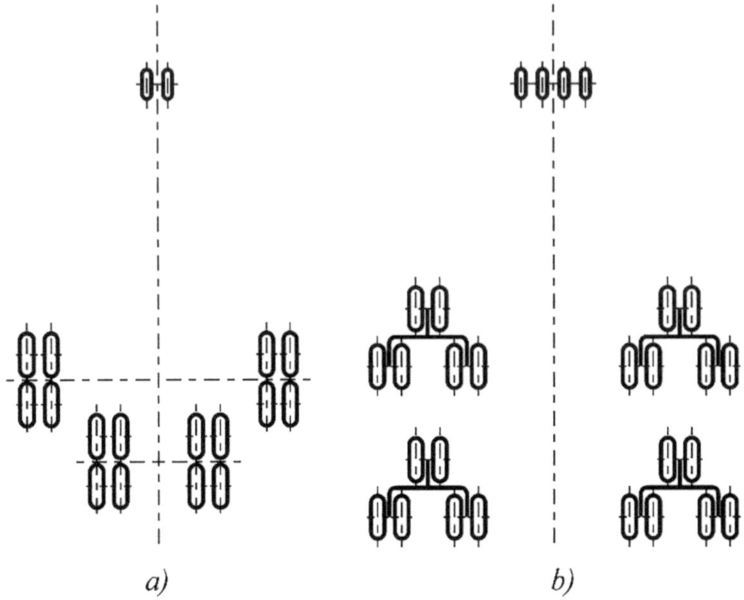

Fig. 21.5 Multiple-wheel landing gear arrangements: **a**—Boeing B-747; **b**—Lockheed C-5A Galaxy

1.4 Multiple-wheel Landing Gear

The development of heavy aircraft has led to the use of the multiple-wheel landing gear arrangement. This arrangement allow a reduction in the load per strut and ensure that the concrete pavement is not damaged during aircraft operation. Multiple-wheel landing gear is used in the Soviet Ilyushin Il-86, which has a take-off mass of 206 tons, and in the Boeing B-747 and Lockheed -5 Galaxy, which have take-off masses of around 350 tons (their landing gear arrangements are shown in Fig. 21.5).

The use of the multiple-wheel landing gear arrangement on such aircraft also creates more favourable conditions for stowing them in the retracted position.

In the aircraft with multiple-wheel landing gear, the nose wheel is located forward of the centre of mass, and, for this reason, they have both the advantages and the shortcomings listed in Sect. 21.1.2, inherent in the tricycle arrangement.

1.5 Aircraft Ground Roll Maneuverability

The aircraft with tricycle, tandem, and multi-wheel landing gear arrangements, have considerably better maneuverability which is ensured by the nose wheel steering control system. Rotation of the nose wheel to angles of 60–80° off the neutral position permit smaller turn radii of the aircraft. At the same time, this system makes it possible to compensate for the disturbances on the runway without resorting to control moments using braking forces. This allows for a reduction of the take-off run and landing roll distances. It is noteworthy that the use of this system makes it possible to accomplish manoeuvring with less thrust and achieve a certain reduction in fuel burn.

2 Selection of the Landing Gear Basic Geometric Parameters

This section deals with selecting parameters for the tricycle landing gear arrangement which is the most common in today's aircraft.

The geometric parameters of a landing gear arrangement determine the location of wheels with respect to the aircraft's centre of mass. Due to this fact, the determination of the landing gear geometric parameters must be preceded by the weight analysis which determines the position of the aircraft's centre of mass and the drawing of its general view in three projections with the indication of the wing MAC position and its incidence angle $\alpha_{\text{incidence}}$ on it (see Fig. 21.1b). The landing gear geometric parameters are selected to ensure the necessary attitude of the aircraft during landing, the minimum take-off run and landing roll distances, and its stability on the RW and during ground manoeuvring [1].

The basic geometric parameters of the tricycle landing gear arrangement are:

- the wheelbase b (the side-view distance between the nose and main wheels' axes);
- the wheel track B (the front-view distance between the points of the main wheels' touchdown on the RW);
- the main landing gear stagger e (the side-view distance between the vertical line passing through the aircraft's centre of mass and the main wheels' axis);
- the nose wheel stagger a (the side-view distance between the vertical line passing through the aircraft's centre of mass and the nose wheel axis);
- the main landing gear stagger angle γ;
- the turnover angle φ (the angle at which the fuselage tail or its skid touches down on the RW);
- the static ground angle ψ (the angle between the fuselage centreline and the RW surface).

Landing gear parameters are determined with unloaded wheels and shock absorbers at the aft take-off or landing center of gravity limit. When multiple-wheel bogeys are installed, the sizing is done relative to the axles on which the bogeys rotate. To ensure the necessary attitude of the aircraft during landing, the landing gear parameters are determined based on the following relationships:

$$\alpha_{land} = \alpha_{C_{L,max}} \Delta\alpha, \tag{21.6}$$

$$\alpha_{\text{land}} = \psi + \varphi + \alpha_{\text{incidence}}, \tag{21.7}$$

where $\Delta\alpha$ is the margin angle guaranteeing that the aircraft will not reach the critical angle of attack at which the wing stall begins.

The static ground angle ψ is usually assumed to be 0° ... 4°. The optimum value of the angle corresponds to the aircraft's minimum drag during take-off run which shortens its distance. During the landing roll, the static ground angle decreases because the braking forces increase the load on the nose wheel and relieve the load on the main gear. If the static ground angle becomes negative, the roll distance decreases.

Knowing the value of the wing incidence angle $\alpha_{\text{incidence}}$, which is assumed to be equal to the angle of attack under the main flight conditions, we can find the aircraft's turnover angle from Eq. (21.7). For various aircraft, $\varphi = 10$ to 18°.

The main wheels stagger angle γ does not allow the aircraft to tip back on the tail during landing, hence its value is assumed to be

$$\gamma = \varphi + \left(1...2^{\circ}\right). \tag{21.8}$$

Using the parameters included in Eqs. (21.6) and (21.7), the stagger of the main gear wheels e and the height H are determined from the side view of the aircraft general arrangement (see Fig. 21.1b).

The magnitude of stagger e is usually

$$e = (0.15 \ldots 0.20)c_{MAC}. \tag{21.9}$$

It should be noted that with a large stagger, it becomes difficult to lift off the nose wheel when the aircraft rotates to the take-off angle of attack. Under these conditions, the lift-off will take place at a higher speed and consequently the ground run distance will increase. Reducing e will ensure an easy nose wheel lift-off; however, with a small stagger the aircraft may tip back on the tail because the aircraft's centre of mass may find itself behind the landing point when $\varphi = 0^{\circ}$.

Parameter e should also be selected to ensure ground operation and parking. The aft center of gravity limit should not be behind the axis of wheels or the bogey installed on the main gear leg. For passenger aircraft, the aft center of gravity position is experienced when a load of 500 kg is applied to the tail, when the servicing personnel with tools work in the area of the empennage, or when, provided there is an aft door, passengers begin to enter the empty aircraft and up to 30% of their number has concentrated in the aft portion of the fuselage. And if the centre of mass does shift behind the main landing gear axis, it is necessary to provide a special tail strut of the type installed on the Ilyushin Il-62.

The nose wheel stagger a is selected such that when the aircraft is parked, the load on it amounts to 6–12% of the aircraft's weight. Hence,

$$a = (0.88 \text{ to } 0.94)\text{b}, \tag{21.10}$$

$$e = (0.06 \text{ to } 0.12)\text{b}, \tag{21.11}$$

where b is the wheelbase.

With a too small load on the nose gear, the aircraft's controllability during taxiing is degraded. With an increase in the load, the mass of the gear and the mass of the fuselage forebody increase.

The wheelbase b affects the ground manoeuvrability of an aircraft. Statistical data on modern aircraft with various arrangements and purposes show that

$$b = (0.3 \text{ to } 0.4)L_{\text{fus}}. \tag{21.12}$$

If the wheelbase is small $(L_{\text{fus}}/4)$, then during taxiing the aircraft encounters significant oscillations in vertical plane.

The landing gear height must ensure a minimum clearance of 200 to 250 mm between the RW surface and the aircraft structure (fuselage, wing, engines, propellers, ventral strakes, etc.) with the tires and shock absorbers of the main and nose gears loaded individually or simultaneously.

This clearance needs to be determined also when the aircraft lands with a bank angle; in this case, the aircraft's pitching angle equals the landing angle, and the bank angle is set at 4°. For an aircraft with a small sweep angle and an anhedral, these conditions usually determine the landing gear height.

Analysis of aircrafts' motion at a 4° bank angle and the practice of operation show that, once one of the gears touches down on the RW, the resultant force on it decreases the bank angle and for this reason the set clearance between the wing tip and the RW is ensured without considering the deflection of the wheels and shock absorber. It permits a significant reduction of the landing gear height. With the unloaded wheels and shock absorbers, the clearance between the ventral tail strakes and the RW also has to be determined because the forces arising on the main gear struts compress the tires and shock absorbers decreasing the pitch angle of the aircraft.

The landing gear height also depends on the aircraft configuration (high-wing monoplane, mid-wing monoplane, low-wing monoplane), the attachment locations of the landing gear and its stowage conditions in the retracted position.

The wheel track B is selected to ensure the aircraft's dynamic stability during take-off, landing, and taxiing. The wheel track mainly depends on the height of the aircraft's centre of mass above ground H, and if the wheel track does not correspond to the height, the aircraft may tip sideways during taxi on the RW. Therefore, the minimum wheel track of the tricycle landing gear must prevent tipping relative to line 1—2, which connects the nose and main landing gear (Fig. 21.6).

Overturning (sideways turnover) will be obviously impossible if the moment created by the aircraft weight with respect to this line will be greater than the moment created by the friction forces:

$$mgc > F_{fric}H, \tag{21.13}$$

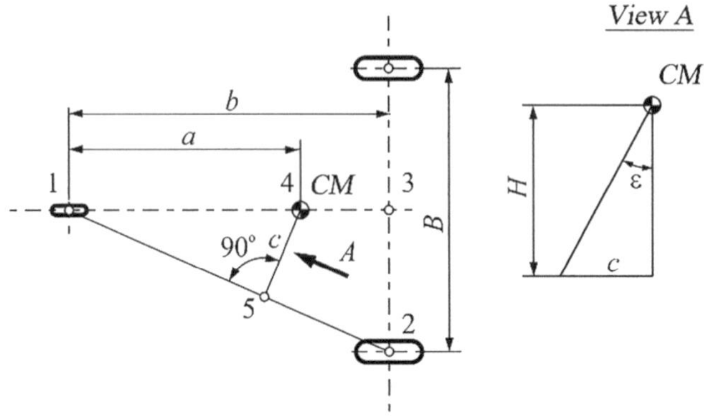

Fig. 21.6 Diagram showing how to determine the minimum wheel track B

$$c > \mu H, \tag{21.14}$$

where F_{fric} *is the* friction force due to the aircraft's sideslip; $\boldsymbol{\mu}$ is the lateral friction coefficient assumed to be 0.85 when determining the wheel track.

The distance between the nose and one of the main landing gears (along diagonal 1—2) equals $\sqrt{b^2+B^2/4}$. Line c is found from the similarity of triangles 1–4–5 and 1–3–2:

$$c = \frac{aB}{2\sqrt{b^2 + B^2/4}}. \tag{21.15}$$

Substituting c in Eq. (21.14), we define the landing gear track B:

$$B = \frac{2Hb\mu}{\sqrt{a^2 - H^2\mu^2}}. \tag{21.16}$$

If in this equation the magnitude $H^2\mu^2$ compared with a^2 is ignored, the wheel track to first approximation may be assumed to be $B \geq 2H$, because $a \approx 0.9b$.

For heavy aircraft, one should try to make the wheel track wider than the width of slabs of the aerodrome concrete pavement, whose size equals seven meters. This reduces the load on the slabs because only one gear leg will be resting on them in that case. The maximum size of the wheel track B should be limited to twelve meters in order to permit aircraft to safely taxi on the taxiways which are not more than 15 m wide.

It is noteworthy that on several aircraft the nose landing gear is not located in the plane of symmetry of the aircraft, but rather in the parallel plane at a short distance from it. This location of the nose landing gear has been used on the Hawker Siddeley Trident, the Fairchild-Republic A-10 bomber and other planes. The nose gear offset with respect to the plane of symmetry permits a more rational configuration, given certain conditions. On the Trident aircraft, a 610 mm offset made it possible to mitigate the reduction of the baggage compartment volume, which is located under the floor aft of the nose gear bay, and on the A-10A aircraft—to place a large-caliber rapid-fire rotary cannon. With a nose wheel steering control system, such a location of the nose landing gear has almost no effect on the aircraft's control on the ground.

When selecting the wheelbase b and wheel track B, it is desirable to determine the possibility for the aircraft to turn by 180° on the runway of the given aerodrome class (the RW dimensions for each class of aerodrome are given in Table 21.1).

It follows from the geometric diagram of the turn performed by an aircraft (Fig. 21.7) that the aircraft may turn by 180° at a given RW width under the following conditions:

$$RWwidth > R_{inboard} + R_{nose} + 2\Delta, \quad (21.17)$$

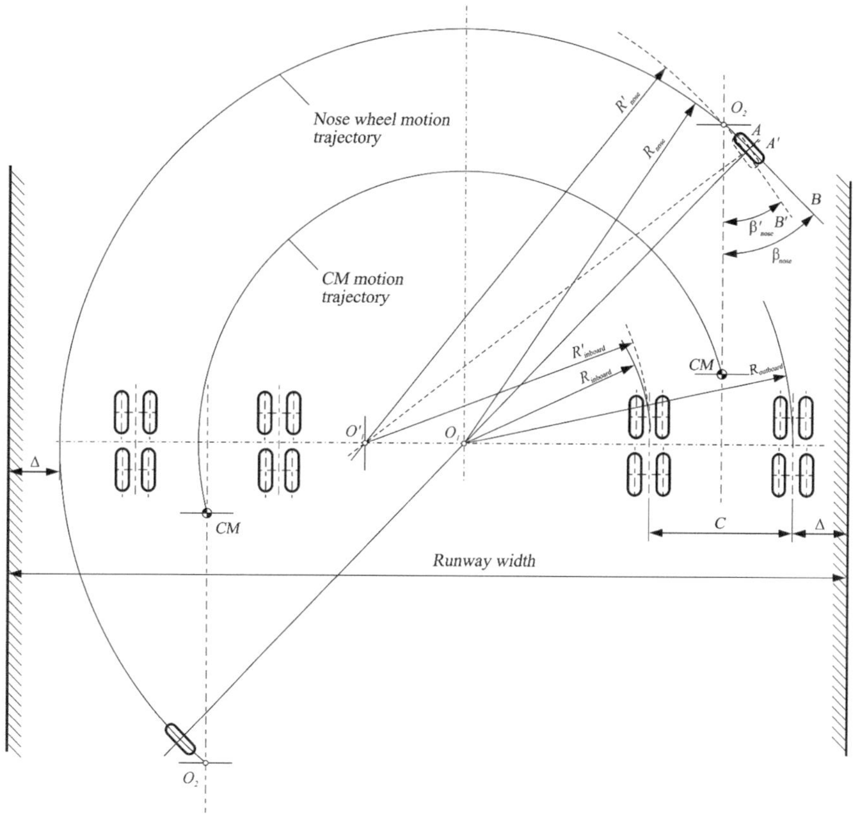

Fig. 21.7 Diagram of turn execution by an aircraft on the RW: O_2—nose wheel axis; $O_1A \perp O_2B$, $O_1'A' \perp O_2B$

where $R_{inboard}$ *is the* radius of the circle along which the inboard main gear moves; R_{nose} is the radius of the circle along which the nose wheel moves; Δ *is the* distance from the RW edge to the wheels, $\Delta = 0.5$ to $1.0\,m$.

Significant influence on the RW width required for turn execution is exerted by the nose wheel rotation angle β_{nose}.

It follows from Fig. 21.7 that a decrease in the maximum value of β_{nose} to β'_{nose} leads to an increase in the RW width due to the increase in $R_{inboard}$ and R_{nose} which are included in the right part of Eq. (21.17). Consequently, the angle β_{nose} should be made as large as possible when designing the nose wheel steering control system. However, in operation, there is some deviation from the above turn execution diagram, which is the consequence of the tire strains arising due to the forces acting on them during a turn. Once the slip emerges, the nose wheels are rolling not in the plane of angle β_{nose} but in a plane that is inclined at an angle δ with respect to it, which is called the slip angle. It diminishes the maximum value of β_{nose} (Fig. 21.8), while the wheel slip on the main landing gear leads to an increase in $R_{inboard}$.

The largest influence on the wheel slip angle is exerted by the aircraft speed during the turn and the tire/runway friction coefficient. Increasing speed and friction coefficient increase the slip angle. As a result, the turning radius of the aircraft's centre of mass, determined by the diagram in Fig. 21.7, should be increased roughly by 20–30%.

The formation of an extra vertical axis, with respect to which the wheels of the multiple wheel bogeys of the main gear are castering, is a design solution that allows the width of the turning strip to be reduced in order to ensure the operation of dimensionally large aircraft at the given class of aerodromes. This design solution is used on the Tupolev Tu-154 and Douglas DC-8. On the DC-8, the aft axis of the two-axle bogey is steerable, while the forward axis of the three-axle bogey is steerable on the Tu-154.

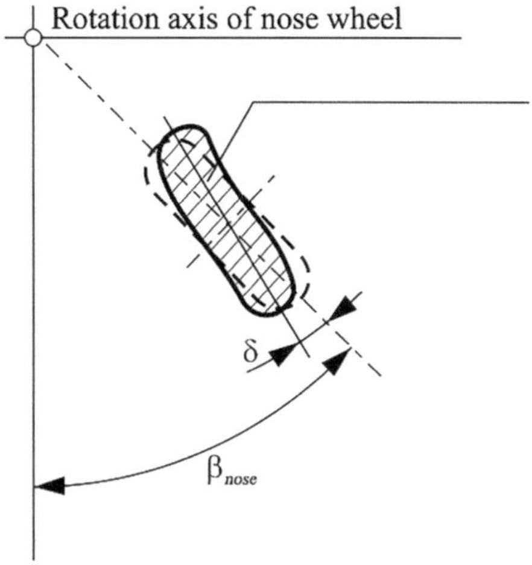

Fig. 21.8 Formation of the wheel slip angle

At the same time, the above design solutions permit a reduction of the magnitude of the torsion that acts on the landing gear as the aircraft accomplishes a turn.

When selecting the geometric parameters of the landing gear, it is always desirable to keep in mind that, despite the need to meet a large number of requirements, the most important of them is to ensure that an aircraft has the required landing angle.

3 Selection of the Number of Gears and Wheels

The selection of the number of gears and the type, size, and the number of wheels mounted on them depends on the requirements for aircraft flotation and its field performance.

Flotation means the possibility of operating an aircraft for an extended period of time on an unpaved or paved aerodrome without damaging its surface.

For the aircraft intended for use on a concrete runway, the necessary number of wheels and their location with respect to one another on the gear to satisfy the flotation requirements (the possibility of operating without damaging the slabs of concrete surfaces) are selected depending on the equivalent single-wheel load P_{ESWL} that corresponds to the set class of aerodrome. The equivalent single wheel load is a load due to a single-wheel gear that is equal, in terms of the force magnitude, to the effect of the real gear on pavement (considering the number, size, and location of wheels on that gear). The concept of the equivalent single wheel load is needed to determine the increase in the total distributed load on the concrete slab in comparison with the rated point load.

Aerodromes with concrete runways are subdivided into several classes based on the length, width, and thickness of the pavement. For each class of aerodrome, the highest value of P_{ESWL} *is set.* In Table 21.1, the characteristics of the classes of aerodromes are given.

It follows from Table 21.1 that, if a tricycle-geared aircraft has one wheel on each gear, then the take-off mass of the aircraft operating on the Class A and Class G aerodromes should not exceed, respectively, 100,000 kg (if the nose wheel is assumed to account for around 10% of the take-off mass) and 4400 kg.

However, the take-off masses of modern aircraft reach 350,000 kg and more, and their operation on the aerodromes with paved runways is ensured by distributing the load by increasing the number of gears and wheels on the gears.

If the landing gear is a four-wheel or eight-wheel bogey, the equivalent single wheel load to first approximation is found by the following formula:

$$P_{\text{ESWL}} = 0.04344 P_{\text{static ground TO}}^{4/3} \left\{ \frac{\left[1 + 0.04(1 - 3 \cdot 10^{-6} P_{\text{static ground TO}})\right](p_0 \cdot 10^{-5} - 4)^{0.81} k_{\text{LG}})}{1 + 0.0534(p_0 \cdot 10^{-5} - 4)^{0.675}} \right\}, \tag{21.18}$$

where:

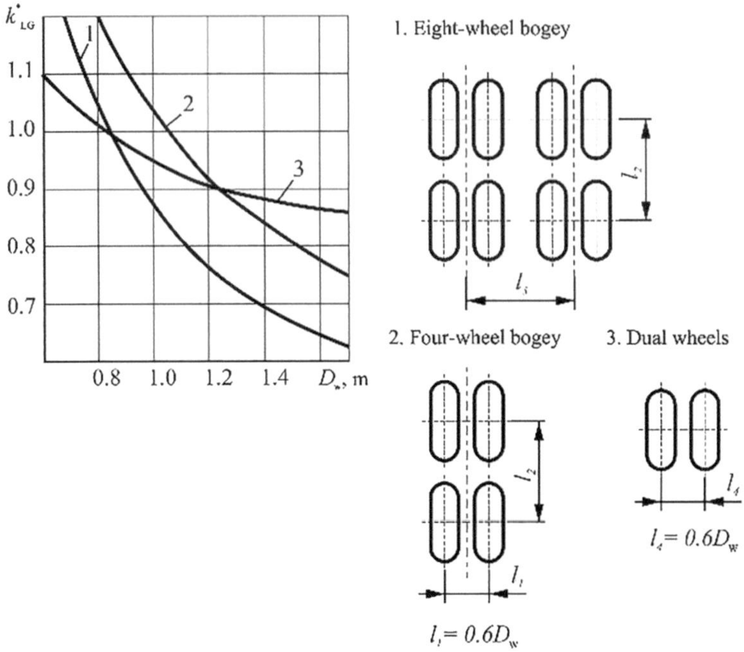

Fig. 21.9 Effect of wheel diameter and gear type on coefficient k_{LG}: l_2= 1.009D_{w}, l_3=D_{w}

$P_{\text{static ground TO}}$ is the static ground load on one wheel of a four-wheel bogey or on two wheels of an eight-wheel bogie, N; p_0 is the pressure in the wheel tire, Pa;

k_{LG} is the factor depending on the gear type and wheel diameter (Fig. 21.9).

For a dual-wheel gear,

$$P_{\mathrm{ESWL}} = 0.2831 P^{1.165}_{\text{static ground TO}} \left\{ \frac{\left[1 + 0.0352(p_0 \cdot 10^{-5} - 4)^{2/3}\right] k_{\mathrm{LG}}}{1 + 0.0488(p_0 \cdot 10^{-5} - 4)^{2/3}} \right\}, \quad (21.19)$$

where $P_{staticgroundTO}$ is the static ground load on one wheel, N; other designations are the same as in Eq. (21.18).

Formulas (21.18) and (21.19) are applicable at $p_0 < 4 \cdot 10^5$ Pa and under the condition that there is only one aircraft gear per RW concrete slab. If two main landing gear legs simultaneously rest on one slab, then the aircraft's flotation sharply decreases. In this case, P_{ESWL}, calculated by formulas (21.18) and (21.19), should be increased by 25–30%.

To avoid degrading aircraft flotation, the distance between the main landing gears at $P_{ESWL} > 0.25MN$ must exceed the slab width equal to 7 m.

Formulas (21.18) and (21.19) show that the use of a large number of wheels of a smaller diameter on an aircraft improves its flotation. The following are more

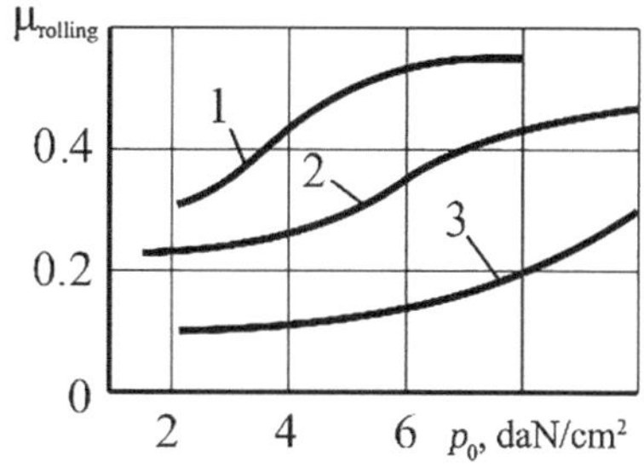

Fig. 21.10 Variation of the ground friction coefficient as a function of the tire pressure: 1—in mud season; 2—in dry period; 3—dry and firm soil

effective methods of increasing the flotation of aircraft with take-off masses in excess of 200,000 kg:

- installation of three or four main landing gears;
- transition to six- and eight-wheel bogeys;
- use of wheel separation on the axles of multiple-wheel bogeys.

When an aircraft is operated on an unpaved aerodrome, its thrust-to-weight ratio must ensure the possibility of starting from rest, maneuvering, and takeoff by using only one power plant thrust, and the track depth on the unpaved surface must not exceed a certain value. The condition of an aircraft's flotation on an unpaved surface is determined by the following formula:

$$\Sigma T > gm_0\mu_{\text{start from rest}}, \tag{21.20}$$

where ΣT is the total thrust of the aircraft's power plants; $\mu_{\text{start from rest}}$ is the ground friction coefficient at the start from rest. One can assume $\mu_{\text{start from rest}} = 1.4\mu_{\text{rolling}}$, where μ_{rolling} is the ground friction coefficient at a speed of 1 m/s. Rough values of μ_{rolling} depending on the tire pressure are given in Fig. 21.10.

The minimum soil strength ensuring flotation may be calculated to first approximation by the following equation:

$$\sigma_{\min} = \frac{P_{\text{static ground TO}}}{\delta_{\text{contact}} B_{\text{wheel}}} \sqrt{\frac{1}{H_{\text{track}} D_{\text{wheel}}}}, \tag{21.21}$$

where $P_{\text{static ground TO}}$ is the static ground load on one wheel, N; H_{track} is the allowable track depth (0.05 m); B_{wheel} is the wheel width, m; D_{wheel} is the wheel diameter, m; $\delta_{contact}$ is the factor taking account of the strain of tire in the area of contact with soil (Fig. 21.11).

σ_{min} found by formula (21.21) must be compared with the value set forth in the mission requirements. It is necessary to meet the following condition:

$$\sigma_{\text{min design}} \le \sigma_{\text{min norm}}. \tag{21.22}$$

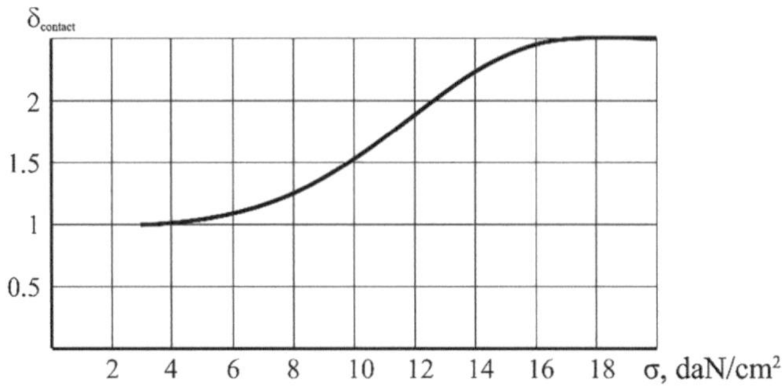

Fig. 21.11 Dependence of $\delta_{contact}$ on σ

It follows from formula (21.21) that, in order to lessen the depth of the track, it is required to decrease the load per wheel and increase its width and diameter.

Flotation on an unpaved surface may be improved if a ski or a wheel-ski undercarriage is used. However, using such landing gear results in the following problems:

- a higher thrust-to-weight ratio (at least 0.6) is required for the aircraft to start from rest;
- the manoeuvrability on the ground is degraded, hence the aircraft virtually loses the ability to move autonomously;
- the landing gear mass increases, which complicates its retraction;
- the power of the landing gear retraction and extension power system increases.

In this context, despite a large number of available designs of ski and wheel-ski undercarriages, they are used mainly on aircraft operated under special conditions.

To determine the required type and size of the wheel, one should refer to the existing wheel nomenclature, which is conventionally divided into four groups (Table 21.2).

The requirements imposed on an aircraft's flotation and field performance help decide which group a wheel belongs to. High- and ultra-high pressure tires are used

Table 21.1 P_{ESWL} for various classes of aerodromes

Aerodrome class	Runway length, m	Runway width, m	P_{ESWL}, MN
A	More than 2550	45	0.45
B	2150–2550	45	0.35
C	1800–2150	45	0.27
D	1500–1800	45	0.20
E	1280–1500	45	0.13
F	1080–1280	30	0.07
G	900–1080	30	0.02

Table 21.2 Nomenclature of landing gear wheels for aircraft

Group number	Group designation	Tire pressure, Pa	Allowable landing and take-off speeds, km/h
I	Ultra low pressure	(2.5–3.5)–106	200
II	Low pressure	(3.5–6.5)–106	250
III	High pressure	(6.5–10)–106	300
IV	Ultra high pressure	10–106	>300

for aircraft that operate off concrete runways. Low and ultra-low pressure tires are used on aircraft that fly off unpaved runways, depending on soil strength.

To ensure reliable and accident-free operation of the aircraft, the parameters and characteristics of the wheels installed on it must not go beyond certain limits. These parameters are:

$P_{\text{static ground land}}$ = wheel load at the aircraft's landing mass;

$P_{\text{static ground TO}}$ = wheel load at the aircraft's take-off mass;

$P_{max\,allowable}$ = maximum allowable wheel load during absorption of operating work A^{oper};

P_{ultimate} = ultimate wheel load during absorption of maximum workA^{max};

P_z = lateral breaking load on wheel;

P_{dynamic} = maximum allowable dynamic load for wheels installed on the nose gear;

V_{land} = aircraft landing speed;

V_{TO} = aircraft take-off speed;

$M_{braking\ torque}$ = maximum braking torque of the wheel;

A_{H} = kinetic energy absorbed by the wheel per one landing;

t = minimum allowable time between landings;

N = guaranteed number of landings.

The minimum allowable time between landings is set to ensure wheel cooling after braking. This is especially important for commuters with a short flight duration and small intervals between take-offs and landings. Non-sealed wheels have a cooling time of 1.5–2.0 h, while sealed wheels have a longer cooling time, limiting aircraft operation.

Dynamic loads for the nose gear wheels are determined by formulas given in Aviation Requirements, chapter 25.

Some of the wheel parameters are determined by the characteristics of the aircraft itself, including its weight, layout and performance. However, other parameters, such as $P_{max\,allowable}$, P_{ultimate}, and P_z, may be finalised only after the landing gear arrangement has been selected and the diagram illustrating the operation of the wheel and shock absorber (specified in Aviation Rules) is prepared.

The wheels on the main landing gear are selected from the range of brake wheels. The brake wheels are also selected for the nose wheel of the tandem arrangement. The wheels intended for installation on the nose and tail gears and the wing outriggers are usually without brake and are selected from the range of brake wheels with arch-shaped tires. To make these wheels lighter, they are used without brakes and brake drums.

When selecting wheels, it is necessary to know the relations between the number of wheels to be installed on the gear and the related weight penalties. Historical dependencies establishing the relations between the mass of the wheels, their number and the aircraft's take-off mass show that an increase in the number of wheels due to an increase in the aircraft's take-off mass leads to an appreciable reduction in the wheel-related weight penalties.

In addition, an increase in the number of wheels on struts helps in most cases to:

- reduce the volume occupied by the wheels in the retracted position;
- reduce the mid-section of the volume occupied;
- reduce cutouts in the airframe and the area of doors;
- enhance the reliability and damage tolerance of the landing gear.

When determining the number of wheels to be installed on a strut, it is desirable to take account of the aircraft layout, which imposes certain limitations on the landing gear stowage in the retracted position.

3.1 ACN—PCN Single Evaluation Method

In accordance with the ICAO (International Civil Aviation Organization) Convention, the bearing capacity of the pavement intended for aircraft with a mass of over 5700 kg is determined by the Aircraft Classification Number—Pavement Classification Number (ACN-PCN) method.

The ACN-PCN method implies comparing the values of the pavement classification numbers (PCN), which characterise the bearing capacity of pavement structures without restricting traffic intensity, with the aircraft classification number (ACN), which characterises their effect on pavements with a specified standard subgrade strength. An aircraft can operate without limitations on pavements if ACN < PCN.

ACN and PCN classification numbers are determined by the following formula:

$PCN(ACN) = 0{,}204P$, where

P is the equivalent single wheel load, kN.

The equivalent single wheel load is such a nominal load that, under a pressure in the wheel tire of 1.25 MPa, creates in the design section of the pavement slab (or in the flexible pavement) a force impact equal to the force impact from a multiple wheel gear load. The equivalence criterion is assumed to be the design condition with respect to tension or strain, which is accepted for a specific type of pavement (rigid, flexible, or composite).

Characteristics of domestically-produced modern wheels—see "Annex 6: Properties of Aircraft Wheels".

3.2 *Structural and Kinematic Diagrams of the Retractable Landing Gear*

The majority of modern aircraft flying at speeds in excess of 250 km/h are made with a landing gear that is retractable in flight. A non-retractable landing gear creates considerable drag and degrades the aircraft's flight performance. There is a wide diversity of landing gear retraction methods [2]. Nose wheels are retracted into the forebody, while main gear wheels can be retracted into the wing, into the wing and fuselage, into the fuselage, into the engine nacelles, or into special nacelles.

The structural and kinematic diagrams of the retractable landing gear have certain distinctive features:

(1) the landing gear is made up of beams;
(2) the landing gear experiencing the loads acting in the retraction and extension planes is, according to the laws of structural mechanics, a statically determinate system;
(3) the landing gear during retraction and extension forms, according to the theory of mechanisms and machines, a four-bar mechanism or a four-link mechanism with the attachment, according to the Assur-Artobolevsky classification, of two-lead groups to it, which serve to ensure the stability of the gear's geometric shape and the retraction/extension process.

The examined retractable landing gear properties make it possible to divide the gears based on their structural design into four groups. In the below diagrams (see Figs. 21.12 through 21.16, 21.19 and 21.22), the landing gear attachment points to the airframe are denoted in Arabic numerals, the retraction and extension mechanisms are denoted in capital letters A, B, and the locks providing stability of geometric shape are denoted by the word "lock".

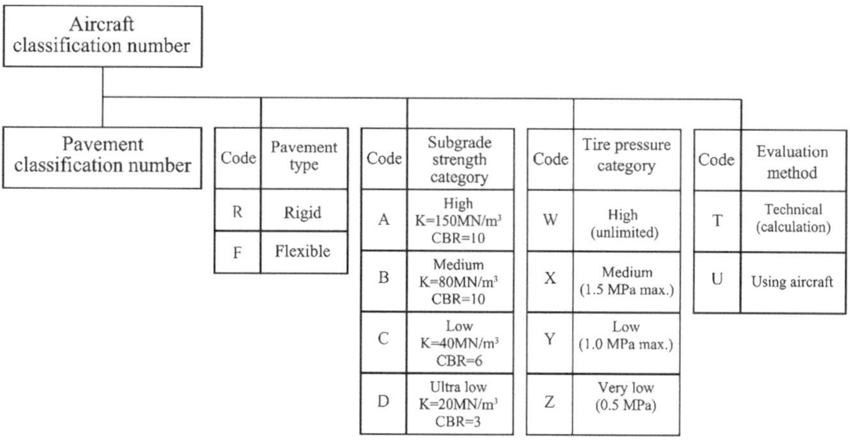

Fig. 21.12 Structure of data presentation by the ACN—PCN method

First group (Fig. 21.13).

1. The stability of the gear's geometric shape in the extended position is ensured by zero bars with a lock, or by the locks substituting for zero bars.
2. Retraction and extension mechanisms are not load-bearing bars of the gear.

Because the locks react forces in the direction of zero bars, the magnitudes of the loads acting on the down locks are small when compared to the loads acting on the folding braces. This group of gears was rarely used in the early days of the retractable landing gear because reliable mechanical locks capable of withstanding high loads had not yet been developed [3].

The gears of this group differ in mass and required volume for the retracted position. The diagrams presented in the middle row of Fig. 21.13 are currently used quite rarely:

- the landing gear mass increases due to a large number of landing gear attachment points to the airframe;
- the volume occupied by the retracted landing gear is larger than the volume occupied by other gears in that group;
- because of the variable force created by the actuating cylinder during retraction, the structure mass increases due to the hydraulic system's oversizing;
- the structure's mass increases due to the complexity of the analysis of the design force experienced by the actuating cylinder when the landing gear is loaded. This force depends on the strain of the structural members.

The most rational diagrams among those given in Fig. 21.13 are those in which:

- the force created by the retraction and extension mechanism varies negligibly (with an increase in the torque created by gravity with respect to the pivot t, the moment arm of the retraction and extension mechanism also increases); the number of structural members is the lowest, which leads to a reduction in mass;
- the landing gear occupies less space in the retracted position.

Second group (Fig. 21.14).

1. The stability of the geometric shape of the gears in the retracted position is ensured by the load-bearing locks, which attach the beams or load-bearing struts of the landing gear directly to the aircraft structure.
2. The retraction and extension mechanisms are not load-bearing bars of the gear.

The diagrams of this group are employed in the design of the nose and main landing gears on several Ilyushin Il and Mikoyan-Gurevich MiG aircraft. These gears occupy small volumes in the airframe.

Third group (Fig. 21.15).

1. The stability of the geometric shape of the gears in the extended position is ensured by mechanical locks installed in the retraction and extension mechanisms.
2. The retraction and extension mechanisms are the load-bearing bars of the gears.

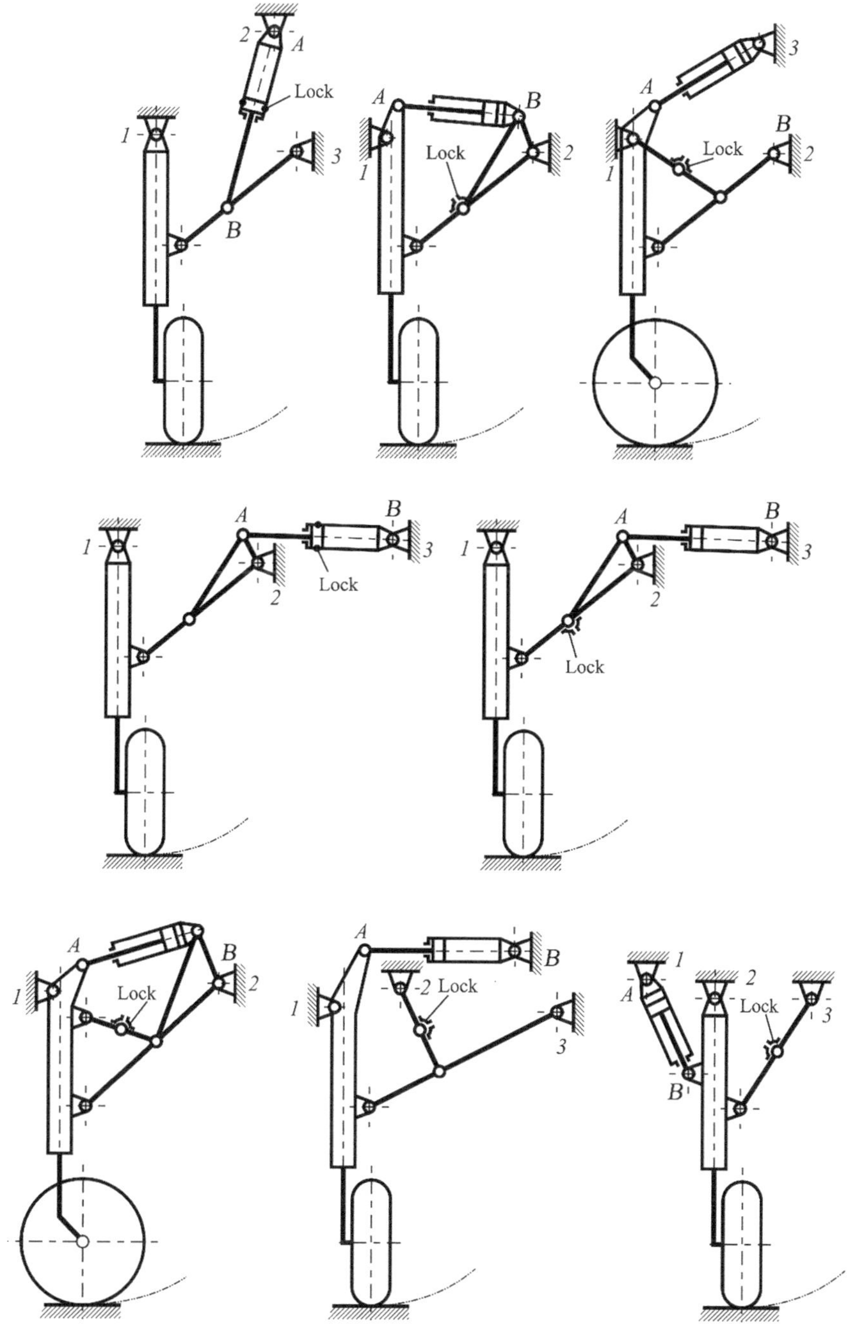

Fig. 21.13 Landing gear diagrams of group I

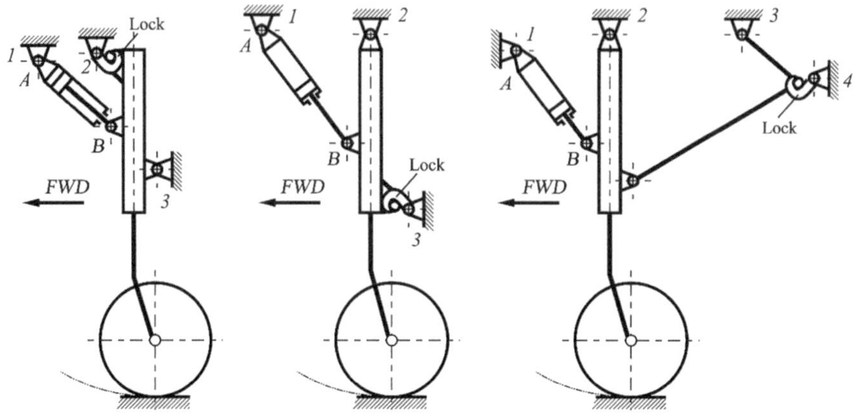

Fig. 21.14 Landing gear diagrams of group II

Fig. 21.15 Landing gear diagrams of group III

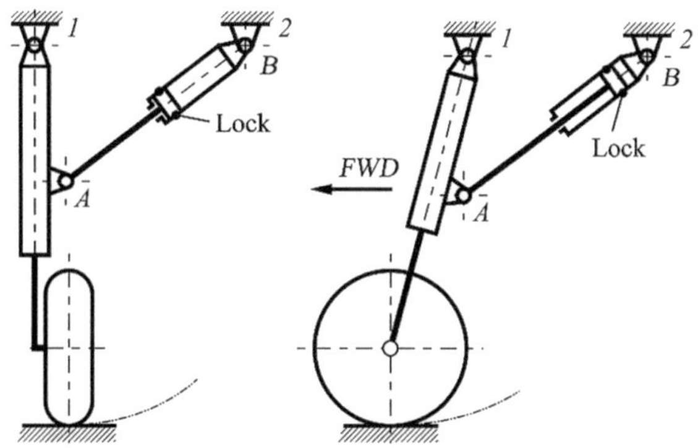

The diagrams of this group have several significant advantages over other groups. These advantages are due to:

- the lowest number of structural elements (only two, the strut and the actuation cylinder);
- the lowest number of pivots (three);
- the lowest number of units (three);
- protection of the load-bearing mechanical locks against external unfavourable factors (sand, dust, water, ice, etc.);
- the fact that strains and clearings in the structure do not affect the gear structural design, which enables a more accurate structural analysis and a reduction in the gear mass;
- a possible redundancy of the load-bearing mechanical lock in the cylinder by using a hydraulic lock;
- a greater structural durability which is attained by reducing the overall number of parts, protecting the mechanical lock, and reducing the maintenance hours.

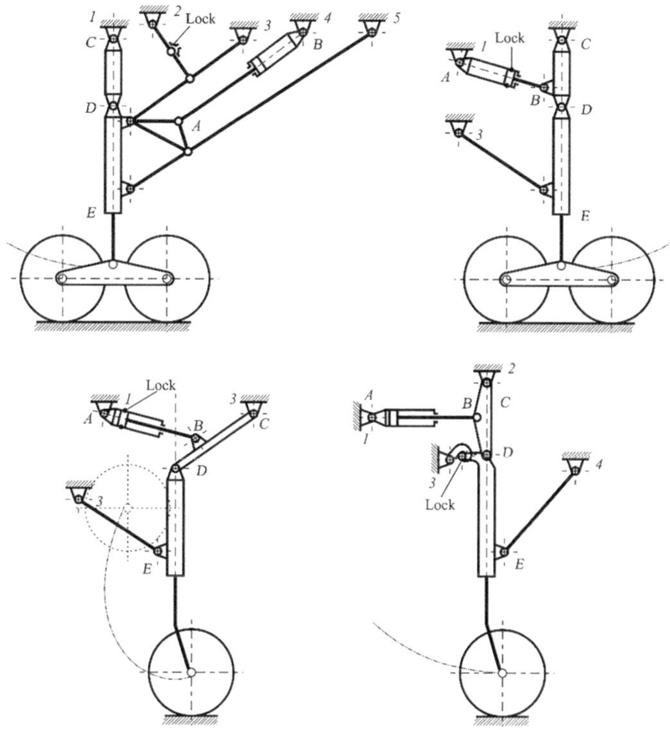

Fig. 21.16 Landing gear diagrams of group IV

The general shortcoming inherent in the third group is the oversizing of the hydraulic system because of the variable force created by the actuating cylinder during retraction.

Fourth group (Fig. 21.16).

1. The stability of the geometric shape of the gears in the extended position is ensured by the methods employed in the designs of the first, second, and third groups.
2. The landing gear beam consists of two parts (links C—D and E—D).

The diagrams of this group are used when the landing gear features a large height; the use of the landing gear of other groups in this case will increase the occupied volumes and penetrate the primary elements of the airframe structure. However, it should be noted that the gain in volume leads to an increased structure mass because the division of the strut into two parts increases the number of pivots, units, and primary structural members.

Today, for the retractable landing gear of any group to be stowed rationally, for the required volume to be reduced, and for the landing gear and airframe structures to be matched, various methods are used. For example, on an airplane, second retraction and extension mechanisms can be installed.

It is also noteworthy that when two retraction and extension mechanisms are used, the reliability and damage tolerance of the main and redundant power systems ensuring the landing gear extension increase appreciably. The increase in reliability and damage tolerance is the result of the complete disconnection of the systems, which eliminates their possible mutual influence during a malfunction and failure. Two retraction and extension mechanisms are used in the main landing gear of the land-based variant of the BAC/Breguet Jaguar, in the nose gear of the A300B passenger airliner, in the main landing gear of the Bristol B175 Britannia, the Handley Page HP.88 Victor, and several other aircraft.

In the selection of the landing gear structural design, much attention should be paid to ensuring the reliability of the landing gear extension and the landing gear downlocking. When the landing gear extension is ensured by the aerodynamic forces and gravity, and the stability of the geometric shape in the extended position is ensured by the acting forces themselves, the reliability will always be significantly higher.

Alongside the landing gear structural diagram, the shock absorber installation diagram must also be considered. If the landing gear has sufficient height and no limitations are imposed on the shock absorber displacement and the initial base between its boxes, the cantilever-telescopic shock absorber allows a reduction in the mass of the landing gear.

The cantilevered-telescopic shock absorbers are not suitable for aircraft operated off unpaved runways because such shock absorbers poorly tolerate direct impact loads on an uneven surface. On aircraft with a limited or small landing gear height, it is desirable to use levered shock absorbers or external shock absorbers. In this case, even with a small height of the landing gear, it is possible to obtain a considerable deflection of the wheels during shock absorption, which results in a reduction of the acting loads and compensates for the increase in the mass of the shock absorbers themselves. Landing gear designs also employ other shock absorber installation diagrams. Double-levered shock absorbers are installed in the main landing gear of the Caravelle and the Convair B-36D. A rocking shock absorber installation is used in the design of the main landing gear of the Tupolev Tu-134 and Tu-154 aircraft. The diagram's strong point is that the cantilevered-telescopic shock absorber absorbs drag loads acting on the landing gear. Attaching a strut to the torque link pivot allows the shock absorber to deflect during compression in the direction of drag loads.

If multiple-wheel bogies are used, they should be arranged such that in the extended position, the initial landing impact is reacted by the rear wheels, which helps decrease the drag loads due to the wheel spin-up.

3.3 Design Solutions to Reduce the Volume of the Landing Gear in the Retracted Position

Recent times have seen the development and application of design solutions to reduce the volume occupied by the landing gear in the retracted position. One such solution is to compress the shock absorber to its maximum displacement, which is accomplished by the following methods.

1. Shock absorber compression to its maximum displacement by the shock absorber itself, which in this case performs the function of the actuating cylinder and to this end is connected with the aircraft hydraulic system (Fig. 21.17). This method of shortening the shock absorber length may be used if the pressure in the gas chamber at the maximum displacement, its geometric dimensions, and the pressure in the hydraulic system, used to accomplish the compression, are related by the following inequality:

$$P_{shock\ absorber\ \max} \pi D^2_{\text{cylinder}}/4 < p_{\text{HS}}(\pi D^2_{\text{cylinder}}/4 - \pi d^2_{\text{piston}}/4)\eta, \quad (21.23)$$

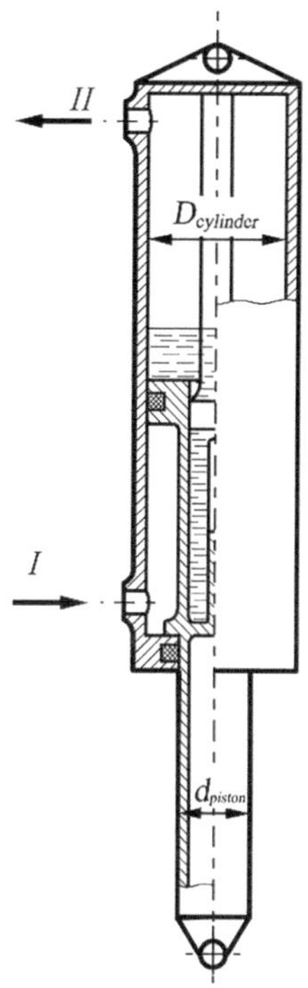

Fig. 21.17 Shock absorber compression during retraction: I—hydraulic fluid supply for shock absorber compression; II—to transfer/separation container

where $P_{shock\ absorber\ \max}$ is the maximum pressure in the shock absorber gas chamber at the maximum displacement; d_{piston} is the outer diameter of the shock absorber piston; D_{cylinder} is the inside diameter of the shock absorber cylinder; p_{HS} is the available pressure in hydraulic system; η is the compression mechanism efficiency.

The reduction in the left term of inequality (21.23) prior to shortening the shock absorber length may be attained by connecting the shock absorber air chamber to transfer/separation containers (Fig. 21.18). This will increase the volume of its air chamber and reduce the pressure. Floating pistons in the transfer/separation container isolate the compressed gas and the fluid contained in the shock absorber from the hydraulic system fluid.

2. Shock absorber compression to some part of its displacement or the maximum displacement with the help of telescopic rod connecting the moving part of the absorber with the aircraft's structure. Such design solution is used in the Sky Hawk A4 D, whose landing gear arrangement is given in Fig. 21.19.

The location of the attachment point of the telescopic rod to the airframe is selected such that, during retraction, the length of the shock absorber is shortened. In the extended position during shock absorber compression, the telescopic rod is also compressed without affecting the operation of the shock absorber.

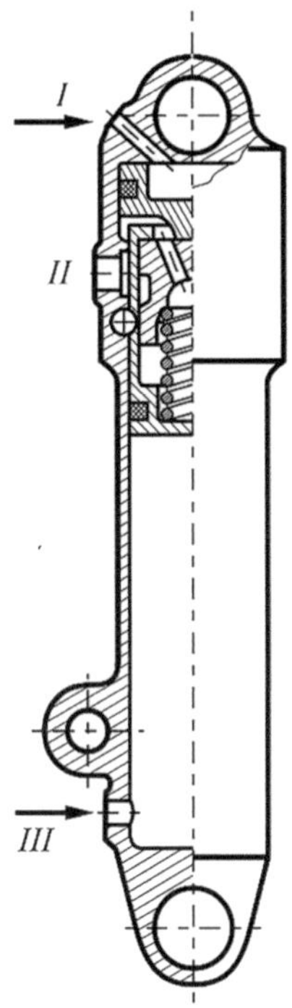

Fig. 21.18 Transfer/separation container: I—hydraulic fluid supply during landing gear retraction; II—supply to shock absorber air chamber; III—hydraulic fluid supply during landing gear extension

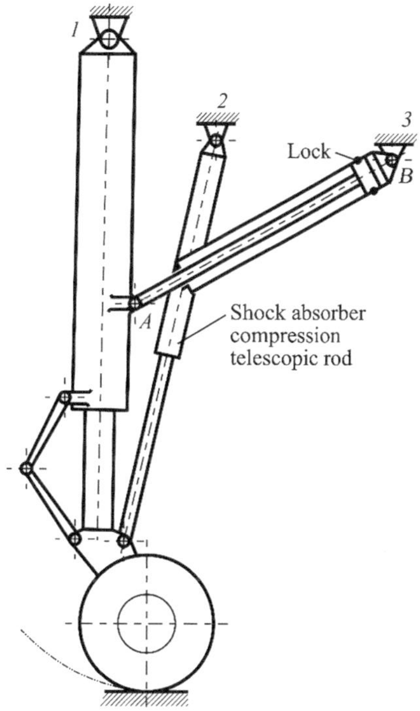

Fig. 21.19 Compression of a shock absorber by a telescopic rod during retraction

3. Displacement of the shock absorber inside the strut with the help of a special drive (Fig. 21.20). This method of shortening the length of a cantilevered-telescopic shock absorber is used in the main landing gear of Concorde.

When an external shock absorber mounted outside the strut is used and the wheels are attached to it with the help of a lever, the volume occupied by the landing gear before retraction may be reduced by:

- pulling the shock absorber and the lever with the wheels attached to it to the strut with the help of a special drive (Fig. 21.21);

- pulling the lever with the wheels attached to it to the strut with the help of a kinematic transmission during landing gear retraction (Fig. 21.22).

In the Hawker Siddeley Harrier STOL, to reduce the volume of the retracted landing gear, a kinematic transmission is used.

To reduce the volume of the landing gear in the retracted position, expanding tires can be used. In the retracted position, such tires are deflated (d is the diameter of a deflated tire); after the landing gear extension and before landing, they are inflated with gas. Below, the characteristics of expanding tires are given:

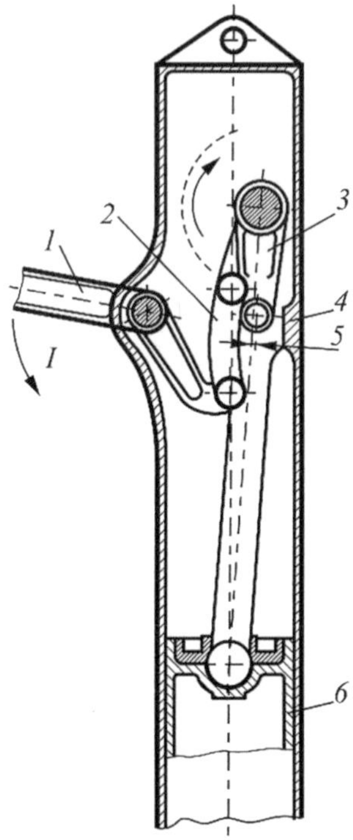

Fig. 21.20 Reduction of the length of the shock absorber during retraction on the concorde airliner: 1—stay; 2—link; 3—two-arm lever; 4—stop; 5—kinematic lock; 6—shock absorber cylinder; I—direction of motion during retraction

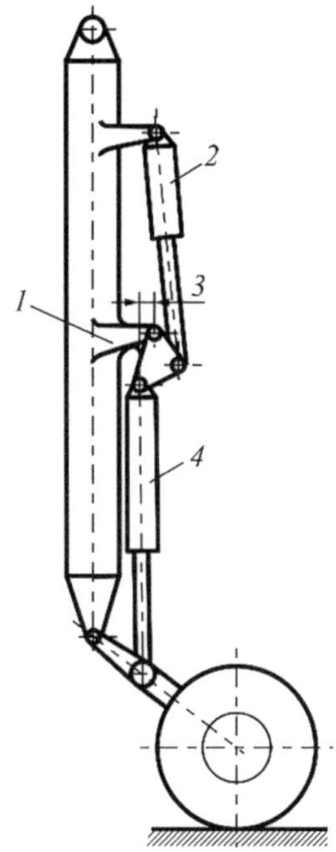

Fig. 21.21 Reduction of the volume occupied by the landing gear during retraction by a special drive: 1–stop; 2–wheel pulling drive; 3–kinematic lock; 4–external shock absorber

Fig. 21.22 Reduction of the volume occupied by a landing gear during retraction with the help of a kinematic transmission

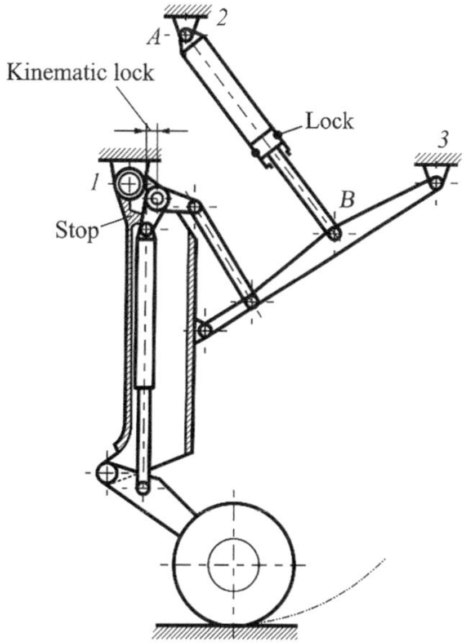

D_{outer}, m	1.45	0.406
d, m	0.965	0.318
$(D - d)/D$	33.5	21.7

The tire with an outer diameter of 1.45 m has been developed for cargo and passenger aircraft and features an inflation pressure of (5 to 6) $\cdot$ 10^5 Pa. The tire with $D_{outer} = 0.406$ m (a high-pressure tire with an inflation pressure of more than 106 Pa) was designed for the C-134 military transport aircraft's main landing gear.

The reduction of the outer diameter of a tire helps reduce the required volume in proportion to the diameter ratio to the second power. For a wheel with $D_{outer} =$ 1.45 m, the occupied volume is reduced 2.2 times, and for a wheel with $D_{outer} =$ 0.406 m—by a factor of 1.6.

Variables		
Symbol	Description	Units
M_0^{avail}	Available stabilizing or control moment	$N \cdot m$
V_{land}	Landing speed	m/s
M_{dest}	Destabilizing moment	N · m
F	Friction force	N
L	Lift	N
μ	Ground roll friction coefficient	–
φ	Planform wheel camber angle	Degrees
b	Wheelbase	m

(continued)

(continued)

Variables		
B	Wheel track	m
e	Main landing gear stagger	m
a	Nose wheel stagger	m
γ	Main landing gear stagger angle	Degrees
φ	Turnover angle	Degrees
ψ	Static ground angle	Degrees
α_{land}	Landing angle of attack	*Degrees*
$\alpha_{C_{L,max}}$	Angle of attack at maximum lift coefficient	Degrees
c_{MAC}	Mean aerodynamic chord	*m*
L_{fus}	Fuselage length	m
H	Height of centre of mass above ground	*m*
β_{nose}	Nose wheel rotation angle	Degrees
P_{ESWL}	Equivalent single-wheel load	N
$R_{inboard}$	Radius of circle along which the inboard main gear moves	m
R_{nose}	Radius of circle along which the nose wheel moves	m
Δ	Distance from RW edge to wheels	m
$P_{staticgroundTO}$	Static ground load on one wheel of four-wheel bogie or on two wheels of eight-wheel bogie	N
p_0	Pressure in wheel tire	Pa
k_{LG}	Factor depending on gear type and wheel diameter	–
H_{track}	Allowable track depth	m
B_{wheel}	Wheel width	m
D_{wheel}	Wheel diameter	m
p_{HS}	Available pressure in hydraulic system	N/m^2
η	Compression mechanism efficiency	–

References

1. Dmitriev AV, Leshchin AV (2003) Vybor skhemy i osnovnyh parametrov opor shassi samolyotov (Selection of the scheme and main parameters of aircraft landing gear supports). MAI, Moscow
2. Kondrashov VA (1991) Proektirovanie ubirayushchihsya shassi samolyotov (Design of retractable landing gear for aircraft). Mashinostroenie, Moscow
3. Dmitriev AV, Leshchin AV (2003) Vybor skhemy i osnovnyh parametrov amortizacionnyh sistem opor shassi samolyotov (Selection of the scheme and main parameters of the shock-absorbing systems of aircraft landing gear supports). MAI, Moscow

Annex 1: Weight Report Example

I. **Airframe**

1. Wings (including fuel compartments sealing).
2. Fuselage (including compartment sealing).
3. Empennage (including dorsal fin fairings and endplates).
4. Landing gear (including doors and fairings).
5. Paint.

II. **Power Plant**

1. Engines (main and auxiliary).
2. Propellers with spinners and mounts (for aircraft with piston and turboprop engines).
3. Means of installing engines:
   (a) engine nacelles with air intakes, air channels and jet nozzles (if the engine is not structurally connected to the fuselage);
   (b) pylons;
   (c) cowlings;
   (d) engines and attachment points;
   (e) engine turning mechanisms (for VTOL aircraft).
4. Engine systems:
   (a) air intake control system;
   (b) exhaust system;
   (c) thrust reversal system;
   (d) oil system;
   (e) cooling system;
   (f) fire extinguishing system;
   (g) engine control system;
   (h) power plant anti-icing system.

M. A. Pogosyan (ed.), *Aircraft Design*, Springer Aerospace Technology,
https://doi.org/10.1007/978-981-96-4599-2

5. Fuel System
   (a) fuel tanks with protectors;
   (b) installation fittings for tanks;
   (c) fuel supply system (fuel lines and pumps);
   (d) pressurization system (with inert gas);
   (e) automatic fuel consumption control system;
   (f) filling system;
   (g) emergency drain system;
   (h) in-flight refueling system.

III. **Equipment and Control System**

1. Hydraulic system:
   (a) sources of energy;
   (b) fittings, tanks, pipelines;
   (c) working fluid;
   (d) fastenings and operational devices.
2. Pneumatic System:
   (a) sources of energy;
   (b) fittings, cylinders, pipelines;
   (c) compressed air;
   (d) attachment parts and operating devices.
3. Electrical Equipment:
   (a) generators and batteries;
   (b) energy converters;
   (c) electrical wiring, accessories, fasteners and operating devices.
4. Aircraft Control:
   (a) radio communication equipment;
   (b) radio navigation equipment;
   (c) radar equipment;
   (d) automatic takeoff and landing systems.

   Note. Radio equipment includes communications, antennas, equipment, and attachments.
5. Air-Navigation Equipment:
   (a) air-navigation instruments;
   (b) devices for monitoring the systems and mechanisms (except for the power plant);
   (c) autopilot;
   (d) computing devices, consoles, dashboards.
6. Anti-icing system.
7. The system of thermal and sound insulation of cabins and instrument compartments.
8. Life support equipment and household equipment:
   (a) crew seats;
   (b) passenger seats;
   (c) air conditioning system;

   (d) oxygen system;
   (e) lavatories;
   (f) galleys and buffets (for passenger aircraft);
   (g) wardrobes and trunks (for passenger aircraft);
   (h) cabin fire protection equipment;
   (i) crew ejection system.
9. Special Equipment, Weapons and Armor:
   (a) a system for automatic control of the operation of equipment and aircraft structure;
   (b) guns, machine guns with installation (without ammunition);
   (c) installation parts and equipment of missile weapons;
   (d) bombing equipment, fastening of active and passive protection devices;
   (e) crew and equipment booking;
   (f) photographic equipment, detection systems, etc.;
   (g) mooring (fixed) equipment.

IV. **Empty aircraft**

1. Aircraft structure.
2. Power plant.
3. Equipment and control system.

V. **Service load**

1. Crew;
   (a) flight crew with personal belongings;
   (b) auxiliary staff (flight attendants, etc.).
2. Water in the buffet and washbasins (passenger aircraft).
3. Unusable fuel.
4. Oil.
5. Technical fluids consumed in flight (for example, anti-icing fluid).
6. Products in the galley of a passenger aircraft (with containers).
7. Literature for passengers, carpets, seat covers, first aid kit, etc.
8. Rescue equipment (boats, emergency ladders, etc.).
9. Spare parts, flight tools, covers for aircraft units, flares.
10. External tanks (without fuel), detachable suspension kits for special cargo, detachable sanitary equipment.
11. Containers for luggage, cargo and mail.

VI. **Empty operational aircraft**

1. Empty aircraft.
2. Service load.

VII. **Fuel**

1. Consumed fuel.
2. Navigation fuel allowance.
3. Fuel in additional and external tanks.

VIII. **Payload**

Items	Aircraft type
1. Bombs, torpedoes, rockets. 2. Ammunition (cartridges, shells). 3. Special cargo. 4. Equipment for photo reconnaissance.	Fighters, bombers, reconnaissance aircraft.
5. Paratroopers with weapons and parachutes. 6. Military equipment, including exhaust and main parachutes, platforms	Military transport aircraft.
7. Passengers. 8. Luggage, cargo, mail.	Passenger aircraft.
9. Chemicals (agricultural aircraft).	

IX. **Full load**

1. Equipment and service load.
2. Fuel.
3. Payload.

X. **Aircraft takeoff mass**

1. Empty aircraft mass.
2. Full load mass.

Annex 2: Sample Project Brief for a Passenger Jet Aircraft

1. **Aircraft mission:**

   1.1 Payload: passengers, their luggage, mail and cargo in containers of a standard type.
   1.2 Aircraft range: midrange.
   1.3 The intensity of passenger—cargo traffic on the serviced airline network: average.
   1.4 Atmospheric conditions must comply with ISA.
   1.5 The weather minimum for landing must correspond to the 2nd category of ICAO.

2. **General requirements:**

   2.1 The aircraft, its systems, engines, equipment, components, ground equipment and repair facilities, as well as operational documentation must comply with:

   - local airworthiness standards;
   - state and industry standards, as well as industry norms.

   2.2 The noise level produced by the aircraft on the ground must ensure that the requirements of the ICAO noise standard are met. Vibration and noise in the passenger cabin and cockpit should not exceed the values regulated by the relevant state and industry standards.

3. **Flight performance:**

   3.1 The maximum passenger capacity is 220.
   3.2 The maximum payload is 22…24 tons.
   3.3 The practical range with a maximum payload is 2500 … 2600 km.
   3.4 Cruise speed should be 810…850 km/h at an altitude of 11,000 m.

M. A. Pogosyan (ed.), *Aircraft Design*, Springer Aerospace Technology,
https://doi.org/10.1007/978-981-96-4599-2

3.5 Flight endurance to an alternate airfield must be 1 h.
3.6 The aircraft must operate off a class B airfield featuring a runway length of not more than 2700 m.

4. **Service life:**

4.1 Aircraft service life must be at least 20 years
4.2 The airframe cumulative life should be 45,000 flight hours or 22,000 flights

5. **Crew**:

5.1 The flight crew must consist of 2 pilots: the pilot-in-command and the co-pilot (+1 checker in the cockpit).
5.2 The number of flight attendants: 6 or 7.

6. **Passenger cabin comfort**:
With the maximum passenger capacity, comfort in passenger cabin must correspond to the economy class.
7. **Aircraft equipment**:
Aircraft equipment should include: flight and navigation equipment, radio communication equipment, anti-icing system, air conditioning and automatic pressure control system in the pressurized cabin, fire fighting equipment, oxygen equipment, emergency equipment, onboard flight data monitoring and recording systems, hydraulic system, control system, power supply system, passenger cabin household equipment.
8. **Power plant and APU**:

8.1 Engine type: high bypass turbofan engine.
8.2 The aircraft must feature an APU.

9. **Maintenance and repair**:

9.1 The aircraft maintenance and repair program must comply with the requirements of state standards.
9.2 The technically possible average annual flight time must be at least 4000 h.

10. **Technical specifications**:
Fuel consumption per 1 passenger-kilometer with a maximum passenger capacity (with 95 kg per passenger) should not exceed 18 grams.

Annex 3: Polars for Aircraft of Selected Classes

Below, an analytical algorithm and typical plots (Figs. A.1, A.2, A.3, A.4, A.5 and A.6) for polars of subsonic and supersonic aircraft are given. These data can be used in graduation projects. Polars for subsonic aircraft are shown for $M_\infty < M_{crit}$. At $M_\infty < M_{crit}$, polars can be represented in the following form (high-lift devices and landing gear are retracted)

$$C_D = C_{D,min} + \frac{(C_L - C_{L,0})^2}{\pi \, AR_{eff}}. \tag{A.1}$$

Here, $C_{L,0}$ is the lift coefficient at $C_{D,min}$. If $C_{L,0} = 0$, then $C_{D,min} = C_{D,0}$, where $C_{D,0}$ corresponds to $C_L = 0$.

To first approximation,

$$C_{D,min} = (0.9 + 0.15\text{M})\left[\begin{array}{l} 3C_f\left(1 + 3.3(t/c)_{average}\right)\cos\Lambda + 0.009\dfrac{S_{tail}}{S} \\ + \left(0.008AR_{fus} + 0.5/AR_{fus}^2\right)\dfrac{S_{Fus.Midsection}}{S} + C_{D,nacelle}\dfrac{S_{Nacelle}}{S} + 0.0002 \end{array}\right] \tag{A.2}$$

Here, $C_f = \frac{0.455}{(\lg \text{Re})^{2.58}}$ is the turbulent wing friction coefficient (according to Schlichting); $C_f = \frac{1.328}{\sqrt{Re}}$ is the laminar wing friction coefficient (according to Blasius; e.g., if boundary layer suction is used); the Reynolds number is

$$Re = \frac{V_{cruise} \cdot 10^5}{1.46 + 0.25 \cdot 10^{-4}H + 1.94 \cdot 10^{-8}H^2} \cdot \sqrt{\frac{S}{AR}}, \tag{A.3}$$

where V_{cruise} is in m/s; H is the flight altitude, m; $S_{Nacelle}$ is the cumulative area of all engine nacelles.

M. A. Pogosyan (ed.), *Aircraft Design*, Springer Aerospace Technology,
https://doi.org/10.1007/978-981-96-4599-2

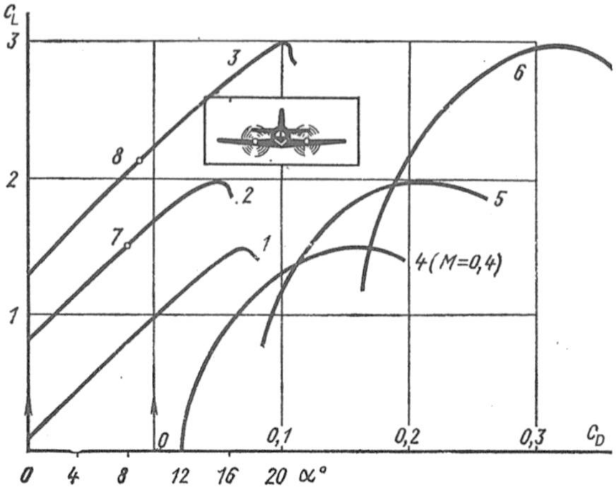

Fig. A.1 Polars for a light aircraft featuring two bypass turbofans: $S = 40m^2$, $AR = 9$; $\Lambda = 0°$; $(t/c)_{root} = 0,17$; $(t/c)_{tip} = 0,12$; $d_{fus} = 2m$; high-lift devices: slats and single-slotted Fowler flaps. **1**—$C_L(\alpha)$, cruise; **2**—$C_L(\alpha)$, take-off ($\delta_{flaps} = 20°$); **3**—$C_L(\alpha)$, landing ($\delta_{flaps} = 40°$); **4**—$C_L(C_D)$, cruise; **5**—$C_L(C_D)$, take-off; **6**—$C_L(C_D)$, landing; **7**—$C_{L,lift-off}$; **8**—$C_{L,land}$

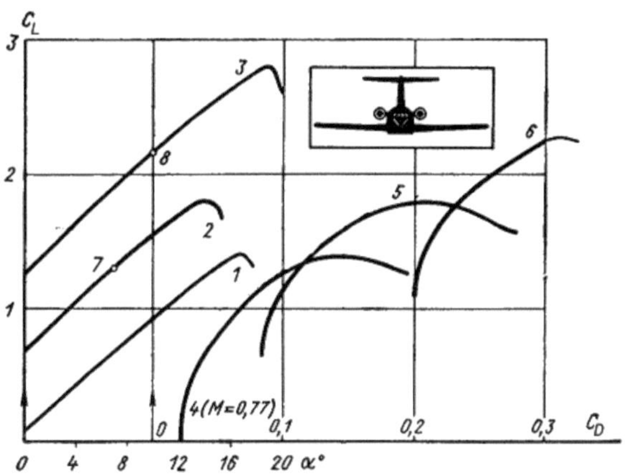

Fig. A.2 Polars for a light subsonic aircraft featuring two bypass turbofans: $S = 32m^2$; $AR = 9$; $\Lambda = 20°$; $(t/c)_{root} = 0{,}14$; $(t/c)_{tip} = 0{,}10$; $d_{fus} = 2m$; high-lift devices: slats and single-slotted fowler flaps. Notes to the curves—see Fig. A.1

In Eq. (A.2), $(t/c)_{average}$ is the average (equivalent) wing thickness ratio (measured parallel to the velocity vector),

$$(t/c)_{average} = \frac{2}{b - d_{fus}} \int_{d_{fus}/2}^{b/2} (t/c)dy \approx \frac{(t/c)_{root} + (t/c)_{tip}}{2}. \qquad \text{(A.4)}$$

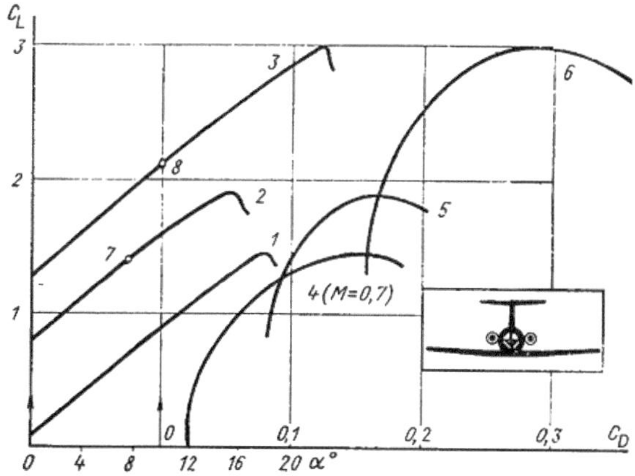

Fig. A.3 Polars for a mid-range airliner featuring two bypass turbofans: $S = 100m^2$; $AR = 8$; $AR = 25°$; $(t/c)_{root} = 0{,}14$; $(t/c)_{tip} = 0, 10$; $d_{fus} = 3, 4m$; high-lift devices as shown in Fig. A.2. Notes to the curves—see Fig. A.1

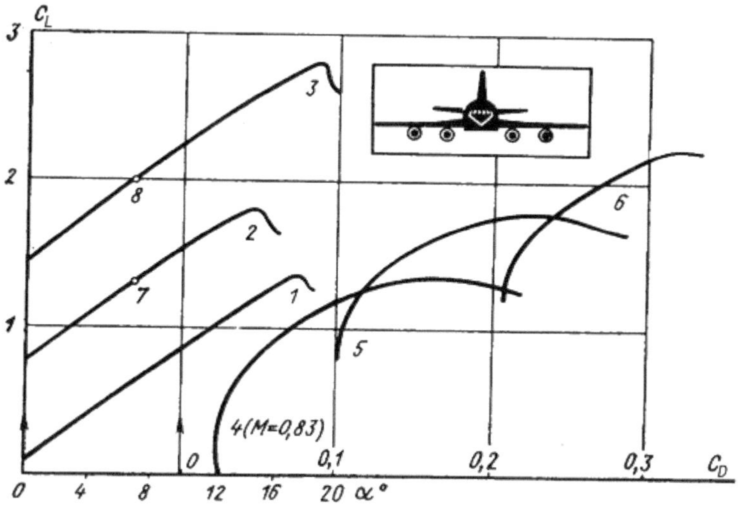

Fig. A.4 Polars for a subsonic airliner featuring four bypass turbofans: S = 350 m^2; AR = 7.5; $\Lambda = 30°$; $t/c_{root} = 0, 14$; $t/c_{tip} = 0, 10$; $d_{fus} = 6m$; high-lift devices: slats and single-slotted Fowler flaps. Notes to the curves—see Fig. A.3

Engine nacelle drag coefficient in Eq. (A.2) depends on the engine bypass ratio (the shape of the nacelles):

m	0	2	4	6
$C_{D,nacelle}$	0.1	0.1	0.085	0.065

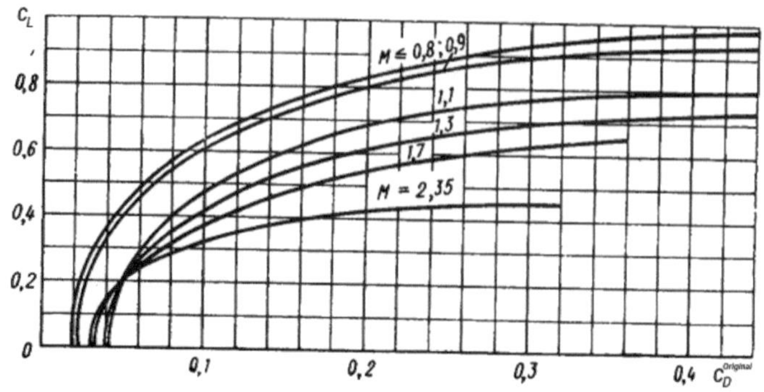

Fig. A.5 Polars for a supersonic maneuvering aircraft (i): original aircraft: $S = 56\text{m}^2$; $AR = 3$; $\Lambda = 40°$; $(t/c)_{average} = 4\%$; $T/W_{Take-off} = 1{,}1$; $C_m^{C_L} = 0,04$; $W_{Take-off}/S = 300\ \frac{daN}{m^2}$. Notes to the curves—see Fig. A.3

Fig. A.6 Drag coefficient of a supersonic maneuvering aircraft as a function of the Mach number

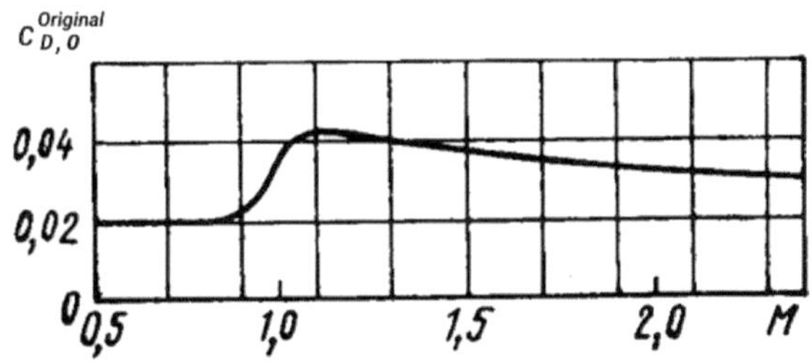

Effective aspect ratio in Eq. (A.1) can be estimated as follows:

$$AR_{eff} = \frac{AR}{1 + 1.53\,\Lambda(-0.4) + (0.018 + 0.0152\,\Lambda) \cdot AR}, \tag{A.5}$$

where AR is the wing geometric aspect ratio ($AR > 5$); Λ is the wing quarter chord sweep angle, radians; M is the Mach number.

To plot the linear portion of $C_L(\alpha)$ (for AR > 3; out of ground effect) the following equation can be used:

$$C_L^\alpha \approx 0,106\left(\frac{1}{\cos\Lambda} + \frac{2}{AR}\right)^{-1}, \tag{A.6}$$

where Λ is the wing quarter chord sweep angle, degrees.

Typical polars for a supersonic maneuvering aircraft (see Chap. 10) are shown in Fig. A.5. To adjust the polars to various values of the thrust-to-weight ratio (which affects the fuselage mid-section) and take-off wing loading (which affects the wing area) the following equation can be used:

$$C_D = C_D^{Original} + \Delta C_D,$$

where

$$\Delta C_D = C_D^{Original}\left(\frac{T/W_0 \cdot (W/S)_0}{700} - 0,5\right). \tag{A.7}$$

$C_D^{Original}$—See Fig. A.5; $C_D^{Original}$—See Fig. A.5; $C_D^{Original}$—See Fig. A.6.

Annex 4: Properties of Aircraft Turbojet Engines

The engine properties shown below (Tables A.1, A.2, Figs. A.7, A.8, A.9, A.10 and A.11) are intended for use at the early stages of aircraft design. Table A.1 presents the main characteristics of afterburning turbojet, bypass turbojet, and afterburning bypass turbojet engines with different thrusts. For the takeoff mode, the maximum thrust, the corresponding specific fuel consumption, the bypass ratio, the compression ratio in the compressor, the gas temperature in front of the turbine, and the air flow rate per second are given. In addition, the value of the specific fuel consumption in cruise flight, engine length, maximum diameter, mass, and specific weight are presented. The aircraft on which these engines are used are specified.

Altitude-airspeed performance. The recommended dependences (see Figs. A.7, A.8, A.9, A.10 and A.11), not being characteristic of any specific engines, show with a sufficient degree of accuracy the influence of flight altitude and speed on thrust and specific fuel consumption of modern aircraft engines.

The specific consumption of cryogenic fuel (liquid hydrogen) can be estimated by a simple recalculation of the specific consumption of carbon fuel (kerosene).

$$c_{,hyd} \approx c_{,ker}/2.5.$$

M. A. Pogosyan (ed.), *Aircraft Design*, Springer Aerospace Technology,
https://doi.org/10.1007/978-981-96-4599-2

Table A.1 Main properties of modern afterburning turbojet, bypass turbojet, and afterburning bypass turbojet engines

Engine thrust range, daN	Engine brand, company, country	Take-off mode						C_p cruise kg/(daN × h)	L_{engine}, m	D_{engine}, m	m_{engine}, kg	γ_{engine}	Airplane, company, country	Purpose
		T_o, daN,	C_{Po}, kg/(daN × H)	m	π^*_K	T_{gas}, ^0K	$m_{air,o}$, kg/s							
0…1000	WR 19–3 Research USA	275	0.69	1	12	1280	6.2	0.69	0.78	0.3	66	0.2	Fockjet Dassault France	C
	Astafan IIG Turbomeca France	620	0.39	7.4	9.1	1244	36	0.67	1.98	0.65	143	0.23	Master Dassault France	M
1000…2500	JT15D-4B Pratt and Whitney Canada	1134	0.56	2.7	10.1	1289	35	0.73	1.6	0.69	253	0.2	Cessna 550 Cessna USA	C
	RD-1700 Soyuz-Klimov Russia	1700	0.65	0.78	14.3	1420	30	0.73	1.98	0.52	300	0.17	Yak-130 "Yakovlev" Russia	M
	AI-222-25F "Motor Sich" Ukraine	4200[a] 2500	1.90[a] 0.67	1.19	15.5	1470	49.2	0.73	3.16	0.63	440	0.22	L-15 Hongdu China	M
	ATF3-6 Garrett USA	2420	0.48	2.8	24	1448	74	0.72	2.34	0.85	430	0.2	Falcon-20 Dassault France	C
2500…5000	AL-55F "Saturn" Russia	3500[a] 2200	1.65[a] 0.65	1.2	22	1445	29.5	0.71	2.59	0.59	445	0.13 0.2	MIG-AT "Mikoyan" Russia	M

(continued)

Table A.1 (continued)

Engine thrust range, daN	Engine brand, company, country	Take-off mode						C_p cruise kg/(daN × h)	L_{engine}, m	D_{engine}, m	m_{engine}, kg	γ_{engine}	Airplane, company, country	Purpose
		T_o, daN,	C_{Po}, kg/(daN × H)	m	$\pi^*_к$	T_{gas}, ^0K	$m_{air,o}$, kg/s							
	M45H-01 Rolls-Royce Great Britain	3570	0.46	3	16.1	1335	101	0.71	2.8	1.09	663	0.19	VFW-614 Fokker Germany	C
	TF-34-CE-400A General Electric USA	4130	0.37	6.2	21	1489	153	0.68	2.54	1.32	670	0.16	S-3A, Lockheed A-10A "Republic" USA	M
5000…10,000	SaM 146 "Saturn-Snecma" Russia-France	6980	0.36	4.43	27.5	1540	238	0.6	3.63	1.22	1450	0.22	SU-100 "Sukhoy" Russia	C
	D-436T2 Motor Sich Ukraine	7200	0.37	4.88	22.2	1550	260	0.63	3.83	1.37	1890	0.26	TU-334 "Tupolev" Russia BE-200 Russia	C
	EJ-200 Eurojet Great Britain	9000[a] 6000	1.70[a] 0.76	0.4	26	1600	77	0.83	4	0.74	1038	0.12[a] 0.17	Eurofighter Eurofighter GmbH Europe	M

(continued)

Table A.1 (continued)

Engine thrust range, daN	Engine brand, company, country	Take-off mode						C_p cruise kg/(daN × h)	L_{engine}, m	D_{engine}, m	m_{engine}, kg	γ_{engine}	Airplane, company, country	Purpose
		T_o, daN,	C_{Po}, kg/(daN × H)	m	π^*_κ	T_{gas}, ^0K	$m_{air,o}$, kg/s							
	RD-33MK Soyuz-Klimov Russia	8800[a] 5400	1.86[a] 0.77	0.55	24	1540	82	0.9	4.23	0.75	1055	0.12[a] 0.19	MiG-29 M "MiG" Russia	M
10,000…20,000	CFM.56 "General Electric" USA	14,240	0.37	4.5	33.4	1560	396	0.55	3.2	1.72	2365	0.17	B-737 Boeing USA A-320,340 Airbus	C
	F-100-229A "Pratt and Whitney" USA	14500[a] 8900	1.91[a] 0.68	0.36	32.4	1590	125	0.82	4.85	1.18	1479	0.10[a] 0.17	F-15E, F-16B, "Boeing" Lockheed USA	M
	AL-31FM "Saturn" Russia	13500[a] 8600	1.96[a] 0.75	0.57	23.2	1527	118	0.9	4.95	0.92	1520	0.11[a] 0.18	Su-27, Su-34 "Sukhoy" Russia	M
10,000…20,000	F119-PW-100 "Pratt and Whitney" USA	15800[a] 11,600	1.94[a] 0.75	0.45	32	1647	125	1.5	5.16	1.17	1775	0.11[a] 0.15	F-22 V Lockheed USA	M

(continued)

Table A.1 (continued)

Engine thrust range, daN	Engine brand, company, country	Take-off mode						C_p cruise kg/(daN × h)	L_{engine}, m	D_{engine}, m	m_{engine}, kg	γ_{engine}	Airplane, company, country	Purpose
		T_o, daN,	C_{Po}, kg/(daN × H)	m	π^*_K	T_{gas}, ^0K	$m_{air,\,o}$, kg/s							
	CF6-32 "General-Electric" USA	16,700	0.37	4.9	23.3	1535	484	0.66	3.73	1.93	3110	0.19	B - 757 Boeing USA	C
	PS-90A2 "Soloviev" Russia	16,800	0.38	4.5	35.5	1640	504	0.6	4.96	1.9	3000	0.18	Il-96 "Ilyushin" TU-214 "Tupolev" Russia	C
	PD - 14 "UEC" Russia	14,000	0.36	8.5	41	1680	480	0.56	4.8	1.9	2870	0.2	MS-21 "Irkut" Russia	C
20,000…40,000	NK-25 "Kuznetsov" Russia	24500[a] 14,500	2.08[a] 0.35	1.45	14	1530	211	0.6	5.2	1.5	3600	0.15[a] 0.25	Tu-160, Tu-22M3 "Tupolev" Russia	M

(continued)

Table A.1 (continued)

Engine thrust range, daN	Engine brand, company, country	Take-off mode						C_p cruise kg/(daN × h)	L_{engine}, m	D_{engine}, m	m_{engine}, kg	γ_{engine}	Airplane, company, country	Purpose
		T_o, daN,	C_{Po}, kg/(daN × H)	m	π^*_K	T_{gas}, ^0K	$m_{air, o}$, kg/s							
	D-18 T Motor Sich Ukraine	23,400	0.35	5.6	28	1630	765	0.6	5.4	2.33	4100	0.18	An-124 "Antonov" Ukraine	C
	CF6-80C "General-Electric" USA	26,300	0.34	4.7	26	1595	754	0.59	4.2	2.5	4600	0.18	B-747-200 V Boeing USA	C

Notes m = bypass ratio; C = civil; M = Military; [a] afterburner on;

[1] $M_{max} = 1, 7$;

[2] $M_{max} = 0, 85$;

[3] $M_{max} = 2, 1$; [4] $M_{max} = 2, 5$ (for F-15);

[5] $M_{max} = 3, 3$;

[6] $M_{max} = 2, 3$

Table A.2 Main properties of turbojet and bypass turbojet engines for VTOL aircraft

Engine thrust range, daN	Engine brand, company, country	Take-off mode						$C_{p\ cruise}$ kg/(daN×h)	L_{engine}, m	D_{engine}, m	m_{engine}, kg	γ_{engine}	Airplane, company, country	Purpose
		T_o, daN	C_{Po},	m	$\pi^*_к$	T_{gas}, ^0K	$m_{air,\ o}$, kg/s							
Lift-cruise turbojet and bypass turbojet engines														
5000…20,000	F402-11-61 "Rolls-Royce" Great Britain	10,400	0.78	1.36	14.7	1453	200	0.9	3.48	1.22	1390	0.13	AV-8B Boeing USA	M
	R-28 V-300 "Soyuz" Russia	6900	0.88	0	10.5	1440	100	–	3.70	1.02	1350	0.2	Yak-38 M "Yakovlev" Russia	M
	P-79B-300 "Soyuz" Russia	15500[a] 10,500	1.6[a] 0.66	0.8	22	1620	180	–	5.23	1.1	1620	0.1 0.15	Yak-141 "Yakovlev" Russia	M
	P135-PW-600 "Pratt and Whitney" USA	19140[a] 12,500	1.75[a] 0.82	0.75	28	1654	159	1.6	5.59	1.29	1654	0.08 0.13	F-35B "Lockheed Martin" USA	M
Lift turbojet engines														
2500…5000	RB 162 Rolls-Royce England	2720	1.72	0	4.5	1300	38.5	–	1.49	0.74	170	0.06	Do 31 Dornier Germany	M
	RD-38 "Kolesov" Russia	3250	1.38	0	5	1270	45.3	–	1.55	0.6	231	0.07	Yak-38 M "Yakovlev" Russia	M
	RD-41 "Kolesov" Russia	4100	1.4	0	6.28	1330	53.5	–	1.6	0.64	290	0.07	Yak-141 "Yakovlev" Russia	M

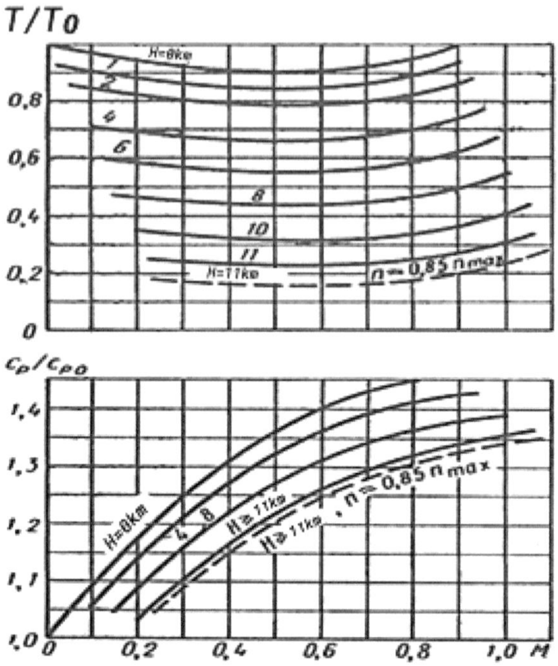

Fig. A.7 Altitude-speed performance of subsonic bypass turbojet engines ($m =$ 0.5…1.5)

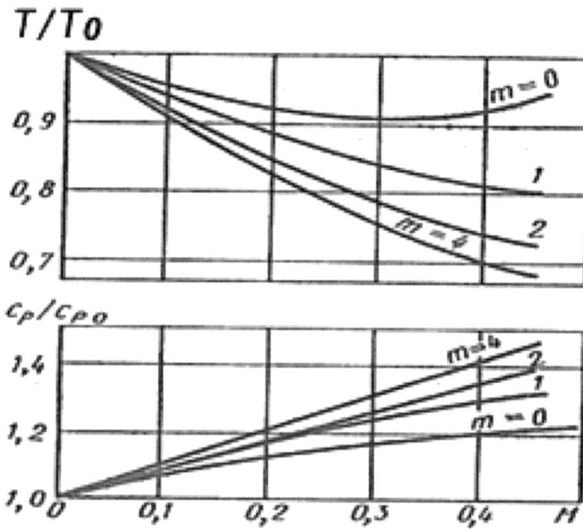

Fig. A.8 Variation of thrust and specific fuel consumption of a bypass turbojet engine as a function of airspeed at $H = 0$

As shown in Chap. 16, the properties of gas turbine engines depend on the type of engine, maximum flight speed, etc., but when performing calculations for the graduation project, their effect on the altitude-airspeed performance can be neglected.

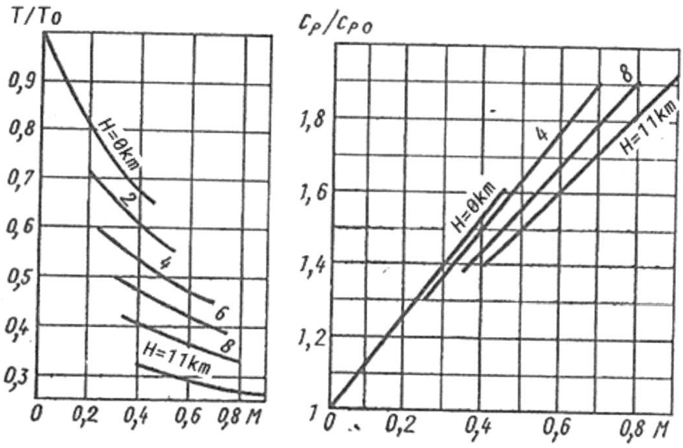

Fig. A.9 Altitude-speed performance of bypass turbojet engines (m = 4…6) estimated at subsonic speeds

Therefore, the above characteristics of supersonic gas turbine engines (designed for M < 3.3) can be applied to turbojet engines (and afterburning turbojet engines), as well as afterburning bypass turbojet engines intended for modern fighters, multi-purpose aircraft, and aircraft with a cruising speed of $M = 2.0...3.0$. The altitude-airspeed performance of subsonic bypass turbojet engines is given separately. The altitude-airspeed performance of vertical lift engines for VTOL and STOL aircraft is identical to the characteristics of subsonic bypass turbojet engines.

The plots show the ratios of thrust and fuel consumption to their take-off values, which enables the determination of T (M, H) и c_p (M, H) for any values of T_0 and $c_{p,0}$.

At H > 11 km, a given engine setting, and M = const, the specific fuel consumption remains constant, and the engine thrust varies linearly with the ambient air density or pressure, so it is determined by a simple recalculation. For example, for any value of M:

$$T_{13} = T_{11}\rho_{13}/\rho_{11} \text{ or } T_{20} = T_{15}p_{20}/p_{15};\; c_{p11} = c_{p13} = c_{p15} = c_{p20}$$

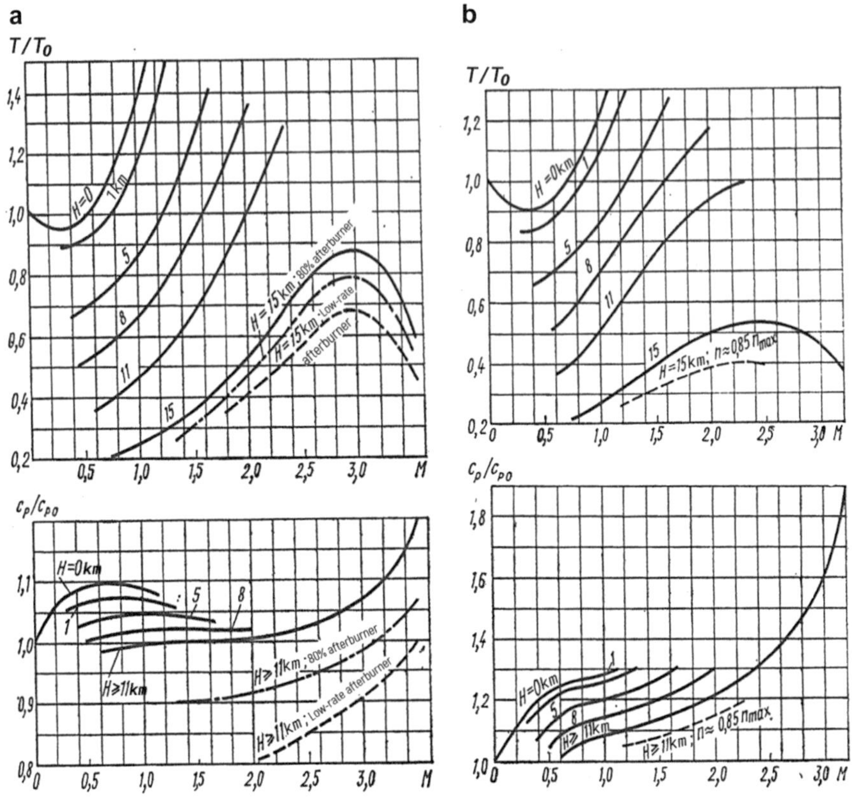

Fig. A.10 **a** Altitude-speed performance of a gas turbine engine designed for $M \leq 3.3$ (afterburner on) **b** Altitude-speed performance of a gas turbine engine designed for $M \leq 3.3$ (maximum thrust, afterburner off)

The transition from the climb mode to the cruise flight mode (see Table 1 for average cruise parameters) is carried out by throttling the engine (reducing the rotor speed). In cruising flight, the number of revolutions of the rotor $n = (0.8...0.9)n_{max}$.

For a supersonic cruise at high altitudes, the partial afterburner mode is usually used. The rotational speed of the rotor is also somewhat reduced.

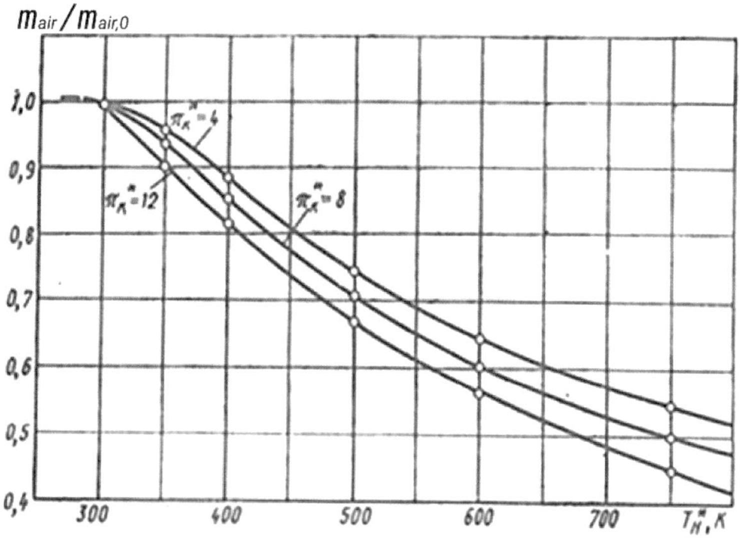

Fig. A.11 Variation of specific air flow for different turbojet engines and bypass turbojet engines as a function of stagnation temperature: $m_{air,0}$ = take off air flow rate per second (See Table A.1)

Annex 5: Properties of the Standard Atmosphere (According to GOST 4401-81)

Geometric altitude H	Temperature T_H	Pressure P_H	Density ρ_H	$\Delta = \rho_H / \rho_0$	$\sqrt{\Delta}$	Speed of sound a_H
km	K	Pa	kg/m^3	–	–	m/s
0	288.15	101,325	1.225	1.000	1.000	340.29
1	281.65	89,876	1.112	0.908	0.953	336.44
2	275.15	79,501	1.007	0.822	0.907	332.53
3	268.66	70,121	0.909	0.742	0.861	328.58
4	262.17	61,660	0.819	0.669	0.818	324.59
5	255.68	54,048	0.736	0.601	0.775	320.55
6	249.19	47,218	0.660	0.539	0.734	316.45
7	242.70	41,105	0.590	0.482	0.694	312.31
8	236.22	35,652	0.526	0.429	0.655	308.11
9	229.73	30,801	0.467	0.381	0.617	303.85
10	223.25	26,500	0.414	0.338	0.581	299.53
11	216.77	22,700	0.365	0.298	0.546	295.15
12	216.65	19,399	0.312	0.255	0.505	295.07
13	216.65	16,580	0.267	0.218	0.467	295.07
14	216.65	14,170	0.228	0.186	0.431	295.07
15	216.65	12,112	0.195	0.159	0.399	295.07
16	216.65	10,353	0.167	0.136	0.369	295.07
17	216.65	8850	0.142	0.116	0.340	295.07
18	216.65	7565	0.122	0.100	0.316	295.07
19	216.65	6468	0.104	0.085	0.291	295.07
20	216.65	5530	0.089	0.073	0.270	295.07

(continued)

M. A. Pogosyan (ed.), *Aircraft Design*, Springer Aerospace Technology,
https://doi.org/10.1007/978-981-96-4599-2

(continued)

Geometric altitude H	Temperature T_H	Pressure P_H	Density ρ_H	$\Delta = \rho_H / \rho_0$	$\sqrt{\Delta}$	Speed of sound a_H
21	217.58	4729	0.076	0.062	0.249	295.70
22	218.57	4048	0.065	0.053	0.230	296.38
23	219.57	3467	0.055	0.045	0.212	297.05
24	220.56	2972	0.047	0.038	0.196	297.72
25	221.55	2549	0.040	0.033	0.181	298.39

Approximate variation of the standard atmosphere as a function of altitude

Temperature T_H [K] at altitude H [m] can be estimated using the following equations:

$$\begin{aligned} T_H &= 288.15 - 0.0065H \quad \text{at } 0\,\text{m} \le \le 11000\,\text{m}; \\ T_H &= 216.65 \quad \text{at } 11000\,\text{m} < \text{H} \le 20000\,\text{m}; \\ T_H &= 196.73 + 0.001H \quad \text{at } 20000\,\text{m} < \text{H} \le 25000\,\text{m}. \end{aligned}$$

Pressure P_H [N/m^2] at altitude H [m] can be estimated using the following equations:

$$\begin{aligned} P_H &= 101325\left(1 - \frac{H}{44308}\right)^{5.2553} \quad \text{at } 0\,\text{m} \le \text{H} < 11000\,\text{m}; \\ P_H &= 22700e^{-\left(\frac{H-11000}{6318}\right)} \quad \text{at } 11000\,\text{m} \le \text{H} \le 25000\,\text{m} \end{aligned}$$

Density ρ_H[kg/m^3] at altitude H [m] can be estimated using the following equations:

$$\begin{aligned} \rho_H &= 1.225\left(1 - \frac{H}{44308}\right)^{4.2553} \quad \text{at } 0 \le \text{H} < 11000\,\text{m}; \\ \rho_H &= 0.365e^{-\left(\frac{H-11000}{6318}\right)} \quad \text{at } 11000\,\text{m} \le \text{H} \le 25000\,\text{m} \end{aligned}$$

Speed of sound a_H at altitude H is calculated using the following equation:

$${}_H = a_0\sqrt{\frac{T_H}{T_0}},$$

where

a_0 = the speed of sound at altitude $H_0 = 0$ m,
T_0 = the temperature at altitude $H_0 = 0$ m,
T_H = the temperature at altitude H.

Annex 6: Properties of Aircraft Wheels

See Tables A.3 and A.4.

M. A. Pogosyan (ed.), *Aircraft Design*, Springer Aerospace Technology,
https://doi.org/10.1007/978-981-96-4599-2

Table A.3 Brake wheels

Product code	Tire size, mm	Operating tire pressure, kgf/cm^2	Parking load due to aircraft takeoff weight, kgf	Allowable takeoff speed, km/h	Allowable landing speed, km/h	Brake operational energy loading, kgf·m	Operational braking torque, kgf·m	Wheel mass, kg	Brake mass, kg
KT239	476 × 178–6	3.5	935	126	123	54,000	52	7.8**	–
KT235	500 × 150–9	7.0	1400	140	126	39,500	30	14.0	7.8
KT236	500 × 150–9	3.0	560	150	130	23,000	30	7.8**	–
KT245	500 × 150–9	4.5	950	126	128	30,500	60	6.3	4.7
KT217	600 × 180	7.0	3210	—	80	46,500	90	12.0**	–
KT254	610 × 185–305	5.5	2080	196	220	294,000	350	28.5**	–
KT176	660 × 200	10.0	2600	290	270	490,000	180	15.8	24.1
KT240	660 × 200–356	12.0	3300	275	240	290,000	265	42.0**	–
KT251	680 × 260	12.0	2730	275	275	420,000	355	58.0**	–
KT228*	680 × 260	10.5	2900	360	310	–	9.9	22.0**	–
KT231	810 × 320	5.5	4300	250	250	275,000	500	20.5	22.0
KT211M	840 × 290	21.0	9300	305	270	1,200,000	1050	43.5	69.5
KT209	840 × 290	15.0	8800	330	310	1,100,000	720	33.0	61.5
KT197*	840 × 290	11.0	6000	320	310	–	–	28.0**	–

(continued)

Table A.3 (continued)

Product code	Tire size, mm	Operating tire pressure, kgf/cm^2	Parking load due to aircraft takeoff weight, kgf	Allowable takeoff speed, km/h	Allowable landing speed, km/h	Brake operational energy loading, kgf·m	Operational braking torque, kgf·m	Wheel mass, kg	Brake mass, kg
KT207	840 × 360	11.0	8800	310	342	1,650,000	650	34.0	75.0
KT163D	840 × 360	9.5	8800	300	265	430,000	650	37.0	72.0
KT218	880 × 315	6.5	5700	255	235	540,000	510	26.0	36.0
KT141E	930 × 305	10.0	8000	325	280	750,000	650	47.0	52.0
KT263	950 × 300	12.0	8400	330	300	1,100,000	1200	40.7	43.0
KT172	950 × 300	12.0	9100	380	320	800,000	750	130.0**	–
KT 175	950 × 300	13.0	10,000	405	350	1,580,000	740	55.8	67.2
KT232	950 × 300P-468	10.0	8370	200	175	652,000	940	45.0	75.0
KT229	950 × 300	6,0	6530	80	80	160,000	265	44,0	19,1
KT206	950 × 400	12,0	11,000	420	300	1,600,000	1000	106,0**	–
KT159Д	1030 × 350	17,0	15,450	400	280	1,650,000	1250	60,9	63,6
KT213	1030 × 350	19,0	19,900	400	308	25,000,000	1700	134,0**	–
KT196	1070 × 390-480P	14,0	12,000	325	280	1,500,000	1550	55,0	68,0
KT186*	1080 × 400	14,5	15,500	410	330	–	–	50,0	10,5

(continued)

Table A.3 (continued)

Product code	Tire size, mm	Operating tire pressure, kgf/cm^2	Parking load due to aircraft takeoff weight, kgf	Allowable takeoff speed, km/h	Allowable landing speed, km/h	Brake operational energy loading, kgf·m	Operational braking torque, kgf·m	Wheel mass, kg	Brake mass, kg
KT159*	1100 × 330	7.0	5830	286	236	–	–	90.0	4.0
KT216	1120 × 450	8.0	9500	300	300	2,200,000	1700	52.0	56.0
KT 166	1270 × 510	11.0	18,375	330	300	3,100,000	2700	80.0	95.0
KT158	1300 × 480	6.5	10,500	330	280	926,000	900	64.3	71.7
KT171	1300 × 480	9.0	16,470	350	300	200,000	2000	90.0	127.0
KT185*	1300 × 480	9.0	14,700	350	300	–	–	80.0	12.0
KT204	1300 × 480–560	11.5	187,000	330	280	2,300,000	2300	83.3	91.0
KT205*	1300 × 480	11.5	16,300	330	280	–	–	90.0**	–
KT106/3*	1450 × 450	11.0	20,000	330	260	2,100,000	2260	140.0	111.0

*– braked nose wheel capable of partial braking

** – wheel and brake mass

Table A.4 Unbraked wheels

Product code	Tire size, mm	Operating tire pressure, kgf/cm^2	Parking load due to aircraft takeoff weight, kgf	Allowable takeoff speed, km/h	Tire mass, kg
KN52	380 × 150–5	3.0	310	126	2.0
KN55	400 × 150–5	3.0	320	125	2.3
KN47	400 × 150–5	4.0	450	165	2.5
KN58	430 × 150–165	4.5	570	196	5.0
KN54	500 × 150–9	5.0	654	230	5.0
KN51	500 × 170–10	7.5	1000	235	7.5
KN36	570 × 140	17.0	2300	365	11.5
KN44	600 × 200	5.0	1600	240	7.8
KN46	620 × 180	7.5	1830	185	13.0
KN35	620 × 180	18–20	3000	335	13.0
KN38	620 × 180	5.0	1280	255	7.3
KN41	680 × 260	14.0	5600	350	25.0
KN39	1120 × 450	8.0	9500	300	36.0

MIX
Papier aus verantwortungsvollen Quellen
Paper from responsible sources
FSC® C105338

If you have any concerns about our products,
you can contact us on
ProductSafety@springernature.com

In case Publisher is established outside the EU,
the EU authorized representative is:
Springer Nature Customer Service Center GmbH
Europaplatz 3, 69115 Heidelberg, Germany

Printed by Libri Plureos GmbH
in Hamburg, Germany